Handbook of Research on Software-Defined and Cognitive Radio Technologies for Dynamic Spectrum Management

Naima Kaabouch
University of North Dakota, USA

Wen-Chen Hu
University of North Dakota, USA

Volume I

A volume in the Advances in Wireless
Technologies and Telecommunication (AWTT)
Book Series

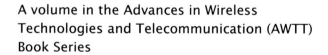

Managing Director:	Lindsay Johnston
Managing Editor:	Austin DeMarco
Director of Intellectual Property & Contracts:	Jan Travers
Acquisitions Editor:	Kayla Wolfe
Production Editor:	Christina Henning
Development Editor:	Rachel Ginder
Typesetter:	Lisandro Gonzalez
Cover Design:	Jason Mull

Published in the United States of America by
Information Science Reference (an imprint of IGI Global)
701 E. Chocolate Avenue
Hershey PA, USA 17033
Tel: 717-533-8845
Fax: 717-533-8661
E-mail: cust@igi-global.com
Web site: http://www.igi-global.com

Library of Congress Cataloging-in-Publication Data

CIP Data
Handbook of research on software-defined and cognitive radio technologies for
dynamic spectrum management / Naima Kaabouch and Wen-Chen Hu, editors.
      volumes cm
  Includes bibliographical references and index.
  ISBN 978-1-4666-6571-2 (hardcover) -- ISBN 978-1-4666-6572-9 (ebook) -- ISBN 978-1-4666-6574-3 (print & perpetual access) 1. Cognitive radio networks. 2. Radio frequency allocation--Management. I. Kaabouch, Naima, 1959- II. Hu, Wen Chen, 1960-
  TK5103.4815.H36 2015
  384.54'524--dc23
                                        2014034430

This book is published in the IGI Global book series Advances in Wireless Technologies and Telecommunication (AWTT) (ISSN: 2327-3305; eISSN: 2327-3313)

British Cataloguing in Publication Data
A Cataloguing in Publication record for this book is available from the British Library.

All work contributed to this book is new, previously-unpublished material. The views expressed in this book are those of the authors, but not necessarily of the publisher.

For electronic access to this publication, please contact: eresources@igi-global.com.

Advances in Wireless Technologies and Telecommunication (AWTT)

Xiaoge Xu
The University of Nottingham Ningbo China

ISSN: 2327-3305
EISSN: 2327-3313

MISSION

The wireless computing industry is constantly evolving, redesigning the ways in which individuals share information. Wireless technology and telecommunication remain one of the most important technologies in business organizations. The utilization of these technologies has enhanced business efficiency by enabling dynamic resources in all aspects of society.

The **Advances in Wireless Technologies and Telecommunication Book Series** aims to provide researchers and academic communities with quality research on the concepts and developments in the wireless technology fields. Developers, engineers, students, research strategists, and IT managers will find this series useful to gain insight into next generation wireless technologies and telecommunication.

COVERAGE

- Network Management
- Radio Communication
- Wireless Sensor Networks
- Wireless Broadband
- Global Telecommunications
- Broadcasting
- Grid Communications
- Mobile Web Services
- Mobile Communications
- Digital Communication

IGI Global is currently accepting manuscripts for publication within this series. To submit a proposal for a volume in this series, please contact our Acquisition Editors at Acquisitions@igi-global.com or visit: http://www.igi-global.com/publish/.

Titles in this Series

For a list of additional titles in this series, please visit: www.igi-global.com

Interdisciplinary Mobile Media and Communications Social, Political, and Economic Implications
Xiaoge Xu (The University of Nottingham Ningbo China, China)
Information Science Reference • copyright 2014 • 409pp • H/C (ISBN: 9781466661660) • US $205.00 (our price)

Cognitive Radio Sensor Networks Applications, Architectures, and Challenges
Mubashir Husain Rehmani (Department of Electrical Engineering, COMSATS Institute of Information Technology, Pakistan) and Yasir Faheem (Department of Computer Science, COMSATS Institute of Information Technology, Pakistan)
Information Science Reference • copyright 2014 • 313pp • H/C (ISBN: 9781466662124) • US $235.00 (our price)

Game Theory Applications in Network Design
Sungwook Kim (Sogang University, South Korea)
Information Science Reference • copyright 2014 • 500pp • H/C (ISBN: 9781466660502) • US $225.00 (our price)

Convergence of Broadband, Broadcast, and Cellular Network Technologies
Ramona Trestian (Middlesex University, UK) and Gabriel-Miro Muntean (Dublin City University, Ireland)
Information Science Reference • copyright 2014 • 333pp • H/C (ISBN: 9781466659780) • US $235.00 (our price)

Handbook of Research on Progressive Trends in Wireless Communications and Networking
M.A. Matin (Institut Teknologi Brunei, Brunei Darussalam)
Information Science Reference • copyright 2014 • 592pp • H/C (ISBN: 9781466651708) • US $380.00 (our price)

Broadband Wireless Access Networks for 4G Theory, Application, and Experimentation
Raul Aquino Santos (University of Colima, Mexico) Victor Rangel Licea (National Autonomous University of Mexico, Mexico) and Arthur Edwards-Block (University of Colima, Mexico)
Information Science Reference • copyright 2014 • 452pp • H/C (ISBN: 9781466648883) • US $235.00 (our price)

Multidisciplinary Perspectives on Telecommunications, Wireless Systems, and Mobile Computing
Wen-Chen Hu (University of North Dakota, USA)
Information Science Reference • copyright 2014 • 305pp • H/C (ISBN: 9781466647152) • US $175.00 (our price)

Mobile Networks and Cloud Computing Convergence for Progressive Services and Applications
Joel J.P.C. Rodrigues (Instituto de Telecomunicações, University of Beira Interior, Portugal) Kai Lin (Dalian University of Technology, China) and Jaime Lloret (Polytechnic University of Valencia, Spain)
Information Science Reference • copyright 2014 • 408pp • H/C (ISBN: 9781466647817) • US $180.00 (our price)

DISSEMINATOR OF KNOWLEDGE

www.igi-global.com

701 E. Chocolate Ave., Hershey, PA 17033
Order online at www.igi-global.com or call 717-533-8845 x100
To place a standing order for titles released in this series, contact: cust@igi-global.com
Mon-Fri 8:00 am - 5:00 pm (est) or fax 24 hours a day 717-533-8661

Editorial Advisory Board

List of Contributors

Table of Contents

Volume I

Section 1
Radio Spectrum Sensing

Raza Umar, King Saud University, Saudi Arabia
Fahham Mohammed, King Fahd University of Petroleum and Minerals, Saudi Arabia
Mohamed Deriche, King Fahd University of Petroleum and Minerals, Saudi Arabia
Asrar U. H. Sheikh, The University of Lahore, Pakistan

Srinivas Nallagonda, National Institute of Technology – Durgapur, India
Sanjay Dhar Roy, National Institute of Technology – Durgapur, India
Sumit Kundu, National Institute of Technology – Durgapur, India
Gianluigi Ferrari, University of Parma, Italy
Riccardo Raheli, University of Parma, Italy

L. Safatly, American University of Beirut, Lebanon
A. H. Ramadan, American University of Beirut, Lebanon
M. Al-Husseini, Lebanese Center for Studies and Research, Lebanon
Y. Nasser, American University of Beirut, Lebanon
K. Y. Kabalan, American University of Beirut, Lebanon
A. El-Hajj, American University of Beirut, Lebanon

Section 2
Radio Spectrum Management and Access

Volume II

Section 3
Software-Defined Radio and Antennas for Cognitive Radio Networks

Detailed Table of Contents

Volume I

Section 1
Radio Spectrum Sensing

Chapter 1

Raza Umar, King Saud University, Saudi Arabia
Fahham Mohammed, King Fahd University of Petroleum and Minerals, Saudi Arabia
Mohamed Deriche, King Fahd University of Petroleum and Minerals, Saudi Arabia
Asrar U. H. Sheikh, The University of Lahore, Pakistan

Cognitive Radio (CR) has emerged as a smart solution to spectrum bottleneck faced by current wireless services under which licensed spectrum is made available to unlicensed Secondary Users (SUs) through robust and efficient Spectrum Sensing (SS). Energy Detection (ED) is the dominantly used SS approach owing to its low computational complexity and ability to identify spectrum holes without requiring a priori knowledge of primary transmission characteristics. In this chapter, the authors present an in-depth analysis of the ED test statistic. Based on the double threshold ED, they analyze the performance of a Hybrid PSO-OR (Particle Swarm Optimization and OR) algorithm for cooperative SS. The sensing decision of "fuzzy" SUs is optimized using PSO and the final collective decision is made based on OR rule. The idea of using two thresholds is introduced to reduce the communication overhead in reporting local data/decision to the fusion center, which also offers reduced energy consumption. The Hybrid PSO-OR algorithm is shown to exhibit significant performance gain over the Hybrid EGC-OR algorithm.

  Srinivas Nallagonda, National Institute of Technology – Durgapur, India
  Sanjay Dhar Roy, National Institute of Technology – Durgapur, India
  Sumit Kundu, National Institute of Technology – Durgapur, India
  Gianluigi Ferrari, University of Parma, Italy
  Riccardo Raheli, University of Parma, Italy

In this chapter, the authors study the performance of Cooperative Spectrum Sensing (CSS) with soft data fusion, given by Maximal Ratio Combining (MRC)-based fusion with Weibull faded channels, and Log-normal shadowed channels. More precisely, they evaluate the performance of a MRC-based CSS with Cognitive Radios (CRs) censored on the basis of the quality of the reporting channels. The performance of CSS with two censoring schemes, namely rank-based and threshold-based, is studied in the presence of Weibull fading, Rayleigh fading, and Log-normal shadowing in the reporting channels, considering MRC fusion. The performance is compared with those of schemes based on hard decision fusion rules. Furthermore, depending on perfect or imperfect Minimum Mean Square Error (MMSE) channel estimation, the authors analyze the impact of channel estimation strategy on the censoring schemes. The performance is studied in terms of missed detection probability as a function of several network, fading, and shadowing parameters.

  L. Safatly, American University of Beirut, Lebanon
  A. H. Ramadan, American University of Beirut, Lebanon
  M. Al-Husseini, Lebanese Center for Studies and Research, Lebanon
  Y. Nasser, American University of Beirut, Lebanon
  K. Y. Kabalan, American University of Beirut, Lebanon
  A. El-Hajj, American University of Beirut, Lebanon

In this chapter, the concepts of Cognitive Radio (CR) and multi-dimensional spectrum sensing are introduced. Spectrum sensing methodologies, energy efficiency consideration, resources scheduling, and self-management and learning mechanisms in cognitive radio networks are also discussed. The entailed challenges of CR RF front-end architectures are looked into. The synthesis and design performance analysis of a tunable RF front-end sensing receiver for CR applications are presented. The chapter also discusses how sensing performance degradation, which is due to RF impairments, is analytically evaluated. Spectrum sensing algorithms that correct imperfect RF issues by compensating induced error effects through digital baseband processing are also illustrated.

    Saud Althunibat, University of Trento, Italy
    Sandeep Narayanan, WEST Aquila s.r.l., Italy & University of L'Aquila, Italy
    Marco Di Renzo, Laboratory of Signals and Systems (L2S), France
    Fabrizio Granelli, University of Trento, Italy

One of the main problems of Cooperative Spectrum Sensing (CSS) in cognitive radio networks is the high energy consumption. Energy is consumed while sensing the spectrum and reporting the results to the fusion centre. In this chapter, a novel partial CSS is proposed. The main concern is to reduce the energy consumption by limiting the number of participating users in CSS. Particularly, each user individually makes the participation decision. The energy consumption in a CSS round is expected by the user itself and compared to a predefined threshold. The corresponding user will participate only if the expected amount of energy consumed is less than the participation threshold. The chapter includes optimizing the participation threshold for energy efficiency maximization. The simulation results show a significant reduction in the energy consumed compared to the conventional CSS approach.

    Mahsa Derakhshani, University of Toronto, Canada
    Tho Le-Ngoc, McGill University, Canada
    Masoumeh Nasiri-Kenari, Sharif University of Technology, Iran

Spectrum sensing is one of the key elements in the establishment of cognitive radio. One of the most effective approaches for spectrum sensing is cyclostationary feature detection. Since modulated signals can be modeled as cyclostationary random signals, this feature can be used to recognize the cyclostationary modulated signal in a background of stationary noise even at low SNR regimes. This chapter reviews non-cooperative cyclostationary sensing approaches and reports recent advances in cooperative cyclostationary sensing algorithms. New results for cooperative cyclostationary spectrum sensing are then presented, which ensure better performance as well as faster and simpler operation. In the proposed schemes, each Secondary User (SU) performs Single-Cycle (SC) cyclostationary detection for fast and simple implementation, while collaboration between SUs in final decision on the presence or absence of the PU is explored to improve its performance. Furthermore, this chapter presents another look at the performance evaluation of cyclostationary detectors in terms of deflection coefficients.

    Ahmed M. Elzanati, Sinai University, Egypt
    Mohamed F. Abdelkader, Port Said University, Egypt
    Karim G. Seddik, American University in Cairo, Egypt

Compressive Sensing (CS) has been proven effective to elevate some of the problems associated with spectrum sensing in wideband Cognitive Radio (CR) networks through efficient sampling and exploiting the underlying sparse structure of the measured frequency spectrum. In this chapter, the authors discuss the motivation and challenges of utilizing collaborative approaches for compressive spectrum sensing. They survey the different approaches and the key published results in this domain. The authors present in detail an approach that utilizes Kronecker sparsifying bases to exploit the two-dimensional sparse structure in the measured spectrum at different, spatially separated cognitive radios. Simulation results show that the presented scheme can substantially reduce the Mean Square Error (MSE) of the recovered power spectrum density over conventional schemes while maintaining the use of a low-rate Sub-Nyquist Analog to Information Converter. It is also shows that one can achieve dramatically lower MSE under low compression ratios using a dense measurement matrix while using Nyquist rate ADC.

Chapter 7

This chapter presents an experimental comparative analysis of the well-known Covariance-Based Detection (CBD) techniques, which include Covariance Absolute Value (CAV), Maximum-Minimum Eigenvalue (MME), Energy with Minimum Eigenvalue (EME), and Maximum Eigenvalue Detection (MED). CBD techniques overcome the noise uncertainty issue of the Energy Detector (ED) and can even outperform ED in the case of correlated signals. They can perform accurate blind detection given sufficient number of signal samples. This chapter also presents a novel CBD algorithm that is based on Principal Component (PC) analysis. A Software-Defined Radio (SDR)-based multiple antenna system is used to evaluate the detection performance of the considered algorithms. The PC algorithm significantly outperforms the MED and EME algorithms and it also outperforms MME and CAV algorithms in certain cases.

In this chapter, the effect of imperfect training data on feature-based signal detection is explored, as it relates to both training time and detection performance in a cognitive radio system. The improved performance of feature-based detection comes at the cost of either having to know in advance the signal features present in primary user transmissions (an unrealistic assumption) or learning them whilst operating "in the field." Such learning, however, necessarily takes place with signal sets which do not perfectly represent the features of the primary users' modulated signals. Using a two-stage detector performing both feature training and sensing functions, it is shown in this chapter that reducing the learning time generally results in poorer detection performance and vice-versa. A suitable trade-off between these two outcomes is obtained by optimizing a cost function that takes both factors into consideration. Cyclostationarity detection is specifically considered.

Chapter 9

In multi-channel Cognitive Radio Networks (CRNs), when the cognitive radio receivers cannot simultaneously sense more than one out of the many possible (groups of) channels, an important challenge is to determine a sensing order for each Cognitive User (CU) so as to optimize a given performance metric. The sensing-order problem is compounded in multi-user CRNs where the multiple users in the network could collide with each other. With the focus on multi-user CRNs, this chapter uses cognitive-throughput maximization as the performance metric and describes how the optimal sensing orders can be computed for different contention management strategies used by the network. In general, the optimal procedures involve a computationally expensive brute-force search, so the chapter also discusses several heuristic-based near-optimal procedures that can be used in practice.

Section 2
Radio Spectrum Management and Access

Chapter 10

Yong Yao, Blekinge Institute of Technology, Sweden
Alexandru Popescu, Blekinge Institute of Technology, Sweden
Adrian Popescu, Blekinge Institute of Technology, Sweden

Cognitive radio networks are a new technology based on which unlicensed users are allowed access to licensed spectrum under the condition that the interference perceived by licensed users is minimal. That means unlicensed users need to learn from environmental changes and to make appropriate decisions regarding the access to the radio channel. This is a process that can be done by unlicensed users in a cooperative or non-cooperative way. Whereas the non-cooperative algorithms are risky with regard to performance, the cooperative algorithms have the capability to provide better performance. This chapter shows a new fuzzy logic-based decision-making algorithm for channel selection. The underlying decision criterion considers statistics of licensed user channel occupancy as well as information about the competition level of unlicensed users. The theoretical studies indicate that the unlicensed users can obtain an efficient sharing of the available channels. Simulation results are reported to demonstrate the performance and effectiveness of the suggested algorithm.

Chapter 11

Yasir Saleem, Sunway University, Malaysia
Farrukh Salim, Technische Universität Ilmenau, Germany & NED University of Engineering
    and Technology, Pakistan
Mubashir Husain Rehmani, COMSATS Institute of Information Technology, Wah Cantt,
    Pakistan
Bushra Rashid, COMSATS Institute of Information Technology, Wah Cantt, Pakistan

In Cognitive Radio Networks (CRNs) there is much dynamicity due to the activities of primary users which results in instability of routes. Therefore, an efficient routing protocol based on good channel assignment strategy is required in CRNs. A good channel selection strategy makes route stable by selecting channels having larger capacity and greater availability time. Therefore, the focus of this chapter is joint channel assignment and routing in CRNs, which provides a comprehensive survey on routing and channel assignment in CRNs. First, the importance of joint channel assignment and routing for successful communication in cognitive radio networks is discussed. Then classification and challenges related to channel assignment and routing are discussed in detail. In order to establish reliable routes in CRNs, some factors are discussed that further enhance the communication in CRNs. Finally, guidelines for the development of efficient routing protocols are discussed.

Dynamic Spectrum Management (DSM) is an effective method for reducing the effect of interference in both wireless and wireline communication systems. This chapter discusses various DSM algorithms, including Optimal Spectrum Balancing (OSB), Iterative Spectrum Balancing (ISB), Iterative Water-Filling (IWF), Selective Iterative Water-filling (SIW), Successive Convex Approximation for Low complExity (SCALE), the Difference of Convex functions Algorithm (DCA), Distributed Spectrum Balancing (DSB), Autonomous Spectrum Balancing (ASB), and Constant Offset ASB using Multiple Reference Users (ASB-MRU). They are compared in terms of their performance (achievable data-rate) by extensive simulation results and their computational complexity.

Chapter 13

This chapter investigates the performance of primary and secondary users in a spectrum-sharing cognitive environment. In this setup, multiple secondary users compete to share a channel dedicated to a primary user in order to transmit their data to a receiver unit. One secondary user is scheduled to share the channel, and to do so, its transmission power should satisfy the outage probability requirement of the primary user. Secondary users are ranked according to their channel strength, and performance measures are derived as a function of a generic channel rank. The performance of different scheduling schemes is also investigated. Further, the performance of the primary user is investigated in this environment. Numerical results are presented to verify the theoretical analysis and investigate the relation between the parameters of the communication environment and the performance measures of the users of the system.

Due to the constraint imposed by the Dynamic Spectrum Access paradigm, Cognitive Radio (CR) networks are entangled in persistent competition for opportunistic access to underutilized spectrum resources. In order to maintain quality of service, each network faces the challenge of acquiring dynamic enough channels to meet channel size requirement. The main goal of every CR network is to minimize the amount of contention experienced during channel acquisition and to maximize the utility derived from acquired channels. This is a major challenge, especially without a global communication protocol that can facilitate communication between the networks. This chapter discusses self-coexistence of CR networks in a decentralized system with no support for coordinated radio transmission activities. Channel acquisition mechanisms that can help networks minimize contention and maximize utility are also discussed. The mechanisms guarantee fast convergence of the system leading to an equilibrium state whereby networks are able to operate on acquired channel with minimal or zero contention.

Chapter 15

Sylwia Romaszko, RWTH Aachen University, Germany
Petri Mähönen, RWTH Aachen University, Germany

In the case of Opportunistic Spectrum Access (OSA), unlicensed secondary users have only limited knowledge of channel parameters or other users' information. Spectral opportunities are asymmetric due to time and space varying channels. Owing to this inherent asymmetry and uncertainty of traffic patterns, secondary users can have trouble detecting properly the real usability of unoccupied channels and as a consequence visiting channels in such a way that they can communicate with each other in a bounded period of time. Therefore, the channel service quality, and the neighborhood discovery (NB) phase are fundamental and challenging due to the dynamics of cognitive radio networks. The authors provide an analysis of these challenges, controversies, and problems, and review the state-of-the-art literature. They show that, although recently there has been a proliferation of NB protocols, there is no optimal solution meeting all required expectations of CR users. In this chapter, the reader also finds possible solutions focusing on an asynchronous channel allocation covering a channel ranking.

Chapter 16

Bin Cao, Jingdezhen Ceramic Institute of Technology, China & Harbin Institute of Technology Shenzhen Graduate School, China
Qinyu Zhang, Harbin Institute of Technology Shenzhen Graduate School, China
Hao Liang, University of Waterloo, Canada & University of Alberta, Canada
Gang Fu, Harbin Institute of Technology Shenzhen Graduate School, China
Jon W. Mark, University of Waterloo, Canada

Radio spectrum underutilization and energy inefficiency become urgent bottleneck problems to the sustainable development of wireless technologies. The research philosophies of wireless communications have been shifted from balancing reliability-efficiency tradeoff in the link level to seeking spectrum-energy efficiency in the network level. Global spectrum-energy efficient designs attract significant attention to improving utilization and efficiency, wherein cognitive radio networks and energy-efficient resource allocation are of particular interests. In this chapter, the authors first provide a systematic study on Cooperative Cognitive Radio Networking (CCRN). As an effort to shed light on addressing spectrum-energy inefficiency at a low complexity, an orthogonal modulation enabled two-phase cooperation framework and an Orthogonally Dual-Polarized Antenna (ODPA) based framework, as well as their resource allocation problems are given and tackled.

Volume II

Raza Umar, King Saud University, Saudi Arabia
Wessam Mesbah, King Fahd University of Petroleum and Minerals, Saudi Arabia

Cognitive radio based on dynamic spectrum access has emerged as a promising technology to meet the insatiable demand for radio spectrum by the emerging wireless applications. In this chapter, the authors address the problem of throughput-efficient spectrum access in Cognitive Radio Networks (CRNs) using Coalitional Game-theoretic framework. They model the problem of joint Coalition Formation (CF) and Bandwidth (BW) allocation as a CF game in partition form with non-transferable utility and present a variety of algorithms to dynamically share the available spectrum resources among competing Secondary Users (SUs). First, the authors present a centralized solution to reach a sum-rate maximizing Nash-stable network partition. Next, a distributed CF algorithm is developed through which SUs may join/leave a coalition based on their individual preferences. Performance analysis shows that the CF algorithms with optimal BW allocation provides a substantial gain in the network throughput over existing coalition formation techniques as well as the simple cases of singleton and grand coalition.

Chungang Yang, Xidian University, China
Jiandong Li, Xidian University, China

The growing demands of radio spectrum urgently require more efficient and effective spectrum exploitation and management technologies. Cognitive radio technology and its networking not only explores the potential white spectrum resources temporally and geographically, but also enables an extensive efficient utilization and optimization of the current allocated spectrum resources. Therefore, rapid progress has been made in the research on cognitive radio and its networking technologies to facilitate more flexibilities in spectrum utilization and management. In this chapter, the authors first summarize the current various advanced and flexible spectrum management schemes, including spectrum trading, leasing, pricing, and harvesting, and analyze their advantages and disadvantages. Then, they take the viewpoints of both the spectrum marketing perspective and spectrum technical perspective, and they propose the centralized and distributed dynamic spectrum sharing schemes, respectively. In particular, the authors introduce many novel advanced spectrum sharing scheme and summarize the open and possible research problems.

Section 3
Software-Defined Radio and Antennas for Cognitive Radio Networks

Chapter 19

Jyoti Sekhar Banerjee, Bengal Institute of Technology, India
Arpita Chakraborty, Bengal Institute of Technology, India

Software Defined Radio (SDR) and Cognitive Radio (CR) are the key enabling technologies to overcome the spectrum scarcity problem a bit, by supporting dynamic spectrum access in which either a network or a wireless node reconfigures its transmission or reception parameters to communicate efficiently, avoiding interference with licensed or unlicensed users. CR senses the environment and enables a secondary system to share the licensed spectrum with the primary system, which usually has exclusive access. The performance of the secondary system could be enhanced by Cooperative Spectrum Sensing (CSS) as it increases the primary detection probability. Again cognitive radio network greatly benefits from a cooperative transmission, employing intermediate nodes as relays. This chapter is focused on software defined radio, its architecture, limitations, then evolution to cognitive radio network, architecture of the CR, and its relevance in the wireless and mobile ad-hoc networks. Additionally, an overview of Cooperative Spectrum Sensing (CSS), its classification, components, challenges, and Cooperative Relay are discussed.

Chapter 20

Erick Gonzalez Rodriguez, Technische Universität Darmstadt, Germany
Yuliang Zheng, Technische Universität Darmstadt, Germany
Holger Maune, Technische Universität Darmstadt, Germany
Rolf Jakoby, Technische Universität Darmstadt, Germany

Cognitive Radio (CR) and Software Defined Radio (SDR), concepts which were mere proposals to solve the population of services over the past two decades, are now enabled by novel materials and components to offer fully reconfigurable devices. Thus, a convergence of services can be attained within a reduced, or even single RF (Radio Frequency) signal path in the device. A solid design of reconfigurable frontends, from the RF part to the digital baseband, should consider different criteria to better exploit the available spectrum. Examples of such criteria are scattering parameters and phase linearity of components at a defined carrier frequency, RF signal bandwidth, and signal quality in terms of Error Vector Magnitude and Bit Error Rate. In this chapter, a general perspective to achieve smarter air interfaces is studied and discussed by setting out strategies based on CR and SDR techniques for the implementation and integration of future reconfigurable RF-Frontends.

In this chapter, the authors discuss the design of precoders in cognitive multi-user multi-way relay systems. When multiple secondary users intend to communicate with each other using the spectrum licensed to the primary user, how to manage the interference between primary and secondary networks as well as among multiple secondary users becomes an important design issue. They discuss one possible solution of using a relay station as well as multiple antenna techniques. Precoding design in such a relay supported multiple antenna secondary network is presented based on the Mean Square Error (MSE) design criterion. The joint design of precoding matrices at all the secondary nodes is a non-convex problem. Therefore, an iterative algorithm is proposed to iteratively optimize the precoding matrices at secondary transmitters, the precoding matrix at the secondary relay, and decoding matrices at secondary receivers. Other non-iterative solutions are also presented to strike a balance between performance and complexity.

Chapter 22

Cognitive Radio presents promising applications in creating a more intelligent and flexible radio and wireless system structures considering the increasing demands faced by spectrum usage and spectrum regulators for specific frequency bands. The flexibility and reconfigurability required by such systems place a significant burden on engineers and radio designers specifically on the radio front-end and antenna elements. This chapter aims to study and analyse the various reconfiguration techniques and types that could be employed in the antenna front-end that will allow cognitive radio to be more flexible and adaptable to different bands and environments. The chapter focuses on theoretical and experimental analyses of novel methods to frequency-reconfigure compact ultra-wideband antennas to work in different bands, and it also explores the possibility of pattern and polarisation reconfiguration of the antenna element. It ultimately shows a method of combining all reconfiguration techniques to realise an original antenna structure capable of adapting the cognitive radio unit to work in congested electromagnetic spectrum bands based on availability of other free gaps, usage rate, and environmental factors. The authors strongly believe the proposed design meets the growing demand of cognitive and smart radio devices for more intelligent and multi-functional antennas.

Chapter 23

This chapter provides a deep insight into multiple antenna eigenvalue-based spectrum sensing algorithms from a complexity perspective. A review of eigenvalue-based spectrum-sensing algorithms is provided. The chapter presents a finite computational complexity analysis in terms of Floating Point Operations (flop) and a comparison of the Maximum-to-Minimum Eigenvalue (MME) detector and a simplified variant of the Multiple Beam forming detector as well as the Approximated MME method. Constant False Alarm Performances (CFAR) are presented to emphasize the complexity-reliability tradeoff within the spectrum-sensing problem, given the strong requirements on the sensing duration and the detection performance.

Chapter 24

Ajay Singh, National Institute of Technology – Raipur, India
Manav R. Bhatnagar, Indian Institute of Technology – Delhi, India
Ranjan K. Mallik, Indian Institute of Technology – Delhi, India

A dual-hop cooperative spectrum sensing approach is studied in detail, where each cooperative Cognitive Radio (CR) makes a binary decision based on the local observation, by using an improved energy detector, and then forwards it to a common receiver. At the common receiver, all binary decisions are fused together. The authors provide an analytical framework for the analysis of performance of the improved energy detector-based cooperative CR network. They discuss how to choose an optimal number of cooperating CRs in order to minimize the total error rate by using an improved energy detector over perfect and imperfect reporting channels. Further, the error performance of dual-hop cooperative spectrum sensing with multiple antennae-based CR is discussed. The authors also exploit the multi-hop cooperative communication approach in an improved energy detector-based CR network for increasing the coverage area of the secondary communication systems with reduced power consumption.

Chapter 25

Mickaël Dardaillon, Université de Lyon, INRIA, INSA-Lyon, France
Kevin Marquet, Université de Lyon, INRIA, INSA-Lyon, France
Tanguy Risset, Université de Lyon, INRIA, INSA-Lyon, France
Jérôme Martin, CEA-Leti – Minatec Campus, France
Henri-Pierre Charles, CEA-List – Minatec Campus, France

Cognitive radio is based on Software Defined Radio (SDR) technology. The commercial success of smart radio applications and cognitive radio networks will be very dependent on cost, performance, and power consumption of SDR hardware platforms. SDR hardware is now available, but many issues have yet to be studied. In this chapter, the authors detail the constraints imposed by recent radio protocols and how hardware architectures support them. Then, they present existing architectures and solutions for SDR programming. Finally, the authors mention challenges related to the programming of future cognitive radio systems.

Chapter 26

Ali H. Mahdi, University of Baghdad, Iraq & Technische Universität Ilmenau, Germany
Mohamed A. Kalil, Suez University, Egypt

Cognitive Radio (CR) systems are smart systems capable of sensing the surrounding radio environment and adapting their operating parameters in order to efficiently utilize the available radio spectrum. To reach this goal, different transmission parameters across the Open Systems Interconnection (OSI) layers, such as transmit power, modulation scheme, and packet length, should be optimized. This chapter discusses the

Adaptive Discrete Particle Swarm Optimization (ADPSO) algorithm as an efficient algorithm for optimizing and adapting CR operating parameters from physical, MAC, and network layers. In addition, the authors present two extensions for the proposed algorithm. The first one is Automatic Repeat reQuest-ADPSO (ARQ-ADPSO) for efficient spectrum utilization. The second one is merging ARQ-ADPSO and Case-Based Reasoning (CBR) algorithms for autonomous link adaptation under dynamic radio environment. The simulation results show improvements in the convergence time, signaling overhead, and spectrum utilization compared to the well-known optimization algorithms such as the Genetic Algorithm (GA).

Chapter 27

    Mahsa Derakhshani, University of Toronto, Canada
    Tho Le-Ngoc, McGill University, Canada

This chapter presents a study on the interference caused by Secondary Users (SUs) due to miss-detection errors and its effects on the capacity-outage performance of the Primary User (PU) in a cognitive radio network assuming Rayleigh and Nakagami fading channels. The effect of beacon transmitter placement on aggregate interference distribution and capacity-outage performance is studied considering two scenarios of beacon transmitter placement: a beacon transmitter located at a PU transmitter or at a PU receiver. Based on the developed statistical models for the interference distribution, closed-form expressions for the capacity-outage probability of the PU are derived to examine the effects of various system parameters on the performance of the PU in the presence of interference from SUs. Furthermore, the model is extended to investigate the cooperative sensing effect on aggregate interference statistical model and capacity-outage performance considering OR (i.e., logical OR operation) and Maximum Likelihood (ML) cooperative detection techniques.

Section 4
Models, Security, and Other Related Topics

Chapter 28

    Yuehong Gao, Beijing University of Posts and Telecommunications, China
    Zhidu Li, Beijing University of Posts and Telecommunications, China
    Guoting Zhang, General Administration of Press and Publication of the People's Republic of China, China
    He Bai, General Administration of Press and Publication of the People's Republic of China, China

Performance evaluation and analysis are of key importance to obtain deep understanding of cognitive radio networks. Some effects have been made to model and analyze the performance of cognitive radio networks. In the literature, there are two methodologies: queuing theory/Markov chain-based analysis and stochastic network calculus-based analysis. These two methodologies rely on different mathematical basics and modeling approaches. Thus, they lead to different output metrics on various viewpoints. This chapter aims to give an overall introduction to both methodologies. First, the fundamental models used in queuing/Markov chain-based analysis are presented, followed by their applications in cognitive radio networks. Then, network calculus basics are introduced with the modeling and application in performance analysis of the cognitive radio network.

Chapter 29

Ju Bin Song, Kyung Hee University, South Korea
Zhu Han, University of Houston, USA

In cognitive radio networks a secondary user needs to estimate the primary users' air traffic patterns so as to optimize its transmission strategy. In this chapter, the authors describe a nonparametric Bayesian method for identifying traffic applications, since the traffic applications have their own distinctive air traffic patterns. In the proposed algorithm, the collapsed Gibbs sampler is applied to cluster the air traffic applications using the infinite Gaussian mixture model over the feature space of the packet length, the packet inter-arrival time, and the variance of packet lengths. The authors analyze the effectiveness of their proposed technique by extensive simulation using the measured data obtained from the WiMax networks.

Chapter 30

Andre Abadie, George Mason University, USA
Damindra Bandara, George Mason University, USA
Duminda Wijesekera, George Mason University, USA

Even though security research in cognitive radio offers specific countermeasures to address known threats, there are a number of unknown conditions or influences that will shape its eventual realization once it reaches capability maturity. To attempt to secure against such unknowns, this chapter describes a risk engine that can incorporate a risk assessment cognition cycle. In various business sectors, risk management is the preferred mechanism to address unknown conditions and therefore offers promise in this context. The chapter describes how the risk engine can potentially address the vulnerabilities inherent to radio operation: in the sensing/perception of spectrum, in the cognition cycle, or in the device infrastructure. It highlights some well-defined threats, their associated countermeasures, and suggests conceptual approaches for a risk engine to intervene in those scenarios. Finally, a case study is introduced to demonstrate an example risk engine's ability to accurately assess particular risks in a given operational environment as well as potentially detect adversarial actions.

Chapter 31

Saed Alrabaee, Concordia University, Canada
Mahmoud Khasawneh, Concordia University, Canada
Anjali Agarwal, Concordia University, Canada

Cognitive radio technology is the vision of pervasive wireless communications that improves the spectrum utilization and offers many social and individual benefits. The objective of the cognitive radio network technology is to use the unutilized spectrum by primary users and fulfill the secondary users' demands irrespective of time and location (any time and any place). Due to their flexibility, the Cognitive Radio Networks (CRNs) are vulnerable to numerous threats and security problems that will affect the performance of the network. Little attention has been given to security aspects in cognitive radio networks. In this chapter, the authors discuss the security issues in cognitive radio networks, and then they present an intensive list of the main known security threats in CRN at various layers and the adverse effects on performance due to such threats, and the current existing paradigms to mitigate such issues and threats. Finally, the authors highlight proposed directions in order to make CRN more authenticated, reliable, and secure.

Chapter 32

Gang Hu, National University of Defense Technology, China
Lixia Liu, National University of Defense Technology, China
Yuxing Peng, National University of Defense Technology, China

Multiple characters of spectrum resource bring many challenges to spectrum trading. The demanders may not find the full-matching spectrum resource. Meanwhile, the optimal matching strategy cannot be determined if the demanders have different matching ratios. This chapter proposes an algorithm called HSO-ST (Heterogeneous Service-Oriented Spectrum Trading) with the target of maximum matching number under the priority restriction. This algorithm can satisfy as many secondary users as possible. Compared with other spectrum trading strategies, HSO-ST can greatly improve the spectrum demand-matching ratio.

Chapter 33

Shree Krishna Sharma, University of Luxembourg, Luxembourg
Symeon Chatzinotas, University of Luxembourg, Luxembourg
Björn Ottersten, University of Luxembourg, Luxembourg

The continuously increasing demand of spectrum and current static spectrum allocation policies are rendering the available radio spectrum scarce. To address the problem of spectrum scarcity in the satellite paradigm, cognitive satellite communications has been considered as a promising technique. In addition to the existing spectrum sharing dimensions such as frequency, time, and space, polarization can be exploited as an additional degree of freedom in order to explore the spectral gaps in the under-utilized licensed spectrum. In this context, this chapter firstly provides an overview of the existing works in polarization-based spectrum sharing. Secondly, it presents the theoretical analysis of energy detection technique for dual polarized Additive White Gaussian Noise and Rayleigh fading channels considering the spectral coexistence scenarios of dual and hybrid satellite systems. Thirdly, it provides the comparison of different combining techniques in terms of the sensing performance. Finally, it provides interesting future research directions in this domain.

 Hailing Zhu, University of Johannesburg, South Africa
 Andre Nel, University of Johannesburg, South Africa
 Hendrik Ferreira, University of Johannesburg, South Africa

Dynamic Spectrum Allocation (DSA) has been viewed as a promising approach to improving spectrum efficiency. With DSA, Wireless Service Providers (WSPs) that operate in fixed spectrum bands allocated through static allocation can solve their short-term spectrum shortage problems resulting from the bursty nature of wireless traffic. Such DSA mechanisms should be coupled with dynamic pricing schemes to achieve the most efficient allocation. This chapter models the DSA problem where a centralized spectrum broker manages "white space" in the spectrum of TV broadcasters and sells the vacant spectrum bands to multiple WSPs, as a multi-stage non-cooperative dynamic game. Furthermore, an economic framework for DSA is presented and a centralized spectrum allocation mechanism is proposed. The simulation results show that the centralized spectrum allocation mechanism with dynamic pricing achieves a DSA implementation that is responsive to market conditions as well as enabling efficient utilization of the available spectrum.

 Daniele Tarchi, University of Bologna, Italy
 Romano Fantacci, University of Firenze, Italy
 Dania Marabissi, University of Firenze, Italy

Machine to Machine (M2M) communications have been recently introduced as a viable paradigm for allowing low cost and efficient communications among devices mainly in an autonomous manner. Even if M2M protocols need dedicated resources, a new paradigm, called Cognitive M2M (CM2M) communications, has been recently considered exploiting cognitive/opportunistic radio communications. After having introduced the problem of applying cognitive techniques in M2M scenarios, the authors focus their attention on the Medium Access Control (MAC) protocols for CM2M scenarios, with a particular attention on the OFDMA-based primary systems. Among other approaches, the authors focus on a data-aided approach for the access of the secondary devices aiming to reduce interference toward the primary system.

Foreword

The development of portable devices and wireless networks has created an increasing demand for radio spectrum channels. This situation has generated a search for more efficient schemes to manage the radio spectrum. Currently, static allocation of frequency bands is the most prevalent way of spectrum management. This strategy has become inefficient, since recent spectrum usage surveys in several locations show that the radio spectrum has been underutilized. Therefore, there is a need for strategies to dynamically manage the access to the radio spectrum. These strategies should take advantage of unused channels without affecting incumbent systems.

A more efficient use of this resource can revolutionize the wireless and mobile communications arena, which will produce not only a technological but also an economic impact on diverse areas that depend on wireless connectivity, such as commerce, education, industry, among others. New technologies that are starving for more spectrum channels will flourish, and operating prices will become more affordable. New wireless applications in markets not currently attractive to conventional operators will also become viable, consequently impacting societies and communities that otherwise would not have that opportunity.

This timely handbook is a collection of current cutting-edge research techniques, trends, and practical applications in the field of radio spectrum. In the book's chapters, you will find descriptions of state-of-the art research projects on the many aspects of cognitive radio and radio spectrum, such as radio spectrum sensing, access, management, security, models, and applications.

This handbook will be a valuable addition to academic and research libraries and hopefully a solid resource for engineers, researchers, scientists, students, and educators involved in information technology, computer science, electrical engineering, and mechanical engineering. It will also be useful reading for anyone interested in learning more about the growing field of radio spectrum management using cognitive radio and software defined radio technologies.

S. Hossein Mousavinezhad
Idaho State University, USA

S. Hossein Mousavinezhad, PhD, Professor, and past Chair, Electrical Engineering, College of Science and Engineering, Idaho State University, is an active member of IEEE and ASEE, having chaired sessions in national/international and regional conferences. He is an ABET Program Evaluator (PEV) for Electrical Engineering and Computer Engineering. He is the Founding General Chair of the IEEE International Electro Information Technology Conferences, www.eit-conference.org. He is IEEE Education Society Vice President and Van Valkenburg Early Career Teaching Award Chair. He was the ECE Program Chair of the 2002 ASEE Annual Conference, Montreal, Quebec, June 16-19 and serves on the division's ExComm. Professor Mousavinezhad received Michigan State University ECE Department's Distinguished Alumni Award, May 2009, ASEE ECE Division's 2007 Meritorious Service Award, ASEE/NCS Distinguished Service Award, April 6, 2002, for significant and sustained leadership. In 2010 and 1994, he received ASEE PNW Section and Zone II Outstanding Campus Representative Awards, respectively. He is also a Senior Member of IEEE, has been a reviewer for IEEE Transactions. His teaching and research interests include Digital Signal Processing (DSP) and Bioelectromagnetics. He has been a reviewer for engineering textbooks including DSP First by McClellan, Schafer, and Yoder, published by Prentice Hall, 1998, and Signal Processing First, Prentice Hall, 2003. He is head of WECEDHA. Hossein is a member of the Editorial Advisory Board of the international research journal Integrated Computer-Aided Engineering. He has published a book chapter, IGI Global. Dr. Mousavinezhad was part of the group promoting economic development in Michigan, MEDC, and was responsible for bringing Innovation Forums to Western Michigan University, January 21, 1999. These forums were a series of meetings and seminars focused on university and industry collaboration initiated by the Michigan Governor. The Forums were sponsored by the Kellogg and Dow Foundations and were designed for finding strategies to create more Hi-Tech jobs in the state. As part of his responsibilities as Professor and Chair of the ECE Department at Western Michigan University, he prepared ABET reports for the two programs offered by the Department (EE and CpE). The graduate programs offered by the ECE Department grew and he was responsible for initiating the first MSEE program in 1987. A new ECE PhD program was offered starting Fall 2002. In addition to administrative responsibilities, he has managed to teach undergraduate/graduate courses in his research area of Digital Signal Processing. He is co-PIs for DSP and globalization and Power/Energy grants funded by NSF. In addition, he received equipment grants from Texas Instruments in support of his teaching/research activities in the DSP field. During May 2009, he received Michigan State University's Electrical and Computer Engineering John D. Ryder Distinguished Alumni Award for contributions in furthering the mission of the department—which is to provide undergraduate and graduate education characterized by quality, access, and relevance, and to develop distinctive research programs in electro-sciences, systems, and computer engineering, with the promise of sustained excellence as measured in scholarship, external investment, reputation, and impact.

Preface

The tremendous increase in portable devices and computers has led to an ever-growing demand for greater data rates for wireless transmission, and thus to an increasing demand for spectrum channels. Conventionally, a licensed spectrum is assigned for comparatively long time spans and projected to be used solely by the license holder (primary user). This static assignment can create both a bottleneck and an under-utilized spectrum. Such inefficient utilization of inadequate wireless spectrum resources has motivated researchers and practitioners to look for advanced and innovative technologies that will enable more efficient use of spectrum resources in both a smarter and a more efficient manner.

INTRODUCTION

To manage the radio spectrum more efficiently, we must realize that since such utilization is dynamic, its management should also be dynamic. In 2003, the Federal Communications Commission (FCC) proposed the development of Dynamic Spectrum Access (DSA), also known as Opportunistic Spectrum Access (OSA). Since then, important efforts have been undertaken to turn this concept into reality. One of these efforts is Cognitive Radio (CR), which has appeared as a supporting platform for DSA. CR combines Artificial Intelligence (AI) with Software-Defined Radio (SDR) technology.

Cognitive Radio is an emerging technology, the primary objective of which is the most efficient utilization of the radio spectrum. A cognitive radio, built on a Software-Defined Radio (SDR), is defined as an intelligent wireless communication system that is aware of the environment, learns from it, and adapts to statistical variations in input stimuli to achieve two main purposes: highly reliable communication and efficient use of radio spectrum.

To accomplish its mission, a cognitive radio executes a series of processes known as a cognitive cycle. This cycle includes three stages: Observing, Decision Making, and Taking Action. Different processes take place at each stage of this cycle. These processes involve techniques drawn from different fields, including digital signal processing, estimation theory, and artificial intelligence. A summary of the main aspects of each of these is given next.

DYNAMIC SPECTRUM MANAGEMENT USING SOFTWARE-DEFINED AND COGNITIVE RADIO TECHNOLOGIES

A cognitive radio is aware of the context wherein it operates. This awareness includes knowledge of the environment, the communication requirements of the users, the regulatory policies, and its own capabilities. Spectrum sensing and channel estimation support the context awareness of a cognitive radio. Spectrum sensing is the process of obtaining awareness about the spectrum usage and the existence of primary users, incumbent users in a determined area, so secondary cognitive users can utilize empty channels, which in turn make the use of the radio spectrum more efficient. Channel estimation is the process of collecting Channel-State Information (CSI) to assess the channel capacity and its characteristics.

Sensing is critical in order to detect when a channel is being used by other users. In terms of spectrum management, sensing allows for the identification of spectrum holes to access the spectrum dynamically and provide dynamic spectrum management. In addition, reliable sensing can prevent interference. SNR and RSS (Received Strength Signal) are basic ways of sensing that help estimate how far apart the nodes are and determine if they are about to lose a connection. Knowing that information allows the cognitive radios to switch to a different frequency channel and either modify the modulation scheme or increase transmission power.

In cognitive radio systems, channel estimation is necessary for the most optimal adjustment of system parameters to changing conditions. In mobile communication systems, such as vehicle networks, the received signal strength oscillates, as the vehicle travels through interference patterns caused by multipath, shadowing due to obstacles, and change in distance between nodes. Generally, CR systems are designed to maximize their throughput and reliability for a given Quality of Service (QoS). This can be accomplished by adapting the system parameters to the fluctuations created by multipaths and shadowing. This process, however, requires estimation, prediction, and tracking of the received signal as accurately as possible. Two of the most popular estimation approaches are Bayesian estimation and maximum likelihood estimation.

The capability of making decisions is what distinguishes a cognitive radio from a conventional radio. It enables the cognitive radio to adapt itself to fulfill the specific requirements of a determinate application. For instance, if the radio starts experiencing problems from interference, the logical move is to switch to another channel. The CR needs a strategy, however, to decide when to switch the channel, determine the best channel to switch to, etc., always keeping in mind the goal. This goal could be maximizing throughput, reliability, or minimizing power consumption and/or delay. It also can be a combination of these features or others. All these features need to be quantifiable in order to formulate a mathematical procedure that can be precisely executed by a computer or computation device.

Making decisions is associated with other processes: orienting, planning, and learning. Orienting establishes priorities based on the observations. If the priority is normal, the next stage is planning, which implies generating and/or evaluating the alternatives. If the priority is high, then the next stage is making a decision on the resources to be allocated. Learning receives information from the other processes to build knowledge, and this knowledge feeds back to the system to refine the deciding process.

The Cognitive Engine (CE) is the entity in the CR that executes orienting, planning, deciding, and learning tasks. The CE takes the stimuli, analyzes them, and classifies the situation. The CE also determines the suitable response to the stimuli and decides how to reconfigure the system along with the SDR forms the CR. In the CR, the CE performs this task by using Artificial Intelligence (AI) techniques.

A CR takes action by configuring its transmission and receiving parameters to obtain a desired behavior that will accomplish a pre-determined goal or a set of goals. The actions that are executed center on two main activities. The first is shaping the transmission profile and configuring any pertinent radio parameters to use the resources given to the CR efficiently and simultaneously not interfere with the resources of other radios. The second action reshapes the transmission profile and reconfigures the parameters when the resources do change. The resources given to a CR are a set of frequencies, a set of time slots, and a set of antennas with beams pointed in different directions or any combination of these resources. The CE is the entity that ultimately decides which actions the CR must take.

ORGANIZATION OF THE BOOK

This handbook is divided into four sections: (1) radio spectrum sensing, (2) radio spectrum access and management, (3) software-defined radio and antennas for cognitive radio networks, and (4) models, security, and other related topics. Each section includes eight or nine chapters that offer basic research and case descriptions, as well as visionary ideas for future applications. The goal is to offer readers new insights on radio spectrum access and management issues and answer a broad array of questions related to these topics. Each section and chapter are briefly introduced next.

Radio Spectrum Sensing

This section describes the techniques related to radio spectrum sensing. Examples of topics covered include energy detection techniques, cooperative spectrum sensing with censoring of cognitive radios, tunable RF front-ends and robust sensing algorithms for cognitive radio receivers, energy-efficient partial-cooperative spectrum sensing in cognitive radio over fading channels, cyclostationary spectrum sensing, collaborative approaches for compressive spectrum sensing, spectrum sensing using principal components, and spectral sensing performance for feature-based signal detection with imperfect training.

Chapter 1: "Hybrid Cooperative Energy Detection Techniques in Cognitive Radio Networks"

An analysis of the Energy Detection (ED) test statistic is presented in this chapter. In addition, it identifies the general structure of ED threshold. Based on the double threshold ED, the authors analyze the performance of a Hybrid PSO-OR (Particle Swarm Optimization and OR) algorithm for cooperative spectrum sensing. The sensing decision of "fuzzy" secondary users is optimized using PSO and the final collective decision is made based on OR rule. The Hybrid PSO-OR algorithm is shown to exhibit significant performance gain over the Hybrid EGC-OR algorithm.

Chapter 2: "Cooperative Spectrum Sensing with Censoring of Cognitive Radios and MRC-Based Fusion in Fading and Shadowing Channels"

This chapter studies the performance of Cooperative Spectrum Sensing (CSS) with soft data fusion, given by Maximal Ratio Combining (MRC)-based fusion with Weibull-faded channels and log-normal shadowed channels. The performance of CSS with two censoring schemes, namely rank-based and threshold-based, is studied in the presence of Weibull-fading and log-normal shadowing in the reporting channels, considering MRC fusion at fusion center.

Chapter 3: "Tunable RF Front-Ends and Robust Sensing Algorithms for Cognitive Radio Receivers"

The concepts of Cognitive Radio (CR) and multi-dimensional spectrum sensing are introduced in this chapter. Spectrum sensing methodologies, energy-efficiency consideration, resources scheduling, and self-management and learning mechanisms in cognitive radio networks are also discussed. The entailed challenges of CR RF front-end architectures are looked into. The synthesis and design performance analysis of a tunable RF front-end sensing receiver for CR applications are presented.

Chapter 4: "Energy-Efficient Cooperative Spectrum Sensing for Cognitive Radio Networks"

Energy efficiency in cooperative spectrum sensing in cognitive radio is investigated in this chapter. The proposed approach aims at reducing the energy consumed in spectrum sensing and improving the resultant energy efficiency of the cognitive transmission. The approach is based on limiting the number of users that participate in the spectrum-sensing task. The participation decision of each user is taken individually by the user itself, where each user compares the expected amount of consumed energy to a predefined threshold. The expected energy consumption is estimated at each user based on its distance from the base station.

Chapter 5: "Cyclostationary Spectrum Sensing in Cognitive Radios at Low SNR Regimes"

This chapter reviews non-cooperative cyclostationary sensing approaches and reports recent advances in cooperative cyclostationary sensing algorithms. New results for cooperative cyclostationary spectrum sensing are then presented, which ensure better performance and faster and simpler operation. In the proposed schemes, each Secondary-User (SU) performs Single-Cycle (SC) cyclostationary detection for fast and simple implementation, while collaboration between SUs in final decision on the presence or absence of the PU is explored to improve its performance. Illustrative and analytical results show that the proposed schemes outperform both SC and Multi-Cycle (MC) cyclostationary detectors, especially in fading channels.

Chapter 6: "A Collaborative Approach for Compressive Spectrum Sensing"

This chapter discusses the motivation and challenges of utilizing collaborative approaches for compressive spectrum sensing. The authors survey the different approaches and the key published results in this domain and present in detail an approach that utilizes Kronecker sparsifying bases to exploit the two-dimensional sparse structure in the measured spectrum at different, spatially separated cognitive radios. Simulation results show that the presented scheme can substantially reduce the Mean Square Error (MSE) of the recovered power spectrum density over conventional schemes while maintaining the use of a low-rate Sub-Nyquist Analog to Information Converter.

Chapter 7: "Spectrum Sensing Using Principal Components for Multiple Antenna Cognitive Radios"

Contrary to the previous work where the main evaluation technique has been theoretical analysis and simulations, this chapter uses Software-Defined Radios (SDRs) with correlated signal reception capability to evaluate the sensing performance of the existing Covariance-Based Detection (CBD) techniques. The existing techniques considered in this work include Covariance Absolute Value (CAV), Maximum-Minimum Eigenvalue (MME), Energy with Minimum Eigenvalue (EME), and Maximum Eigenvalue Detection (MED). This chapter also presents a novel technique for blind signal detection that uses Principal Component (PC) analysis.

Chapter 8: "Spectral Sensing Performance for Feature-Based Signal Detection with Imperfect Training"

Cognitive radio is a technique proposed to overcome the problem of high-required transmission data rate through wireless networks with limited available radio spectrum. Low latency and accurate spectral sensing by secondary users is crucial for effective cognitive radio implementation. This work characterizes the uncertainty of the imperfect training data in terms of the effect on training time and detection performance. The trade-off between training time and detection performance is determined. Spectrum sensing is implemented in a two-stage detector, which performs both feature training and sensing functions.

Chapter 9: "Sensing Orders in Multi-User Cognitive Radio Networks"

In multi-channel Cognitive Radio Networks (CRNs), an important challenge is to determine a *sensing order* for each Cognitive User (CU) so as to optimize a given performance metric. The sensing-order problem is compounded in multi-user CRNs where the multiple users in the network could collide with each other. With the focus on multi-user CRNs, this chapter uses cognitive-throughput maximization as the performance metric and describes how the optimal sensing orders can be computed for different contention management strategies used by the network.

Radio Spectrum Management and Access

This section relates to radio spectrum access and management. Topics covered include competition-based channel selection, routing through efficient channel assignment, spectrum management through trading, cooperative and non-cooperative access techniques, sensing order techniques in multi-user environment, competition-based channel selection methods, asynchronous channel allocation, distributed mechanisms for multiple-channel acquisition, channel and performance studies for spectrum-sharing, and game theoretic approaches.

Chapter 10: "On Fuzzy Logic-Based Channel Selection in Cognitive Radio Networks"

This chapter is reporting a fuzzy logic-based decision-making algorithm for competition-based channel selection. The underlying decision criterion integrates both statistics of licensed users' channel occupancy and information about the competition level of unlicensed users. By using such an algorithm, the unlicensed user competitors can achieve an efficient sharing of the available channels. Simulation results are reported to demonstrate the performance and the effectiveness of the suggested algorithm.

Chapter 11: "Routing through Efficient Channel Assignment in Cognitive Radio Networks"

In this chapter, the authors focus on joint channel assignment and routing in cognitive radio networks, providing a comprehensive survey on routing and channel assignment in CRNs. First, the importance of joint channel assignment and routing for successful communication in cognitive radio networks is discussed. Then, classification and challenges related to channel assignment and routing are discussed in detail.

Chapter 12: "Dynamic Spectrum Management Algorithms for Multiuser Communication Systems"

This chapter discusses various Dynamic Spectrum Management (DSM) algorithms, including Optimal Spectrum Balancing (OSB), Iterative Spectrum Balancing (ISB), Iterative Water-Filling (IWF), Selective Iterative Water-filling (SIW), Successive Convex Approximation for Low complExity (SCALE), the Difference of Convex functions Algorithm (DCA), Distributed Spectrum Balancing (DSB), Autonomous Spectrum Balancing (ASB), and Constant Offset ASB using Multiple Reference Users (ASB-MRU). They are compared in terms of their performance (achievable data-rate) by extensive simulation results and their computational complexity.

Chapter 13: "Performance Studies for Spectrum-Sharing Cognitive Radios under Outage Probability Constraint"

This chapter investigates the performance of primary and secondary users in a spectrum-sharing cognitive environment. In this environment, multiple secondary users compete to share a channel dedicated to a primary user in order to transmit their data to a receiver unit. Only one secondary user is scheduled to share the channel, and to do so, the transmission power of the scheduled secondary user should satisfy the outage probability requirement of the primary user. Secondary users are ranked according to their channel strength, and performance measures are derived as a function of a generic channel rank.

Chapter 14: "Distributed Mechanisms for Multiple Channel Acquisition in a System of Uncoordinated Cognitive Radio Networks"

In this chapter, the authors consider a system of Cognitive Radio (CR) networks, where networks cannot communicate with one another and are incapable of implementing a specific and global communication protocol. They are concerned with the issue of coexistence in a decentralized system of CR networks.

The authors discuss channel acquisition mechanisms that can help CR networks maximize utility and minimize contention. The channel acquisition mechanisms are well suited for an uncoordinated system of CR networks. The mechanisms discussed here ensure fast convergence of the system by minimizing the contention experienced until a stable state of equilibrium is attained.

Chapter 15: "Asynchronous Channel Allocation in Opportunistic Cognitive Radio Networks"

The channel service quality, and the neighborhood discovery (NB) phase are fundamental and challenging due to the dynamics of cognitive radio networks. The authors provide an analysis of these challenges, controversies, and problems, and review the state-of-the-art literature. They show that, although recently there has been a proliferation of NB protocols, there is no optimal solution meeting all required expectations of cognitive radio users.

Chapter 16: "Cooperative Cognitive Radio Networking: Towards a New Paradigm for Dynamic Spectrum Access"

In this chapter, the authors first provide a systematic study on cooperative cognitive radio networking. As an effort to shed light on addressing spectrum-energy inefficiency at a low complexity, an orthogonal modulation-enabled two-phase cooperation framework and an orthogonally dual-polarized antenna-based framework, as well as their resource allocation problems, are given and tackled.

Chapter 17: "Throughput-Efficient Spectrum Access in Cognitive Radio Networks: A Coalitional Game Theoretic Approach"

This chapter addresses the problem of throughput-efficient spectrum access in cognitive radio networks using coalitional game-theoretic framework. The authors model the problem of joint Coalition Formation (CF) and Bandwidth (BW) allocation as a CF game in partition form with non-transferable utility and present a variety of algorithms to dynamically share the available spectrum resources among competing Secondary Users (SUs). Performance analysis shows that the CF algorithms with optimal BW allocation provides a substantial gain in the network throughput over existing coalition formation techniques as well as the simple cases of singleton and grand coalition.

Chapter 18: "Advanced Cognitive Radio-Enabled Spectrum Management"

In this chapter, motivated by the recent advanced cognitive radio-enabled spectrum management schemes, the authors first summarize the current various advanced and flexible spectrum management schemes, including spectrum trading, leasing, pricing, and harvesting, and then analyze their advantages and disadvantages. Then, they take the viewpoints of both the spectrum marketing perspective and spectrum technical perspective, and they propose the centralized and distributed dynamic spectrum-sharing schemes, respectively.

Software-Defined Radio and Antennas for Cognitive Radio Networks

This third section describes novel antennas for cognitive radio networks as well as cognitive radio techniques and their implementation, using Software-Defined Radio technology. The topics covered include fundamentals of Software-Defined Radio and cooperative spectrum sensing, future reconfigurable radio front-ends for cognitive radio and Software-Defined Radio, precoder design for cognitive multiuser multi-way relay systems, reconfigurable antennas for flexible radio front-end, and complexity issues related to eigenvalue-based, multi-antenna spectrum sensing.

Chapter 19: "Fundamentals of Software-Defined Radio and Cooperative Spectrum Sensing: A Step Ahead of Cognitive Radio Networks"

Cognitive Radio (CR) adapts itself to the newer environment on the basis of its intelligent sensing and captures the best available spectrum to meet user communication requirements. The performance of secondary systems could be enhanced by a Cooperative Spectrum Sensing (CSS) approach, as it increases the probability of detection of primary activities. This chapter is focused on software-defined radio, its architecture, limitations, then evolution to cognitive radio network, architecture of the CR, and its relevance in the wireless and mobile Ad-hoc networks.

Chapter 20: "Future Reconfigurable Radio Frontends for Cognitive Radio and Software-Defined Radio: From Functional Materials to Spectrum Management"

A solid design of reconfigurable frontends, from the RF part to the digital baseband, should take into account different criteria to better exploit the available spectrum. In this chapter, architectures for the implementation and integration of future reconfigurable RF-frontends are presented. Furthermore, a general perspective to achieve smarter air interfaces is studied and discussed by setting out different strategies based on CR and SDR techniques.

Chapter 21: "Precoder Design for Cognitive Multiuser Multi-Way Relay Systems Using MSE Criterion"

This chapter discusses the design of precoders in cognitive multi-user multi-way relay systems. The authors discuss one possible solution of using a relay station as well as multiple antenna techniques. Precoding design in such a relay-supported multiple antenna secondary network is presented based on the Mean Square Error (MSE) design criterion. An iterative algorithm is proposed to iteratively optimize the precoding matrices at secondary transmitters, the precoding matrix at the secondary relay, and decoding matrices at secondary receivers. The design objective is to minimize the sum MSE of all received signals under transmit power constraints at each secondary transmitter as well as the relay station, while the interference to primary network is nulled out or kept under a certain level.

Chapter 22: "Compact and Efficient Reconfigurable Antennas for Flexible Radio Front-End in Cognitive Radio Systems"

The aim of this chapter is to investigate possible roles that different categories of reconfigurable antennas can play in cognitive and smart radio. Hence, this chapter focuses on investigating some novel methods to frequency-reconfigure compact ultra-wideband antennas to work in different bands; this will offer additional filtering to the radio front-end. Furthermore, the design of novel pattern and polarization reconfigurable antennas will also be investigated to assist cognitive radio through spatial rather than frequency means.

Chapter 23: "Complexity Issues within Eigenvalue-Based Multi-Antenna Spectrum Sensing"

This chapter provides deep insight into multi-antenna eigenvalue-based spectrum-sensing algorithms from a complexity point of view. A review of eigenvalue-based spectrum-sensing algorithms is provided. The chapter presents a finite computational complexity analysis and a comparison of the Maximum to Minimum Eigenvalue (MME) detector and a simplified variant of the Multiple Beamforming detector as well as the approximated MME method. It is shown that the complexity/reliability tradeoff is a difficult challenge within spectrum sensing, given the strong requirements on sensing duration and detection performance.

Chapter 24: "Dual-Hop and Multi-Hop Cooperative Spectrum Sensing with an Improved Energy Detector and Multiple Antennae-Based Secondary Users"

A dual-hop cooperative spectrum-sensing approach is studied in detail, where each cooperative Cognitive Radio (CR) makes a binary decision based on the local observation, by using an improved energy detector, and then forwards it to a common receiver. At the common receiver, all binary decisions are fused together. The authors provide an analytical framework for the analysis of performance of the improved energy detector-based cooperative CR network.

Chapter 25: "Cognitive Radio Programming Survey"

Based on the authors' analysis, the success of cognitive radio heavily depends on Software-Defined Radio (SDR). The cost, performance, and power consumption of SDR hardware platforms will enable (or forbid) smart radio applications and cognitive radio networks. SDR has evolved rapidly and is now reaching market maturity, but many issues have yet to be studied. In this chapter, the authors highlight how hardware architectures fulfill the constraints imposed by recent radio protocols, and they present current architectures and solutions for programming SDR. The authors also list the challenges to overcome in order to program future cognitive radio systems.

Chapter 26: "Cross-Layer Optimization and Link Adaptation in Cognitive Radios"

This chapter discusses the Adaptive Discrete Particle Swarm Optimization (ADPSO) algorithm as an efficient algorithm for optimizing and adapting CR operating parameters from physical, MAC, and network layers. In addition, the authors present two extensions for the proposed algorithm. The first one is Automatic Repeat reQuest-ADPSO (ARQ-ADPSO) for efficient spectrum utilization. The second one is merging ARQ-ADPSO and Case-Based Reasoning (CBR) algorithms for autonomous link adaptation under dynamic radio environment. The simulation results show improvements in the convergence time, signaling overhead, and spectrum utilization compared to the well-known optimization algorithms such as the Genetic Algorithm (GA).

Chapter 27: "Interference Statistics and Capacity-Outage Analysis in Cognitive Radio Networks"

This chapter presents a study on the interference caused by Secondary Users (SUs) due to miss-detection errors and its effects on the capacity-outage performance of the Primary User (PU) in a cognitive radio network assuming Rayleigh and Nakagami fading channels. The effect of beacon transmitter placement on aggregate interference distribution and capacity-outage performance is studied considering two scenarios of beacon transmitter placement: a beacon transmitter located at a PU transmitter and at a PU receiver. It is shown that the beacon transmitter at the PU receiver imposes less interference and, hence, better capacity-outage probability to the PU than the beacon transmitter at the PU transmitter.

Models, Security, and Other Related Topics

This section includes the chapters on modeling, security, pricing, and applications. The topics covered include modeling and performance evaluations, nonparametric Bayesian prediction of primary user air traffics, interacting particle system approaches, analysis of multiple cognitive radio networks and their coexistence, analysis of security issues and solutions, spectrum trading, competitive spectrum pricing under a centralized dynamic spectrum allocation, and applications of cognitive radio networks.

Chapter 28: "Modeling and Performance Evaluation"

In the literature, there are two methodologies: queuing theory/Markov chain-based analysis and stochastic network calculus-based analysis. These two methodologies rely on different mathematical basics and modeling approaches. Thus, they lead to different output metrics on various viewpoints. This chapter aims to give an overall introduction to both methodologies. First, the fundamental models used in queuing/Markov chain-based analysis will be presented, followed by its applications in cognitive radio networks. Then, network calculus basics are introduced with the modeling and application in performance analysis of the cognitive radio network.

Chapter 29: "Nonparametric Bayesian Prediction of Primary Users' Air Traffics in Cognitive Radio Networks"

In cognitive radio networks, a secondary user needs to estimate the primary users' air traffic patterns to optimize its transmission strategy. In this chapter, the authors describe a nonparametric Bayesian method for identifying and clustering traffic applications. In the proposed algorithm, the collapsed Gibbs sampler is applied to cluster the air traffic applications using the infinite Gaussian mixture model over the feature space of the packet length, the packet inter-arrival time, and the variance of packet lengths. The authors analyze the effectiveness of their proposed technique by extensive simulation using the measured data obtained from the WiMax networks.

Chapter 30: "Risk Engine Design as a Key Security Enhancement to the Standard Architecture for Cognitive Radio"

This chapter describes a *risk engine* that can incorporate a risk assessment cognition cycle. In various business sectors, risk management is the preferred mechanism to address unknown conditions and therefore offers promise in this context. The chapter describes how the risk engine can potentially address the vulnerabilities inherent to radio operation in the sensing/perception of spectrum, in the cognition cycle, or in the device infrastructure. It highlights some well-defined threats, their associated countermeasures, and suggests conceptual approaches for a risk engine to intervene in those scenarios.

Chapter 31: "Towards Security Issues and Solutions in Cognitive Radio Networks"

This chapter discusses the security issues in cognitive radio networks, and then it presents an intensive list of main known security threats in Cognitive Radio Networks (CRN) at various layers and the adverse effects on performance due to such threats, and the current existing paradigms to mitigate such issues and threats. Finally, the authors highlight proposed directions in order to make CRN more authenticated, reliable, and secure.

Chapter 32: "Heterogeneous Service-Oriented Spectrum Trading"

This chapter proposes an algorithm called HSO-ST (Heterogeneous Service-Oriented Spectrum Trading) with the target of maximum matching number under the priority restriction. This algorithm can satisfy more secondary users. Compared with other spectrum-trading strategies, HSO-ST can improve the spectrum demand-matching ratio greatly.

Chapter 33: "Exploiting Polarization for Spectrum Awareness in Cognitive Satellite Communications"

The authors firstly provide an overview of the existing works in polarization-based spectrum sharing. Secondly, they present the theoretical analysis of Energy Detection technique for dual polarized Additive White Gaussian Noise (AWGN) and Rayleigh fading channels considering the spectral coexistence scenarios of dual and hybrid satellite systems. Finally, the authors provide the comparison of different combining techniques in terms of the sensing performance in the considered dual polarized channels with the help of theoretical analysis and numerical results.

Chapter 34: "Competitive Spectrum Pricing under Centralized Dynamic Spectrum Allocation"

This chapter models the dynamic spectrum allocation problem in wireless networks with a centralized spectrum broker, who manages "white space" in the spectrum of TV broadcasters in a given area and sells the vacant spectrum bands for revenue to multiple WSPs, as a multi-stage non-cooperative dynamic game. The simulation results show that the centralized spectrum allocation mechanism with dynamic pricing achieves a dynamic spectrum allocation implementation that is responsive to market conditions as well as enabling efficient utilization of the available spectrum.

Chapter 35: "Cognitive Radio Techniques for M2M Environments"

A new paradigm, called Cognitive Machine to Machine (CM2M) communications, has been recently considered to exploit cognitive/opportunistic radio communications. After having introduced the problem of applying cognitive techniques in M2M scenarios, the authors focus their attention on the Medium Access Control protocols for CM2M scenarios, with a particular attention on the OFDMA-based primary systems. Among other approaches, the authors focus on a data-aided approach for the access of the secondary devices, aiming to reduce the interference toward the primary system.

SUMMARY

Spectrum management and cognitive radio are new areas that span several fields, including information technology, computer science, computer engineering, and electrical engineering. Only a few books related to these areas have been published thus far. This book will be the first that covers all the concepts related to spectrum management and cognitive radio that are supported by the software that defines radio technology. It will provide timely, important technologies and methods of spectrum management and cognitive radio and be a valuable reference book for educators, researchers, practitioners, and graduate students.

The handbook covers a wide range of topics, including channel estimation and characterization, spectrum sensing, decision-making, antenna design, security in cognitive radio networks, and models. Books that cover all these topics are not currently available. This handbook offers a collection of engineering and computer science articles written by well-established researchers and industrials who have considerable expertise in cognitive radio, wireless communications, and electromagnetics.

Finally, this handbook is a timely contribution to assist researchers, students, engineers, educators, indeed anyone interested in radio spectrum, its management, and working in information technology, computer science, electrical engineering, and/or mechanical engineering. It offers chapters on current cutting-edge research into techniques, trends, and practical applications in the growing field of cognitive radio and radio spectrum management. It also discusses the most up-to-date research on radio spectrum sensing, access, management, security, models, antennas for cognitive radio networks, and their applications.

Naima Kaabouch
University of North Dakota, USA

Wen-Chen Hu
University of North Dakota, USA

Acknowledgment

Dynamic spectrum management is critical for the mobile age because of limited radio frequencies. This handbook discusses various issues related to dynamic spectrum management using software-defined and cognitive radio technologies. It was a substantial project and took us more than one year to finish, from July 2013 to August 2014. Owing to the overwhelming responses and submissions, the project became a handbook. The editors spent a great deal of time communicating with (potential) authors and reviewers via numerous emails, organizing, and managing this handbook. The successful accomplishment of this handbook is a credit to many people. It consists of 4 sections and 35 chapters contributed by more than 100 world-renowned authors. The editors thank the authors for their quality work and great effort to revise their chapters based on the reviewers' comments. The reviewers, who provided such helpful feedback and detailed comments, are particularly appreciated. Special thanks go to the staff at IGI Global, especially to Allison McGinniss, Mehdi Khosrow-Pour, and Jan Travers. Finally, the biggest thanks goes to our family members for their love and support throughout this project.

Naima Kaabouch
University of North Dakota, USA

Wen-Chen Hu
University of North Dakota, USA

Section 1
Radio Spectrum Sensing

Chapter 1
Hybrid Cooperative Energy Detection Techniques in Cognitive Radio Networks

Raza Umar
King Saud University, Saudi Arabia

Mohamed Deriche
King Fahd University of Petroleum and Minerals, Saudi Arabia

Fahham Mohammed
King Fahd University of Petroleum and Minerals, Saudi Arabia

Asrar U. H. Sheikh
The University of Lahore, Pakistan

ABSTRACT

Cognitive Radio (CR) has emerged as a smart solution to spectrum bottleneck faced by current wireless services under which licensed spectrum is made available to unlicensed Secondary Users (SUs) through robust and efficient Spectrum Sensing (SS). Energy Detection (ED) is the dominantly used SS approach owing to its low computational complexity and ability to identify spectrum holes without requiring a priori knowledge of primary transmission characteristics. In this chapter, the authors present an in-depth analysis of the ED test statistic. Based on the double threshold ED, they analyze the performance of a Hybrid PSO-OR (Particle Swarm Optimization and OR) algorithm for cooperative SS. The sensing decision of "fuzzy" SUs is optimized using PSO and the final collective decision is made based on OR rule. The idea of using two thresholds is introduced to reduce the communication overhead in reporting local data/decision to the fusion center, which also offers reduced energy consumption. The Hybrid PSO-OR algorithm is shown to exhibit significant performance gain over the Hybrid EGC-OR algorithm.

INTRODUCTION

Current spectrum allocation policy adopted by the government spectrum regulatory agencies provides each wireless service provider with a license to operate within a particular frequency band in one geographical location. With the focus shifting to new multimedia services, demand for higher bandwidth allocation has increased. As the radio spectrum is limited, the present scenario doesn't allow the wireless systems to adapt to these fast changing demands. As a result, Federal Communications Commission (FCC) car-

DOI: 10.4018/978-1-4666-6571-2.ch001

ried out a number of studies to investigate current spectrum scarcity with the goal to optimally manage available radio resources. Recent measurements have revealed that a large portion of assigned spectrum is sporadically utilized. According to FCC (2003a) notice of proposed rulemaking and order, spectrum utilization varies from 15% to 85% with wide variance in time and space. This suggests that the root cause of current spectrum scarcity is not the physical shortage of spectrum rather it is inefficient fixed spectrum allocation. This fact questioned the effectiveness of traditional spectrum policies and opened doors to a new communication paradigm to exploit radio resources dynamically and opportunistically.

Dynamic and Opportunistic Spectrum Access (DOSA) is proposed to be the solution for inefficient spectrum utilization wherein unlicensed users are allowed to opportunistically access the un-used licensed spectrum under the stringent requirement of avoiding interference to the licensed users of that spectrum. In this way, the NeXt Generation (xG) wireless networks based on DOSA techniques are proposed to meet the requirements of wireless users over heterogeneous wireless architectures by making them intelligently interact with their radio environment. Cognitive Radio, which was first discussed in (Mitola & Maguire, 1999) is seen as an important technology that enables xG network to use the spectrum dynamically and opportunistically. The key component of CR technology is the ability to measure, sense and ultimately adapt to the radio's operating environment. In CR terminology, the users with legacy rights on the usage of specific part of the spectrum are called *primary users* (PU) while the term *secondary users* (SU) is reserved for low-priority un-licensed users which are equipped with a cognitive capability to exploit this spectrum without being noticed by PU. Therefore, as identified by Raza Umar & Sheikh (2012), the fundamental task of SU (also termed as simply CR in literature) is to reliably sense the spectrum with an objective to identify a vacant band and to update its transmission parameters to exploit the unused part of the spectrum in such a way that it does not interfere with PU.

Being the focus of this chapter, we identify *Spectrum sensing* (SS) as the key cognitive functionality. Spectrum sensing in essence is the task of obtaining awareness about the spectrum usage at a specific time in a given geographical region. Intuitively this awareness can be obtained by using beacons or geo-location and database. These approaches though appear simple but are practically infeasible because of prohibitively large infrastructure requirements and implementation complexity. In this article we focus on local spectrum sensing at CR based on primary transmitter detection. To achieve spectrum efficiency, spectrum sensing must be performed by secondary users continuously to use the licensed band whenever the primary user is absent, however; it should also minimize interference with the PU (Raza Umar & Sheikh (2012b).

A review on spectrum sensing techniques for Cognitive Radio Networks (CRNs), with a special focus on sensing methods that need little or no information about the primary signals and the propagation channels, is presented in (Yonghong Zeng, Liang, Hoang, & Zhang, 2010). A comparative study of these schemes (Raza Umar & Sheikh, 2012a) reveals that energy detection (ED) is the most widely used spectrum sensing approach since it does not need any a priori information about the primary transmission characteristics, is easy to implement, and has low computational complexity while being optimal for detecting independent and identically distributed (IID) primary signals. Furthermore, the importance of ED based sensing is evident from the fact that most of the cooperative sensing techniques reported in the recent literature (Arshad, Imran, & Moessner, 2010; Saad, Han, Basar, Debbah, & Hjorungnes, 2011) employ ED for local detection of primary transmissions in addition to a variety of two-stage sensing schemes which perform at the first stage coarse detection based on ED (Maleki, Pandharipande, & Leus, 2010).

In this chapter, we review energy detection based spectrum sensing and present an in depth analysis of the test statistic for energy detector. General structure of the test statistic and corresponding threshold are presented to address existing ambiguities in the literature. The derivation of exact distribution of the test statistic, reported in the literature, is revisited and hidden assumptions on the primary user signal model are unveiled. In our sensing model, we focus on cooperative energy detection. Energy detectors measure the energy of the signal by sensing the targeted frequency band for a fixed amount of sensing time. The received energy is compared with a fixed threshold λ. This threshold value depends mainly on the noise floor (Urkowitz, 1967). If the received energy is more than the threshold value, it indicates that PU is present, otherwise PU is considered to be absent. However, due to noise uncertainties and other environmental effects, the observed energy values in the vicinity of the threshold λ are not quite reliable. Decisions from such CRs become unreliable, and hence, they are referred as Fuzzy CRs. Here, the double threshold energy detector proposed earlier in (Chunhua Sun, Zhang, & Ben Letaief, 2007) is implemented. The reliable CRs report their local binary decisions, while the Fuzzy CRs directly report their energy observations to the fusion centre. The Particle Swarm Optimization (PSO) technique is then used to optimally combine the received energy values from the Fuzzy CRs. In this fashion, the number of sensing bits over the reporting channel as well as the energy consumed to transmit these bits is reduced at the cost of negligible performance degradation compared to conventional data fusion technique using PSO (Zheng, Lou, & Yang, 2010).

BACKGROUND

Transmitter Detection

ED was first discussed by Urkowitz in his classic paper (Urkowitz, 1967) as binary hypothesis testing problem for the detection of deterministic signals in white Gaussian noise which was further investigated by (Fadel F Digham, Alouini, & Simon, 2007) for unknown deterministic signals operating over Rayleigh and Nakagami fading channels. ED based spectrum sensing model in CRN aims at deciding between two hypotheses (PU signal absent (H_0) and PU signal present (H_1)) and is based on the fundamental results presented by Urkowitz and Digham et al. with three characteristic features: (1) The source (primary) signal to be detected is unknown narrowband deterministic. Although the spectral region, specified by central frequency (f_c) and bandwidth (W), to which the primary transmissions are confined is assumed to be known a priori but its distinguishing parameters are not available. (2) The received waveform is constructed from the limited number of samples (N) of band limited signal observed during the sensing duration time (T). Energy in the received waveform is normalized by the two-sided (receiver) noise power spectral density ($N_0 / 2$) to provide a dimensionless test metric. This metric is then compared with a threshold to decide about the primary signal presence. In discrete domain, this translates to defining test statistic as sum of the squares of N received samples scaled by noise variance (σ_n^2). (3) Noise is considered to be a white Gaussian process and the exact noise power spectral density or variance (power) is known a priori. The validity of the above mentioned fundamental assumptions has been challenged in recent literature from different perspectives. In the following, we briefly describe some of the ambiguities/conflicts in the reported research activities and highlight critical hidden assumptions in ED based spectrum sensing.

Ambiguities in ED Test Statistic

The decision metric for ED, in principle, is the energy content in the received signal at CR. However, there exists a noticeable ambiguity in defining the exact test statistic for ED in the literature. The classical work of (Fadel F Digham et al., 2007; Urkowitz, 1967) and some recent publications like (A. Ghasemi & Sousa, 2005; Amir Ghasemi & Sousa, 2007) belong to class of techniques that normalize energy in the received samples by noise variance to get the test statistic. Whereas, other authors like (Ye, Memik, & Grosspietsch, 2008; Y. Zeng, Liang, Hoang, & Peh, 2009) define the average energy in the received samples as the decision metric i.e. they scale the energy in the received samples by the number of observed samples to make a decision on the presence/absence of primary signal which, in fact, becomes the measure of power in the received signal. On the other hand, authors like (Sonnenschein & Fishman, 1992) consider unscaled version of energy content in the received samples as the test statistic. As a result of different scaling factors employed in test statistics, we will see that the probability of detection P_d and the probability of false alarm P_f are different across various approaches yielding to a source of confusion for novice researchers in the field of spectrum sensing.

Hidden Assumptions on PU Signal Model

The exact distribution of the ED test statistic depends on the nature of noise and PU signal. The classical results on ED by (Urkowitz, 1967) were developed for radar applications where the deterministic source signal is to be detected in the presence of white Gaussian noise. Many authors (Fadel F Digham et al., 2007; A. Ghasemi & Sousa, 2005; Amir Ghasemi & Sousa, 2007) and others used the results reported by Urkowitz under the assumption that the probability of detection can be considered as a conditional probability. However, we will show that this is only possible when the unknown PU signal originates from a circular constellation as illustrated later in the chapter.

Cooperative Spectrum Sensing

The most serious limitation of transmitter detection approach is its degraded performance in multi-path fading and shadowed environments. This problem can be solved by exploiting the inherent spatial diversity in a multi-user environment resulting from the fact that if some SUs are in deep fade or observe severe shadowing, there might be other SUs in the network that may have a good line-of-sight to Primary transmitter. Cooperative detection results in much improved sensing performance of CR network by improving the detection probability while simultaneously decreasing the probabilities of miss-detection and false alarms. The improvement in performance as a result of cooperation is referred to as *cooperative gain* (Akyildiz, Lo, & Balakrishnan, 2011). Cooperative gain is not only about the improvement in detection performance, it can also be seen in terms of sensing hardware. In addition, cooperative detection solves the hidden PU problem. This result in decreased sensitivity requirements for CR added with improved agility. Consequently, combining the sensing information from different CRs, results in more reliable spectrum awareness. This gives rise to the concept of cooperative spectrum sensing wherein CRs employing different technologies, exchange information about the time and frequency usage of spectrum to exploit any vacant spectrum opportunity efficiently. Researchers, in (F.F. Digham, Alouini, & Simon, 2003), (Amir Ghasemi & Sousa, 2007) and (Ben Letaief, 2009) have proposed different cooperative spectrum sensing approaches to improve PU detection.

Cooperation among CRs can be implemented in either centralized or distributed manner. In distributed cooperation, each cooperating CR decides about the spectrum opportunity locally after gathering spectrum observations from other CRs in the network. The shared spectrum observations may be in the form of *soft sensing* results or *quantized* (*binary/hard*) version of local decisions about spectrum hole availability. On the other hand, in centralized cooperation, each CR performs local spectrum sensing, and sends its local binary decision (1 for hypothesis H_1 and 0 for hypothesis H_0) or its energy value to the fusion centre. The fusion centre would then fuse all the local decisions or energies and make a final decision on whether the PU is present or not. Here, the central unit, also called the *fusion center*, decides about the spectrum hole after collecting local sensing results from cooperating SUs. This spectrum opportunity is then either broadcasted to all CRs or central unit itself controls the CR traffic by managing the detected spectrum usage opportunity in an optimum fashion. This central node is an *Access Point* (*AP*) in a Wireless Local Area Network (WLAN) or a *Base Station* (*BS*) in a cellular network. In cooperative sensing, CRs need a control channel to share local spectrum sensing results and frequency band allocation information with each other. This control channel, depending upon system requirements, can be implemented using a dedicated band, an un-licensed band such as ISM or an underlay system such as Ultra Wide Band (UWB) (Cabric, Mishra, Willkomm, Brodersen, & Wolisz, 2005). In this chapter, we will use the OR-rule for decision fusion in centralized sensing, as it is very conservative for accessing the targeted band, and minimizes the interference caused to the PU (Ben Letaief, 2009).

ENERGY DETECTION BASED SPECTRUM SENSING

Spectrum sensing is the task of obtaining awareness of spectrum usage. Practically, CR bases its decision about the presence or absence of PU on locally observed signal samples which can be formulated as a binary hypothesis testing problem as follows (R Umar & Sheikh, 2012).

$$x(k) = \begin{cases} n(k), & H_0 \\ hs(k) + n(k), & H_1, \quad k = 1, \cdots, N \end{cases} \tag{1}$$

Here, $x(k)$ is the received signal sample at CR, $n(k)$ represents the additive white Gaussian noise (AWGN) with zero mean and variance σ_n^2, $s(k)$ represents the PU transmitted signal sample which is to be detected and h is the channel gain. N denotes the number of samples of observed signal of bandwidth W for T seconds, mathematically given by

$$N = 2TW = 2m\,,$$

where m represents the time-bandwidth product. This is a binary signal detection problem in which CR has to decide between two hypotheses, H_0 (band vacant) and H_1 (band occupied).

Appropriate PU Signal Model and Optimal Signal Detection

The classical Neyman-Pearson (NP) approach (Kay, 1998) to hypothesis testing defines the test statistic as likelihood ratio test (LRT) given by:

$$u_{LRT} = \frac{p(\boldsymbol{x} \mid H_1)}{p(\boldsymbol{x} \mid H_0)} \tag{2}$$

Here,

$$\boldsymbol{x} = \left[x(1), x(2) \ldots x(N) \right]$$

represents the set of received signal samples at CR and $p(.)$ denotes the pdf. According to the NP theorem (Kay, 1998), LRT maximizes the probability of detection for a given false alarm rate provided the exact distribution of $\boldsymbol{x}$ under both H_0 and H_1 is known *a priori*. As evident from (1), the distribution of $\boldsymbol{x}$ under H_0 depends on noise distribution, while the distribution of $\boldsymbol{x}$ under H_1 is related to the wireless channel, PU transmitted signal distribution and the noise distribution.

If we assume that the channel is AWGN ($h = constant = 1$) and the primary signal samples ($s(k) \forall k$) are independent, the pdf in (2) can be decoupled as:

$$\begin{aligned} p(\boldsymbol{x} \mid H_0) &= \prod_{k=1}^{N} p\left(x(k) \mid H_0 \right) \\ p(\boldsymbol{x} \mid H_1) &= \prod_{k=1}^{N} p\left(x(k) \mid H_1 \right) \end{aligned} \tag{3}$$

The noise samples are typically assumed to be IID with Gaussian distribution (resulting from central limit theorem owing to various independent sources of noise) i.e.

$$n(k) \sim N\left(0, \sigma_n^2\right).$$

If we further assume that the signal samples are also Gaussian (Yonghong Zeng et al., 2010) i.e. $s(k) \sim N\left(0, \sigma_s^2\right)$, the LRT becomes the estimator-correlator (EC) detector (Kay, 1998), wherein the energy content in the received samples is used as the test statistic for detecting any PU activity with maximum probability of detection i.e.

$$u = \sum_{k=1}^{N} \boldsymbol{x}^T(k)\boldsymbol{x}(k) = \sum_{k=1}^{N} x^2(k) \qquad (4)$$

Hence, energy detection is optimal for detecting IID signals.

It is important to highlight here that, in general, an EC detector requires the source signal covariance matrix (related to unknown channel) and noise power (Kay, 1998). However, when the PU signal presence is unknown yet, it is unrealistic to rely on signal covariance matrix for its detection (Yonghong Zeng et al., 2010), and hence, PU signal is usually modeled as an IID Gaussian random process and ED is widely applied for its detection.

Energy Detector

The block diagram of energy detector is depicted in Figure 1.

The input band-pass filter removes the out of band signals based on spectrum of interest, known to be centered around f_c and spanning over bandwidth W. The filtered received signal $x(t)$ is digitized by an analog to digital converter (ADC) and a simple squaring device followed by an accumulator gives the energy content in N samples of $x(k)$, which acts as the test statistic for ED. This decision metric, u, is then compared with a threshold, λ, to decide if the scanned band is vacant (H_0) or occupied (H_1).

It is important to point out that in classical literature (Fadel F Digham et al., 2007; Urkowitz, 1967) etc., the energy detector measures the energy in the band limited ($bandwidth = W\,Hz$) received waveform $x(t)$ over sensing duration T and approximates this measure by the sum of squares of limited number of N received samples (see [11], p.524, Figure 1). However, in this chapter, the received signal is digitized (by sampling at the Nyquist rate i.e. $f_s = 2W$ samples/sec, where f_s is the sampling frequency which yields $N = 2WT$ samples for the observation window of T sec) after pre-filtering it with a band-pass filter of bandwidth W and instead of reconstructing the signal waveform to find its energy, the energy content in N received samples is used as the test statistic for ED as shown in Figure 1.1.

The performance of ED is gagged with its *specificity* and *sensitivity* (Yucek & Arslan, 2009) which are measured by two metrics: probability of false alarm P_f, which denotes the probability that the detection algorithm falsely decides that PU is present in the considered frequency band when it actually is absent, and probability of detection P_d, which represents the probability of correctly detecting the PU signal in the scanned frequency band. Mathematically, P_f and P_d are given by:

Figure 1. Block diagram of an energy detector

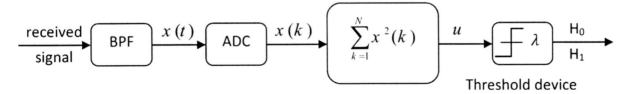

$$P_f = P(signal\ is\ detected\mid H_0) = P(u > \lambda \mid H_0)$$

$$= \int_\lambda^\infty f(u\mid H_0)du \qquad (5)$$

$$P_d = P(signal\ is\ detected\mid H_1) = P(u > \lambda \mid H_1)$$

$$= \int_\lambda^\infty f(u\mid H_1)du \qquad (6)$$

Here, $f(u\mid H_i)$ represents the pdf of test statistic under hypothesis H_i with $i = 0,1$. The exact and approximate distribution of test statistic is discussed in detail later in the chapter.

Thus, we target at maximizing P_d while minimizing P_f. The $P_d\ vs\ P_f$ plot depicts *Receiver Operating Characteristics* (ROC) and is considered as an important performance indicator. Achievable ROC curve strongly depends upon threshold λ. If λ is kept high, targeting minimum P_f, we end up in much decreased P_d, on the other hand, if value of λ is kept low, with an aim to get maximum P_d, resulting P_f exceeds the acceptable limits (Raza Umar & Sheikh, 2012). Hence, a careful tradeoff has to be considered while setting the ED threshold. In practice, if a certain spectrum re-uses probability of unused spectrum is targeted, P_f is fixed to a small value (e.g. $\leq 5\%$) and P_d is maximized. This is referred to as constant false alarm rate (CFAR) detection principle (Raza Umar & Sheikh, 2012a) in which λ is calculated using (5). However, if the CRN is required to guarantee a given non-interference probability P_d is fixed to a high value (e.g. $\geq 95\%$) and P_f is minimized. This requirement is met by evaluating λ based on (6) and this approach is known as constant detection rate (CDR) principle. As evident from (5) and (6), the derivations of λ are very similar for CFAR and CDR, so the analytical results derived under the assumption of CFAR can be applied to CDR based detection with minor modifications and vice versa (Raza Umar & Sheikh, 2012). In this chapter, the analysis is based on CFAR and the results can be applied to CDR based detection with minor modifications.

EXACT TEST STATISTIC DISTRIBUTION FOR ENERGY DETECTION

As indicated in the previous section, probability of false alarm and detection depend on the pdf of the test statistic under H_0 and H_1 respectively. Hence, accurate evaluation of P_f and P_d depend on the exact test statistic distribution. In the following, we will explore the exact expressions of P_f and P_d for deterministic PU signal model.

Exact P_f under AWGN

Under H_0,

$$x(k) = n(k) \sim N\left(0, \sigma_n^2\right)$$

where, without loss of generality, $n(k)$ is assumed to be Gaussian with zero mean and σ_n^2 variance. The test statistic u, is simply the sum of squares of N Gaussian random variables, each with zero mean and σ_n^2 variance. Hence, u normalized with σ_n^2 is said to have a *central Chi-square* distribution with N degrees of freedom:

$$H_0 : \frac{1}{\sigma_n^2} u = \sum_{k=1}^{N} \left(\frac{1}{\sigma_n} n(k) \right)^2$$

$$= \sum_{k=1}^{N} \left(y(k) \right)^2 \; where \;\; y(k) \sim N(0,1) \tag{7}$$

$$\sim \chi_N^2 \tag{8}$$

Using (5) and the fact that

$$f\left(\frac{1}{\sigma_n^2} u | H_0 \right) = \chi_N^2$$

and λ represents the threshold, P_f can be expressed as:

$$P_f = P\left(\frac{1}{\sigma_n^2} u > \frac{1}{\sigma_n^2} \lambda \mid H_0 \right)$$

$$= \int_{\frac{\lambda}{\sigma_n^2}}^{\infty} f(\frac{1}{\sigma_n^2} u \mid H_0) du \tag{9}$$

$$= Q_{\chi_N^2}\left(\frac{\lambda}{\sigma_n^2} \right) \tag{10}$$

Here,

$$Q_{\chi^2_N}\left(\frac{\lambda}{\sigma^2_n}\right)$$

indicates the right-tail probability for a χ^2_N random variable as given by ((Kay, 1998), pg. 52). Using the pdf of χ^2_N, the exact closed-form expression for P_f is:

$$P_f = \int_{\frac{1}{\sigma^2_n}\lambda}^{\infty} \frac{x^{\frac{N}{2}-1} e^{-\frac{x}{2}}}{2^{\frac{N}{2}} \Gamma(\frac{N}{2})} dx$$

With a change of variables $x/2 = y$, putting $N/2 = m$ and using the definition of incomplete Gamma function,

$$\Gamma\left(\alpha,\beta\right) = \int_{\beta}^{\infty} x^{\alpha-1} e^{-x} dx ,$$

we get:

$$P_f = \frac{1}{\Gamma(m)} \int_{\frac{1}{\sigma^2_n}\frac{\lambda}{2}}^{\infty} y^{m-1} e^{-y} dy = \frac{\Gamma(m,\frac{\lambda}{2\sigma^2_n})}{\Gamma(m)} \triangleq F_m(\frac{\lambda}{2\sigma^2_n}) \tag{11}$$

This result matches with the one derived by ((Fadel F Digham et al., 2007), (Equation (4)) and by ((Amir Ghasemi & Sousa, 2007), Equation (4) for $\sigma^2_n = 1$). The only difference is that they defined the scaled ED test statistic as $u_{scl} = u / \sigma^2_n$.

General Structure of ED Threshold and Resolving the Energy Scaling Conflict

The ED threshold for constant false alarm constraint can be derived from (11) as:

$$\lambda = 2\sigma^2_n F_m^{-1}(P_f) \tag{12}$$

This clearly indicates that the threshold depends on noise variance σ^2_n, number of observed samples N, and targeted constant false alarm probability. It can be represented in a generic form as:

$$\lambda = \sigma^2_n \varepsilon(N, P_f) \tag{13}$$

Here, ε is a constant related to the number of sample N and target P_f.

Therefore, in the general form, the test statistic given by (4) is compared to a threshold of the form indicated in (13). A careful look at these equations reveal that all prior reported ED algorithms are special cases of the general form of the energy metric u, and threshold $\sigma_n^2 \varepsilon$. For example, (Urkowitz, 1967), (Fadel F Digham et al., 2007), (A. Ghasemi & Sousa, 2005; Amir Ghasemi & Sousa, 2007) suggested to compare u / σ_n^2 with ε, (Y. Zeng et al., 2009) and (Ye et al., 2008) proposed the comparison between u / N and $\sigma_n^2 \dfrac{\varepsilon}{N}$ to identify holes in the scanned frequency spectrum.

Exact P_d under AWGN (no fading)

In a non-fading environment where the channel gain h is deterministic and can be considered as *unity* without loss of generality ($h = 1$), the received signal under H_1 is given by:

$$x(k) = s(k) + n(k)$$

with,

$$n(k) \sim N\left(0, \sigma_n^2\right).$$

Thus, the test statistic u, depends on the statistics of $s(k)$.

Deterministic PU Signals (Detection of Equal Energy PU Signals)

The simplest signals to be detected under AWGN environment belong to the class of unknown deterministic signals. This case was analyzed by (Urkowitz, 1967), where it was shown that the assumption of unknown deterministic signal results in Gaussian received signal $x(k)$, similar to noise, with same variance σ_n^2 but with non-zero mean. Following the work of Urkowitz, exact closed-form expression for P_d was obtained by (Fadel F Digham et al., 2007). Most of the literature on ED based spectrum sensing refers to these fundamental works to identify the presence or absence of primary signal in the scanned frequency band. In general, PU signal contains information for its intended primary receiver and hence it is random in nature and cannot be treated as deterministic. However, as reported by (Urkowitz, 1967), detection probability results of unknown deterministic signal are applicable to random signal model provided the probability of detection is considered a conditional probability of detection where the condition is that the unknown signal to be detected (PU signal in case of spectrum sensing) has a certain amount of energy. This suggests that PU signal must not contain any information in its amplitude resulting in underlying assumption that PU signal must have deterministic, although unknown, energy. Only in that case, detection probability of PU signal is given by the classical results reported in (Fadel F Digham et al., 2007; Urkowitz, 1967).

For example, if $s(k)$ belongs to an *M-ary Phase Shift Keying* (PSK) signaling, all PU signal points lie on a circle of radius, say A, and have equal power A^2. The symmetry of the constellation indicates that the detection probability of the system is equal to the detection probability when any one signal point is transmitted. This is similar to evaluating the detection probability by assuming unknown PU signal to be a deterministic signal with $s(k) = A$. It is noteworthy that for a deterministic PU signal of finite duration T with (unknown) constant amplitude A, A^2 represents the power of the PU signal point while TA^2 is the measure of total energy content E_s, of the unknown signal $s(k)$.

Mathematically, for

$$s(k) \in \{A_i\} \quad \text{for} \quad i = 1, 2, \cdots, M$$

with

$$|A_i| = A \quad \forall i \quad \text{and} \quad \sum_{i=1}^{M} P(s(k) = A_i) = 1 \tag{14}$$

The detection probability based on the energy content of the received signal is given by:

$$P_d = \sum_{i=1}^{M} P_d \mid \left(s(k) = A_i\right) P(s(k) = A_i) \tag{15}$$

$$= P_d \mid \left(s(k) = A\right) \tag{16}$$

$$(\because P_d \mid \left(s(k) = A_i\right) = P_d \mid \left(s(k) = A\right) \ \forall \ i)$$

Hence, this case simplifies into the detection of the unknown deterministic signals $x(k) = A + n(k)$. Under H_1, with

$$x(k) \sim N\left(A, \sigma_n^2\right),$$

the test statistic is simply the sum of squares of N Gaussian random variables, each with mean A and variance σ_n^2. Hence, u normalized with σ_n^2 is said to have a *non-central Chi-square* distribution with N degrees of freedom:

$$H_1 : \frac{1}{\sigma_n^2} u = \sum_{k=1}^{N} \left(\frac{1}{\sigma_n}(A + n(k))\right)^2$$

$$= \sum_{k=1}^{N} \big(y(k)\big)^{2} : y(k) \sim N\left(\frac{A}{\sigma_{n}}, 1\right) \tag{17}$$

$$\sim \chi_{N}^{2}(\Omega) \left(deterministic \ \ s(k)\right) \tag{18}$$

Here, Ω is the non-centrality parameter given by:

$$\Omega = \sum_{k=1}^{N}\left(\frac{A}{\sigma_{n}}\right)^{2} = \frac{NA^{2}}{\sigma_{n}^{2}} \tag{19}$$

$$= 2\gamma \tag{20}$$

where,

$$\gamma = \frac{E_{s}}{N_{0}}$$

which represents SNR.

It is important to point out here that this definition of SNR is consistent with the works of some well-known authors like (Fadel F Digham et al., 2007; Urkowitz, 1967) and also (A. Ghasemi & Sousa, 2005) . The non-centrality parameter, given by (19) is evaluated as:

$$\Omega = \frac{NA^{2}}{\sigma_{n}^{2}} = \frac{2TWA^{2}}{2WN_{0}/2} = \frac{TA^{2}}{N_{0}/2} = \frac{2E_{s}}{N_{0}} = 2\gamma$$

Using (6), and the fact that

$$f\left(\frac{1}{\sigma_{n}^{2}}u\Big|H_{0}\right) = \chi_{N}^{2} ,$$

detection probability for deterministic PU signal, P_{d} can be found as:

$$P_{d} = P\left(\frac{1}{\sigma_{n}^{2}}u > \frac{1}{\sigma_{n}^{2}}\lambda \mid H_{1}\right)$$

$$= \int_{\frac{\lambda}{\sigma_n^2}}^{\infty} f(\frac{1}{\sigma_n^2} u \mid H_1) du \tag{21}$$

$$= Q_{\chi_N^2(\Omega)}\left(\frac{\lambda}{\sigma_n^2}\right) \tag{22}$$

Here, $Q_{\chi_N^2(\Omega)}\left(\frac{\lambda}{\sigma_n^2}\right)$ indicates the right-tail probability for a $\chi_N^2(\Omega)$ random variable as given by ((Kay, 1998)pg. 52). Using the pdf of $\chi_N^2(\Omega)$, the exact closed-form expression for P_d is:

$$P_d = \int_{\frac{1}{\sigma_n^2}\lambda}^{\infty} \frac{1}{2} e^{-\frac{x+\Omega}{2}} \left(\frac{x}{\Omega}\right)^{\frac{m-1}{2}} I_{m-1}(\sqrt{\Omega x}) dx$$

Here, $I_a(y)$ is the ath order modified Bessel function of the first kind:

$$I_a(y) = \left(\frac{y}{2}\right)^a \sum_{j=1}^{\infty} \frac{(y^2/4)^j}{j!\Gamma(a+j+1)}$$

With change of variables $x = y^2, \Omega = a^2$ and using definition of generalized Marcum Q-function incomplete gamma function:

$$Q_\delta(\alpha, \beta) = \int_j^\alpha \frac{x^\delta}{\alpha^{\delta-1}} e^{-\frac{x^2+\alpha^2}{2}} I_{\delta-1}(\alpha x) dx$$

$$P_d = \int_{\sqrt{\frac{\lambda}{\sigma_n^2}}}^{\infty} \frac{y^m}{a^{m-1}} e^{-\frac{y^2+a^2}{2}} I_{m-1}(ay) dy$$

$$= Q_m(a, \sqrt{\frac{\lambda}{\sigma_n^2}}) = Q_m(\sqrt{\Omega}, \sqrt{\frac{\lambda}{\sigma_n^2}}) = Q_m(\sqrt{2\gamma}, \sqrt{\frac{\lambda}{\sigma_n^2}}) \tag{23}$$

Here $m = N/2$ is the time-bandwidth product, assumed to be an integer number. This result matches with the one derived by (Fadel F Digham et al., 2007), eq.(5) for $a = 2$, $\sigma_n^2 = 1$. Similar results were shown by (Amir Ghasemi & Sousa, 2007),eq.(3) for $\sigma_n^2 = 1$).

The key point to highlight here is the exact closed-form expression of P_d, (23), reported extensively in the literature based on (Fadel F Digham et al., 2007; A. Ghasemi & Sousa, 2005; Amir Ghasemi & Sousa, 2007; Urkowitz, 1967) etc., caters only the cases in which the total detection probability can be considered as a conditional probability of detection which is true only for the PU signals originating from circular constellations. To the best of our knowledge, this hidden assumption has never been pointed out in the literature though equal energy restriction on PU signal points is not always true in practice. Furthermore, if PU signal constellation points belong to non-circular constellation, the simplification of Equations (15) to (16) does not remain valid and hence total detection probability needs to be evaluated in accordance with Equation (15). This means that the transmission probabilities of possible PU signal points would also become critical and need to be known *a priori* to evaluate the weighted summation encountered in (15).

The role of SNR in determining the ED performance is highlighted in Figure 2. Exact ROC comparison for deterministic PU signal model for N=10 at -5dB, 0 dB and 5 dB SNR2 by fixing the observed number of samples, $N = 10$.

DETECTION UNDER CHANNEL FADING/SHADOWING

The above analysis ignores any possible uncertainties in the signal model. For example, the results presented above assume that the noise variance is known *a priori* and also the channel gain remains constant (non-fading environment) throughout the sensing duration. In practice, however, these uncertainties are unavoidable and hence must be incorporated in the above derived P_d and P_f expressions.

Figure 2. Exact ROC comparison for deterministic PU signal model for N=10 at -5dB, 0 dB and 5 dB SNR2

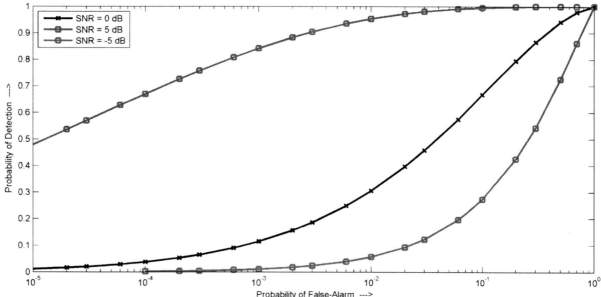

If there exists any uncertainty (θ) in the signal model, the performance metric probabilities ($P_d \rightarrow P_d(\theta)$ and $P_f \rightarrow P_f(\theta)$) give the *instantaneous* probability measure as a function of uncertain statistic (originating either from the noise variance estimate or the channel fading coefficient). Thus, in this case, average probabilities are needed to be evaluated by averaging already derived probabilities over the varying statistic. Hence,

$$\overline{P_d} = \int_x P_d(\theta) f_\theta(x) dx \tag{24}$$

$$\overline{P_f} = \int_x P_f(\theta) f_\theta(x) dx \tag{25}$$

Here, $f_\theta(x)$ represents the pdf of the uncertain statistic θ. For the noise uncertainty case, θ is replaced by the *noise uncertainty factor*, α (Y. Zeng et al., 2009), while for the fading environment, θ represents the *SNR*, γ (Amir Ghasemi & Sousa, 2007).

For the case of Rayleigh fading channel, the SNR γ follows the exponential distribution (F.F. Digham et al., 2003). The probability of detection for the case of Rayleigh fading is given by (Fadel F Digham et al., 2007):

$$P_{d\,Ray} = e^{-\frac{\lambda}{2\sigma^2}} \sum_{n=0}^{u-2} \frac{1}{n!}\left(\frac{\lambda}{2}\right)^n + \left(\frac{1+\overline{\gamma}}{\overline{\gamma}}\right)^{u-1}\left[e^{-\frac{\lambda}{2\sigma^2(1+\overline{\gamma})}} - e^{-\frac{\lambda}{2\sigma^2}} \sum_{n=0}^{u-2} \frac{1}{n!}\frac{\lambda\overline{\gamma}}{2\sigma^2(1+\overline{\gamma})}\right] \tag{26}$$

In the above equation, $\overline{\gamma}$ is the average received SNR. The expression for P_f is the same as shown in (11), i.e. without fading, as it is evaluated when the received signal is absent and hence, doesn't depend on the SNR.

ENERGY DETECTOR BASED COOPERATIVE SPECTRUM SENSING

Cooperative detection is popularly used as an effective method to improve the sensing performance in uncertain conditions. A comprehensive survey of cooperative spectrum sensing (CSS) schemes is provided in (Akyildiz et al., 2011) which identify ED as the most popular sensing technique owing to its simplicity and non-coherent/blind nature. The significance of ED is also evident from the fact that most of the cooperative sensing techniques reported in recent literature (Arshad et al., 2010; Saad et al., 2011) employ ED for local detection of primary transmissions. In cooperative detection, combining the local observations from spatially distributed CRs provides a more reliable spectrum awareness but requires additional processing (which takes time and consumes energy) of the sensing information to reach a unified global decision on PU activity.

ED, being computationally very simple, proves to be a suitable building block of a cooperative detection framework as it offers significant *cooperative gain* with minimum *cooperation overhead*. This is evident from the Figure 3. Sensing performance of ED based cooperative spectrum sensing using decision and data fusion3 which compares the sensing performance of ED based CSS employing *decision fusion* and *data fusion*. For decision fusion, we reach a global decision by combining 1-bit local sensing results using OR-rule. In comparison, all the locally observed energies are added up equally (Equal Gain Combining (EGC)), in the data fusion approach, to form the global test statistic which is then compared with global threshold. The performance comparison is based on given constraint on false-alarm rate at 10 dB SNR for the time-bandwidth product of 5 for 20 cooperating CRs. Results in Figure 3. Sensing performance of ED based cooperative spectrum sensing using decision and data fusion3 show that even 1-bit decision fusion (incurring minimum cooperation overhead in terms of control channel bandwidth) of ED based local sensing provides miss-detection rate below 0.00001 at $P_f = 0.1$ and it remains < 0.01 (detection rate $> 99\%$) even when P_f is decreased up to 0.0001. This indicates that the ED based CSS can yield high throughput efficient cognitive radio networks by significantly decreasing the average false alarm rate per CR for the given probability of detection constraint.

In a typical CRN, due to spatial diversity, it is very unlikely for all CRs to experience the same amount of fading/shadowing, and noise uncertainty. Hence, each CR will have a different local SNR. While performing data combining for such a network, it is very important to smartly combine the energy values from various CRs. This can be achieved by weighing each CR differently, depending on the amount of fading it undergoes and its noise variance. In other words, the observation value from a CR having a higher SNR should be given more weightage. Moreover, weights for CRs should be assigned such that it should optimize the overall detection performance. Researchers, in (Zheng et al., 2010), have used Particle Swarm Optimization technique to find the optimum weights for a CRN, in a manner which optimizes the detection performance. Data combining using PSO performs better than Equal Gain

Figure 3. Sensing performance of ED based cooperative spectrum sensing using decision and data fusion

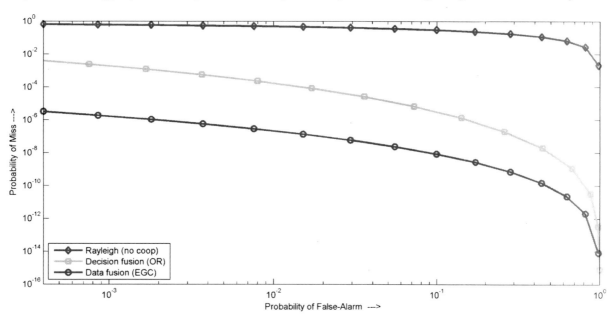

Combining (EGC) and Maximal Ratio Combining (MRC) schemes (Zheng et al., 2010). Data fusion gives a much better detection performance, but requires more bits to report the observed energy values compared to the decision fusion, where only local binary decisions are reported (Yucek & Arslan, 2009).

HYBRID DOUBLE THRESHOLD COOPERATIVE SPECTRUM SENSING ALGORITHM USING SWARM INTELLIGENCE

In this section, a new hybrid approach for cooperative spectrum sensing in CRNs is presented, wherein the unreliable CRs are identified, and the reliable and the unreliable CRs are treated separately. Generally, using a single threshold approach, due to noise uncertainties and other environmental effects, the observed energy values from certain CRs are very close to the prefixed threshold value. Decisions from such CRs are unreliable, hence, such CRs can be considered as *Fuzzy* (unreliable). In the presented approach, a double threshold energy detector is used at each cooperating CR to check its reliability. The reliable CRs report their local binary decision to the fusion centre, while the unreliable CRs report their observed energy values. The fusion centre uses Particle Swarm Optimization to optimally combine the reported energy values from the Fuzzy CRs. Since the hard and soft decisions are fused at the fusion centre, the hybrid technique is termed as the Hybrid PSO-OR technique. It is shown that the presented hybrid approach considerably reduces the number of reporting bits and the energy consumption at the expense of a negligible performance loss compared to the conventional single threshold data fusion technique using PSO. The Hybrid PSO-OR algorithm is compared to other hybrid approaches, which combines energies from Fuzzy CRs using EGC (Hybrid EGC-OR), and also with the conventional OR rule. The expressions for the cooperative probabilities of detection and false alarm, and expression for the normalized average number of bits over the reporting channel are derived for the Hybrid PSO-OR algorithm.

System Model

As cooperative spectrum sensing is widely used for its power in detecting the PU, we will be using a centralized cooperative sensing scheme in our model and each cooperating CR would perform double threshold energy detection (Chunhua Sun et al., 2007) as shown in Figure 4. The reliability of the observed energy values from each CR is measured using the two thresholds λ_1 and λ_2.

If the observed energy E_i is more than λ_2, decision H_1 is sent, and if E_i is more than λ_1, decision H_0 is sent to the fusion centre. If the energy value E_i for any CR falls in the region between the two thresholds (fuzzy region), it is considered unreliable to send its local decision, and hence, it directly reports its energy value to the fusion centre. The fusion centre would then combine the hard decisions (local results) and soft decisions (combining energy values according to weights obtained from PSO) to make a final decision on whether the PU is present or not. The Hybrid PSO-OR framework is shown in Figure 5 wherein PSO is used at the fusion centre to optimally combine the received energy values from all the Fuzzy CRs. The fusion centre finally combines the hard and the soft decisions, to give a final decision on whether the primary user is present or absent.

Figure 4. Double Threshold Energy Detector for a typical CR

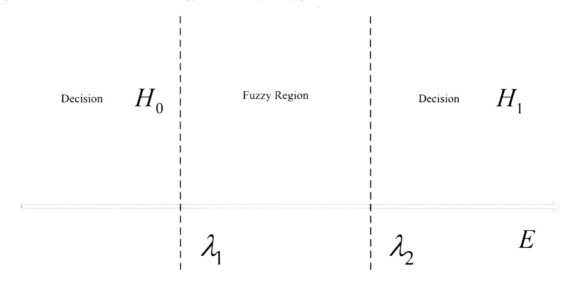

Figure 5. Framework for the Hybrid PSO-OR algorithm

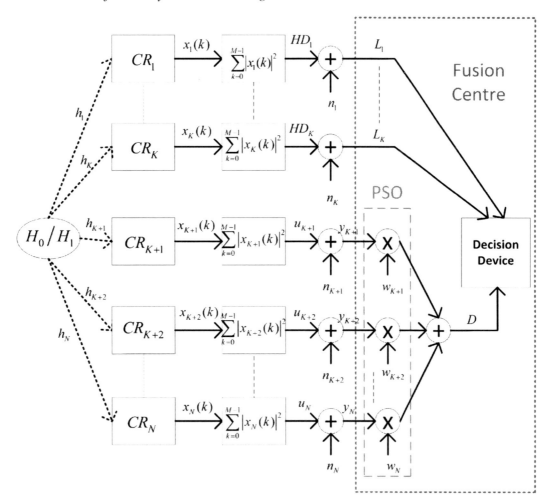

1. Every i^{th} cooperating CR, for $i = 1, 2, \ldots N$, performs local energy detection spectrum sensing from observing energy value E_i. Depending on the observed energy value, each CR sends its local decision (H_0 or H_1) or the observed energy value to the fusion centre (FC) as mentioned below:

$$FC_i = \begin{cases} E_i, & \lambda_1 < E_i \leq \lambda_2 \\ L_i, & otherwise \end{cases}, \text{ for } i = 1,2,\ldots N \tag{27}$$

In the above equation, the hard decision L_i is either a binary 1 (if local decision is H_1) or a binary 0 (if local decision is H_0), given by:

$$L_i = \begin{cases} 0, & 0 \leq E_i \leq \lambda_1 \\ 1, & E_i > \lambda_2 \end{cases} \tag{28}$$

2. It is assumed that among the N cooperating CRs, the fusion centre receives K hard decisions and $(N - K)$ observed energy values. The fusion centre first makes a soft decision based on the received energy values. PSO is employed at the fusion centre (Zheng et al., 2010), to assign an optimal weight vector w with $(N - K)$ dimensions, for optimally combining the $(N - K)$ received energies. The soft decision is given as:

$$D = \begin{cases} 0, & 0 \leq \sum_{i=1}^{N-K} w_i E_i \leq \lambda \\ 1, & \sum_{i=1}^{N-K} w_i E_i > \lambda \end{cases} \tag{29}$$

The threshold value λ is discussed in the next section.

3. Finally, the fusion centre makes an overall decision in the following manner:

$$F = \begin{cases} 1, & D + \sum_{i=1}^{K} L_i \geq 1 \\ 0, & otherwise \end{cases} \tag{30}$$

Thus, only when the combination of all the hard decisions and the soft decision is zero, the fusion centre decides that the target frequency band is vacant and ready for opportunistic access.

Particle Swarm Optimization for the Fuzzy CRs

Particle Swarm Optimization (PSO) is a population based, stochastic optimization technique, which is modelled after the flocking and swarming behaviour in birds and animals (Kennedy & Eberhart, 1995). In PSO, each particle is a solution to the optimization problem and these candidate solutions are referred to as swarm of particles (Trelea, 2003). As opposed to other population based algorithms, PSO maintains a static population whose particles are adjusted according to the new discoveries about the space. We can see the unique behaviour of PSO where each particle in a population moves towards the best optimal solution depending upon its past experiences as well as its neighbours (Kennedy & Eberhart, 2001). The fitness function determines the performance of each particle (Rashid et al., 2011). PSO is governed by two basic equations representing the position and velocity of each particle at a given time. The position and velocity of a particle are updated at each time step until a termination condition is reached. The search process is stopped automatically once the predetermined output or the maximum number of iterations are achieved.

$$v_{id}^{t} = \omega v_{id}^{t-1} + c_{1}\zeta\left(p_{id}^{t-1} - x_{id}^{t-1}\right) + c_{2}\eta\left(p_{gd}^{t-1} - x_{id}^{t-1}\right) \tag{31}$$

$$x_{id}^{t} = x_{id}^{t-1} + v_{id}^{t} \tag{32}$$

Here,

$$\boldsymbol{v}_{i}^{t} = \left[v_{i1}^{t}, v_{i2}^{t}, \ldots, v_{iD}^{t}\right]$$

and

$$\boldsymbol{x}_{i}^{t} = \left[x_{i1}^{t}, x_{i2}^{t}, \ldots, x_{iD}^{t}\right]$$

are the velocity and position vectors for particle

$$i\left(1 \leq i \leq S\right),$$

at the time iteration t, with dimension

$$d\left(1 \leq d \leq D\right).$$

We define

$$\boldsymbol{p}_i^t = \left[p_{i1}^t, p_{i2}^t, \ldots, p_{iD}^t \right]$$

as the best solution of particle i until time t and

$$\boldsymbol{p}_g^t = \left[p_{g1}^t, p_{g2}^t, \ldots, p_{gD}^t \right]$$

as the global best solution until time t. The parameters c_1 and c_2 are positive constants, ω is the weight of inertia, ζ and η are uniform random variables in the range $\left[0,1\right]$. The flowchart for the basic PSO algorithm is shown in Figure 6. Flowchart for basic PSO algorithm6.

In the analyzed framework, PSO is performed to optimally combine the energy values of the $N-K$ fuzzy CRs in the double threshold energy detection algorithm. In this framework, the probabilities of detection and false alarm for the fuzzy CRs are expressed as follows (Quan, Cui, & Sayed, 2008):

$$P_{d.PSO} = Q\left(\frac{Q^{-1}\left(P_f\right)\sqrt{\boldsymbol{w}^T A \boldsymbol{w}} - E_s \boldsymbol{h}^T \boldsymbol{w}}{\sqrt{\boldsymbol{w}^T B \boldsymbol{w}}} \right) \tag{33}$$

Figure 6. Flowchart for basic PSO algorithm

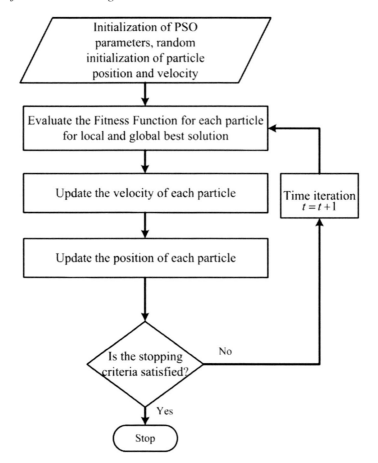

$$P_f = Q \left(\frac{\lambda - M\boldsymbol{\sigma}^T \boldsymbol{w}}{\sqrt{\boldsymbol{w}^T \boldsymbol{A} \boldsymbol{w}}} \right) \tag{34}$$

For a given target probability of false alarm, the threshold value λ for the fuzzy CRs is given by:

$$\lambda = M\boldsymbol{\sigma}^T \boldsymbol{w} + Q^{-1}\left(P_f\right)\sqrt{\boldsymbol{w}^T \boldsymbol{A} \boldsymbol{w}} \tag{35}$$

The terms $\boldsymbol{w}^T \boldsymbol{A} \boldsymbol{w}$ and $\boldsymbol{w}^T \boldsymbol{B} \boldsymbol{w}$ are the received global test statistic variances under hypothesis H_0 and H_1 respectively. The expression for matrices $\boldsymbol{A}$ and $\boldsymbol{B}$ are given as:

$$\boldsymbol{A} = 2M \begin{bmatrix} \sigma_1^4 & \cdots & 0 \\ \vdots & \ddots & \vdots \\ 0 & \cdots & \sigma_{N-K}^4 \end{bmatrix} + \begin{bmatrix} \delta_1^2 & \cdots & 0 \\ \vdots & \ddots & \vdots \\ 0 & \cdots & \delta_{N-K}^2 \end{bmatrix} \tag{36}$$

$$\boldsymbol{B} = 2M \begin{bmatrix} \sigma_1^4 & \cdots & 0 \\ \vdots & \ddots & \vdots \\ 0 & \cdots & \sigma_{N-K}^4 \end{bmatrix} + \begin{bmatrix} \delta_1^2 & \cdots & 0 \\ \vdots & \ddots & \vdots \\ 0 & \cdots & \delta_{N-K}^2 \end{bmatrix} + 4E_s \begin{bmatrix} h_1^2 & \cdots & 0 \\ \vdots & \ddots & \vdots \\ 0 & \cdots & h_{N-K}^2 \end{bmatrix} \begin{bmatrix} \sigma_1^2 & \cdots & 0 \\ \vdots & \ddots & \vdots \\ 0 & \cdots & \sigma_{N-K}^2 \end{bmatrix} \tag{37}$$

Here $\boldsymbol{w}$ is the weight vector, used for controlling the global spectrum detector,

$$E_s = \sum_{k=0}^{M-1} \left| s(k) \right|^2$$

is the transmitted signal energy for M samples in each detection interval, and

$$\gamma = \frac{E_s \left| h_i \right|^2}{\sigma_i^2}$$

is the local SNR, which is M times the average SNR at the output of local energy detector. The squared amplitude channel gains for all CRs are represented in a vector form as $\boldsymbol{h} = \left[\left| h_1 \right|^2, \left| h_2 \right|^2, \ldots, \left| h_{N-K} \right|^2 \right]^T$ and the corresponding sensing noise (which is zero-mean AWGN) variances are given by

$$\boldsymbol{\sigma} = \left[\left| \sigma_1 \right|^2, \left| \sigma_2 \right|^2, \ldots, \left| \sigma_{N-K} \right|^2 \right]^T.$$

The reporting channel noises are spatially uncorrelated Gaussian variables with zero-mean, and variances:

$$\delta = \left[\left|\delta_1\right|^2, \left|\delta_2\right|^2, \ldots, \left|\delta_{N-K}\right|^2 \right]^T$$ (Quan et al., 2008).

In our framework, optimizing the cooperative sensing from the fuzzy CRs becomes a problem of optimization of weight vector w to maximize the probability of detection P_d. From the expression of $P_{d,PSO}$, it is clear that since the Q function is a monotonically decreasing function, maximizing $P_{d,PSO}$ is same as minimizing the following expression (Quan et al., 2008):

$$f(w) = \frac{Q^{-1}(P_f)\sqrt{w^T A w} - E_s h^T w}{\sqrt{w^T B w}}$$ (38)

If w is the optimal weight vector which minimizes $f(w)$, then its scaled version αw, where α is a positive real number, is also an optimal vector which minimizes $f(w)$. Hence, to limit the number of optimal solutions, extra constraints are introduced to reduce the optimization problem as:

$$\min_{w} f(w) \, st. \, \sum_{l=1}^{N-K} w_l = 1, \, 0 \leq w_l \leq 1, \, for \, l = 1, 2, \ldots, N - K$$ (39)

Here, the dimension of the weight vector w is equal to the number of fuzzy CRs, i.e. $N - K$. The PSO algorithm for minimizing $f(w)$ to get the weight vector solution is explained in flowchart in Figure 6. Flowchart for basic PSO algorithm6. The PSO algorithm provides a weight vector w, which will be the solution for optimizing the probability of detection $P_{d,PSO}$.

Performance Analysis of the Hybrid PSO-OR Algorithm

In this section, the expressions for the cooperative probability of detection (Q_d), probability of missed detection (Q_m), and probability of false alarm (Q_f) are derived. The probability that a particular CR's observed energy value falls in the Fuzzy region, under hypothesis H_0 and H_1 respectively, is defined as follows:

$$\Delta_{0,i} = P\left(\lambda_1 < E_i \leq \lambda_2 \middle| H_0\right) = F\left(\lambda_2\right) - F\left(\lambda_1\right)$$ (40)

$$\Delta_{1,i} = P\left(\lambda_1 < E_i \leq \lambda_2 \middle| H_1\right) = G\left(\lambda_2\right) - G\left(\lambda_1\right)$$ (41)

The functions $F(.)$ and $G(.)$ are the CDFs of the observed energy value E, under hypothesis H_0 and H_1 respectively, and are defined as:

$$F(\lambda_2) = P\left(E < \lambda_2 | H_0\right) \tag{42}$$

$$G(\lambda_2) = P\left(E < \lambda_2 | H_1\right) \tag{43}$$

For a typical CR with local SNR γ, and local sensing noise variance $\sigma_n^2 = \sigma_i^2$, the probability of detection ($P_{d,i}$), the probability of missed detection ($P_{m,i}$), and the probability of false alarm ($P_{f,i}$) are defined as follows:

a. Under AWGN, the probability of detection is obtained from (23), by using $\lambda = \lambda_2$ and $\sigma_n^2 = \sigma_i^2$ as:

$$P_{d,i} = P\left(E_i > \lambda_2 | H_1\right) = \left(\sqrt{2\gamma}, \sqrt{\frac{\lambda_2}{\sigma_i^2}}\right) \tag{44}$$

b. Under Rayleigh fading, the probability of detection is obtained from (26), by using $\lambda = \lambda_2$ and $\sigma_n^2 = \sigma_i^2$ as:

$$P_{d,i} = P\left(E_i > \lambda_2 | H_1\right) = e^{-\frac{\lambda_2}{2\sigma_i^2}} \sum_{k=0}^{u-2} \frac{1}{k!}\left(\frac{\lambda_2}{2\sigma_i^2}\right)^k + \left(\frac{1+\bar{\gamma}}{\bar{\gamma}}\right)^{u-1}\left[e^{-\frac{\lambda_2}{2(1+\bar{\gamma})}} - e^{-\frac{\lambda_2}{2\sigma_i^2}} \sum_{k=0}^{u-2} \frac{1}{k!}\left(\frac{\lambda_2\bar{\gamma}}{2\sigma_i^2\left(1+\bar{\gamma}\right)}\right)^k\right] \tag{45}$$

The probability of miss detection $P_{m,i}$ can be derived from the above probability of detection as:

$$P_{m,i} = P(E_i \le \lambda_1 | H_1) = 1 - \Delta_{1,i} - P_{d,i} \tag{46}$$

Similarly, the probability of false alarm $P_{f,i}$ can be obtained from (11), by using $\lambda = \lambda_2$ and $\sigma_n^2 = \sigma_i^2$ as:

$$P_{f,i} = P\left(E_i > \lambda_2 | H_0\right) = \frac{\Gamma\left(m, \frac{\lambda_2}{2\sigma_i^2}\right)}{\Gamma(m)} \tag{47}$$

The cooperative probability of missed detection (Q_m) is defined as follows:

$$Q_m = P\left(F = 0 \mid H_1\right) \tag{48}$$

Using the total probability theorem, Q_m can be expressed as follows:

$$Q_m = P\left(F = 0, K \neq N \mid H_1\right) + P(F = 0, K = N \mid H_1)$$

$$Q_m = P\left(\sum_{i=1}^{K} L_i = 0 \cap D = 0, K \neq N \mid H_1\right) + P\left(\sum_{i=1}^{K} L_i = 0, K = N \mid H_1\right)$$

After some mathematical manipulations, Q_m can finally be expressed as follows:

$$Q_m = \sum_{K=0}^{N} \binom{N}{K} \prod_{i=1}^{K} \{P_{m,i}\} \prod_{i=1}^{N-K} \{\Delta_{1,i}\} \left[1 - P_{d,PSO(N-K)}\right] \tag{49}$$

Here, $P_{d,PSO(N-K)}$ is probability of detection for the $N - K$ fuzzy CRs, which can be expressed as:

$$P_{d,PSO(N-K)} = P\left(\sum_{i=1}^{N-K} w_i E_i > \lambda \mid H_1\right) \tag{50}$$

The cooperative probability of detection (Q_d) is obtained using:

$$Q_d = 1 - Q_m \tag{51}$$

The cooperative probability of false alarm (Q_f) is simply defined as:

$$Q_f = P\left(F = 1 \mid H_0\right) = 1 - P(F = 0 \mid H_0) \tag{52}$$

Using the total probability theorem again, Q_f can be expressed as:

$$Q_f = 1 - \left[\begin{array}{l} P\left(F = 0, K \neq N \mid H_0\right) \\ +P(F = 0, K = N \mid H_0) \end{array}\right]$$

Again, Q_f can be expressed as:

$$Q_f = 1 - \left[\sum_{K=0}^{N} \binom{N}{K} \prod_{i=1}^{K} \{F(\lambda_{1,i})\} \prod_{i=1}^{N-K} \{\Delta_{0,i}\} \left[1 - P_{f.PSO(N-K)} \right] \right] \qquad (53)$$

Here, $F(\lambda_{1,i})$ is the CDF for i^{th} CR, and $P_{f.PSO(N-K)}$ is the probability of false alarm for the $N - K$ fuzzy CRs, which can be expressed as:

$$P_{f.PSO(N-K)} = P\left(\sum_{i=1}^{N-K} w_i E_i > \lambda \mid H_0 \right) \qquad (54)$$

In the analysis above, if the energies from the $N - K$ fuzzy CRs are combined using the EGC technique, the cooperative probabilities of missed detection and false alarm become:

$$Q_m = \sum_{K=0}^{N} \binom{N}{K} \prod_{i=1}^{K} \{P_{m,i}\} \prod_{i=1}^{N-K} \{\Delta_{1,i}\} \left[1 - P_{d.EGC(N-K)} \right] \qquad (55)$$

$$Q_f = 1 - \left[\sum_{K=0}^{N} \binom{N}{K} \prod_{i=1}^{K} \{F(\lambda_{1,i})\} \prod_{i=1}^{N-K} \{\Delta_{0,i}\} \left[1 - P_{f.EGC(N-K)} \right] \right] \qquad (56)$$

The expression for $P_{d.EGC(N-K)}$ in the equations depends on the underlying channel characteristics. For the case of AWGN, the exact closed-form expression of P_d as derived in (23) is used, whereas for Rayleigh fading, (26) is used for $P_{d.EGC(N-K)}$.

The expression for $P_{f.EGC(N-K)}$ is independent of the channel behavior. Hence, for both the cases of AWGN and Rayleigh fading, the exact closed-form expression of P_f, given by (11) is used for $P_{f.EGC(N-K)}$.

On the other hand, if the CRs whose observed energy values fall in the fuzzy region are neglected, the final decision will then be based only on the local hard decisions reported by the reliable CRs (Chunhua Sun et al., 2007). Under such a scenario, the cooperative probabilities of detection, missed detection and false alarm become:

$$Q_d = P\left(F = 1, K \geq 1 \mid H_1 \right) = 1 - \prod_{i=1}^{N} \{G(\lambda_{2,i})\} \qquad (57)$$

$$Q_m = P\left(F = 0, K \geq 1 | H_1\right) =$$

$$1 - \prod_{i=1}^{N} \Delta_{1,i} - Q_d = \prod_{i=1}^{N} \left\{ G\left(\lambda_{2,i}\right) \right\} \tag{58}$$

$$-\prod_{i=1}^{N} \Delta_{1,i} = \prod_{i=1}^{N} \left\{ G\left(\lambda_{1,i}\right) \right\}$$

$$Q_f = P\left(F = 1, K \geq 1 | H_0\right) = 1 - \prod_{i=1}^{N} \left\{ F\left(\lambda_{2,i}\right) \right\} \tag{59}$$

Here, $F\left(\lambda_i\right)$ and $G\left(\lambda_i\right)$ are the CDFs for i^{th} CR, under H_0 and H_1 respectively.

Bit Savings over the Reporting Channel for the Hybrid PSO-OR Algorithm

The data fusion schemes such as the conventional EGC (Fadel F Digham et al., 2007) and conventional PSO (i.e. single threshold gain combining using PSO (Zheng et al., 2010)) achieve higher detection performance compared to the decision fusion techniques, but at the expense of high communication burden over the reporting channel. Hence, due to the bandwidth constraints over the reporting channel, there is a need to reduce the number of reporting bits, with preferably a negligible loss in performance.

In this section, the average communication overhead over the reporting channel is evaluated. In the Hybrid PSO-OR algorithm, during each spectrum sensing cycle, the fusion centre receives K hard decisions and $\left(N - K\right)$ observation values (energies, local noise variances and channel gains). Each CR uses 1 bit to report its hard decision, and b bits to report its observation values. The total number of bits received by the fusion centre is given as:

$$B = K + b\left(N - K\right) \tag{60}$$

Firstly, let us define event V_K, as the case in which K CRs report their hard decisions to the fusion centre. Thus, the probability of event V_K is given by:

$$P\left(V_K\right) = \left[1 - P\left(\lambda_1 < E \leq \lambda_2\right)\right]^K \tag{61}$$

The other event W_{N-K} is defined, when $\left(N - K\right)$ CRs report their observed information to the fusion centre. The probability of event W_{N-K} is given by:

$$P\left(W_{N-K}\right) = \left[P\left(\lambda_1 < E \leq \lambda_2\right)\right]^{N-K} \tag{62}$$

Also, P_1 and P_0 are denoted as the probabilities of a primary user being present and absent respectively. Using these equations, the expression for the average number of CRs reporting their observed information is given as:

$$K_{soft} = \sum_{K=0}^{N}(N-K)\left[\begin{array}{c}\binom{N}{K}P\left(W_{N-K}\mid H_0\right)P\left(V_K\mid H_0\right)P_0 \\ +\binom{N}{K}P\left(W_{N-K}\mid H_1\right)P\left(V_K\mid H_1\right)P\left(H_1\right)P_1\end{array}\right]$$

After some mathematical manipulations, the normalized average number of bits over the reporting channel is given as:

$$\bar{B} = \frac{B_{avg}}{N} = b - (b-1)\bar{K} \tag{63}$$

Here, $\bar{K} = \dfrac{K_{hard}}{N}$ is the normalized average number of hard decisions (Chunhua Sun et al., 2007).

From equation, it is observed that the normalized average number of bits over the reporting channel is obviously less than b bits. Hence, the normalized average number of reporting bits for the presented algorithm is low in comparison to the conventional single threshold PSO algorithm, wherein each CR uses b bits to report its observed information. For the case of $\Delta_0 = 0$, the two thresholds λ_1 and λ_2 coincide, and the normalized average number of reporting bits $\bar{B} = 1$ (corresponding to the conventional OR rule). On the other hand, for $\Delta_0 = 1$, the two thresholds λ_1 and λ_2 are infinitely apart, and the normalized average number of reporting bits $\bar{B} = b$ (corresponding to the conventional PSO algorithm).

Energy Savings for the Hybrid PSO-OR Algorithm

The energy savings for the Hybrid PSO-OR algorithm, wherein K out of N cooperating CRs transmit one bit hard sensing decision, while $N - K$ fuzzy CRs transmit their observation values (energies, local noise variances and channel gains) to the fusion center using b bits each, is evident from the bit savings achieved in the Hybrid PSO-OR algorithm, as given by (63). Under the assumption that one unit of energy is utilized to transmit one bit, the conventional single threshold PSO requires b units of normalized energy while Hybrid PSO-OR algorithm utilizes $b - (b-1)\bar{K}$ units of normalized energy. Hence, $(b-1)\bar{K}$ units of normalized energy are saved using Hybrid PSO-OR algorithm.

Simulation Results

In this section, the performance analysis is carried for the hybrid cooperative spectrum sensing algorithm (Hybrid PSO-OR). The Hybrid PSO-OR algorithm is compared to the Hybrid EGC-OR algorithm, the conventional single threshold PSO algorithm, and the conventional OR rule algorithm.

The time-bandwidth product is taken as $u = TW = 5$, and the number of received signal samples is $M = 2u = 10$. The parameters used for PSO are: number of particles $S = 30$, maximum number of iterations $t = 50$, inertia weight $\omega = 1$, and the learning factors $c_1 = c_2 = 2$.

The performance analysis is carried for 10 cooperating CRs, for different average SNR values under both AWGN and Rayleigh fading, for a particular value of Δ_0. For simplicity, the reporting channels are assumed to be perfect, i.e. $\delta = 0$.

Case 1: Under AWGN

The performance of the Hybrid PSO-OR algorithm is compared for an average SNR $= 4\,\mathrm{dB}$, and $\Delta_0 = 0.2$ in Figure 7.

The quantitative analysis is performed for $\Delta_0 = 0.2$, where the Hybrid PSO-OR algorithm is compared with the Hybrid EGC-OR algorithm, the conventional PSO data fusion technique and the conventional OR rule (decision fusion). The performance of the Hybrid PSO-OR algorithm, for any value of Δ_0 between 0 and 1, lies in between the conventional OR rule and the conventional PSO data fusion technique. For $Q_f = 10^{-3}$ and $\Delta_0 = 0.2$, the Hybrid PSO-OR algorithm gives 49% missed detection as opposed to the Hybrid EGC-OR algorithm which gives 66% missed detection.

Figure 7. C-ROC – Comparison of Hybrid PSO-OR algorithm with existing techniques (10 CRs, Avg. SNR = 4 dB, AWGN Case)

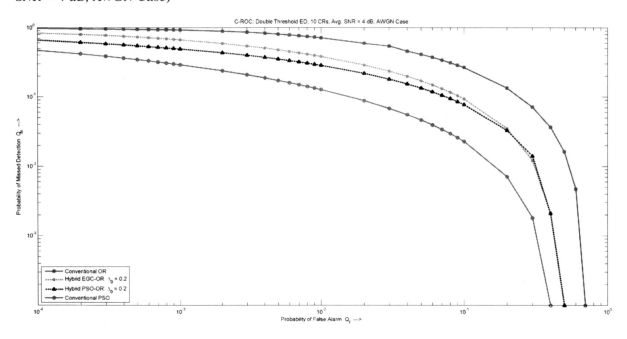

For the Hybrid PSO-OR algorithm, $\Delta_0 = 0$ corresponds to the conventional OR rule, where the two thresholds λ_1 and λ_2 coincide. On the other hand, $\Delta_0 = 1$ corresponds to the conventional PSO algorithm, where all the CRs fall in fuzzy region, and the received energies are combined using the optimal weight vector calculated using PSO.

Case 2: Under Rayleigh Fading

The performance of the Hybrid PSO-OR algorithm is analyzed for an average SNR = 5 dB, and $\Delta_0 = 0.05$ in Figure 8.

The sensing channel gains follow a Rayleigh distribution.

Similar to the AWGN case, the performance of the Hybrid PSO-OR algorithm, for different values of Δ_0, lies in between the conventional OR rule and the conventional PSO technique. Here, the case of $\Delta_0 = 0.05$ is used to show the performance of the Hybrid PSO-OR algorithm in comparison with other existing techniques. For $Q_f = 10^{-3}$ and $\Delta_0 = 0.05$, the Hybrid PSO-OR algorithm gives 30% missed detection as opposed to Hybrid EGC-OR algorithm which gives 39% missed detection.

Summary of Simulation Results

The performance analysis of the Hybrid PSO-OR algorithm under the case of AWGN and Rayleigh fading is summarized in Table 1 and Table 2 respectively. All the quantitative analysis values mentioned in the tables are for $Q_f = 10^{-3}$.

Figure 8. ROC – Comparison of Hybrid PSO-OR algorithm with existing techniques (10 CRs, Avg. SNR = 5 dB, Rayleigh Fading)

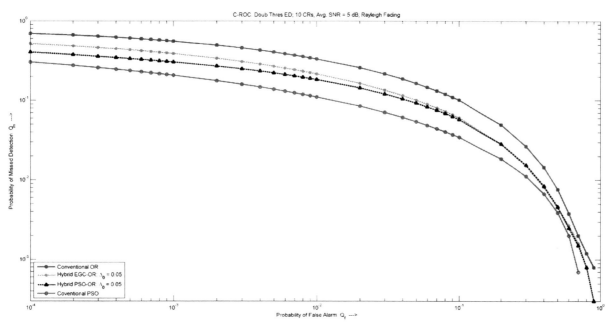

Table 1. Performance analysis summary of the Hybrid PSO-OR algorithm under AWGN

SNR (dB)	Δ_0	Improvement in P_d Compared to Hybrid EGC-OR	Loss in P_d Compared to Conventional PSO
4	0.01	103.3%	64.4%
	0.2	52.1%	28.2%
6	0.01	25.7%	45.7%
	0.2	**8.4%**	**8%**

Table 2. Performance analysis summary of the Hybrid PSO-OR algorithm under Rayleigh Fading

SNR (dB)	Δ_0	Improvement in P_d Compared to Hybrid EGC-OR	Loss in P_d Compared to Conventional PSO
5	0.01	15.2%	20.2%
	0.1	13.4%	10.7%
7	0.01	3.1%	9.7%
	0.1	**2.9%**	**5.4%**

The above tables indicate that at low SNR for a given cooperative false alarm probability and a particular value of Δ_0, the Hybrid PSO-OR algorithm achieves a much better detection performance compared to the Hybrid EGC-OR algorithm. Under AWGN, for the case of $\Delta_0 = 0.2$, the Hybrid PSO-OR algorithm achieves 52.1% detection performance improvement compared to Hybrid EGC-OR algorithm, whereas it show 28.2% loss compared to the conventional PSO technique. Under Rayleigh fading, for the case of $\Delta_0 = 0.1$, the Hybrid PSO-OR algorithm achieves 13.4% improvement in detection compared to Hybrid EGC-OR algorithm, whereas it loses only 10.7% compared to the conventional PSO technique.

FUTURE RESEARCH DIRECTIONS

The spectrum sensing problem was addressed in time domain using the double threshold energy detector. As an extension to our work, the double threshold energy detector can be explored further. Some of the recommended research directives are briefly outlined below:

Maximum Eigenvalue Based Detection for Fuzzy CRs

For the CRs falling in the fuzzy region, the received signal samples can be collected to estimate the covariance matrix. The fusion centre can make a soft decision based on the maximum eigenvalue of the covariance matrix. This soft decision can be combined with the other hard decisions from the reliable CRs, to make a final decision on the presence or absence of the primary user.

Cluster-Based Two-Threshold Cooperative Spectrum Sensing

In the presented Hybrid PSO-OR algorithm, the reporting channels were assumed to be perfect. However, when the cooperating CRs report their local sensing observations to the fusion centre through fading channels, the sensing performance can seriously degrade. Under such scenarios, cluster based cooperative sensing, proposed in (C. Sun, Zhang, & Letaief, 2007), can be used to enhance sensing performance. Each cooperating CR would use a double threshold energy detector locally. All the cooperating CRs are divided into clusters (Younis & Fahmy, 2004). For each cluster, a CR with the maximum instantaneous reporting channel gain is selected as the cluster head, whose task is to collect local sensing information from the other CRs in the cluster, then forward it to the fusion centre (C. Sun et al., 2007).

Based on the double threshold energy detector, each CR in the cluster either sends its local hard decision or the observed energy value to the cluster head. The cluster head can then use an EGC scheme to combine the received energies and make a soft decision. Finally, the cluster head combines the hard and soft decisions to make a final decision, and reports it to the fusion centre. Hence, this would reduce the reporting errors due to the fading channels (C. Sun et al., 2007).

CONCLUSION

Cognitive Radio has emerged as a promising technology to solve the current spectrum scarcity problem by opening secondary access to the licensed spectrum under stringent interference avoidance conditions. Spectrum sensing lies at the heart of CR. In this chapter, we focused on cooperative spectrum sensing based on energy detection approach. We started by deriving the exact distribution of test statistics for the energy detection of unknown primary signal under both AWGN and fading environment. We showed the ROC curves to analyze the impact of varying SNR and sensing performance constraints in terms of P_f and P_d. We then presented a hybrid cooperative spectrum sensing algorithm using a double threshold energy detector and a Particle Swarm Optimization technique. The algorithm combines decision fusion and data fusion. The simulation results showed that the Hybrid PSO-OR algorithm significantly reduces the communication overhead in terms of average number of reporting bits and the energy consumption, at the expense of a negligible loss in performance as compared to the conventional single threshold data fusion technique using PSO. It was also shown that the Hybrid PSO-OR algorithm outperforms the conventional hybrid approach, called the Hybrid EGC-OR technique, which uses Equal Gain Combining for data fusion.

ACKNOWLEDGMENT

The authors thank KFUPM for the support provided to carry out this work.

REFERENCES

Akyildiz, I. F., Lo, B. F., & Balakrishnan, R. (2011). Cooperative spectrum sensing in cognitive radio networks: A survey. *Physical Communication, 4*(1), 40–62. doi:10.1016/j.phycom.2010.12.003

Arshad, K., Imran, M. A., & Moessner, K. (2010). Collaborative Spectrum Sensing Optimisation Algorithms for Cognitive Radio Networks. *International Journal of Digital Multimedia Broadcasting, 2010,* 1–20. doi:10.1155/2010/424036

Ben Letaief, K., & Wei Zhang, . (2009). Cooperative Communications for Cognitive Radio Networks. *Proceedings of the IEEE, 97*(5), 878–893. doi:10.1109/JPROC.2009.2015716

Cabric, D., Mishra, S. M., Willkomm, D., Brodersen, R., & Wolisz, A. (2005). A Cognitive Radio Approach for Usage of Virtual Unlicensed Spectrum. In Proceedings of 14th IST Mobile Wireless Communications Summit. Dresden, Germany: IST.

Digham, F. F., Alouini, M. S., & Simon, M. K. (2003). On the energy detection of unknown signals over fading channels. In *Proceedings of IEEE International Conference on Communications, (Vol. 5,* pp. 3575–3579). IEEE. doi:10.1109/ICC.2003.1204119

Digham, F. F., Alouini, M.-S., & Simon, M. K. (2007). On the Energy Detection of Unknown Signals Over Fading Channels. *IEEE Transactions on Communications, 55*(1), 21–24. doi:10.1109/TCOMM.2006.887483

Ghasemi, A., & Sousa, E. S. (2007). Opportunistic Spectrum Access in Fading Channels Through Collaborative Sensing. *Journal of Communication, 2*(2), 71–82. doi:10.4304/jcm.2.2.71-82

Ghasemi, A., & Sousa, E. S. (2005). Collaborative spectrum sensing for opportunistic access in fading environments. In *Proceedings of First IEEE International Symposium on New Frontiers in Dynamic Spectrum Access Networks,* (pp. 131–136). IEEE. doi:10.1109/DYSPAN.2005.1542627

Kay, S. M. (1998). Fundamentals of Statistical Signal Processing, Volume II: Detection Theory. Prentice Hall.

Kennedy, J., & Eberhart, R. (1995). Particle swarm optimization. In *Proceedings of ICNN'95 - International Conference on Neural Networks* (Vol. 4, pp. 1942–1948). IEEE. doi:10.1109/ICNN.1995.488968

Kennedy, J., & Eberhart, R. C. (2001). *Swarm Intelligence.* Morgan Kauffman Publishers.

Maleki, S., Pandharipande, A., & Leus, G. (2010). Two-stage spectrum sensing for cognitive radios. In *Proceedings of 2010 IEEE International Conference on Acoustics, Speech and Signal Processing* (pp. 2946–2949). IEEE. doi:10.1109/ICASSP.2010.5496149

Mitola, J., & Maguire, G. Q. (1999). Cognitive radio: Making software radios more personal. *IEEE Personal Communications, 6*(4), 13–18. doi:10.1109/98.788210

Quan, Z., Cui, S., & Sayed, A. H. (2008). Optimal Linear Cooperation for Spectrum Sensing in Cognitive Radio Networks. *IEEE Journal of Selected Topics in Signal Processing, 2*(1), 28–40. doi:10.1109/JSTSP.2007.914882

Rashid, R. A., Baguda, Y. S., Fisal, N., Sarijari, M. A., Yusof, S. K. S., Ariffin, S. H. S., & Mohd, A. (2011). Optimizing Achievable Throughput for Cognitive Radio Network using. *Swarm Intelligence*, (October): 354–359.

Saad, W., Han, Z., Basar, T., Debbah, M., & Hjorungnes, A. (2011). Coalition Formation Games for Collaborative Spectrum Sensing. *IEEE Transactions on Vehicular Technology, 60*(1), 276–297. doi:10.1109/TVT.2010.2089477

Sonnenschein, A., & Fishman, P. M. (1992). Radiometric detection of spread-spectrum signals in noise of uncertain power. *IEEE Transactions on Aerospace and Electronic Systems, 28*(3), 654–660. doi:10.1109/7.256287

Sun, C., Zhang, W., & Ben Letaief, K. (2007). Cooperative Spectrum Sensing for Cognitive Radios under Bandwidth Constraints. In *Proceedings of 2007 IEEE Wireless Communications and Networking Conference*, (pp. 1–5). IEEE. doi:10.1109/WCNC.2007.6

Sun, C., Zhang, W., & Letaief, K. B. (2007). Cluster-Based Cooperative Spectrum Sensing in Cognitive Radio Systems. In *Proceedings of 2007 IEEE International Conference on Communications*, (pp. 2511–2515). IEEE. doi:10.1109/ICC.2007.415

Trelea, I. C. (2003). The particle swarm optimization algorithm: Convergence analysis and parameter selection. *Information Processing Letters, 85*(6), 317–325. doi:10.1016/S0020-0190(02)00447-7

Umar, R., & Sheikh, A. U. H. (2012a). Spectrum Access and Sharing for Cognitive Radio. In Developments in Wireless Network Prototyping, Design, and Deployment: Future Generations (pp. 241–271). Academic Press.

Umar, R., & Sheikh, A. U. H. (2012b). A comparative study of spectrum awareness techniques for cognitive radio oriented wireless networks. *Physical Communication*. doi:10.1016/j.phycom.2012.07.005

Umar, R., & Sheikh, A. U. H. (2012c). Cognitive Radio oriented wireless networks: Challenges and solutions. In *Proceedings of 2012 International Conference on Multimedia Computing and Systems* (pp. 992–997). IEEE. doi:10.1109/ICMCS.2012.6320105

Urkowitz, H. (1967). Energy detection of unknown deterministic signals. *Proceedings of the IEEE, 55*(4), 523–531. doi:10.1109/PROC.1967.5573

Ye, Z., Memik, G., & Grosspietsch, J. (2008). Energy Detection Using Estimated Noise Variance for Spectrum Sensing in Cognitive Radio Networks. In *Proceedings of 2008 IEEE Wireless Communications and Networking Conference* (pp. 711–716). IEEE. doi:10.1109/WCNC.2008.131

Younis, O., & Fahmy, S. (2004). *Distributed clustering in ad-hoc sensor networks: a hybrid, energy-efficient approach. In Proceedings of IEEE INFOCOM 2004* (Vol. 1, pp. 629–640). IEEE. doi:10.1109/INFCOM.2004.1354534

Yucek, T., & Arslan, H. (2009). A survey of spectrum sensing algorithms for cognitive radio applications. *IEEE Communications Surveys and Tutorials, 11*(1), 116–130. doi:10.1109/SURV.2009.090109

Zeng, Y., Liang, Y.-C., Hoang, A. T., & Peh, E. C. Y. (2009). Reliability of Spectrum Sensing Under Noise and Interference Uncertainty. In *Proceedings of 2009 IEEE International Conference on Communications Workshops* (pp. 1–5). IEEE. doi:10.1109/ICC.2009.5199287

Zeng, Y., Liang, Y.-C., Hoang, A. T., & Zhang, R. (2010). A Review on Spectrum Sensing for Cognitive Radio: Challenges and Solutions. *EURASIP Journal on Advances in Signal Processing, 2010*, 1–16. doi:10.1155/2010/381465

Zheng, S., Lou, C., & Yang, X. (2010). Cooperative spectrum sensing using particle swarm optimisation. *Electronics Letters, 46*(22), 1525. doi:10.1049/el.2010.2115

ADDITIONAL READING

Akyildiz, I. F., Lee, W.-Y., Vuran, M. C., & Mohanty, S. (2006). NeXt generation/dynamic spectrum access/cognitive radio wireless networks: A survey. *Computer Networks, 50*(13), 2127–2159. doi:10.1016/j.comnet.2006.05.001

Ghasemi, A., & Sousa, E. S. (2007). Spectrum sensing in cognitive radio networks: The cooperation-processing tradeoff. *Wireless Communications and Mobile Computing, 7*(9), 1049–1060. doi:10.1002/wcm.480

Herath, S. P., Member, S., Rajatheva, N., Member, S., & Tellambura, C. (2011).. . *Energy Detection of Unknown Signals in Fading and Diversity Reception, 59*(9), 2443–2453.

Joshi, D. R., Popescu, D. C., & Dobre, O. A. (2011). Gradient-Based Threshold Adaptation for Energy Detector in Cognitive Radio Systems. *IEEE Communications Letters, 15*(1), 19–21. doi:10.1109/LCOMM.2010.11.100654

Kulkarni, R. V., & Venayagamoorthy, G. K. (2011). Particle Swarm Optimization in Wireless-Sensor Networks: A Brief Survey. *IEEE Transactions on Systems, Man and Cybernetics. Part C, Applications and Reviews, 41*(2), 262–267. doi:10.1109/TSMCC.2010.2054080

Maleki, S., Pandharipande, A., & Leus, G. (2010). Two-stage spectrum sensing for cognitive radios. *2010 IEEE International Conference on Acoustics, Speech and Signal Processing* (pp. 2946–2949). IEEE. doi:10.1109/ICASSP.2010.5496149

Mishra, S., Sahai, A., & Brodersen, R. (2006). Cooperative Sensing among Cognitive Radios. *2006 IEEE International Conference on Communications* (Vol. 00, pp. 1658–1663). IEEE. doi:10.1109/ICC.2006.254957

Nuttall, A. (1975). Some integrals involving the<tex>Q_M</tex>function (Corresp.). *IEEE Transactions on Information Theory, 21*(1), 95–96. doi:10.1109/TIT.1975.1055327

Rashid, R. A., Baguda, Y. S., Fisal, N., Sarijari, M. A., Yusof, S. K. S., Ariffin, S. H. S., & Mohd, A. (2011). Optimizing Achievable Throughput for Cognitive Radio Network using. *Swarm Intelligence*, (October): 354–359.

Tandra, R., & Sahai, A. (2008). SNR Walls for Signal Detection. *IEEE Journal of Selected Topics in Signal Processing, 2*(1), 4–17. doi:10.1109/JSTSP.2007.914879

Umar, R., Sheikh, A. U. H., & Deriche, M. (2014). Unveiling the hidden assumptions of energy detector based spectrum sensing for cognitive radios. *IEEE Communications Surveys and Tutorials, 16*(2), 713–728. doi:10.1109/SURV.2013.081313.00054

Wang, Y., Tian, Z., & Feng, C. (2010). A Two-Step Compressed Spectrum Sensing Scheme for Wideband Cognitive Radios. *2010 IEEE Global Telecommunications Conference GLOBECOM 2010* (pp. 1–5). IEEE. doi:10.1109/GLOCOM.2010.5683246

Zhou, X., Li, G. Y., Li, D., Wang, D., & Soong, A. C. K. (2010). Bandwidth efficient combination for cooperative spectrum sensing in cognitive radio networks. *2010 IEEE International Conference on Acoustics, Speech and Signal Processing* (pp. 3126–3129). IEEE. doi:10.1109/ICASSP.2010.5496092

KEY TERMS AND DEFINITIONS

Cognitive Radio (CR): An intelligent reconfigurable device that can interact with its radio environment and update its transmission parameters (in software) on the run to optimally benefit from the available radio resources.

Fuzzy CR: A CR with the received energy value falling in the fuzzy region.

Fuzzy Region: A region in between the two threshold values of energy detector.

Primary User (PU): High priority licensed user with legacy rights on usage of specific part of frequency spectrum.

Probability of Detection (P_d): It is the probability of correctly detecting the primary signal when the primary user is actually active in the scanned frequency band.

Probability of False Alarm (P_f): It is the probability of falsely detecting the primary signal when the primary user is actually silent in the scanned frequency band.

Secondary User (SU): It is a low priority un-licensed user equipped with cognitive capability to exploit any vacant spectrum band without interfering with the primary user.

Spectrum Sensing (SS): It is the ability to determine which portion(s) of the scanned frequency band are vacant and identify any primary user activity.

Chapter 2
Cooperative Spectrum Sensing with Censoring of Cognitive Radios and MRC–Based Fusion in Fading and Shadowing Channels

Srinivas Nallagonda
National Institute of Technology – Durgapur, India

Sumit Kundu
National Institute of Technology – Durgapur, India

Sanjay Dhar Roy
National Institute of Technology – Durgapur, India

Gianluigi Ferrari
University of Parma, Italy

Riccardo Raheli
University of Parma, Italy

ABSTRACT

In this chapter, the authors study the performance of Cooperative Spectrum Sensing (CSS) with soft data fusion, given by Maximal Ratio Combining (MRC)-based fusion with Weibull faded channels, and Log-normal shadowed channels. More precisely, they evaluate the performance of a MRC-based CSS with Cognitive Radios (CRs) censored on the basis of the quality of the reporting channels. The performance of CSS with two censoring schemes, namely rank-based and threshold-based, is studied in the presence of Weibull fading, Rayleigh fading, and Log-normal shadowing in the reporting channels, considering MRC fusion. The performance is compared with those of schemes based on hard decision fusion rules. Furthermore, depending on perfect or imperfect Minimum Mean Square Error (MMSE) channel estimation, the authors analyze the impact of channel estimation strategy on the censoring schemes. The performance is studied in terms of missed detection probability as a function of several network, fading, and shadowing parameters.

DOI: 10.4018/978-1-4666-6571-2.ch002

INTRODUCTION

There is huge demand of spectrum in this decade due to increase in internet traffic and other new services However, good part of the spectrum has already been licensed/ leased to different government, semi government and private organizations. Although, it can be noticed that the licensed and leased spectrums are not efficiently used and hence efficient spectrum allocation and utilization policies are required from spectrum traders and researchers.In order to deal with this conflict between spectrum scarcity and spectrum under-utilization, cognitive radio (CR) has been proposed as a revolutionary technology for the next generation of wireless communication networks (Mitola, 1999), (Haykin, 2005). In order to guarantee that the operation of the primary users (PUs) is not affected, the secondary users (SUs) need to sense the presence of active PUs: this process is referred to as *spectrum sensing*.

It is necessary to detect the presence of PUs accurately and quickly in order to find available unused spectrum, which is called *spectrum holes*. This is done by "Spectrum sensing" an important feature of CR technology. Accurate sensing of spectrum holes is a hard task because of the time-varying nature of wireless channels (Cabric, 2004), including fading and shadowing. Presence of multi-path fading or shadowing in the sensing (S) channel (S-channel) between a PU and a SU, may limit the successful detection of the PU by a single SU (Digham, 2003). The detection/sensing performance can be improved, by limiting the negative impact of fading, if different SUs are allowed to cooperate by sharing their local sensing information on the activity status of PUs: this is the essence of *cooperative spectrum sensing* (CSS) (Akyildiz, 2006), (Ghasemi, 2005), (Zhang, 2008). More precisely, CR systems allow the CR users[1] to sense the spectrum of PUs opportunistically without creating any intolerable interference to PUs. In many wireless applications, it is of great interest to check the presence and availability of an active communication link when the PU signal is unknown. In such scenarios, one appropriate choice consists in using an energy detector (ED) which measures the energy in the received waveform over a proper observation time window (Urkowitz, 1967), (Digham, 2003). Therefore, CSS using EDs improves the detection performance when all CR users sense the PU individually and send their sensing information via reporting (R) channels (R-channels) to a fusion center (FC). In CSS systems, the sensing information on the PUs's activity status sent by several CR users is combined at the FC to obtain a global decision. In general, the sensing information reported to the FC by several CR users can be combined in two different ways: through (i) soft combining (Teguig, 2012), (Sun, 2011), (Nallagonda, 2013a) or (ii) hard combining (Choudhari, 2012(a)), (Choudhari, 2012(b)), (Choudhari, 2013), (Nallagonda, 2011b), (Nallagonda, 2012). In this book chapter, we focus on soft combining of spectrum sensing decisions from several CR users when the S- and R-channels are affected by fading and shadowing. Specifically, we study the impact of Weibull fading as well as Lognormal shadowing in the R-channels and the benefits of censoring the CR users on the basis of quality of the R-channels.

The rest of the chapter is organized as follows. Initially, we discuss on the background of this chapter: in particular, the motivation of the present work and the basics of CSS, along with existing works, are introduced. Next, we evaluate performance of CSS in faded environments (Rayleigh fading, Weibull fading, and Log-normal shadowing) under several hard and soft data fusion rules. We introduce the concept of censoring on the basis of the quality of R-channels, which is then incorporated into CSS systems. Specifically, two different censoring methods, such as rank-based and threshold-based censoring, have been analyzed under both perfect and imperfect channel estimation schemes. Finally, we conclude this chapter. The logical structure of the work presented in this chapter is shown in Figure 1.

Figure 1. Logical structure of the work in this chapter

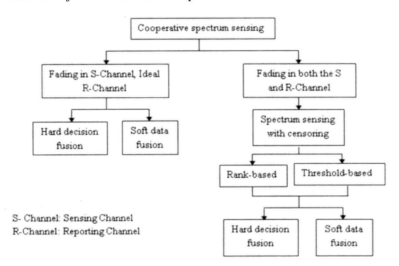

BACKGROUND

Hard decision combining fusion rules, such as OR-logic, AND-logic, and Majority-logic, can be implemented at FC to make the final decision on the presence or on the absence of a PU (Ghasemi, 2007), (Zhang, 2008). The existing literature on ED, using a single CR user (Nallagonda, 2011a) or cooperative CR users (Nallagonda, 2011b), (Nallagonda, 2012) performing spectrum sensing with hard decision combining fusion rules, typically assumes the following models for the channels Rayleigh/ Rician/ Nakagami-*m*/ Weibull fading and Log-normal shadowing. In contrast to hard decision combining fusion rules, where the FC receives a 1-bit binary local decision from a CR, it is possible to apply soft data fusion at FC to improve the performance of CSS. According to a soft combining approach, CR users transmit the entire local sensing samples or the complete local test statistics (instead of sending just 1-bit binary decision), which are combined using any one of possible diversity combining techniques.

Our present study focuses not only on hard decision fusion rules but also on the soft data fusion rule assuming Weibull fading and log-normal shadowing environments. Weibull fading has been proved to exhibit an excellent fitting for indoor (Hashemi, 1993) and outdoor (Adawi, 1988) environments. The Weibull distribution reduces to Rayleigh distribution for a certain value of the fading parameter (Ismail, 2006). In (Fathi, 2012), the authors mentioned that existing receiver diversity techniques, such as equal gain combining (EGC) and maximal ratio combining (MRC), can be utilized for soft combining of local observations or test statistics at the FC. This has motivated us to evaluate the performance of CSS under soft data fusion in Weibull fading channels. Existing works mostly examine the CSS under various soft data combining fusion schemes, such as square law selection (SLS), MRC, square law combining (SLC), selection combining (SC) using ED in AWGN, Nakagami-*m* fading, Log-normal shadowing, and Rician fading channels (Teguig, 2012), (Sun, 2011), (Nallagonda, 2013a). For example, in (Niu, 2003), the likelihood ratio test (LRT) fusion is discussed in the case of wireless sensor networks. In (Ma, 2008), an optimal soft combination scheme based on Neyman-pearson (NP) criterion is proposed to combine the weighted local observations. The proposed scheme reduces to EGC at high signal-to-noise ratio (SNR) and reduces to MRC at low SNR. In (Choudhari, 2012(a)), the authors provide the performance analysis and comparison of hard decision (HD) and soft decision (SD)- based CSS in the presence of R-channel errors. The effects of channel errors are incorporated in the analysis through the bit error probability (BEP). A general expression for the detection probability with K-out-of-N fusion rule has been derived for HD fusion in the presence of R-channel error. While, for SD-based CSS, an optimal fusion rule with R-channel errors is derived and its distribution is established, SD-based CSS has been found to yield a significant performance gain with regard to conventional counting rule-based HD (i.e. K-out-of-N) even in the presence of channel BEP. If the BEP is above a certain value, then regardless of the received signal strength of a PU, the constraints on the probabilities of correct detection and false alarm cannot be met. Furthermore, it is shown that SD-based CSS is more robust in terms of channel BEP (Choudhari, 2012(a). In (Choudhari, 2013), the impact of error detection and error correction coding for cooperative sequential sensing by CRs has been shown. In particular, a distributed parallel detection network is considered, where each SU sends a soft decision in the form of quantized local LLRs to the FC. At each SU, the quantized LLRs are converted to bits using Gray mapping. These bits are then channel encoded and transmitted, using BPSK, to the FC. The reporting channels between the SUs and the FC induce errors in the decision statistics. Performance with a simple error correcting code yields a significant improvement (Choudhari, 2013). In Chaudhari (2012b), the authors evaluate the performance of sequential detection scheme for CSS in CRs with R-channel errors. The sequential local LLRs are transmitted through erroneous R-channels. The R-channels are modeled as a simple binary symmetric channel (BSC) that induces errors with a certain bit error probability (BEP). However, in (Choudhari, 2012a, Choudhari, 2013, Chaudhari, 2012b) CSS with distributed detection approach is mainly considered. In (Chen, 2014), the authors provide a tutorial on various cooperative techniques in cognitive networks, with emphasis on spectrum sensing and access based cooperation, interference constraint-based adaptive cooperative feedback, rate less network coding-based cooperative transmission.

In (Han, 2013), the authors propose two novel quantization schemes to improve sensing performance considering soft decision fusion rule in CSS. In (Atlay, 2012), the authors consider the effect of imperfect R-channels on decision logics. In (Hamza, 2014), the authors find the global detection probabilities and secondary throughput through moment generating function (MGF)-based approach for the case of sensing with equal gain combining (EGC) in CSS. In particular, EGC is compared to MRC and situations where the former outperforms the latter are also identified. Sensing based on EGC always outperforms the sensing using orthogonal R-channels---such as TDMA---in terms of secondary throughput. Further, the effects of phase and synchronization errors on sensing performance are also evaluated in (Hamza, 2014). In (Abdi, 2013), a simulation study is conducted to optimize the linear SC scheme at the FC jointly with as a function of following these parameters, the number of bits used by each node to quantize the local sensing outcomes and the power levels at which each node reports its sensing outcome to the FC. Thus, the authors propose to optimize the linear SC scheme at the FC jointly with two significant mechanisms at reporting phase. It is demonstrated that by joint consideration of reporting and fusion phases the error detection performance can be improved because of better exploitation of spatial/user diversities. In (Zhao, 2013(b)), the authors propose a new soft fusion scheme based on LLRs, which simplifies the analysis of the data at the FC, and allows to derive exact closed-form expressions for the probabilities of missed detection and false alarm. In (Cui, 2013), the authors consider a relay-based dual-stage collaborative spectrum sensing (DCSS) model that combines distributed and centralized approaches. Furthermore, an efficient fast sensing algorithm for a large scale CR network is derived which requires the smallest number of CR users for DCSS while satisfying the target detection error rate bound (Cui, 2013). In (Paula, 2014), the authors consider a CSS scheme using a distributed approach with a

FC in presence of unreliable R-channel. The impact of errors introduced by R-channels on decision rule is analyzed following the Bayesian risk criterion. However cooperative sensing can incur cooperation overhead in terms of extra sensing time, delay, energy, and operations devoted to cooperative sensing. A detailed survey on these issues has been presented in (Akyildiz, 2011), which specifically addresses the issues of cooperation method, cooperative gain, and cooperation overhead.

In existing ED-based CSS systems, i.e., CSS with hard decision combining (Nallagonda, 2011b), (Nallagonda, 2012) or CSS with soft data combining fusions (Nallagonda, 2013a), R-channels are assumed to be ideal and S-channels are considered as noisy and faded / shadowed channels. However, in many practical situations R-channels may also be affected by noise and fading / shadowing channel (Zhao, 2013), (Zou, 2011), (Ferrari, 2006). Although most works on spectrum sensing assume ideal R-channels (Nallagonda, 2011b), (Nallagonda, 2012), (Nallagonda, 2013a), the presence of fading or shadowing in R-channels is likely to affect the sensing information sent by CR users where the FC is far from them (Choudhari, 2012(a)). If the R-channel is heavily faded or shadowed, the sensing information received at the FC is likely to be erroneous with respect to that transmitted by the CR user. If this is the case, it is better to interrupt the transmission of sensing information from such CR user and, thus, the use of censoring is *expedient*. The CR users whose R-channels are estimated as reliable by the FC are censored, i.e., they are allowed to transmit. The CR users which are not participating in improving the detection performance may be stopped, so that the system complexity can be reduced and the detection performance can be improved. This has the additional benefit of reducing the energy consumption in energy constrained network. Therefore, censoring of CR users is necessary to improve the performance of CSS. In (Zhao, 2013(a)), under the assumption of both S and R-channels are noisy and faded, a filter bank-based soft decision fusion (SDF) CSS system is proposed. In (Zou, 2011), a selective-relay based CSS scheme, assuming noisy and Rayleigh faded channels in both S- and R-channels, is proposed. Similar fading or shadowing conditions is considered in S and R-channels. For example, both S-channel and R-channel are noisy and Weibull faded or noisy and Log-normal shadowed. Although all the CR users detect PUs using EDs, only the CR users censored on the basis of the R-channel quality are allowed to transmit. The censoring decision is taken by the FC on the basis of estimation of R-channel. Using minimum mean square estimation (MMSE)-based estimation of the R-channels, the FC selects a subset of CR users among all the available ones (say P out of N) which have the highest channel coefficients, i.e. the CR users associated with best estimated channel coefficients are selected --- this approach is referred to as rank-based censoring (Nallagonda, 2013b). However, an alternative censoring scheme, based on channel thresholding (denoted also as threshold-based censoring and such that a CR user is selected to transmit its decision if the estimated R-channel fading coefficient exceeds a given threshold), is considered and analyzed in (Nallagonda, 2013c) (Nallagonda, 2013d). In (Nallagonda, 2014), the performance of CSS with both rank-based and threshold-based censoring is evaluated only with Rayleigh and Nakagami-*m* fading channels, considering majority-logic only at FC.

Channel estimation can be either perfect (no estimation error) or imperfect (with estimation errors). Accordingly, for each censoring strategy, there are two possibilities, namely perfect or imperfect channel estimation. The FC employs coherent reception to fuse the binary local decisions received from the censored CR users, in order to obtain a global decision regarding the presence or the absence of PUs. In (Nallagonda, 2014), a probabilistic model for CR user selection is proposed for Rayleigh, and Nakagami-*m* faded channels. In the current chapter, we consider the same censoring concept of (Nallagonda, 2013b), (Nallagonda, 2014) and evaluate the performance of CSS with rank-based censoring of CR users, based on R-channels quality. Similarly, considering the concept of channel thresholding of (Nallagonda, 2013c),

(Nallagonda, 2013d), (Nallagonda, 2014), we evaluate the performance of CSS with threshold-based censoring in the presence of Weibull fading, Log-normal shadowing in R-channels, with MRC and majority logic fusions at the FC. The investigation of majority-logic and MRC fusion schemes where both S- and R-channels are Weibull faded or Log-normal shadowed is an interesting research extension. Moreover, we develop the required analytical and simulation testbed for CSS, with both rank-based and threshold-based in the presence of Weibull faded, and Log-normal shadowed channels. In order to reach this goal, we develop closed-form expressions for the estimation error variance for Weibull fading channels. These expressions are useful to evaluate the performance of CSS with both censoring strategies in the presence of imperfect channel estimation. To the best of our knowledge, these expressions are not readily available in the literature. Along the way, we also develop probabilistic models of CR selection for Weibull faded, and Log-normal shadowed channels.

Our main contributions in the present chapter can be summarized as follows.

We carryout performance evaluation of soft data fusion (MRC-based) in Weibull faded channels as well as Log-normal shadowed channels. Furthermore, the performance of SD-based fusion is compared with that of HD-based fusion (typically; OR-logic, AND-logic, and majority-logic) and the impact of fading and shadowing on soft data fusion is investigated.

The performances of several hard-decision fusion rules are also evaluated and compared with each other in the presence of Weibull fading and Log-normal shadowing conditions.

A closed-form expression for the estimation error variance with Weibull fading is presented. This expression is expedient to evaluate the performance of CSS with censoring based on imperfect channel estimation. The impact of the R-channel estimation error on the detection performance in the considered fading scenario is evaluated. Direct performance comparisons between perfect and imperfect channel estimation cases, for various values of the main channel and network parameters, are carried out.

The performance, in terms of missed detection probability, under both perfect and imperfect channel estimation conditions, is investigated. The effects of Weibull fading and Log-normal shadowing in S- and R-channel and the impact of channel SNRs on the performance of the considered CSS schemes are investigated. In threshold-based censoring scenarios, novel analytical expressions, as functions of the censoring threshold C_{th}, for the selection of CRs are derived in Weibull fading and Log-normal shadowing channels. In particular, the probability mass function (PMF) of the number of censored CRs is analyzed. The impact of the number of available CRs and the average R-channel SNRs on the average missed detection probability of CSS is investigated. In threshold-based censoring schemes, the impact of the censoring threshold on the average missed detection is discussed. Finally, the performance comparison between MRC fusion and majority logic fusion, for various network parameters, as well as the effect of imperfect channel estimation is also highlighted.

COOPERATIVE SPECTRUM SENSING

We consider a CSS network of N CRs, one PU and one FC, as shown in Figure 2. Each CR senses the PU individually using energy detector and sends its sensing information/data to the FC. Depending on the transmitted sensing information from each CR to FC, the FC employs HD combining fusion if it receives one bit binary values (1/0) from the CRs or soft data combining fusion if it receives energy values (E) from the CRs. Particularly, OR-logic, AND logic, majority-logic, and MRC fusion are performed separately at FC to make the global decision about the presence or the absence of the PU. In this chapter,

Figure 2. Cooperative spectrum sensing (CSS) system

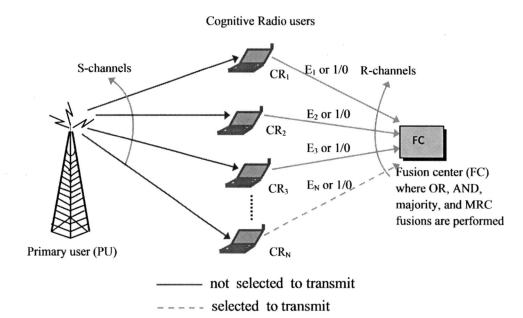

we consider additive noise and Weibull fading or additive noise and shadowing in the S-channels, while R-channels are assumed as either (i) ideal channels (i.e., noiseless channels) or (ii) Weibull faded or Log-normal shadowed channels. Based on these two cases of interest, a CSS system can be classified as follows. In the first scenario, we assume that the R-channels are ideal and we evaluate the performance of CSS. In the second scenario, we consider fading or shadowing in the R-channels.

The received signal $x(t)$ at k-th CR can be represented as:

$$x_k(t) = \begin{cases} n_k(t) & H_0 \\ h_k s(t) + n_k(t) & H_1 \end{cases} \tag{1}$$

where $s(t)$ is the PU signal with energy E_s and $n(t)$ is the noise waveform. The noise $n(t)$ is modeled as a zero-mean white Gaussian random process. The S-channel fading coefficient for the k-th CR is denoted as h_k. H_1 and H_0 are the two hypotheses associated with the presence and absence of a PU, respectively. When the PU is absent i.e., under hypotheses H_0, each CR receives only the noise signal at the input of the ED and the noise energy at k-th CR can be approximated over the time interval $(0, T)$, as (Urkowitz, 1967), (Digham, 2003):

$$\int_0^T n_k^2(t)dt = \frac{1}{2W} \sum_{i=1}^{2u} n_{k_i}^2, \tag{2}$$

where u $(=TW)$ is the time-bandwidth product, T is the observation time, and W is the one-sided bandwidth. Furthermore,

$$n_{k_i} \sim N(0, N_{01}W), \ \forall \ i. \tag{3}$$

where N_{01} is the one-sided noise power spectral density, and $N(\mu, \sigma^2)$ is a Gaussian random variable with mean μ and variance σ^2. The received signal energy under hypothesis H_0 at k-th CR, denoted as E_k, can be written as:

$$E_k = \sum_{i=1}^{2u} n_{k_i}'^2 \ ; \ n_{k_i}' = \frac{n_{k_i}}{\sqrt{N_{01}W}}, \tag{4}$$

The same approach can be followed to evaluate the received signal energy under hypothesis H_1 at the k-th CR by replacing n_{k_i} with $n_{k_i} + s_i$, where $s_i = s(\frac{i}{2W})$.

In a non-faded environment (i.e., $h_k = 1$) i.e., with only additive white Gaussian noise, the detection and false alarm probabilities for the k-th CR can be expressed as follows (Digham, 2003), (Nallagonda, 2011b), (Nallagonda, 2012):

$$P_{d,k} = P(E_{k1} > \lambda | H_1) = Q_u(\sqrt{2\gamma_{s,k}}, \ \sqrt{\lambda}) \tag{5}$$

$$P_{f,k} = P(E_{k0} > \lambda | H_0) = \Gamma(u, \lambda/2) / \Gamma(u) \tag{6}$$

$$P_{m,k} = 1 - P_{d,k} \tag{7}$$

where $\gamma_{s,k}$ is the instantaneous SNR of k-th S-channel, $\Gamma(.,.)$ is the incomplete gamma function, and $Q_u(.,.)$ is the generalized Marcum Q-function. The expression for $P_{f,k}$ for the k-th CR, as given in (6), remains the same when fading is considered in the S-channel due to independence of $P_{f,k}$ from SNR. The detection threshold λ can be set for a chosen $P_{f,k}$ following (6). Equation (5) gives the probability of detection as a function of $\gamma_{s,k}$. However when h_k is varying due to fading, the average detection probability at k-th CR ($\bar{P}_{d,k}$) may be obtained by averaging (5) over fading or shadowing statistics (Ghasemi, 2005), (Nallagonda, 2011b), (Nallagonda, 2012)

$$\bar{P}_{d,k} = \int_0^\infty Q_u(\sqrt{2\gamma_s}, \ \sqrt{\lambda}) f_{\gamma_s}(\gamma_s) d\gamma_s \tag{8}$$

where $f_\gamma(x)$ is the probability density function (PDF) of SNR under fading or shadowing.

RAYLEIGH FADING

In the Rayleigh faded scenario, the k-th S-channel fading coefficient h_k can be expressed as a function of the Gaussian in-phase X_1 and quadrature X_2 elements of the multipath components (Simon, 2014), (Nallagonda, 2011a, 2012):

$$| h_k |=| X_1 + jX_2 |= \sqrt{X_1^2 + X_2^2} \; ; \; X_{1,2} \sim N(0, \; 1/2) . \tag{9}$$

If the signal amplitude follows a Rayleigh distribution, then the SNR γ_s follows an exponential PDF given by (Simon, 2014):

$$f_{\gamma_s}(\gamma_s) = \tfrac{1}{\bar{\gamma}_s} \exp\left(-\tfrac{\gamma_s}{\bar{\gamma}_s}\right) \qquad \gamma_s \geq 0, \tag{10}$$

where $\bar{\gamma}_s$ is the average S-channel SNR. The average P_d in this case, $\overline{P}_{dRay}$, can be evaluated by substituting (10) in (8) (Nallagonda, 2014), (Digham, 2003), as

$$\overline{P}_{d,k,Ray} = \exp(-\tfrac{\lambda}{2})\sum_{k=0}^{m-2} \tfrac{1}{k!} (\tfrac{\lambda}{2})^k + (\tfrac{1+\bar{\gamma}_s}{\bar{\gamma}_s})^{m-1}$$

$$\times \left(\exp(-\tfrac{\lambda}{2(1+\bar{\gamma}_s)}) - \exp(-\tfrac{\lambda}{2})\sum_{k=0}^{m-2} \tfrac{1}{k!} (\tfrac{\lambda\bar{\gamma}_s}{2(1+\bar{\gamma}_s)})^k \right) \tag{11}$$

WEIBULL FADING

In the Weibull fading model, the k-th S-channel fading coefficient h_k can be expressed as a function of the Gaussian in-phase X_1 and quadrature X_2 elements of the multipath components (Hashemi, 1993) (Adawi, 1988), (Nallagonda, 2011b):

$$h_k = (X_1 + jX_2)^{2/v} \tag{12}$$

where v is the Weibull fading parameter and $j = \sqrt{-1}$. Let X be the magnitude of h_k, i.e., $X = |h_k|$. If $R = |X_1 + jX_2|$ is a Rayleigh distributed random variable, a Weibull distributed random variable can be obtained by transforming R and using (12) as

$$X = R^{2/v} \tag{13}$$

The PDF of the SNR (γ_s) in a Weibull faded channel is given by (Sagias, 2004) i.e.,

$$f_{\gamma_s}(\gamma_s) = c \left[\frac{\Gamma(p)}{\overline{\gamma}_s} \right]^c \gamma_s^{c-1} \exp\left[-\left\{ \frac{\gamma_s \Gamma(p)}{\overline{\gamma}_s} \right\}^c \right] \tag{14}$$

where $c = v/2$ and $p = 1 + 1/c$. Furthermore, the average probability of detection $\overline{P}_{d,k,we}$ can be obtained by substituting (14) in (8).

LOG-NORMAL SHADOWING

In this case, we assume that the S-channel between PU and CR is shadow faded. The linear channel gain, h_k may be modeled by a log-normal random variable i.e. $h_k = e^X$ where X is a zero-mean Gaussian random variable with variance σ^2. Log-normal shadowing is usually characterized in terms of its dB-spread, σ_{dB} which is related to σ by $\sigma = 0.1\ln(10) \sigma_{dB}$. The PDF of SNR, γ_s, in Log-normal shadowing channel is given by (Simon, 2004)

$$f_{\gamma_s}(\gamma_s) = \frac{10}{\ln(10)\sqrt{2\pi}\sigma\gamma_s} \exp\left[-\frac{(10\log_{10}\gamma_s - \overline{\gamma}_s)^2}{2\sigma^2} \right] \tag{15}$$

Further, the average probability of detection $\overline{P}_{d,k,\ln}$ can be obtained by substituting (15) in (8).

The FC employs different hard decision and soft data combining fusion rules to attain a final decision from the received signals. Several types of hard decision fusion algorithms such as OR-logic, AND-logic, and majority-logic and soft data fusion such as MRC fusion are considered in the current text as described in the next section.

HARD DECISION FUSION RULES

In the case of HD, the CRs make the one-bit binary decision called local decision by comparing the received energy with a detection threshold λ. Several HD fusion rules, such as OR-rule, AND-rule, and majority-rules, can be implemented at the FC. Assuming independent decisions, the fusion problem, using k-out of-N CRs for decision, is described by a binomial distribution or is derived from Bernoulli trials where each trial represents the decision of each CR. The generalized formula for overall probability of detection Q_d at FC for the k-out of-N fusion rule is given by (Nallagonda, 2011b, 2012, 2014), (Ghasemi, 2007), (Zhang, 2008)

$$Q_d = \sum_{l=k}^{N} \binom{N}{l} \overline{P}_d^l \left(1 - \overline{P}_d \right)^{N-l}. \tag{16}$$

where $\overline{P}_d$ is the probability of detection for each individual CR as defined by (8).

The OR-logic fusion rule (i.e., 1-out of-*N* rule) can be evaluated by setting *k*=1 in (16)

$$Q_{d,OR} = \sum_{l=1}^{N} \binom{N}{l} \bar{P}_d^l \left(1 - \bar{P}_d\right)^{N-l} = 1 - (1 - \bar{P}_d)^N .$$

(17)

The AND-logic fusion rule (i.e., *N*-out of-*N* rule) can be evaluated by setting *k*=*N* in (16)

$$Q_{d,AND} = \sum_{l=N}^{N} \binom{N}{l} \bar{P}_d^l \left(1 - \bar{P}_d\right)^{N-l} = (\bar{P}_d)^N .$$

(18)

For the case of majority-rule (i.e., *N*/2 out of-*N* rule) the $Q_{d,MAJ}$ is evaluated by setting $k = \lfloor N/2 \rfloor$ in (16). Similarly, the overall probability of false alarm Q_f for the generalized *k*-out of-*N* fusion rule can be evaluated by replacing $\bar{P}_d$ with P_f in (16).

SOFT DATA FUSION RULE

In the case of SD combining fusion, each CR forwards the entire sensing information (energy value, *E*) to the FC without performing any local decision. The FC employs soft data combining fusion and takes the final decision about the PU. The MRC fusion is considered in this chapter. First the energy value of the received signal at each CR is obtained using an ED. Next, all CRs send their respective energy values, with appropriate weighting, to the FC. Then the FC gathers all the data (energy values with appropriate weights) from all CRs, combines them, and makes a global decision by comparing the obtained value with a detection threshold, λ. The MRC fusion requires sensing channel state information to amplify the energy values. It may be observed that as we assume R-channels to be ideal, the MRC fusion rule at FC simply becomes the sum of the weighted signals received from all the CRs over their ideal R-channels. We also assume that signals sent by CRs over their R-channels are orthogonal, i.e., there is no interference amongst the signals sent by different CRs to the FC. Over AWGN channels, the probabilities of false alarm and correct detection under MRC fusion, are given by (Teguig, 2012), (Sun, 2011), (Nallagonda, 2013a)

$$Q_{f,MRC} = \frac{\Gamma(u, \lambda/2)}{\Gamma(u)}$$

(19)

$$Q_{d,MRC} = Q_u(\sqrt{2\gamma_{s,MRC}}, \ \sqrt{\lambda})$$

(20)

where $\gamma_{s,MRC} = \sum_{i=1}^{N} \gamma_{s,i}$, denotes the instantaneous SNR at the output of the MRC combiner. However, in the presence of fading, the probability of correct detection can be found by averaging $Q_{d,MRC}$ given by (20) with respect to fading statistical distribution or, equivalently, over the SNR distribution as follows:

$$\bar{Q}_{d,MRC} = \int_0^\infty Q_{d,MRC}(\gamma_{s,MRC}, \lambda) f(\gamma_{s,MRC}) \, d\gamma_{s,MRC} \qquad (21)$$

where $f(\gamma_{s,MRC})$ is the PDF of S-channel SNR under fading or shadowing.

Results

The following results are obtained using MATLAB-based simulations. The performance of CSS is evaluated in the presence of Weibull fading and Log-normal shadowing environments. The R-channels are assumed ideal. In particular, a performance comparison between HD and SD fusion rules at the FC is investigated.

In Figure 3 and Figure 4, the performance of CSS with hard decision and soft data fusion rules, respectively, is investigated. In particular, the performance is investigated through the use of complementary ROC curves, i.e., Q_m versus Q_f. In all cases, 3 CRs are considered and the performance is evaluated in Weibull fading environment (Figure 3) and Log-normal shadowing environments (Figure 4). In these figures, for comparison purposes, the curve for non-cooperation case ($N=1$) is also shown. It is obvious that the system with CSS among multiple CRs outperforms the system with a single CR (no-cooperation case), except for the AND rule. This is due to the fact that cooperation among the CRs cancels the negative effects of fading or shadowing in the S-channels as compared to the single CR case. It can be observed from both figures, that, for a particular value of Q_f, the performance with OR-rule is better than those with majority and AND-rules. It can also be seen, in both Weibull faded and Log-normal shadowed environments, that the schemes with MRC fusion outperforms all HD fusion rules. However, these benefits are obtained at the cost of a larger bandwidth for the reporting channel. On the other hand, HD fusion rules have a lower complexity. The impacts of fading and shadowing on the detection performance of MRC fusion are also investigated in Figure 3 and Figure 4, respectively. Sensing performance improves significantly for higher values of the Weibull fading parameter and lower values of the shadowing parameter. This is due to the fact that higher values of the fading parameter or lower values of the shadowing parameter reduce the severity of fading and shadowing in the S-channel.

Spectrum Sensing with Censoring

As already discussed in above sections, in the first considered scenario, the performance of CSS, considering S-channels as noisy-faded and R-channels as ideal, has been well studied. However, in a more realistic scenario, R-channels may not be noiseless (ideal) channels. Though most works on spectrum sensing assume noiseless R-channels (Nallagonda, 2013b, 2013c, 2013d, 2014), the presence of fading or shadowing in R-channels is likely to affect the sensing information sent by CRs to FC. If the R-channel is heavily faded or shadowed, the sensing information received at the FC is likely to be an erroneous version of that transmitted by the CR. Under such conditions, it is wise to stop transmitting sensing

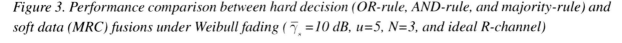

Figure 3. Performance comparison between hard decision (OR-rule, AND-rule, and majority-rule) and soft data (MRC) fusions under Weibull fading ($\bar{\gamma}_s = 10$ dB, u=5, N=3, and ideal R-channel)

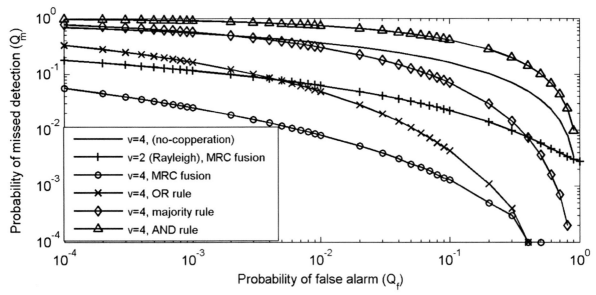

Figure 4. Performance comparison between hard decision (OR-rule, AND-rule, and majority-rule) and soft data (MRC) fusions under Log-normal shadowing ($\bar{\gamma}_s = 10$ dB, u=5, N=3, and ideal R-channel)

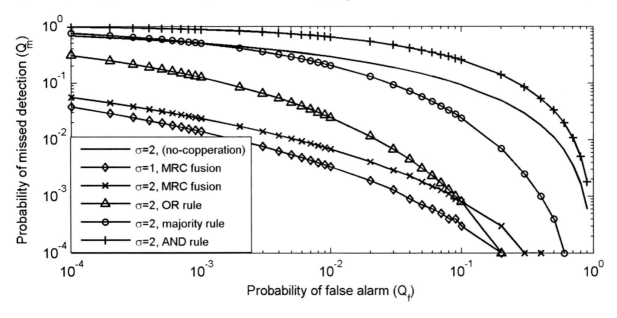

information from these CRs: censoring is thus expedient in such scenarios. The CRs whose R-channels are estimated as reliable by the FC are censored, i.e., they are allowed to transmit their sensing information, while the CRs with poor R-channels are stopped. This helps to reduce system complexity and to improve detection performance. Therefore, censoring of CRs is necessary to improve the performance of CSS. The CSS system with censoring is also shown in Figure 2.

In censoring-based CSS systems, we assume that both S-and R-channels are modeled as noisy and faded or noisy with shadowing. In this section, we study the performance of CSS with two censoring schemes, namely: (i) rank-based censoring and (ii) threshold-based censoring in the presence of Weibull fading and Log-normal shadowing. MRC and majority-logic fusion rules are considered at the FC on the reception of sensing information, received from censored CRs to obtain a global decision regarding the presence or the absence of PUs. The overall probability of missed detection is selected as the key performance metric and is evaluated, through simulations, under several channel and network conditions. A CR has its individual sensing information and, if censored, transmits its information, using binary phase shift keying (BPSK) as modulation format, to the FC over the corresponding faded or shadowed R-channel.

Transmissions between the CRs and the FC are carried out in two phases. In the first transmission phase, each CR sends one training symbol to enable the FC to estimate all fading or shadowing channel coefficients between the FC and N participating CRs. MMSE-estimation of the R-channel coefficients is carried out at the FC using training symbols sent by the CRs to the FC. The signal from the k-th CR received at the FC can be expressed as (Nallagonda, 2013b, 2013c, 2013d, 2014):

$$y_k = s_k h_k + n_k \quad k \in \left\{1, 2, ..., N\right\} \tag{22}$$

where $s_k = {}^{+}\!\!/_{-}\sqrt{E_b}$ is a BPSK signal corresponding to H_1 / H_0, respectively. The R-channel fading or shadowing coefficient is denoted as h_k and the R-channel noise is $n_k \sim CN\left(0, \tilde{A}_n^2\right)$. It should be observed that identical notation of h_k for both S- and R-channel coefficients is indicated which means that the same fading/shadowing parameter is considered in both S-and R-channels. The Gaussian channel noise samples $\left\{n_k\right\}$ and R-channel coefficients $\left\{h_k\right\}$ are mutually independent. We assume that the FC estimates the k-th CR's R-channel fading or shadowing coefficient h_k according to an MMSE estimation strategy on the basis of the observable y_k as follows (Nallagonda, 2013b, 2013c, 2013d, 2014):

$$\hat{h}_k = E[h_k \mid y_k] = \frac{\sqrt{E_b}}{E_b + \sigma_n^2} y_k$$

$$= \frac{E_b}{E_b + \sigma_n^2} h_k + \frac{\sqrt{E_b}}{E_b + \sigma_n^2} n_k. \tag{23}$$

The estimation error for k-th R-channel co-efficient can be expressed as $\tilde{h}_k = h_k - \hat{h}_k$

$$\tilde{h}_k = h_k \left(1 - \frac{E_b}{E_b + \sigma_n^2}\right) - \frac{\sqrt{E_b}}{E_b + \sigma_n^2} n_k$$

$$= h_k \frac{\sigma_n^2}{E_b + \sigma_n^2} - \frac{\sqrt{E_b}}{E_b + \sigma_n^2} n_k . \tag{24}$$

The channel estimation is either perfect $\left(\hat{h}_k = h_k, \tilde{h}_k = 0\right)$ or imperfect $\left(\hat{h}_k = h_k - \tilde{h}_k, \tilde{h}_k \neq 0\right)$. After the first phase of transmission, 'Z' $\left(Z \leq N\right)$ CRs, out of 'N' available CRs, are selected using any one of the two possible censoring schemes (i) *rank-based* ($Z=P$; the selected CRs are associated with the best P estimated channel coefficients, where $P \leq N$), and (ii) *threshold-based* ($Z=K$; the selected K CRs have estimated channel coefficients exceeding a predefined threshold C_{th} and the value of 'K'. The FC informs the selected CRs via one-bit feedback (we assume that feedback channels are error-free).

In the second transmission phase, the selected CRs send their local binary BPSK modulated sensing information to the FC over the corresponding R-channels. The fading or shadowing coefficients of R-channels are assumed to be fixed over a symbol transmission time, as the channel is assumed to be slowly faded or shadowed. The signal, received from the k-th selected CR, at the FC is (Nallagonda, 2013b, 2013c, 2013d, 2014)

$$y_{k,d} = m_k h_k + n_{k,d}; \quad k \in \left\{1, 2, ..., Z\right\} \tag{25}$$

where channel noise $n_{k,d} \sim CN\left(0, \sigma_n^2\right)$ and $m_k \in \left\{+\sqrt{E_b}, -\sqrt{E_b}\right\}$ is the BPSK modulated binary decisions.

The MRC fusion rule depends on the R-channel estimates, the CR's performance indices, and incorporates the effect of channel estimation error. Assuming that the CRs have identical local performance indices and BPSK is the used modulation format, the MRC fusion rule can be obtained by simplifying the LRT fusion. In the case of ideal R-channel, as described earlier, the weights used for MRC depend on S-channel state information, while in present case the weights used for MRC fusion depend on R-channel state information. The LRT fusion can be written as follows (Nallagonda, 2013b, 2013d)

$$\Lambda = \prod_{k=1}^{N} \frac{f(y_{k,d} \mid H_1)}{f(y_{k,d} \mid H_0)}$$

$$= \prod_{k=1}^{N} \frac{\bar{P}_{d_k} + (1 - \bar{P}_{d_k})e^{\frac{-4\sqrt{E_b}}{\sigma_w^2} \text{Re}(y_{k,d}\hat{h}_k^*)}}{P_{f_k} + (1 - P_{f_k})e^{\frac{-4\sqrt{E_b}}{\sigma_w^2} \text{Re}(y_{k,d}\hat{h}_k^*)}} \tag{26}$$

where

$$\sigma_w^2 = E_b \sigma_{\tilde{h}}^2 + \sigma_n^2 = \frac{E_b \sigma_n^2}{E_b + \sigma_n^2} + \sigma_n^2, \ \bar{P}_{d_k} = \bar{P}_d, \ P_{f_k} = P_f \quad \forall \ k \tag{27}$$

In the case of perfect channel estimation, σ_w^2 is equal to σ_n^2 when the estimation error variance $\tilde{A}_h^2$ is zero. At very low R-channel SNRs both σ_n^2 and σ_w^2 tend to be very large. By taking a logarithm of both sides in (26) and using the approximations $e^{-x} \approx 1 - x$ and $\log(1 + x) \approx x$ for small values of x, we can simplify the LRT rule as:

$$\Lambda_1 = \log(\Lambda) = \frac{2\sqrt{E_b}}{\sigma_w^2} \sum_{k=1}^{N} (\overline{P}_{d_k} - P_{f_k}) \operatorname{Re}(y_{k,d} \hat{h}_k^*). \tag{28}$$

Under the assumption that the CRs have identical local performance indices, Λ_1 can be simplified further as follows (Nallagonda, 2013b, 2013d):

$$\Lambda_{MRC} = \sum_{k=1}^{N} \operatorname{Re}\left[y_{k,d} \, \hat{h}_k^* \right] \tag{29}$$

where $\hat{h}_k^*$ is the complex conjugate of the estimated channel coefficient $\hat{h}_k$ and the signal at the FC received from k-th selected CR is $y_{k,d}$ (given by Equation (25)). Given $\hat{h}_k$, one can observe from (29) that Λ_{MRC} is a linear combination of Gaussian random variables and, therefore, has a Gaussian distribution. The FC can then take a decision in favor of H_1 or H_0 simply by comparing Λ_{MRC} with the threshold zero.

Generation of Estimated and Estimated Error Coefficients for *k*-th R-Channel

In this section, we discuss methods for generating the estimated R-channel coefficient ($\hat{h}_k$) and estimated error coefficient ($\tilde{h}_k$) for Rayleigh, Weibull faded, and Log-normal shadowed R-channels. To the best of our knowledge, the error variance expression for Weibull fading channel is not readily available in the literature.

Estimation Error in Rayleigh Channel

For k-th Rayleigh faded R-channel, the estimation error is a zero-mean complex Gaussian random variable with the following variance (Nallagonda, 2013b, 2013d, 2014):

$$\sigma_{\tilde{h}.R}^2 = \left(\frac{E_b}{\sigma_n^2} + 1 \right)^{-1} \tag{30}$$

At this point, the k-th Rayleigh faded R-channel estimation error coefficient can be generated according to the following equations:

$$| \tilde{h}_k | = \sqrt{\tilde{h}_{kI}^2 + \tilde{h}_{kQ}^2} \; ; \; \tilde{h}_{kI} \sim N\left(0, \; \frac{\sigma_{\tilde{h},R}^2}{2}\right), \; \tilde{h}_{kQ} \sim N\left(0, \; \frac{\sigma_{\tilde{h},R}^2}{2}\right). \tag{31}$$

Estimation Error in Weibull Fading Channel

For the k-th Weibull faded R-channel, the actual fading coefficient h_k can be expressed, in terms of in-phase (h_{kI}) and quadrature (h_{kQ}) components as

$$h_k = (h_{kI} + jh_{kQ})^{2/v} \tag{32}$$

where v is the Weibull fading parameter (also denoted as shape parameter). The amplitude or envelope $| h_k |$ is Weibull distributed only when $h_{k,I,Q} \sim N(0, \sigma^2 / 2); \; \sigma^2 = 1$. The mean and variance of h_k are

$$\mu_{h,we} = E[h_k]_{we} = w\Gamma(1 + 1 / v) \tag{33}$$

$$\sigma_{h,we}^2 = w^2\Gamma(1 + 2 / v) - w^2\Gamma(1 + 1 / v)^2 \tag{34}$$

where w is the scale parameter. The estimated k-th Weibull faded R-channel coefficient ($\hat{h}_k$) can be generated by substituting (32) in (23) as

$$\hat{h}_k = \frac{E_b}{E_b + \sigma_n^2} (h_{kI} + jh_{kQ})^{2/v} + \frac{\sqrt{E_b}}{E_b + \sigma_n^2} n_k . \tag{35}$$

Finally, the estimated Weibull fading coefficient can be generated by substituting $h_{kI}, h_{kQ} \sim N(0, 1 / 2)$ in (35) and $n_k = n_{kI} + jn_{kQ}$, where $n_{kI}, n_{kQ} \sim N(0, \sigma_n^2 / 2)$. From (35), the estimation error coefficient for the k-th Weibull faded R-channel $\tilde{h}_k (= h_k - \hat{h}_k)$ can be generated by substituting (32) in (24) as

$$\tilde{h}_k = \frac{\sigma_n^2}{E_b + \sigma_n^2} (h_{kI} + jh_{kQ})^{2/v} - \frac{\sqrt{E_b}}{E_b + \sigma_n^2} n_k . \tag{36}$$

The mean and variance of $\tilde{h}_k$ can be evaluated by using (33), and (34) as

$$E[\tilde{h}_k]_{we} = \frac{1}{1 + \bar{\gamma}_r} w\Gamma(1 + 1 / v) \tag{37}$$

$$\sigma^2_{\hat{h},we} = \frac{1}{(1+\bar{\gamma}_r)^2}\left[\bar{\gamma}_r + w^2\Gamma(1+2/v) - w^2\Gamma(1+1/v)^2\right]$$

$$= \frac{1}{(1+\bar{\gamma}_r)^2}\left[\bar{\gamma}_r + \sigma^2_{h,we}\right] \tag{38}$$

For *v*=2, the term $\sigma^2_{h,we}$ in (38) is equal to the variance of a Rayleigh distributed random variable, which is assumed to be equal to one and (38) gives an alternative expression for estimated error variance in Rayleigh channel ($\sigma^2_{\hat{h},R}$), which matches with (30).

Estimation Error in Log-Normal Shadowing Channel

For the *k*-th Log-normal shadowing R-channel, the actual Log-normal coefficient h_k can be expressed as follows: if *X* is a zero-mean Gaussian RV with variance σ^2, i.e., $X \sim N(0,\sigma^2)$, then a log-normal coefficient can be modeled as $h_k = \exp(X)$. Finally, the Log-normal shadowing coefficient can be generated by using σ of $X \sim N(0,\sigma^2)$ as $\sigma = 0.1\ln(10)\sigma_{dB}$, where σ_{dB} is the shadowing parameter, generally expressed in terms of its dB-spread. The estimated shadowing coefficient for the *k*-th R-channel ($\hat{h}_k$) can be generated by substituting $h_k = \exp(X)$, where $X \sim N(0,(0.1\ln(10)\sigma_{dB})^2)$, and $n_k = n_{kI} + jn_{kQ}$, where $n_{kI}, n_{kQ} \sim N(0, \sigma^2_n / 2)$ in (23), as follows:

$$\hat{h}_k = \frac{E_b}{E_b + \sigma^2_n}\exp(X) + \frac{\sqrt{E_b}}{E_b + \sigma^2_n}(n_{kI} + jn_{kQ}) \tag{39}$$

From (39), the estimation error $\tilde{h}_k$ for *k*-th R-channel can be generated by substituting the shadowing coefficient h_k and the complex Gaussian noise coefficient n_k in (24) as follows:

$$\tilde{h}_k = \frac{\sigma^2_n}{E_b + \sigma^2_n}\exp(X) - \frac{\sqrt{E_b}}{E_b + \sigma^2_n}(n_{kI} + jn_{kQ}) \tag{40}$$

Due to analytical complexity, it may not be possible to simplify further the Equations (35), (36), (39), and (40).

Rank-Based Censoring

According to this censoring scheme (Nallagonda, 2013b, 2013d, 2014), *P* (out of *N*) CRs—those with the best estimated channel coefficients (i.e., the highest ones)—are selected. More precisely, using MMSE-based estimation of the R-channels, the amplitudes of the *N* estimated R-channel coefficients are sorted in decreasing order and the FC selects the *P* CRs ($P \leq N$) which have highest estimated R-channel coefficients i.e., CRs associated with best estimated R-channel coefficients are selected. The FC fuses the sensing information, received as BPSK signal from the P selected CRs, to obtain a final decision on the presence or absence of the PU. The overall probability of missed detection can be evaluated by using MRC fusion.

The following results are obtained using MATLAB based simulations. The performance of CSS with rank-based censoring for both perfect and imperfect channel estimation cases has been evaluated in Weibull faded and Log-normal shadowed environments. Particularly, the overall missed detection probability (Q_m) is evaluated as function of number (P) of selected CRs, considering the impact of average S-channel SNR ($\overline{\gamma}_s$) and the average R-channel SNRs ($\overline{\gamma}_r$). The results for majority-logic fusion are also shown for comparison purpose.

In Figure 5 and Figure 6, the performance of CSS with rank-based censoring is evaluated in the presence of Weibull fading and Log-normal shadowing, respectively. In both figures, Q_m is shown as a function of number(P) of selected CRs. The results are shown for both the cases with perfect and imperfect channel estimations by performing MRC fusion at FC in both the figures. For comparison purposes, the curves for majority-logic fusion are shown. In Figure 5, the impact of $\overline{\gamma}_r$ and $\overline{\gamma}_s$ on missed detection performanceis investigated. Two values of $\overline{\gamma}_r$ (-8 dB, -6 dB) and two values of $\overline{\gamma}_s$ (15 dB, 20 dB) are considered. When any one of the parameters P, $\overline{\gamma}_r$ and $\overline{\gamma}_s$ increases, Q_m reduces with both perfect and imperfect channel estimations. The probability of incorrect reception from CRs at FC reduces with higher $\overline{\gamma}_r$. As expected, for a given value of the $\overline{\gamma}_r$, Q_m is higher with imperfect channel estimation, as channel-based censoring leads to the selection of a group of CRs which may not be the best ones due to channel estimation errors. Furthermore, according to (38), an increase in $\overline{\gamma}_r$ leads to a decrease in estimation error variance $\sigma^2_{\hat{h},weib}$ and this, in turns, reduces the average estimation error. A reduced estima-

Figure 5. Performance of CSS in terms of Q_m under perfect and imperfect channel estimation for different values of $\overline{\gamma}_r$, and $\overline{\gamma}_s$ in Weibull faded (v=4) environment (u=5, P_f=0.05, N=30, and rank-based censoring).

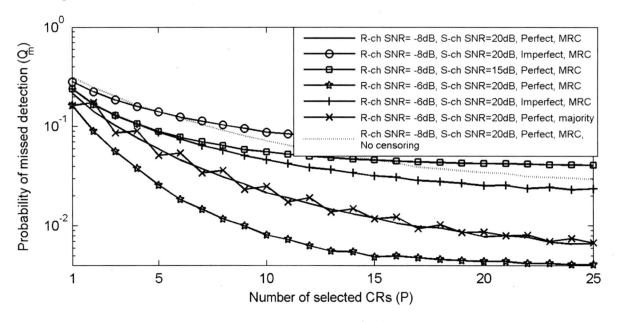

Figure 6. Performance of CSS in terms of Q_m under perfect and imperfect channel estimation for different values of σ, and $\bar{\gamma}_r$ in Log-normal shadowing environment (u=5, P_f=0.05, N=30, and rank-based censoring).

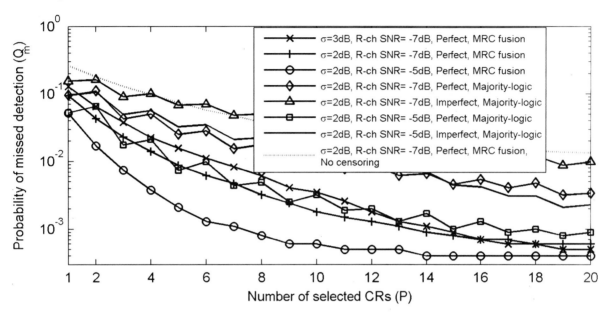

tion error leads to a further reduction of Q_m. In particular, in case of imperfect channel estimation with P=15, Q_m decreases by 59.16% when $\bar{\gamma}_r$ increases from -8 dB to -6 dB. Similarly, in the case of perfect channel estimation, Q_m decreases by 53.76% for the same values of P and $\bar{\gamma}_r$. Higher $\bar{\gamma}_s$ improves the detection of the PU at the CR. For example, in the case of perfect channel estimation with P=12 and $\bar{\gamma}_r$ =-8 dB, as $\bar{\gamma}_s$ increases from 15 dB to 20 dB, Q_m decreases by 73.96%. We observe from Figure 5 that missed detection performance with MRC fusion is better than the performance with majority-logic fusion in both the cases of channel estimations. For example, in case of perfect channel estimation, for P=12, $\bar{\gamma}_r$ = -6 dB and $\bar{\gamma}_s$ =20 dB, Q_m with MRC is 58.47% lower than Q_m with majority-logic fusion. In Figure 6, the missed detection is evaluated for various values of shadowing parameter (σ in dB) and $\bar{\gamma}_r$. It is seen that there is a significant impact of shadowing parameter on missed detection performance. More precisely, the performance degrades for higher value of σ due to increase in value of σ, S/R-channels undergo severe effect of shadowing so that quality of S/R-channels become very poor and erroneous transmission/reception occurs. The Q_m reduces with increase in the value of $\bar{\gamma}_r$ due to reduction of noise effect in the R-channel. Further, MRC fusion provides better performance than majority-logic fusion, though MRC fusion depends on channel estimation. From both Figure 5 and Figure 6, it is observed that in case of majority logic fusion, zigzag nature of the curve may be attributed due to occurrence of tie at FC in the case of even number of selected CRs. However in the case of no censoring, the missed detection probability is seen to be high.

Threshold Based Censoring

In this censoring scheme (Nallagonda, 2013c, 2013d, 2014), a CR (say the k-th) is selected for transmission if the amplitude of the corresponding estimated R-channel coefficient $\hat{h}_k$ is above a censoring threshold (C_{th}). This approach involves two transmission phases: in the first phase, the FC estimates the R-channel corresponding to each CR using MMSE; in the second phase, the FC censors a CR if the corresponding estimated channel coefficient exceeds a chosen threshold C_{th}. In this section, novel analytical expressions, as functions of C_{th}, for the selection of CRs are derived in different channels such as Weibull fading, and Log-normal shadowing channels. The threshold based selection of CRs in the case of Rayleigh faded R-channels is analyzed in (Nallagonda, 2013c, 2013d, 2014). In particular, the probability mass functions (PMF) of the number of censored CRs; the expression for the average missed detection ($\bar{Q}_m$) and false alarm ($\bar{Q}_f$) probabilities are analyzed in the context of Rayleigh fading in R-channel.

Threshold Based Censoring in Weibull Fading and Log-Normal Shadowing Channels

If the amplitude of estimated R-channel fading coefficient is a Weibull distributed random variable, the cumulative distribution function (CDF) of Weibull distributed random variable in terms of C_{th} can be derived as

$$F_{we}(C_{th}) = 1 - \exp(-\{C_{th} / \Omega\}^v); \quad C_{th} \geq 0 \tag{41}$$

where v is the Weibull fading parameter which ranges from 1 to ∞ and $\Omega = 1$. Now the probability of selecting a CR in Weibull faded R-channel can be derived as

$$
\begin{aligned}
p_{we} &= \Pr(|\hat{h}_k| > C_{th}) \\
&= 1 - F_{we}(C_{th}) \\
&= \exp(-\{C_{th} / \Omega\}^v)
\end{aligned}
\tag{42}
$$

Similarly, in case of Log-normal shadowing, amplitude of estimated R-channel coefficient is a Log-normal distributed random variable, so the CDF of Log-normal distributed RV in terms of C_{th} can be derived as

$$F_{we}(C_{th}) = \frac{1}{2} erfc\left(-\frac{\ln C_{th}}{\sigma\sqrt{2}}\right); \quad C_{th} \geq 0 \tag{43}$$

where erfc (.) is complementary error function, σ is the Log-normal shadowing parameter generally expressed in dB. Then the probability of selecting a CR for Log-normal shadowing channel can be derived as

$$p_{\ln} = \text{Pr}\left(\ |\hat{h}_k|\ > C_{th}\right) = 1 - \frac{1}{2}erfc\left(-\frac{\ln C_{th}}{\sigma\sqrt{2}}\right) \tag{44}$$

The probability of selecting K number of CRs from N available CRs can now be expressed by utilizing the binomial distribution function as follows (Nallagonda, 2013c, 2013d, 2014)

$$P(K) = \binom{N}{K} p^K (1-p)^{(N_1 - K)} \tag{45}$$

where p is p_{we} or $p_{\ln}$ depends on censoring in Weibull or log-normal channels.

Let $P_m(\text{error} \mid K)$ indicates the conditional missed detection probability when sensing information from K number of CRs are fused using MRC fusion. Given $P(K)$, the probability of selecting K number of CRs in (45), the average probabilities of missed detection and false alarm can be expressed following (Nallagonda, 2013c, 2013d, 2014) as:

$$\bar{Q}_m = P(\text{missed detection}) = \sum_{K=0}^{N_1} P_m(\text{error} \mid K)\ P(K) \tag{46}$$

$$\bar{Q}_f = P(\text{false alarm}) = \sum_{K=0}^{N_1} P_f(\text{error} \mid K)\ P(K) \tag{47}$$

The $\bar{Q}_m$ and $\bar{Q}_f$ are the function of chosen C_{th}, as the probability of mass function (pmf) $\{P(K)\}$ of the number of censored CRs depends on C_{th}. It may be noted that $P_m(\text{error} \mid K)$ and $P_f(\text{error} \mid K)$ are evaluated here considering sensing information from K selected CRs.

Results

The following results are obtained, as in the previous sections, using MATLAB based simulations. The performance of CSS with threshold based censoring for both the perfect and imperfect channel estimation cases has been evaluated in Weibull fading and Log-normal shadowing environments. The average missed detection probability is evaluated as a function of C_{th} considering the impact of various network parameters, such as the Weibull fading parameter (v), shadowing parameter (σ in dB), the number of available CR users (N), and the average R-channel SNRs ($\bar{\gamma}_r$).

In Figure 7, the binomially distributed probability mass fusion (PMF) of the number of selected CRs is shown for different values of v and C_{th} under perfect channel estimation. We observe that as v increases from 2 to 4, the binomially-distributed PMF is shifted right side i.e. more number of CRs is selected by reducing the fading severity in the R-channel. The binomially distributed PMF of the number of selected CR users as obtained for $v=2$ matches exactly with result obtained for Rayleigh (Figure 7.7, Nallagonda, 2014) under perfect channel estimation. In the same figure, the effects of C_{th} on the probability of select-

Figure 7. PMF of the number of selected CR users for different values of v, and C_{th}, under perfect channel estimation in Weibull faded environment (N=30)

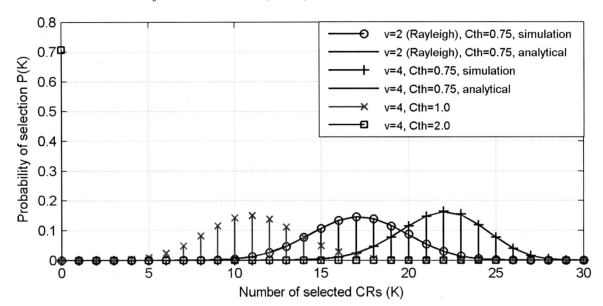

ing K number of CRs, i.e., on the value of $P(K)$, is also shown. It can be observed that for small values of C_{th}, a larger number of CRs is likely to be selected, while the PMF tends to concentrate around small values for higher values of C_{th}. For example, for $v=4$, and $C_{th}=0.75$, it is seen that $K=22$ CRs have highest probability (0.17) of being selected. One can also observe that as C_{th} increases, the PMF moves towards origin. This is due to the fact that a larger value of C_{th} decreases the number of selected CRs. If the value of C_{th} is increased to a very high level (say $C_{th}=3.5$), no CR is selected to transmit, i.e., the probability of selecting no CR is equal to 1 ($P(0) = 1$). The binomially distributed PMF of the number of selected CRs obtained with our simulation test bed confirms that obtained based on the analytical expression (Equation 42 followed by Equation 45). This validates our analytical framework.

In Figure 8, $\overline{Q}_m$ is shown as a function of the C_{th} in presence of Weibull fading. Various values of N (i.e. $N=10$, and 30) and of $\overline{\gamma}_r$ ($\overline{\gamma}_r=-8$ dB, -6 dB) are considered. Two fusion rules such as majority-logic fusion and MRC fusion have been performed separately at FC. The performance comparison between these two fusions is evaluated considering perfect channel estimation case. It can be seen from this figure that as C_{th} increases $\overline{Q}_m$ attains a minimum value for an 'optimal' C_{th} level and thereafter increases with further increase in C_{th} to finally attain a value of 0.5. This behavior of $\overline{Q}_m$ is due to the changing PMF of the number of censored CRs for various values of C_{th}. For very small values of the C_{th}, even unreliable links tend to be selected, and $\overline{Q}_m$ is rather high. On the other hand, as C_{th} is increased to a very high level, no CR is selected to transmit, i.e. $P(0)=1$, and the FC takes a decision by flipping a fair coin resulting in $\overline{Q}_m$ of 0.5. Therefore, there exists an optimal value of C_{th}, in correspondence to which $\overline{Q}_m$ is minimized. It can be seen that a larger value of N leads to a reduced $\overline{Q}_m$ in correspondence to the optimized value of C_{th}. In particular, for $C_{th}=0.5$ and $\overline{\gamma}_r=-6$ dB, $\overline{Q}_m$ decreases by 81.81% when N increases from 10 to 30. When $\overline{\gamma}_r$ increases, the FC receives a larger number of correct decisions (large

Figure 8. Average missed detection probability ($\overline{Q}_m$) as a function of C_{th} for various values of v, N, and $\overline{\gamma}_r$ under perfect channel estimation in Weibull fading ($\overline{\gamma}_s$ =20 dB, P_f=0.05, and u=5)

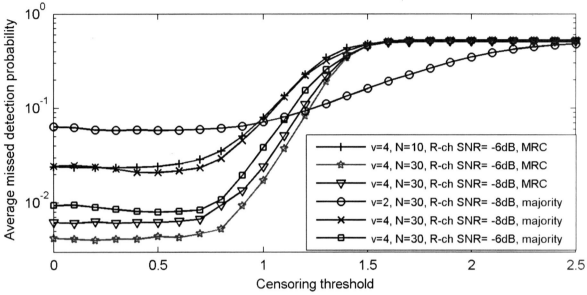

number of CRs selected) and this, in turn, leads to a reduction in $\overline{Q}_m$. The optimum value of C_{th} depends on network parameters. Furthermore, the MRC fusion-based CSS outperforms the majority-logic fusion-based CSS for the same values of network parameters.

Figure 9. and Figure 10 show the impacts of σ on binomially-distributed PMF of the number of selected CRs, and on the average missed detection performance, respectively. Different values of σ (σ

Figure 9. PMF of the number of selected CR users for different values of σ and C_{th} under perfect channel estimation in Log-normal shadowing environment (N=30).

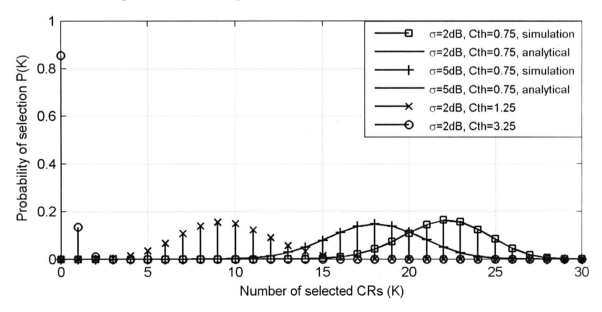

Figure 10. Average missed detection probability ($\overline{Q}_m$) as a function of C_{th} for various values of N, and σ under perfect and imperfect channel estimations in Log-normal shadowing ($\overline{\gamma}_s$ =20 dB, $\overline{\gamma}_r$ =-8 dB P_f=0.05, and u=5).

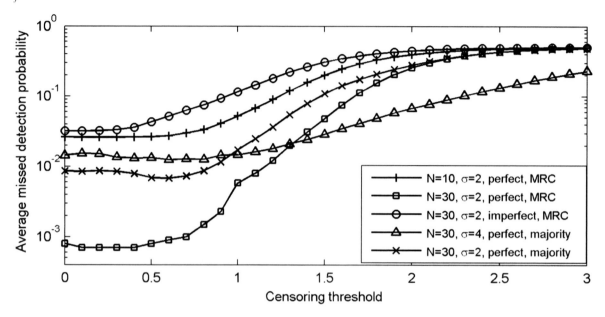

=2 dB, 4 dB, and 5 dB) are considered for these figures. In Figure 9, it is seen that when σ increases from 2 to 5, the PMF is shifted towards origin, it means that probability of selecting K number of CRs decreases due to increase in severity of shadowing effect in the R-channel. Due to this reason, the missed detection performance degrades with increase in parameter σ from 2 to 4 as shown in Figure 10. In Figure 10, the performance is evaluated by performing majority-logic, and MRC fusions individually at FC. We observe that there is an improvement in performance of CSS by increasing number of CRs in the network. The binomially distributed PMF of the number of selected CR users as obtained based on our simulation test bed matches exactly with result obtained based on the analytical expression given in equations (Equation 44 followed by Equation 45), which validates our simulation test bed. Furthermore, in presence of shadowing also, the MRC fusion provides better results as compared to the results with majority-logic fusion.

The considered framework can be applied to a spectrum overlaid cognitive radio network where secondary users need to identify spectrum holes. Our censoring scheme saves energy by not allowing transmissions from CRs whose R-channels are deeply faded. The scheme is energy efficient—even though the FC, which is not energy constrained, needs to spend some energy for MRC operation.

FUTURE RESEARCH DIRECTIONS

Future work might be to perform censoring not only on basis of R-channel quality but also to consider reliability of CR's decisions along with R-channel quality. Thus censoring could be done on the basis of some combined metric assigning appropriate weight to each of R-channel quality and decision reliability of CR. Further censoring could be pursued to achieve certain target objectives such as energy efficiency, a given level of sensing throughput, agility etc.

CONCLUSION

In this chapter, the performance of cooperative spectrum sensing (CSS) using energy detection with and without censoring in Weibull fading and Log-normal shadowing channels has been investigated. The performance of a few hard decision fusion rules (OR-rule, AND-rule, and majority-rule) and a soft data fusion rule (MRC fusion) has been analyzed in a comparative way, considering meaningful performance metrics and evaluating the impact of several system parameters. Our results show that in case of CSS with CRs using ED achieves the lowest probability of missed detection with MRC fusion, as compared to OR-rule, AND-rule, majority-rule fusions, under the same conditions in both the Weibull and Log-normal shadowing channels. We have also investigated the performance of CSS with CRs censored on the basis of the quality of the R-channels, considering both Weibull and Log-normal shadowing channels. The performance with perfect and imperfect channel estimation has been analyzed, in a comparative way, under MRC fusion. Our results show that missed detection probability reduces for increasing values of the number of selected CRs, regardless of the channel estimation quality (perfect or imperfect). However, in the presence of perfect channel estimation no further improvement, in terms of missed detection probability, is obtained by increasing the number of selected CRs beyond a given limit. The fading parameter and the R-/S-channel SNRs have a significant impact on the missed detection probability. The censoring threshold for the selection of CRs has a significant impact on the average missed detection probability. Depending on the configuration of relevant network parameters, such as the available number of CRs, fading/shadowing parameters, and the average R-channel SNRs, there exists an optimal censoring threshold, which corresponds to the minimum average missed detection probability, for both the perfect and imperfect channel estimation. The framework presented in this chapter is useful in designing a cooperative spectrum sensing scheme able to prolong the lifetime of an energy-constrained cognitive radio network by minimizing the number of less useful transmissions.

REFERENCES

Abdi, Y., & Ristaniemi, T. (2013). Joint reporting and linear fusion optimization in Collaborative Spectrum Sensing for cognitive radio networks. In *Proceedings of 9th International Conference on Information, Communications and Signal Processing (ICICS)*. Academic Press. doi:10.1109/ICICS.2013.6782816

Adawi, N. S. (1988). Coverage prediction for mobile radio systems operating in the 800/900 MHz frequency range. *Transactions on Vehicular Technology, 37*(1), 3–72. doi:10.1109/25.42678

Akyildiz, I. F., Lee, W. Y., Vuran, M. C., & Mohanty, S. (2006). Next generation/dynamic spectrum access/cognitive radio wireless networks: A survey. *Computer Networks, 50*(13), 2127–2159. doi:10.1016/j.comnet.2006.05.001

Altay, C., Yilmaz, H. B., & Tugcu, T. (2012). Cooperative sensing analysis under imperfect reporting channel. In *Proceedings of IEEE Symposium on Computers and Communications (ISCC)*. IEEE. doi:10.1109/ISCC.2012.6249392

Cabric, S. D., Mishra, S. M., & Brodersen, R. W. (2004). Implementation issues in spectrum sensing for cognitive radios. In *Proceedings of the Asilomar Conference on Signals, Systems, and Computers*, (*Vol. 1*, pp. 772-776). Academic Press. doi:10.1109/ACSSC.2004.1399240

Chaudhari, S., & Koivunen, V. (2013). Impact of reporting-channel coding on the performance of distributed sequential sensing. In *Proceeding of IEEE 14th Workshop on Signal Processing Advances in Wireless Communications (SPAWC)*. IEEE.

Chaudhari, S., & Lunden, J., & Koivunen, V. (2012a). Cooperative Sensing with Imperfect Reporting Channels: Hard Decisions or Soft Decisions. *IEEE Transactions on Signal Processing*. doi:10.1109/TSP.2011.2170978

Chaudhari, S., Lunden, J., & Koivunen, V. (2012b). Effects of quantization and channel errors on sequential detection in cognitive radios. In *Proceeding of 46th Annual Conference on Information Sciences and Systems (CISS)*. CISS.

Cui, C., & Wang, Y. (2013, October). Analysis and Optimization of Sensing Reliability for Relay-Based Dual-Stage Collaborative Spectrum Sensing in Cognitive Radio Networks. *Wireless Personal Communications*, *72*(4), 2321–2337. doi:10.1007/s11277-013-1152-6

Digham, F. F., Alouini, M. S., & Simon, M. K. (2003). On the energy detection of unknown signals overfading channels. In *Proceedings of the International conference on Communications*, (*Vol. 5*, pp. 3575-3579). Academic Press.

Fathi, Y., & Tawfik, M. H. (2012). Versatile performance expression for energy detector over á-μ generalized fading channels. *Electronics Letters*, *48*(17), 1081–1082. doi:10.1049/el.2012.1744

Ferrari, G., & Pagliari, R. (2006). Decentralized binary detection with noisy communication links. *Transactions on Aerospace Electronic Systems*, *42*(4), 1554–1563. doi:10.1109/TAES.2006.314597

Ghasemi, A., & Sousa, E. S. (2005). Collaborative spectrum sensing for opportunistic access in fading environments. In *Proceedings of the International Symposium on Dynamic Spectrum Access Networks* (pp. 131-136). Baltimore, MD: Academic Press. doi:10.1109/DYSPAN.2005.1542627

Ghasemi, A., & Sousa, E. S. (2007). Opportunistic spectrum access in fading channels through collaborative sensing. *Journal on Selected Areas in Communications*, *2*(2), 71-82.

Hamza Mohamed, D., Aissa, S., & Aniba, G. (2014). Equal Gain Combining for Cooperative Spectrum Sensing in Cognitive Radio Networks. *IEEE Transactions on Wireless Communications*. doi:10.1109/TWC.2014.2317788

Han, W., Li, J., Li, Z., Si, J., Zhang, Y. (2013). Efficient Soft Decision Fusion Rule in Cooperative Spectrum Sensing. *IEEE Transactions on Signal Processing*, *61*(8), 1931-1943. doi:10.1109/TSP.2013.2245659

Hashemi, H. (1993). The indoor radio propagation channel. *Proceedings of the IEEE*, *81*(7), 943–968. doi:10.1109/5.231342

Haykin, S. (2005). Cognitive radio: Brain-empowered wireless communications. *Journal on Selected Areas in Communications, 23*(2), 201–220. doi:10.1109/JSAC.2004.839380

Ismail, M. H., & Matalgah, M. M. (2006). BER analysis of BPSK modulation over the Weibull fading channel with CCI. In *Proceedings of the Vehicular Technology Conference* (pp. 1-5). Montreal, Canada: Academic Press. doi:10.1109/VTCF.2006.334

Ma, J., Zhao, G., & Li, Y. (2008). Soft combination and detection for cooperative spectrum sensing in cognitive radio networks. *Transactions on Wireless Communications, 7*(11), 4502–4507. doi:10.1109/T-WC.2008.070941

Mitola, J., & Maguire, G. (1999). Cognitive radio: Making software radios more personal. *Personal Communications, 6*(4), 13–18. doi:10.1109/98.788210

Nallagonda, S., Shravan kumar, B., Roy, S. D., & Kundu, S. (2013a). Performance of cooperative spectrum sensing with soft data fusion schemes in fading channels. In *Proceedings of the India Conference* (pp. 1-5). IIT Kanpur. doi:10.1109/INDCON.2013.6725924

Nallagonda, S., Roy, S. D., & Kundu, S. (2011a). Performance of energy detection based spectrum sensing in fading channels. In *Proceedings of the Computer & Communication Technology* (pp. 575-580). Allahabad, India: Academic Press. doi:10.1109/ICCCT.2011.6075107

Nallagonda, S., Roy, S. D., & Kundu, S. (2011b). Performance of Cooperative Spectrum Sensing in Rician andWeibull Fading Channels. In *Proceedings of the India Conference* (pp. 1-5). Hyderabad, India: Academic Press.

Nallagonda, S., Roy, S. D., & Kundu, S. (2012). Performance of cooperative spectrum sensing in Fading Channels. In *Proceedings of the Recent Advances in Information Technology* (pp. 1-6). ISM Dhanbad. doi:10.1109/RAIT.2012.6194506

Nallagonda, S., Roy, S. D., & Kundu, S. (2013b). Performance evaluation of cooperative spectrum sensing with censoring of cognitive radios in Rayleigh fading channel. *Wireless Personal Communications, 70*(4), 1409–1424. doi:10.1007/s11277-012-0756-6

Nallagonda, S., Roy, S. D., Kundu, S., Ferrari, G., & Raheli, R. (2013c). Cooperative spectrum sensing with censoring of cognitive radios in Rayleigh fading under majority logic fusion. In *Proceedings of the National Conference on Communications* (pp. 1-5). IIT Delhi. doi:10.1109/NCC.2013.6487926

Nallagonda, S., Roy, S. D., Kundu, S., Ferrari, G., & Raheli, R. (2013d). Performance of MRC fusion based cooperative spectrum sensing with censoring of cognitive radios in Rayleigh fading channel. In *Proceedings of the Wireless Communications and Mobile Computing Conference* (pp. 30-35). Sardinia, Italy: Academic Press. doi:10.1109/IWCMC.2013.6583530

Nallagonda, S., Roy, S. D., Kundu, S., Ferrari, G., & Raheli, R. (2014). Cooperative spectrum sensing with censoring of cognitive radios in fading channel under majority logic fusion. In F. Bader & M. G. Di Benedetto (Eds.), Cognitive Communications and Cooperative HetNet Coexistence (pp. 133-161), Springer International Publishing. doi:10.1007/978-3-319-01402-9_7

Niu, R., Chen, B., & Varshney, P. K. (2003). Decision fusion rules in wireless sensor networks using fading statistics. In *Proceedings of the Annual Conference on Information Sciences and Systems* (pp. 1-6). The Johns Hopkins University.

Paula, A. D., & Panazio, C. (2014). Cooperative spectrum sensing under unreliable reporting channels. Elsevier.

Sagias, N. C., Karagiannidis, G. K., & Tombras, G. S. (2004). Error-rate analysis of switched diversity receivers in Weibull fading. *Electronics Letters, 40*(11), 681–682. doi:10.1049/el:20040479

Simon, M. K., & Alouini, M. S. (2004). *Digital Communication over Fading Channels* (2nd ed.). John Wiley and Sons. doi:10.1002/0471715220

Sun, H., Nallanathan, A., Jiang, J., & Wang, C. X. (2011). Cooperative spectrum sensing with diversity reception in cognitive radios. In *Proceedings of the International Conference on Communications and Networking in China* (216-220). Harbin, China: Academic Press. doi:10.1109/ChinaCom.2011.6158151

Teguig, D., Scheers, B., & Nir, V. L. (2012). Data fusion schemes for cooperative spectrum sensing in cognitive radio networks. In *Proceedings of the Military Communications and Information Systems Conference* (pp. 1-7). Gdansk, Poland: Academic Press.

Urkowitz, H. (1967). Energy detection of unknown deterministic signals. *Proceedings of the IEEE, 55*(4), 523–531. doi:10.1109/PROC.1967.5573

Zhang, W., Mallik, R. K., & Letaief, K. B. (2008). Cooperative spectrum sensing optimization in cognitive radio networks. In *Proceedings of the International conference on Communications* (pp. 3411-3415). Beijing, China: Academic Press. doi:10.1109/ICC.2008.641

Zhao, N., Pu, F., Xu, X., & Chen, N. (2013). Optimization of multi-channel cooperative sensing in cognitive radio networks. *IET Communications, 7*(12), 1177–1190. doi:10.1049/iet-com.2012.0748

Zhao, Y., Kang, G., Wang, J., Liang, X., & Liu, Y. (2013). A soft fusion scheme for cooperative spectrum sensing based on the log-likelihood ratio. In *Proceedings of IEEE 24th International Symposium on Personal Indoor and Mobile Radio Communications (PIMRC)*. IEEE. doi:10.1109/PIMRC.2013.6666688

Zou, Y., Yao, Y. D., & Zheng, B. (2011). A selective-relay based cooperative spectrum sensing scheme without dedicated reporting channels in cognitive radio networks. *Transactions on Wireless Communications, 10*(4), 1188–1198. doi:10.1109/TWC.2011.021611.100913

KEY TERMS AND DEFINITIONS

Cooperative Spectrum Sensing: A technique where the Cognitive radios share their individual sensing information to improve the over all sensing information about the primary user.

Fading Channels: A wireless communication channel undergoing fading which may either be due to multipath propagation, referred to as multipath induced fading, or due to shadowing from obstacles affecting the wave propagation, sometimes referred to as shadow fading.

Fusion Rules: Schemes for combining local decisions to obtain a global decision.

Censoring: Allowing to send information.

Missed Detection: The situation of not detecting the PU/ target.

MRC (Maximal Ratio Combining): Is a diversity combining technique in which signals from several branch or channel are added together with the gain of each channel is made proportional to the rms signal level and inversely proportional to the mean square noise in that channel.

MMSE: In signal processing, a minimum mean square error (MMSE) estimator is an estimation method which minimizes the mean square error (MSE) of the estimated value and the actual value of a desired random variable. This is a common measure of estimator quality.

ENDNOTE

[1]   Note that with the generic term CR we also refer to a secondary (cognitive) user (SU). The context eliminates any ambiguity.

Chapter 3
Tunable RF Front-Ends and Robust Sensing Algorithms for Cognitive Radio Receivers

L. Safatly
American University of Beirut, Lebanon

Y. Nasser
American University of Beirut, Lebanon

A. H. Ramadan
American University of Beirut, Lebanon

K. Y. Kabalan
American University of Beirut, Lebanon

M. Al-Husseini
Lebanese Center for Studies and Research, Lebanon

A. El-Hajj
American University of Beirut, Lebanon

ABSTRACT

In this chapter, the concepts of Cognitive Radio (CR) and multi-dimensional spectrum sensing are introduced. Spectrum sensing methodologies, energy efficiency consideration, resources scheduling, and self-management and learning mechanisms in cognitive radio networks are also discussed. The entailed challenges of CR RF front-end architectures are looked into. The synthesis and design performance analysis of a tunable RF front-end sensing receiver for CR applications are presented. The chapter also discusses how sensing performance degradation, which is due to RF impairments, is analytically evaluated. Spectrum sensing algorithms that correct imperfect RF issues by compensating induced error effects through digital baseband processing are also illustrated.

INTRODUCTION

Cognitive Radio (CR), by which a wireless device can sense the radio environment and use unoccupied frequency bands, is thought of as a drastic solution to the increasing demand for efficient radio frequency (RF) spectrum management, and the use of this spectrum by RF devices with limited resources, such as energy and access rights. The solution offered by cognitive radio relies on its capability to observe whether a specific frequency band is occupied or not, and to use an unoccupied band, called a white space

DOI: 10.4018/978-1-4666-6571-2.ch003

(WS), without interfering with the operation of other wireless devices, especially those of authorized or primary users (PUs). Furthermore, if an authorized terminal starts transmission in a WS occupied by a secondary user, the terminal of the secondary user (SU) jumps into a new WS, or stays in the same WS but alters its transmission power level or modulation scheme, to suppress interference.

A survey of spectrum sensing methodologies for cognitive radio is presented by (Yucek, 2009). Besides studying the various aspects of spectrum sensing problem, and their associated challenges, multi-dimensional spectrum sensing concept is introduced. Herein, the spectrum sensing term is declared as a general term that involves obtaining the spectrum usage characteristics across multiple dimensions such as time, space, frequency, and code. It is not only based on measuring the spectral content, or measuring the radio frequency energy over the spectrum, as traditionally understood. The conventional definition of the spectrum opportunity, which is often defined as a band of frequencies that are not being used by the primary user of that band at a particular time in a particular geographic area, only exploits the frequency, time, and space dimensions of the spectrum. However, other dimensions, such as location, angle of arrival, and code need to be explored. For the location dimension, the spectrum can be available in some parts of the geographical area while it is occupied in others. With the knowledge of the location or direction of primary users, secondary ones can alter their transmission direction without creating any interference. Simultaneous transmission without interfering with primary terminals would be possible in code domain upon using orthogonal coding schemes. The radio space with the introduced dimensions can be defined as a theoretical hyperspace occupied by radio signals, which has dimensions of location, angle of arrival, frequency, time, and possibly others.

The most common spectrum sensing techniques in the cognitive radio literature are given by (Axell 2010, Kwan 2012, Weifang 2009, Yucek, 2009). These include energy detector, waveform-, cyclo-stationarity-, radio identification, matched-filtering, and sub-sampling based sensing techniques. The selection of a specific sensing method depends on several factors such as required accuracy, sensing duration, computational complexity, and network requirements, thus tradeoffs should be considered. Cooperative sensing, whether it is centralized or distributed, is also proposed by (Yucek, 2009, Axell, 2010), as a solution to problems that arise in spectrum sensing due to noise uncertainty, fading, and shadowing. However, energy efficiency in cooperative cognitive radio networks has to be considered. The energy consumption of such networks increases as the number of cooperating users grows. Hence, techniques such as on-off sensing and censoring have been developed to improve the energy efficiency in cognitive radio networks. The implementation of a reconfigurable cognitive sensing methodology is exemplified by (Weingart, 2007). It may adjust itself in a manner that highly minimizes the probabilities of misdetection and false alarm to meet performance goals. Although this would be an optimum solution of the spectrum sensing problem, it is worth mentioning here that other challenges such as high sampling rate, high resolution analogue to digital converters (ADCs) with large dynamic range, and high speed signal processors are demanding associated requirements with spectrum sensing. On the other hand, RF receivers are expected to process narrowband baseband signals of the wide frequency spectrum with reasonably low complexity and low power processors. In other words, the RF components such as antennas, amplifiers, mixers and oscillators are expected to operate over a wide range of frequencies, as reported by (Yucek, 2009, Cabric, 2005).

Spectrum mobility, channel sensing, resource allocation, and spectrum sharing are important functional cognitive radio stimuli, which are exploited by Medium Access Control (MAC) protocols. The design of cognitive MAC protocols is still open to research and investigation according to (De Domenico, 2012). It aims to build up a spectrum opportunity map, schedule available resources, improve coexistence between users that belong to heterogeneous systems, and allow cognitive users to vacate selected channels when their quality becomes unacceptable. Self-management and learning mechanisms have been devised, by (Bantouna, 2012), as solution for reconfigurable cognitive systems to respectively identify opportunities to improve their performance without the need of human intervention and to increase the reliability of their decision making. An autonomous cognitive radio architecture, which goes beyond adaptive radio systems to exploit both the self-learning and self-reconfiguration ingredients of cognition without any prior knowledge of the RF environment, is discussed by (Bkassiny, 2012). Architectures for future dynamic spectrum access, with the emphasis on the key research challenges associated with the autonomous paradigm, have been reported by (Nekovee, 2006).

RF CR receivers are expected to process narrowband baseband signals received over a wide frequency spectrum with reasonably low-complexity and low-power processors. In other words, the RF components such as antennas, amplifiers, mixers and oscillators are expected to operate over a wide range of frequencies. This necessitates the move towards reconfigurable and tunable RF CR front-end architectures. A tunable RF front-end receiver should function in the presence of an efficient software-defined engine. Software defined engines are not only expected to sense, process, learn and tune/control to guarantee an acceptable performance of the receiver, but are also required to account for RF impairments. This helps to avoid performance degradation due to the nonlinearity or phase noise of the spectrally-agile components in the RF receiver. The deterioration in a down-converted signal could be due to any of several factors, including interference, nonlinearities, phase noise or mixing spurs. These always-present issues influence the level and quality of the down-converted signals, which dictate how the software defined engine, performs its sense-process-learn-tune and control tasks. As a result, it is crucial in a CR design to consider and account for RF impairments to efficiently sense the frequency spectrum.

In real CR scenarios, the performance of traditional spectrum sensing techniques shows degradation if additional spectrum components are introduced due to nonlinearities in the PU's front-end equipment, especially when the PU operates with higher output powers. Such unwanted frequency components will overlap with weak SU transmissions, thus possibly rendering them undetectable and decreasing the opportunity of finding a WS. This scenario, where the spectrum is crowded by distortion products of a strong interferer, is called "blocking". To face such an issue and other harmful RF impairments, the software defined engine should be able to digitally mitigate the effects of nonlinearities by using "Dirty RF". As described by (Fettweis, 2005), the idea behind "Dirty RF" is to design and implement digital signal processing algorithms to compensate and mitigate RF distortions. In other words, it is the correction of hardware impairments using software techniques. Such optimized solution could improve the scalability, enhance power-saving, reveal the dynamic aspect of SU front-ends, and allow relaxation of the complex analogue design of CR tunable receivers. In modern software-defined radios, several digital approaches were devised to cleanse the received signal from RF induced parasites. These are feed-forward techniques with reference nonlinearity, feed-back equalization, and training symbols based equalization.

Based on the above, a cognitive receiver should digitally enhance RF front-ends by performing digital post-distortion of the received signals. Accordingly, aforementioned spectrum sensing techniques are not only urged to scan, sense, and detect a WS but also to mitigate the harmful impact of RF imperfections. For that, the analysis of the sensitivity of various detectors against RF impairments will lead to a selection of optimal and robust detectors. Then, a post distortion capability will be added to spectrum sensing algorithms to compensate the RF impairments. Resulted robust dual stage algorithms, performing sensing and mitigation, are studied and analyzed using the Universal Software Radio Peripheral (USRP).

In this chapter, robust sensing algorithms for cognitive radio receivers are presented. Adaptive compensation techniques, which blindly compensate the main RF imperfections to guarantee a reliable sensing, are detailed herein.

The chapter is organized as follows: Section II reviews cognitive radio from an RF engineering perspective, and highlights its entailed challenges. In Section III, the synthesis and design performance analysis of a tunable RF CR front-end sensing receiver addresses the RF impairments impact on spectrum sensing. Solutions that contribute perfect functioning of the dynamic terminal, and suppress to the main harmful RF impairments are presented in Section IV. Section V discusses a future assessment plan to pragmatically test the presented mitigation and sensing algorithms. Conclusions are presented in Section VI.

COGNITIVE RADIO RF ENGINEERING: DESIGN AND CHALLENGES

Unlike software-defined radios (SDRs), their CR counterparts are expected to sense the occupancy or target any channel in the entire spectrum, and tolerate interferers at any frequency as well. These requirements constrain stringent issues on antenna design, low–noise amplification, frequency synthesizers that provide a carrier frequency from tens of megahertz to about 10 GHz, mixing spurs, and spectrum sensing, as given by (Razavi, 2009). A feasibility study on software-defined cognitive radio equipment is carried out by (Harada, 2008A). Herein, it is reported that broadband and tunable antennas, multi-band amplifiers, RF filters, broadband direct conversion mixers, baseband filters, ADCs/DACs are needed to realize software-defined cognitive radio equipment. As for the signal processing unit, low-power consumption reconfigurable baseband signal processor is needed. A small-sized software-defined radio unit that consists of a hardware platform, which includes a multi-band/tunable RF unit with multiple switching circuits and FPGA-based signal processing unit, has been introduced by (Harada, 2008B). The software platform of the proposed radio is built to manage the spectrum sensing task and reconfigure the communications schemes. A wideband RF front-end architecture for software defined radio is proposed by (Tyagi, 2010). Starting from the original idea, which is based on digitizing the signal directly after the antenna of the SDR, four RF front–end sensing receiver architectures are analyzed in terms of performance, power consumption, cost and size. As illustrated in Figure 1, these are multiple narrowband, wideband with few medium bands, wideband, and tunable RF front-end architectures. The latter is chosen as the best architecture to overcome the performance limitations due to IM products in a wideband environment. Both the tunable and multiple narrowband architectures are discounted because tunable preselect filters suffer from high insertion loss and of few standards support, respectively. However, the adopted architecture is still tunable, but in a different fashion.

An architecture for a software-defined radio is reported by (Dutta, 2010). Herein, the bare bones of an OFDM-based physical layer, which can adapt to perform various tasks in different radio networks, are provided. A flexible universal radio platform to receive and transmit, which can be programmed to steer to any frequency band, tune to a channel of any bandwidth, and adapt to any modulation scheme-all within reasonable constraints is crucial. Ultimately, the receiver itself is only half of a radio communication device, which is not complete without an efficient software defined engine, as reported by (Abidi, 2007). According to (Carey-Smith, 2005), employing tunable narrowband circuits rather than pursuing a truly wideband approach, the required flexibility of software-defined radio front-end may be met with better overall performance. Although having attractive features, a wideband front-end may

Figure 1. RF front-end sensing receiver architectures

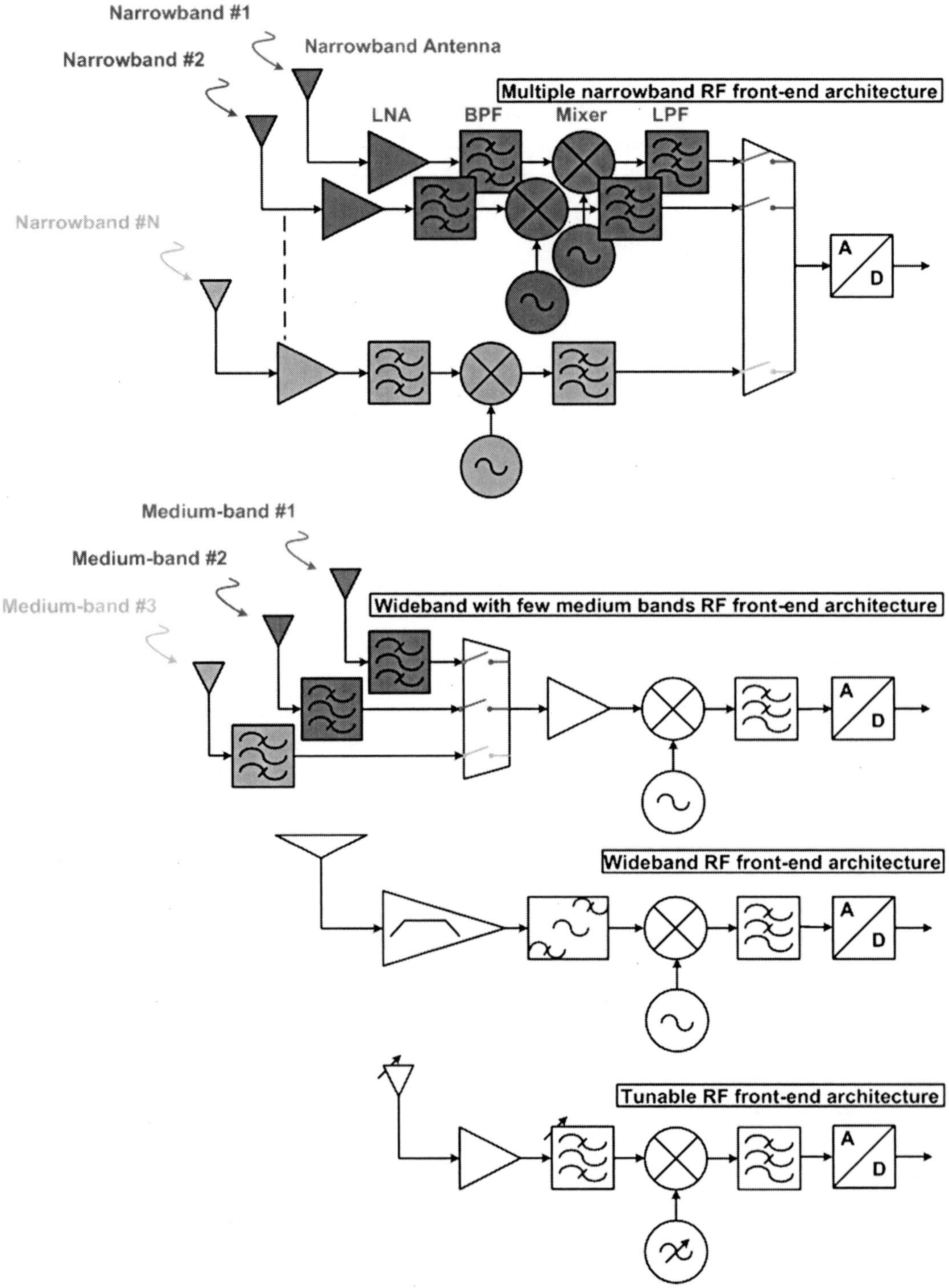

not necessarily be the optimum solution for software-defined radio as it leads to a compromise in the transceiver performance, caused by limitations in the front-end components. Therefore, techniques of introducing flexible frequency discrimination, which include tunable bandpass, tunable bandstop filters, and tunable narrowband antennas, help reducing spurious spectral content in the transmitter and limiting out-of-band interference in the receiver.

The broad frequency allocation with the variety of the existing standards, calls for reconfigurable and frequency-agile, with reduced number of functional blocks, microwave circuits to pave the way toward reconfigurable radio front-end architectures, as discussed by (Lourandakis, 2012). According to (Acampora, 2010), The "spectrum overcrowding" issue arising in cognitive radio, which is due to the coexisting telecommunication standards in overlapping frequency bands, calls for flexible receiver architectures to efficiently share wireless resources. The ability to design linear and spectrally-agile components and architectures in the RF front-end of the transceiver is considered a primary technological concern in cognitive radio architectures, as given by (Hayar, 2007). An adaptive weaver architecture radio with spectrum sensing capabilities to relax RF component requirements is presented by (Lessing, 2007). Therein, large interferers are avoided by sensing the environment, and learning/adapting capabilities are enabled by means of a flexible receiver architecture, which employs variable local oscillators at RF and intermediate frequency. The design of a reconfigurable RF Front-end, along with the system analysis, for wireless multi-standard terminals is discussed by (Agnelli, 2006). Herein, the global transceiver requirements in terms of noise, linearity, dynamic range and bandwidth are verified. Additionally, the characterization of each single block in the transceiver chain is reported. A reconfigurable RF front-end, for frequency agile direct conversion receivers and cognitive radio system applications, is reported in (Djoumessi, 2010). A tutorial on highly integrated and tunable RF front-ends for reconfigurable multi-band transceivers is presented by (Darabi, 2010). High dynamic range RF front-ends from multi-band multi-standard to cognitive radio have been discussed by (Heinen, 2011). Besides the proposed RF spectrum sensing circuit, (Jongsik, 2011) categorizes harmonic problems in the up/down frequency conversion and ultra-wideband frequencies synthesis as two critical arising issues to deal with in the design considerations of cognitive radio-based TV white space transceivers.

Emerging opportunities of RF IC/system for future cognitive radio wireless communications have been reported by (Kyutae, 2008). High performance silicon-based RF front-end design techniques for adaptive and cognitive radios have been discussed by (Larson, 2012). Herein, tunable filters based on N-path techniques have demonstrated excellent noise and linearity properties. Techniques to implement frequency-agile RF front-end ICs and modules for application in intelligent cognitive radios have been proposed by (Mukhopadhyay, 2005). A 2.4 GHz image rejection (IR) RF front-end receiver is presented by (Fayrouz, 2010). The proposed IR front-end is implemented in 130 nm standard CMOS technology. A 65 nm technology-based 0.7-2.6 GHz high-linearity RF front-end for cognitive radio spectrum sensing is reported by (Stadius, 2011). A 30 MHz-2.4 GHz 90 nm technology-based CMOS receiver with integrated RF filter and scalable dynamic range energy detector for cognitive radio systems has been discussed by (Kitsunezuka, 2012). A 600 MHz-3.4 GHz flexible spectrum sensing receiver, which consists of a wideband front-end, spectrum-adaptive filtering, switched-capacitor amplifiers, and a filtering spectrum-adaptive receiver ADC, in 65 nm CMOS technology is proposed by (Lin, 2012). A wideband 2 to 6 GHz RF front-end with blocker filtering is introduced by (Kaltiokallio, 2012). A 300-

800 MHz tunable filter and linearized Low-noise amplifier applied in a low-noise harmonic-rejection RF sampling receiver, which is based on 65 nm CMOS technology, is proposed by (Ru, 2010). A 3.1-10.6 GHz RF receiver front-end in 0.18 μm CMOS technology for ultra–wideband applications is reported by (Bonghyuk, 2010).

Alternative approaches, which aim at integrating into the digital front-end and the CR-enabled user equipment for efficient interference detection, rapid spectrum evaluation, and multi-standard access, are reported by (Mayer, 2007). According to (Balasubramanian, 2012), it is seen that the design of a multi-band cognitive radio systems require significant analog and RF processing, which is the major cause of nonlinearities, regardless of the proposed incorporated circuit techniques to alleviate them. Therefore, different design approaches, such as RF DACs and digital PAs, have been reported to address those perennial nonlinearities.

In conventional systems, the wireless data transfer between the transmitter and the receiver is assumed to be linear. However, several analog parts of the transceiver RF front-end exhibit impairments such as the analogue-to-digital (ADC) and digital-to- analogue (DAC) converters, the mixer, the local oscillator (LO) and the amplifiers, i.e. the power amplifier (PA) in the transmitter and the low-noise amplifier (LNA) in the receiver.

At the receiver side, these imperfection generators become more and more crucial with the increased dynamic range of the received signal. The challenge becomes harder when the receiver operates over a wide bandwidth since nonlinear distortions will show up as unwanted signals in free bands and can hit the target band.

On the other hand, emerging wireless systems push towards designing flexible and reconfigurable receivers able to operate over multiple frequency bands and supporting multiple standards. A fundamental key in building such cognitive radios is the integration of sophisticated digital signal processing (DSP) techniques. Recently, extensive research work has focused on developing such techniques by presenting the algorithmic and the implementation levels.

The main functionality of a cognitive receiver is to sense the highly dynamic band of interest and locate the primary users in order to transmit in the detected white spaces. DSP techniques were applied to enhance the classical sensing performance by using adaptive algorithms that extract special features of the signal. To ameliorate the detection results, researchers have studied the efficiency of recent spectrum sensing techniques in a nonlinear environment.

In fact, the nonlinear characteristics of the receiver analog front-end were proved to be a critical source of degradation of the sensing performance. Unwanted frequency components resulting from such imperfections are clearly detected as real signals since they carry the special feature of the strong signal.

Recently, many papers proposed analog improvement of the front-end stages to enhance the receiver performance. However, compensating the RF impairments via advanced DSP techniques, by benefiting from the software reconfigurability of the device, has not been studied efficiently. Since the design of a reconfigurable RF front-end is considered nowadays the bottleneck of the synthesis of a fully adaptive engine, linearizing the hardware in dynamic scenarios is currently beyond of the state-of-the-art in electronics. Thus, mitigation DSP techniques are urged not only to enhance the detection efficiency but also to relax the delicate conception of the cognitive hardware. Accordingly, the sensitivity of advanced spectrum sensing techniques against RF impairments is highlighted and recently proposed compensation techniques are described.

RF IMPAIRMENTS IMPACT ON RF CR FRONT-ENDS

A Tunable RF CR Front-end Sensing Receiver Scenario

A tunable RF front-end sensing receiver for cognitive radio applications is presented in this subsection. It targets the WLAN IEEE 802.11 g commercial wireless standard key specifications. It incorporates a tunable filter-antenna, low-noise amplifier, tunable bandpass filter, single-ended balanced mixer, voltage-controlled oscillator, and variable gain IF amplifier. The reception of the tunable RF front-end sensing receiver is then tested to assess its performance.

A typical block diagram of the proposed receiver, which incorporates crucial components for the reception, low-noise amplification, bandpass filtering, down-conversion, and proper detection of an RF signal in the 2.4-2.485 GHz frequency range, is given in Figure 2. Each component has a unit number, ranging from (U1) to (U8). The role of each unit is discussed herein.

Besides amplifying the received signal via (U1), the LNA unit (U2) is required to make the receiver noise figure as small as possible, thus making the receiver very sensitive. The tunable bandpass filter unit (U3) is required to reject all spurious-signal creating frequencies, including the image frequency, while simultaneously letting the desired RF bandwidth pass to the mixer. The mixer unit (U4) is required along with a tunable local oscillator (U5) for the down-conversion process to be successfully completed. The lowpass filter (U6) is required to suppress all the mixed output signals except the desired IF signal to detect. The AGC (Automatic Gain Control) unit (U7) is required to control the overall gain of the receiver. The AGC unit is followed by unit (U8), which detects the down-converted signals. The sense, process, learn, tune/control loop, which is driven by a specific sensing algorithm, is responsible for taking care of tuning the receiver's blocks operations. To synthesize the illustrated tunable RF-frontend sensing receiver and analyze its performance, the key specifications of the adopted commercial wireless standard have to be targeted from the very beginning.

The reception of the tunable RF front-end sensing receiver is tested for two different carrier frequencies n the 2.4-2.485 GHz vicinity. Figure 3 illustrates two examples of down-converted signals in the megahertz vicinity. Compared to the 20-MHz one, the 19.22-MHz down-converted signal shows a higher noise level besides the frequency shift. This deterioration in the down-converted signal could be

Figure 2. Block diagram of the tunable RF front-end sensing receiver

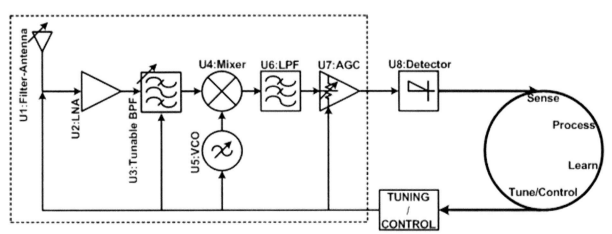

Figure 3. Down-converted Signals in the Megahertz Vicinity

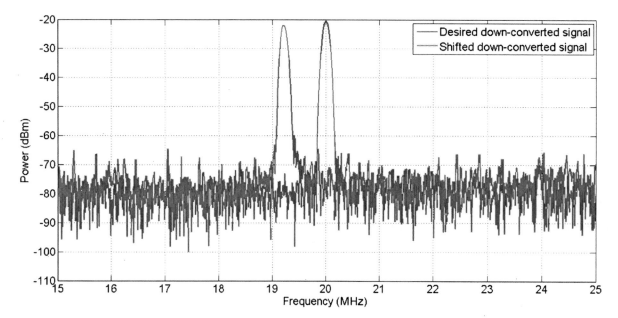

due to any of several factors, including interference, nonlinearities, phase noise or mixing spurs. These always-present issues influence the level and quality of the down-converted signals, which dictate how the software defined engine performs its sense, process, learn, tune and control tasks. As a result of this discussion, it is crucial in a CR design to consider and account for RF impairments to efficiently sense the frequency spectrum.

A Spectrum Sensing Overview

A CR transceiver scans its environment through spectrum sensing and adapts itself to be able to utilize vacant spectrum bands. One key feature of CR is the capability of fast adaptation with highly changing conditions. Thus, the main feature of CR devices is to perform a reliable spectrum sensing in a dynamic environment to be able to ensure an efficient utilization of the spectrum resource. A secondary user is required to identify white spaces through direct sensing of the licensed bands. In this case, it monitors the availability of a licensed frequency band and transmits when the band is vacant. This section discusses spectrum sensing methodologies for cognitive radio to ensure reliable identification of white spaces. A sorted list (in ascending order of complexity) of well-known spectrum sensing techniques is: energy detection (ED), matched filter (MF) based sensing, cyclostationarity based sensing (CD), covariance based sensing, and Eigen value based sensing.

The spectrum sensor essentially performs a binary hypothesis test on whether or not there are primary users in a particular channel. The channel is idle under the hypothesis H_0 and busy under the hypothesis H_1. It is obvious that under H_0 the received signal is only the ambient noise while it consists of the PU's signal and the ambient noise under H_1.

$$H_0 : y_0(k) = w(k)$$
$$H_1 : y_1(k) = s(k) + w(k)$$

For k = 1, …, n where n is the number of received samples, w(k) represents ambient noise, and s(k) is the PU signal. It is evident that the received signal will have more energy when the channel is busy than when it is idle. An energy detector is based on this concept to detect the presence of a PU. When aspects of the PU signal are known, one can exploit the feature of the received signal and identify the PU; a special case leads to the cyclostationarity detector. Regardless of the used sensing algorithm, sensing errors are inevitable due to additive noise, limited observations, and inherent randomness of the observed data, and RF impairments.

As a proof of concept, a study of the main RF impairments effects on the performance of two widely used spectrum sensing techniques i.e. the energy detector and a blind cyclostationarity based detector is presented below. The results prove the critical impact caused by hardware non-idealities on the efficiency of these techniques. After introducing this crucial problem, an overview on main RF mitigation techniques is presented in the next section.

1. Energy Detector

In many cases, the signaling scheme of the PU may be unknown to the SU. The energy detector is based on the idea that with the presence of a signal in the channel, there would be significantly more energy than if there was no signal present. This concept is applicable for any PU without knowing any of its characteristics. Energy detection simply involves the application of a threshold τ on the collected energy from the channel. Since the detection is only based on the received amount of energy, the signal can be simply modeled as a zero-mean stationary white Gaussian process, independent of the white Gaussian noise. The spectrum sensing problem is to distinguish between two mutually independent Gaussian sequences: $y_0(k)$ and $y_1(k)$. w(k) and s(k) are zero-mean complex Gaussian random variables with respective variances σ_w^2 and σ_s^2. The variance of the received sequence **y** of the n observed samples, under the two hypotheses, could be written as:

$$H_0 : \sigma_0^1 = \sigma_w^2$$
$$H_1 : \sigma_1^2 = \sigma_w^2 + \sigma_s^2$$

The test statistic of the ED is to decide H_1 when:

$$z = \frac{1}{2N\sigma_w^2}\sum_{k=1}^{n}|y(k)|^2 > \tau$$

2. Blind Cyclostationarity Based Detector

To modulate signals, transmitters couple the base band signal with sine wave carriers, pulse trains, repeating spreading, hoping sequences, or cyclic prefixes which result in built-in periodicity. Thus, a modulated signal is characterized as cyclostationary since its statistics, mean and autocorrelation, exhibit periodicity. This feature can be extracted and analyzed using Fourier analysis. In communications, this periodicity is introduced intentionally so a receiver can estimate several parameters such as carrier phase, pulse timing, or direction of arrival. In CR networks, this periodicity is the feature for detecting a random signal with a particular modulation type in a background of noise and other modulated signals.

Instead of studying the traditional cyclostationarity based detector, a recent version of these feature based schemes is adopted. The algorithm was selected because of its reduced complexity, complete blindness to noise and channel variations, and accurate results (Safatly, 2014). Its hypothesis selection is based on the estimation of the Cyclic Autocorrelation Function (CAF) formulated below:

$$R_{yy}^{\alpha}(\tau) = \lim_{T \to \infty} \frac{1}{T} \sum_{t=0}^{T-1} y(t)y*(t+\tau)e^{-j2\pi\alpha t}$$

where τ is the time delay and α is the fundamental cyclic frequency. The extraction of the built-in periodicity of signals is performed by observing the symmetry of the estimation of the CAF. The idea was proposed by Khalaf et al. in (Khalaf, 2012), where they recommended using the sparse property of the CAF to reduce the complexity of the cyclostationarity feature detector. An estimation method that utilizes the sparseness constraint is applied to estimate the CAF. Finally, to ensure the blindness of the algorithm, a symmetry test is checked on the estimated version of the CAF. In summary, the steps of the detection algorithm based on the symmetry property of the cyclic autocorrelation function SP-CAF are shown in Table 1.

RF Impairments Impact on Spectrum Sensing

In this section, the modeling and analysis of main RF impairments are highlighted and simulated results of the degradation of ED and SP-CAF are shown and discussed. In communications systems, the most crucial RF impairments are the Carrier Frequency Offset CFO, Phase Noise, IQ Mismatch, and nonlinearities induced by the non-ideal stages of the RF front-end depicted in Figure 4. Let $\mathbf{r}$ be the received sequence while taking into account the imperfections of the hardware, while $\mathbf{y}$ is the theoretically ideal received sequence. The goal of the following sections is to present the formulation of the main RF impairments and study their impact of the spectrum sensing techniques test statistics.

1. Carrier Frequency Offset Impact

During the up and down conversion at the OFDM transceiver, the local oscillators should produce an ideal sine wave at the standard RF carrier frequency. In practice, the produced carrier frequency may differ from one to another LO, which causes CFO between the transmitter and receiver (Tandur, 2007). Let Δf be the carrier frequency offset between the transmitter and receiver and T_s is the sampling period. In this case, the baseband received signal could be written as:

$$H_0 : r(k) = y(k).e^{2\pi\Delta fkT_s} = w(k).e^{2\pi\Delta fkT_s}$$
$$H_1 : r(k) = y(k).e^{2\pi\Delta fkT_s} = s(k).e^{2\pi\Delta fkT_s} + w(k).e^{2\pi\Delta fkT_s}$$

Table 1. SP-CAF based Detection Algorithm

Initialize:
    a) Acquire n data samples
    b) Set L=3 the number of OMP iterations and M=5 the number of delays τ ($\tau = 1...M$).

For M different values of τ:

a) Calculate the autocorrelation vector $\overline{f}_\tau$:

$$\overline{f}_\tau = \left[f_\tau(0), f_\tau(1), ..., f_\tau(N-1) \right]^T \text{ where } f_\tau = y(t)\,y(t+\tau)$$

b) Calculate the elements of the matrix A(n,N) performing the IDFT transform:

$$a_{pq} = e^{2i\pi(p-1)(q-1)/N}$$

c) Estimate the cyclic autocorrelation vector $\overline{r}_\tau$ by solving the underdetermined system:

$$A\overline{r}_\tau = \overline{f}_\tau$$

The calculation of the approximated solution $\hat{r}_\tau$ is done using an iterative optimization technique called Orthogonal Matching Pursuit (OMP) (Davis, 1997).

d) Run the OMP algorithm L times.

e) Calculate the symmetry index for this value of τ by measuring the mean value of the abscissa of the non-zero elements in $\hat{r}_\tau$

$$IND_{SYM}^\tau = \frac{1}{L-1} \sum_{i=1}^{n} (\hat{r}_\tau)_i$$

End For

Equivalent decision:
a) Calculate the equivalent symmetry index resulting from M symmetry indexes obtained with different values of delays τ

$$IND_{SYM}^{EQU} = \sum_{i=1}^{M} IND_{SYM}^{\tau,i}$$

b) The test statistic of the SP-CAF is to decide H1 when:

$$IND_{SYM}^{EQU} > 0$$

Figure 4. Stages of an RF front-end receiver

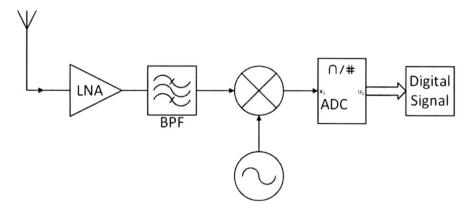

Obviously, since $|r(k)| = |y(k).e^{2\pi\Delta fkT_s}| = |y(k)|$, the test statistic of ED is not sensitive to the CFO. In the case of the SP-CAF, the discrete CAF of the distorted received signal yields to:

$$R_{rr}^{\alpha}(\tau) = \frac{1}{M}\sum_{k=1}^{M} y(k).e^{j2\pi\Delta fk}y*(k+\tau)e^{-j2\pi\Delta f(k+\tau)T_s}e^{-j2\pi\alpha k} = e^{-j\Delta f\tau T_s}R_{yy}^{\alpha}(\tau)$$

This demonstrates that, after the CFO impact, $|R_{rr}^{\alpha}(\tau)| = |R_{yy}^{\alpha}(\tau)|$. This means that the SP-CAF detector is not sensitive to the CFO.

To validate these conclusions, the Receiver Operational Characteristic (ROC), i.e. probability of detection P_D versus probability of false alarm P_{FA} of these detectors is simulated and plotted. Figure 5 and Figure 6 show that the ED and SP-CAF detectors maintain their efficiencies independently of the CFO value $\varepsilon=N\Delta fT_S$.

Figure 5. ROC of the ED at different values of CFO $\varepsilon=N\Delta fT_S$.

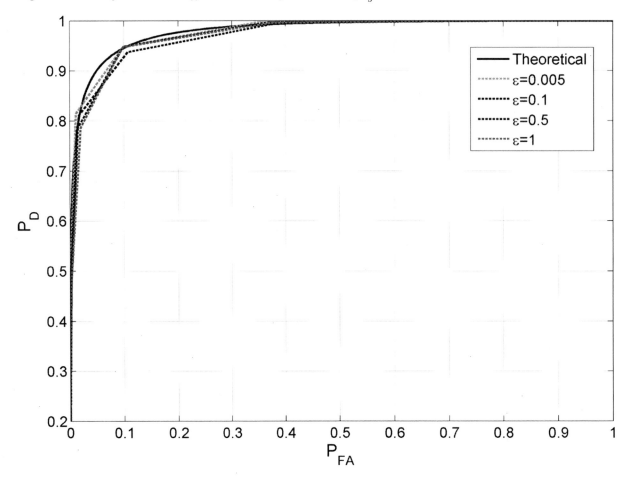

Figure 6. ROC of the SP-CAF at different values of CFO $\varepsilon=N\Delta f T_s$.

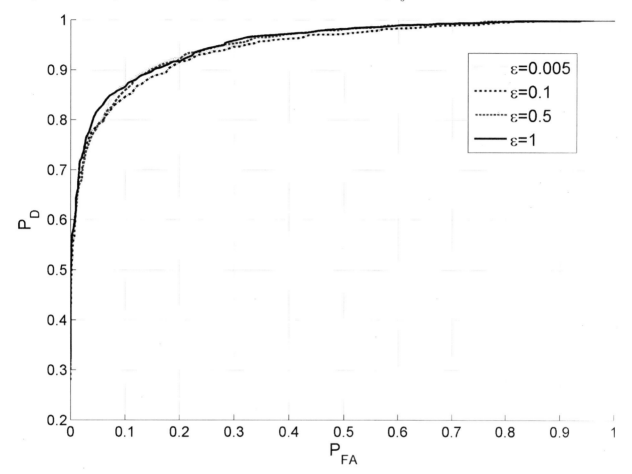

2. Phase Noise Impact

Phase noise is introduced by the local oscillator at both the transmitter and the receiver. It is a random process caused by the frequency fluctuation of the LO. Phase noise is the difference between the phase of the receiver oscillator and the phase of the carrier of the received signal. The distortion could be modeled as:

$$r(k) = y(k).e^{j\phi_k}$$

ϕ_k is the phase rotation of the received signal due to the phase noise. It is usually modeled as a Wiener random process or a wide-sense stationary (WSS) random process as:

$$\phi_k = \phi_{k-1} + \varphi_k$$

φ_k is assumed to be an independent identically distributed real Gaussian process with zero mean and a variance of $\sigma_\tau^2 = 2\pi\beta T_s$, where β is called the Full-Width at Half-Maximum (FWHM) or the diffusion factor.

The ED, being dependent on the calculation of the amplitude of the signal, is robust against phase noise. For the SP-CAF, the CAF will be affected as demonstrated below:

$$R_{rr}^a(\tau) = \frac{1}{M}\sum_{k=1}^{M} y(k)e^{j\varphi_k}y*(k+\tau)e^{-j\varphi_{k-\tau}}e^{-j2\pi\alpha k}$$

$$R_{rr}^a(\tau) = \frac{1}{M}\sum_{k=1}^{M} y(k)y*(k+\tau)e^{j(\varphi_k-\varphi_{k-\tau})}e^{-j2\pi\alpha k}$$

$$R_{rr}^a(\tau) = \frac{1}{M}\sum_{k=1}^{M} y(k)y*(k+\tau)e^{-j2\pi\alpha k}e^{-j\sum_{i=1}^{\tau}\varphi_{k-i}}$$

$$R_{rr}^a(\tau) = \frac{1}{M}\sum_{k=1}^{M} y(k)y*(k+\tau)e^{-j(2\pi\alpha k+\varphi_k')}$$

The CAF of the distorted signal r(k) is clearly affected by a phase component φ_k'. To observe the impact of this phase factor on the test statistic and consequently on the detection performance, simulation were driven and results are shown below. In the simulation, the ROC curve is simulated for the SP-CAF while taking several values of the factor βT_s. As demonstrated in Figure 7, increasing βT_s dramatically degrades the performance of the SP-CAF. For $\beta T_s=5.10^{-4}$, the probability of detection is equivalent to the probability of false alarm, which corresponds to the worst detector performance.

3. I/Q Imbalance Impact

Another major source of impairments in wireless communications system is the mismatch between the I and Q branches or, equivalently between the real and imaginary parts of the complex signal at both the transmitter (during up-conversion) and receiver (during down-conversion). In reality, matching between the I and Q branches is not perfect, consequently the phase and the amplitude of the Q-path are altered.

In this case, the baseband distorted received signal r(k) could be written as a linear expression of the ideal signal y(k) and its conjugate y*(k) as follows:

$$r(k) = \mu.y(k) + \upsilon.y*(k)$$

The distortion parameters μ and υ depends on the mismatch amplitude g and phase θ between the I and Q branches in the analog stages. μ and υ could be formulated as:

$$\mu = \cos(\theta/2) + jg\sin(\theta/2)$$
$$\upsilon = g\cos(\theta/2) - j\sin(\theta/2)$$

Figure 7. ROC of the SP-CAF detector at different values of .

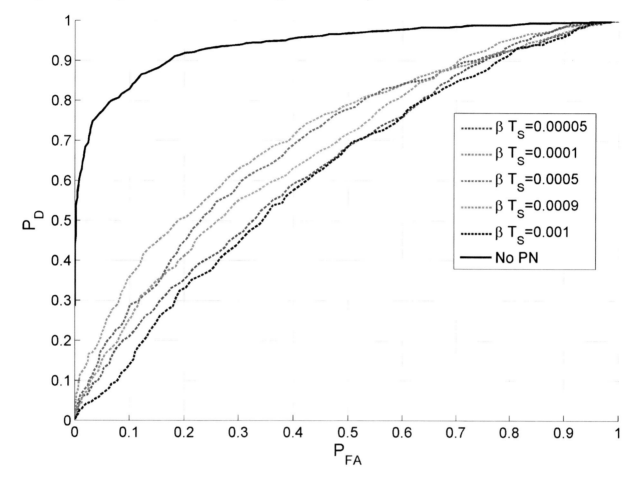

To analyze the impact of the I/Q imbalance on the energy detector, the distortion between $|r(k)|^2$ and $|y(k)|^2$ is observed.

$$| r(k) |^2 = | \mu.y(k) + \upsilon.y*(k) |^2$$
$$| r(k) |^2 = | \mu.y(k) |^2 + | \upsilon.y*(k) |^2 + 2\Re\{\mu.y(k).(\upsilon.y*(k))*\}$$
$$| r(k) |^2 = | \mu |^2 . | y(k) |^2 + | \upsilon |^2 . | y(k) |^2 + 2\Re\{\mu.\upsilon*.y(k)^2\}$$
$$| r(k) |^2 = (| \mu |^2 + | \upsilon |^2). | y(k) |^2 + 2\Re\{\mu.\upsilon*.y(k)^2\}$$
$$| r(k) |^2 = (1 + g^2). | y(k) |^2 + 2\Re\{\mu.\upsilon*.y(k)^2\}$$

It is obvious that the mismatch amplitude g and phase θ could alter the performance of the ED. To observe this modification, the ROC of the ED is plotted for several values of g and θ in Figure 8 and Figure 9. The degraded performance is obvious in the simulated results which demonstrated the high sensitivity of the ED to I/Q imbalance.

The CAF is distorted due to I/Q imbalance as formulated below:

Figure 8. ROC of the ED detector at different values of θ

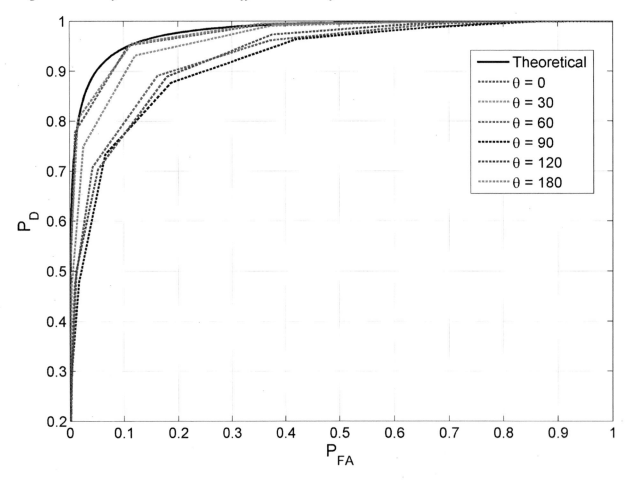

$$R_{rr}^{\alpha}(\tau) = \frac{1}{M}\sum_{k=1}^{M} r(k)r*(k+\tau)e^{-j2\pi\alpha k}$$

$$R_{rr}^{\alpha}(\tau) = \frac{1}{M}\sum_{k=1}^{M} [\mu.y(k) + v.y*(k)][\mu.y(k+\tau) + v.y*(k+\tau)]* e^{-j2\pi\alpha k}$$

$$R_{rr}^{\alpha}(\tau) = \frac{1}{M}\sum_{k=1}^{M} [\mu.y(k) + v.y*(k)][\mu*.y*(k+\tau) + v*.y(k+\tau)]e^{-j2\pi\alpha k}$$

$$R_{rr}^{\alpha}(\tau) = \frac{1}{M}\sum_{k=1}^{M} [|\mu|^2.y(k).y*(k+\tau) + v.\mu*.y*(k).y*(k+\tau) \\ + v*.\mu.y(k).y(k+\tau) + |v|^2 y*(k).y(k+\tau)]e^{-j2\pi\alpha k}$$

$$R_{rr}^{\alpha}(\tau) = (|\mu|^2 + |v|^2)R_{yy}^{\alpha}(\tau) + 2\Re(v.\mu*.R_{yy}^{'\alpha}(\tau))$$

where $R_{yy}^{'\alpha}(\tau) = \frac{1}{M}\sum_{k=1}^{M} y(k)y(k+\tau)e^{-j2\pi\alpha k}$

As concluded from the formulation above, cyclostationarity based detectors are dramatically harmed by I/Q imbalance effects. However, this is not true for the SP-CAF detector since it is sufficient to ob-

Figure 9. ROC of the ED detector at different values of g

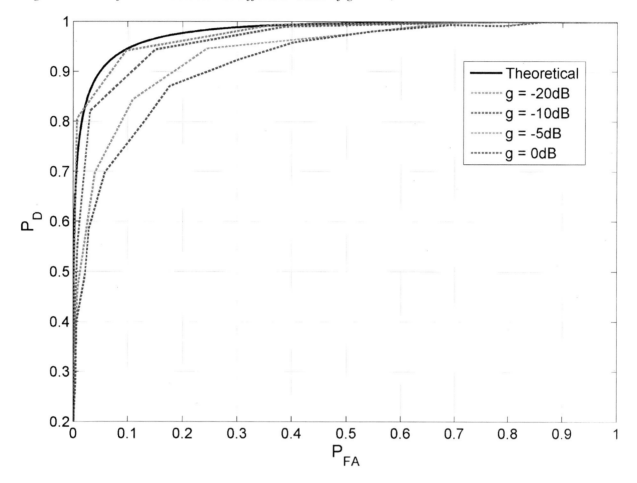

serve only the real part of the received signal (I branch) and estimate its CAF in order to decide on the presence of a PU. Results plotted in Figure 10 and Figure 11 show the robustness on SP-CAF against I/Q imbalance effects.

4. Nonlinearities Impact

RF designers are struggling to design linear analog components for CR systems. However, the linearity constraint is becoming more and more challenging especially when building a dynamic and flexible receiver with tunable stages. Thus, RF nonlinearities are considered the most serious impairment caused by a reconfigurable hardware chain. In the case of nonlinearities, the distorted signal could be written as:

$$r(k) = \alpha_1 y(k) + \alpha_2 (y(k))^2 + \alpha_3 (y(k))^3$$

where $\alpha_1 y(k)$, $\alpha_2 (y(k))^2$, and $\alpha_3 (y(k))^3$ represent the linear component, the second order nonlinear component, and the third order nonlinear component. α_1, α_2, and α_3, are real constants and their values depend

Figure 10. ROC of the SP-CAF detector at different values of θ

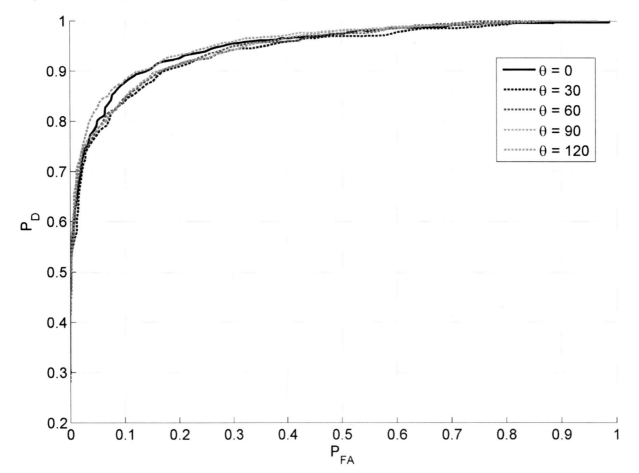

on the hardware characteristics of the board such as the gain and the measured input-referred intercept points (IIP). Superior nonlinear components are omitted in this equation since it is only a proof of concept.

It is clear that the nonlinearities components will expand in the expression of $r(k) = w(k)$ in H_0 and $r(k) = w(k) + y(k)$ in H_1. This will create the same (P_{FA}, P_D) couple but for higher threshold since the noise level is also increased by the nonlinear effect. Thus, the ROC curve of the ED will remain unmodified as illustrated in Figure 12. Also for the SP-CAF detector, the symmetry (cyclostationary feature) will be identically detected with or without nonlinearities since the symmetry is observed with low or high levels. In H_0 scenarios, inducing nonlinear components of the noise will not add a cyclostationary feature to its randomness. The ROC of the SP-CAF simulated with the nonlinear components induced by a real board is shown in Figure 13. The harmful effect of nonlinearly induced components is observed while sensing a wide band, since these unwanted harmonics can virtually occupy free bands and mislead the detector as shown in Figure 14.

Figure 11. ROC of the SP-CAF detector at different values of g

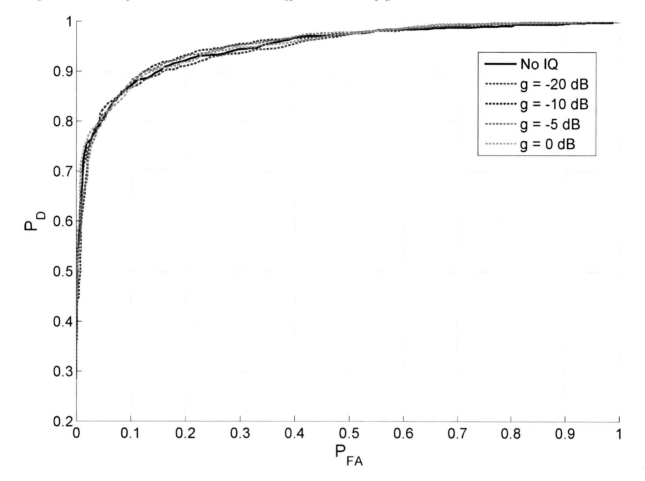

Proposed Algorithms for RF Impairments Compensation

"Dirty RF" is an adequate solution for the RF impairments problem in CR receivers as proposed by (Fettweis, 2005). It will tolerate some RF impairments and mitigate them digitally in order to give some relaxation to the analog design of such devices. This is an optimized solution that could improve the scalability, enhance power-saving, and reveal the dynamic aspect of SU front-ends. Several pre-distortion techniques were presented to mitigate these impairments on the transmitter side. However, in the critical design of a cognitive receiver, these DSP techniques should blindly compensate the main RF imperfections to guarantee a reliable sensing, thus a perfect functioning of the dynamic terminal. In this section, three adaptive compensation techniques, that address the main harmful impairments, are described.

1. Phase Noise Impact Compensation

In the previous subsection, the destructive impact of the phase noise caused by LO is illustrated. It is shown that every second-order cyclostationarity-based sensing technique will lose its performance in the

Figure 12. ROC of the ED detector with real nonlinear components

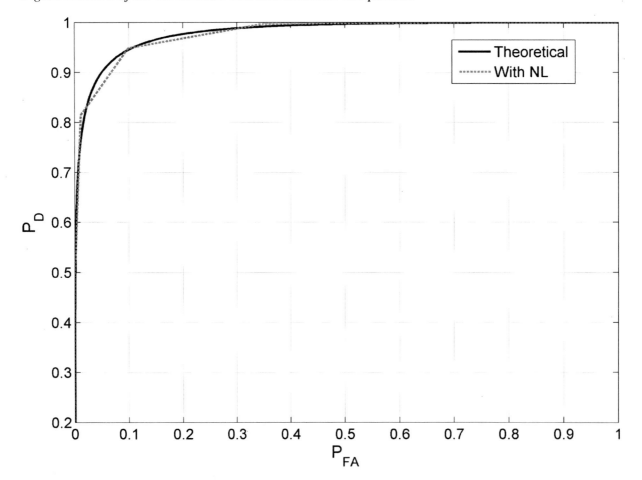

presence of such impairment. A blind compensation technique was reported by (Lee, 2012) to lower the noise effect. The blindness criteria of this scheme make it suitable for CR receivers since it conserves its flexibility feature. To combat the phase noise degradations, the algorithm estimates and compensates two major effects of phase noise, i.e. the inter-carrier interference (ICI) and the common phase error (CPE), while assuming an OFDM based signal. For that, the received signal is partitioned into sub-blocks and time-averages of the phase noise is calculated over each block. A reduction of the differences between obtained time-averages is applied as a first stage of the algorithm to mitigate the ICI resulting from phase noise. Then, an estimation the CPE is applied by using the discrete Fourier transform of the corrected signal resulting from the ICI compensation stage. Finally, the CPE is compensated by computing the minimum mean square error (MMSE) filter coefficient of the one-tap equalizer. A block diagram of the algorithm is shown in Figure 15.

2. I/Q Imbalance Impact Compensation

Another RF impairment that harms the sensing efficiency of a cognitive terminal is the I/Q imbalance. For that, a blind compensation scheme should be added before the sensing technique to cleanse the sig-

Figure 13. ROC of the SP-CAF detector with real nonlinear components

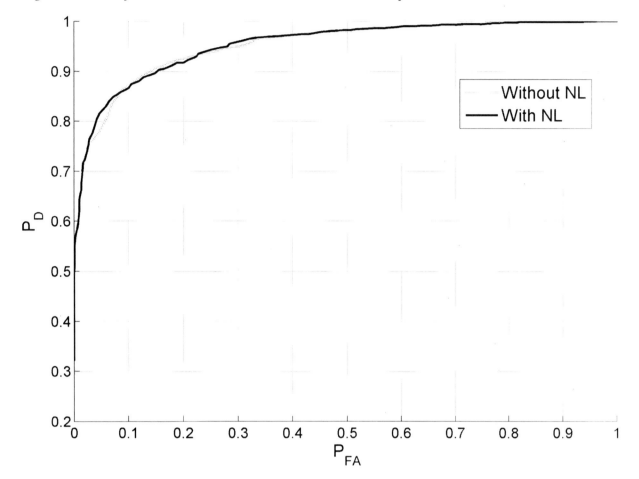

nal from damaging distortions. Such scheme was lately discussed by (Matsui, 2011) et al. using 2-tap adaptive filters. Tap filters weights are computed using the constant modulus algorithm (CMA) and the algorithm is applied subcarrier by subcarrier in an OFDM based signal. To adaptively compute the weights of the compensator, CMA exploits the constant modularity of the transmitted signal. Since the CMA does not require any training signals, the algorithm is considered as blind and could be used for CR scenarios. The block diagram of this scheme is illustrated in Figure 16.

3. Nonlinearities Impact Compensation

As demonstrated above, strong primary users could harm the performance of traditional spectrum sensing techniques by inducing additional spectrum components via frontend's nonlinearity. This is the most crucial source of impairments that could totally harm the efficiency of any sensing technique. Unfortunately, unwanted frequency components could overlap with weak secondary users thus degrading the reliability

Figure 14. Spectrum Density of a received signal distorted by hardware imperfections

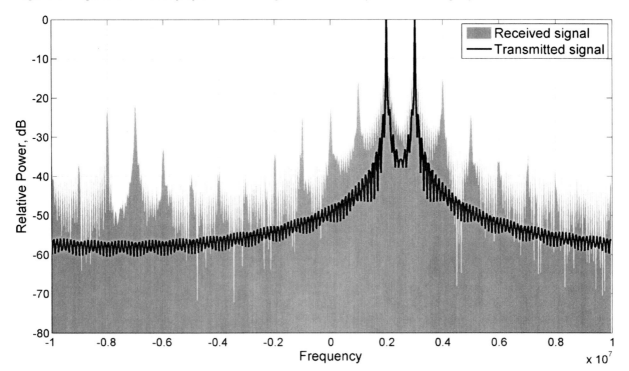

Figure 15. Block diagram of the blind phase noise compensator scheme

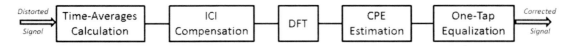

of SU transmissions, or virtually occupy the spectrum and decrease the opportunity to find a vacant transmission band. To mitigate these effects, a smart DSP should be able to digitally lower the effects of non-linearities, not only near SU transmissions but also along the whole wideband. (Valkama, 2006) devised the adaptive interference cancellation (AIC) algorithm, which is a feed-forward algorithm for mitigation of second, third and fifth order intermodulation distortion. The idea is to model the distortion caused by the interferer and then subtract them from the received signal. A mathematical formulation of the distortion model and order is studied before implementing algorithm which is based on imitating the distortion products and adaptively adjust their levels to compensate them from the distorted signal.

A block diagram of an enhanced version of the AIC algorithm, as presented in (Grimm, 2014), is illustrated in Figure 17. The algorithm starts by splitting the band of the received signal in order to differentiate between the strong PU and other frequency components (SUs + distortions). The band splitting

Figure 16. Block diagram of the blind I/Q Imbalance compensator

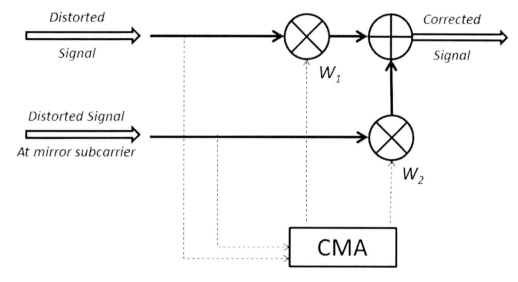

Figure 17. Block diagram of the AIC algorithm

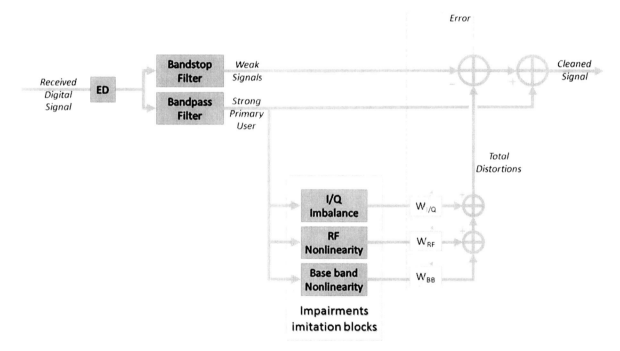

is accompanied by a coarse energy detector used to locate the strong interferer. Then a parallel block of reference non-linearities is used to extract potential distortion products from the strong interferer. An adaptive filter, the Least Mean Square (LMS), is used to adjust digitally-created distortions levels. The adaptive filter utilizes the distorted signal resulted from the band splitter as an input parameter and minimizes the common error signal e(t):

$$e(k) = r_{des}(k) - r_{ref}(k)$$

The adjusted non linearities are finally subtracted from the received signal to cleanse the band from non-linearity distortions.

It is shown by (Grimm, 2012) that the application of the AIC algorithm before the spectrum sensing algorithm increases the detection reliability in CR devices. Our future work focuses on implementing a dual-stage algorithm based on the ideas in this document to be able to alleviate the RF frontend difficulties in CR wideband receivers.

CONCLUSION AND FUTURE RESEARCH PLAN

This chapter highlights the main RF impairments that degrade the spectrum sensing performance in RF CR front-ends. The carrier frequency offset, phase noise, I/Q imbalance, and nonlinearities RF impairments impact on spectrum sensing is analytically evaluated. Robust sensing algorithms for the above RF impairments compensation are proposed. The discussed spectrum sensing algorithms blindly correct imperfect RF issues by compensating induced error effects through digital baseband processing. The illustrated solutions are found to contribute better functioning of the dynamic terminal, and suppress the main harmful RF impairments. A future assessment plan to pragmatically test the proposed mitigation and sensing algorithms is presented.

Among state-of-the-art sensing and mitigation techniques overviewed in this chapter, very few schemes were implemented on real test-beds. Unfortunately, a large number of the devised algorithms for CR systems fail to achieve the expected performance in real-time measurements due to the highly dynamic environment. For this reason, the main future research plan is to extend the work, presented in this chapter, to test the mitigation and sensing algorithms on a practical platform. This is considered the main focus of CR researchers nowadays after spending more than ten years on devising and proving theoretical techniques.

A suggested platform is the Universal Software Radio Peripheral (USRP) which is a software defined radio transceiver where real RF signal could be manipulated via a hosting computer. An RF frontend, consisting of a daughter board and an antenna, is added to the USRP to acquire and deliver RF signals by performing RF filtering, amplifying, mixing and digitizing. The baseband digital operations, i.e. coding/decoding, interleaving, frequency hopping, equalization, and compression, are done by using an embedded FPGA. The acquired signal is accessible on the hosting computer by using GNU Radio or Matlab, where DSP techniques could be implemented and real-time measurements could be collected.

Having such practical environments, the implementation of a fast, blind, and robust sensing algorithm is required to meet the requirements of highly adaptive cognitive receivers. The degradation observed in the performance of the detectors is a motivation to devise a sensing scheme aware of the hardware impairments to be perfectly deployable in real-world scenarios. For that, presented mitigation techniques should assist the fast and blind sensing schemes to alleviate the hardware harmful effects by cleaning the received band before searching for potential spectrum holes. Such dual stage architectures are expected to be a modern candidate for efficient cognitive sensing.

ACKNOWLEDGMENT

The authors gratefully acknowledge the support of the Lebanese National Council for Scientific Research (LNCSR) to the Electromagnetic and Radio Frequency Group (EMRF) at the American University of Beirut (AUB).

REFERENCES

Abidi, A. A. (2007). The path to the software-defined radio receiver. *IEEE Journal of Solid-State Circuits*, *42*(5), 954–966. doi:10.1109/JSSC.2007.894307

Acampora, A., Collado, A., & Georgiadis, A. (2010). Nonlinear analysis and optimization of a distributed voltage controlled oscillator for cognitive radio. In *Proceedings of IEEE International Microwave Workshop Series on RF Front-Ends for Software Defined and Cognitive Radio Solutions (IMWS)*, (pp. 1-4). IEEE. doi:10.1109/IMWS.2010.5440987

Agnelli, F., Albasini, G., Bietti, I., Gnudi, A., Lacaita, A., & Manstretta, D. et al. (2006). Wireless multi-standard terminals: System analysis and design of a reconfigurable RF front-end. *IEEE Circuits and Systems Magazine*, *6*(1), 38–59. doi:10.1109/MCAS.2006.1607637

Axell, E., Leus, G., & Larsson, E. G. (2010). Overview of spectrum sensing for cognitive radio. In *Proceedings of 2nd International Workshop on Cognitive Information Processing (CIP)*, (pp. 322-327). Academic Press. doi:10.1109/CIP.2010.5604136

Balasubramanian, S., Boumaiza, S., Sarbishaei, H., Quach, T., Orlando, P., & Volakis, J. et al. (2012). Ultimate transmission. *IEEE Microwave Magazine*, *13*(1), 64–82. doi:10.1109/MMM.2011.2173983

Bantouna, A., Stavroulaki, V., Kritikou, Y., Tsagkaris, K., Demestichas, P., & Moessner, K. (2012). An overview of learning mechanisms for cognitive systems. *EURASIP Journal on Wireless Communications and Networking, 22*.

Bkassiny, M., Jayaweera, S. K., Li, Y., & Avery, K. A. (2012). Wideband spectrum sensing and non-parametric signal classification for autonomous self-learning cognitive radios. *IEEE Transactions on Wireless Communications*, *11*(7), 2596–2605. doi:10.1109/TWC.2012.051512.111504

Cabric, D., & Brodersen, R. W. (2005). Physical layer design issues unique to cognitive radio systems. In *Proceedings of IEEE 16th International Symposium on Personal, Indoor and Mobile Radio Communications (PIMRC)*, (vol. 2, pp. 759-763). IEEE.

Carey-Smith, B. E., Warr, P. A., Rogers, P. R., Beach, M. A., & Hilton, G. S. (2005). Flexible frequency discrimination subsystems for reconfigurable radio front-ends. *EURASIP Journal on Wireless Communications and Networking*, *2005*(3), 372196. doi:10.1155/WCN.2005.354

Darabi, H. (2010). Highly integrated and tunable RF front-ends for reconfigurable multi-band transceivers. In *Proceedings of IEEE Custom Integrated Circuits Conference (CICC)*, (pp. 1-8). IEEE. doi:10.1109/CICC.2010.5617622

De Domenico, A., Strinati, E. C., & Di Benedetto, M. (2012). A survey on MAC strategies for cognitive radio networks. *IEEE Communications Surveys and Tutorials*, *14*(1), 21–44. doi:10.1109/SURV.2011.111510.00108

Djoumessi, E. E., & Ke Wu. (2010). Reconfigurable RF front-end for frequency-agile direct conversion receivers and cognitive radio system applications. In *Proceedings of IEEE Radio and Wireless Symposium (RWS)*, (pp. 272-275). IEEE. doi:10.1109/RWS.2010.5434205

Dutta, A., Saha, D., Grunwald, D., & Sicker, D. (2010). An architecture for software defined cognitive radio. In *Proceedings of ACM/IEEE Symposium on Architectures for Networking and Communications Systems (ANCS)*, (pp. 1-12). ACM/IEEE. doi:10.1145/1872007.1872014

Fayrouz, H., Wenceslas, R., Lakhdar, Z., & Oussama, F. (2010). Design of fully-integrated RF front-end for large image rejection and wireless communication applications. In *Proceedings of 17th IEEE International Conference on Electronics, Circuits, and Systems (ICECS)*, (pp. 902-905). IEEE.

Fettweis, G., Lohning, M., Petrovic, D., Windisch, M., Zillmann, P., & Tave, W. (2005). Dirty RF: A New Paradigm. In *Proceedings of IEEE 16th International Symposium on Personal, Indoor and Mobile Radio Communications (PIMRC)*, (pp. 2347–2355). IEEE.

Grimm, M., Allen, M., Marttila, J., Valkama, M., & Thoma, R. (2014). Joint Mitigation of Nonlinear RF and Baseband Distortions in Wideband Direct-Conversion Receivers. *IEEE Transactions on Microwave Theory and Techniques*, *62*(1), 166–182. doi:10.1109/TMTT.2013.2292603

Grimm, M., Sharma, R. K., Hein, M. A., & Thomä, R. S. (2012). DSP-based Mitigation of RF Front-end Non-linearity in Cognitive Wideband Receivers. *Frequenz.*, *6*(9-10), 303–310.

Harada, H. (2008A). A feasibility study on software defined cognitive radio equipment. In *Proceedings of 3rd IEEE Symposium on New Frontiers in Dynamic Spectrum Access Networks (DySPAN)*, (pp. 1-12). IEEE. doi:10.1109/DYSPAN.2008.8

Harada, H. (2008B). A small-size software defined cognitive radio prototype. In *Proceedings of IEEE 19th International Symposium on Personal, Indoor and Mobile Radio Communications (PIMRC)* (pp. 1-5). IEEE.

Hayar, A. M., Pacalet, R., & Knopp, R. (2007). Cognitive radio research and implementation challenges. In *Proceedings of Conference Record of the Forty-First Asilomar Conference on Signals, Systems and Computers (ACSSC)*, (pp. 782-786). ACSSC.

Heinen, S., & Wunderlich, R. (2011). *High dynamic range RF frontends from multiband multistandard to cognitive radio*. In *Proceedings of Semiconductor Conference Dresden* (pp. 1–8). SCD.

Kaltiokallio, M., Saari, V., Kallioinen, S., Parssinen, A., & Ryynnen, J. (2012). Wideband 2 to 6 GHz RF front-end with blocker filtering. *IEEE Journal of Solid-State Circuits*, *47*(7), 1636–1645. doi:10.1109/JSSC.2012.2191348

Khalaf, Z., Nafkha, A., & Palicot, J. (2012, June). Blind spectrum detector for cognitive radio using compressed sensing and symmetry property of the second order cyclic autocorrelation. In *Proceedings of 7th International ICST Conference on Cognitive Radio Oriented Wireless Networks and Communications (CROWNCOM)*, (pp. 291-296). IEEE.

Kim, J., & Shin, H. (2011). Design considerations for cognitive radio based CMOS TV white space transceivers. In *Proceedings of International SoC Design Conference (ISOCC)*, (pp. 238-241). ISOCC.

Kitsunezuka, M., Kodama, H., Oshima, N., Kunihiro, K., Maeda, T., & Fukaishi, M. (2012). A 30-MHz–2.4-GHz CMOS receiver with integrated RF filter and dynamic-range-scalable energy detector for cognitive radio systems. *IEEE Journal of Solid-State Circuits, 47*(5), 1084–1093. doi:10.1109/JSSC.2012.2185531

Kwan, A., Bassam, S. A., & Ghannouchi, F. M. (2012). Sub-sampling technique for spectrum sensing in cognitive radio systems. In *Proceedings of IEEE Radio and Wireless Symposium (RWS)*, (pp. 347-350). IEEE. doi:10.1109/RWS.2012.6175350

Larson, L., Abdelhalem, S., Thomas, C., & Gudem, P. (2012). High-performance silicon-based RF front-end design techniques for adaptive and cognitive radios. In *Proceedings of IEEE Compound Semiconductor Integrated Circuit Symposium (CSICS)*, (pp. 1-4). IEEE. doi:10.1109/CSICS.2012.6340060

Lee, M. K., Lim, S. C., & Yang, K. (2012). Blind Compensation for Phase Noise in OFDM Systems over Constant Modulus Modulation. *IEEE Transactions on Communications, 60*(3), 620–625. doi:10.1109/TCOMM.2011.121511.110088

Lim, K., & Laskar, J. (2008). Emerging opportunities of RF IC/system for future cognitive radio wireless communications. In *Proceedings of IEEE Radio and Wireless Symposium*, (pp. 703-706). IEEE. doi:10.1109/RWS.2008.4463589

Lin, D. T., Hyungil Chae, L. L., & Flynn, M. P. (2012). A 600MHz to 3.4GHz flexible spectrum-sensing receiver with spectrum-adaptive reconfigurable DT filtering. In *Proceedings of IEEE Radio Frequency Integrated Circuits Symposium (RFIC)*, (pp. 269-272). IEEE. doi:10.1109/RFIC.2012.6242279

Lourandakis, E., Weigel, R., Mextorf, H., & Knoechel, R. (2012). Circuit agility. *IEEE Microwave Magazine, 13*(1), 111–121. doi:10.1109/MMM.2011.2173987

Luu, L., & Daneshrad, B. (2007). An adaptive weaver architecture radio with spectrum sensing capabilities to relax RF component requirements. *IEEE Journal on Selected Areas in Communications, 25*(3), 538–545. doi:10.1109/JSAC.2007.070404

Matsui, M., Nakagawa, T., Kudo, R., Ishihara, K., & Mizoguchi, M. (2011). Blind frequency-dependent IQ imbalance compensation scheme using CMA for OFDM system. In *Proceedings of IEEE 22nd International Symposium on Personal, Indoor and Mobile Radio Communications (PIMRC)*, (pp. 1386-1390). IEEE.

Mayer, A., Maurer, L., Hueber, G., Dellsperger, T., Christen, T., Burger, T., & Zhiheng, C. (2007). RF front-end architecture for cognitive radios. In *Proceedings of IEEE 18th International Symposium on Personal, Indoor and Mobile Radio Communications (PIMRC)*, (pp. 1-5). IEEE.

Mukhopadhyay, R., Yunseo Park, , Sen, P., Srirattana, N., Jongsoo Lee, , & Chang-Ho Lee, et al. (2005). Reconfigurable RFICs in si-based technologies for a compact intelligent RF front-end. *IEEE Transactions on Microwave Theory and Techniques, 53*(1), 81–93. doi:10.1109/TMTT.2004.839352

Nekovee, M. (2006). Dynamic spectrum access with cognitive radios: Future architectures and research challenges. In *Proceedings of 1ˢᵗ International Conference on Cognitive Radio Oriented Wireless Networks and Communications*, (pp. 1-5). Academic Press. doi:10.1109/CROWNCOM.2006.363464

Park, B., Lee, K., Choi, S., & Hong, S. (2010). A 3.1–10.6 GHz RF receiver front-end in 0.18 μm CMOS for ultra-wideband applications. In *Proceedings of IEEE MTT-S International Microwave Symposium Digest (MTT)*, (pp. 1616-1619). IEEE.

Razavi, B. (2009). Challenges in the design of cognitive radios. In *Proceedings of IEEE Custom Integrated Circuits Conference (CICC)*, (pp. 391-398). IEEE.

Ru, Z., Klumperink, E. A. M., Saavedra, C. E., & Nauta, B. (2010). A 300–800 MHz tunable filter and linearized LNA applied in a low-noise harmonic-rejection RF-sampling receiver. *IEEE Journal of Solid-State Circuits, 45*(5), 967–978. doi:10.1109/JSSC.2010.2041403

Safatly, L., Aziz, B., Nafkha, A., Louet, Y., Nasser, Y., El-Hajj, A., & Kabalan, K. Y. (2014). A Blind Spectrum Sensing Using Symmetry Property of Cyclic Autocorrelation Function: From Theory to Practice. *EURASIP Journal on Wireless Communications and Networking, 2014*(26).

Stadius, K., Kaltiokallio, M., Ollikainen, J., Parnanen, T., Saari, V., & Ryynanen, J. (2011). A 0.7 – 2.6 GHz high-linearity rf front-end for cognitive radio spectrum sensing. In *Proceedings of IEEE International Symposium on Circuits and Systems (ISCAS)*, (pp. 2181-2184). IEEE. doi:10.1109/ISCAS.2011.5938032

Tandur, D., & Moonen, M. (2007). Joint adaptive compensation of transmitter and receiver IQ imbalance under carrier frequency offset in OFDM-based systems. *IEEE Transactions on Signal Processing, 55*(11), 5246–5252. doi:10.1109/TSP.2007.898788

Tyagi, A. K., & Rajakumar, R. V. (2010). A wideband RF frontend architecture for software defined radio. In *Proceedings of 2010 International Conference on Industrial and Information Systems (ICIIS)*, (pp. 25-30). Academic Press. doi:10.1109/ICIINFS.2010.5578742

Valkama, M., Shahed Hagh Ghadam, A., Anttila, L., & Renfors, M. (2006). Advanced Digital Signal Processing Techniques for Compensation of Nonlinear Distortion in Wideband Multicarrier Radio Receivers. *IEEE Transactions on Microwave Theory and Techniques, 54*(6), 2356–2366. doi:10.1109/TMTT.2006.875274

Wang, W. (2009). Spectrum sensing for cognitive radio. In *Proceedings of Third International Symposium on Intelligent Information Technology Application Workshops (IITAW)*, (pp. 410-412). Academic Press. doi:10.1109/IITAW.2009.49

Weingart, T., Yee, G. V., Sicker, D. C., & Grunwald, D. (2007). Implementation of a reconfiguration algorithm for cognitive radio. In *Proceedings of 2ⁿᵈ International Conference on Cognitive Radio Oriented Wireless Networks and Communications (CrownCom)*, (pp. 171-180). Academic Press. doi:10.1109/CROWNCOM.2007.4549792

Yucek, T., & Arslan, H. (2009). A survey of spectrum sensing algorithms for cognitive radio applications. *IEEE Communications Surveys and Tutorials, 11*(1), 116–130. doi:10.1109/SURV.2009.090109

ADDITIONAL READING

Al-Hussein, M., Ramadan, A., Zamudio, M. E., Christodoulou, C. G., El-Hajj, A., & Kabalan, K. Y. (2011). A UWB Antenna Combined with a Reconfigurable Bandpass Filter for Cognitive Radio Applications. *IEEE-APS Topical Conference on Antennas and Propagation in Wireless Communications (APWC)*, pp. 902-904.

Al-Husseini, M., El-Hajj, A., Tawk, Y., Kabalan, K. Y., & Christodoulou, C. G. (2010). A simple dual-port antenna system for cognitive radio applications. *International Conference on High Performance Computing and Simulation (HPCS)*, pp. 549-552.

Al-Husseini, M., Kabalan, K. Y., El-Hajj, A., & Christodoulou, C. G. (2010). Cognitive Radio: UWB Integration and Related Antenna Design. Meng Joo Er (Ed.), New Trends in Technologies: Control, Management, Computational Intelligence and Network Systems (pp. 395-412). InTech.

Al-Husseini, M., Ramadan, A., El-Hajj, A., & Kabalan, K. Y. (2012). A Tunable Filter Antenna for Cognitive Radio Systems. *IEEE International Symposium on Antennas and Propagation and USNC/URSI National Radio Science Meeting*, pp. 1-5.

Al-Husseini, M., Ramadan, A., El-Hajj, A., Kabalan, K. Y., Tawk, Y., & Christodoulou, C. G. (2011). Design Based on Complementary Split-ring Resonators of an Antenna with Controllable Band Notches for UWB Cognitive Radio Applications. *IEEE International Symposium on Antennas and Propagation and USNC/URSI National Radio Science Meeting*, pp. 1120-1122.

Al-Husseini, M., Safatly, L., Ramadan, A., El-Hajj, A., Kabalan, K. Y., & Christodoulou, C. G. (2012). Reconfigurable Filter Antennas or Pulse Adaptation in UWB Cognetive Radio Systems. *Progress In Electromagnetics Research B, 37*, 327–342.

Al-Husseini, M., Tawk, Y., Christodoulou, C. G., Kabalan, K. Y., & El-Hajj, A. (2010). A reconfigurable cognitive radio antenna design. *IEEE International Symposium on Antennas and Propagation and USNC/URSI National Radio Science Meeting*, pp. 11-17.

Allén, M., Marttila, J., & Valkama, M. (2010). Modeling and mitigation of nonlinear distortion in wideband A/D converters for cognitive radio receivers. *International Journal of Microwave and Wireless Technologies, 2*(02), 183–192.

Arslan, H. (Ed.). (2007). *Cognitive radio, software defined radio, and adaptive wireless systems* (Vol. 10). Berlin: Springer.

Blossom, E. (2004). GNU radio: tools for exploring the radio frequency spectrum. *Linux journal, 2004*(122), 4.

Davis, G., Mallat, S., & Avellaneda, M. (1997). Adaptive Greedy Approximations. *Constructive Approximation, 13*(1), 57–98.

Ettus, M. (2009). *Universal software radio peripheral.* Mountain View, CA: Ettus Research.

Gardner, W. A. (1991). Exploitation of spectral redundancy in cyclostationary signals. *IEEE Signal Processing Magazine, 8*(2), 14–36.

Gardner, W. A., Napolitano, A., & Paura, L. (2006). Cyclostationarity: Half a century of research. *Signal Processing, 86*(4), 639–697.

Kiayani, A., Anttila, L., Zou, Y., & Valkama, M. (2012). Advanced receiver design for mitigating multiple RF impairments in OFDM systems: Algorithms and RF measurements. *Journal of Electrical and Computer Engineering, 2012,* 3.

Mitola, J. (1999). Cognitive radio for flexible mobile multimedia communications. *IEEE International Workshop on Mobile Multimedia Communications (MoMuC),* pp. 3-10.

Ramadan, A., Al-Husseini, M., Kabalan, K. Y., El-Hajj, A., Tawk, Y., Christodoulou, C. G., & Costantine, J. (2012). A Narrowband Frequency-tunable Antenna for Cognitive Radio Applications. *6th European Conference on Antennas and Propagation (EUCAP),* pp. 3273-3277.

Ramadan, A. H., Costantine, J., Al-Husseini, M., Kabalan, K. Y., Tawk, Y., & Christodoulou, C. G. (2014). Tunable filter-antennas for cognitive radio applications. *Progress In Electromagnetics Research B, 57,* 253–265.

Safatly, L., Al-Husseini, M., El-Hajj, A., & Kabalan, K. Y. (2012). Advanced Techniques for Pulse Shaping in UWB Cognitive Radio. *International Journal of Antennas and Propagation, 2012.* doi:10.1155/2012/390280

Schenk, T. (2008). RF imperfections in high-rate wireless systems. Springer, 9, 2033-2043.

Tawk, Y., Christodoulou, C. G., Costantine, J., & Barbin, S. E. (2012). A frequency and radiation pattern reconfigurable antenna system with sensing capabilities for cognitive radio. *IEEE International Symposium on Antennas and Propagation and USNC/URSI National Radio Science Meeting,* pp. 8-14.

Tawk, Y., Costantine, J., Avery, K., & Christodoulou, C. G. (2011). Implementation of a Cognitive Radio Front-End Using Rotatable Controlled Reconfigurable Antennas. *IEEE Transactions on Antennas and Propagation, 59*(5), 1773–1778.

Tawk, Y., Hemmady, S., Christodoulou, C. G., Costantine, J., & Balakrishnan, G. (2011). A cognitive radio antenna design based on optically pumped reconfigurable antenna system (OPRAS). *IEEE International Symposium on Antennas and Propagation and USNC/URSI National Radio Science Meeting,* pp. 1116-1119.

Valkama, M., Renfors, M., & Koivunen, V. (2001). Advanced methods for I/Q imbalance compensation in communication receivers. *IEEE Transactions on Signal Processing, 49*(10), 2335–2344.

Valkama, M., Springer, A., & Hueber, G. (2010, May). Digital signal processing for reducing the effects of RF imperfections in radio devices-An overview. *Proceedings of IEEE International Symposium on Circuits and Systems (ISCAS),* pp. 813-816.

Wyglinski, A. M., Nekovee, M., & Hou, T. (Eds.). (2009). *Cognitive radio communications and networks: principles and practice*. Academic Press.

Zheng, C. (2009). OFDM Interference Analysis with 'Dirty RF'.

KEY TERMS AND DEFINITIONS

Cognitive Radio: A fully reconfigurable wireless transceiver which automatically adapts its communication parameters to network and user demands.

Dirty RF: Design and implement digital signal processing algorithms to compensate and mitigate RF distortions.

Licensed-Band Cognitive Radio: Users are capable of using bands assigned by licensed users.

RF Impairment: Radio frequency impediments (or harms) produced by the RF devices causing signals distortion.

Sensing-Based Spectrum Sharing: Users first listen to the spectrum allocated to the licensed users to detect the state of the licensed users. Based on the detection results, cognitive radio users decide their transmission strategies. If the licensed users are not using the bands, cognitive radio users will transmit over those bands. If the licensed users are using the bands, cognitive radio users share the spectrum bands with the licensed users by restricting their transmit power.

Spectrum Mobility: Process by which a cognitive-radio user changes its frequency of operation.

Spectrum Sharing: Spectrum sharing cognitive radio networks allows cognitive radio users to share the spectrum bands of the licensed-band users.

Unlicensed-Band Cognitive Radio: Users can only utilize unlicensed parts of the radio frequency (RF) spectrum when it is not utilized by a licensed user.

Software Defined Radio: A flexible radio based on a domination of software modules and capable to host multiple standards.

Chapter 4
Energy-Efficient Cooperative Spectrum Sensing for Cognitive Radio Networks

Saud Althunibat
University of Trento, Italy

Marco Di Renzo
Laboratory of Signals and Systems (L2S), France

Sandeep Narayanan
WEST Aquila s.r.l., Italy & University of L'Aquila, Italy

Fabrizio Granelli
University of Trento, Italy

ABSTRACT

One of the main problems of Cooperative Spectrum Sensing (CSS) in cognitive radio networks is the high energy consumption. Energy is consumed while sensing the spectrum and reporting the results to the fusion centre. In this chapter, a novel partial CSS is proposed. The main concern is to reduce the energy consumption by limiting the number of participating users in CSS. Particularly, each user individually makes the participation decision. The energy consumption in a CSS round is expected by the user itself and compared to a predefined threshold. The corresponding user will participate only if the expected amount of energy consumed is less than the participation threshold. The chapter includes optimizing the participation threshold for energy efficiency maximization. The simulation results show a significant reduction in the energy consumed compared to the conventional CSS approach.

1. INTRODUCTION

Recently, energy efficiency in wireless networks has received a significant amount of research. This is because mobile users are usually battery-powered. The limited energy resources represent a challenge hindering wide implementation of some recent technologies (Fettweis & Zimmermann, 2008). Some wireless systems, such as cognitive radio (CR), implies more energy consumption than other systems. Cognitive radios in general require more energy to operate, as compared to the conventional transceivers due to the additional tasks required to perform cognitive transmission. In CR, a licensed spectrum can be exploited by unlicensed users when it is (temporarily and spatially) unused by licensed users. This requires awareness of spectrum status, which is performed by a process termed as spectrum sensing (Mitola & Maguire, 1999), (Haykin, 2005).

DOI: 10.4018/978-1-4666-6571-2.ch004

In order to identify the unused spectrum portions, the unlicensed users, also called cognitive users (CUs), are enforced to sense it for specific period, inducing energy consumption which does not exist in the typical wireless systems. Moreover, aiming at improving the reliability of spectrum sensing, cooperative spectrum sensing (CSS) is proposed (Mishra, Sahai & Brodersen, 2006), (Di Renzo, Imbriglio, Graziosi & Santucci, 2009), (Ghasemi & Sousa, 2007). In CSS, the local sensing results are reported to a central entity, called fusion centre (FC). The FC is in charge of making a global decision regarding the spectrum occupancy by applying a specific fusion rule (FR). Although CSS decreases the probability of erroneous decision considerably by mitigating the effects of multipath fading and shadowing, it causes extra delay, security risks (I.F. Akyildiz, Lo & Balakrishnan, 2011) (Di Renzo Graziosi & Santucci, April 2009) and more energy consumption.

The high energy expenditure in CSS is caused by the individual sensing and reporting the sensing results to the FC. In case of large number of CUs and/or large number of sensed channels, energy-efficient CSS becomes a pressing need for CR systems. Aiming at reducing energy consumption, limiting the amount of reported results has been widely investigated. In general, two well-known schemes for results' reporting (S. Chaudhari, Lunden, Koivunen & Poor, 2012) (Viswanathan & Varshney, 1997), soft scheme (SS) and hard scheme(HS). In SS, the sensing result of each CU is quantized locally by a multiple number of bits and sent to the FC. On the other hand, the result is quantized by only one bit in HS. As a CU employing HS reports only one bit, it is clear that the energy consumption is lower than if SS is employed (Maleki, Chepuri, & Leus, 2011). Thus, in this work, we consider only HS.

Many works have investigated the reduction of energy consumption in CSS. These works can be classified into four different approaches: (i) Reducing the number of sensing users, (ii) Reducing the sensing time, (iii) Reducing the reported sensing data, and (ii) Optimizing the decision-making rule.

In the first approach, (Maleki et al., 2011), and (Pham, Zhang, Engelstad, Skeie & Eliassen, 2010) have proposed algorithms that use the minimum number of sensing users based on different setups while satisfying predefined thresholds on the detection accuracy. In (Cheng et al.,2012), the CUs are divided into non-disjoint subsets such that only one subset senses the spectrum while the other subsets enter a low power mode. The energy minimization problem is formulated as a network lifetime maximization problem with constraints the detection accuracy. An algorithm for user selection is proposed in (Najimi et. al., 2013), where the user subset that has the lowest cost function and guarantees the desired detection accuracy is selected. The cost function is related to the system energy consumption.

The works in (Zhao et al., 2012), (Feng et al., 2012) and (Pham, Zhang, Engelstad, Skeie & Eliassen, 2010) consider the sensing time as a possible approach to reduce the energy consumption. In (Zhao et al., 2012), the CUs perform an initial sensing stage and report their local decisions to the FC. If a global decision cannot be made (no majority exists), a longer sensing stage is used. In (Feng et al., 2012), a utility function that consists of the difference between the achievable throughput (revenue) and the consumed energy (cost) is maximized by optimizing the sensing time with a constraint on the detection probability. The optimal sensing time that maximizes the energy efficiency with constraints based on the detection accuracy is obtained in (Pham, Zhang, Engelstad, Skeie & Eliassen, 2010). A joint optimization of the number of sensing users, the sensing time and the local detection threshold is presented in (Gao et al., 2013), aiming to maximize the energy efficiency by imposing a constraint on the detection accuracy.

Following the third approach, censoring, confidence voting and clustering have been proposed. Censoring is a promising technique that can significantly reduce the reporting CUs. In censoring, a CU does not report its sensing result unless it lies outside a specific range (Sun et al., March 2007), (Lunden et al.,

2007), (Appadwedula et al., 2008). The censoring thresholds are optimized for minimizing the energy consumption with constraints on the detection accuracy (Maleki et al., 2011). In (Maleki et al., 2013), truncated sequential sensing and censoring are combined in order to reduce the energy consumption in CSS. Specifically, the spectrum is sequentially sensed, and once the accumulated energy of the sensed samples lies outside a certain region, the sensing is stopped and a binary decision is sent to the FC. If the sequential sensing process continues until a timeout, censoring is applied an no decision is sent. The thresholds of the censoring region are optimized in order to minimize the maximum energy consumption per CU subject to a constraint on the detection accuracy. In (Lee et al., 2008), a confidence voting scheme is presented. It works as follows: if the spectrum sensing of a specific CU agrees with the global decision, it gains its confidence; otherwise, it loses its confidence. When a user's confidence level drops below a threshold, it considers itself as unreliable and stops sending its results. But it still keeps sensing the spectrum and tracking the global decision. As long as the result matches, it gains its confidence. Once its confidence level passes beyond the threshold, it rejoins the voting. The energy saving and the detection accuracy of this approach are investigated in (Lee et al., 2008).

Clustering is a popular approach to reduce the overhead load between the CUs and the FC. In clustering, CUs are separated into clusters and one from each cluster is nominated as cluster-head, which is in charge of collecting sensing results from cluster-members and reporting a cluster-decision to the FC on behalf of the cluster-members (Sun et al., June 2007). The cluster-head is changed randomly in each CSS round. The energy saving and the accuracy loss are investigated in (Lee et al., 2008). In addition to energy consumption analysis, time delay is conducted in (Xia et al., 2009). In (Wei et al., 2010) and (Khasawneh et al., 2012) clustering and censoring approaches are combined in one energy-efficient algorithm considering the noisy reporting channels.

Another simple algorithm for reducing the number of reporting CUs without affecting the detection accuracy can be found in (Althunibat & Granelli, 2013 June). The idea is based on an instantaneous processing of the received results at the FC. Whenever a global decision can be made, the reporting process is terminated and the rest of the CUs do not report their local sensing results. Despite its simplicity, this approach does not impact the detection accuracy and it offers a higher energy efficiency than other algorithms.

Following the approach of optimizing decision-making rule, in (Peh et al., 2011), the fusion threshold of the K-out-of-N rule is optimized for maximizing energy efficiency without constraints, while a constraint on resulting interference represented by the missed detection probability is set in (Althunibat et al., 2013 December). In (Maleki et al., 2012), the optimal fusion threshold that maximizes the throughput of CRN is obtained with constraints on the consumed energy per CU and the overall detection probability.

In this chapter, a novel proposal for improving energy efficiency in CSS is presented. The proposed algorithm is based on limiting the number of participating CUs in CSS. However, the selection of the participating CUs is not random. Instead, those CUs that consume higher energy are prevented from participation in CSS. The participation decision of each CU is taken individually by the CU itself. In detail, each CU estimates the expected amount of energy that will be consumed if it participates, and compares it to a predefined threshold, called participation threshold. Accordingly, the CU will participate only if its expected energy consumption is less than the participation threshold. Using this approach, the users that will greatly increase the energy consumption will be prevented from participating, resulting in lower energy consumption. It is worth mentioning that the proposed approach improves the energy efficiency not only by reducing energy consumption but also by increasing the amount of successfully transmitted data. The increase in amount of successfully transmitted data is due to the decrease in the overall false alarm probability, as the number of involving users in CSS decreases. As the number of the involving CUs depends on the predefined threshold, an optimization of this threshold is carried out to maximize energy efficiency.

The rest of this chapter is organized as follows; system model is described in Section II, where in-detail discussion of energy consumption during CSS is carried out. In Section III, the proposed algorithm is presented along with its mathematical formulas followed by performance evaluation through computer simulations in section IV. Conclusions are drawn in Section V.

2. SYSTEM MODEL

A cognitive network consisting of N CUs is considered. All CUs try to access a target spectrum, called primary spectrum. The primary spectrum is licensed for a set of users called primary users (PUs). All the stages of the cognitive transmission is organized by a central FC located at the base station. The channels between the CUs and the licensed users (sensing channels) and the channels between the CUs and the FC (reporting channels) are modelled as narrow-band Rayleigh fading with additive white Gaussian noise (AWGN). The channel variance between any CU and the target spectrum is denoted by μ^2, while the channel variance between any CU and the FC is denoted as σ_i^2. The CUs are distributed randomly around the FC. The distance between i^{th} CU and the FC, denoted as (d_i), is uniformly distributed $d_i \sim U[d^{min}, d^{max}]$, where d^{min} and d^{max} are the minimum and the maximum distances, respectively.

As depicted in Figure 1, during spectrum sensing, the target spectrum is sensed for a specific time, denoted by T_s. The optimal method for spectrum sensing is energy detection method especially when no prior information is available (Cabric, Mishra & Brodersen, 2004). This method implies collecting a number of samples and computing the average energy contained in these samples. According to the HS, the resultant average is compared to a predefined threshold, and a local binary decision $u_i\{1,0\}$ about spectrum status is made. If $u_i = 1$ then the ith CU decides that the spectrum is being used. Otherwise, the spectrum is identified as unused by the ith CU.

Figure 1. The general description of cooperative spectrum sensing

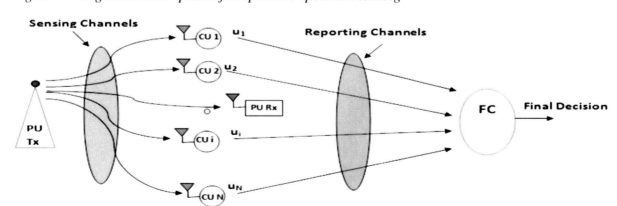

The local sensing performance is measured by two probabilities, namely, the detection probability ($P_{d,i}$) and the false-alarm probability ($P_{f,i}$). The detection probability is the probability of identifying a channel as used given that it is actually used. In other words, making a local decision of $u_i = 1$, when the channel is used. The false alarm probability is the probability of identifying a channel as used channel given that it is unused, which means making a local decision of $u_i = 1$ when the channel is unused. For simplicity, we assume an identical performance among the CUs, and hence $P_{d,1} = P_{d,2} = ... = P_d$ and $P_{f,1} = P_{f,2} = ... = P_f$. P_d and P_f for Rayleigh fading channels are given as (Ghasemi & Sousa, 2007), (Digham, Alouini & Simon, 2007):

$$P_d = \int_\rho Q_m(\sqrt{(2mx)}, \sqrt{(\lambda)}) f_\rho(x) \ dx^{\,1}$$

(1)

$$P_f = \frac{\Gamma\left(m, \dfrac{\lambda}{2}\right)}{\Gamma(m)}$$

(2)

where $Q_m(.,.)$ is the generalized Marcum Q-function (Nuttall, 1975), m is the time-bandwidth product, ρ is the signal-to-noise ratio, $f_n(x)$ is the probability density function (pdf) of the Rayleigh fadingchannel. λ is the energy threshold used by the energy detector and $\Gamma(.,.)$ is the incomplete gamma function (Ryzhik, 1994).

3. CONVENTIONAL APPROACH OF COOPERATIVE SPECTRUM SENSING

In the conventional approach of CSS, all CUs should participate in the spectrum sensing process. Therefore, after a local decision is issued individually by each CU, all local decisions should be reported to the FC. The general FR to process the received local decisions in HS is called $K - out - of - N$ rule (Althunibat, Di Renzo & Granelli, 2013), where N denotes the total number of reporting users and K represents the number of users who detect a signal in the target spectrum, i.e., have obtained a local decision of 1. $K - out - of - N$ rule implies comparing K with a predefined threshold (K'). If   , then the spectrum is identified as used. Otherwise, the spectrum is identified as unused. Mathematically, the function of $K - out - of - N$ rule is written as follows:

$$FinalDecision = \begin{cases} used & \text{if } K \geq K' \\ unused & \text{if } K < K' \end{cases}.$$

(3)

Some popular rules are derived from this rule like OR-rule ($K' = 1$), AND-rule ($K' = N$). Without loss of generality, we consider only OR-rule ($K' = 1$). The overall performance is measured by the overall detection probability and the overall false alarm probability, which are given for OR-rule as (Viswanathan & Varshney, 1997) (Ghasemi & Sousa, 2007):

$$P_D = 1 - (1 - P_d)^N \tag{4}$$

$$P_F = 1 - (1 - P_f)^N . \tag{5}$$

Regarding the total energy consumed in this approach, if we denote the energy consumed by the ith CU during sensing and reporting by $E_{s,i}$, $E_{r,i}$, respectively, and the energy consumed by the scheduled user is E_t, the total energy consumed is given as:

$$E_{tot} = \sum_{i=1}^{N} E_{s,i} + \sum_{i}^{N} E_{r,i} + P_{unused} E_t \tag{6}$$

where P_{unused} is the probability of identifying the spectrum as *unused*, and is given as:

$$P_{unused} = 1 - P_0 P_F - P_1 P_D \tag{7}$$

where, P_0 and P_1 are the probabilities that the spectrum is actually unused and used, respectively.

Notice that the sensing energy is identical for all CUs and equal to Es, thus, (6) can be simplified as:

$$E_{tot} = NE_s + \sum_{i}^{N} E_{r,i} + P_{unused} E_t . \tag{8}$$

As the energy is defined as the consumed power multiplied by the time, (8) can be rewritten as:

$$E_{tot} = N\alpha_s T_s + \sum_{i=1}^{N} \alpha_{r,i} T_r + P_{unused} \alpha_t T_t \tag{9}$$

where: i) T_s, T_r, and T_t are the time consumed by a CU in sensing, reporting and transmission, respectively; ii) α_s, α_r and α_t are the consumed power during sensing, reporting and transmission, respectively.

Another important quantity that should be defined is the amount of the successfully transmitted data (D) measured in bits. Notice that D depends on the correct identification of the unused spectrum. D is given as:

$$D = P_0(1 - P_F)RT_t \tag{10}$$

where R is the data rate in bps, and the factor $P_0(1 - P_F)$ represents the probability of the correct identification of the unused spectrum. From (10), it is also clear that the D increases as P_F decreases.

Finally, for the purpose of assessing the energy efficiency in $[Joule / bit]$, we define the consumed energy per bit (EpB) as follows:

$$EpB = \frac{E_{tot}}{D} \tag{11}$$

4. THE PROPOSED APPROACH

Motivated by improving the energy efficiency in cognitive radio systems, we propose a novel approach for spectrum sensing which reduces energy consumption during this process with a constraint on the achievable detection accuracy. The idea is to reduce the number of users participating in spectrum sensing, which results in a partial cooperative spectrum sensing. The novelty of our proposal is that the participation decision is taken individually by each CU, and on a base of expected energy consumption. In other words, each CU calculates its expected energy consumption in case of participating in spectrum sensing, and compares it to a predefined threshold (γ). If it is lower than γ, the CU will participate. Otherwise, the CU will not participate. The participation threshold is identical for all CUs, and it is decided at the FC and broadcasted to CUs. By such mechanism, we try to reduce energy consumption by an effective way that implies preventing the CUs who will consume large amount of their energy in spectrum sensing from participation.

If we denote the estimated energy consumed during spectrum sensing by the ith CU by E_i, the following equation describes the participation decision (S_i):

$$S_i = \begin{cases} 1 \ (Participate) & \text{if} \quad E_i < \gamma \\ 0 \ (Don't \ participate) & \text{if} \quad E_i \geq \gamma \end{cases} . \tag{12}$$

Next, we discuss the calculation of E_i, the resulting performance based on our proposal, and finally, we address the energy efficiency improvement achieved by the proposed approach.

4.1 Calculation of E_i

E_i includes the energy consumed during local sensing and decision reporting by the CU. Thus, E_i is given as:

$$E_i = E_s + E_{r,i} \tag{13}$$

As E_s is identical for all CUs, then the determinant factor in E_i is $E_{r,i}$ that can be written as a product of the reporting time T_r and the power consumed during reporting $\alpha_{r,i}$, as follows:

$$E_{r,i} = \alpha_{r,i} T_r \tag{14}$$

In results' reporting, the user is in transmission status, and hence, $\alpha_{r,i}$ mainly depends on the distance from the FC and the desired bite error rate. $\alpha_{r,i}$ is given as (S. Cui, 2004):

$$\alpha_{r,i} = \alpha^c + \alpha_i^{PA} \tag{15}$$

where α_i^{PA} is the power consumed in the power amplifier stage of the ith user, and α^c is the power consumed by the other circuit elements. α^c is identical in all users and can be modelled as:

$$\alpha^c = \alpha^{DAC} + \alpha^{filt} + \alpha^{mix} + \alpha^{syn} \tag{16}$$

where α^{DAC}, α^{filt}, α^{mix}, and α^{syn} are the power consumption at the digital-to-analog converter (DAC), the transmit filters, the mixer, and the frequency synthesizer, respectively. α^{filt}, α^{mix}, and α^{syn} can be modelled as constants, while α^{DAC} can be approximated as:

$$\alpha^{DAC} = \left(\frac{1}{2} V_{dd} I_0 (2^{n_1} - 1) + n_1 C_p (2B + f_{cor}) V_{dd}^2 \right) \tag{17}$$

where I_0 is the current supply, n_1 is the number of bits in the DAC, C_p is the parasitic capacitance, V_{dd} is the voltage supply, f_{cor} is the corner frequency, and B is the symbol bandwidth. The second part of (15), α_i^{PA} is given as:

$$\alpha_i^{PA} = \frac{\zeta}{\delta} \alpha_i^{out} \tag{18}$$

where δ is the drain efficiency of the RF power amplifier, ζ is the Peak-to-Average Ratio (PAR) which is dependent on the modulation scheme and the constellation size, and α_i^{out} is the transmitted power from the amplifier. When the channel only experiences a square-law path loss we have:

$$\alpha_i^{out} = \bar{E}^b R_b \tag{19}$$

where $\bar{E}^b$ is the required energy per bit at the receiver for a given BER requirement, and R_b is the bit rate. Under Rayleigh fading, $\bar{E}^b$ in (19) for BPSK modulation can be given as follows:

$$\bar{E}^b = \frac{N_o(1 - 2P_e)^2}{4\sigma_i^2 P_e(1 - P_e)} \tag{20}$$

where P_e is the BER and σ_i^2 is channel variance that is given as:

$$\sigma_i^2 = \frac{(4\pi d_i)^2}{G_t G_r \lambda^2} M_l N_f \tag{21}$$

where G_t is the transmitter antenna gain, G_r is the receiver antenna gain, λ is the carrier wavelength, M_l is the link margin compensating the hardware process variations and other additive background noise or interference. N_f is the receiver noise figure defined as $N_f = \frac{N_r}{N_o}$, with $N_o = -171\ dBm\ /\ Hz$ the single-sided thermal noise Power Spectral Density (PSD) at room temperature. N_r is the PSD of the total effective noise at the receiver input.

In addition to multipath fading, the performance of cooperative wireless networks is also affected by shadowing in realistic operating conditions. Recently several studies have been conducted to analyze the impact of shadowing in CSS (Di Renzo, Imbriglio, Graziosi & Santucci, 2009), (Di Renzo Graziosi & Santucci, April 2009), (A. Ghasemi, 2007 January). In general, all these studies have concluded that shadowing significantly reduces the performance of CSS. As such, our proposed partial CSS approach is also not free from the impacts of shadowing. Usually, shadowing is modelled using log-normal distribution (Alouini, 2005).

A composite shadowing/multipath fading environment consists of shadowing superimposed on multipath fading environment. In the case of a composite log-normal shadowing/Rayleigh fading channel, with BPSK modulation, the BER can be obtained from (Eq. 5.27, Alouini, 2005) as follows:

$$P_e = \frac{1}{\sqrt{2\pi}} \sum_{n=1}^{N} w_n \left[1 - \sqrt{\frac{c_n}{1 + c_n}} \right] \tag{22}$$

where $c_n = 10^{\left(x_n \sqrt{2}\Omega + 10\log_{10}\bar{\gamma}\right)/10}$, $\bar{\gamma} = \left(\bar{E}^b / N_o\right)\sigma_i^2$, Ω is the logarithmic standard deviation of shadowing, and $\{x_n\}$ and $\{w_n\}$, with $n = 1, 2, ..., N$, are the zeros and weights of the N th-order Hermite polynomial, respectively (Table 25.10, Stegun, 1972). For the ease of illustration of the optimization scheme, in the subsequent analysis, the affects of shadowing have not been not considered. More specifically, we only assume independent and identically distributed Rayleigh fading channels. However, the interested readers may extend our proposed approach to include shadowing by taking into account that the expression for $\bar{E}^b$ in (20) must be replaced by $\bar{E}^b$ obtained from (22).

4.2 The Achievable Performance

Let us consider the estimated energy of each CU (E_i) as a random variable with a Probability Density Function (pdf), f , and Cumulative Distribution Function (CDF), F_E . Therefore, for any CU, the probability of participation in the spectrum sensing equals to $F_E(\gamma)$. Also, the number of CUs who have decided to participate (N*) follows a binomial distribution described as:

$$Prob.(N^* = n) = \binom{N}{n}(F_E(\gamma))^n \left(1 - F_E(\gamma)\right)^{N-n} \tag{23}$$

where the average number of sensing users $\overline{N^*}$ is given by:

$$\overline{N^*} = NF_E(\gamma) \tag{24}$$

After reporting the local decisions made by $\overline{N^*}$ CUs, OR-rule is applied and a final decision is made. In case of $N^* = 0$, i.e., no users have participated, a random final decision is made at the FR. Therefore, the average overall detection probability (P_D^*) and the average overall false alarm probability (P_F^*) can be written as:

$$P_D^* = \begin{cases} 1-(1-P_d)^{N^*} & \text{if} \quad N^* \geq 1 \\ 0.5 & \text{if} \quad N^*=0 \end{cases} \tag{25}$$

$$P_F^* = \begin{cases} 1-(1-P_f)^{N^*} & \text{if} \quad N^* \geq 1 \\ 0.5 & \text{if} \quad N^*=0 \end{cases} \tag{26}$$

where $N^* = 1, 2, ..., N$.

4.3 Energy Efficiency Optimization

The total energy consumed by the whole system by following the proposed approach (E_{tot}^*) can be written as follows:

$$E_{tot}^* = \sum_{i=1}^{N} S_i E_i + P_{unused}^* E_t^* \tag{27}$$

where the first term represents the consumed energy during spectrum sensing process, which equals to 0 for the CUs who have not participated because it is multiplied by $S_i = 0$. The second term represents the energy consumed during data transmission (E_t^*) which is conditioned by P_{unused}^*. P_{unused}^* is the probability of identifying the spectrum as unused in our approach, which can be obtained by substituting P_D^* and P_F^* instead of P_D and P_F in (7).

Regarding the calculation of E_t^*, we assume that a CU is randomly scheduled for data transmission. Therefore, the calculation of E_t^* follows the same procedure as E_r with a proper substitution of the values of f_{cor}, B, and P_c.

The amount of successfully transmitted data in bits, D^* depends mainly on the performance of the spectrum sensing, and can be given as:

$$D^* = RT_t\left(P_0(1 - P_F^*)\right) \tag{28}$$

Hence, the total energy consumed per successfully transmitted bit based on the proposed approach (EpB^*) is given as:

$$EpB^* = \frac{E_{tot}^*}{D^*} \tag{29}$$

Remember that the resulting EpB depends mainly on the number of participating users which is a function of γ. Therefore, in order to minimize EpB^*, an optimization of γ is required.

5. SIMULATION RESULTS

In this section, we present some simulation results in order to illustrate the advantage of the proposed partial-cooperative spectrum sensing scheme. The considered N CUs are randomly distributed around the FC, each at a distance that is uniformly distributed between 0.1 Km to 7 Km. In particular, we are interested in finding an optimal value of the energy threshold, γ, which minimizes the total energy consumed per successfully transmitted bit in partial CSS, EpB^*. Table 1 lists the simulation parameters used in this section.

Figure 2 shows the average number of participating CUs in CSS versus the participation threshold γ. The x-axis is shown in terms of E_{min}, E_{max} and Δ, where E_{min} and E_{max} are the energy consumed in spectrum sensing by a CU at a distance equals to d^{min} and d^{max}, respectively. Δ is the step between each two consecutive lines equals to 1×10^{-11}. Obviously, as γ increases, the probability of participating in CSS for each CU increases as well, and hence, the number of participating will increase. Different numbers of available CUs in the network are shown in Figure 2. However, such change in the participating CUs will undoubtedly influence the transmitted data, energy consumption and energy efficiency, as we will see in the next results.

Table 1. Simulation parameters

Parameter	Value	Parameter	Value
N	10	P_0	0.5
P_d	0.8	P_f	0.2
T_s	$3\,ms$	T_r	$0.01\,ms$
T_t	$40\,ms$	d^{min}	$100\,m$
d^{max}	$7\,Km$	α_s	$106\,mW$
α^{syn}	$50.0\,mW$	α^{filt}	$2.5\,mW$
α^{mix}	$30.3\,mW$	I_0	$3\,\mu A$
V_{dd}	$3\,V$	f_r	$2.5\,GHz$
R_b	$10\,Kbps$	n_1	10
C_p	$1\,pF$	$G_t G_r$	$5\,dBi$
N_f	$10\,dB$	M_l	$40\,dB$
δ	0.35	f_{cor}	$1\,kHz$
P_e	10^{-5}	ζ	514×10^{-3}

As the number of participating CUs is variable depending on the participation threshold, it is expected that both the false alarm and detection probabilities will be affected. In order to show the effect of the participation threshold on the sensing accuracy, we use the false-decision probability (δ) as an evaluation metric. δ is defined as the weighted sum of false alarm probability and the missed detection probability as follows:

$$\delta = P_0 P_F + P_1 \left(1 - P_D\right) \tag{30}$$

Figure 3 plots the false-decision probability versus participation threshold at N=10. At low values of the participation threshold, δ is equal to 0.5 since the decision is random when no CUs participate in the CSS process. Optimizing the participation threshold yields in a minimum false-decision probability as shown in Figure 3. Notice that the minimum false-decision probability attained by our approach is much less than the attained by the conventional approach.

Figure 2. The average number of participating CUs versus γ *(* $\Delta = 1 \times 10^{-14}$ *)*

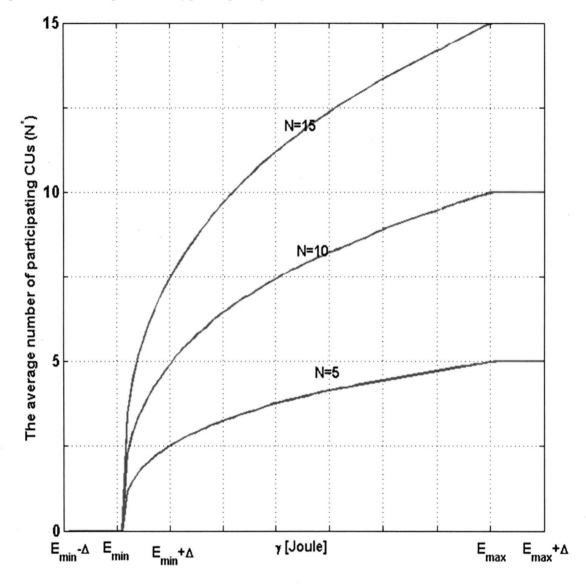

The achievable amount of successfully transmitted data versus the threshold γ is shown in Figure 4. For low values of γ, all CUs will not participate in CSS since they have E_i larger than γ, which results in $P_F^* = 0.5$, according to (26). Hence, D^* is constant since it depends mainly on P_F^*, as stated in (28).

When γ increases so that the number of sensing users equals [1], P_F^* improves, and consequently, the transmitted data increases. As γ increases, D^* decreases since P_F^* increases. For comparative purposes, Figure 4 also shows the plot for the achievable amount of successfully transmitted data using the conventional approach, D, where all the users take part in CSS. Since in conventional approach, D is independent of the value of γ, the plot will be a constant with respect to γ.

Figure 3. The false decision probability versus γ ($\Delta = 1 \times 10^{-14}$)

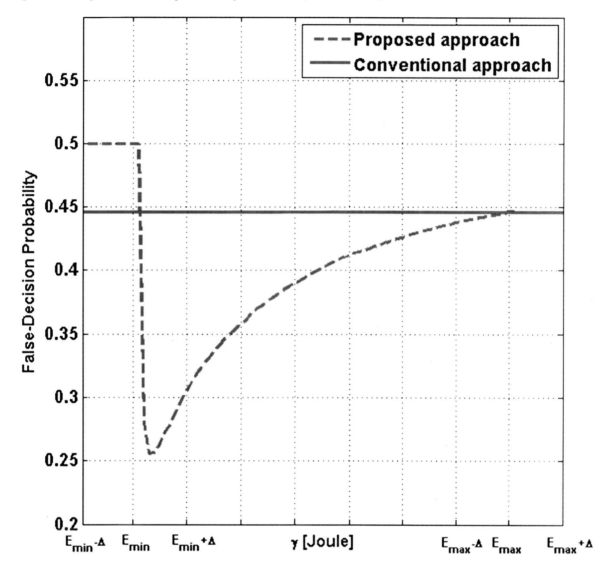

In Figure 5, the total energy consumed by the system in partial CSS, E_{tot}^* over different values of γ is plotted. We can see that, as γ increases, E_{tot}^* first remains the same, but then decreases and then gradually becomes stable for larger values of γ. The initial flat region in the plot is due to the fact the estimated energy, E_i of all the CU's is above γ. Hence, all the CU's will not participate in spectrum sensing, and energy is consumed only in transmission. As γ is increased, E_{tot}^* decreases even though more CUs participate in CSS. This is due to the decrease in P_{unused}. The plot for total energy consumed by the system in conventional approach is also shown in Figure 5.

Figure 4. The amount of transmitted data versus γ ($\Delta = 1 \times 10^{-14}$)

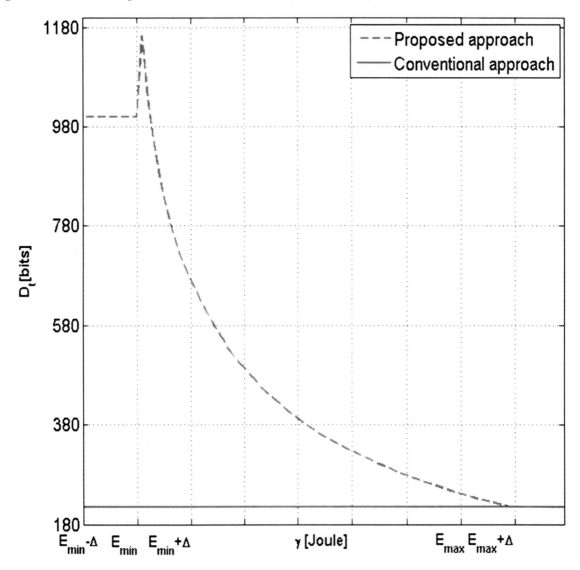

From the previous figures, it is clear that increasing γ lowers the energy consumption but with lower transmitted data. Thus, in order to find the optimal value of γ that balances the two contrasting effects, the total energy consumed per successfully transmitted bit in partial CSS, EpB^* versus different values of γ is plotted in Figure 6. As γ increases, EpB^* first remains the same, but then decreases and then increases after a particular value of γ. The value of γ, where EpB^* is minimum gives the optimal value of γ. The plot for total energy consumed per successfully transmitted bit is also shown in Figure 6. The results in Figure 6 clearly show the potential gain of using the proposed partial-CSS scheme over the conventional approach. More precisely, when the optimal value of γ is used, partial CSS provides a Relative Average Energy Reduction (RAER) per successfully transmitted bit of approximately 78% with respect to the conventional approach.

Figure 5. Total consumed energy of the whole system versus γ *(* $\Delta = 1{\times}10^{-14}$ *)*

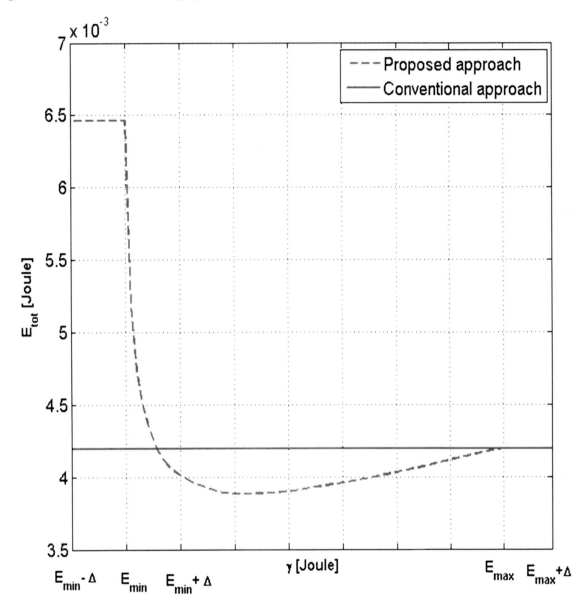

The threshold γ plays a key factor in the performance of the proposed approach. In Figure 7, we plot the percentage of RAER compared to the conventional approach versus the total number of CUs, where RAER is expressed as follows:

$$\mathrm{RAER}[\%] = \frac{EpB_{conventional} - EpB_{proposed}}{EpB_{conventional}} \times 100 \tag{30}$$

Figure 6. Total energy consumed per successfully transmitted bit versus γ ($\Delta = 1 \times 10^{-14}$)

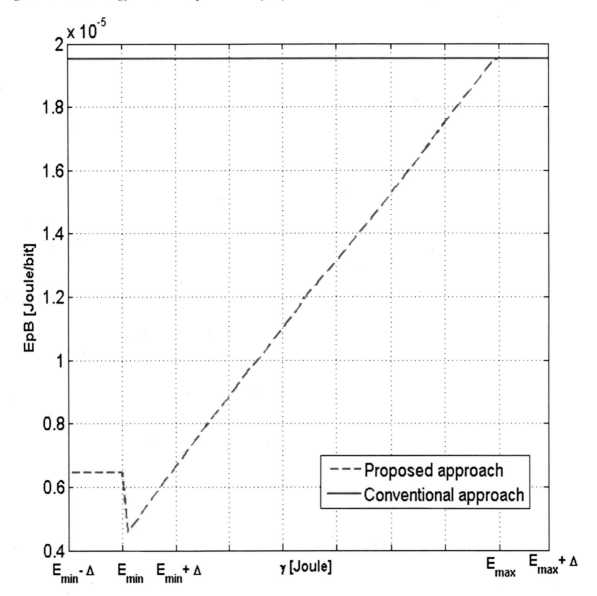

6. CONCLUSION

A partial cooperative spectrum sensing approach is presented in this chapter, which aims at reducing the energy consumption in cognitive radio. The proposed approach is based on reducing the number of sensing users. Each user decides to participate in spectrum sensing if its expected energy consumption during this process is less than a threshold. The participation threshold is optimized to minimize the energy consumption through computer simulations.

As a future work, the performance of the proposed approach can be investigated considering non-identical local sensing performance and channel conditions among users. Also, the influence of shadowing

Figure 7. The Relative Average Energy Reduction per bit versus the total number of CUs

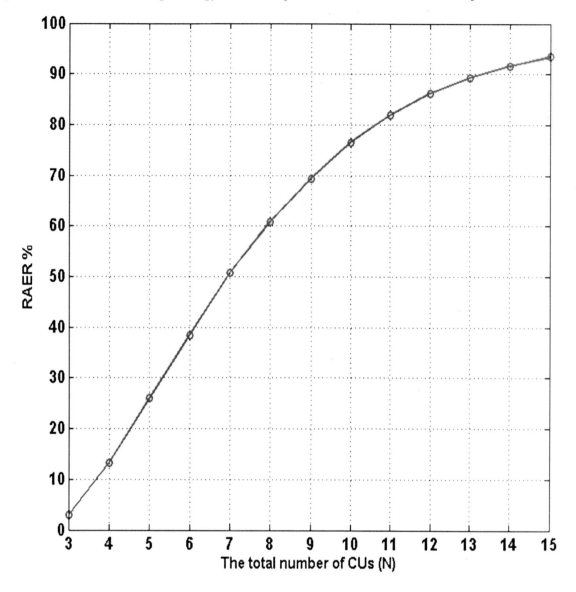

in sensing and reporting channels on the overall performance is worth to be studied. Another possible future work is to extend the proposed approach to a fully-distributed approach. The fully-distributed approach implies that the participation threshold is locally decided at each CU considering the instantaneous battery status.

ACKNOWLEDGMENT

This work was supported by the European Commission under the auspices of the FP7–PEOPLE MITN–GREENET project (grant 264759).

REFERENCES

Akyildiz, I. F., Lo, B. F., & Balakrishnan, R. (2011). Cooperative spectrum sensing in cognitive radio networks: A survey. *Physical Communication*, *4*(1), 40–62. doi:10.1016/j.phycom.2010.12.003

Alouini, M. K. (2005). *Digital Communication over Fading Channels* (2nd ed.). John Wiley & Sons.

Alouini, M. S. (2005). *Digital Communication over Fading Channels* (2nd ed.). John Wiley & Sons.

Althunibat, S., Di Renzo, M., & Granelli, F. (2013, Decmebr). Optimizing the K-out-of-N Fusion rule for Cooperative Spectrum Sensing in Cognitive Radio Networks. In *Proceedings of Global Telecommunications Conference (GLOBECOM 2013)*, (pp. 1-5). IEEE.

Althunibat, S., & Granelli, F. (2013, June). Novel energy-efficient reporting scheme for spectrum sensing results in cognitive radio. In *Proceedings of Communications (ICC)*, (pp. 2438-2442). IEEE. doi:10.1109/ICC.2013.6654897

Althunibat, S., Narayanan, S., Di Renzo, M., & Granelli, F. (2012, September). On the Energy Consumption of the Decision-Fusion Rules in Cognitive Radio Networks. In *Proceedings of Computer Aided Modeling and Design of Communication Links and Networks (CAMAD)*, (pp. 125-129). IEEE. doi:10.1109/CAMAD.2012.6335312

Appadwedula, S., Veeravalli, V. V., & Jones, D. L. (2008). Decentralized detection with censoring sensors. *IEEE Transactions on* Signal Processing, *56*(4), 1362–1373. doi:10.1109/TSP.2007.909355

Cabric, D., Mishra, S. M., & Brodersen, R. W. (2004, November). Implementation issues in spectrum sensing for cognitive radios. In *Proceedings of Signals, Systems and Computers*, (Vol. 1, pp. 772-776). IEEE. doi:10.1109/ACSSC.2004.1399240

Chaudhari, S., Lunden, J., Koivunen, V., & Poor, H. V. (2012). Cooperative sensing with imperfect reporting channels: Hard decisions or soft decisions?. *IEEE Transactions on* Signal Processing, *60*(1), 18–28. doi:10.1109/TSP.2011.2170978

Cheng, P., Deng, R., & Chen, J. (2012). Energy-efficient cooperative spectrum sensing in sensor-aided cognitive radio networks. *IEEE Wireless Communications*, *19*(6), 100–105. doi:10.1109/MWC.2012.6393524

Cui, S., Goldsmith, A. J., & Bahai, A. (2004). Energy-efficiency of MIMO and cooperative MIMO techniques in sensor networks. *IEEE Journal on* Selected Areas in Communications, *22*(6), 1089–1098.

Di Renzo, M., Graziosi, F., & Santucci, F. (2009, April). Cooperative spectrum sensing in cognitive radio networks over correlated log-normal shadowing. In *Proceedings of Vehicular Technology Conference*, (pp. 1-5). IEEE. doi:10.1109/VETECS.2009.5073470

Di Renzo, M., Imbriglio, L., Graziosi, F., & Santucci, F. (2009). Distributed data fusion over correlated log-normal sensing and reporting channels: Application to cognitive radio networks. *IEEE Transactions on* Wireless Communications, *8*(12), 5813–5821.

Digham, F. F., Alouini, M. S., & Simon, M. K. (2007). On the energy detection of unknown signals over fading channels. *IEEE Transactions on* Communications, *55*(1), 21–24. doi:10.1109/TCOMM.2006.887483

Feng, X., Gan, X., & Wang, X. (2011). Energy-constrained cooperative spectrum sensing in cognitive radio networks. In *Proceedings of Global Telecommunications Conference (GLOBECOM 2011)*. IEEE.

Fettweis, G., & Zimmermann, E. (2008, September). ICT energy consumption-trends and challenges. In *Proceedings of the 11th International Symposium on Wireless Personal Multimedia Communications* (*Vol. 2*, p. 6). Academic Press.

Gao, Y., Xu, W., Yang, K., Niu, K., & Lin, J. (2013). Energy-efficient transmission with cooperative spectrum sensing in cognitive radio networks. In *Proceedings of Wireless Communications and Networking Conference (WCNC)*, (pp. 7-12). IEEE.

Ghasemi, A., E. S. (2007). Asymptotic performance of collaborative spectrum sensing under correlated log-normal shadowing. *IEEE Communications Letters*, 34–36.

Ghasemi, A., & Sousa, E. S. (2007, January). Opportunistic spectrum access in fading channels through collaborative sensing. *Journal of Communication*, *2*(2), 71–82.

Ghasemi, A., & Sousa, E. S. (2007). Asymptotic performance of collaborative spectrum sensing under correlated log-normal shadowing. *IEEE Communications Letters*, *11*(1), 34–36. doi:10.1109/LCOMM.2007.060662

Haykin, S. (2005). Cognitive radio: Brain-empowered wireless communications. *IEEE Journal on* Selected Areas in Communications, *23*(2), 201–220.

Khasawneh, M., Agarwal, A., Goel, N., Zaman, M., & Alrabaee, S. (2012, July). Sureness efficient energy technique for cooperative spectrum sensing in cognitive radios. In *Proceedings of Telecommunications and Multimedia (TEMU)*, (pp. 25-30). IEEE. doi:10.1109/TEMU.2012.6294727

Lee, C. H., & Wolf, W. (2008). Energy efficient techniques for cooperative spectrum sensing in cognitive radios. In *Proceedings of Consumer Communications and Networking Conference*, (pp. 968-972). IEEE. doi:10.1109/ccnc08.2007.223

Lunden, J., Koivunen, V., Huttunen, A., & Poor, H. V. (2007, November). Censoring for collaborative spectrum sensing in cognitive radios. In *Proceedings of Signals, Systems and Computers*, (pp. 772-776). IEEE. doi:10.1109/ACSSC.2007.4487321

Maleki, S., Chepuri, S. P., & Leus, G. (2011, June). Energy and throughput efficient strategies for cooperative spectrum sensing in cognitive radios. In *Proceedings of Signal Processing Advances in Wireless Communications (SPAWC)*, (pp. 71-75). IEEE. doi:10.1109/SPAWC.2011.5990482

Maleki, S., Chepuri, S. P., & Leus, G. (2013). Optimization of hard fusion based spectrum sensing for energy-constrained cognitive radio networks. *Physical Communication*, *9*, 193–198. doi:10.1016/j.phycom.2012.07.003

Maleki, S., & Leus, G. (2013). Censored truncated sequential spectrum sensing for cognitive radio networks. *IEEE Journal on* Selected Areas in Communications, *31*(3), 364–378.

Maleki, S., Pandharipande, A., & Leus, G. (2011). Energy-efficient distributed spectrum sensing for cognitive sensor networks. *IEEE Sensors Journal*, *11*(3), 565–573. doi:10.1109/JSEN.2010.2051327

Mishra, S. M., Sahai, A., & Brodersen, R. W. (2006, June). Cooperative sensing among cognitive radios. In *Proceedings of Communications,* (Vol. 4, pp. 1658-1663). IEEE. doi:10.1109/ICC.2006.254957

Mitola, J. III, & Maguire, G. Q. Jr. (1999). Cognitive radio: Making software radios more personal. *IEEE Personal Communications*, *6*(4), 13–18. doi:10.1109/98.788210

Najimi, M., Ebrahimzadeh, A., Andargoli, S. M. H., & Fallahi, A. (2013). A Novel Sensing Nodes and Decision Node Selection Method for Energy Efficiency of Cooperative Spectrum Sensing in Cognitive Sensor Networks. *IEEE Sensors Journal*, *1*(5), 1610–1621. doi:10.1109/JSEN.2013.2240900

Nuttall, A. (1975). Some integrals involving the QM function. *IEEE Transactions on* Information Theory, *21*(1), 95–96.

Peh, E. C. Y., Liang, Y. C., Guan, Y. L., & Pei, Y. (2011, Dec.). Energy-Efficient Cooperative Spectrum Sensing in Cognitive Radio Networks. In *Proceedings of Global Telecommunications Conference (GLOBECOM 2011)*, (pp. 1-5). IEEE. doi:10.1109/GLOCOM.2011.6134342

Pham, H. N., Zhang, Y., Engelstad, P. E., Skeie, T., & Eliassen, F. (2010, March). Energy minimization approach for optimal cooperative spectrum sensing in sensor-aided cognitive radio networks. In *Proceedings of Wireless Internet Conference (WICON)*, (pp. 1-9). IEEE. doi:10.4108/ICST.WICON2010.8531

Ryzhik, I. S. (1994). Table of Integrals, Series, and Products (5th ed.). Academic Press.

Stegun, M. A. (1972). Handbook of Mathematical Functions with Formulas, Graphs, and Mathematical Tables (9th ed.). New York: Dover Press.

Stegun, M. A. (1972). Handbook of Mathematical Functions with Formulas, Graphs, and Mathematical Tables (9th ed. ed.). New York: Dover Press.

Sun, C., Zhang, W., & Ben, K. (2007, June). Cluster-based cooperative spectrum sensing in cognitive radio systems. In *Proceedings of Communications,* (pp. 2511-2515). IEEE. doi:10.1109/ICC.2007.415

Sun, C., Zhang, W., & Letaief, K. (2007, March). Cooperative spectrum sensing for cognitive radios under bandwidth constraints. In *Proceedings of Wireless Communications and Networking Conference,* (pp. 1-5). IEEE. doi:10.1109/WCNC.2007.6

Viswanathan, R., & Varshney, P. K. (1997). Distributed detection with multiple sensors I. Fundamentals. *Proceedings of the IEEE*, *85*(1), 54–63. doi:10.1109/5.554208

Wei, J., & Zhang, X. (2010, March). Energy-efficient distributed spectrum sensing for wireless cognitive radio networks. In *Proceedings of INFOCOM IEEE Conference on Computer Communications Workshops,* (pp. 1-6). IEEE. doi:10.1109/INFCOMW.2010.5466680

Xia, W., Wang, S., Liu, W., & Chen, W. (2009, September). Cluster-based energy efficient cooperative spectrum sensing in cognitive radios. In *Proceedings of Wireless Communications, Networking and Mobile Computing,* (pp. 1-4). Academic Press. doi:10.1109/WICOM.2009.5301709

Zhao, N., Yu, F. R., Sun, H., & Nallanathan, A. (2012). An energy-efficient cooperative spectrum sensing scheme for cognitive radio networks. In *Proceedings of Global Communications Conference (GLOBECOM)*, (pp. 3600-3604). IEEE. doi:10.1109/GLOCOM.2012.6503674

KEY TERMS AND DEFINITIONS

Cognitive Radio Networks: A type of wireless networks that consists a set of users cooperating to use a specific spectrum based on cognitive radio technology.

Cognitive Radio: A wireless technology proposed to improve spectrum efficiency. It is based on exploiting the temporarily unused portions of the spectrum by unlicensed users.

Cooperative Spectrum Sensing: An approach proposed to enhance the reliability of the spectrum sensing process. It implies sharing the local sensing results of several users at a central entity, aiming at improving the reliability of the process decision.

Energy Efficiency: The ratio of the average successfully transmitted data in bits to the average consumed energy in Joule.

Spectrum Sensing: A method used to identify the spectrum status either used or unused.

Wireless Communications: The transfer of data between two or more devices that are not electrically connected to each other.

Wireless Networks: A type of communication networks that uses wireless data transfer techniques.

ENDNOTES

[1]   A closed form expression of the integral in (1) can be found in (Digham et al., 2007)

Chapter 5
Cyclostationary Spectrum Sensing in Cognitive Radios at Low SNR Regimes

Mahsa Derakhshani
University of Toronto, Canada

Tho Le-Ngoc
McGill University, Canada

Masoumeh Nasiri-Kenari
Sharif University of Technology, Iran

ABSTRACT

Spectrum sensing is one of the key elements in the establishment of cognitive radio. One of the most effective approaches for spectrum sensing is cyclostationary feature detection. Since modulated signals can be modeled as cyclostationary random signals, this feature can be used to recognize the cyclostationary modulated signal in a background of stationary noise even at low SNR regimes. This chapter reviews non-cooperative cyclostationary sensing approaches and reports recent advances in cooperative cyclostationary sensing algorithms. New results for cooperative cyclostationary spectrum sensing are then presented, which ensure better performance as well as faster and simpler operation. In the proposed schemes, each Secondary User (SU) performs Single-Cycle (SC) cyclostationary detection for fast and simple implementation, while collaboration between SUs in final decision on the presence or absence of the PU is explored to improve its performance. Furthermore, this chapter presents another look at the performance evaluation of cyclostationary detectors in terms of deflection coefficients.

INTRODUCTION

It has been confirmed that the conventional fixed spectrum allocation strategies lead to spectrum underutilization. Despite the activity of licensed users, measurements reveal that there still exists a plenty of instantaneous spectrum availabilities in the licensed spectrum (Harrison, Mishra, & Sahai, 2010; Van de Beek, Riihija, Achtzehn, & Mahonen, 2012). This motivates the idea of opportunistic spectrum ac-

DOI: 10.4018/978-1-4666-6571-2.ch005

cess (OSA) that allows SUs to utilize temporal spectrum opportunities in the licensed spectrum. Under this model, SUs need to sense the spectrum and identify spectrum holes for their transmissions, while limiting the level of interference to the licensed primary users (PUs) (Akyildiz, Lee, Vuran, & Mohanty, 2006; Goldsmith, Jafar, Maric, & Srinivasa, 2009; Haykin, 2005; Mitola, 2000; Zhao & Sadler, 2007).

Spectrum sensing, the detection of presence or absence of PUs in a specific frequency band, is one of the key elements in the establishment of cognitive radio based on OSA. This is because its accuracy, computational complexity and response time directly affect the efficiency of SU spectrum usage and the PU protection (Zhao & Sadler, 2007). The problem of spectrum sensing can be modeled as a binary hypothesis testing problem with null hypothesis that the PU is absent, and the alternative hypothesis that the PU is present (Axell, Leus, Larsson, & Poor, 2012; Yucek & Arslan, 2009).

Different approaches have been proposed for spectrum sensing in cognitive radio (Axell et al., 2012; Yucek & Arslan, 2009). Energy detection (radiometry) has been widely known for its simplicity. It measures the energy contained within the band of interest and compares it with a threshold to detect the presence of a PU signal. Though it is very well-known and uncomplicated to implement, there are several drawbacks for energy detection that might alleviate its benefits in some cases, especially at low SNR regimes. Energy detectors cannot differentiate between the energy of signal of interest and noise, thus it has a low performance at low SNR regimes. Furthermore, it is sensitive to uncertainty in noise statistics (Cabric, Mishra, & Brodersen, 2004; Digham, Alouini, & Simon, 2007; Sonnenschein & Fishman, 1992; Tandra & Sahai, 2005).

Since modulated signals exhibit cyclostationarity or spectral coherence, this feature can be used to recognize cyclostationary modulated signal in a background of stationary noise, even at low SNR regimes. Distinctive characteristics of cyclostationary signal from stationary noise make signal detection possible at low SNR regimes and in the presence of noise uncertainty, by measuring signal power at non-zero CFs (Gardner, Napolitano, & Paura, 2006; Gardner & Spooner, 1992; Gardner, 1986, 1988; Yeung & Gardner, 1996). In the cyclostationary detection, the values of either spectral correlation function or cyclic autocorrelation function at different non-zero CFs are used as test statistics for detecting signals, which improve the detection performance compared to energy detection. In the cognitive radio application, SUs are not informed of the PU signal. Thus, they cannot identify the proper CF at which signal has power. As a result, one drawback of cyclostationary non-cooperative detectors is the need for a time-consuming and computationally complex search over all possible CFs to assure presence or absence of the PU (Yeung & Gardner, 1996).

Furthermore, against severe shadowing and fading effects, excessively long detection time is required to enable reliable detection, and hence, guarantee a limited level of interference to the PUs. Cooperative sensing is proposed as a solution to enhance the detection performance without increasing the detection time. In the cooperative spectrum sensing, SUs collaborate with each other to sense the spectrum to find the spectrum holes. In other words, they share their sensing information, and then each of them decides about absence or presence of the PU in a specific bandwidth based on the shared knowledge. Since SUs sense different shadowing and fading, cooperative spectrum sensing can improve the detection performance by multi-user diversity (Akyildiz, Lo, & Balakrishnan, 2011).

Thus, in this chapter, OR (i.e., logical OR operation) and maximum-likelihood (ML) cooperative cyclostationary techniques are presented to improve the detection performance and reduce the complexity comparing with non-cooperative cyclostationary detectors. In the proposed OR and ML cooperative cyclostationary methods, each user only measures signal power at a single CF and exchanges its information with the others for further decision. The use of parallel SC detectors at different CFs by different SUs can reduce the computational complexity as compared to the use of MC detector in each SU as in (Lunden, Koivunen, Huttunen, & Poor, 2009; Sadeghi & Azmi, 2008). Since each SU sees independent noise and fading, effectively, decision based on the information related to the signal powers simultaneously measured at different CFs by the SUs can offer a better performance via multi-user and space diversity, as compared to non-cooperative MC detectors.

Furthermore, performance of non-cooperative and the proposed cooperative cyclostationary techniques is evaluated from another perspective in terms of deflection coefficients in AWGN channels. In (B. Shen, Ullah, & Kwak, 2010), deflection coefficient maximization criterion is used to propose an optimal cooperative energy detector. Then, outage probability of deflection coefficient is defined as a measure to evaluate the performance of different cyclostationary detectors in a fading channel. In addition, the detection and false-alarm probabilities of the SC detector and presented cooperative schemes in AWGN channels are derived. Simulation results are also provided in AWGN and fading channels to confirm the analytical results.

BACKGROUND

In order to detect spectrum holes, each SU decides between two following hypotheses that represent the presence or absence of the PU signal (Taherpour, Norouzi, Nasiri-Kenari, Jamshidi, & Zeinalpour-Yazdi, 2007).

$$\begin{cases} H_0 : x_i(t) = n_i(t) \\ H_1 : x_i(t) = h_i s(t) + n_i(t), i = 1, ..., N_u \end{cases} \tag{1}$$

where $x_i(t)$ is the i-th SU's received signal, N_u is the number of SUs, $s(t)$ is the PU's transmitted signal, $n_i(t)$ is the additive white Gaussian noise (AWGN) and h_i is the gain of channel between the PU and the i-th SU. H_0 denotes the null hypothesis that the PU is absent, and H_1 is the alternative hypothesis that the PU is present (Taherpour, Norouzi, et al., 2007). In the spectrum sensing, SUs measure the test statistics at first, and then compare it with a threshold which is determined with a desirable false alarm probability in order to decide between two hypotheses.

Due to modulation or/and channel coding, communication signals can be modeled as cyclostationary random processes. A random process $s(t)$ is said to be cyclostationary if its mean and autocorrelation are both periodic with the same period $T_c \neq 0$. More specifically, $E\{s(t+T_c)\} = E\{s(t)\}=\mu$ and $E\{s(t)s*(t+\tau)\} = E\{s(t+T_c)s*(t+T_c+\tau)\} = R_s(t, \tau)$ if $s(t)$ is cyclostationary. For cyclostationary signals, the periodic auto-correlation function $R_s(t, \tau)$ accepts the Fourier-series expansion as follows

$$R_s\left(t,\tau\right) = \sum_k R_s^{\alpha_k}\left(\tau\right) e^{j2\pi\alpha_k t}, \alpha_k = k/T_c, \ k \in \mathbb{Z} \tag{2}$$

where α_k is the k-th CF and Fourier coefficient $R_s^{\alpha k}(\tau)$ (known as cyclic autocorrelation function (CAF)) is the projection of $R_s(t, \tau)$ onto the exponential basis function with the CF α_k. For a cyclostationary signal,

CAF $R_s^{\alpha k}(\tau)$ is non-zero for some non-zero CF α_k and some time lag τ, while CAF of the background stationary noise is zero for a non-zero CF α_k, i.e., $R_n^{\alpha k}(\tau) = 0$. Thus, the value of CAF of the received signal (i.e., $x(t)$) can be used as a test statistic as follows

$$T_{\mathrm{CAF}}\left(\alpha_k, \tau\right) = \left|R_x^{\alpha_k}\left(\tau\right)\right|^2. \tag{3}$$

For this cyclostationary detection technique, the priori information of signal parameters such as the signal's bit or chip rate, and carrier frequency are required to know the exact CFs of $s(t)$. The CAF test statistic (i.e., T_{CAF}) is sub-optimal since it examines cyclostationary features only at a single CF and a single time lag. This test statistic to detect the presence of cyclostationarity can be extended to multiple CFs and multiple time lags. For a single CF and multiple time lags, Dandawate & Giannakis, 1994 developed a generalized likelihood-ratio test (GLRT) and proposed cyclostationary detectors based on the estimates of CAF at different time lags. Subsequently, the criterion to choose the optimal time lags is studied by Feng, Wang, & Li, 2008 and Lunden, Koivunen, Huttunen, & Poor, 2007.

To improve the sensing performance, simultaneous use of multiple CFs are considered by Kim, Kimtho, & Takada, 2010; Lunden, Koivunen, Huttunen, & Poor, 2007. For a given set of lags, Lunden et al., 2007 presented two test statistics based on maximum and sum of CAF test statistics over CFs of interest. In addition, Kim et al., 2010 proposed a maximum ratio combining (MRC) test based on transmit signal characteristics. Exploiting multiple CFs, Sadeghi, Azmi, & Arezumand, 2011 presented another test statistic which is a linear combination of test statistics over multiple CFs, while optimizing the weights in terms of the deflection coefficient criterion. To jointly use multiple CFs and multiple time lags for a more reliable spectrum sensing performance, J. Shen & Alsusa, 2013 presented an optimal scheme to select test points in CF and time lag domains to maximize asymptotic detection performance.

Another test statistic that can be used for cyclostationary detection is based on using the spectral correlation function (SCF) which is the Fourier transform of CAF with respect to the lag variable τ. The test statistic is computed as follows

$$T_{\mathrm{SCF}}\left(\alpha_k, f\right) = \left|S_x^{\alpha_k}\left(f_l\right)\right|^2 \tag{4}$$

where $S_x^{\alpha k}(f_l)$ is SCF of $x(t)$ at the k-th CF, $\alpha_k = k / T_c$, and the frequency f_l. This test statistic involves two dimensions CF and frequency. By integrating the SCF over all frequencies, another test statistic can be obtained as follows

$$T_{\mathrm{SC}} = \left|\int_{-\frac{f_s}{2}}^{\frac{f_s}{2}} S_x^{\alpha_k}(f)\mathrm{d}f\right|^2 \tag{5}$$

where f_s is the sampling rate. Since this detector measures the signal power only for one specific CF, it is called the SC cyclostationary detector. The test statistic of the SC cyclostationary detector is also given by (Gardner & Spooner, 1992)

$$T_{\mathrm{SC}} = \left| R_x^{\alpha_k} \left(\tau = 0 \right) \right|^2 . \tag{6}$$

Generally, cyclostationary detectors need long detection time to guarantee low false-alarm and miss-detection probabilities. In order to reduce the average detection time, Choi, Jeon, & Jeong, 2009 proposed a sequential detection framework to apply for a SC cyclostationary detector.

As a cyclostationary detector, the energy detector can be interpreted as a SC detector at zero CF, i.e., $\alpha_k = 0$, and its test statistic can be represented as

$$T_{\mathrm{ED}} = \left| R_x^{\alpha_k = 0} \left(\tau = 0 \right) \right|^2 . \tag{7}$$

In other words, the energy detector measures the received signal power at CF equal to zero. Since the background stationary noise also has energy at CF equal to zero, the energy detection suffers from a poor performance under low SNR regimes or/and noise uncertainty.

Another cyclostationary detection scheme is called MC cyclostationary detection in which the test statistic is the sum of the received signal powers at all CFs as follows

$$T_{\mathrm{MC}} = \sum_{k=1}^{N_\alpha} \left| \int_{-\frac{f_s}{2}}^{\frac{f_s}{2}} S_x^{\alpha_k} (f) \mathrm{d}f \right|^2 \tag{8}$$

where N_α is the number of CFs. Moreover, the total power can be calculated as

$$T_{\mathrm{MC}} = \sum_{k=1}^{N_\alpha} \left| R_x^{\alpha_k} \left(\tau = 0 \right) \right|^2 \tag{9}$$

where $|R_x^{\alpha_k}(\tau = 0)|$ represents the cyclostationary signal power at the k-th CF. In the literature, MC cyclostationary detection has received considerable attention (Gardner, 1988; Izzo, Paura, & Tanda, 1992). For the cognitive radio application, a non-cooperative MC detector using cyclostationary signatures for OFDM-based waveforms was proposed by Sutton, Nolan, & Doyle, 2008. Furthermore, Xiangzhen & Xuping, 2008 proposed a non-cooperative MC detector in which the detector combines distinct SC detectors for different CFs and the final decision is obtained by ORing the primitive decisions of the SC detectors.

The performance of cyclostationary spectrum sensing can be further enhanced by using cooperation among SUs. By exploiting spatial diversity, cooperative sensing can be an effective solution to mitigate the detection performance degradations due to multipath fading, shadowing and receiver uncertainty issues. Recently, different cooperative methods are considered for spectrum sensing based on energy detection in order to have a better performance, especially in fading channels (Ganesan & Li, 2007a, 2007b; Ghasemi & Sousa, 2005; Ma, Zhao, & Li, 2008; Mishra, Sahai, & Brodersen, 2006; Quan, Cui, Poor, & Sayed, 2008; Taherpour, Nasiri-Kenari, & Jamshidi, 2007; Taherpour, Norouzi, et al., 2007; Unnikrishnan & Veeravalli, 2008).

However, studies of cooperative spectrum sensing based on cyclostationary detection for cognitive radios have been rather limited. Lunden et al., 2009; Sadeghi & Azmi, 2008 studied cooperative sensing schemes with cyclostationary feature detection in which each SU uses a MC cyclostationary detector. To alleviate the processing requirement at each SU, Derakhshani, Le-Ngoc, & Nasiri-Kenari, 2011 proposed cooperative cyclostationary schemes in which each SU performs SC cyclostationary detection at different CFs. This enables computationally efficient parallel searching in the CF domain as compared to a MC detector.

NON-COOPERATIVE SINGLE-CYCLE CYCLOSTATIONARY DETECTION

The performance of a detection scheme can be evaluated with two probabilities, including detection probability (P_d) and false alarm probability (P_{fa}). Generally, SUs need a small false alarm probability and a large detection probability to sufficiently protect PUs and efficiently use spectrum opportunities. In this section, the detection probability and false alarm probability of the SC detector in an AWGN channel (i.e., $h_i = 1$), are derived.

If $x(t)$ is assumed as a cyclo-ergodic process, its CAF is (Gardner et al., 2006; Gardner & Spooner, 1992; Gardner, 1986, 1988; Yeung & Gardner, 1996)

$$R_x^{\alpha_k}(\tau) = \lim_{T_s \to \infty} \int_{t_0 - \frac{T_s}{2}}^{t_0 + \frac{T_s}{2}} x(t) x^*(t + \tau) e^{-j2\pi\alpha_k t} dt \tag{10}$$

where $\alpha_k = k / T_c$ is the k-th CF and T_s is the observation interval. Given the time and energy constraints, we consider a limited observation interval of length $T_s = N_s \Delta t$ (Δt is the sampling period) where N_s is the total number of received samples. Assuming $t = m\Delta t$ and $T_c = M\Delta t$ (where T_c is the chip (or bit) time interval), $Y_{SC} = R_x^{\alpha k}(\tau = 0)$ can be estimated by

$$Y_{SC} \simeq \frac{1}{N_s} \sum_{m=1}^{N_s} |x[m]|^2 \, e^{-j2\pi\alpha'_k m} \tag{11}$$

where $\alpha'_k = k / M$. Hence, Y_{SC} in different hypotheses (H_0 and H_1) according to the binary hypotheses test (1) can be represented as

$$\begin{cases} H_0 : Y_{SC} \simeq \dfrac{1}{N_s} \sum_{m=1}^{N_s} |n[m]|^2 \, e^{-j2\pi\alpha'_k m} \\ H_1 : Y_{SC} \simeq \dfrac{1}{N_s} \sum_{m=1}^{N_s} |s[m] + n[m]|^2 \, e^{-j2\pi\alpha'_k m} \end{cases} \tag{12}$$

where $n[m] = n_r[m] + j\, n_i[m]$ is a complex AWGN sample with zero mean and variance σ^2, and $N_s = N_c M$ where N_c is the number of chips (or bits) and M is the number of samples in each chip (or bit) time interval. The structure of the SC detector can be defined as

$$T_{SC} = |Y_{SC}|^2 \underset{H_0}{\overset{H_1}{\gtrless}} \eta \tag{13}$$

where η denotes detection threshold. Consequently,

$$P_{fa,\alpha} = \Pr\left(T_{SC} > \eta \mid H_0\right) = \int_{\eta}^{+\infty} \Pr\left(T_{SC} \mid H_0\right) dT_{SC} \tag{14}$$

$$P_{d,\alpha} = \Pr\left(T_{SC} > \eta \mid H_1\right) = \int_{\eta}^{+\infty} \Pr\left(T_{SC} \mid H_1\right) dT_{SC} \tag{15}$$

where the conditional probability density functions (PDF), $\Pr(T_{SC} \mid H_0)$ and $\Pr(T_{SC} \mid H_1)$, of $T_{SC} = |Y_{SC}|^2$ given H_0 and H_1 can be derived from those of Y_{SC}, $\Pr(Y_{SC} \mid H_0)$ and $\Pr(Y_{SC} \mid H_1)$. Based on the central limit theorem for large $N_s \gg 1$, Y_{SC} is approximately Gaussian. For $\alpha'_k = k / M$, $\sum_{m=1}^{M} e^{-j2\pi(k/M)m} = 0$ and $N_s = N_c M$ (i.e., including several periods of the exponential function), since $E\{|n[m]|^2\} = \sigma^2$ and $\text{var}\{|n[m]|^2\} = \text{var}\{n_r^2[m] + n_i^2[m]\} = \sigma^4$, the mean and variance of Y_{SC} given H_0 are

$$E\left\{Y_{SC} \mid H_0\right\} = \frac{\sigma^2}{N_s} \sum_{m=1}^{N_s} e^{-j2\pi\alpha'_k m} = 0 \tag{16}$$

$$\text{var}\left\{Y_{SC} \mid H_0\right\} = \frac{1}{N_s^2} \sum_{m=1}^{N_s} \left|e^{-j2\pi\alpha'_k m}\right|^2 \text{var}\left\{|n[m]|^2\right\} = \frac{\sigma^4}{N_s}. \tag{17}$$

Since $\{Y_{SC} \mid H_0\} = Y_0^r + jY_0^i$ is a complex Gaussian random variable, it follows that $\{T_{SC} \mid H_0\} = |(Y_{SC} \mid H_0)|^2 = (Y_0^r)^2 + (Y_0^i)^2$ is central chi-square distributed with two degrees of freedom, i.e.,

$$\Pr\left(T_{SC} \mid H_0\right) = \frac{1}{2\sigma_0^2} e^{-\frac{T_{SC}}{2\sigma_0^2}}, \sigma_0^2 = \frac{\sigma^4}{N_s}. \tag{18}$$

The false-alarm probability is

$$P_{fa,\alpha'_k} = \int_{\eta}^{+\infty} \frac{1}{2\sigma_0^2} e^{-\frac{T_{SC}}{2\sigma_0^2}} dT_{SC} = e^{-\frac{\eta}{2\sigma_0^2}}. \tag{19}$$

Similarly, since $E\{|s[m]+n[m]|^2\} = |s[m]|^2 + \sigma^2$ and $\sum_{m=1}^{M} e^{-j2\pi(k/M)m} = 0$, the mean of $\{Y_{SC} \mid H_1\}$ is

$$\mathrm{E}\left\{Y_{\mathrm{SC}} \mid H_1\right\} = \frac{1}{N_s} \sum_{m=1}^{N_s} \left(\left|s[m]\right|^2 + \sigma^2\right) e^{-j2\pi\alpha'_k m} = \frac{1}{N_s} \sum_{m=1}^{N_s} \left|s[m]\right|^2 e^{-j2\pi\alpha'_k m} = P_{\alpha'_k} \tag{20}$$

where the complex-valued $P_{\alpha'k}$ represents the signal power at the CF α'_k. Since noise samples are statistically independent, the variance of $\{Y_{\mathrm{SC}} \mid H_1\}$ is

$$\mathrm{var}\left\{Y_{\mathrm{SC}} \mid H_1\right\} = \frac{1}{N_s^2} \sum_{m=1}^{N_s} \left|e^{-j2\pi\alpha'_k m}\right|^2 \mathrm{var}\left\{\left|s[m] + n[m]\right|^2\right\} = \frac{2\sigma^2 P}{N_s} + \frac{\sigma^4}{N_s} \tag{21}$$

where $P = (1 / N_s) \sum_{m=1}^{Ns} |s[m]|^2$. In other words, $(Y_{\mathrm{SC}} \mid H_1) = Y_1^{\mathrm{r}} + jY_1^{\mathrm{i}}$ is a complex Gaussian random variable, and, hence, $\{T_{\mathrm{SC}} \mid H_1\} = |(Y_{\mathrm{SC}} \mid H_1)|^2 = (Y_1^{\mathrm{r}})^2 + (Y_1^{\mathrm{i}})^2$ has a non-central chi-square distribution with two degrees of freedom, i.e., its PDF is given by

$$\mathrm{Pr}\left(T_{\mathrm{SC}} \mid H_1\right) = \frac{1}{2\sigma_1^2} e^{-\frac{\left(s_1^2 + T_{\mathrm{SC}}\right)}{2\sigma_1^2}} I_0\left(\sqrt{T_{\mathrm{SC}}} \frac{s_1}{\sigma_1^2}\right) \tag{22}$$

where $s_1^2 = (\mathrm{E}\{Y_1^{\mathrm{r}} \mid H_1\})^2 + (\mathrm{E}\{Y_1^{\mathrm{i}} \mid H_1\})^2 = |P_{\alpha'k}|^2$ and $\sigma_1^2 = (2\sigma^2 P) / N_s + \sigma^4 / N_s$. The detection probability P_{d} is

$$P_{\mathrm{d},\alpha'_k} = \int_\eta^{+\infty} \mathrm{P}\left(T_{\mathrm{SC}} \mid H_1\right) \mathrm{d}T_{\mathrm{SC}} = Q_1\left(\frac{s_1}{\sigma_1}, \frac{\sqrt{\eta}}{\sigma_1}\right) \tag{23}$$

where $Q_1(a, b)$ is the first-degree Marcum Q function. Figure 1 shows that the receiver operating characteristic (ROC) curves (i.e., plots of detection probability versus false alarm probability) obtained by the above analysis are in good agreement with simulation results for various SNRs.

All simulations are implemented in Matlab. In the simulation results, the considered PU signal is the direct-sequence (DS) spread-spectrum BPSK signal using Walsh-Hadamard code with length 16 (i.e., processing gain of 16). It is also assumed that the SUs experience independent and statistically identical channel fading and noises. Furthermore, N_b denotes the number of bits and $N_c = N_b \times 16$. In the simulation results, ROC curves are generated by plotting the fraction of true detections out of the total actual PU signal presences versus the fraction of false alarms out of the total actual PU signal absences, at various threshold settings.

COOPERATIVE CYCLOSTATIONARY DETECTION

In cooperative spectrum sensing, the SUs collaborate with each other to sense the spectrum to find the spectrum holes. In other words, they share their sensing information, and then each of them decides about absence or presence of the PU in a specific bandwidth based on the shared knowledge. Since SUs sense different noise and fading, cooperative spectrum sensing can improve the performance by multi-

Figure 1. ROC curves for different SNRs

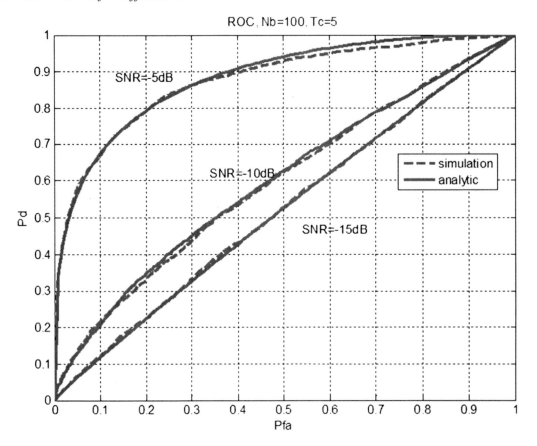

user diversity. In particular, we present efficient cooperative cyclostationary spectrum sensing schemes in which each SU performs SC cyclostationary detection for fast and simple implementation, while collaboration between SUs in final decision on the presence or absence of the PU is explored to improve its performance. As the SUs simultaneously measure the spectral correlation functions at different CFs and exchange their information regarding the measured results, a sufficient number of CFs are effectively examined in a short period of time because of parallel searching, which makes the proposed cooperative spectrum sensing more reliable and faster.

Furthermore, we assume that a SU combines the local sensing information from only proper close SUs in order to alleviate complexity and error probability of information exchange among cooperating SUs (Sun, Zhang, & Letaief, 2007). Therefore, since the cooperating SUs are expected to be located close to each other and are monitoring the same frequency band, their average received powers can be modeled as being identical.

OR Detector

In OR cooperative detection(Ghasemi & Sousa, 2005; Mishra et al., 2006), each SU first informs all other cooperating SUs only about its preliminary decision regarding the presence/absence of the PU by one bit (0 for absence or 1 for presence) and, subsequently, makes its final decision by ORing the preliminary decision bits from itself and all others, i.e., the PU is declared to be present if at least one preliminary decision bit is 1. Therefore, the detection probability $P_{d,OR}$ and false alarm probability $P_{fa,OR}$ of this OR cooperative detector can be determined by

$$P_{d,OR} = 1 - \prod_{k=1}^{N_u} \left(1 - P_{d,\alpha'_k}\right) \tag{24}$$

$$P_{fa,OR} = 1 - \prod_{k=1}^{N_u} \left(1 - P_{fa,\alpha'_k}\right) \tag{25}$$

where N_u denotes the number of SUs, and $P_{d,\alpha'k}$ and $P_{fa,\alpha'k}$, given in (19) and (23), are, respectively, the detection probability and false alarm probability of the k-th SU for CF α'_k.

ML Detector

The proposed ML cooperative cyclostationary detector is based on the idea of MC detection. Based on (9), the test statistic for MC cyclostationary detector can be given by

$$T_{MC} = \sum_{k=1}^{N_u} \left|R_x^{\alpha'_k}(\tau = 0)\right|^2, \quad \alpha'_k = \frac{k}{M}. \tag{26}$$

Thus, in the proposed ML cooperative cyclostationary detection, each SU measures the received signal power at different α'_k and sends it to all other cooperating SUs. Subsequently, each SU uses all the measured signal powers at different CFs to compute the following test statistic

$$T_{ML-CO} = \sum_{k=1}^{N_u} \left|R_{x_k}^{\alpha'_k}(\tau = 0)\right|^2, \quad \alpha'_k = \frac{k}{M} \tag{27}$$

where $R_{xk}^{\alpha'k}(\tau = 0)$ represents the cyclic autocorrelation for the k-th SU at CF α'_k. From the derived distributions for the test statistic of SC detection in (18) and (22), it can be concluded that $\{T_{ML-CO} \mid H_0\}$ has a central chi-square distribution with $2N_u$ degrees of freedom and $\{T_{ML-CO} \mid H_1\}$ has a non-central chi-square distribution with $2N_u$ degrees of freedom, i.e., their conditional PDFs are

$$\Pr\left(T_{ML-CO} \mid H_0\right) = \frac{1}{\sigma_0^{2N_u} 2^{N_u} \Gamma\left(N_u\right)} T_{ML-CO}^{N_u-1} e^{-\frac{T_{ML-CO}}{2\sigma_0^2}}, \quad \sigma_0^2 = \frac{\sigma^1}{N_s} \tag{28}$$

$$\Pr\left(T_{\text{ML-CO}} \mid H_1\right) = \frac{1}{2\sigma_1^2} \left(\frac{T_{\text{ML-CO}}}{s_1^2}\right)^{\frac{N_u-1}{2}} \times e^{-\frac{\left(s_1^2 + T_{\text{ML-CO}}\right)}{2\sigma_1^2}} I_{N_u-1}\left(\sqrt{T_{\text{ML-CO}}} \frac{s_1}{\sigma_1^2}\right) \tag{29}$$

with $s_1^2 = \sum_{k=1}^{N_u} |P_{a\dot{k}}|^2$ and $\sigma_1^2 = (2\sigma^2 P) / N_s + \sigma^4 / N_s$. The corresponding false-alarm and detection probabilities are, respectively,

$$P_{\text{fa,ML-CO}} = \int_{\eta}^{+\infty} \mathrm{P}\left(T_{\text{ML-CO}} \mid H_0\right) \mathrm{d} T_{\text{ML-CO}} = \int_{\eta}^{+\infty} \frac{1}{\sigma_0^{2N_u} 2^{N_u} \Gamma\left(N_u\right)} T_{\text{ML-CO}}^{N_u-1} e^{-\frac{T_{\text{ML-CO}}}{2\sigma_0^2}} \mathrm{d} T_{\text{ML-CO}}, \tag{30}$$

and

$$P_{\text{d,ML-CO}} = \int_{\eta}^{+\infty} \mathrm{P}\left(T_{\text{ML-CO}} \mid H_1\right) \mathrm{d} T_{\text{ML-CO}} = Q_{N_u}\left(\frac{s_1}{\sigma_1}, \frac{\sqrt{\eta}}{\sigma_1}\right), \tag{31}$$

where $Q_{N_u}(a, b)$ is the Marcum function with N_u degrees. The ROC curves for SNR = -15dB and different numbers of SUs in Figure 2 show a good agreement between simulation and analytical results.

Quantization Effect

In the ML cooperative detector, it is assumed that measured power of each SU can be presented with infinite precision. Therefore, it is needed to consider the quantization effect on the performance of the ML detector by assuming a finite number of bits to present the measured power of each SU, and examine the performance for different numbers of bits. Figure 3 and Figure 4 illustrate the performance of the ML detector for different numbers of quantization bits in an AWGN channel with SNR = -12dB, and in a Rayleigh fading channel with SNR = -10dB, respectively. The results indicate that, in both channels, for 6 or more bits, the performance is extremely close to that for unquantized measurements.

DEFLECTION COEFFICIENTS OF DIFFERENT DETECTORS IN AWGN CHANNELS

Performance of different cyclostationary detectors under consideration can be compared in terms of their ROC curves and deflection coefficient that is defined as (Poor, 1994)

$$D = \frac{\left|\mathrm{E}(\theta \mid H_1) - \mathrm{E}(\theta \mid H_0)\right|}{\sqrt{\mathrm{var}(\theta \mid H_0)}} \tag{32}$$

Figure 2. ROC curves for different numbers of SUs

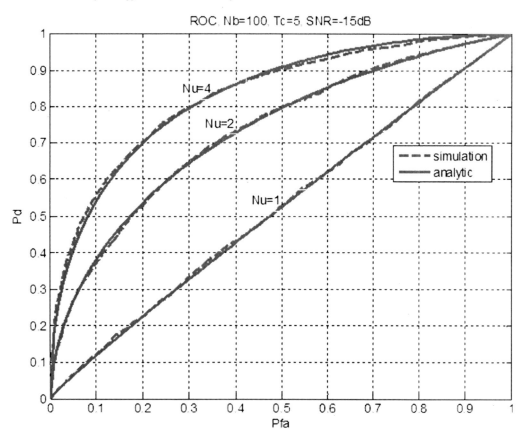

where θ is the decision coefficient of the detector. A larger deflection coefficient indicates a larger difference between the two conditional means of θ given H_0 and H_1 as compared to the conditional standard of deviation given H_0, and hence, a better discrimination between two hypotheses or a better detection performance.

In this section, deflection coefficients of different cyclostationary detectors under consideration are derived and compared in AWGN channels.

Single-Cycle Detector

As it is discussed before, $\{T_{\mathrm{SC}} | H_0\}$ has a central chi-square distribution with two degrees of freedom and $\{T_{\mathrm{SC}} | H_1\}$ has a non-central chi-square distribution with two degrees of freedom for SC detector, therefore the mean and variance of $\{T_{\mathrm{SC}} | H_0\}$ and the mean of $\{T_{\mathrm{SC}} | H_1\}$ according to (18) and (22) are given by

$$\mathrm{E}\left\{T_{\mathrm{SC}} \mid H_0\right\} = \sigma_0^2 = \frac{\sigma^4}{N_s} \tag{33}$$

Figure 3. ROC curves of ML cooperative cyclostationary detector in an AWGN channel for different numbers of quantization bits (N)

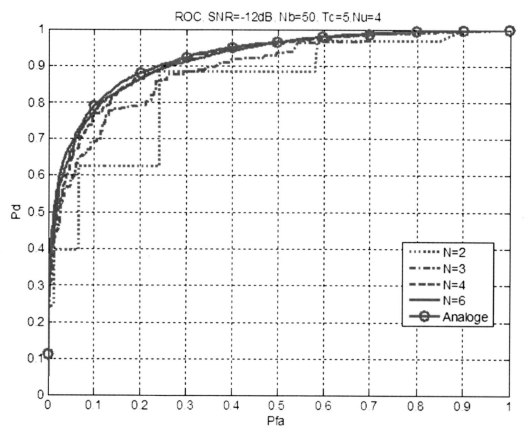

$$E\left\{T_{SC} \mid H_1\right\} = 2\sigma_1^2 + s_1^2 = 2\frac{2P\sigma^2 + \sigma^4}{N_s} + \left|P_{\alpha_k'}\right|^2 \tag{34}$$

$$\operatorname{var}\left\{T_{SC} \mid H_0\right\} = 2\sigma_0^4 = 2\left(\frac{\sigma^4}{N_s}\right)^2. \tag{35}$$

Consequently, the deflection coefficient of the SC detector can be derived as

$$D_{SC}^{\alpha_k'} = \frac{\left|P_{\alpha_k'}\right|^2 + \dfrac{4P\sigma^2 + \sigma^4}{N_s}}{\sqrt{2}\sigma^4 / N_s} \tag{36}$$

Figure 4. ROC curves of ML cooperative cyclostationary detector in a Rayleigh fading channel for different numbers of quantization bits (N)

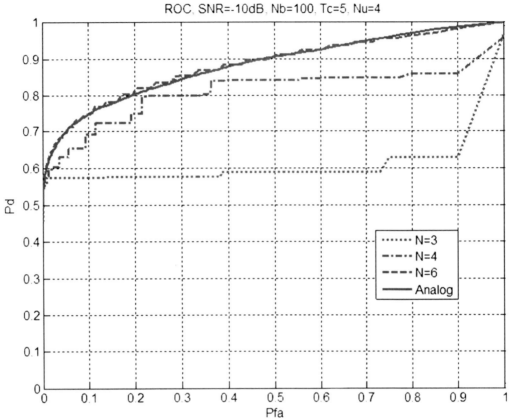

If $N_s \gg 1$, then $(4P\sigma^2 + \sigma^4) / N_s$ is negligible in comparison with $|P_{\alpha'k}|^2$. Thus, the deflection coefficient reduces to

$$D_{SC}^{\alpha'_i} \approx \frac{1}{\sqrt{2}} \frac{|P_{\alpha'_i}|^2}{\sigma^1 / N_s}. \tag{37}$$

OR Cooperative Cyclostationary Detector

The final decision in OR detection indicates presence of the PU's signal (H_1 hypothesis) when at least one of the primitive decisions of the SUs is 1 (H_1 hypothesis). Hence, effectively, it is based on the primitive decision of the SU that has the best performance. Thus, the deflection coefficient for the OR cooperative cyclostationary detector corresponds to the maximum deflection coefficient of different SUs which search in different CFs, i.e.,

$$D_{\text{OR-CO}} = \max_{\alpha_k'} \left(D_{\text{SC}}^{\alpha_k'} \right) = \frac{\max_{\alpha_k'} \left(\left| P_{\alpha_k'} \right|^2 \right) + \dfrac{4P\sigma^2 + \sigma^4}{N_s}}{\sqrt{2}\sigma^4 / N_s}. \tag{38}$$

If $N_s \gg 1$, then $(4P\sigma^2 + \sigma^4) / N_s$ is negligible in comparison with $\max_{\alpha'k} (|P_{\alpha'k}|^2)$. Thus, deflection coefficient reduces to

$$D_{\text{OR-CO}} \approx \frac{1}{\sqrt{2}} \frac{\max_{\alpha_k'} \left(\left| P_{\alpha_k'} \right|^2 \right)}{\sigma^4 / N_s}. \tag{39}$$

ML Cooperative Cyclostationary Detector

Based on (28), (29), and the characteristics of the mean and variance of the central and non-central chi-square random variables, we can obtain the following mean and variance of $\{T_{\text{ML-CO}} | H_0\}$ and $\{T_{\text{ML-CO}} | H_1\}$

$$\text{E}\left\{ T_{\text{ML-CO}} \mid H_0 \right\} = N_u \sigma_0^2 = N_u \frac{\sigma^4}{N_s} \tag{40}$$

$$\text{E}\left\{ T_{\text{ML-CO}} \mid H_1 \right\} = 2N_u \sigma_1^2 + s_1^2 = 2N_u \frac{2P\sigma^2 + \sigma^4}{N_s} + \sum_{k=1}^{N_u} \left| P_{\alpha_k'} \right|^2 \tag{41}$$

$$\text{var}\left\{ T_{\text{ML-CO}} \mid H_0 \right\} = 2N_u \sigma_0^4 = 2N_u \left(\frac{\sigma^4}{N_s} \right)^2. \tag{42}$$

Therefore, the deflection coefficient of the ML cooperative detector is

$$D_{\text{ML-CO}} = \frac{\displaystyle\sum_{k=1}^{N_u} \left| P_{\alpha_k'} \right|^2 + N_u \frac{4P\sigma^2 + \sigma^4}{N_s}}{\sqrt{2N_u}\sigma^4 / N_s} \tag{43}$$

For $N_u \ll N_s$, in the numerator, the second term is negligible as compared to the first term, and

$$D_{\text{ML-CO}} \approx \frac{1}{\sqrt{2N_{\text{u}}}} \frac{\sum_{k=1}^{N_{\text{u}}} \left| P_{\alpha_k'} \right|^2}{\sigma^4 / N_{\text{s}}} = \frac{1}{\sqrt{N_{\text{u}}}} \sum_{k=1}^{N_{\text{u}}} D_{\text{SC}}^{\alpha_k'} \ . \tag{44}$$

It indicates that the deflection coefficient of the ML cooperative detector is proportional to the sum of the deflection coefficients of all SUs in the network for $N_{\text{u}} \ll N_{\text{s}}$.

Multi-Cycle Detector

Assuming $N_{\alpha} = N_{\text{u}}$, since

$$T_{\text{MC}} \simeq \sum_{k=1}^{N_{\text{u}}} \left| \frac{1}{N_{\text{s}}} \sum_{m=1}^{N_{\text{s}}} |x[m]|^2 e^{-j2\pi\alpha_k' m} \right|^2$$

and

$$T_{\text{ML-CO}} \simeq \sum_{k=1}^{N_{\text{u}}} \left| \frac{1}{N_{\text{s}}} \sum_{m=1}^{N_{\text{s}}} |x_k[m]|^2 e^{-j2\pi\alpha_k' m} \right|^2,$$

it can be concluded that $E\{T_{\text{MC}}\} = E\{T_{\text{ML-CO}}\}$ and $\text{var}\{T_{\text{MC}}\} = \text{var}\{T_{\text{ML-CO}}\} + 2\sum_{i<j} \text{cov}\{X_i, X_j\}$ where

$$X_i = \left| \frac{1}{N_{\text{s}}} \sum_{m=1}^{N_{\text{s}}} |x[m]|^2 e^{-j2\pi\alpha_i' m} \right|^2 .$$

In order to find $\text{var}\{T_{\text{MC}} \mid H_0\}$, $\text{cov}(X_i, X_j \mid H_0)$ need to be calculated.

$$\begin{aligned}
\text{cov}\left(X_i, X_j \mid H_0\right) &= E\left(X_i X_j \mid H_0\right) - E\left(X_i \mid H_0\right) E\left(X_j \mid H_0\right) \\
&= \frac{[14 + 3(N_{\text{s}} - 1)]\sigma^8}{N_{\text{s}}^3} - \frac{\sigma^8}{N_{\text{s}}^2} \simeq \frac{2\sigma^8}{N_{\text{s}}^2}
\end{aligned} \tag{45}$$

Then, $\text{var}\{T_{\text{MC}} \mid H_0\} = \text{var}\{T_{\text{ML-CO}} \mid H_0\} + N_{\text{u}}(N_{\text{u}}-1)(2\sigma^8) / N_{\text{s}}^2$. Therefore, the deflection coefficient of the MC detector can be derived as

$$D_{\text{MC}} = \frac{\left| E(T_{\text{MC}} \mid H_1) - E(T_{\text{MC}} \mid H_0) \right|}{\sqrt{\text{var}(T_{\text{MC}} \mid H_0)}} = \frac{\left| E(T_{\text{ML-CO}} \mid H_1) - E(T_{\text{ML-CO}} \mid H_0) \right|}{\sqrt{\text{var}(T_{\text{ML-CO}} \mid H_0) + N_{\text{u}}(N_{\text{u}} - 1)\dfrac{2\sigma^8}{N_{\text{s}}^2}}}$$

$$= \frac{\sum_{k=1}^{N_u} \left| P_{\alpha_k'} \right|^2 + N_u \dfrac{4P\sigma^2 + \sigma^4}{N_s}}{\sqrt{\dfrac{2N_u \sigma^8}{N_s^2} + \dfrac{2(N_u^2 - N_u)}{N_s^2}\sigma^8}} = \frac{\sum_{k=1}^{N_u} \left| P_{\alpha_L'} \right|^2 + N_u \dfrac{4P\sigma^2 + \sigma^4}{N_s}}{\sqrt{2}N_u \sigma^4 / N_s}. \tag{46}$$

For $N_u \ll N_s$, in the numerator, the second term is negligible as compared to the first term, and

$$D_{MC} \approx \frac{\sum_{k=1}^{N_u} \left| P_{\alpha_k'} \right|^2}{\sqrt{2}N_u \sigma^4 / N_s} = \frac{1}{N_u}\sum_{k=1}^{N_u} D_{SC}^{\alpha_k'} = \frac{D_{ML\text{-}CO}}{\sqrt{N_u}}. \tag{47}$$

Performance Comparison of Different Detectors in AWGN Channels

Obviously, the OR detector with $D_{OR\text{-}CO} = \max_{\alpha'k}(D_{SC}^{\alpha'k})$ outperforms the SC detector with $D_{SC}^{\alpha'k}$ because the OR detector examines different CFs contrary to the SC detector which measures power only at α'_k. Moreover, from (39) and (47), the OR detector with $D_{OR\text{-}CO} = \max_{\alpha'k}(D_{SC}^{\alpha'k})$ outperforms the MC detector with $D_{MC} \approx (1/N_u)\sum_{k=1}^{N_u} D_{SC}^{\alpha'k}$ in AWGN channels.

Comparing the ML cooperative detector with $D_{ML\text{-}CO} = (1/\sqrt{(N_u)})\sum_{k=1}^{N_u} D_{SC}^{\alpha'k}$ and the OR cooperative detector with $D_{OR\text{-}CO} = \max_{\alpha'k}(D_{SC}^{\alpha'k})$, it can be concluded that the ML cooperative detector outperforms the OR cooperative detector whenever $D_{ML\text{-}CO}/D_{OR\text{-}CO} = ((1/\sqrt{(N_u)})\sum_{k=1}^{N_u} D_{SC}^{\alpha'k})/\max_{\alpha'k}(D_{SC}^{\alpha'k}) > 1$. According to (37), this condition can be reduced to $((1/\sqrt{(N_u)})\sum_{k=1}^{N_u} |P_{\alpha'k}|^2)/\max_{\alpha'k}(|P_{\alpha'k}|^2) = \sqrt{(N_u)}[\text{mean}_{\alpha'k}(|P_{\alpha'k}|^2)/\max_{\alpha'k}(|P_{\alpha'k}|^2)] > 1$. This condition shows that for large number of cooperating users, N_u, the ML detector outperforms the OR detector. The plot of $D_{ML\text{-}CO}/D_{OR\text{-}CO}$ versus N_u based on numerical results in Figure 5 indicates that, for $N_u > 2$, the ML cooperative detector outperforms the OR cooperative detector in an AWGN channel.

From (44) and (47), it is clear that the ML cooperative detector also has a larger deflection coefficient (and hence better performance) than the MC detector. This can be explained by the fact that, in the proposed ML cooperative detection, different SUs observe independent white noise samples. Thus, it can help to improve the performance by concurrently examining different CFs by different SUs. Figure 6 and Figure 7 illustrate the deflection coefficients of cooperative and non-cooperative cyclostationary detectors versus SNR for $N_u = 4$ and $N_u = 10$ based on numerical results in an AWGN channel.

Figure 8. presents the ROC curves based on simulation results in an AWGN channel of the proposed ML and OR cooperative SC-based cyclostationary detectors and existing non-cooperative and ML cooperative MC-based cyclostationary, and ML cooperative energy detectors. The MC-based ML cooperative cyclostationary detector has the best ROC curve since it benefits from both multi-user cooperative diversity and MC detection. However, the performance gap between the MC-based and proposed SC-based ML cooperative cyclostationary detectors is not large while the proposed detector is much simpler since each SU only needs to search one CF, contrary to a MC-based detector in which each SU should examine all

Figure 5. $D_{ML\text{-}CO} / D_{OR\text{-}CO}$ versus number of cooperating users, N_u

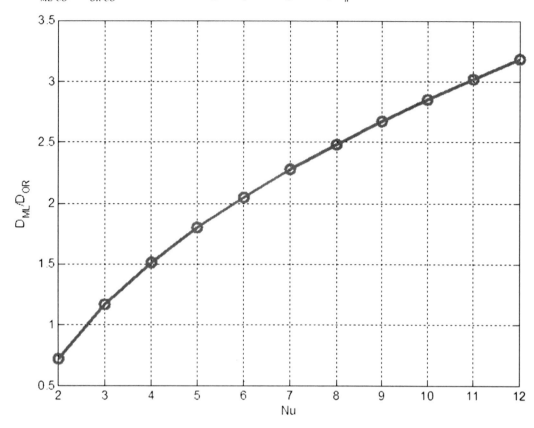

CFs. Performance of the SC-based ML cooperative cyclostationary detector with identical CF strongly depends on the selected CF as shown by a large difference in ROC between the best and worst selected CFs. If SUs have full knowledge about primary-user signal to choose the best CF, monitoring best CF by all SUs is the most favorable option among SC-based cooperative detectors. However, in general, SUs may not have sufficient knowledge about the primary-user signal to choose the best CF. Therefore, it is more beneficial for the cooperating SUs to use different CFs as in the proposed SC-based ML cooperative detection scheme. The MC cyclostationary detector (without multi-user cooperation) performs slightly better than the ML-cooperative energy detector for false-alarm probability lower than 0.4 but the situation is reversed when the false-alarm exceeds 0.4. The results in Figure 8 confirm the benefits of the proposed schemes in combining the advantages of multi-user cooperation and SC cyclostationary spectrum-sensing with different CFs for low SNR operation with relatively low complexity.

DEFLECTION COEFFICIENTS OF DIFFERENT DETECTORS IN RAYLEIGH FADING CHANNELS

In this section, the outage probability (i.e., the probability that the deflection coefficient of each detector falls below a certain threshold in a fading channel) for each cyclostationary detector is derived. By comparing different outage probabilities, the improvement which cooperative detectors provide due to

Figure 6. Deflection coefficient versus SNR in an AWGN channel ($N_u = 4$)

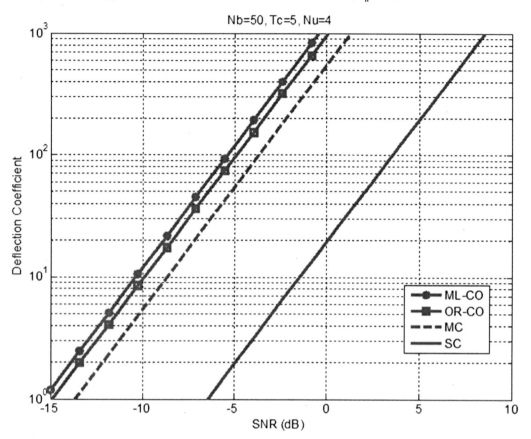

the diversity can be measured. As discussed, detectors with larger deflection coefficient have better performance. Therefore, for the same threshold, so the smaller outage probability is more desirable. In the cooperative techniques, the outage probability decreases exponentially with N_u. In this section, outage probabilities are calculated in a Rayleigh fading channel in which $a = |h|^2$ follows an exponential distribution with parameter 1.

Single-Cycle Detector

According to (37), in a fading channel, the deflection coefficient of SC detector given the channel power a is $(D_{SC}^{\alpha'k} \mid a) \approx (1/\sqrt{2}) a^2 |P_{\alpha'k}|^2 / (\sigma^4 / N_s) = a^2 D_{SC}^{\alpha'k}$ where $D_{SC}^{\alpha'k}$ denotes the deflection coefficient of SC detector in AWGN channel. In order to have a better understanding of the performance in a fading channel, we define the outage probability for the deflection coefficient as

$$\Pr\left(\left(D_{SC}^{\alpha'_k} \mid a\right) < D_{th}\right) = \Pr\left(a^2 D_{SC}^{\alpha'_k} < D_{th}\right) = 1 - e^{-\sqrt{\frac{D_{th}}{D_{SC}^{\alpha'_k}}}}. \tag{48}$$

Figure 7. Deflection coefficient versus SNR in an AWGN channel ($N_u = 10$)

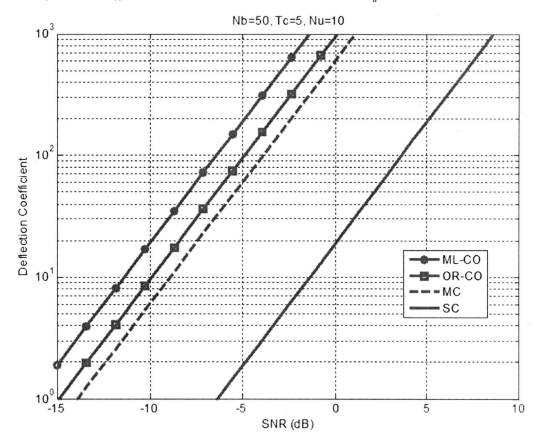

Note that for $1-e^{-x} \leq 0.03$ the linear approximation $1-e^{-x} \approx x$ is very accurate (less than 1%). In other words, for $D_{th} < 0.001\, D_{SC}^{\alpha'k}$, (48) can be approximated as

$$\Pr\left(\left(D_{SC}^{\alpha'_k} \mid a\right) < D_{th}\right) = \sqrt{\frac{D_{th}}{D_{SC}^{\alpha'_k}}}. \tag{49}$$

Multi-Cycle Detector

According to (47), the conditional deflection coefficient in a fading channel can be written as $D_{MC} = (1 / N_u)\, a^2 \sum_{k=1}^{N_u} D_{SC}^{\alpha'k}$. Hence, the outage probability can be derived as

$$\Pr\left(\left(D_{MC} \mid a\right) < D_{th}\right) = \Pr\left(\frac{1}{N_u} a^2 \sum_{k=1}^{N_u} D_{SC}^{\alpha'_k} < D_{th}\right) = 1 - e^{\sqrt{\frac{N_u D_{th}}{\sum_{k=1}^{N_u} D_{SC}^{\alpha'_k}}}}. \tag{50}$$

Similarly, for $D_{th} < (0.001 / N_u) \sum_{k=1}^{N_u} D_{SC}^{\alpha'k}$, (50) can be approximated as

Figure 8. ROC Curves of ML cooperative cyclostationary, OR cooperative cyclostationary, multi-cycle and cooperative energy detectors in an AWGN channel

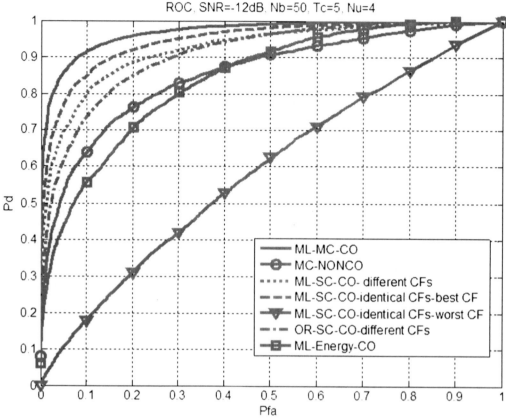

$$\Pr\left(\left(D_{\mathrm{MC}}\mid a\right)<D_{\mathrm{th}}\right)=\sqrt{\frac{N_{\mathrm{u}}D_{\mathrm{th}}}{\sum_{k=1}^{N_{\mathrm{u}}}D_{\mathrm{SC}}^{\alpha_{k}}}}. \tag{51}$$

OR Cooperative Cyclostationary Detector

The deflection coefficient for the OR cooperative cyclostationary detector in an AWGN channel corresponds to the maximum deflection coefficient of different SUs who search in different CFs according to (39). Therefore, the outage probability in a fading channel for the OR detector can be derived as

$$\Pr\left(\left(D_{\mathrm{OR\text{-}CO}}\mid \mathbf{a}\right)<D_{\mathrm{th}}\right)=\Pr\left(a_{1}^{2}D_{\mathrm{SC}}^{\alpha_{1}}<D_{\mathrm{th}},...,a_{N_{\mathrm{u}}}^{2}D_{\mathrm{SC}}^{\alpha_{N}}<D_{\mathrm{th}}\right)$$
$$=\prod_{k=1}^{N_{\mathrm{u}}}\Pr\left(a_{k}^{2}D_{\mathrm{SC}}^{\alpha_{k}}<D_{\mathrm{th}}\right)=\prod_{k=1}^{N_{\mathrm{u}}}\left(1-e^{-\sqrt{\frac{D_{\mathrm{th}}}{D_{\mathrm{SC}}^{\alpha_{k}}}}}\right) \tag{52}$$

where $a=\{a_k, k = 1, 2, .., N_u\}$. Similarly, for $D_{th} < 0.001 \min_{a'k} D_{SC}^{a'k}$, (52) can be approximated as

$$\Pr\left(\left(D_{OR\text{-}CO} \mid \mathbf{a}\right) < D_{th}\right) = \prod_{k=1}^{N_u}\left(\sqrt{\frac{D_{th}}{D_{SC}^{\alpha'_k}}}\right) = \sqrt{\frac{D_{th}^{N_u}}{\prod_{k=1}^{N_u} D_{SC}^{\alpha'_k}}}. \tag{53}$$

ML Cooperative Cyclostationary Detector

According to (44), the deflection coefficient of the ML cooperative cyclostationary detector in a fading channel becomes

$$\left(D_{ML\text{-}CO} \mid \mathbf{a}\right) = \frac{1}{\sqrt{N_u}} \sum_{k=1}^{N_u} a_k^2 D_{SC}^{\alpha'_k}. \tag{54}$$

Since a_k is an exponential random variable with parameter 1 in a Rayleigh fading channel, a_k^2 has the Weibull distribution with scale parameter $\lambda = 1$ and shape parameter $\zeta = 0.5$. The sum of independent Weibull random variables does not have a closed-form distribution, i.e.,

$$\Pr\left(\left(D_{ML\text{-}CO} \mid \mathbf{a}\right) < D_{th}\right) = \Pr\left(\sum_{k=1}^{N_u} a_k^2 D_{SC}^{\alpha'_k} < \sqrt{N_u} D_{th}\right). \tag{55}$$

Therefore, we present a lower-bound for the outage probability in this case to measure how diversity can help to improve the performance of the detector, i.e., since $\sum_{k=1}^{N_u} a_k^2 D_{SC}^{\alpha'k} < (\sum_{k=1}^{N_u} a_k)^2 \max_{a'k}(D_{SC}^{\alpha'k})$,

$$\Pr\left(\left(D_{ML\text{-}CO} \mid \mathbf{a}\right) < D_{th}\right) > \Pr\left(\left(\sum_{k=1}^{N_u} a_k\right)^2 \max_{\alpha'_k}\left(D_{SC}^{\alpha'_k}\right) < \sqrt{N_u} D_{th}\right)$$
$$= \Pr\left(\sum_{k=1}^{N_u} a_k < \sqrt{\frac{\sqrt{N_u} D_{th}}{\max_{\alpha'_k}\left(D_{SC}^{\alpha'_k}\right)}}\right). \tag{56}$$

Since a_k's are independent exponential random variables with parameter 1, $\sum_{k=1}^{N_u} a_k$ has a Gamma distribution $G(N_u, 1)$. Hence, (56) can be computed as

$$\Pr\left(\left(D_{ML\text{-}CO} \mid \bar{a}\right) < D_{th}\right) > \frac{\gamma\left(N_u, \sqrt{\dfrac{\sqrt{N_u} D_{th}}{\max_{\alpha'_k}\left(D_{SC}^{\alpha'_k}\right)}}\right)}{\Gamma\left(N_u\right)} \tag{57}$$

where $\Gamma(s) = \int_0^{+\infty} t^{s-1} e^{-t} dt$ is the Gamma function and $\gamma(s, x) = \int_0^x t^{s-1} e^{-t} dt$ is the lower incomplete Gamma function. Since $\gamma(s, x) \simeq (1/s) x^s$, $x \rightarrow 0$, assuming $D_{th} < (0.001 / \sqrt{N_u}) \max_{\alpha' k} (D_{SC}^{\alpha' k})$, (57) can be approximated as

$$\Pr\left(\left(D_{\text{ML-CO}} \mid a\right) < D_{th}\right) > \frac{N_u^{\frac{N_u}{4}}}{N_u \Gamma(N_u)} \sqrt{\frac{D_{th}^{N_u}}{\left(\max_{\alpha'_k} D_{SC}^{\alpha'_k}\right)^{N_u}}}. \tag{58}$$

Performance Comparison of Different Detectors in Rayleigh Fading Channels

Equations (49) and (51) indicate the outage probabilities of both SC and MC detectors are of $\mathcal{O}(D_{th}^{1/2})$ for $D_{th} \ll \sum_{k=1}^{N_u} D_{SC}^{\alpha k}$. On the other hand, Equations (53) and (58) show that the outage probabilities of both cooperative OR and ML cyclostationary detectors are of $\mathcal{O}(D_{th}^{N_u/2})$ for $D_{th} \ll \sum_{k=1}^{N_u} D_{SC}^{\alpha k}$. In other words, in fading channels, cooperative sensing exploits the multi-user diversity of order Nu to provide much more performance gain as their outage probabilities decrease exponentially with N_u.

Figure 9 and Figure 10 show the deflection outage probability versus deflection coefficient threshold in Rayleigh fading channels for $N_u = 4$ and $N_u = 10$ based on simulation results. It is shown that the

Figure 9. Deflection coefficient outage probability versus D_{th} in a Rayleigh fading channel ($N_u = 4$)

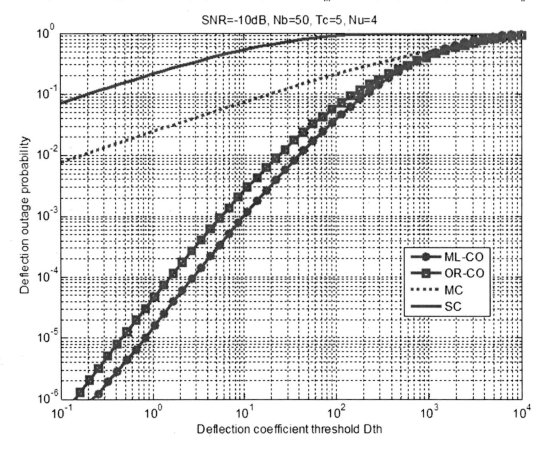

Figure 10. Deflection coefficient outage probability versus D_{th} in a Rayleigh fading channel ($N_u = 10$)

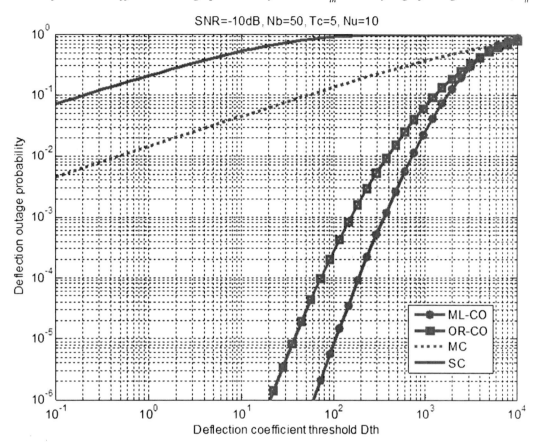

cooperative techniques including OR and ML have smaller outage probability in comparison with the non-cooperative methods. Figure 9 and Figure 10 confirm that the deflection outage probability is $\mathcal{O}(D_{th}^{1/2})$ in the non-cooperative methods and $\mathcal{O}(D_{th}^{Nu/2})$ in the cooperative methods. It indicates that the ML cooperative detector outperforms the OR cooperative detector in a Rayleigh fading channel. It is obvious that the gap between cooperative and non-cooperative methods becomes exponentially larger by increasing N_u.

Figure 11 shows the ROC curves of the ML cooperative, OR cooperative and MC detectors in a Rayleigh fading channel based on simulation results. The simulation results in agreement with the analytical results confirm that the performance gain offered by the cooperative detectors in the Rayleigh fading channel is larger than that in an AWGN channel. It is also clear from Figure 11 that the performance gain of the cooperative methods increases in the fading channels due to the diversity in comparison with non-cooperative methods.

In order to show the benefits of cooperation to solve hidden terminal problem caused by severe shadowing, and hence, improve detection performance, we consider another case in which SUs cooperate

Figure 11. ROC curves of ML cooperative cyclostationary, OR cooperative cyclostationary and multi-cycle detectors in a Rayleigh fading channel

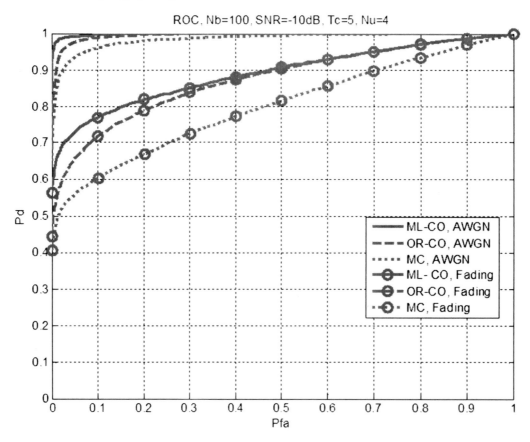

over a wider area. If the SUs span a distance that is larger than the correlation distance of the shadow fading, cooperating SUs experience independent shadowing and it is unlikely that all of them are under a deep shadow simultaneously. Figure 12 illustrates the ROC curves of ML cooperative and MC detectors in a log-normal shadowing channel where cooperating SUs sense independent shadowing based on simulation results, and indicates that the performance gain offered by the ML cooperative detector increases as the channel variance increases.

CONCLUSION

This chapter presents an overview of recent advances in cyclostationary spectrum sensing, including non-cooperative and cooperative approaches, for cognitive radio systems. In particular, two cooperative cyclostationary spectrum sensing schemes are introduced. In these methods, distinctive CFs are exam-

Figure 12. ROC curves of ML cooperative cyclostationary and multi-cycle detectors in a shadowing channel

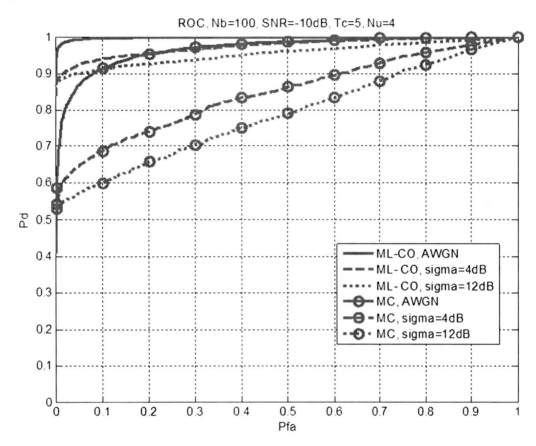

ined by different SUs. While each SU performs SC detection for low complexity, the performance of the proposed cooperative decision is even better than that of an optimal MC detector because the available power at all CFs is measured by different SUs which face independent noise and fading. Performance of the proposed schemes in terms of false-alarm and detection probabilities and deflection coefficients is evaluated by analysis and simulation in AWGN and fading channels. Comparing the outage probability of deflection coefficient of different cyclostationary detectors shows that the detection performance improves exponentially with number of cooperating users in the cooperative methods in a fading channel. Illustrative and analytical results show that the proposed schemes outperform both SC and MC cyclostationary detectors, especially in fading channels. Among the proposed cyclostationary cooperative detection schemes, ML detector offers further performance improvement at the costs of more signaling information as compared to the OR detector.

REFERENCES

Akyildiz, I. F., Lee, W. Y., Vuran, M. C., & Mohanty, S. (2006). NeXt generation/dynamic spectrum access/cognitive radio wireless networks: A survey. *Computer Networks*, 50(13), 2127–2159. doi:10.1016/j.comnet.2006.05.001

Akyildiz, I. F., Lo, B. F., & Balakrishnan, R. (2011). Cooperative spectrum sensing in cognitive radio networks: A survey. *Physical Communication, 4*(1), 40–62. doi:10.1016/j.phycom.2010.12.003

Axell, E., Leus, G., Larsson, E., & Poor, H. (2012). Spectrum sensing for cognitive radio: State-of-the-art and recent advances. *IEEE Signal Processing Magazine, 29*(3), 101–116. doi:10.1109/MSP.2012.2183771

Cabric, D., Mishra, S. M., & Brodersen, R. W. (2004). Implementation issues in spectrum sensing for cognitive radios. In *Proceedings of Conference Record of the Asilomar Conference on Signals, Systems and Computers (Vol. 1*, pp. 772–776). IEEE. doi:10.1109/ACSSC.2004.1399240

Choi, K. W., Jeon, W. S., & Jeong, D. G. (2009). Sequential detection of cyclostationary signal for cognitive radio systems. *IEEE Transactions on Wireless Communications, 8*(9), 4480–4485. doi:10.1109/TWC.2009.090288

Dandawate, A. V., & Giannakis, G. B. (1994). Statistical tests for presence of cyclostationarity. *IEEE Transactions on Signal Processing, 42*(9), 2355–2369. doi:10.1109/78.317857

Derakhshani, M., Le-Ngoc, T., & Nasiri-Kenari, M. (2011). Efficient cooperative cyclostationary spectrum sensing in cognitive radios at low SNR regimes. *IEEE Transactions on Wireless Communications, 10*(11), 3754–3764. doi:10.1109/TWC.2011.080611.101580

Digham, F. F., Alouini, M.-S., & Simon, M. K. (2007). On the energy detection of unknown signals over fading channels. *IEEE Transactions on Communications, 55*(1), 21–24. doi:10.1109/TCOMM.2006.887483

Feng, H., Wang, Y., & Li, S. L. S. (2008). Statistical test based on finding the optimum lag in cyclic autocorrelation for detecting free bands in cognitive radios. In *Proceedings of International Conference on Cognitive Radio Oriented Wireless Networks and Communications* (pp. 1–6). Academic Press. doi:10.1109/CROWNCOM.2008.4562528

Ganesan, G., & Li, Y. (2007a). Cooperative spectrum sensing in cognitive radio, part I: Two user networks. *IEEE Transactions on Communications, 6*(6), 2204–2213.

Ganesan, G., & Li, Y. (2007b). Cooperative spectrum sensing in cognitive radio, part II: Multiuser networks. *IEEE Transactions on Communications, 6*(6), 2214–2222.

Gardner, W. A. (1986). Measurement of spectral correlation. *IEEE Transactions on Acoustics, Speech, and Signal Processing, 34*(5), 1111–1123. doi:10.1109/TASSP.1986.1164951

Gardner, W. A. (1988). Signal interception: A unifying theoretical framework for feature detection. *IEEE Transactions on Communications, 36*(8), 897–906. doi:10.1109/26.3769

Gardner, W. A., Napolitano, A., & Paura, L. (2006). Cyclostationarity: Half a century of research. *Signal Processing, 86*(4), 639–697. doi:10.1016/j.sigpro.2005.06.016

Gardner, W. A., & Spooner, C. M. (1992). Signal interception: Performance advantages of cyclic-feature detectors. *IEEE Transactions on Communications, 40*(1), 149–159. doi:10.1109/26.126716

Ghasemi, A., & Sousa, E. S. (2005). Collaborative spectrum sensing for opportunistic access in fading environments. In *Proceedings of IEEE International Symposium on New Frontiers in Dynamic Spectrum Access Networks* (pp. 131–136). IEEE. doi:10.1109/DYSPAN.2005.1542627

Goldsmith, A., Jafar, S. A., Maric, I., & Srinivasa, S. (2009). Breaking spectrum gridlock with cognitive radios: An information theoretic perspective. *Proceedings of the IEEE, 97*(5), 894–914. doi:10.1109/JPROC.2009.2015717

Harrison, K., Mishra, S. M., & Sahai, A. (2010). How much white-space capacity is there? In *Proceedings of IEEE Symposium on New Frontiers in Dynamic Spectrum* (pp. 1–10). IEEE. doi:10.1109/DYSPAN.2010.5457914

Haykin, S. (2005). Cognitive radio: Brain-empowered wireless communications. *IEEE Journal on Selected Areas in Communications, 23*(2), 201–220. doi:10.1109/JSAC.2004.839380

Izzo, L., Paura, L., & Tanda, M. (1992). Signal interception in non-Gaussian noise. *IEEE Transactions on Communications, 40*(6), 1030–1037. doi:10.1109/26.142793

Kim, M., Kimtho, P., & Takada, J.-i. (2010). Performance enhancement of cyclostationarity detector by utilizing multiple cyclic frequencies of OFDM signals. In *Proceedings of IEEE Symposium on New Frontiers in Dynamic Spectrum* (pp. 1–8). IEEE. doi:10.1109/DYSPAN.2010.5457876

Lunden, J., Koivunen, V., Huttunen, A., & Poor, H. V. (2007). Spectrum sensing in cognitive radios based on multiple cyclic frequencies. In *Proceedings of International Conference on Cognitive Radio Oriented Wireless Networks and Communications* (pp. 37–43). IEEE. doi:10.1109/CROWNCOM.2007.4549769

Lunden, J., Koivunen, V., Huttunen, A., & Poor, H. V. (2009). Collaborative cyclostationary spectrum sensing for cognitive radio systems. *IEEE Transactions on Signal Processing, 57*(11), 4182–4195. doi:10.1109/TSP.2009.2025152

Ma, J., Zhao, G., & Li, Y. (2008). Soft combination and detection for cooperative spectrum sensing in cognitive radio networks. *IEEE Transactions on Wireless Communications, 7*(11), 4502–4507. doi:10.1109/T-WC.2008.070941

Mishra, S., Sahai, A., & Brodersen, R. (2006). Cooperative sensing among cognitive radios. In *Proceedings of IEEE International Conference on Communications* (pp. 1658–1663). IEEE.

Mitola, J. (2000). *Cognitive radio: An integrated agent architecture for software defined radio*. (Ph.D. dissertation). KTH Royal Inst. Technol.

Poor, H. V. (1994). *An introduction to signal detection and estimation. Book*. Springer-Verlag. doi:10.1007/978-1-4757-2341-0

Quan, Z. Q. Z., Cui, S. C. S., Poor, H., & Sayed, A. (2008). Collaborative wideband sensing for cognitive radios. *IEEE Signal Processing Magazine, 25*(6), 60–73. doi:10.1109/MSP.2008.929296

Sadeghi, H., & Azmi, P. (2008). Cyclostationarity-based cooperative spectrum sensing for cognitive radio networks. In *Proceedings of International Symposium on Telecommunications* (pp. 429–434). Academic Press. doi:10.1109/ISTEL.2008.4651341

Sadeghi, H., Azmi, P., & Arezumand, H. (2011). Optimal multi-cycle cyclostationarity-based spectrum sensing for cognitive radio networks. In *Proceedings of Iranian Conference on Electrical Engineering* (pp. 1–6). IEEE.

Shen, B., Ullah, S., & Kwak, K. (2010). Deflection coefficient maximization criterion based optimal cooperative spectrum sensing. *AEÜ: International Journal of Electronics and Communications, 64*(9), 819–827. doi:10.1016/j.aeue.2009.06.006

Shen, J., & Alsusa, E. (2013). Joint cycle frequencies and lags utilization in cyclostationary feature spectrum sensing. *IEEE Transactions on Signal Processing, 61*(21), 5337–5346. doi:10.1109/TSP.2013.2278810

Sonnenschein, A., & Fishman, P. M. (1992). Radiometric detection of spread-spectrum signals in noise of uncertain power. *IEEE Transactions on Aerospace and Electronic Systems, 28*(3), 654–660. doi:10.1109/7.256287

Sun, C., Zhang, W., & Letaief, K. B. (2007). Cluster-based cooperative spectrum sensing in cognitive radio systems. In *Proceedings of IEEE International Conference on Communications* (pp. 2511–2515). IEEE. doi:10.1109/ICC.2007.415

Sutton, P. D., Nolan, K. E., & Doyle, L. E. (2008). Cyclostationary signatures in practical cognitive radio applications. *IEEE Journal on Selected Areas in Communications, 26*(1), 13–24. doi:10.1109/JSAC.2008.080103

Taherpour, A., Nasiri-Kenari, M., & Jamshidi, A. (2007). Efficient cooperative spectrum sensing in cognitive radio networks. In *Proceedings of IEEE International Symposium on Personal, Indoor and Mobile Radio Communications* (pp. 1–6). IEEE. doi:10.1109/PIMRC.2007.4394424

Taherpour, A., Norouzi, Y., Nasiri-Kenari, M., Jamshidi, A., & Zeinalpour-Yazdi, Z. (2007). Asymptotically optimum detection of primary user in cognitive radio networks. *IET Communications, 1*(6), 1138–1145. doi:10.1049/iet-com:20060645

Tandra, R., & Sahai, A. (2005). Fundamental limits on detection in low SNR under noise uncertainty. In *Proceedings of International Conference on Wireless Networks, Communications and Mobile Computing* (*Vol. 1*, pp. 464–469). IEEE. doi:10.1109/WIRLES.2005.1549453

Unnikrishnan, J., & Veeravalli, V. V. (2008). Cooperative sensing for primary detection in cognitive radio. *IEEE Journal of Selected Topics in Signal Processing, 2*(1), 18–27. doi:10.1109/JSTSP.2007.914880

Van de Beek, J., Riihija, J., Achtzehn, A., & Mahonen, P. (2012). TV white space in Europe. *IEEE Transactions on Mobile Computing, 11*(2), 178–188. doi:10.1109/TMC.2011.203

Xiangzhen, L., & Xuping, Z. (2008). Feature Detection Based on Multiple Cyclic Frequencies in Cognitive Radios. In *Proceedings of China-Japan Joint Microwave Conference* (pp. 290 – 293). IEEE. doi:10.1109/CJMW.2008.4772428

Yeung, G. K., & Gardner, W. A. (1996). Search-efficient methods of detection of cyclostationary signals. *IEEE Transactions on Signal Processing, 44*(5), 1214–1223. doi:10.1109/78.502333

Yucek, T., & Arslan, H. (2009). A survey of spectrum sensing algorithms for cognitive radio applications. *IEEE Communications Surveys and Tutorials, 11*(1), 116–130. doi:10.1109/SURV.2009.090109

Zhao, Q., & Sadler, B. M. (2007). A survey of dynamic spectrum access: Signal processing, networking, and regulatory Policy. *IEEE Signal Processing Magazine, 24*(3), 79–89. doi:10.1109/MSP.2007.361604

ADDITIONAL READING

Bhargavi, D., & Murthy, C. R. (2010). Performance comparison of energy, matched-filter and cyclo-stationarity-based spectrum sensing. In *IEEE Workshop on Signal Processing Advances in Wireless Communications (SPAWC)*. (pp. 1-5). IEEE. doi:10.1109/SPAWC.2010.5670882

Cabric, D., & Brodersen, R. W. (2005). Physical layer design issues unique to cognitive radio systems. In *IEEE Symposium on Personal, Indoor and Mobile Radio Communications*. (Vol. 2, pp. 759 – 763). IEEE. doi:10.1109/PIMRC.2005.1651545

Chen, H. S., Gao, W., & Daut, D. G. (2007). Spectrum sensing using cyclostationary properties and application to IEEE 802.22 WRAN. In *Global Telecommunications Conference*. (pp. 3133-3138). IEEE. doi:10.1109/GLOCOM.2007.593

Fehske, A., Gaeddert, J., & Reed, J. (2005). A new approach to signal classification using spectral correlation and neural networks. In *IEEE Symposium on New Frontiers in Dynamic Spectrum Access Networks*. (pp. 144 – 150). IEEE. doi:10.1109/DYSPAN.2005.1542629

Ghozzi, M., Marx, F., Dohler, M., & Palicot, J. (2006). Cyclostatilonarilty-based test for detection of vacant frequency bands. In *Conference on Cognitive Radio Oriented Wireless Networks and Communications*, 2006. (pp. 1 – 5). IEEE. doi:10.1109/CROWNCOM.2006.363454

Han, N., Shon, S., Chung, J. H., & Kim, J. M. (2006). Spectral correlation based signal detection method for spectrum sensing in IEEE 802.22 WRAN systems. In *Conference on Advanced Communication Technology*. (Vol. 3, pp. 6 – pp). IEEE.

Haykin, S., Thomson, D. J., & Reed, J. H. (2009). Spectrum sensing for cognitive radio. *Proceedings of the IEEE*, *97*(5), 849–877. doi:10.1109/JPROC.2009.2015711

Khambekar, N., Dong, L., & Chaudhary, V. (2007). Utilizing OFDM guard interval for spectrum sensing. In *IEEE Wireless Communications and Networking Conference*. (pp. 38 – 42). IEEE. doi:10.1109/WCNC.2007.13

Kim, K., Akbar, I. A., Bae, K. K., Um, J. S., Spooner, C. M., & Reed, J. H. (2007). Cyclostationary approaches to signal detection and classification in cognitive radio. In *IEEE symposium on New Frontiers in Dynamic Spectrum Access Networks* (pp. 212–215). IEEE. doi:10.1109/DYSPAN.2007.35

Lundén, J., Kassam, S. A., & Koivunen, V. (2010). Robust nonparametric cyclic correlation-based spectrum sensing for cognitive radio. *Signal Processing. IEEE Transactions on*, *58*(1), 38–52. doi:10.1109/TSP.2009.2029790

Oner, M., & Jondral, F. (2004). Cyclostationarity based air interface recognition for software radio systems. In *IEEE Radio and Wireless Conference*. (pp. 263 – 266). IEEE. doi:10.1109/RAWCON.2004.1389125

Oner, M., & Jondral, F. (2004). Cyclostationarity-based methods for the extraction of the channel allocation information in a spectrum pooling system. In *IEEE Radio and Wireless Conference*. (pp. 279 – 282). IEEE. doi:10.1109/RAWCON.2004.1389129

Qihang, P., Kun, Z., Jun, W., & Shaoqian, L. (2006). A distributed spectrum sensing scheme based on credibility and evidence theory in cognitive radio context. In *international symposium on Personal, indoor and mobile radio communications*. (pp. 1 – 5). IEEE. doi:10.1109/PIMRC.2006.254365

Shankar, N. S., Cordeiro, C., & Challapali, K. (2005). Spectrum agile radios: utilization and sensing architectures. In *IEEE Symposium on New Frontiers in Dynamic Spectrum*. (pp. 160 – 169). IEEE.

Tian, Z., Tafesse, Y., & Sadler, B. M. (2012). Cyclic feature detection with sub-Nyquist sampling for wideband spectrum sensing. *Selected topics in signal processing. IEEE Journal of*, 6(1), 58–69.

Ye, Z., Grosspietsch, J., & Memik, G. (2007). Spectrum sensing using cyclostationary spectrum density for cognitive radios. In *IEEE Workshop on Signal Processing Systems*. (pp. 1-6). IEEE. doi:10.1109/SIPS.2007.4387507

KEY TERMS AND DEFINITIONS

Cognitive Radio: A cognitive radio is an intelligent and reconfigurable wireless communication system that enables monitoring the radio environment, learning, and accordingly, adapting transmission parameters in order to achieve the optimal spectrum utilization.

Cooperative Spectrum Sensing: Cooperative sensing is a solution to enhance the detection performance, in which secondary users collaborate with each other to sense the spectrum to find the spectrum holes.

Cyclostationary Detection: Cyclostationary detection is one of the most effective approaches for spectrum sensing, which uses cyclostaionary feature to recognize the cyclostationary modulated signal in a background of stationary noise even at low SNR regimes.

Deflection Coefficient: Deflection coefficient of a detector is a measure to characterize the detection performance. The larger deflection coefficient indicates the easier differentiation between two hypotheses, and thus the better the detection performance.

Multi-Cycle Cyclostationary Detector: Multi-cycle detector is a cyclostationary detector, in which the test statistic is the sum of the received signal powers at all cycle-frequencies.

Single-Cycle Cyclostationary Detector: Single-cycle detector is a cyclostationary detector which measures the signal power only for one specific cycle-frequency.

Spectrum Sensing: Spectrum sensing is the process of periodically monitoring a specific frequency band, aiming to identify presence or absence of primary users.

Chapter 6
A Collaborative Approach for Compressive Spectrum Sensing

Ahmed M. Elzanati
Sinai University, Egypt

Mohamed F. Abdelkader
Port Said University, Egypt

Karim G. Seddik
American University in Cairo, Egypt

ABSTRACT

Compressive Sensing (CS) has been proven effective to elevate some of the problems associated with spectrum sensing in wideband Cognitive Radio (CR) networks through efficient sampling and exploiting the underlying sparse structure of the measured frequency spectrum. In this chapter, the authors discuss the motivation and challenges of utilizing collaborative approaches for compressive spectrum sensing. They survey the different approaches and the key published results in this domain. The authors present in detail an approach that utilizes Kronecker sparsifying bases to exploit the two-dimensional sparse structure in the measured spectrum at different, spatially separated cognitive radios. Simulation results show that the presented scheme can substantially reduce the Mean Square Error (MSE) of the recovered power spectrum density over conventional schemes while maintaining the use of a low-rate Sub-Nyquist Analog to Information Converter. It is also shows that one can achieve dramatically lower MSE under low compression ratios using a dense measurement matrix while using Nyquist rate ADC.

INTRODUCTION: CONCEPTS AND CHALLENGES

CR technology is gaining more ground each day as a promising technology to mitigate the limited availability of radio spectrum resources. Spectrum Sensing is considered one of the most important tools in opportunistic CR systems, where the secondary user should efficiently detect the signals from PUs, so that it does not cause harmful interference to them.

In spite of the tremendous evolution that occurs in CR related technologies, spectrum sensing in wideband CR networks remains one of the most challenging issues facing the widespread of this technol-

DOI: 10.4018/978-1-4666-6571-2.ch006

ogy. The CRs need to sense the wide-band spectrum in order to detect the unoccupied channels available for opportunistic use. For example, digital TV signals with a power above the threshold of −116 dBm should be detected with a probability of at least 0.9 and with 0.1 maximum probability of a false alarm. Several issues that challenge spectrum sensing performance could be summarized as follows;

- **Hardware Requirements:** One of the hardware challenges relates to the need to sample the signal at a very high sampling rate especially in wide-band networks. As, according to Shannon theorem, the sampling rate must be at least twice the highest frequency component of the signal to avoid aliasing, which is called the Nyquist sampling rate. This requires expensive, complicated Analog to Digital Converter (ADC) with high resolution, wide dynamic range, and low power consumption. The current technology forms an obstacle to design such a high sampling rate with wide dynamic range ADC (Yucek & Arslan, 2009). Novel approaches are required to simultaneously sense wide-band multiple channels using a limited number of RF interfaces.
- **Noise Uncertainty:** Energy detection based spectrum sensing requires a precise estimation for the environmental noise statistics, which are used to obtain an accurate calculation of the threshold needed to differentiate between free and busy channels. The main problem is that such a precise estimation of noise is not always attainable. In general, energy detection performance deteriorates as noise uncertainty increases. This behavior is continued till the performance decreases significantly and the system fails at specific SNR no matter the number of measurements. This phenomenon is called SNR wall.
- **Hidden Primary User:** The Hidden Primary User problem arises when the cognitive radio device fails to detect the PU transmitted signal, causing unwanted interference at the primary user receiver. This problem occurs due to many factors, such as the relative locations of devices and severe multipath fading and shadowing. Cooperative sensing is proposed as a solution for handling this problem, where exploiting spatial diversity among several collaborating CRs had proven to be an effective method to improve the detection performance, but at the expense of cooperation overhead on the network resources. Researches try to reduce the amount of the overload data. Various Fusion techniques such as soft decision methods or hard decision methods can be used for combining information from different cognitive radios.

The enormous technological progress in digital signal processing enhanced by the engagement the wireless communications and signal processing has paved the way for mitigating some of these issues. Specifically, CS has been proposed as an effective technique to alleviate some of these problems through efficient sampling that exploits the underlying sparse structure of the measured frequency spectrum.

The hardware requirements issue for example can be mitigated through CS based spectrum sensing approaches; the signal is captured using low speed Analog to Information Converter (AIC) rather than conventional high speed Nyquist ADC. The AIC is known for its ability to capture the signal using its information rate rather than its symbol rate. A practical realization of it was given in (Kirolos, Ragheb, Laska, Duarte, Massoud, & Baraniuk, 2006) where; the analog signal is multiplied by a pseudo-random maximal length PN sequence of {1,-1}. The rate of change of this PN sequence is higher than the Nyquist rate. The signal is integrated then sampled at its information rate using low speed ADC. The only problem with this scheme is that it is subject to model mismatches and non-linearity inherent in pseudo random generator, multiplier and integrator.

In this chapter, we investigate the use of CS approaches to reduce both the sampling rate and the cooperation overhead problem while using collaborative spectrum sensing. This can be achieved by efficient compression that significantly reduces the amount of data sent from CRs to FC and vice versa. Therefore, the overhead data in cooperative schemes can be diminished, allowing more frequency spectrum for opportunistic use.

A comprehensive survey on compressive spectrum sensing is presented. Particularly, the major research approaches in this area are presented and compared to each other in terms of their complexity, efficiency, and underlying assumptions. Then, an approach for collaborative compressive spectrum sensing is presented. The presented approach achieves improved sensing performance through utilizing Kronecker sparsifying bases to exploit the two dimensional sparse structure in the measured spectrum at different, spatially separated cognitive radios. A recovery approach that utilizes Kronecker spectrum sensing is built to recover the spectrum with a fewer number of samples and improved performance under noise.

The remainder of this chapter is organized as follows: First, an overview of different trends in Compressive Sensing (CS) is illustrated, followed by a literature review on CSS approaches. Then a Kronecker approach for compressive collaborative spectrum sensing is presented. Result section presents the simulation results that show the effectiveness of the presented compressed sensing approach.

PRINCIPLES OF COMPRESSIVE SENSING

Traditional approach of signal acquisition follows Shannon Nyquist theorem, which states that the signal should be captured using analog to digital converter with sampling rate higher than or equal Nyquist rate. In order to compress the signal for storage or transmission purposes, we first capture the signal at the Nyquist rate, and then compress it. The question that arises, can we directly capture the signal in a compressed representation?

CS Data Acquisition and Representation

CS attempts to provide an answer to that question. It is an acquisition technique in which the signal can be recovered from fewer number of measurements compared to the number of measurements needed by conventional sampling method. We also do not have to sample the signal then compress it as we utilize the sparsity if the signal to capture only the important information directly.

In order to perfectly recover the signal, there are two major conditions that should be satisfied. The first condition is related to the signal itself. The signal should be sparse in some sparsifying basis. Sparse signals are those signals with a small number of non-zero coefficients compared to their dimension. The information rate in these signals is much less than its bandwidth. Therefore, CS techniques can capture those signals with a sampling rate that is higher than its information rate and much lower than its Nyquist sampling rate. The second important condition is related to the sampling system. Mainly, the measurement matrix used to capture this important information from the signal must be incoherent with the sparsifying basis (Donoho, 2006).

Assume a signal $\mathbf{x} \in \mathbb{R}^N$ that has a sparse representation in some domain Ψ such as Fourier Domain, Wavelet, or Discrete Cosine Transform (DCT), such that

$$\mathbf{x} = \mathbf{\Psi}\mathbf{s}, \tag{1.1}$$

where the N dimensional vector $\mathbf{s}$ is a sparse and compressible vector. It is said that $\mathbf{s}$ is K sparse, if it has at most K non-zero entries where $K = \|\mathbf{s}\|_0$ is the $L0$ norm of the sparse vector $\mathbf{s}$, and $K << N$.

The signal $\mathbf{x}$ can be acquired through $M \geqslant cKlog(N \,/\, K)$ linear measurements of $\mathbf{x}$ where $M < N$, and c is small constant that depends on the recovery algorithm. The measurement vector $\mathbf{y} \in \mathbb{R}^M$ can be written as:

$$\mathbf{y} = \mathbf{\Phi}\mathbf{x} = \mathbf{\Phi}\mathbf{\Psi}\mathbf{s} = \mathbf{\Theta}\mathbf{s}, \tag{1.2}$$

where $\mathbf{\Phi}$ is $M \times N$ measurement matrix.

The measurement matrix $\mathbf{\Phi}$ must preserve the values of non-zero components in $\mathbf{s}$, Therefore; it should be carefully designed so that $\mathbf{\Theta}$ satisfies the following condition

$$1 - \epsilon \leqslant \frac{\|\mathbf{\Theta}\mathbf{v}\|_2}{\|\mathbf{v}\|_2} \leqslant 1 + \epsilon \tag{1.3}$$

where $\mathbf{v}$ is a $3K$ sparse vector. This condition is referred to as Restricted Isometry Property (RIP). There is another related important condition that must be taken into consideration while designing $\mathbf{\Phi}$, which is the incoherency. In this condition, the basis $\mathbf{\Psi}$ must not lie in the null space of $\mathbf{\Phi}$. Thus, the rows of $\mathbf{\Phi}$ do not sparsify the columns of $\mathbf{\Psi}$. An indicator of coherency between both the measurement and the sparsifying bases matrices is given by

$$\mu(\mathbf{\Phi}, \mathbf{\Psi}) = \sqrt{N} \max_{(1 \leq i \leq M, 1 \leq j \leq N)} |\langle \phi_i \psi_j \rangle|, \tag{1.4}$$

where $\mu(\mathbf{\Phi}, \mathbf{\Psi}) \in [1, \sqrt{N}]$ indicates the maximal correlation between both measurement and sparsifying matrices elements. Many measurements matrices satisfy both the conditions of RIP and Incoherency such as:

- Gaussian measurement matrix: The elements of this matrix are independent and identically distributed (i.i.d) from a Gaussian probability density function with zero mean and $1\,/\,M$ variance (Baraniuk, Compressive Sensing [Lecture Notes], 2007).
- Bernoulli measurement matrix: The elements of this matrix are i.i.d with symmetric Bernoulli distribution.

CS Recovery Algorithms

In the traditional linear algebra theorem, the system presented by Equation (1.2) is considered as an under-determined system, where the dimension of the measurement vector M is less than the signal dimension N. Therefore, there is an infinite number of solutions to such system. On the other hand, under the sparsity constraint on the recovered signal, exact and unique solution can be retrieved.

The signal could be recovered exactly by solving the following ℓ_0 minimization program, where the ℓ_0 norm counts the number of non-zero coefficient, so it is optimum to recover the sparse signal captured by the compressive sensing paradigm. Furthermore, perfect signal reconstruction could be achieved from just $K+1$ linear measurement by solving the following optimization problem

$$\hat{\mathbf{s}} = \arg\min \|\mathbf{s}\|_0 \quad \text{such that} \quad \mathbf{\Theta s} = \mathbf{y}. \tag{1.5}$$

Unfortunately, the problem with this reconstruction method is that ℓ_0 norm minimization is an NP-Hard (Non-deterministic Polynomial time Hard) problems. Hence, the ℓ_0 algorithms exhibit high computational complexity and instability.

Another simple approximation of the solution can be recovered by solving the following ℓ_2 norm minimization problem.

$$\hat{\mathbf{s}} = \arg\min \|\mathbf{s}\|_2 \quad \text{such that} \quad \mathbf{\Theta s} = \mathbf{y}. \tag{1.6}$$

This estimates vector $\hat{s}$ with the smallest ℓ_2 energy. There is also a closed form solution to this program as $\hat{\mathbf{s}} = \mathbf{\Theta}^T(\mathbf{\Theta\Theta}^T)^{-1}\mathbf{y}$. Although ℓ_2 norm involves the lowest computational complexity, there is no guarantee to exactly recover the signal as it does not recover the sparsest solution.

Another reconstruction method is ℓ_1 norm minimization. It was shown that by solving the ℓ_1 minimization program, we could recover the signal perfectly using only $M = cK\log(N/K)$ Gaussian linear measurements. This could be achieved by solving the following program:

$$\hat{s} = \arg\min \|s\|_1 \quad \text{such that} \quad \mathbf{\Theta} s = y \tag{1.7}$$

This minimization problem is a convex program and could be solved efficiently using different approaches such as Basis Pursuit (BP) and Matching Pursuit (MP) (Mallat & Zhifeng, 1993).

Recent Trends in CS

Compressive sensing paradigm is featured by its non-adaptive behavior, where it does not depend on the measured signal as long as it is sparse. However, exploiting the structure of the signal to increase the performance of compressed signals recovery and recover the signal with fewer numbers of the measurements was the subject of a lot of work. This work could be classified as;

- **Model based CS:** The model based CS recovery algorithms exploit additional prior structure for the sparse signals (Baraniuk, Cevher, Duarte, & Hegde, 2010). There are different types of these algorithms depending on the structure of the signal such as tree structure and block sparsity algorithms.

- **Bayesian CS:** The Bayesian CS recovery algorithms are based on Bayesian estimation. Prior information about the signal is assumed like the probability distribution of the recovered signal, which tends to be Bernoulli distributed in sparse signals (Baron, Sarvotham, & Baraniuk, 2010).

- **Adaptive Compressive Sensing:** While traditional compressive sensing is basically superior by being non-adaptive recovery method, adaptive CS recovery algorithms can significantly enhance performance at the expense of complexity involved in the adaptive acquisition of the signals (Haupt, Baraniuk, Castro, & Nowak, 2012).

- **Dynamic Compressive Sensing:** Dynamic CS algorithms are effective for sparse signals that possess a slow varying pattern (support and values). Numerous algorithms have exploited this feature. The simplest intuitive dynamic algorithm relies on recovering only the difference between the current and previous signals (CS-Diff). If the signal is slowly varying in both support and values, the number of non-zeros coefficients in the difference will decrease causing the number of required measurements needed for perfect reconstruction to decrease.

- **Distributed Compressive Sensing (DCS):** DCS is a distributed coding algorithm for acquiring and recovering multiple signals which share the same sparsity order and locations of non-zero components simultaneously. The DCS theory relies on a new notion that exploits the joint sparsity of a signal ensemble. The multiple measurements matrix can be represented by the following equation

$$\mathbf{Y}_2 = \mathbf{\Phi}\mathbf{X}_2 = \mathbf{\Phi}\mathbf{\Psi}\mathbf{S}_2, \tag{1.8}$$

where $\mathbf{Y}_2$ is an $M \times J$ matrix, J is the number of signals in the ensemble, and $\mathbf{S}_2$ is $N \times J$ sparse matrix with a common sparse support and different coefficient values.

We can recover the signals by using the simultaneous orthogonal matching pursuit (SOMP) algorithm presented in (Duarte, Sarvotham, Baron, Wakin, & Baraniuk, 2005). DCS-SOMP is an efficient greedy algorithm for joint signal recovery based on the SOMP algorithm for simultaneous sparse approximation with only K measurements not M like the conventional ℓ_1 algorithm.

- **Kronecker Compressive Sensing (KCS):** In contrast to DCS, which exploits joint sparsity in one dimension, KCS exploits the structure of a multidimensional signal in every dimension (Duarte & Baraniuk, Kronecker Compressive Sensing, 2012). Kronecker product bases are suitable for CS applications concerning multidimensional signals. These bases can be used both to obtain sparse or compressible representations of many real signals. KCS depends on the concept that every dimension has its own sparsifying basis, so we can jointly apply these sparsifying bases by getting the Kronecker product of all sparsifying matrices.

We assume that a three dimensional signal is represented by a 3-D matrix $\mathbf{X}_3$, where $\mathbf{X}_3 \in \mathbb{R}^{N1 \times N2 \times N3}$. This signal can be captured using the Kronecker product measurement matrix $\mathbf{\Phi}_k = \mathbf{\Phi}_{k1} \otimes \mathbf{\Phi}_{k2} \otimes \mathbf{\Phi}_{k3}$, where $\mathbf{\Phi}_{k1}, \mathbf{\Phi}_{k2}$, and $\mathbf{\Phi}_{k3}$ are the measurement matrices that operate individually on portions of the multidimensional signal and $\otimes$ denotes the Kronecker product. The measurement vector $\mathbf{y}_3$ can be written as

$$y_3 = \Phi_k \bar{x}_3 = \Phi_k \Psi_k \bar{s}_3 = \Theta \bar{s}_3. \tag{1.9}$$

The sparsifying basic $\Psi_k = \Psi_{k1} \otimes \Psi_{k2} \otimes \Psi_{k3}$ is the Kronecker product of all individual bases, where Ψ_{k1}, Ψ_{k2}, and Ψ_{k3} are the sparsifying bases for the first, second, and third dimensions respectively, and $\bar{x}_3$, $\bar{s}_3$ are the column vector-reshaped representation of the matrix X_3 and the sparse coefficient matrix S_3, respectively. We may then recover $\bar{s}_3$ by solving ℓ_1 minimization program using Basis Pursuit.

COMPRESSIVE SPECTRUM SENSING

There are many surveys covering different aspects of wideband spectrum sensing in cognitive radio networks such as the work in (Hongjian Sun, Nallanathan, Cheng-Xiang Wang, & Yunfei Chen, 2013). Sun *et. al* classified the work in wideband spectrum sensing into two categories based on Nyquist sampling and sub-Nyquist sampling.

At Nyquist sampling, the signal is sampled at Nyquist rate such as the approaches in wavelet detection (Zhi Tian & Giannakis, 2007) and filter bank sampling (Farhang-Boroujeny, 2008). The former scheme still has the problem of requiring high speed ADC, while in the later scheme some constrains put on the ADC are relaxed on the expense of requiring large number of high frequency radio components.

Therefore, the need for Sub-Nyquist solutions has aroused. There are two major frameworks developed in parallel; sub-Nyquist multi-channel based spectrum sensing approaches and Compressive sensing based spectrum sensing approaches. In sub-Nyquist Multi-channel Spectrum Sensing (MSS), multi sub-nyquist ADCs are used. The authors in (Mishali & Eldar, 2009) presented Multicast and Modulated Wideband Converter (MWC) approaches. These schemes tend to mitigate the effect of modeling mismatch mentioned in AIC mentioned before.

In the rest of this section, we present a literature review of some of the work in CS based wideband spectrum sensing approaches. We provide a novel classification of the research in Compressive Spectrum Sensing (CSS) based on; sparsity bases selection, cooperation schemes, recovery algorithms, and dynamic compressive spectrum sensing. The latest trends in CSS are illustrated in Figure 1.

Sparsity Bases Selection

The sparsity of the spectrum determines how efficiently the signal can be recovered. In the spectrum sensing problem, the spectrum is usually under-utilized, so the signal is highly sparse in the frequency domain. The exact choice of the sparsity basis classifies the CCS approaches into four main sets:

- Sparsity in the PSD edges: one of the first attempts to use compressive spectrum sensing was (Zhi Tian & Giannakis, 2007). In this work, the signal is captured using low rate AIC, coarsely recovered by CS recovering algorithm, then the edges of the spectrum were extracted by wavelet based edge detector. Finally, a fine estimation of the spectrum is obtained by averaging the recovered spectrum over the spectrum edges. The main underlying assumptions are that, the spectrum

Figure 1. Classification of compressive spectrum sensing schemes

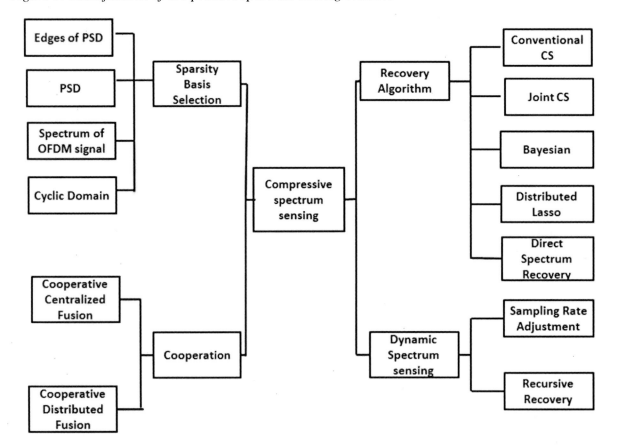

is under-utilized and the PSD of the measured wide-band signal is smooth within each sub-band, but exhibits a discontinuous change between adjacent sub-bands. Hence, there is small number of edges in the frequency domain.

- In practice, this approach requires the calculation of PSD from the signal autocorrelation function rather estimating the spectrum. Hence the signal has to be sampled at the Nyquist rate then its autocorrelation function is calculated. A following approach was proposed in (Polo, Ying Wang, andharipande, & Leus, 2009), where the wideband analog signal is directly captured by analog to information converter, by proposing a matrix which relates the autocorrelation functions of both the compressed and uncompressed signals solving the bottleneck in the sampling rate presented in the previous scheme.

- Sparsity in the PSD: Instead of representing the spectrum with the edges of PSD, The authors of (Bazerque & Giannakis, 2010) exploited sparsity in both the PSD and the locations of PUs to construct PSD map in both space and frequency. They provided a salient model for the power spectrum sensing for each PU as a combination of non-overlapping rectangular basis of unit height or any suitable basis like Gaussian bell-shaped with different coefficients. These coefficients represented the power of this sub-band for specific PU. This leads again to piecewise constant modeling. The Weighted Moving Average Periodogram (WMAP) is used to estimate the PSD. The reconstructed PSD map is useful especially in wide area ad-hoc networks. Hence, the CRs can adjust its transmitting power to avoid interfering with the PUs signals. Unfortunately, the signal must be captured using high sampling rate ADC.

- Another promising approach, which combines spectrum sensing and primary user localization, was proposed in (Li, l, Han, Chakravarthy, & Wu, 2010). This was achieved by combining the received signals from all possible discrete PUs locations then these measurements are combined and stacked in a vector at the FC. Finally, compressive sensing recovery algorithm was applied to recover the information needed to get the power and the location of the PUs. A following robust approach in (Hu, et al., 2012) used weighted centroid method rather than discretization step in (Li, l, Han, Chakravarthy, & Wu, 2010) to identify the location of PU, and non-negative matrix factorization was proposed as a dimension reduction method.

- Sparsity in the spectrum of OFDM signals: unlike the aforementioned method, the framework presented in (Zeng, Li, & Tian, 2011) adopts sparsity in the instantaneous spectrum not PSD to alleviate one of the limitations of the PSD calculation using Periodogram. The signal is sampled at time domain and assumed to be an OFDM signal with predefined channels locations and unknown power spectrum density level for the primary user (PU) in each channel. The spectrum is underutilized with few active channels. Hence, the signal is sparse in the frequency domain.

- Cyclic spectrum domain: The cyclostationary feature detection is a method for detecting primary user transmissions by exploiting the cyclostationary feature of the received signal. This feature detection alleviates the limitations in the energy detector schemes such as; noise uncertainty especially at low SNR and inability to differentiate different types of PU signals. Tian et al. in (Tian, Tafesse, & Sadler, Cyclic feature detection with sub-Nyquist sampling for wideband spectrum sensing, 2012) introduced CS in Cyclostationary spectrum detection, as they have tried to alleviate the sampling requirements of cyclic detectors by utilizing the compressive sampling principle and exploiting the unique sparsity structure in the two-dimensional cyclic spectrum domain.

In general, the signal in the cyclic domain is sparser than it is in the Fourier basis even when the spectrum is not underutilized. Hence this scheme allowed significant reduction in sampling rate compared to Energy detection schemes Furthermore; it can differentiate between the PUs signals and the noise, so it is more robust to noise.

Cooperation

Cooperation is one of the proposed solutions to handle the problem of Hidden Primary User by utilizing spatial diversity gain at the expense of increased overload data on the CR network. Applying compressive sensing in cooperative spectrum sensing can reduce that overload on the network while addressing the ADC rate limitations and hidden primary user problems. In cooperative spectrum sensing, each CR senses the spectrum compressed then sends either its compressed measurements, recovered spectrum or final occupancy decision vector to the Fusion Center (FC) or to the neighboring CRs. These three fusion techniques are known as measurement fusion, data fusion and decision fusion, respectively. Cooperative schemes could be classified into two major subsets; Cooperative centralized fusion techniques and Cooperative distributed fusion techniques.

In centralized fusion cooperative techniques, CRs share information with the FC, which in turn takes the final decision and sends it to the CRs via control channels. Such as the work in (Wang, Pandharipande,

Polo, & Leus, 2009), which adopted a centralized measurement fusion technique, and developed a joint recovery algorithm. This algorithm made use of the common spectrum support sensed at different CRs. This scheme evinced a performance gain due to diversity over the single CR scheme presented in (Polo, Ying Wang, andharipande, & Leus, 2009). The work in (Wang, Tian, & Feng, Cooperative spectrum sensing based on matrix rank minimization, 2011) proposed an approach where a tradeoff between channel diversity and sampling cost is applied.

On the other hand, the CRs in distributed fusion cooperative techniques share information among each other and reach consensus on their own decisions without the aid of any FC. Distributed Sensing is more reliable than centralized sensing, as there is does not need a backbone infrastructure, it offers reduced cost, and more robustness to node failure.

The authors in (Bazerque & Giannakis, 2010) introduced a distributed collaborative CSS approach where each CR shares its measured spectrum with the neighboring CRs within a single hop. Each CR reaches a global decision by recovering the common spectrum using its local data and using all the measurements from other CRs in the secondary network as extra constraints. In contrast, (Zeng, Li, & Tian, 2011) presented an iterative distributed collaborative CSS approach. In this approach, the signal is captured compressed in the time domain rather than in the autocorrelation function where each CR reaches the consensus decision by exchanging the low power recovered spectrum or decision vector between the neighboring CRs within a single hop. Different from the model in (Bazerque & Giannakis, 2010), the CR recovered the spectrum by using its local measurements besides, the recovered spectrum from just its neighbor within a single hop not from the entire network, which reduces the computational cost.

Compressive Spectrum Sensing in Dynamic Environments

The dynamic nature of spectrum sensing in cognitive radio network has motivated researchers to explore the dynamic behavior of the sparse signal over time. The motivation was to provide more efficient compressive approaches through tracking the dynamic behavior of the spectrum signals. The work in this direction is split into two main directions, dynamic sampling rate adjustment and dynamic recursive recovery.

Dynamic Sampling Rate Adjustment: These approaches estimate the needed number of compressed measurements at each instant of time through efficient tracking of the sparsity order of the spectrum signal.. In (Wang, Tian, & Feng, A two-step compressed spectrum sensing scheme for wideband cognitive radios, 2010) the overall sampling rate is minimized adaptively, where a two-step compressed spectrum sensing (TS-CSS) scheme for efficient wideband sensing is applied. The first step quickly estimates the actual sparsity order of the wide spectrum of interest using a small number of samples. The second step adjusts the total number of samples collected according to the estimated signal sparsity order. In contrast, the work in (Li, Hong, Han, & Wu, 2011) proposed a probabilistic model for obtaining the minimum number of samples and, they made use of error bars recovered from Bayesian compressive sensing recovery algorithm to judge the suitability of the current sampling rate. Although this framework of dynamic sampling rate adjustment adapts the sampling rate efficiently to decrease the sensing resources, it doesn't utilize the temporal correlation between different instants for recursively decreasing the sparsity order of the recovered vector.

Dynamic Recursive Recovery: In these algorithms, the slow time varying pattern of spectrum signal is exploited. The work in (Yin, Wen, Li, Meng, & Han, 2011) exploited the dynamic behavior of the spectrum in the framework of collaborative compressive spectrum sensing. An approach for recovering only the recent change of the spectrum using least square based method was proposed. Although this algorithm has low complexity compared to BP or greedy algorithms, it fails if more than one change in the spectrum occurs simultaneously. This was extended in (Liu, Han, Wu, & Qian, 2011) where two types of sparsity are utilized, sparsity in locations of primary users and in the spectrum. It also utilized the sparsity in the locations of the primary users to refine the search space in the dynamic spectrum sensing using least square.

Recovery Algorithm

Several recovery algorithms can be used for spectrum recovery from the compressed measurements. These algorithms offer different tradeoffs in recovery complexity, performance, and robustness to noise, besides the allowable compression ratios. Some of the recovery techniques are easy to implement, but require a large number of samples in order to reach a desired performance level. Energy saving in computation is offset by the sampling cost. Recovery algorithms also differ from each other based on the required data to recover the signal like sparsity order, probability distribution, the number of the clusters, and so on. Different recovery algorithms were used for compressive spectrum sensing as mentioned below.

- **Conventional CS Recovery Algorithms:** the spectrum could be recovered using traditional ℓ_1 minimization CS recovery algorithms, such as linear programming and Orthogonal Matching Pursuit which is greedy search algorithm, that is not as complex as ℓ_1 but it needs more measurements to get the same MSE performance ℓ_1 norm based reconstruction algorithms.
- **Joint Compressive Sensing (JCS):** In collaborative spectrum sensing a variety of joint recovery algorithms were proposed. These algorithms utilize the fact that the sensed spectrums at adjacent CRs share common sparse support, like the proposed algorithms in (Wang, Pandharipande, Polo, & Leus, 2009).
- **Distributed LASSO:** It is an online version of LASSO where, the signal is recovered by distributed processing at CRs not at FC (Zeng, Li, & Tian, 2011).
- **Bayesian CS:** The spectrum holes can be detected using Bayesian CS. Frequency domain sparsity if used as prior information about the signal as in (Havary-Nassab, Hassan, & Valaee, 2010). BCS algorithm inspired by Relevance Vector Machine (RVM) is presented in (Huang, Wu, & Wang, 2010), where the signal parameters such as the existence of primary user, the carrier frequency, and the power could be estimated without recovering the entire signal. In addition, an extension to this work is provided in (Hong, 2010) where a multi-resolution wavelet Bayesian recovery algorithm is implemented. This algorithm is superior due to its low complexity. Moreover, the performance of Bayesian CS could be enhanced when more prior information is available about the sensed spectrum. Meanwhile, the error bars generated during recovery could be used to determine if any further measurements are required.
- **Algebraic Detector:** Unlike previous methods (except the Bayesian based approaches) that recover the signal then detect the holes, (Guibene, Moussavinik, & Hayar, 2011) Detect the location of holes without recovering the signal just from the compressed vector under certain linearity condition on measurement matrix using algebraic detector.

A comparison among some of the key mentioned approaches is provided in Table 1 and Table 2 to get an overview of each approach with its merits and limitations. In conclusion, compressive sensing proved to be an effective tool to solve the issues related to wideband spectrum sensing such as high speed ADC and high overload data in cooperative CR network. Unfortunately, the complexity of CS algorithms limits the practical implementation of compressive spectrum sensing, but the rapid development in efficient CS algorithms and software-defined radios (SDR) could mitigate these limitations.

Table 1.Comparsion between diffferent compressive spectrum sensing approaches

Approach	Objective	Sparsity	Cooperation	Fusion Techniques
(Zhi Tian & Giannakis, 2007)	Coarse Spectrum estimation	Edges of PSD	Local	X
(Wang, Pandharipande, Polo, & Leus, 2009)	Spectrum estimation, detection	Edges of PSD	Centralized	Measurement Fusion
(Bazerque & Giannakis, 2010)	Spectrum estimation and PU localization	PSD and locations of PU	Distributed	Data Fusion
(Li, I, Han, Chakravarthy, & Wu, 2010)	Spectrum estimation and PU localization	PSD and locations of PU	Centralized	Measurement Fusion
(Hu, et al., 2012)	Spectrum estimation and PU localization	PSD and locations of PU	Centralized	Measurement Fusion
(Zhi Tian & Giannakis, 2007)	Coarse Spectrum estimation	Edges of PSD	Local	X

Table 2.Comparsion between diffferent compressive spectrum sensing approaches (continued)

Approach	Recovery Algorithm	Sampling Rate	Hardware Requirements	Computational complexity	Dynamic
(Zhi Tian & Giannakis, 2007)	OMP	Nyquist	Moderate	low	no
(Wang, Pandharipande, Polo, & Leus, 2009)	DOMP	Sub-Nyquist	Moderate	Moderate	no
(Bazerque & Giannakis, 2010)	D Lasso	Nyquist	Moderate	high	no
(Li, I, Han, Chakravarthy, & Wu, 2010)	Weighted Lp	Nyquist	Moderate	high	no
(Hu, et al., 2012)	Non-negative matrix factorization	Nyquist	Moderate	High	no
(Zhi Tian & Giannakis, 2007)	OMP	Nyquist	Moderate	low	no

COLLABORATIVE KRONECKER COMPRESSIVE SPECTRUM SENSING APPROACH

Most of the mentioned collaborative approaches only consider the signal structure at each CR for compression. In order to overcome the communication overhead that results from cooperation, utilizing different sparse structures to reduce the number of the measurements is needed. In this section, we present an approach to this problem that makes use of the correlation between the measurements of different cognitive radios in addition to the special structure of the spectrum at each CR. This is achieved by using a Kronecker product matrix as a sparsifying basis, which enables us to jointly model the different types of structures presented in the signal and to exploit the two dimensional correlation nature of the collaborative observations.

We experiment this approach under a modified signal ensemble model that address the problem of hidden primary user. We propose several different measurements matrices, and show through simulation results that the proposed approach can substantially reduce the mean square error of the recovered power spectrum density by about 50% over conventional schemes at low compression ratio (around 0.2) while maintaining the use of low-rate ADC. We also show that using a dense measurement matrix we can achieve a dramatically low MSE under low compression ratios. However, such matrix can only be used under Nyquist rate ADC. This can help in special cases where resources are of utmost importance as compared to radio complexity.

Signal Model and Problem Statement

We consider a network of K cognitive radio terminals, distributed randomly in a certain geographic area and performing collaborative spectrum sensing through a centralized fusion center (FC), as shown in Figure 2.

Each CR locally monitors P non-overlapped channels, where each channel is either occupied by a primary user or unoccupied. We assume a low ratio of occupied channel U / P due to the low percentage of spectrum occupancy by active radios.

Each CR captures a wideband analog N dimensional signal $\mathbf{x}_k \in \mathbb{R}^N$, where the subscript k denotes a specific CR terminal. The signal $\mathbf{x}_k$ is captured using the pseudo random demodulation scheme for compressive sampling presented in (Kirolos, Ragheb, Laska, Duarte, Massoud, & Baraniuk, 2006). The autocorrelation of the measurements are sent synchronously to the fusion center over Additive White Gaussian Noise (AWGN) wireless channel. The recovered signals at FC can be modeled as $N \times K$ matrix, which can be written as:

$$\mathbf{X} = \left[\mathbf{x}_1, \mathbf{x}_2, \cdots, \mathbf{x}_K\right] = \begin{bmatrix} x_1^1 & x_2^1 & \cdots & x_K^1 \\ x_1^2 & x_2^2 & \cdots & x_K^2 \\ \vdots & \vdots & \ddots & \vdots \\ x_1^N & x_2^N & \cdots & x_K^N \end{bmatrix}, \quad (1.10)$$

where x_i^j represents the j-th measurement of the i-th CR.

Figure 2. System model

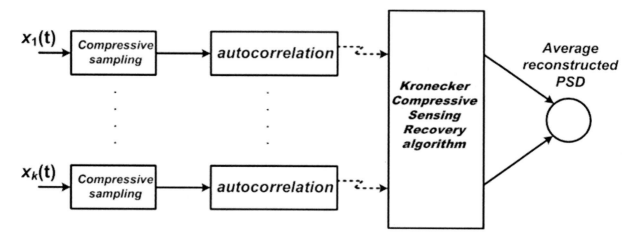

Each columns of the matrix represent the individual signals of each CR corresponding to different snapshots, while the rows of the matrix represent the same snapshot of the signal at different cognitive radio terminals.

We use a signal ensemble that is a modified version of the joint sparsity model (JSM2) presented in (Duarte, Sarvotham, Baron, Wakin, & Baraniuk, 2005). The original (JSM2) model assumes only common sparse support, neglecting the possible effect of hidden users, while the (JSM1) model assumes common signal amplitude which is valid only if CRs are distributed in very small area. The proposed modified model (JSM2M) relaxes these assumptions and generates signals, which have a common sparse support in frequency domain with different amplitudes plus innovations due to hidden primary user problem.

Kronecker Compressive Collaborative Spectrum Sensing

In our proposed approach, we utilize the Kronecker CS technique to exploit the different structures embedded in the signal, which allows us to recover the spectrum measurements with a better accuracy using fewer numbers of samples. We use a sparsifying basis which make use of the piecewise constant structure of the power spectrum density across different frequency channels, and another sparsifying basis which exploits the correlation structure of the measurements from different CR terminals. Finally, we propose a combined structure which exploits both dimensions for signal recovery.

The signal is captured at each CR terminal using the method stated in (Wang, Pandharipande, Polo, & Leus, 2009), the signal x_k is sampled at sub-Nyquist rate by analog to information converter according to

$$y_k = \Phi_A x_k, \tag{1.11}$$

where $\mathbf{y_k}$ is the measurement vector and $\mathbf{\Phi_A}$ is $M \times N$ random measurement matrix with i.i.d Gaussian entities. Denote the autocorrelation at lag j by $r_x(j) = E(x_n x_{j-n}^*)$. Denote the $2N \times 1$ autocorrelation vector of $\mathbf{x_k}$ by $\mathbf{r_{kx}} = [0, r_x(-N+1), \ldots, r_x(N-1)]^T$. The Nyquist rate autocorrelation vector and the compressed autocorrelation vector are related by the following equation

$$\mathbf{r_{ky}} = \mathbf{\Phi_{II}} \mathbf{r_{kx}}, \tag{1.12}$$

where $\mathbf{r_{kx}}$ and $\mathbf{r_{ky}}$ denote the autocorrelation vectors for $\mathbf{x_k}$ and $\mathbf{y_k}$, respectively. The matrix $\mathbf{\Phi_{II}}$ is formed as follows:

$$\mathbf{\Phi_{II}} = \begin{bmatrix} \overline{\mathbf{\Phi}}_A \mathbf{\Phi_{11}} & \overline{\mathbf{\Phi}}_A \mathbf{\Phi_{22}} \\ \overline{\mathbf{\Phi}}_A \mathbf{\Phi_{33}} & \overline{\mathbf{\Phi}}_A \mathbf{\Phi_{44}} \end{bmatrix}. \tag{1.13}$$

Let $\phi_{i,j}^*$ denote the (i, j)-th element of $\mathbf{\Phi_A}$. The $M \times N$ matrix $\overline{\mathbf{\Phi}}_A$ has its (i, j)-th elements given by

$$[\overline{\mathbf{\Phi}}_A]_{i,i} = \begin{cases} 0 & i = 1, j = 1, \cdots, N, \\ \phi_{M+2-i} & i \neq 1, j = 1, \cdots, N, \end{cases}$$

and the $N \times N$ matrices $\mathbf{\Phi_{11}}, \mathbf{\Phi_{22}}, \mathbf{\Phi_{33}}, \mathbf{\Phi_{44}}$ are

$$\mathbf{\Phi_{11}} = hankel([0_{N \times 1}], [0 \; \dot{\phi}_{1,1} \cdots \phi_{1,N}^*]),$$

$$\mathbf{\Phi_{22}} = hankel([\phi_{1,1}^* \cdots \phi_{1,N}^*], [\phi_{1,N}^* \; 0_{1 \times (N-1)}]),$$

$$\mathbf{\Phi_{33}} = teoplitz([0_{N \times 1}], [0 \; \phi_{1,N} \cdots \phi_{1,2}]),$$

$$\mathbf{\Phi_{44}} = teoplitz([\phi_{1,1} \cdots \phi_{1,N}], [\phi_{1,1} \; 0_{1 \times (N-1)}]),$$

where $hankel(c, r)$ is a hankel matrix (i.e., symmetric and constant across the anti-diagonals) whose first column is c and last row is r, $toeplitz(c, r)$ is a toeplitz matrix (i.e., symmetric and constant across the diagonals) whose first column is c and first row is r, $0_{N \times 1}$ is a column vector of N zeros, and $0_{1 \times (N-1)}$ is a row vector of $N-1$ zeros.

The power spectrum density (PSD) is the Fourier transform of the autocorrelation function as

$$\mathbf{S_k}(f) = \mathbf{F} \mathbf{r_{kx}}, \tag{1.14}$$

where $\mathbf{F}$ denotes $2N \times 2N$ discrete Fourier transform matrix. The edges of the spectrum are sparse in frequency domain and can be approximated using the differentiation of PSD as follows

$$\mathbf{z}_{k} = \mathbf{\Gamma S}_{k}(f) = \mathbf{\Gamma F r}_{kx}, \tag{1.15}$$

where $\mathbf{\Gamma}$ is the first order difference matrix given by

$$\mathbf{\Gamma} = \begin{bmatrix} 1 & 0 & 0 & \cdots & 0 \\ -1 & 1 & 0 & \cdots & 0 \\ 0 & -1 & 1 & \ddots & 0 \\ \vdots & \ddots & \ddots & \ddots & \vdots \\ 0 & \cdots & 0 & -1 & 1 \end{bmatrix}.$$

The structure of the signals $\mathbf{x}_{1}, \mathbf{x}_{2}, \cdots, \mathbf{x}_{K}$ which is observable on each columns of the matrix could be sparsified by the sparsifying matrix $\mathbf{\Psi}_{1}$ as

$$\mathbf{r}_{kx} = (\mathbf{\Gamma F})^{-1}\mathbf{z}_{k} = \mathbf{\Psi}_{1}\mathbf{z}_{k}. \tag{1.16}$$

We assume that the values of the measurements which span all the nearby cognitive radio stations $\mathbf{x}^{1}, \mathbf{x}^{2}, \cdots, \mathbf{x}^{N}$ are expected to be highly correlated (Pradhan, Kusuma, & Ramchandran, 2002). Therefore, it is quite reasonable to assume that $\mathbf{x}^{j}$ is compressible, where the piecewise smooth signals tend to be compressible in wavelet basis

$$\mathbf{x}^{j} = \mathbf{\Psi}_{2}\mathbf{s}^{j}, \tag{1.17}$$

where $\mathbf{x}^{j}$ is a $K \times 1$ column vector, $\mathbf{\Psi}_{2}$ is a $K \times K$ wavelet sparsifying basis matrix and $\mathbf{s}^{j}$ is a $K \times 1$ column vector represents the sparse coefficient of vector $\mathbf{x}^{j}$ in the basis $\mathbf{\Psi}_{2}$.

The Kronecker product is used not only for the generation of sparsifying basis that combines both structures presented in the signal, but also for the formation of the measurement matrix used in compressive sensing. Assuming that we use the same Analog to Information Converter with the same measurement matrix at each CR terminal $\mathbf{\Phi}_{A}$, then from Equation (1.12) we can find the compressed autocorrelation matrix for all CRs through the following equation

$$\mathbf{R}_{y} = \mathbf{\Phi}_{II}\mathbf{R}_{x}, \tag{1.18}$$

where $\mathbf{R}_{x}$ and $\mathbf{R}_{y}$ matrices are given by

$$\mathbf{R}_{x} = \left[\mathbf{r}_{1x}, \mathbf{r}_{2x}, \cdots, \mathbf{r}_{Kx}\right]_{2N \times K}, \tag{1.19}$$

$$\mathbf{R}_y = \left[\mathbf{r}_{1y}, \mathbf{r}_{2y}, \cdots, \mathbf{r}_{Ky} \right]_{2M \times K}.$$

The joint $2MK \times 2NK$ measurement matrix $\boldsymbol{\Phi}_1$ can be given as follows:

$$\boldsymbol{\Phi}_1 = \mathbf{I}_K \otimes \boldsymbol{\Phi}_{\mathrm{II}}, \boldsymbol{\Phi}_1 = \mathbf{I}_K \otimes \boldsymbol{\Phi}_{\mathrm{II}}, \tag{1.20}$$

$$\boldsymbol{\Phi}_1 = \begin{bmatrix} \boldsymbol{\Phi}_{\mathrm{II}} & \mathbf{0}_{2M \times 2N} & \cdots & \mathbf{0}_{2M \times 2N} \\ \mathbf{0}_{2M \times 2N} & \boldsymbol{\Phi}_{\mathrm{II}} & \ddots & \vdots \\ \vdots & \ddots & \ddots & \vdots \\ \mathbf{0}_{2M \times 2N} & \cdots & \cdots & \boldsymbol{\Phi}_{\mathrm{II}} \end{bmatrix}, \tag{1.21}$$

where $\mathbf{I}_K$ denotes the $K \times K$ identity matrix.

From Equation (1.18) we can express the reshaped vector of $\mathbf{R}_y$ in the following form

$$\bar{\mathbf{r}}_y = \boldsymbol{\Phi}_1 \bar{\mathbf{r}}_x, \tag{1.22}$$

where $\bar{\mathbf{r}}_x$ and $\bar{\mathbf{r}}_y$ denote the reshaped vectors of $\mathbf{R}_x$ and $\mathbf{R}_x$, respectively, as shown below

$$\bar{\mathbf{r}}_y = [\mathbf{R}_y^T(:,1), \mathbf{R}_y^T(:,2), \cdots, \mathbf{R}_y^T(:,K)]^T. $$

We can now generate the sparsifying matrix $\boldsymbol{\Psi}$, which combines the structures of both the rows and columns of the matrix $\mathbf{X}$. The global sparsifying basis is the Kronecker product of both sparsifying bases presented in Equations (1.16) and (1.17) given by

$$\boldsymbol{\Psi} = \boldsymbol{\Psi}_1 \otimes \boldsymbol{\Psi}_2. \tag{1.23}$$

The reshaped vector of autocorrelation matrix $\mathbf{R}_x$ can be viewed in $\boldsymbol{\Psi}$ domain as follows

$$\bar{\mathbf{r}}_x = \boldsymbol{\Psi}\bar{\mathbf{e}}, \tag{1.24}$$

where $\bar{\mathbf{e}}$ is the sparse coefficient vector in the basis $\boldsymbol{\Psi}$, also the reshaped vector of autocorrelation matrix $\mathbf{R}_y$ can be viewed in $\boldsymbol{\Psi}$ domain from Equation (1.22) as

$$\bar{\mathbf{r}}_y = \boldsymbol{\Phi}_1 \boldsymbol{\Psi}\bar{\mathbf{e}}. \tag{1.25}$$

The edges of the spectrum $\bar{e}$ can be recovered by the Basis Pursuit algorithm for the following ℓ_1 minimization,

$$\hat{\bar{\mathbf{e}}} = \arg\min \left\| \bar{\mathbf{e}} \right\|_1 \text{ subject to } \boldsymbol{\Phi}_1 \boldsymbol{\Psi} \bar{\mathbf{e}} = \bar{\mathbf{r}}_{//}. \tag{1.26}$$

The reshaped recovered vector can be estimated by

$$\hat{\bar{\mathbf{r}}}_x = \boldsymbol{\Psi} \hat{\bar{\mathbf{e}}}. \tag{1.27}$$

The recovered power spectrum densities seen by the K cognitive radio terminals at FC $\hat{\mathbf{S}}_{N \times K}$ can be estimated by

$$\hat{\mathbf{S}} = \mathbf{F} \hat{\mathbf{R}}_x, \tag{1.28}$$

where $\hat{\mathbf{R}}_x$ is the reshaped matrix from the vector $\hat{\bar{\mathbf{r}}}_x$.

Simulation Results

In this section, we illustrate through numerical simulations the performance of the presented Kronecker approach. In all of the experiments, we consider a spectrum of interest with a bandwidth in the range [50 150] MHz centered on a carrier frequency f_c. The PSD is smooth within each subband, but exhibits a discontinuous change between adjacent subbands similar to (Zhi Tian & Giannakis, 2007), with number of samples $N = 128$. The network consists of $k = 8$ CRs connected to a FC through error free channels. The total number of PUs $p = 6$, and the number of active PUs $U = 3$. Hence the spectrum utilization equals 0.5. The Daubechies wavelet "db4" is used in forming the sparsifying basis Ψ_2.

The received signal at CR receiver is corrupted by additive white Gaussian noise. It is further assumed that all CR receivers experience fading/shadowing. Where, each subband received at each CR is subjected to different gain due to fading and hidden primary user problem. The signal to noise ratio is considered as the inverse of the noise variance, where $SNR = 3$ dB.

In order to compare the performance of our approach with current CS approaches like (Wang, Pandharipande, Polo, & Leus, 2009), we use the normalized mean square error between the average of the recovered spectrum and the average of the noise free spectrum versus the Compression Ratio $(_{M/N})$ as our performance criteria.

In the first experiment, we examine the effect of ensemble model selection and recovery algorithm on the sparsity order K (the number of non-zero elements). The sparsity order of the different ensembles models under both Kronecker and traditional sparsifying bases are shown in Table 3. It is obvious that the signal has the sparsest representation (lowest sparsity order) under the Kronecker basis Ψ. This proves that the presented approach can exploit more underlying sparsity than 1-D approach since it

Table 3. The sparsity order of different signal models under different sparsifying basis

Signal Model	JSM2	JSM2M
Kronecker basis Ψ	7	17
Traditional basis Ψ_1	47	47

exploits different structure presented in the ensemble (like the strong correlation between the measurements at different CRs), as compared to the traditional sparsifying basis Ψ_1 which only works on a single dimension of the received signals (intra-signal).

In the second experiment, we compare the performance of different recovery algorithms using the same Kronecker measurement matrix presented in (1.21). Figure 3 elucidates a noticeable performance improvement while using a very low rate compressive sampler, as we can achieve a lower NMSE using the Kronecker approach specially under low compression ratio (about 50% reduction at $M/N = 0.2$) as compared to both the Simultaneous Orthogonal Matching Pursuit (SOMP) presented in (Wang, Pandharipande, Polo, & Leus, 2009) with Ψ_1 as a sparsifying basis, and the Independent Recovery using Basic Pursuit (BP) algorithm (Donoho, 2006) where each CR independently recovers the spectrum then send it to the FC for averaging the recovered spectrum from all CRs.

Figure 3. NMSE performance versus compression ratio for different recovery algorithms using Kronecker Gaussian measurment matrix

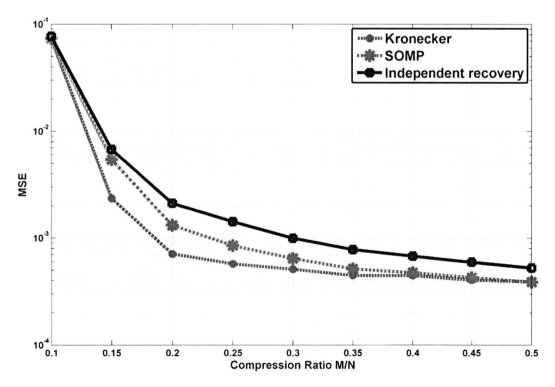

In the third experiment, we evaluate the performance of collaborative compressive sensing under different measurement matrices such as:

Kronecker measurement matrix in the presented approach $\Phi_1 = I_K \otimes \Phi_{II}$ where Φ_{II} is the matrix given in (1.21).

Kronecker measurement matrix $\Phi_2 = I_K \otimes \Phi_{BB}$ where Φ_{BB} is a matrix formed in the same way as Φ_{II} but by substituting for Φ_A by a Bernoulli measurement matrix rather than a Gaussian matrix as in (1.21).

Global random dense measurement $2KM \times 2KN$ matrix Φ_3 with i.i.d Gaussian entries.

The performance results for these different measurement matrices are shown in Figure 4. It is evident from the figure that the dense matrix Φ_3 achieves a significant improvement in the MSE performance as compared to the Kronecker measurement matrices, especially at very low compression ratios (less than 0.2). However, we cannot theoretically find a dense matrix that relates the Nyquist autocorrelation vector and the compressed autocorrelation vectors in the way shown in Equation(1.18). These results in a computational bottleneck, as we have to sample the signals at the Nyquist rate, find the autocorrelation function, compress it using compressive sampler, and then send it compressed to the FC. While this may deprive our system from the advantage of using a low-rate ADC, it can significantly reduce the cooperation overload on some of the radio resources required to achieve communication between CRs and

Figure 4. NMSE versus compression ratio for different measurement matrices using Kronecker sparsifying basis

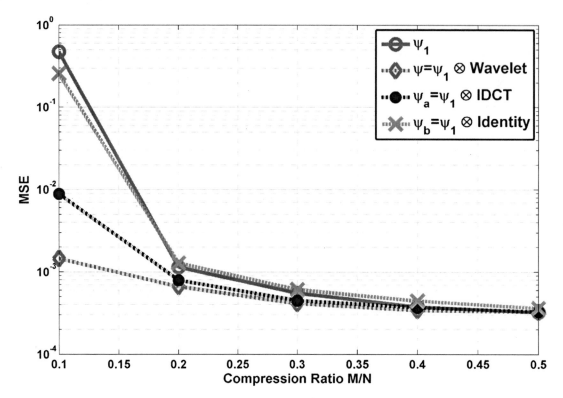

the FC like the number of multiple access channels. A similar observation can be made for the Bernoulli measurement matrix Φ_2, which results in the worst MSE performance. However, it has the merit of putting some CRs at sleep mode, which saves both power and bandwidth.

In the fourth experiment, we evaluate the performance of collaborative compressive sensing using the dense measurement matrix Φ_3 under different sparsifying bases. Four different sparsifying bases used in our simulation are as follows.

SOMP algorithm with sparsifying basis Ψ_1 used in (Wang, Pandharipande, Polo, & Leus, 2009).

Kronecker CS with $\Psi = \Psi_1 \otimes \Psi_2$.

1. $\Psi_a = \Psi_1 \otimes IDCT$.
2. $\Psi_b = \Psi_1 \otimes I_K$.

Figure 5 depicts how effective is our choice of sparsifying basis in exploring the underlying structure of the measurements. The Kronecker basis Ψ performs significantly better than all other bases. Both the sparsifying bases Ψ_1, Ψ_b have the largest MSE. Since, these matrices exploit the sparsity only in one dimension. On the other hand, the sparsifying basis Ψ_a exploits sparsity of 2-D signals, the first dimension is the ordinary sparsifying matrix Ψ_1 of the individual signals and the second is the DCT sparsifying basis. The signal in the DCT domain is sparse only under the assumption that the measure-

Figure 5. NMSE versus compression ratio for different sparsifying basis using dense measurement matrix

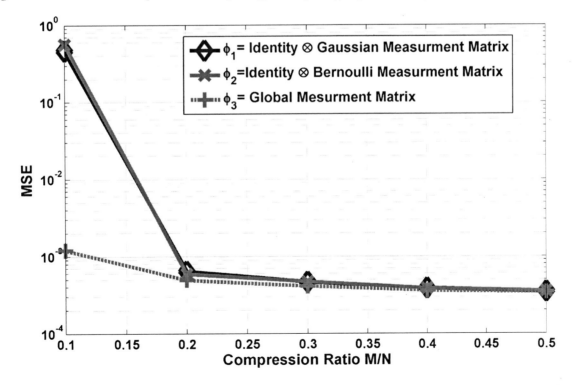

ments have strong correlation without sudden peaks. But, this is not the case when there is a hidden primary user, since there will be abrupt peaks which would destroy the sparsity. Therefore, wavelet basis is more suitable for that case and evinced the lowest MSE at a very low compression ratio.

The computational complexity of the proposed algorithm is compared to the other approaches is shown in Table 4. Although the complexity of the proposed algorithm is proportional to K^3 and the former algorithms keeps complexity in order of K, the overall complexity is still comparable. This is attributed to the fact that the number of the measurements M needed to achieve a specific MSE is the proposed approach is much lower than in the DSC approach. If we take into account that the number of CRs K is much lower than the number of measurement M, we find the computational complexity and its associated power consumption are comparable.

Other aspects of power consumption such as signal acquisition are not of great concern in our approach. Through utilizing a conventional AIC front end, the power consumption is analogous to the other compressive schemes. Which is much lower than the Nyquist ADCs as shown in (Bellasi, Bettini, Benkeser, Burger, Qiuting, & Studer, 2013).

FUTURE RESEARCH DIRECTIONS

Compressive sensing in wireless communication is still in its early stages. However, its widespread success indicates that it might be the key to making fundamental advances in next generation communication systems. Furthermore, the scalability of wireless communication especially cognitive radios associated with novel signal processing techniques will continue to pose theoretical and practical challenges. Bringing Cognitive-based technologies into the mainstream of commination systems will require a multidisciplinary effort, integrating experts with knowledge about diverse application domains.

One of the open potential research projects in is Dynamic Compressive Spectrum Sensing. In this approach, not only the sparsity in the frequency domain is utilized but also the dynamic behavior inherent in the channels transition between the free and busy states may be exploited. In addition, the spatial correlation between the channels measured at neighboring CRs should be taken into consideration in the model. These schemes try to exploit the known parameters of the statistical model to increase the recovery performance. This could be done using a low complexity Bayesian compressive sensing recovery algorithm tailored to that model.

Table 4. Computational complexity

Algorithm	Complexity
Distributed Compressive Sensing - Simultaneous Orthogonal Matching Pursuit.	$O(NKM^2)$ (Duarte, Sarvotham, Baron, Wakin, & Baraniuk, 2005)
Independent recovery using OMP	$O(NKM^2)$
Kronecker recovery using OMP	$O((KN)*(KM)^2)=O(NK^3M^2)$

CONCLUSION

The rapid growth in wireless applications leads to spectrum congestion. Static spectrum allocation schemes avoid interference at the expense of low spectrum utilization. The cognitive radio technique promises to increase the utilization of the radio spectrum by utilizing dynamic opportunistic spectrum access schemes.

The under-utilization results in signals that are sparse in the frequency domain. Such sparsity allowed the use of compressive sensing in capturing and reconstructing the spectrum with far-less time samples than Nyquist theorem imposes. Moreover, compressive wide-band spectrum sensing techniques allows sensing multiple frequency bands, simultaneously rather than channel-by channel scanning.

The main objectives of this chapter are the efficient sensing of the wideband spectrum and the accurate detection of the spectrum holes. This is achieved through utilizing one of the latest trends in both compressive and spectrum sensing. In particular, the sensing performance is enhanced by exploiting the different structures embedded in the signal. This is achieved by formulating the problem as a Kronecker compressive sensing recovery problem, and carefully designing suitable measurement and sparsifying bases. We present a modified signals ensemble model that accounts for the scenario of hidden primary user problem. The performance using MSE under different sparsifying bases and measurement matrices evinced a significant improvement, as the Kronecker sparsifying basis exploits different structures presented in the signal which allows for significant reduction in sampling rate, relaxes constrains put on (ADCs), and finally reduces the amount of radio resources needed for the communication between CRs and FC.

REFERENCES

Baraniuk, R. G. (2007, July). Compressive Sensing. *Signal Processing Magazine, IEEE, 24*(4), 118–121. doi:10.1109/MSP.2007.4286571

Baraniuk, R. G., Cevher, V., Duarte, M. F., & Hegde, C. (2010, April). Model-Based Compressive Sensing. *IEEE Transactions on* Information Theory, *56*, 1982–2001.

Baron, D., Sarvotham, S., & Baraniuk, R. G. (2010, January). Bayesian Compressive Sensing Via Belief Propagation. *IEEE Transactions on* Signal Processing, *58*(1), 269–280. doi:10.1109/TSP.2009.2027773

Bazerque, J. A., & Giannakis, G. B. (2010). Distributed spectrum sensing for cognitive radio networks by exploiting sparsity. *IEEE Transactions on* Signal Processing, *58*(3), 1847–1862. doi:10.1109/TSP.2009.2038417

Bellasi, D. E., Bettini, L., Benkeser, C., Burger, T., Qiuting, H., & Studer, C. (2013). VLSI Design of a Monolithic Compressive-Sensing Wideband Analog-to-Information Converter. *IEEE Journal on* Emerging and Selected Topics in Circuits and Systems, *3*(4), 552–565.

Donoho, D. L. (2006, April). Compressed sensing. *IEEE Transactions on* Information Theory, *52*, 1289–1306.

Duarte, M. F., & Baraniuk, R. G. (2012, February). Kronecker Compressive Sensing. *IEEE Transactions on* Image Processing, *21*(2), 494–504.

Duarte, M. F., Sarvotham, S., Baron, D., Wakin, M. B., & Baraniuk, R. G. (2005). Distributed Compressed Sensing of Jointly Sparse Signals. In *Proceedings of the Asilomar Conference on Signals, Systems and Computers*. Academic Press. doi:10.1109/ACSSC.2005.1600024

Elzanati, A. M., Abdelkader, M. F., Seddik, K. G., & Ghuniem, A. M. (2013). Collaborative compressive spectrum sensing using kronecker sparsifying basis. In *Proceedings of Wireless Communications and Networking Conference (WCNC)* (pp. 2902-2907). IEEE. doi:10.1109/WCNC.2013.6555022

Farhang-Boroujeny, B. (2008, May). Filter bank spectrum sensing for cognitive radios. *IEEE Transactions on* Signal Processing, *56*(5), 1801–1811. doi:10.1109/TSP.2007.911490

Guibene, W., Moussavinik, H., & Hayar, A. (2011). Combined compressive sampling and distribution discontinuities detection approach to wideband spectrum sensing for cognitive radios. In *Proceedings of Ultra Modern Telecommunications and Control Systems and Workshops (ICUMT)*, (pp. 1--7). IEEE.

Haupt, J., Baraniuk, R., Castro, R., & Nowak, R. (2012). Sequentially designed compressed sensing. In *Proceedings of Statistical Signal Processing Workshop (SSP)*, (pp. 401-404). IEEE.

Havary-Nassab, V., Hassan, S., & Valaee, S. (2010). Compressive detection for wide-band spectrum sensing. In *Proceedings of Acoustics Speech and Signal Processing (ICASSP)*, (pp. 3094--3097). IEEE.

Hong, S. (2010). Multi-resolution bayesian compressive sensing for cognitive radio primary user detection. In *Proceedings of Global Telecommunications Conference (GLOBECOM 2010)*, (pp. 1-6). IEEE. doi:10.1109/GLOCOM.2010.5683291

Hu, Z., Ranganathan, R., Zhang, C., Qiu, R., Bryant, M., Wick, M., et al. (2012). Robust non-negative matrix factorization for joint spectrum sensing and primary user localization in cognitive radio networks. In *Proceedings of IEEE Waveform Diversity and Design Conference*. IEEE.

Huang, C. C., & Wang, L. C. (2012). Dynamic sampling rate adjustment for compressive spectrum sensing over cognitive radio network. *IEEE Wireless Communications Letters*, *1*(2), 57–60. doi:10.1109/WCL.2012.010912.110136

Huang, D., Wu, S., & Wang, P. (2010). Cooperative spectrum sensing and locationing: A sparse Bayesian learning approach. In *Proceedings of Global Telecommunications Conference (GLOBECOM 2010)*, (pp. 1-5). IEEE. doi:10.1109/GLOCOM.2010.5684081

Kirolos, S., Ragheb, T., Laska, J., Duarte, M. E., Massoud, Y., & Baraniuk, R. G. (2006). Practical Issues in Implementing Analog-to-Information Converters. In *Proceedings of the 6th International Workshop on System-on-Chip for Real-Time Applications*, (pp. 141-146). Cairo: Academic Press.

Li, S., Hong, S., Han, Z., & Wu, Z. (2011). Bayesian Compressed Sensing based Dynamic Joint Spectrum Sensing and Primary User Localization for Dynamic Spectrum Access. In *Proceedings of Global Telecommunications Conference (GLOBECOM 2011)*, (pp. 1-5). IEEE.

Li, X., Han, Q., Chakravarthy, V., & Wu, Z. (2010). Joint spectrum sensing and primary user localization for cognitive radio via compressed sensing. In Proceedings of Military Communications Conference, (pp. 329--334). IEEE.

Liu, L., Han, Z., Wu, Z., & Qian, L. (2011). Collaborative Compressive Sensing based Dynamic Spectrum Sensing and Mobile Primary User Localization in Cognitive Radio Networks. In *Proceedings of Global Telecommunications Conference (GLOBECOM 2011)*, (pp. 1-5). IEEE.

Mallat, S. G., & Zhifeng, Z. (1993). Matching pursuits with time-frequency dictionaries. *IEEE Transactions on* Signal Processing, *41*(12), 3397–3415. doi:10.1109/78.258082

Mishali, M., & Eldar, Y. C. (2009). Blind multiband signal reconstruction: Compressed sensing for analog signals. *IEEE Transactions on* Signal Processing, *57*(3), 993–1009. doi:10.1109/TSP.2009.2012791

New, M. F., Sarvotham, S., Baron, D., Wakin, M. B., & Baraniuk, R. G. (2005). Distributed Compressed Sensing of Jointly Sparse Signals. In *Proceedings of Signals, Systems and Computers,* (pp. 1537-1541). Academic Press.

Polo, Y. L., Wang, &., Andharipande, A., & Leus, G. (2009). Compressive wide-band spectrum sensing. In *Proceedings of IEEE International Conference on Acoustics, Speech and Signal Processing (ICASSP)*, (pp. 2337-2340). Taipei, Taiwan: IEEE.

Pradhan, S. S., Kusuma, J., & Ramchandran, K. (2002). Distributed compression in a dense microsensor network. *IEEE Signal Processing Magazine, 19*(2), 51–60. doi:10.1109/79.985684

Sun, H., Nallan, A., Wang, C-X., & Chen, Y. (2013, April). Wideband spectrum sensing for cognitive radio networks: A survey. *IEEE Wireless Communications, 20*(2), 74-81.

Sun, H., Chiu, W., & Nallanathan, A. (2012, November). Adaptive Compressive Spectrum Sensing for Wideband Cognitive Radios. *IEEE Communications Letters*, *16*(11), 1812–1815. doi:10.1109/LCOMM.2012.092812.121648

Tian, Z., & Giannakis, G. G. (2007). Compressed Sensing for Wideband Cognitive Radios. In *Proceedings of Acoustics, Speech and Signal Processing*. Honolulu, HI: IEEE.

Tian, Z. (2008). Compressed Wideband Sensing in Cooperative Cognitive Radio Networks. In *Proceedings of IEEE Global Telecommunications Conference (GLOBECOM)*. IEEE. doi:10.1109/GLOCOM.2008.ECP.721

Tian, Z., Tafesse, Y., & Sadler, B. M. (2012). Cyclic feature detection with sub-Nyquist sampling for wideband spectrum sensing. *IEEE Journal of Selected Topics in Signal Processing*, 58-69.

Wang, X., Guo, W., Lu, W., & Wang, W. (2011). Adaptive compressive sampling for wideband signals. In *Proceedings of Vehicular Technology Conference (VTC Spring)*, (pp. 1-5). IEEE.

Wang, Y., Pandharipande, A., Polo, Y. L., & Leus, G. (2009). Distributed compressive wide-band spectrum sensing. In *Proceedings of Information Theory and Applications Workshop*. Academic Press.

Wang, Y., Tian, Z., & Feng, C. (2010). A two-step compressed spectrum sensing scheme for wideband cognitive radios. In *Proceedings of Global Telecommunications Conference (GLOBECOM 2010)*, (pp. 1-5). IEEE. doi:10.1109/GLOCOM.2010.5683246

Wang, Y., Tian, Z., & Feng, C. (2011). Cooperative spectrum sensing based on matrix rank minimization. In *Proceedings of Acoustics, Speech and Signal Processing (ICASSP)*, (pp. 3000-3003). IEEE.

Yin, W., Wen, Z., Li, S., Meng, J., & Han, Z. (2011). Dynamic compressive spectrum sensing for cognitive radio networks. In *Proceedings of Information Sciences and Systems (CISS)*, (pp. 1-6). IEEE.

Yucek, T., & Arslan, H. (2009). first quarter). A survey of spectrum sensing algorithms for cognitive radio applications. *IEEE Communications Surveys and Tutorials*, *11*(1), 116–130. doi:10.1109/SURV.2009.090109

Zeng, F., Li, C., & Tian, Z. (2011). Distributed compressive spectrum sensing in cooperative multihop cognitive networks. *IEEE Journal of Selected Topics in Signal Processing*, 37-48.

KEY TERMS AND DEFINITIONS

Collaborative Compressive Spectrum Sensing: Is a spectrum sensing that exploits the presence of multiple secondary users distributed in the space to increase the detection performance and alleviate the effect of hidden primary user problem.

Compressive Sensing: CS is a signal processing technique by which the signal can be captured and recovered using only a small number of sub Nyquist measurements under certain conditions.

Compressive Spectrum Sensing: Is a new paradigm which uses compressive sampling at the front end of the secondary user system to determine the vacant PUs' channels using a realizable low speed AIC.

Kronecker Compressive Sensing: Is a collaborative compressive spectrum sensing technique used to increase the detection performance by exploiting the different sparse structures of the measured spectrum at neighboring secondary users.

Sparse Signal: Is a signal which contains only a small number of non-zero elements compared to its dimension. Analog to Information Converter: AIC is the front end of compressive sampling systems that is able to capture linear combinations of signal measurements at sub Nyquist rate. Wide-band Spectrum Sensing: is an important stage in the cognitive radio technology at which the PU shall detect a wideband spectrum to identify vacant channels for opportunistic use.

Chapter 7
Spectrum Sensing Using Principal Components for Multiple Antenna Cognitive Radios

Farrukh A. Bhatti
Institute of Space Technology, Pakistan

Gerard B. Rowe
The University of Auckland, New Zealand

Kevin W. Sowerby
The University of Auckland, New Zealand

ABSTRACT

This chapter presents an experimental comparative analysis of the well-known Covariance-Based Detection (CBD) techniques, which include Covariance Absolute Value (CAV), Maximum-Minimum Eigenvalue (MME), Energy with Minimum Eigenvalue (EME), and Maximum Eigenvalue Detection (MED). CBD techniques overcome the noise uncertainty issue of the Energy Detector (ED) and can even outperform ED in the case of correlated signals. They can perform accurate blind detection given sufficient number of signal samples. This chapter also presents a novel CBD algorithm that is based on Principal Component (PC) analysis. A Software-Defined Radio (SDR)-based multiple antenna system is used to evaluate the detection performance of the considered algorithms. The PC algorithm significantly outperforms the MED and EME algorithms and it also outperforms MME and CAV algorithms in certain cases.

INTRODUCTION

Spectrum sensing is an enabling technology for a cognitive radio (CR) system. Extensive research has been performed in the past decade in devising accurate and robust wireless signal detection methods for CR applications. Spectrum sensing techniques can be classified into two main types. The first type relies on a priori knowledge of the channel and/or licensed user (LU) signal characteristics for signal

DOI: 10.4018/978-1-4666-6571-2.ch007

detection, and includes cyclostationarity-based feature detection, matched filter detection and waveform-based sensing (Yucek & Arslan, 2009). The second type of sensing techniques do not require prior knowledge of channel or PU signal characteristics, and includes energy detection, wavelet based sensing and covariance-based sensing (Yucek & Arslan, 2009). Techniques of the second type are often typically referred to as blind signal detectors. The first type of techniques generally offer a higher level of detection accuracy but their application is limited to a specific type of known wireless signal and might require the use of complex estimation techniques. On the other hand, blind signal detection techniques can be applied to most types of signals which make them a suitable choice for heterogeneous wireless environments. However, they do have some performance limitations.

Due to their flexibility and broad scope of application, blind signal detection techniques have drawn significant research interest in recent years. Energy detection (ED) is the simplest of all detection techniques, however, it requires knowledge of noise variance to correctly set the detection threshold. In practice, the noise variance is not constant and is influenced by various factors, such as temperature, humidity, device aging, radio interference etc. The performance of ED is adversely affected by even the slightest variation in the noise variance (Tandra & Sahai, 2005), which makes it undesirable for implementation in real systems. Moreover, ED is optimal for detecting independent and identically distributed (i.i.d.) Gaussian signals (Kay, 1998) but it is not optimal for detecting correlated signals, which is the case in most practical wireless systems. These limitations of ED can be overcome by covariance-based detection techniques (Zeng & Liang, 2009b; Zeng & Liang, 2009a; Zeng, Koh, & Liang, 2008; Zeng & Liang, 2007b; Zeng & Liang, 2007a; Penna, Garello, Figlioli, & Spirito, 2009; Kortun, Ratnarajah, Sellathurai, & Zhong, 2010; Zeng & Liang, 2010) that exploit the structure of the covariance matrix of the received signals and, in general, do not require knowledge of the noise variance. (An exception is the MED method that does require knowledge of the noise variance (Zeng, Koh, & Liang, 2008). CBD algorithms can accurately detect a signal at low SNR, provided a sufficient number of signal samples are used. This was the motivation for further exploring a new detection algorithm that relied on the calculation of the covariance matrix. Most of the work done on CBD methods has focused on theoretical analysis and simulations. In (Oh, et al., 2008), a hardware implementation has been performed using the CAV algorithm only; however, no comparison of results for all the CBD algorithms is presented. There are two main contributions of this chapter.

- Firstly we present a performance comparison of the existing CBD algorithms using a multi antenna SDR-based signal acquisition system. To the author's knowledge, this is the first time that a comprehensive performance comparison of the CBD algorithms is presented while using actual wireless signals. In the previous related works, individual CBD algorithms have been reported separately in various studies but no overview is available to give a clear picture of the relative performance of different CBD algorithms in real systems. Such an overview is a major contribution of this chapter.
- The second contribution of this chapter is that, an innovative blind signal detection technique is proposed, that is based on principal component analysis (PCA). The PCA based algorithm (or PC algorithm for brevity) is very promising in comparison with the other CBD algorithms. The initial processing step required of the PC algorithm is similar to the other CBD algorithms i.e. calculation of sample covariance matrix. Due to this similarity the new technique is presented along with the CBD algorithms to give a performance comparison.

BACKGROUND

This section describes the system model used in this work. Some of the detection performance metrics are also described here. This is followed by a review of the existing CBD algorithms.

System Model

We consider a multiple receive antenna system, where each antenna is connected to an independent radio frequency (RF) front-end and the inter-antenna distance is less than half of the wavelength of the LU's center frequency being received. This enables correlated signal reception at each antenna. The signal is sampled at $f_s \geq W$ where f_s is the sampling frequency and W is the bandwidth of the received signal. Usually the signal is oversampled at $f_s \gg W$ to get a higher temporal correlation between the samples. Let $T_s = 1/f_s$ be the sampling period. We define $z_i(n) \triangleq z_i(nT_s)$, $s_i(n) \triangleq s_i(nT_s)$ and $w_i(n) \triangleq w_i(nT_s)$, where $z_i(n)$ is the received signal at the ith front-end, $s_i(n)$ is the (noiseless) sample of the LU's signal and $w_i(n)$ is the white noise (AWGN) added at the ith front-end[1] and it follows a normal distribution $w_i(n) \sim \mathcal{N}(0, \sigma_w^2)$ where σ_w^2 is the noise variance. The process of signal detection requires distinguishing between the following two hypotheses:

$$z_i(n) = \begin{cases} w_i(n), & \mathcal{H}_0 \\ h_i s_i(n) + w_i(n), & \mathcal{H}_1. \end{cases} \tag{1}$$

$\mathcal{H}_0$ is the null hypothesis that indicates the absence of the LU signal, and $\mathcal{H}_1$ is the alternative hypothesis that indicates the presence of the LU signal. h_i is the complex channel gain between the source and the ith antenna, and it represents multipath fading and path-loss effects. Each antenna is connected with an independent RF front-end and the RF front-ends are synchronized for synchronous reception of the LU signals. The receiver multiplexes the complex baseband samples received from K front-ends to make a single stream of S complex samples that can be expressed as

$$\mathbf{z}(n) = \left[z_1(n), \cdots, z_K(n), z_1(n-1), \cdots, z_K(n-1), \cdots, z_1\left(n - \frac{S}{K} + 1\right), \cdots, z_K\left(n - \frac{S}{K} + 1\right) \right]. \tag{2}$$

S samples are used to make a single sensing decision about the presence of an LU signal where each RF front-end contributes S/K samples. Two probabilities characterize the performance of the spectrum sensing process: 1) probability of detection $P_d = \Pr(\text{decide } \mathcal{H}_1 | \mathcal{H}_1)$ and 2) probability of false alarm $P_{fa} = \Pr(\text{decide } \mathcal{H}_1 | \mathcal{H}_0)$. It is desired to maximize P_d while minimizing P_{fa}. In addition, the probability of missed detection is defined as $P_{md} = 1 - P_d$.

Existing CBD Algorithms

The CBD algorithms exploit the covariance matrix of the received signal, for signal detection. The performance of these methods depends upon the correlation of the received signals. Therefore, we use correlated signal reception as described above to increase the spatial correlation and do over-sampling to increase the temporal correlation in the received signal samples. Expressing $\mathbf{z}(n)$ as $\mathbf{z}$, we define the statistical covariance matrix of the received signal as $\mathbf{R} = E\left[\mathbf{z}\mathbf{z}^{\dagger}\right]$, where $[.]^{\dagger}$ denotes the transpose conjugate. In practice, a sample covariance matrix is used as an approximation of $\mathbf{R}$ and is defined as

$$\hat{\mathbf{R}}(N) = \frac{1}{N}\sum_{i=1}^{N} \overline{\mathbf{z}}(n)\overline{\mathbf{z}}^{\dagger}(n). \tag{3}$$

N is the number of times averaging is done to compute $\hat{\mathbf{R}}(N)$. $\hat{\mathbf{R}}(N)$ is Hermitian and Toeplitz, where the diagonal entries are the variances and the non-diagonal entries are the covariances of the signal samples. $\overline{\mathbf{z}}(n)$ is a column vector of length $L = S/N$ (L is also called the smoothing factor) and it satisfies the following conditions: 1) $\mathbf{z} = \left[\overline{\mathbf{z}}(1), \overline{\mathbf{z}}(2), \cdots, \overline{\mathbf{z}}(N)\right]$ 2) $L = \omega K$ where ω is a non-zero positive integer. Increasing N improves the performance of CBD algorithms, but it also entails more samples, which in turn increases the sensing time. The total number of samples required for making a sensing decision is $S = NL$. The CBD algorithms considered in this chapter are CAV, MME, MED and EME. These algorithms are briefly discussed here and later their performance is compared with the PC algorithm.

Covariance Absolute Value

Let $r_{mn}(N)$ be the elements of the matrix $\hat{\mathbf{R}}(N)$. Then, compute the following (Zeng & Liang, 2007b):

$$T_1(N) = \frac{1}{L}\sum_{n=1}^{L}\sum_{m=1}^{L}\left|r_{mn}(N)\right| \tag{4}$$

$$T_2(N) = \frac{1}{L}\sum_{n=1}^{L}\left|r_{nn}(N)\right|. \tag{5}$$

Under $\mathcal{H}_0$, the non-diagonal elements $r_{mn}(N)$, for $m \neq n$, approach zero, and the diagonal elements $r_{nn}(N)$ approach σ_w^2. Thus, in this case, $T_1(N)/T_2(N) \approx 1$. Under $\mathcal{H}_1$, the received signal correlation is higher, which results in $r_{mn}(N) > 0$ and $r_{nn}(N) \approx \sigma_s^2$, where σ_s^2 is the signal variance. Thus, in this case, $T_1(N)/T_2(N) > 1$. The signal exists if $T_1(N)/T_2(N) > \psi_{CAV}$; otherwise, the signal does not exist. Here, ψ_{CAV} is the threshold for detection.

Maximum Minimum Eigenvalue

The MME detection algorithm calculates the eigenvalues (λ) of $\hat{\mathbf{R}}(N)$ and then finds the ratio of the highest and lowest eigenvalues (Zeng & Liang, 2007a). If $\lambda_{\max} > \psi_{MME}\lambda_{\min}$, then the signal exists; otherwise, the signal does not exist. This method relies on the fact that, in the case of noise, all the eigenvalues are approximately equal, i.e. $\lambda_{\max}/\lambda_{\min} \approx 1$, whereas in the case that a signal is present, at least a few eigenvalues are larger than the others.

Maximum Eigenvalue Detection

The MED algorithm compares the highest eigenvalue $\lambda_{\max}$ of $\hat{\mathbf{R}}(N)$ with the noise variance σ_w^2. If $\lambda_{\max} > \psi_{MED}\sigma_w^2$, then the signal exists otherwise it does not exist (Zeng, Koh, & Liang, 2008). However, a major drawback in this method is that similar to ED, it also requires exact knowledge of the noise variance. Since noise variance cannot be determined accurately in advance and is likely to vary over time, this method is unreliable for implementation in a real system.

Energy with Minimum Eigenvalue

The EME algorithm compares the received signal power $p = \mathrm{tr}\left(\hat{\mathbf{R}}(N)\right)/L$ with the minimum eigenvalue $\lambda_{\min}$ of $\hat{\mathbf{R}}(N)$ (Zeng & Liang, 2009a). Here, $\mathrm{tr}(.)$ denotes the trace of a matrix. If $p > \psi_{EME}\lambda_{\min}$, the signal exists; otherwise it does not exist.

PC ALGORITHM

Principal Component Analysis in Statistics

Principal component analysis (PCA) is a multivariate analysis technique which is used to reduce the dimensionality of a data set consisting of a large number of interrelated variables, while preserving as much as possible of the variation present in the data set (Jolliffe, 1986). This reduction is attained through a linear transformation to a new set of variables, the principal components (PCs), which are uncorrelated and ordered such that the first few PCs retain most of the variation present in the original data set.

Although related work on singular value decomposition and factor analysis was undertaken in the late nineteenth century, it is generally agreed that PCA was invented by Pearson and Hotelling. Pearson presented his work on PCA in 1901 where he was concerned with finding lines and planes that best fit a set of points in p-dimensional space and he also discussed a geometric optimization problem that lead to PCA (K. Pearson, 1901). In his work, Pearson observed that his methods could be easily applied to numerical problems, although the calculations could become cumbersome for four or more variables, but he suggested that they were still quite feasible.

Hotelling presented his pioneering work on PCA in 1933. His approach was different from Pearson and he argued that there could be a smaller fundamental set of independent variables that could determine the values of the original variables. He observed that such variables had been called 'factors' in the psychological literature, and he introduced the alternative term 'components' to maintain distinction with the term 'factor'. Hotelling defined his 'components' in such a way so as to maximize their contributions to the total of the variances of the original variables (Hotelling, 1933). Hotelling's work is closer to the modern description of PCA and therefore PCA is sometimes also referred to as the Hotelling Transform. In engineering, PCA is also known as the Karhunen-Loeve expansion (Fukunaga, 1990).

Application of PCA in Different Fields

Due to the computational complexity involved in performing manual calculations, the use of PCA was very limited until the dawn of the computer era, in the latter half of the twentieth century. Since then PCA has been applied to a wide variety of multivariate problems related to different fields such as, neuroscience, computer vision, meteorology, oceanography, gene expression, material science, agricultural studies, pharmaceutics, ecological studies etc.

In computer vision, principal components have been used for face recognition (Turk & Pentland, 1991). The input digital image is a U×U vector of *l* bit numbers. This vector is represented using a small set of features, by using principal component analysis. This representation is called the eigenface. The eigenface approach is fast and relatively simple, as it does not require the exploitation of the physical facial features, such as geometry of nose, length of ears, distance between eyes etc. The eigenface approach, while relying on the principal component analysis, can accurately match an eigenface of an input image with a database of the eigenfaces stored in the system.

PCA has also been used extensively in climate studies (Storch, 1999). Climatic data typically ignores the short term variability in time, e.g. a series of values can represent mean air temperature in the month of January, over several years, for a particular location. The climatic series formed in this way shows a statistical behavior and is similar to white or red noise stochastic processes. The main aim of statistical climatology analysis is to extract useful properties from climatic series data that has low signal to noise ratio. PCA is used in understanding the principal modes of climatic variability of an atmospheric variable. PCA serves as a compression tool that reduces the variability of the data to a small number of modes, which explain a considerable part of the overall variance of the data.

Signal Detection Using Principal Components

In this section we present an innovative technique for blind signal detection that is based on principal component analysis (Bhatti, Rowe, & Sowerby, 2012). The computation of PCs involves the eigenvalue decomposition of a positive semi-definite symmetric matrix, which is the covariance matrix in our case. For the application of PCA, the data should have a zero mean. The data considered here is composed of time domain, modulated[2], complex baseband samples, and each of the I and Q components of the complex samples follow a symmetric distribution (i.e. $p(x) = p(-x)$) with zero mean. The signal samples exhibit these characteristics both under $\mathcal{H}_0$ and $\mathcal{H}_1$. The complex baseband samples from each antenna source are first decomposed into their constituent I and Q components and are expressed as

$$\mathbf{x}(n) = \left[x_{1I}, x_{1Q}, x_{2I}, x_{2Q}, \cdots, x_{KI}, x_{KQ}\right]^T, \tag{6}$$

where K is the number of receive antennas and $[.]^T$ denotes the transpose operation. Due to synchronous reception, all the decomposed complex baseband samples in the column vector correspond to the same time instant. Multiplexing received samples from different antenna sources increases correlation in the data. S complex samples, used in making a single sensing decision, form a *2K* dimensional data set that is expressed in the form of a matrix as

$$\mathbf{X} = \left[\mathbf{x}(n), \mathbf{x}(n-1), \cdots, \mathbf{x}(n-M+1)\right], \tag{7}$$

where $M = S/K$. For convenience of notation we express $\mathbf{x}(n-i)$ as $\mathbf{x}_i$. We define a sample covariance matrix of the data matrix $\mathbf{X}$ as

$$\hat{\mathbf{R}}(M) = \frac{1}{M} \sum_{n=1}^{M} \mathbf{x}(n)\mathbf{x}^T(n). \tag{8}$$

We assume that $\mathbf{X}$ has a zero mean vector. In general, wireless communication is an AC-coupled phenomenon (McClaning, 2012), therefore, both under $\mathcal{H}_0$ and $\mathcal{H}_1$, the mean of the received data samples is zero. Let $\lambda_1 \geq \lambda_2 \cdots \geq \lambda_{2K}$ be the ordered eigenvalues (characteristic roots) of $\hat{\mathbf{R}}(M)$, such that the following condition is satisfied (Srivastava & Khatri, 1979)

$$\left|\hat{\mathbf{R}}(M) - \lambda\mathbf{I}\right| = 0, \tag{9}$$

where $\mathbf{I}$ is the identity matrix having the same dimensions as $\hat{\mathbf{R}}(M)$. Let $\gamma_1, \gamma_2, \cdots, \gamma_{2K}$ be the normalized eigenvectors (characteristic vectors) of $\hat{\mathbf{R}}(M)$, i.e. the vectors satisfying

$$\hat{\mathbf{R}}(M)\gamma_i = \lambda_i\gamma_i, \tag{10}$$

$$\gamma_i^T\gamma_j = \delta_{ij}, \tag{11}$$

where δ_{ij} is the Kronecker delta. Let $\mathbf{F}$ be a feature matrix, which is composed of the k most significant eigenvectors, where $1 \leqslant k \leqslant 2K$, i.e.

$$\mathbf{F} = [\gamma_1, \cdots, \gamma_k]. \tag{12}$$

The k most significant eigenvectors correspond to the k highest ordered eigenvalues. The new transformation of the original data set to the principal components can now be defined as

$$\mathbf{p}_i = \mathbf{F}^T \mathbf{x}_i, i = 1, 2, \cdots, M. \tag{13}$$

These PCs give an orthogonal linear transformation of the original data set. The complete set of PCs can be expressed as $\mathbf{P} = [\mathbf{p}_1, \mathbf{p}_2, \cdots, \mathbf{p}_M]$. However, it is more useful to express $\mathbf{P}$ in terms of row vectors $\mathbf{y}_j$, where $\mathbf{y}_j = [p(j)_1, p(j)_2, \cdots, p(j)_M]$ and where $p(j)_i$ is the jth element of $\mathbf{p}_i$ for $j = 1 \cdots k$. We can now represent $\mathbf{P}$ in terms of k row vectors as

$$\mathbf{P} = \begin{bmatrix} \mathbf{y}_1 \\ \vdots \\ \mathbf{y}_k \end{bmatrix}. \tag{14}$$

$\mathbf{P}$ has k rows and M columns, or equivalently it can be said that it has k principal components. In $\mathbf{P}$ the variate $\mathbf{y}_1$ is the one with the largest variance λ_1 and is uncorrelated with all the remaining variates $\mathbf{y}_2, \cdots, \mathbf{y}_k$. Similarly the variate $\mathbf{y}_2$ is the one with the second largest variance and is uncorrelated with $\mathbf{y}_1$ and all the other variates, and so on (Srivastava & Khatri, 1979). Representing $p(j)_i$ as α_{ji} for convenience of notation, we have $\mathbf{y}_j = [\alpha_{j1}, \alpha_{j2}, \cdots, \alpha_{jM}]$, then

$$\sum_{i=1}^{M} \alpha_{ji}^2 = \lambda_j. \tag{15}$$

It follows from (15) that the jth PC gives the distribution of the total energy given by the jth eigenvalue (Mardia, Kent, & Bibby, 1979). Since:

$$\sum_{j=1}^{2K} \lambda_j = tr\mathbf{R},$$

the sum of the variances of the original variables is the same as the sum of the variances of the PCs. Therefore the variables with smaller variances can be ignored without significantly effecting the total variance, thereby reducing the number of variables. For this the last eigenvector in (12) is removed first and then so on. While removing the least significant eigenvectors from the feature matrix a tradeoff has to be made between dimension reduction and the information loss that occurs. This tradeoff depends upon the problem type. In the context of signal detection, dimensionality reduction saves computation time which in turn also reduces the sensing time. The error induced due to dimension reduction has been discussed in detail in (Fukunaga, 1990). Choosing an appropriate value of k gives the optimum tradeoff between the dimension reduction and the information loss, as is discussed in the next section.

Rejection of Insignificant Principal Components

In practice, only the first few (k) principal components account for the maximum variation of the data, and so the remaining principal components can be rejected. There is no universally accepted method for choosing k, rather it depends more upon the specific problem. Several procedures have been suggested to determine k.

1.   **Average Eigenvalue:** A simple criteria to determine the number of useful principal components (k) is the Guttman-Kaiser criterion (Jackson, 1993). Principal components associated with eigen-values (λ_k), which are larger in magnitude than the average of the eigenvalues, are retained. While this method works well in general, when it identifies k incorrectly, it is likely to retain too many components (Jolliffe, 1986). Having a k higher than the optimum value results in additional components that do not contribute any significant information but they do cause a significant increase in computational complexity.

2.   **Scree Plot:** This is a graphical approach that can be used to find k, and it was suggested by Cattell (Cattell, 1966). A scree plot is a plot of eigenvalues versus the index of the eigenvalues (or the components). The eigenvalues obtained from the covariance matrix are plotted in successive order of their extraction, i.e. in ascending order. The plot is observed for randomness, like rocks falling on a scree down a hill. The line corresponding to this scree is found, and then k is determined from the number of eigenvalues preceding this scree. On many occasions, this scree appears where the slope of the plot changes drastically to generate the scree. The point where the slope changes drastically is also termed an elbow.

3.   **Proportion of the Total Variance Explained:** In PCA, each eigenvalue represents the variation associated with the corresponding principal component. One of the criteria for determining k is based on the proportion of the total variance explained by the k principal components that are retained. For k principal components that are retained out of a total of P components, the proportion of variance explained by the k components can be represented as

$$v_k = \frac{\sum_{i=1}^{k} \lambda_i}{\sum_{i=1}^{P} \lambda_i}. \qquad (16)$$

A satisfactory value of v_k is determined, which in turn determines the value of k. The obvious problem with this approach is to decide on an appropriate value of v_k. In practice it is common to select v_k between 70% to 90% (Jolliffe, 1986). However, this approach has been criticized for its subjectivity (Kim, 1978).

In general, applying an appropriate criterion can help in finding an appropriate value of k for a particular problem. In our case, finding k was rather simple due to a relatively smaller number of variables considered i.e. $P = 2K$, and also due to the specific pattern of the eigenvalues. Figure 1 shows the plot of the magnitude of the eigenvalues versus their index. Clearly, it can be observed that the first eigen-

Figure 1. Eigenvalues magnitude versus the eigenvalues index, obtained from the covariance matrix that is calculated from the signals acquired from the USRP based receiver at different SNRs (dB)

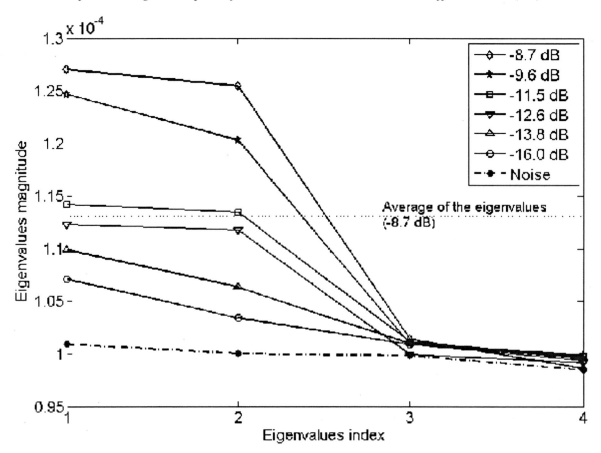

value has the highest magnitude whereas the fourth eigenvalue has the lowest magnitude, under $\mathcal{H}_1$. These eigenvalue plots have been obtained for the same signals that are captured using the USRP (Universal Software Radio Peripheral) based system, and are also used for the performance evaluation of the signal detection algorithms discussed in this chapter. The different curves represent the eigenvalues obtained from the covariance matrix that is calculated from signals at different SNRs, as shown in the legend, in Figure 1. For the eigenvalues that are computed from the signal at relatively higher SNR, the range[3] of the eigenvalues is higher, whereas as the SNR decreases the range of the eigenvalues also decreases. In the case of noise the magnitude of the eigenvalues is almost equal, i.e. the range ≈ 0.

The number of principal components to retain was found by applying the Guttman-Kaiser criterion. The eigenvalues were compared with the average eigenvalue. The first two eigenvalues were much higher than the average whereas the last two eigenvalues were much lower than the average value. Figure 1 shows the average eigenvalue represented by the horizontal dotted line for the case of a signal at -8.6 dB SNR. For the PC algorithm, k is determined once, while considering the total number of antennas K connected to the receiver. Therefore finding the value of k should not be considered as an additional complexity of this algorithm.

In our case the variables (data set) are the synchronized signals (I and Q channels) from different RF front-ends. For the case of two RF front-ends (the case considered in this chapter), there are four variables i.e. a pair of I and Q channels from each RF front-end. The reason for treating I and Q channels from the same antenna as different variables (or dimensions) is that in the case of noise, the I and Q channels are uncorrelated, i.i.d. variables. However in the case a signal is present, the I and Q channels are correlated as they have a phase difference of $\pi / 2$ radians. In the case of noise, the eigenvalues of the covariance matrix are (approximately) equal to the noise variance (for a sufficiently large N). Thus the corresponding eigenvectors have equal significance and the resulting principal components have equal variation or scatter. Figure 2 shows a 2-dimensional plot of PCs generated from noise, for $k=2$. The arrangement of points on the scatterplot is regular, with fixed spacing in between. This spacing corresponds to the quantization step of the ADC in the USRP. In the case a signal is present, the orthonormal transformation to PCs results in a circle (Figure 3) because the I and Q channels are orthogonal to each other. Their transformation to an orthonormal basis results in an equal amount of spread along both the basis, thus resulting in a circle whose radius is proportional to the received signal power. For the same

Figure 2. Scatter plot of PCs obtained from noise for k=2 and M=10000

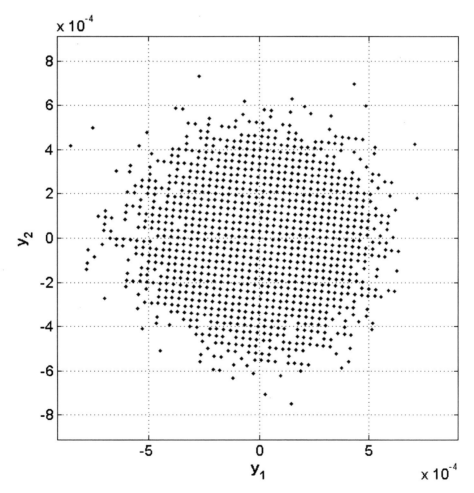

Figure 3. Scatter plot of PCs obtained from wireless microphone signals at 10 dB SNR for k=2 and M=10000

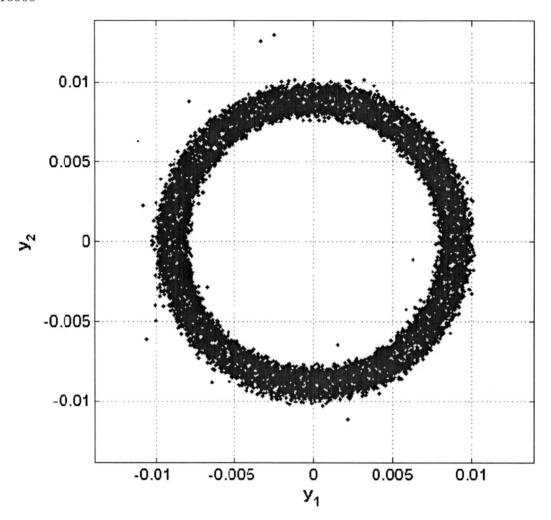

reason we observe the specific pattern in the Figure 1, where the magnitude of the first two and the last two eigenvalues appears to be quite similar. The pair of similar eigenvalues represent the variation in the I and Q channels, that have a similar magnitude as well. Figure 3 shows a 2-dimensional plot of PCs generated from a wireless microphone signal.

Detection Criteria for PC Algorithm

We can now present a test statistic for distinguishing between signal and noise based on the above discussion for k PCs as

$$\psi_{PC} = \frac{1}{M} \sum_{j=1}^{k} (\alpha_{j1}^2 + \alpha_{j2}^2 + \cdots + \alpha_{jM}^2) \overset{\mathcal{H}_0}{\underset{\mathcal{H}_1}{\lessgtr}} \tau, \tag{17}$$

where α_{ji} is the *i*th variable of the *j*th PC and τ is the detection threshold. In this work, τ was determined empirically according to the desired P_{fa}. Since $\sum_{i=1}^{M} \alpha_{ji}^2 = \lambda_j$, therefore the test statistic ψ_{PC} is merely a sum of the *k* eigenvalues. Under $\mathcal{H}_0$ and $\mathcal{H}_1$, ψ_{PC} follows an asymptotic and normal distribution, as $M \to \infty$ (Mardia, Kent, & Bibby, 1979), i.e.

$$\psi_{PC} \sim \mathcal{N}\left(\sum_{j=1}^{k} \lambda_j, 2\sum_{j=1}^{k} \frac{\lambda_j^2}{M} \right). \tag{18}$$

In the case of AWGN that follows a standard normal distribution, $\lambda_j \approx 1$, for $j = 1 \ldots k$. In this case $\psi_{PC} \sim \mathcal{N}\left(k, 2k/M \right)$. For a given probability of false alarm P_{fa}, the detection threshold τ can be computed in terms of the mean *k* and variance $2k/M$. The cumulative density function (cdf) of ψ_{PC} is expressed as

$$\Theta(\psi_{PC}) = \frac{1}{2}[1 + erf(\frac{\tau - k}{2^{3/2} k/M})]. \tag{19}$$

The area in the tail of a normal probability density function (pdf) is given by the Q function, where $Q(\tau) = 1 - \Theta(\tau)$. Therefore

$$P_{fa} = 1 - \frac{1}{2}[1 + erf(\frac{\tau - k}{2^{3/2} k/M})]. \tag{20}$$

By performing some mathematical manipulation and solving for τ, we get

$$\tau = k + \frac{2^{3/2} k}{M}[erf^{-1}(1 - 2P_{fa})], \tag{21}$$

where $erf^{-1}(.)$ is the inverse error function, which can be easily computed. Under $\mathcal{H}_1$, $\lambda_j \neq 1$, for $j = 1 \ldots k$. In this case finding the mean and variance for (18) becomes more complicated. This task will be addressed in the future work.

DETECTION PERFORMANCE ANALYSIS

The existing CBD techniques and the proposed PC algorithm were tested with actual wireless signals, while using the SDRs. The experimental setup comprising of the SDRs is described next.

Experimental Setup

For correlated reception of wireless signals, a receiver system with two RF front-ends was setup using two USRPs as shown in Figure 4. The USRP (model N210) is a high performance SDR that has been developed by Ettus Research (Ettus Research, 2013).

The USRP consists of two main parts: a motherboard and a daughterboard. The motherboard, is the heart of the USRP, that has a Xilinx Spartan XC3SD3400A FPGA, that performs the core digital signal processing operations. It has two pairs of ADCs (Analog to Digital Converters) and DACs (Digital to Analog Converters). The motherboard has a Gigabit Ethernet interface that allows high speed communication with the PC. A WBX daughterboard is mounted on the motherboard and contains two RF front-ends for simultaneous transmission and reception. A WBX daughterboard can only receive one signal at a time using its single receive chain. The WBX uses a direct-conversion receiver architecture in which the local oscillator (LO) frequency is set to the same frequency as the desired RF center frequency. It pre-filters, amplifies, mixes and low pass filters the passband signals. An omnidirectional vertical an-

Figure 4. Wireless signals acquisition system comprising of two USRPs

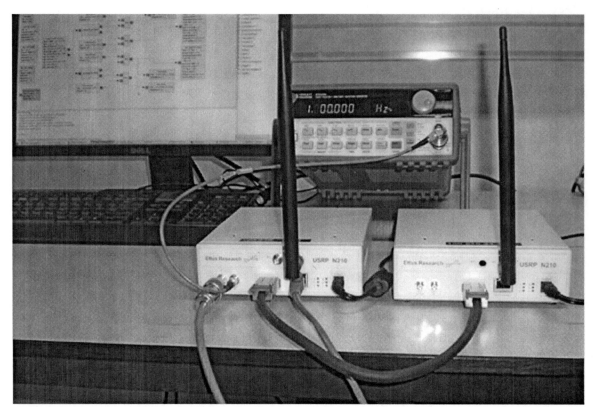

tenna was connected to the daughterboard for wireless signal reception. The two USRPs were connected in a master/slave configuration, where the master USRP was connected with the PC via the Ethernet interface. The two were inter-linked with a MIMO cable that ensured fully coherent signal reception at the two USRPs. The MIMO cable also transferred complex baseband data from slave to master that was eventually routed through the gigabit Ethernet to the PC along with the complex baseband stream of the master USRP. To synchronize the USRPs two reference signals were fed to the master USRP. A 10 MHz signal to provide a single frequency reference and a 1 pulse per second (PPS) signal to synchronize the sample time across the devices.

To represent an LU, transmissions representing wireless microphone signals were used in these experiments. Agilent E4438C signal generator was used as a transmitter. The LU signal was an FM signal with a bandwidth of 200 kHz, and it was transmitted at a center frequency of 410 MHz. The received baseband signal was sampled at 6.25 Mega samples per second. The signal was significantly oversampled in order to ensure higher correlation between the samples. Oversampling requires more energy consumption. Therefore, this approach is more suitable for devices that do not have a major power constraint, such as cognitive base-stations in IEEE 802.22. Low-power devices can adopt this approach by reducing the oversampling factor, thereby, doing a tradeoff between sensing accuracy and power consumption. The received signals were stored in a computer, where the signal detection algorithms were applied. This ensured identical operational conditions for all the algorithms to establish a fair performance comparison.

Figure 5. shows the schematic diagram of the correlated reception system that was used for capturing the signals. These signals were later used for the performance evaluation of the detection algorithms.

Figure 5. Block diagram of USRP based signal acquisition system

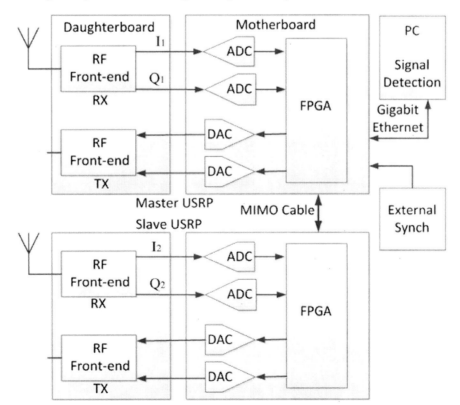

The antenna spacing was kept less than $\lambda_c / 2$, where λ_c is the wavelength of the carrier frequency $\lambda_c = 0.731$ meters. Small antenna spacing results in higher correlation between the signals received at the two antennas. The SNR at the receiver was varied by adjusting the transmit power of the signal generator.

Results and Discussion

In this section we present the results of our experiments, in which the performance of the proposed PC algorithm is compared with the existing CBD algorithms. All the results presented here for the PC algorithm are for $k=2$. Figure 6 shows P_d of all the algorithms obtained for different values of mean SNR at 10% P_{fa}. Since the two receive antennas are positioned at a distance less than half the wavelength of the carrier frequency, therefore the SNRs observed at the two antennas are fairly close. The SNR used in the results is the mean of the SNRs achieved at the two front-ends. The number of samples used in performing a single sensing decision (S) is kept constant for all the algorithms. For CBD algorithms a tradeoff has to be done between choosing L (i.e. the length of the covariance matrix) and M (the number of averages done for calculating the sample covariance matrix) for a fixed S. The PC algorithm gives the best sensing performance and is followed by the CAV and the MME algorithms, e.g. at -22 dB SNR, the P_d of the PC algorithm is 10% better than the CAV algorithm, which is the second best. The EME algorithm remains significantly poor in performance e.g. at -20 dB SNR its P_d is about 40% below the PC, CAV and MME algorithms. The MED algorithm gives the worst performance even with an optimistic value of $v=0.5$dB, where v is the noise uncertainty factor defined as

Figure 6. P_d for wireless microphone signals for 10% P_{fa} and S=60000

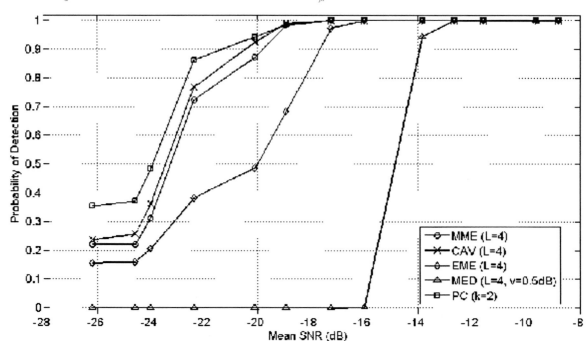

$$v = \sup \left\{ 10 \log_{10} \left(\frac{\hat{\sigma}_w^2}{\sigma_w^2} \right) \right\}. \tag{22}$$

Here $\hat{\sigma}_w^2$ is the estimated noise variance. In practice, the noise uncertainty factor of a receiving device ranges from 1 to 2 dB (Shellhammer & Tandra, 2006). The sensing performance of MED is dependent on having an exact knowledge of noise variance. This is the same requirement that limits the application of the energy detection method in real systems where noise variance cannot be estimated exactly due to the variation in conditions over time. In all of the results given here for MED, we consider $v=0.5$ dB, which is an optimistic choice and the sensing performance degrades sharply for higher values of v.

Figure 7. shows the sensing results for 1% P_{fa}. The relative performance of all the algorithms remains the same in this case, except for the PC algorithm that is marginally outperformed by the CAV algorithm. The performance of the PC and MME algorithms is similar above -19 dB SNR, but for lower SNRs the PC algorithm performs better than the MME algorithm, with CAV outperforming both. In Figure 6 and Figure 7, $L=4$ so that the size of covariance matrix in the case of CBD algorithms remains equal to the size of covariance matrix of the PC algorithm. This allows a comparison of the CBD and PC algorithms for same size of sample covariance matrix.

Figure 8. shows the effect of increasing S on the P_d, for fixed values of P_{fa} and SNR and different values of L. The MME and CAV algorithms perform better than the PC algorithm for a larger covariance matrix i.e. for $L=20$. However, for $L=4$ the performance of the PC algorithm is better than the perfor-

Figure 7. P_d for wireless microphone signals for 1% P_{fa} and S=60000

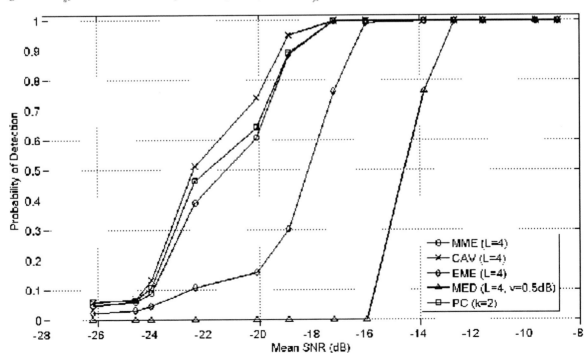

Figure 8. P_d vs sample size for microphone signal at -20 dB SNR for 10% P_{fa}

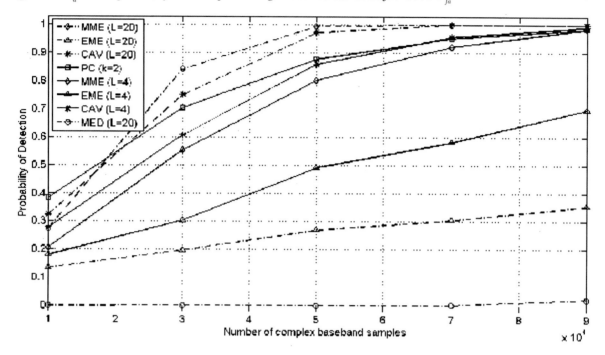

mance of the MME and CAV algorithms. For CBD algorithms, using a larger value of L also increases the computational complexity. In the case of the EME algorithm, the performance improves for smaller values of L, however it remains significantly lower than the PC, MME and CAV algorithms. The MED algorithm for v=0.5 dB performs the worst of all at -20 dB SNR. Increasing S, while keeping the size of the covariance matrix constant, in effect increases N and M for CBD and PC algorithms, respectively. It can be observed from Figure 8, that increasing S does not increase P_d linearly. E.g. in the case of PC, for $S = 9 \times 10^4$, P_d is almost 100%, therefore, increasing S any further does not give additional improvement in performance. Overall the sensing performance of all the algorithms except MED, significantly exceeds the FCC's sensing requirement of achieving 90% P_d for 10% P_{fa} at -12 dB SNR in the case of wireless microphone signals (Shellhammer S. J., 2008).

FUTURE RESEARCH DIRECTIONS

In this research work the USRP based system was used for signal acquisition and the signal detection was performed on a computer. As an extension to this work, the existing CBD algorithms and the PC algorithm can be implemented on a standalone hardware platform that performs the complete sensing process. A USRP E100 or E110 can be used for prototyping. Although, the PC algorithm has a higher complexity than the existing algorithms, given the high computational capacity of many available hardware platforms, their implementation is still possible. Implementing these algorithms on a prototype platform may lead to greater understanding about the real time performance of these algorithms.

In covariance based algorithms, for a given number of samples S, a tradeoff has to be made between choosing the smoothing factor L and the number of averages N performed for calculating the covariance matrix. Increasing L improves the detection performance, as shown in results, at the same time, increasing N also improves the detection performance as it smoothens the covariance matrix and the resulting eigenvalues approach their asymptotic limit. As a future task, the optimum tradeoff between L and N can be found analytically and empirically.

CONCLUSION

In this chapter, a novel blind signal detection algorithm based on principal component analysis has been presented. Under realistic channel conditions, the PC algorithm has been compared with four other existing covariance based algorithms. The PC algorithm performs better than the EME algorithm under all conditions and it also performs better than the MED algorithm for even very small variations in noise variance, which is the case in real systems. The MME and CAV algorithms perform better than the PC algorithm for larger sizes of covariance matrices, but for equal sizes of covariance matrices the PC algorithm outperforms MME and CAV. The PC algorithm performs better than all of the algorithms if a smaller number of samples is used for sensing. Overall, the PC algorithm is much better than the EME and MED algorithms and is better than the MME and CAV algorithms in certain conditions for implementation in real systems. The PC algorithm can be used effectively for the detection of very low power signals without any knowledge of the source signal and the channel.

REFERENCES

Bhatti, F. A., Rowe, G. B., & Sowerby, K. W. (2012). Spectrum sensing using principal component analysis. In *Proceedings of IEEE Wireless Communications and Networking Conference*, (pp. 725 –730). IEEE.

Cattell, R. (1966). The scree test for the number of factors. *Multivariate Behavioral Research, 1*(2), 245–276. doi:10.1207/s15327906mbr0102_10

Ettus Research. (2013, Nov). Retrieved from Ettus Research: http://www.ettus.com/

Fukunaga, K. (1990). *Introduction to Statistical Pattern Recognition*. Academic Press.

Hotelling, H. (1933). Analysis of a complex of statistical variables into principal components. *Journal of Educational Psychology, 24*(6), 417–441. doi:10.1037/h0071325

Jackson, D. A. (1993). Stopping rules in principal components analysis: A comparison of heuristical and statistical approaches. *Ecology, 74*(8), 2204–2214. doi:10.2307/1939574

Jolliffe, I. (1986). *Principal Component Analysis*. Springer-Verlag. doi:10.1007/978-1-4757-1904-8

Kay, S. M. (1998). *Fundamentals of Statistical Signal Processing: Detection theory*. Prentice Hall.

Kim, J. O. (1978). *Factor Analysis: Statistical Methods and Practical Issues*. Beverly Hills, CA: Sage.

Kortun, A., Ratnarajah, T., Sellathurai, M., & Zhong, C. (2010). On the Performance of Eigenvalue-Based Spectrum Sensing for Cognitive Radio. In *Proceedings of IEEE Symposium on New Frontiers in Dynamic Spectrum*, (pp. 1-6). IEEE. doi:10.1109/DYSPAN.2010.5457844

Mardia, K. V., Kent, J. T., & Bibby, J. M. (1979). *Multivariate Analysis*. Academic Press.

McClaning, K. (2012). *Wireless Receiver Design for Digital Communications*. Scitech Publishing.

Oh, S. W., Le, T. P., Zhang, W., Ahmed, S. N., Zeng, Y., & Kua, K. J. (2008). *TV white-space sensing prototype*. Wireless Communications and Mobile Computing.

Pearson, K. (1901). LIll. on lines and planes of closest fit to systems of points in space. Retrieved from Philosophical Magazine Series.

Penna, F., Garello, R., Figlioli, D., & Spirito, M. A. (2009). Exact non-asymptotic threshold for eigenvalue-based spectrum sensing. In *Proceedings of 4th International Conference on Cognitive Radio Oriented Wireless Networks and Communications*, (pp. 1-5). Academic Press. doi:10.1109/CROWN-COM.2009.5189008

Shellhammer, S., & Tandra, R. (2006). Performance of the Power Detector with Noise Uncertainty. *IEEE 802.22-06/0134r0 Std.*

Shellhammer, S. J. (2008). Spectrum sensing in IEEE 802.22. In *Proceedings of Cognitive Information Processing Workshop*. IEEE.

Sonnenschein, A., & Fishman, P. M. (1992). Radiometric detection of spread-spectrum signals in noise of uncertain power. *IEEE Transactions on Aerospace and Electronic Systems*, 28(3), 654–660. doi:10.1109/7.256287

Srivastava, M. S., & Khatri, C. G. (1979). *An Introduction to Multivariate Statistics*. Elsevier North Holland.

Storch, F. W. (1999). *Statistical Analysis in Climate Research*. Cambridge University Press.

Tandra, R., & Sahai, A. (2005). Fundamental limits on detection in low SNR under noise uncertainty. In *Proceedings of International Conference on Wireless Networks, Communications and Mobile Computing*, (pp. 464–469). Academic Press. doi:10.1109/WIRLES.2005.1549453

Turk, M., & Pentland, A. (1991). Eigenfaces for recognition. *Journal of Cognitive Neuroscience*, 3(1), 71–86. doi:10.1162/jocn.1991.3.1.71 PMID:23964806

Yucek, T., & Arslan, H. (2009). A survey of spectrum sensing algorithms for cognitive radio applications. *IEEE Communications Surveys and Tutorials*, 11(1), 116–130. doi:10.1109/SURV.2009.090109

Zeng, Y., Koh, L. C., & Liang, Y. C. (2008). Maximum Eigenvalue Detection: Theory and Application. Academic Press.

Zeng, Y., & Liang, Y. C. (2007a). Maximum-Minimum Eigenvalue Detection for Cognitive Radio. In *Proceedings of IEEE 18th International Symposium on Personal, Indoor and Mobile Radio Communications*, (pp. 1-5). IEEE.

Zeng, Y., & Liang, Y. C. (2007b). Covariance Based Signal Detections for Cognitive Radio. In *Proceedings of 2nd IEEE International Symposium on New Frontiers in Dynamic Spectrum Access Networks*, (pp. 202-207). IEEE. doi:10.1109/DYSPAN.2007.33

Zeng, Y., & Liang, Y. C. (2009a). Eigenvalue-based spectrum sensing algorithms for cognitive radio. *IEEE Transactions on Communications, 57*(6), 1784–1793. doi:10.1109/TCOMM.2009.06.070402

Zeng, Y., & Liang, Y. C. (2009b). Spectrum-Sensing Algorithms for Cognitive Radio Based on Statistical Covariances. *IEEE Transactions on Vehicular Technology, 58*(4), 1804–1815. doi:10.1109/TVT.2008.2005267

Zeng, Y., & Liang, Y. C. (2010). Robust spectrum sensing in cognitive radio. In *Proceedings of IEEE 21st International Symposium on Personal, Indoor and Mobile Radio Communications Workshops*, (pp. 1-8). IEEE.

KEY TERMS AND DEFINITIONS

Blind Detection: Sensing or detecting the presence of a signal without prior knowledge of the signal characteristics or channel information.

Cognitive Radio: An intelligent wireless communication system that is aware of its radio environment and that adapts its operating parameters such that it facilitate secondary communication while co-existing with the licensed users.

Covariance-Based Sensing: Spectrum sensing based on the calculation of the covariance matrix from the received signal samples.

Probability of Detection: The probability of correctly detecting a signal when the signal is actually present.

Probability of False Alarm: The probability of incorrectly detecting a signal when it is actually not present.

Software Radio: A radio in which some or all of the physical layer functions are implemented in a software.

Spectrum Sensing: The process of detecting the presence of a licensed wireless transmission, with the intention of locating an idle frequency band for self-usage.

ENDNOTES

[1]   In this system model, the terms antenna and RF front-end have similar meanings and so they are used interchangeably.

[2]   The signal samples meet the condition required for PCA, as they are modulated and downconverted samples. The frequency carrier present in the modulated signals makes it a zero mean random variable that is symmetrically distributed. The samples may not follow a symmetric distribution under, if they are demodulated.

[3]   Range is defined as the difference between the highest and the lowest value in a given data.

Chapter 8

Spectral Sensing Performance for Feature-Based Signal Detection with Imperfect Training

Quang Thai
Macquarie University, Australia

Sam Reisenfeld
Macquarie University, Australia

ABSTRACT

In this chapter, the effect of imperfect training data on feature-based signal detection is explored, as it relates to both training time and detection performance in a cognitive radio system. The improved performance of feature-based detection comes at the cost of either having to know in advance the signal features present in primary user transmissions (an unrealistic assumption) or learning them whilst operating "in the field." Such learning, however, necessarily takes place with signal sets which do not perfectly represent the features of the primary users' modulated signals. Using a two-stage detector performing both feature training and sensing functions, it is shown in this chapter that reducing the learning time generally results in poorer detection performance and vice-versa. A suitable trade-off between these two outcomes is obtained by optimizing a cost function that takes both factors into consideration. Cyclostationarity detection is specifically considered.

INTRODUCTION

The flexible and on-demand nature of wireless communications has seen it become increasingly ubiquitous in recent years, and it has become the focus of research and development activities that seek out new ways in which it can improve and enrich quality of life. The proliferation of wireless applications has also required a rethink of how the radio environment is used and how radio spectrum is planned and managed. In order to support both increased demand for existing wireless applications as well as new ones yet to be foreseen, either new exploitable radio spectrum must be found, or the existing spectrum

DOI: 10.4018/978-1-4666-6571-2.ch008

must be used more efficiently. It has been proposed that the latter measure be met in the near future by equipping wireless devices with cognitive radio capabilities. Such devices would be able to recognize and exploit radio channels that were unutilized or under-utilized by existing *primary* networks for their own communication needs via a *secondary* network, effectively allowing channels to be multiplexed amongst these networks.

To be able to recognize such channels, cognitive radios must be able to perform spectral sensing (or spectrum sensing) - the process of determining whether or not a spectrally defined channel is in use at a particular time. A cognitive radio, as a secondary user, must perform spectral sensing for a number of spectrally separated channels to determine which of the channels are being used by primary users and which channels are available for exploitation. The reliable detection of the *absence* of primary user transmission on a channel establishes an opportunity for secondary user transmission on that channel. Correspondingly, the reliable detection of the *presence* of primary user transmission on the channel identifies the secondary user requirement to stop transmission on that channel and change to another channel for subsequent transmissions. This use of spectral sensing by the secondary user is referred to as "detect and avoid" (Reisenfeld, 2009).

Many spectral sensing algorithms have been considered for implementation in cognitive radio (Yucek & Arslan, 2009). Good detection performance has been obtained from detectors which use known patterns, or features, of the primary user modulated signal. However, cognitive radio transmission is over dynamic, fading channels with rapidly changing characteristics. It would not be reasonable to expect that a spectral sensing algorithm could be pre-programmed with all the known patterns or features of the modulated signals for all classes of primary users. Furthermore, the patterns may be modified by the specific channel characteristics which may be encountered. Therefore, it is extremely advantageous for feature-based detectors to train themselves to recognize the signal features of primary user transmissions using channel output observations.

A feature-based approach is used in cyclostationarity detection (Yucek & Arslan, 2009). A cyclostationarity detector has been described where the detector was trained to recognize features for the recognition of various classes of modulated signals (Thai & Reisenfeld, 2011). The training set consisted of signal observations where the primary user is *known to be transmitting*, and this was used to obtain the required features. This training set was 'perfect' because it was known that all observations contain primary user features. In an operational environment, this perfect training set is unavailable. The same operational environment must then be used to conduct training for spectral sensing. Supervised learning is achieved with an 'imperfect' training set - so-called because it may contain observations where the primary user signal is, in fact, absent.

In the remainder of this chapter, a general approach for training feature-based detectors in the field will be described. This approach is applicable to any algorithm which relies on a training set. The degradation in performance due uncertainty in the imperfect training set will be described.

BACKGROUND: A MODEL FOR TRAINING USING FIELD CHARACTERISTICS

In supervised learning, learning algorithms require a training set to determine the specific features which may be used for discrimination and classification. In cognitive radio applications, as mentioned previously, it may be required to obtain the training set from field observations. In previous work, training was done for a cyclostationarity detector with a training set containing perfect feature information (Thai & Reisenfeld, 2011). In an operational scenario, a training set would need to be experimentally obtained from channel output signals.

For the purposes of obtaining a training set during field operation, a *screening detector* may be used. The function of the screening detector is to train a feature-based detector. The screening detector must be non-parametric and should also be computationally efficient. The energy detector is well suited to meet the needs of the screening detector, and it will be assumed in the remainder of this chapter that the screening detector is an energy detector. After the screening detector isolates a training set of appropriate size, this training set is used by the learning algorithm. The learning algorithm identifies the features in the primary user modulated signal which can then be used by the feature-based detector for spectrum sensing. The block diagram of the feature-based detector using imperfect training is shown in Figure 1.

For each *candidate* training observation, the screening detector makes a binary decision on whether the primary user is absent, which is the null hypothesis labelled H_0, or, alternatively, on whether the primary user is currently present, which is the alternative hypothesis, labelled H_1. If the screening detector decides H_1, then the observation is added to the training set. If the screening detector decides H_0, then the observation is discarded. The screening detector performance is imperfect because the detector makes some decision errors. This model represents a learning environment subject to an imperfect teacher (the screening detector) where the student is the learning algorithm. Discussion of imperfect training has previously been presented (Shanmugam & Breipohl, 1971).

While the energy detector is used as the screening detector in this work, any reasonably performing detection algorithm could be used. This model is agnostic to the specific details of the training algorithm and the feature-based detector.

Two-stage detectors have been previously used to achieve previous goals related to spectrum sensing. Coarse/refined detection for improved performance (Zhang & Sanada, 2010) and coarse/fine resolution of occupancy sensing across a wide bandwidth (Luo & Roy, 2008) have been discussed. In this work, a two-stage detector is used to address the problem of detector training and operation, in which both functions are implemented with field observations of channel outputs. The goal of this approach is the creation of a feature-based detector which, after training, will be able to use what it has learned in relation to the features of the primary user signal in its detection algorithm. This leads to the following benefits:

- Superior detection error rate performance relative to the performance of the (blind) screening detector can be achieved by exploiting the learned features; and
- In operation, the feature-based detector has the capacity to *reduce the amount of computation required* by the detection algorithm, by only calculating the detection decision test statistic based on the relevant features only. In the context of cyclostationarity detection, this means that only the subset of the spectral correlation density that changes depending on the presence or absence of the primary user signal needs to be computed, rather than the entire function.

IMPLICATIONS OF IMPERFECT TRAINING ON TRAINING TIME

Quantifying the Training Time

Ideally, the learning algorithm would:

Figure 1. A model for imperfect training of a feature-based detector for cognitive radio, using obser-vations taken from the field and a simple training algorithm to identify cyclic spectral features in the primary user signal

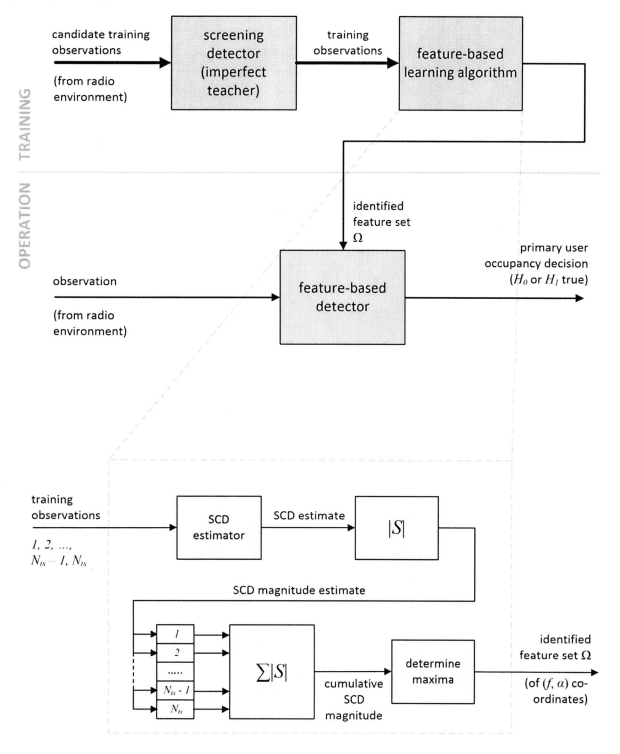

- Have fast learning time, and
- Identify signal features robustly.

Time spent in training is an opportunity lost in secondary user transmission and requires energy. In addition, accurate feature identification is important to optimize the error rate performance of the feature-based detector. However, these outcomes tend to be mutually exclusive, and it shall be seen that a trade-off between these two needs to be made in a feature detector. It shall also be shown that imperfect training may result in degradations in both of the desired outcomes compared to the idealized, perfect scenario of the screening detector making error free decisions.

The training time may be defined in terms of the *average* of candidate training observations encountered, N_{obs}. It may be assumed that some constant time (or known average time) elapses between each such observation. N_{ts} may be defined as the number of observations in the training set deemed to be required to acceptably identify the signal features needed by the feature-based detector during the training process.

If the a-priori probability of the primary user transmitting on a channel is P_1, and the probability of not transmitting on a channel is $P_0 = 1 - P_1$, in the ideal case where a perfect screening detector exists, N_{obs} is given by $N_{obs,ideal}$ whose *average* value is,

$$N_{obs,ideal} = \frac{N_{ts}}{P_1} \tag{1}$$

Under imperfect training, N_{obs} might be greater than $N_{obs,ideal}$. The imperfect screening detector will have some probability of missed detection, P_{md}. Missed detections by the screening detector erroneously exclude from the training set observations for which the primary user is present. This represents a missed training opportunity or *false exclusion* of training observations from the training set.

The imperfect screening detector may also admit into the training set signal observations for which the primary user is absent. This represents false alarms and the probability of a false alarm for the training detector is P_{fa}. False alarms may weaken the identification of features by the learning algorithm due to the erroneous information present in the training set, and may therefore degrade the performance of the feature based detector. Such false alarm events of the screen detector are called *false inclusions* to the training set.

In summary, both false exclusions and false inclusions impinge upon the desirable outcomes of reduced training time and increased reliability in primary user signal feature identification. Where the screening detector's false exclusions exceeds its false inclusions over time, training time is adversely increased. On the other hand, where the screening detector's false inclusions exceeds its false exclusions over time, training time is reduced, but the signal features in the training set will not be as pronounced when considered in aggregate, being diluted by noisy observations (where the primary user's signal is absent) which have no features.

Analysis on Required Training Times Under Imperfect Training

An imperfect detector will always be characterized by a non-zero value of P_{fa} and a non-zero value of P_{md}. Suppose the screening detector calculates a test statistic λ from each of the candidate training observations. The statistic λ is compared to a decision threshold, λ_T. The resulting binary hypothesis test is used to decide whether or not this observation should be included into the training set: if $\lambda \geq \lambda_T$, include it; if $\lambda < \lambda_T$, exclude it. Therefore, under imperfect training,

$$
\begin{aligned}
N_{obs} &= \frac{N_{ts}}{P(\lambda \geq \lambda_T)} \\
&= \frac{N_{ts}}{P_0 P(\lambda \geq \lambda_T \mid H_0) + P_1 P(\lambda \geq \lambda_T \mid H_1)} \\
&= \frac{N_{ts}}{P_0 P_{fa,sc} + P_1(1 - P_{md,sc})}
\end{aligned}
\tag{2}
$$

where $P_{fa,sc}$ is the false alarm probability of the screening detector and $P_{md,sc}$ is the missed detection probability of the screening detector. $P(A)$ represents the probability of event A occurring.

The training time degradation factor, D, is then defined as,

$$
D = \frac{N_{obs}}{N_{obs,ideal}} = \frac{P_1}{P_0 P_{fa,sc} + P_1(1 - P_{md,sc})} = \frac{P_1}{(1 - P_1)P_{fa,sc} + P_1(1 - P_{md,sc})}
\tag{3}
$$

D is actually a normalisation of the total number of observations encountered during the training process, as given by (2), by the number of observations that would be encountered on average if the screening detector was a perfect detector, as given by (1). If $D > 1$, the training time has increased due to the screening detector (but the training outcome in terms of feature identification has been improved), and if $D < 1$, the training time has decreased due to the screening detector (but the training outcome in terms of feature identification has been degraded).

Screening Detector Based on Energy Detection

For the energy detector used as the screening detector, a normalized test statistic may be used given by (Urkowitz, 1967),

$$
\lambda = \frac{2}{\sigma_N^2} \sum_{n=0}^{L-1} |x[n]|^2
\tag{4}
$$

where, $x[n]$ is the discrete-time, complex candidate training signal,

L is the signal observation length in samples, and σ_N^2 is the complex noise power. $\frac{\sigma_N^2}{2}$ is the noise power in each of the in-phase (real) and quadrature (imaginary) parts of $x[n]$. The normalization by $\frac{\sigma_N^2}{2}$ ensures that the test statistic λ represents the sum of the squares of $2L$ Gaussian random variables with unity variance.

The statistical distribution of λ has been investigated (Urkowitz, 1967). Under H_0, λ has a central χ^2 distribution with $k = 2L$ degrees of freedom. The cumulative distribution function (CDF) of λ conditioned on H_0 being true may be denoted as $F_0(\lambda, k)$. Under H_1, λ has a non-central χ^2 distribution with $k = 2L$ degrees of freedom and non-centrality parameter $\Lambda = 2L \times \text{SNR}$. The SNR is the signal to noise ratio which is defined as the ratio between the power of the primary user's transmission and the noise power, both as seen by the cognitive radio receiver within the bandwidth of the channel on which the signal detection is being made. The cumulative distribution function of λ condition upon H_1 being true may be denoted as $F_0(\lambda, k, \Lambda)$.

Then,

$$D = \frac{P_1}{(1 - P_1)(1 - F_0(\lambda_T, k)) + P_1(1 - F_1(\lambda_T, k, \Lambda))} \tag{5}$$

where, $k = 2L$, and, $\Lambda = 2L \times \text{SNR}$.

Asymptotic Behavior for Increasing SNR

Suppose that λ_T is chosen corresponding to a particular value of $P_{fa,sc}$, which is an acceptable percentage of false inclusions. For high SNR, the mean values of λ under H_0 and H_1 will be very disparate. For this case, with a fixed λ_T, as $\text{SNR} \to \infty$, $P_{md,sc} \to 0$. Then, D has an asymptotic approximation given by,

$$D \to \frac{P_1}{(1 - P_1)P_{fa,sc} + P_1}, \text{ for high SNR} \tag{6}$$

Visualization

Figure 2 shows a plot of the training time degradation factor, D, as a function of $P_{fa,sc}$, for various low SNR scenarios with $P_0 = P_1 = 0.5$. For very low SNR values (-9, -7.5, -6 dB), $D > 1$ for low $P_{fa,sc}$ (i.e. large λ_T). While large λ_T is desirable in that it reduces the number of false inclusions, it is undesirable because it causes a large number of false exclusions. Since at low SNR, the probability density functions of λ under H_0 and H_1HHh are relatively close together, and given that $P_0 = P_1$, the missed training opportunities become dominant for large λ_T. The result is $D > 1$. As λ_T decreases, $P_{fa,sc}$ increases and the erroneous inclusions into the training set starts to dominate, such that $D < 1$.

Figure 3 shows corresponding information to Figure 2, but for varying values of P_1 and with SNR fixed at -3 dB. As the primary user channel occupancy probability, P_1, decreases, false exclusions have a diminishing effect (and false inclusions have a dominating effect) upon training time which causes D to decrease more rapidly as a function of $P_{fa,sc}$.

Figure 4 shows D for a low occupancy channel, with $P_1 = 0.1$. This figure also illustrates how the asymptotic expression for D because more accurate as SNR increases. In this scenario, it was found that

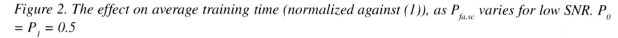

Figure 2. The effect on average training time (normalized against (1)), as $P_{fa,sc}$ varies for low SNR. $P_0 = P_1 = 0.5$

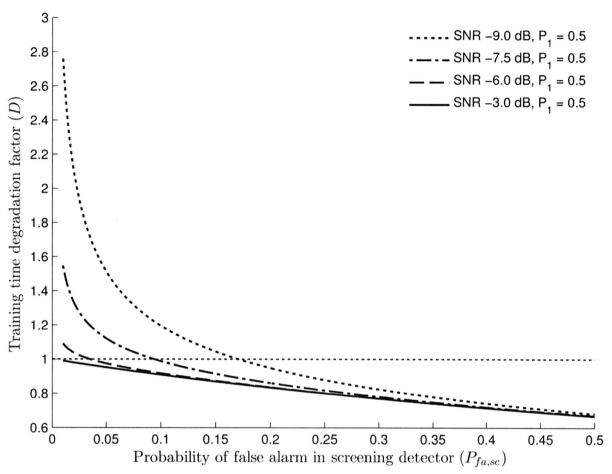

the asymptotic expression, (6), becomes a good approximation for expression (5) when SNR = −6 dB, for $P_{fa,sc} > 0.05$. For larger values of SNR, the agreement between (5) and (6) would be even closer still over all values of $P_{fa,sc}$.

IMPLICATIONS OF IMPERFECT TRAINING ON DETECTION PERFORMANCE

It is intuitive that as $P_{fa,sc}$ increases, the feature-based detector's performance will deteriorate with more false inclusions during training. A MATLAB simulation can verify this empirically, whereby a simple

Figure 3. The effect on average training time (normalized against (1)), as P_1 varies, for SNR = −3 dB

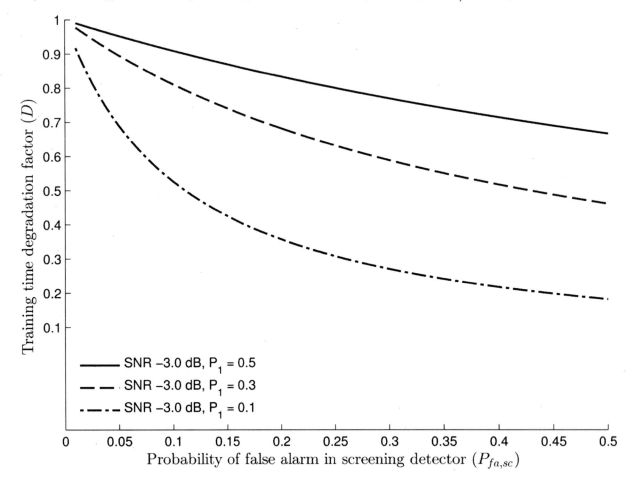

learning algorithm identified the features present in the primary users signal's spectral correlation density, which is denoted $S(f,\alpha)$. These features in $S(f,\alpha)$ are used in cyclostationarity detection.

The learning algorithm is outlined in Figure 1. For every signal observation ultimately admitted into the training set by the screening detector, its spectral correlation density was estimated using the FFT accumulation method (FAM) (Roberts, Brown, & Loomis, 1991). The magnitude of the estimated spectral correlation density was obtained for each training observation, and the magnitudes from all training observations were summed together. From this summed discrete function, the points with the M largest values on the (f,α) cyclic spectral plane were identified using a sorting operation. The coordinates of these M points were then used to form a set of points, Ω, for the cyclostationarity detector to use. It should be noted that other training algorithms may be used besides a simple maximal feature magnitude approach, and these may offer better feature identification performance or efficiency. This is beyond the scope of this work.

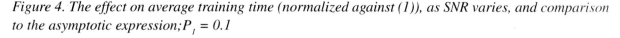

Figure 4. The effect on average training time (normalized against (1)), as SNR varies, and comparison to the asymptotic expression; $P_1 = 0.1$

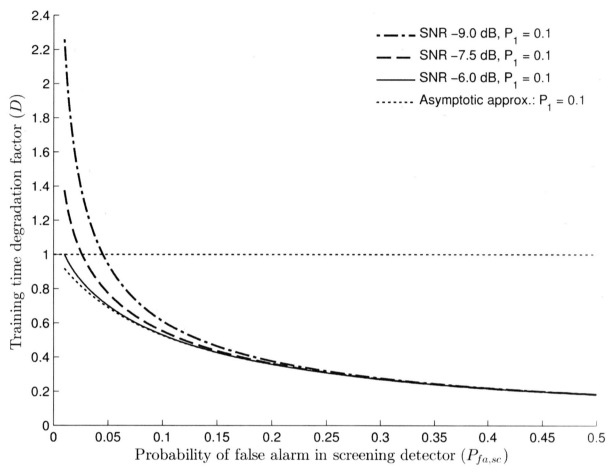

Figure 5 shows an example of identified features obtained from simulation for the spectral correlation density training process described. The simulation was repeated with $P_{fa,sc} = 0.05$ and 0.5. It was assumed that the primary user transmitted using OFDM with equally-spaced pilot subcarriers. The M feature points identified by the algorithm as being elements of Ω are shown as black dots. For $P_{fa,sc} = 0.05$, the selected feature points are closer to the feature points that would be expected for an OFDM signal with pilot subcarriers (i.e. peaks in the spectral correlation density would be expected at the points of intersection on the grid pattern) than when $P_{fa,sc} = 0.5$. For $P_{fa,sc} = 0.05$, there are a reduced number of false inclusions in the determination of Ω, which explains this observation.

After training is complete, the feature detector enters its operational phase where it tries to identify whether or not the primary user is transmitting on a channel during an observation time. The feature detector operates by calculating a test statistic, γ, which is obtained by summing the magnitude of an observation's spectral correlation density at all (f,α) co-ordinates in the feature set Ω, as well as all co-ordinates where $\alpha = 0$. This resulting test statistic is then given by,

Figure 5. Plots of the primary user modulated signal features selected using the aggregate spectral correlation density magnitude, resulting from training. The primary user signal was OFDM with pilot subcarriers. Under perfect screening, the aggregate spectral correlation density magnitude would have peaks located at the points of intersection in a diagonal grid pattern. Results are shown for two values of $P_{fa,sc}$. The identified features are less accurate for $P_{fa,sc} = 0.5$ compared to $P_{fa,sc} = 0.05$. As $P_{fa,sc}$ increases, the larger number of false inclusions during training decrease the accuracy of feature identification.

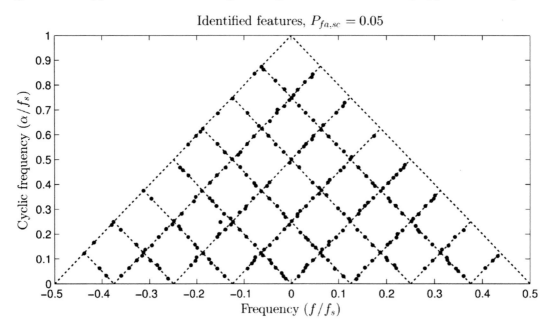

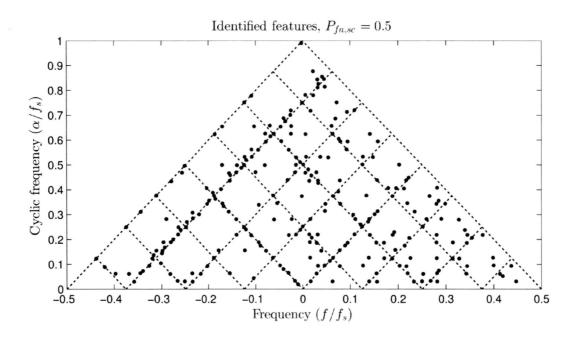

$$\gamma = \sum_{\substack{((f,\alpha)\in\Omega) \\ \cup((f,\alpha):\alpha=0)}} |S(f,\alpha)| \qquad\qquad (7)$$

The rationale for including all points where $\alpha = 0$ in the calculation of γ is the observation that *for these points, the spectral correlation density of a signal is equivalent to its power spectral density* (Gardner, 1991). Consequently, a signal's spectral correlation density gives at least as much information as its power spectral density. Therefore, *if only the points where $\alpha = 0$ were to be included in the summation, then the detector would be equivalent to an energy detector*. By also including the points in Ω, additional signal feature information is exploited – hence, the cyclostationarity detector is able to achieve detection error rate performance that will, *by definition*, be at least as good as that of an energy detector.

A detection decision is determined by comparing γ to the threshold, γ_T. The simulation determined the detection error rate performance for a range of values of γ_T - *these are points on a single detection error performance curve*. The simulation also measured the performance of the feature-based detector for various values of $P_{fa,sc}$ employed during training - *these resulted in different error detection performance curves*. Some parameters of the detection simulation are listed in Table 1, and Figure 6 shows the resultant error detection performance curves. A low value of P_1 was selected because this is representative of the under-utilized channel situations which is most useful for cognitive radio.

The simulation approach can be described in more detail as follows. Firstly the cognitive radio undergoes a training phase:

1. A value of $P_{fa,sc}$ is selected, and the corresponding screening detector threshold λ_T determined. Since the screening detector is an energy detector, this can be calculated by:

$$\lambda_T = F_0^{-1}(1 - P_{fa,sc}) \qquad\qquad (8)$$

2. The screening detector is presented with signal observations from the radio environment, of length L samples, that randomly represent either the primary user signal (seen by the cognitive radio with

Table 1. Training and detection simulation parameters

Parameter	Description
L (signal observation length)	256
N_{ts} (training set size – see Figure 1)	96
Screening Detector Type	Energy Detector
Primary user signal	OFDM, 1024 carriers per symbol, unmodulated pilot carriers with amplitude 7 repeated every 128 carriers, guard interval 1/8 symbol
SNR (as seen by the cognitive radio with respect to the primary user's signal)	-3 dB
P_0, P_1	0.9, 0.1
M (no. cyclostationarity feature points identified during training relating to the primary user signal)	256

Figure 6. Cyclostationarity detector performance for different $P_{fa,sc}$, as indicated by detection error performance curves

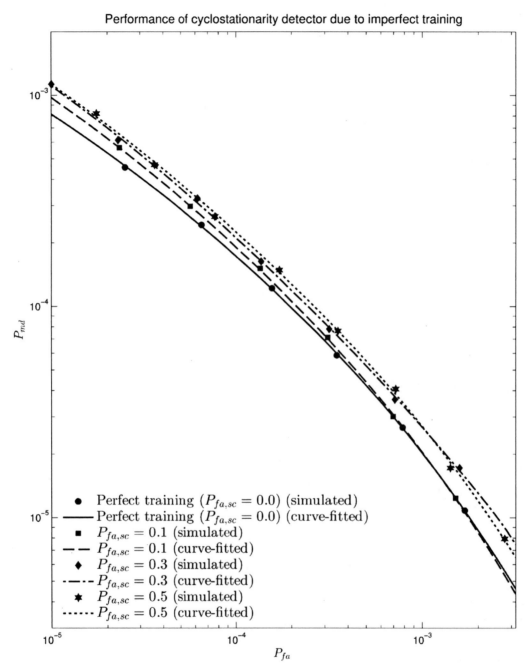

the specified SNR) or additive white Gaussian noise (with the same noise power), according to probabilities P_0 and P_1.

3. The training set is comprised of the first N_{ts} signal observations whose energy detector test statistic, as described by (4), exceeds λ_T. For each observation here, the spectral correlation density was estimated, and the magnitude of each estimate was summed together.

4. From the cumulative spectral correlation density estimate, the cyclostationarity features of the primary user was deemed to be the locations on the (f,α) plane corresponding to the M points with the largest magnitude.

5. The complete set of features used by the cyclostationarity detector, Ω, is then the union of these M points with all points where $\alpha = 0$ (the latter defining the power spectral density of an observed signal).

Then, the operational phase of the cognitive radio is simulated as follows:

1. The cyclostationarity detector is presented with a large number of signal observations from the radio environment, of length L samples, representing additive white Gaussian noise only. For each observation, the detection test statistic γ is calculated according to (7).

2. Using this set of test statistics, an estimate of the probability of false alarm for various values of detection threshold γ_T is obtained, by determining the fraction of test statistics whose values lie above γ_T.

3. The cyclostationarity detector is presented with a large number of signal observations from the radio environment, of length L samples, representing the primary user signal (seen by the cognitive radio with the specified SNR). For each, the detection test statistic γ is calculated according to (7).

4. Using this set of test statistics, and the same detection thresholds as in step 2, an estimate of the *corresponding* probability of missed detection is obtained, by determining the fraction of test statistics whose value lies below each threshold.

After the empirical detection error performance points were determined, a smooth curve fit was applied using an exponential model to obtain an approximate relationship between P_{fa} and P_{md}, described by:

$$\log_{10}(P_{md}) = ae^{b(\log_{10}(P_{fa}))} + ce^{d(\log_{10}(P_{fa}))} \tag{9}$$

where a, b, c and d are fitted constants.

As indicated in Figure 6, there is perceptible performance degradation due to imperfect training as $P_{fa,sc}$ increases. The sensitivity of the detection performance curves to $P_{fa,sc}$ is dependent upon the relative strength of the features selected in the cyclostationarity analysis compared to noise.

To summarize, this section has demonstrated through a comprehensive simulation that a high value of $P_{fa,sc}$ used in the screening detector tends to result in a trained cyclostationarity detector with a poorer error detection performance curve. This is due to the increased prevalence of false inclusions into the training set.

TRADING OFF DETECTION TIME AND DETECTION PERFORMANCE

The preceding discussions have confirmed the trade-off between shortened training time and accurate detection performance, as $P_{fa,sc}$ varies. This trade-off may be formulated in an optimization problem, whose solution is an appropriate choice of $P_{fa,sc}$ and the corresponding appropriate value of the screening detector threshold, λ_T. The function to optimize is a cost function. The cognitive radio will incur several costs during its operation. These costs include:

- Cost of training, C_{train}. C_{train} is the opportunity cost on data transmission.
- Cost of false alarm, C_{fa}. C_{fa} is the cost of not exploiting an available spectral opportunity. This is an opportunity cost on data transmission.
- Cost of missed detection, C_{md}. C_{md} is the cost of interference by the secondary user to the primary user. This is a penalty, which may be financial, for the creation of the interference.

Let τ_{train} be the average time spent on training. τ_{train} is a function of λ_T, which in turn is a function of $P_{fa,sc}$. It is possible to specify D as a function of $P_{fa,sc}$, denoted by $D(P_{fa,sc})$.

Referring to (5) and (8),

$$D(P_{fa,sc}) = \frac{P_1}{(1 - P_1)P_{fa,sc} + P_1(1 - F_1(F_0^{-1}(1 - P_{fa,sc})))} \tag{10}$$

Then, with the assumption that the cost of training is directly proportional to training time,

$$C_{train} \propto \tau_{train} = k_1 N_{obs} = k_1 D(P_{fa,sc})\left(\frac{N_{ts}}{P_1}\right) \tag{11}$$

where, k_1 is a weighting constant.

After the completion of training, the cognitive radio may operate for some time, τ_{op}, before being required to retrain for a new primary user modulation or a new geographical location. During this time, the channel will be occupied by the primary user for average time $P_1\tau_{op}$ and vacant for average time $P_0\tau_{op}$. Under the assumption that the costs of false alarm and missed detection are both directly proportional to the average operational times,

$$C_{fa} = k_2 P_{fa} P_0 \tau_{op} = k_2 P_{fa}(1 - P_1)\tau_{op}$$
$$C_{md} = k_3 P_{md} P_1 \tau_{op} \tag{12}$$

where k_2 and k_3 are also weighting constants. Then, the total cost function is,

$$C = C_{train} + C_{fa} + C_{md}$$

$$C(P_{fa,sc}, P_{fa}, P_{md}) = k_1 D(P_{fa,sc}) \left(\frac{N_{ts}}{P_1} \right) + k_2 P_{fa}(1 - P_1)\tau_{op} + k_3 P_{md} P_1 \tau_{op} \tag{13}$$

A cognitive radio will set its threshold, γ_T, for the feature-based detector in operation. This value of γ_T will determine the P_{fa} and P_{md} values that characterize the feature-based detector's operation - characterized by a single point on the error detection performance curve. Furthermore, it has been seen that the detection error rate performance curve that defines the relationship between P_{fa} and P_{md} varies with $P_{fa,sc}$. From (13), the cost C is a function of these three variables, so the optimization problem may then be stated as,

$$\arg\min_{P_{fa,sc}, P_{fa}, P_{md}} \left[k_1 D(P_{fa,sc}) \left(\frac{N_{ts}}{P_1} \right) + k_2 P_{fa}(1 - P_1)\tau_{op} + k_3 P_{md} P_1 \tau_{op} \right] \tag{14}$$

In other words, the 'optimal' point is defined by the value of $P_{fa,sc}$ (training), P_{fa} and P_{md} (operation) that gives the minimum combined cost C. $P_{fa,sc}$ determines which error detection performance curve the feature-based detector is subject to, whereas P_{fa} and P_{md} define the 'operating point' on this curve. With detection error rate performance curves determined for a wide variety of values of $P_{fa,sc}$, it is possible to use numerical techniques to obtain an approximate solution to this optimization problem, described as follows:

1. Select a value of $P_{fa,sc}$ for which the feature-based detector's corresponding detection error rate performance curve has been characterized (according to simulation and subsequent curve-fitting procedure indicated by (9)).
2. For the minimum value of P_{fa} on the error detection performance curve, find the corresponding value of P_{md}.
3. Evaluate the cost C using (14).
4. Repeat steps 2 - 3 for $P_{fa} + \Delta P_{fa}, P_{fa} + 2\Delta P_{fa}, P_{fa} + 3\Delta P_{fa}, \dots$, where ΔP_{fa} is small, stopping at the maximum value of P_{fa} on the error detection performance curve.
5. Repeat from step 1, for all values of $P_{fa,sc}$ for which the feature-based detector's detection error rate performance curve has been characterized (according to (9)).

The results of carrying out this procedure can be illustrated with the detection error rate performance curves shown in Figure 6. In Figure 7, the cost function is evaluated and graphed as one 'traverses along' (changes the operating point on) the detection error performance curve, for $P_{fa,sc} = 0.1, 0.3$ and 0.5. ($P_{fa,sc} = 0$ was ignored, as this is not possible to achieve in reality.) In this case, the optimal solution is to set the threshold on the screening detector during training such that $P_{fa,sc} = 0.5$. Then, for the corresponding detection error rate performance curve, the detection threshold λ_T of the feature-based detector is set to give values of P_{fa} and P_{md} that result in $C = 0.392$.

Similarly, in Figure 8, the approach is repeated for a different set of weighting constants, leading to a different solution. In this case, the optimal solution is to set the threshold on the screening detector

Figure 7. The cost function, C, for $P_{fa,sc}$ = 0.1, 0.3 and 0.5, when $k_1 = \dfrac{P_1}{N_{ts}}$, k_2 = 80, k_3 = 640, τ_{op} = 10, P_1 = 0.1 . The value of k_1 was chosen to give a training cost of C_{train} = 1 when the training time degradation factor D = 1.

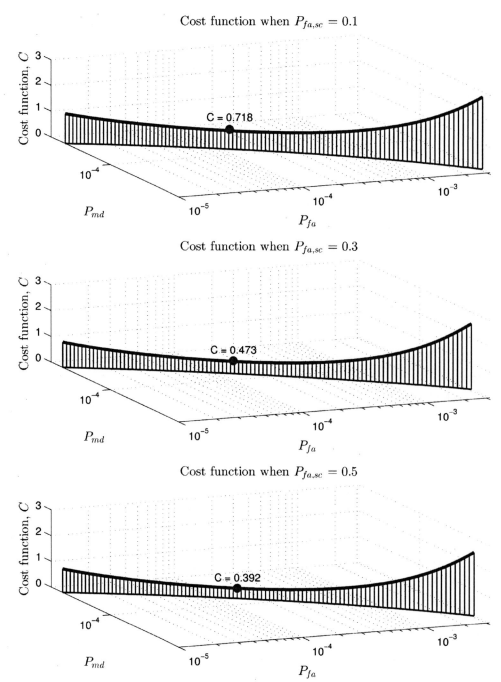

Figure 8. The cost function, C, for $P_{fa,sc}$ = 0.1, 0.3 and 0.5, when $k_1 = \dfrac{P_1}{N_{ts}}$, $k_2 = 2000$, $k_3 = 8000$, $\tau_{op} = 10$, $P_1 = 0.1$. The value of k_1 was chosen to give a training cost of $C_{train} = 1$ when the training time degradation factor D = 1.

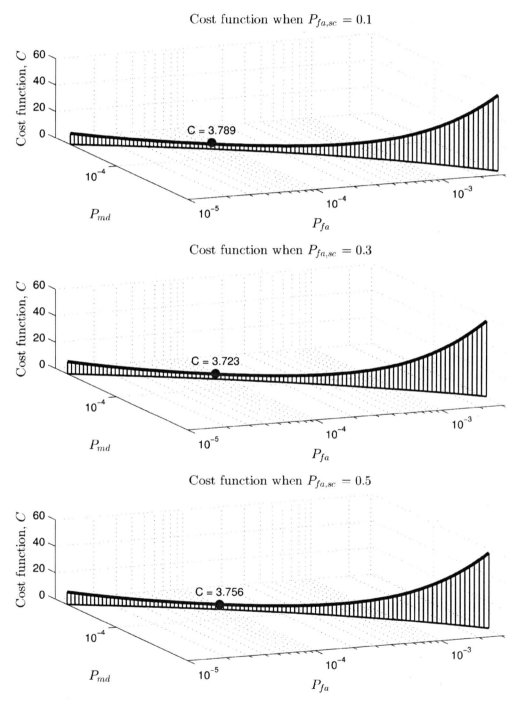

during training such that $P_{fa,sc} = 0.3$. Then, for the corresponding detection error rate performance curve, the detection threshold λ_T of the feature-based detector is set to give values of P_{fa} and P_{md} that result in $C = 3.723$.

It is apparent that the values of the weighting constants play a role in the solution to the optimization problem. It should also be noted that the cost function has only been evaluated in these examples for three values of $P_{fa,sc}$, and thus the solution is only optimal subject to this limited search space. By conducting simulations at other values of $P_{fa,sc}$, a larger ensemble of error detection performance curves can be obtained, and the search for the optimal solution would then proceed to take this new information into account.

FUTURE RESEARCH DIRECTIONS

In this work, it was assumed that the only channel perturbation between the primary user and the cognitive radio was additive Gaussian noise. A next step for further investigation would be to consider the effect of channel fading on both the training and operational stages. Also, this work illustrates how the performance of the feature-based detector can vary depend upon the value of the detection threshold applied to signal observations taken by the screening detector during training. However, the detection error rate performance curves were established empirically, and will change for different detection algorithms and primary user signal characteristics. Consideration of how to derive suitable expressions for, or approximations to, these functions in a more general way may lead to useful outcomes. Finally, the numerical optimization illustrated in Figures 7 and 8 only consider a fairly limited set of detection error performance curves at different threshold values $P_{fa,sc}$ during training. Further simulations can be carried out so that the same procedure is carried out on a wider gamut of known detection error rate performance curves.

CONCLUSION

A model for a two-stage detector structure has been described. This detector has been applied to cognitive radios employing a feature-based signal detector for spectrum sensing. The first stage employs a screening detector, whose function is to identify a suitable training set from radio observations taken in the field and to identify and learn the primary user's modulated signal features. The second stage is a feature-based detector used for spectral sensing. This second stage detector exhibits superior detection error rate performance compared to that achievable by the screening detector, due to the utilization of signal features learned by the screening detector. The screening detector itself is not perfect. This leads to deteriorative effects such as lengthened training time, in cases such as low signal to noise ratio, and worse feature detection performance compared to detection with perfect feature training. There is a trade-off between the desirable detector characteristics of short training time and accurate detection performance. This trade-off function depends upon the decision threshold used in the screening detector. An optimization has been formulated for the selection of the decision threshold of the screening detector, in which a cost function based upon the expected training time and detection error rates is minimized.

ACKNOWLEDGMENT

The authors would like to thank the Commonwealth Government of Australia for their support of this research through the Australian Postgraduate Awards Program.

REFERENCES

Gardner, W. A. (1991). Exploitation of spectral redundancy in cyclostationary signals. *IEEE Signal Processing Magazine, 8*(2), 14–36. doi:10.1109/79.81007

Luo, L., & Roy, S. (2008). *A two-stage sensing technique for dynamic spectrum access.* Paper presented at the IEEE International Conference on Communications (ICC '08). Beijing, China. doi:10.1109/ICC.2008.785

Reisenfeld, S. (2009). *Performance bounds for detect and avoid signal sensing.* Paper presented at the Second International Workshop on Cognitive Radio and Advanced Spectrum Management (CogART 2009). Aalborg, Germany. doi:10.1109/COGART.2009.5167248

Roberts, R. S., Brown, W. A., & Loomis, H. H. Jr. (1991). Computationally efficient algorithms for cyclic spectral analysis. *IEEE Signal Processing Magazine, 8*(2), 38–49. doi:10.1109/79.81008

Shanmugam, K., & Breipohl, A. M. (1971). An error correcting procedure for learning with an imperfect teacher. *IEEE Transactions on Systems, Man, and Cybernetics, 1*(3), 223–229. doi:10.1109/TSMC.1971.4308289

Thai, Q., & Reisenfeld, S. (2011). *Adaptive exploitation of cyclostationarity features for detecting modulated signals.* Paper presented at the International Conference on Advanced Technologies for Communications (ATC). New York, NY. doi:10.1109/ATC.2011.6027450

Urkowitz, H. (1967). Energy detection of unknown deterministic signals. *Proceedings of the IEEE, 55*(4), 523–531. doi:10.1109/PROC.1967.5573

Yucek, T., & Arslan, H. (2009). A survey of spectrum sensing algorithms for cognitive radio applications. *IEEE Communications Surveys and Tutorials, 11*(1), 116–130. doi:10.1109/SURV.2009.090109

Zhang, W., & Sanada, Y. (2010). *Dual-stage detection scheme for detect and avoid.* Paper presented at the IEEE International Conference on Ultra-Wideband (ICUWB). Nanjing, China. doi:10.1109/ICUWB.2010.5615282

ADDITIONAL READING

Bhargava, D., & Murthy, C. R. (2010). *Performance comparison of energy, matched-filter and cyclostationarity-based spectrum sensing.* Paper presented at the IEEE 11th International Workshop on Signal Processing Advances in Wireless Communications, Marrakech. doi:10.1109/SPAWC.2010.5670882

Bourgeois, T., & Sanada, Y. (2011). *Cyclostationarity-based detector for OFDM signals with correlated pilot subcarriers.* Paper presented at the 14th International Symposium on Wireless Personal Multimedia Communications (WPMC), Brest.

Dudley, S. M., Headley, W. C., Lichtman, M., Imana, E., Ma, X., & Abdelbar, M. et al. (2014). Practical issues for spectrum management with cognitive radios. *Proceedings of the IEEE, 102*(3), 242–264. doi:10.1109/JPROC.2014.2298437

Fazeli-Dehkordy, S., Konstantinos, N. P., & Pasupathy, S. (2010). *Two-stage spectrum detection in cognitive radio networks.* Paper presented at the IEEE International Conference on Acoustics, Speech, and Signal Processing, Texas. doi:10.1109/ICASSP.2010.5496090

Gardner, W. A. (1987). Spectral correlation of modulated signals. *IEEE Transactions on Communications, 35*(6), 584–601. doi:10.1109/TCOM.1987.1096820

Gardner, W. A. (1988). Signal interception: A unifying theoretical framework for feature detection. *IEEE Transactions on Communications, 36*(8), 897–906. doi:10.1109/26.3769

Ghasemi, A., & Sousa, E. S. (2008). Spectrum sensing in cognitive radio networks: Requirements, challenges and design trade-offs. *IEEE Communications Magazine, 46*(4), 32–39. doi:10.1109/MCOM.2008.4481338

Islam, M. H., Koh, C. L., & Oh, S. W., & al., e. (2008). *Spectrum survey in Singapore: occupancy measurements and analysis.* Paper presented at the Third International Conference on CROWNCOM, Singapore. doi:10.1109/CROWNCOM.2008.4562457

Li, Z., Wang, H., & Kuang, J. (2011). *A two-step spectrum sensing scheme for cognitive radio networks.* Paper presented at the International Conference on Information Science and Technology, Nanjing, China.

Luo, L., Neihart, N. M., & Roy, S. (2009). A two-stage sensing technique for dynamic spectrum access. *IEEE Transactions on Wireless Communications, 8*(6), 3028–3037. doi:10.1109/TWC.2009.080482

Maeda, K., Benjebbour, A., & Asai, T. (2007). *Recognition among OFDM based systems utilizing cyclostationary inducing transmission.* Paper presented at the Second IEEE International Symposium on New Frontiers in Dynamic Spectrum Access Networks, Dublin, Ireland. doi:10.1109/DYSPAN.2007.74

Maleki, S., Pandaripande, A., & Leus, G. (2010). *Two-stage spectrum sensing for cognitive radios.* Paper presented at the IEEE International Conference on Acoustics, Speech, and Signal Processing, Texas.

Nair, P. R., Vinod, A. P., & Krishna, A. K. (2011). *A fast two-stage detector for spectrum sensing in cognitive radios.* Paper presented at the IEEE Vehicular Technology Conference, San Francisco. doi:10.1109/VETECF.2011.6092897

Nair, P. R., Vinod, A. P., Smitha, K. G., & Krishna, A. K. (2012). Fast two-stage spectrum detector for cognitive radios in uncertain noise channels. *Communications, IET, 6*(11), 1341–1348. doi:10.1049/iet-com.2011.0751

Peh, E. C., & Liang, Y. C. (2007). *Optimization for cooperative sensing in cognitive radio networks.* Paper presented at the IEEE Wireless Communications and Networking Conference, Kowloon. doi:10.1109/WCNC.2007.11

Schunur, S. R. (2009). *Identification and classification of OFDM based signals using preamble correlation and cyclostationary feature extraction.* (Masters Thesis), Naval Postgraduate School, Monterey, California.

Scudder, H. I. (1965). Adaptive communication receivers. *IEEE Transactions on Information Theory, 11*(2), 167–174. doi:10.1109/TIT.1965.1053752

Shanmugam, K. (1972). A parametric procedure for learning with an imperfect teacher. *IEEE Transactions on Information Theory, 18*(2), 300–302. doi:10.1109/TIT.1972.1054780

Sohn, S. H., Han, N. K., & Jae, M., & al., e. (2007). *OFDM signal sensing method based on cyclostationary detection.* Paper presented at the Second International Conference on Cognitive Radio Oriented Wireless Networks and Communications, Orlando, Florida. doi:10.1109/CROWNCOM.2007.4549773

Sonnenschien, A., & Fishman, P. M. (1992). Radiometric detection of spread spectrum signals in noise of uncertain power. *IEEE Transactions on Aerospace and Electronic Systems, 28*(3), 654–660. doi:10.1109/7.256287

Sutton, P. D., Nolan, K. E., & Doyle, L. (2008). Cyclostationary signatures in practical cognitive radio applications. *IEEE Journal on Selected Areas in Communications, 26*(1), 13–24. doi:10.1109/JSAC.2008.080103

Tandra, R., & Sahai, A. (2008). SNR walls for signal detection. *IEEE Journal of Selected Topics in Signal Processing, 2*(1), 4–17. doi:10.1109/JSTSP.2007.914879

Wang, B., & Liu, K. (2011). Advances in cognitive radio networks: A survey. *IEEE Journal of Selected Topics in Signal Processing, 5*(1), 5–23. doi:10.1109/JSTSP.2010.2093210

Zahedi-Ghasabeth, A., Tarighat, A., & Daneshrad, B. (2010). *Spectrum sensing of OFDM waveforms using embedded pilot carriers.* Paper presented at the IEEE International Conference on Communications, Cape Town.

Zeng, Y., Koh, C. L., & Liang, Y. C. (2008). *Maximum eigenvalue detection: theory and application.* Paper presented at the IEEE International Conference on Communications (ICC), Beijing.

Zhang, Q., Kokkeler, A. B. J., & Smit, G. J. M., & al., e. (2008). *An efficient multi-resolution spectrum sensing method for cognitive radio.* Paper presented at the Third International Conference on Communications and Networking, Hangzhou, China.

Zheng, Y., Luang, Y. C., & Hoang, A. I. (2010). A review on spectrum sensing for cognitive radio: Challenges and solutions. *EURASIP Journal on Advances in Signal Processing, 2010*, 1–15. doi:10.1155/2010/381465

Zhou, Y., & Tian, F. (2010). *A spectrum sensing algorithm based on random matrix theory in cognitive radio networks.* Paper presented at the IEEE International Conference on Wireless Communication Systems. doi:10.1109/ISWCS.2010.5624506

KEY TERMS AND DEFINITIONS

Cyclostationarity Detection: A category of signal detection approaches exploiting the presence of certain periodicities which tend to be present in an information-bearing signal, but absent in a signal comprised of only noise. In this chapter, cyclostationarity detection involves exploiting information in a signal observation's spectral correlation density estimate. It is a form of feature-based detection.

Detection Error Performance Curve: A locus of points that describes how the detector's *probability of false alarm* and *probability of missed detection* vary, parameterized by the detection threshold used. In other words, every point on the curve (an *operating point*) represents a different detection threshold.

Energy Detection: A signal detection approach that use the energy in a signal observation as the sole discriminant to determine whether or not an information-bearing signal is present.

Feature-Based Detection: A category of signal detection approaches that exploit signal *features* (beyond its energy) to determine whether or not it is present in a signal observation. Unlike energy detection which makes use of a signal characteristic (i.e. its energy) that is applicable to all signals, different signals may have different features. A feature-based detector must *know* or *learn* what these features are, in order to operate without performing redundant computation. Cyclostationarity detection is an example of this type of detection.

Imperfect Training: Describes a training process for a learning algorithm where there is some uncertainty as to whether or not all of the elements in the training set are truly representative of the features that need to be learned. For example, a feature-based detector that is learning to identify the features in a signal may be presented with observations in the training set where the signal is, in fact, absent.

Screening (Training) Detector: A signal detector that is responsible for isolating signal observations which are suitable for use by a learning algorithm. Since no detector is perfect to begin with, this leads to imperfect training.

Signal Detection: The process of *deciding* whether or not a signal of interest is present in a signal observation. This decision is often made via a hypothesis test, and gathering evidence via a signal detection algorithm to either support the null hypothesis H_0 (hence deciding the signal is absent) or reject it (hence deciding the signal is present, which is the alternative hypothesis H_1). In the context of cognitive radios, the 'signal of interest' is the transmission of a primary user.

Signal-to-Noise Ratio (SNR): In the context of cognitive radios, this is the ratio between the primary user's signal power *as observed by the cognitive radio's receive antenna* during spectrum sensing, and the noise power similarly observed, within a defined frequency band.

Spectral (Spectrum) Sensing: The task of determining which frequency bands in a cognitive radio's environment is vacant at a given time. This is generally accomplished by employing signal detection.

Spectral Correlation Density: A two-dimensional function (in frequency (f) and *cyclic frequency* (α)) of a signal. Conceptually, it provides an indication of the average amount of correlation between the signal's components at frequencies $f + 0.5\alpha$, $f - 0.5\alpha$. Cyclostationary signals have some non-zero values for this function, for $\alpha \neq 0$.

Chapter 9
Sensing Orders in Multi-User Cognitive Radio Networks

Rakesh Misra
Stanford University, USA

Arun Pachai Kannu
Indian Institute of Technology Madras, India

ABSTRACT

In multi-channel Cognitive Radio Networks (CRNs), when the cognitive radio receivers cannot simultaneously sense more than one out of the many possible (groups of) channels, an important challenge is to determine a sensing order for each Cognitive User (CU) so as to optimize a given performance metric. The sensing-order problem is compounded in multi-user CRNs where the multiple users in the network could collide with each other. With the focus on multi-user CRNs, this chapter uses cognitive-throughput maximization as the performance metric and describes how the optimal sensing orders can be computed for different contention management strategies used by the network. In general, the optimal procedures involve a computationally expensive brute-force search, so the chapter also discusses several heuristic-based near-optimal procedures that can be used in practice.

INTRODUCTION

In an age where the wireless RF spectrum is becoming increasingly congested and scarcer than ever, the concept of *cognitive radios* has attracted significant attention as a means for enhancing the efficiency of spectrum usage (Hossain, Niyato & Han, 2009). A cognitive radio is a smart radio that can intelligently detect available wireless channels in its vicinity and dynamically configure its transmission or reception parameters so as to make the best use of these *spectrum holes*. Cognitive radios are capable of communicating on licensed spectrum without causing interference to the incumbent or *primary* users of the licensed bands, and therefore hold great potential for improving the efficiency of the usage of licensed spectrum that is otherwise poorly utilized due to static frequency allocations.

However, the spectrum that a cognitive radio would be allowed to operate on can be expected to be scattered and heterogeneous in general. In other words, a cognitive radio would need to search over

DOI: 10.4018/978-1-4666-6571-2.ch009

multiple portions of the licensed spectrum, possibly having different bandwidths and primary user characteristics, in order to select the best free channel for its use. Also, these radios are expected to be small devices, often hand-held with limited footprint, and therefore cognitive radios that can simultaneously sense more than one portion of the spectrum quickly, efficiently and reliably would be expensive to build in practice. As a result, each cognitive user (CU) in a cognitive radio network (CRN) would need to have a *sensing-order* i.e., an order in which it will sequentially sense the different channels until it finds a suitable channel for its communication.

The *sensing-order problem* in cognitive radio networks relates to finding the sensing orders for the cognitive users in the network that jointly optimize a chosen objective function. For example, if the objective is to minimize the expected sum of sensing times or equivalently, the expected sum of times to transmission, the sensing orders need to be jointly selected in a way that lets the CUs find disjoint and free channels as early as possible in their respective sensing orders. A more popular objective, which is also the objective of interest throughout this chapter, is the expected sum throughput of all CUs, also referred to as the *cognitive throughput*. When cognitive throughput is used as the objective, each CU needs to find not just a free channel but also a good quality channel i.e., if a channel it senses to be free has a very bad signal-to-noise ratio (SNR), it can skip the channel and continue sensing according to its sensing order with the good hope of finding a better quality channel in the future that will result in a higher effective throughput. Therefore, with cognitive throughput maximization, the sensing-order problem also encompasses determining *optimal stopping rules* for each CU i.e., a rule based on its instantaneous channel conditions that it can use whenever it finds a free channel to decide whether to stop sensing or continue sensing with the hope of finding a better channel in the future.

One might wonder, "If the different channels are identical except for their *primary-free probabilities*, isn't it optimal to just sense the channels in the descending order of their primary-free probabilities?" Jiang, Lai, Fan and Poor (2009) showed that when there is a single CU in the network, this intuitive sensing order is indeed optimal when the CU does not use rate adaptation i.e., when the CU transmits at a constant rate irrespective of the channel quality. However, if the CU uses rate adaptation, the intuitive sensing order is not optimal in general. Finding the optimal sensing order involves computing the cognitive throughput as a function of the sensing order and using a dynamic programming approach to search for the optimal sensing order. The section titled *Background* briefly discusses these and other results for a single-user CRN that will set the background for the material that follows in this chapter.

This chapter focuses on multi-user CRNs. In multi-user CRNs, expressing the cognitive throughput itself becomes challenging because in addition to the complexities in a single-user CRN, the cognitive throughput also depends on how contentions are managed among CUs, or in other words, how multiple CUs are allowed to access a common set of channels. Based on earlier work (Misra & Kannu, 2012a; Misra & Kannu, 2012b), this chapter considers the following two broad lines of contention management strategies and studies the optimal sensing order separately for each of them.

1. *Contention avoidance strategies* (studied in detail in the section titled *Centralized Sensing Order*), where a central coordinator in the network makes all spectrum access decisions and ensures that two or more CUs never contend for the same channel. In such networks, the coordinator has a sensing order, referred to as a *centralized sensing order*, which it uses to sequentially decide which CU, if any, should be allowed to access each channel. Note that the spectrum sensing is still performed locally by the individual CUs who report their measurements to the coordinator, so the centralized sensing order is really just an order for making spectrum access decisions.

2. *Contention resolution strategies* (studied in detail in the section titled *Decentralized Sensing Orders*), where each CU makes its own spectrum access decisions and the network uses a pre-defined mechanism, for e.g. exponential back off, to resolve contentions whenever two or more CUs request access to the same channel simultaneously. In such networks, the CUs have their own sensing orders, referred to as *decentralized sensing orders*, which they use to search for suitable channels for their communication.

This chapter presents a systematic study of the sensing-order problem for several scenarios that fall within the purview of the above-mentioned strategies. In general, determining the optimal sensing order involves a prohibitively complex brute-force search across all possible sensing orders – a space that is huge even for a moderately large number of channels and CUs. Hence, this chapter also discusses heuristic-based sub-optimal procedures for different scenarios that are shown to achieve near-optimal performance with significantly lesser complexities.

MODEL: COGNITIVE RADIO NETWORKS

A general cognitive radio network (CRN) comprises of multiple channels, multiple cognitive users (CUs) that can access the channels in a way that does not interfere with the primary users of the channels, and a coordinator that facilitates this spectrum access. Figure 1 shows a CRN with M CUs, each trying to access one of N channels via a central coordinator. The coordinator is a centrally located node that can sense the traffic patterns of the primary users within the geographical confines of the CRN.

In cognitive radio literature, a commonly used traffic model for primary users for studying the sensing-order problem is a slotted-time arrival model (Jiang et al., 2009; Fan & Jiang, 2009). In this model, time is partitioned into slots of fixed duration T and the primary users can start their transmissions only at the beginning of a time slot. In other words, if a channel is sensed to be free of primary traffic after the beginning of a slot, it can be assumed to remain free for the remainder of the slot. Such a slotted-time model allows us to find the optimal sensing order(s) at the beginning of a time slot with the aim of maximizing the cognitive throughput in the given slot. Note that T is a parameter of this model - the

Figure 1. Illustration of a general cognitive radio network. The coordinator continually estimates the primary-free probabilities of the different channels and computes the sensing order(s) in the network

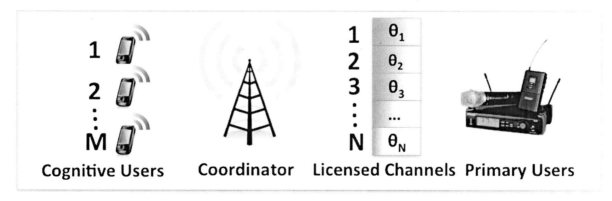

smaller T is chosen to be, the closer this discrete-time arrival process can be made to approximate any continuous-time arrival process; although, as described later, the sensing duration imposes a limit on how small T can be.

With slotted-time operation, the network is assumed to be synchronous. In any time slot, channel i is modeled to be either free of primary traffic with probability θ_i or busy with probability $(1 - \theta_i)$. In other words, θ_i is the *primary-free probability* of channel i and provides an abstraction for the activities of the primary users. It can be assumed with only a slight loss of generality that $\theta_1 > \theta_2 > ... > \theta_N$. The coordinator is responsible for continually estimating the primary-free probabilities of the different channels, and triggering a re-computation of sensing orders whenever the probabilities change significantly. In this chapter, however, our focus is on *per-slot* optimization, so how the primary traffic patterns change across time slots will not be a matter of concern. In other words, this chapter focuses on the procedure(s) the coordinator would use to compute the sensing orders every time it is provided with a new set of primary-traffic statistics for the different channels.

Due to the fading nature of the wireless channels, the received signal strengths are random. So, the CUs use rate adaptation in general, based on the instantaneous channel quality, in order to exploit the time-varying features of the wireless channel. Specifically, the transmission rate used by CU m on channel i will be denoted as $r(\gamma_i^m)$, where γ_i^m is the instantaneous SNR seen on channel i by CU m to its receiver and $r(\cdot)$ is a non-decreasing function. We will also consider the special case when the CUs do not use rate adaptation i.e., transmit at a constant rate $r(\gamma_i^m) = R$ irrespective of the channel quality. If τ is the time taken to sense a channel with the desired accuracy and if k channels need to be sensed in a slot before a CU starts transmission, then the *effectiveness* of transmission in that slot will be denoted by $c_k = 1 - k\tau / T$ (see Figure 2).

The instantaneous SNRs can be assumed to be independent across CUs and across channels. In addition, if T is chosen to be smaller than T_c, the coherence time of the channels, the instantaneous SNRs can be assumed to remain constant over the duration of a time slot T. Also, T needs to be at least as large as $N\tau$ so that a maximum of N channels can be sensed sequentially in any time slot in the worst case.

Figure 2. Slot structure. The network operates in slots of duration T. The time for sensing one channel reliably is τ. If k channels need to be sensed before transmission, the data transmission time is $T - k\tau$.

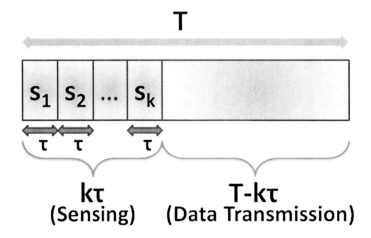

Thus, the parameter T of the slotted-time operation model needs to be chosen such that $N\tau < T < T_c$, and in turn the sensing algorithm needs to be designed in a way that the duration τ for reliable sensing satisfies $N\tau < T_c$.

Contention Avoidance (Centralized Sensing Order)

Since the coordinator enjoys a centralized view of the network, one way of avoiding contentions altogether when there are multiple CUs is to let the coordinator make all spectrum access decisions in the network based on local spectrum sensing reports from the individual CUs. But since the CUs themselves cannot sense more than one channel at a time, the coordinator needs to have a centralized sensing-order $s = (s_1, s_2, \ldots, s_N)$ that it would follow to collect local sensing reports from the CUs and assign channels to them in a sequential manner.

When the network uses this contention avoidance strategy, the coordinator senses the availability of the different channels at the beginning of each time slot according to s using a *cooperative centralized sensing* procedure as illustrated in Figure 3. Note that the sensing is actually performed by the individual CUs, so 'coordinator senses a channel' throughout the chapter would actually mean 'coordinator requests for sensing reports for a channel'. It continues sensing until it has allotted a channel to each of the M CUs or exhausted sensing all N channels. Whenever it senses a channel to be primary free, it also decides which CU, if any, should get to access this channel. The problem that we address in Section 4

Figure 3. Illustration of cooperative centralized sensing. Spectrum sensing is performed locally by the individual cognitive users. The coordinator fuses the local sensing reports to make a centralized decision on whether the channel being sensed is free to be used by the cognitive users.

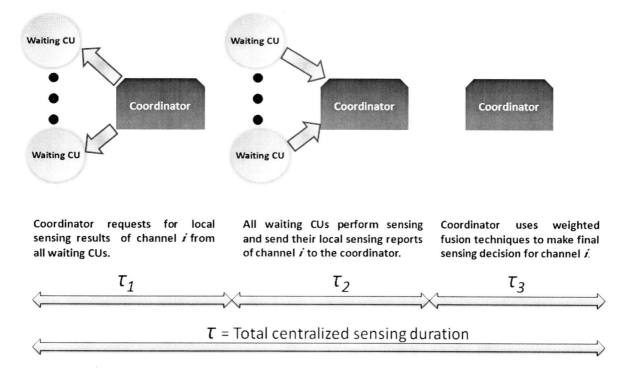

| Coordinator requests for local sensing results of channel *i* from all waiting CUs. | All waiting CUs perform sensing and send their local sensing reports of channel *i* to the coordinator. | Coordinator uses weighted fusion techniques to make final sensing decision for channel *i*. |

τ_1 τ_2 τ_3

τ = Total centralized sensing duration

is two-fold: (1) How does the coordinator decide whether or not to allow access to a channel that is found to be primary free, and if yes, to which CU? (2) How does the coordinator determine the *optimal centralized sensing order* given the primary-free probabilities and the channel fading statistics?

Contention Resolution (Decentralized Sensing Orders)

On second thoughts, it might seem wasteful to assign channels sequentially to the CUs when all of them could simultaneously sense and find a free channel for their respective transmissions. In other words, every CU *j* could have a decentralized sensing-order $\mathbf{s}^j = (s_1^j, s_2^j, \ldots, s_N^j)$ which it could follow to sense the availability of channels one after the other until it finds a free and *good* channel for its transmission. However, since the spectrum access decisions are made by the CUs individually, two or more CUs can often end up trying to access the same channel simultaneously and therefore, the network should pre-define a mechanism for resolving contentions.

When the network uses a contention resolution strategy, each CU *j* sequentially senses the channels according to its decentralized sensing-order $\mathbf{s}^j$ at the beginning of each time slot. Whenever it senses a channel to be primary free, it also decides whether to access the channel or skip it with the hope of finding a better channel in its subsequent sensing. The problem that we address in Section 5 is again two-fold: (1) How do we decide whether a CU should access a primary-free channel or skip it for a better *future* channel? (2) How do we determine the *optimal decentralized sensing orders* for the CUs given the primary probabilities, the channel fading statistics and the contention resolution mechanism?

In all scenarios discussed in this chapter, it is assumed that no *recall* or *guess* is allowed i.e., the CUs/ coordinator cannot go back in their sensing orders to sense a previously sensed channel nor can they proceed to access an un-sensed channel.

BACKGROUND

Single-User CRNs

Jiang et al. (2009) have studied the optimal sensing-order problem for single-user CRNs in detail. Their main results are as follows.

- When the CUs do not use rate adaptation, the intuitive sensing order i.e., channels arranged in the descending order of their primary-free probabilities, is optimal.
- When the CUs use rate adaptation, the intuitive sensing order is not optimal in general. The cognitive throughput however can be expressed as a function of the sensing order in the form of a recursion, and a dynamic programming approach can be formulated to find the optimal sensing order with much lesser complexity than a brute-force search.

The above-mentioned work also relaxes several assumptions in the network model and studies the sensing-order problem for each of them, for e.g. when the primary-free probabilities are not known, when there are channel sensing errors, when the sensing capacity in each slot is limited (i.e., $N\tau > T$) etc.

The problem of jointly optimizing the sensing order and the sensing duration to maximize cognitive throughput has been addressed by Hamza and Aissa (2013). They have also studied the effect of sensing errors in their optimization problem. Apart from maximization of throughput, other metrics such as minimal opportunity-discovery delay have also been considered in the literature (Kim & Shin, 2013; Kim & Shin, 2008) to determine the optimal sensing order.

In the absence of a priori knowledge of the primary-free probabilities, Cheng and Zhuang (2011) have formulated a simple sensing order based on the descending order of achievable rates. In the absence of a priori statistics, the CUs may learn this statistical information through spectrum sensing. Here a tradeoff exists between learning the statistics (exploration) and accessing a free channel (exploitation). Several studies on these using connections to the multi-armed bandit problem have been done in the literature (Zhang, Jiang, Tan & Slevinsky, 2013; Lai, El Gamal, Jiang & Poor, 2011; Lai, Jiang & Poor, 2008).

Multi-User CRNs

Spectrum access of cognitive users in multi-user CRNs faces additional challenges since collision among CUs can occur if multiple CUs sense and make access decisions on the same channel at the same time. In addition to the channel sensing order, multi-user CRNs need to have a contention-resolution mechanism for handling collisions. An overview of various tools used to study dynamic spectrum access in multi-user CRNs is presented in Xu et al. (2013). The cognitive throughput depends on the primary-free probabilities, collision probabilities among CUs that depend on the contention-resolution mechanism, and the statistics of randomly fading channels.

In Fan & Jiang (2009), the sensing order problem to maximize cognitive throughput was studied for a two-user CRN with three different contention resolution mechanisms. Their main contributions were as follows.

- When the CUs do not use rate adaptation, the optimal sensing order for each user can be obtained through a brute-force search. Suboptimal algorithms were proposed and their performance was studied through numerical simulations.
- When the CUs use rate adaptation, some findings were reported on how the optimal stopping rule and sensing order settings should be jointly considered. However, the optimal solution was not found.

In this chapter, however, the sensing order problem is studied (1) for a general *M-user* CRN (2) with rate adaptation, which forms a much more general setting of the problem.

In the following works, the multi-user sensing-order problem is addressed with the assumption that the data rates supported by channels are not random. Zhao and Wang (2012) have addressed the multi-user sensing order problem by optimizing a metric that involves the primary-free probabilities of channels, the transmission rates on those channels and the probability of collision among CUs. Shokri-Ghadikolaei, Sheikholeslami and Nasiri-Kenari (2013) have developed modified random access protocols along with the channel sensing orders to minimize collisions among CUs and to balance the load of CUs across all channels. Khan, Lehtomaki, Mustonen and Matinmikko (2011) have developed adaptive CU strategies to find the sensing orders that minimize the probability of collision. This chapter, however, focuses on randomly fading channels and allows the CUs to adapt their data rates based on instantaneous channel conditions, which in turn poses additional challenges to the sensing-order problem.

CENTRALIZED SENSING ORDER

In this section, we consider multi-user cognitive radio networks where spectrum access decisions are made by a central coordinator and study how to determine the optimal centralized sensing order for the coordinator. The rest of this section is organized as follows.

1. In the subsection titled *Mathematical formulation of cognitive throughput*, we describe how to compute the cognitive throughput as a function of the centralized sensing order $\mathbf{s} = (s_1, s_2, \ldots, s_N)$ of the coordinator. The problem that remains then is to determine which sensing order results in the maximum cognitive throughput.
2. In the subsection titled *Optimal centralized sensing order without rate adaptation*, we show that when the CUs do not rate adaptation, the optimal centralized sensing order and the optimal stopping rule are neat generalizations of the corresponding results for a single-user CRN.
3. In the subsection titled *Optimal centralized sensing order with rate adaptation,* we demonstrate via an illustrative example how the procedure described in (1) could be used to compute the cognitive throughput when the CUs do use rate adaptation, and also discuss how the optimal centralized sensing order can be determined with a complexity much lesser than that of a brute-force search.

Mathematical Formulation of Cognitive Throughput

Let v_i^j denote the instantaneous throughput that can be obtained in the i-th sensing slot given that channel s_i is sensed to be primary free, where j denotes the number of *waiting CUs* i.e., CUs who in the current sensing slot are yet to be assigned a channel for their transmissions. Let V_{i+1}^j denote the expected sum throughput of j waiting CUs in the network if the coordinator proceeds to sense channel s_{i+1}. If $j = 0$, $V_{i+1}^j = 0$. For $j > 0$, we can write V_{i+1}^j as follows.

$$V_{i+1}^j = \begin{cases} \theta_{s_{i+1}} E[v_{i+1}^j], & i+1 = N \\ \theta_{s_{i+1}} E[v_{i+1}^j] + (1 - \theta_{s_{i+1}})V_{i+2}^j, & \text{otherwise} \end{cases} \tag{4.1}$$

Recall that $\theta_{s_{i+1}}$ refers to the primary-free probability of channel s_{i+1}, which is the probability that s_{i+1} is free of primary traffic in the time slot under consideration and is assumed to be known a priori to the coordinator. In order to write an expression for v_i^j, we make use of the following observations. If $j = 0$, then $v_i^j = 0$. Otherwise, in the i-th sensing slot with j (> 0) waiting CUs when the coordinator senses channel s_i to be primary-free, it can choose to either

1. let one of the waiting CUs *access* this channel, or
2. unless $i = N$, *skip* the channel altogether and proceed to sense s_{i+1},

depending on which choice results in a higher reward. A key observation here is that if the coordinator chooses to do (1), it should let the waiting CU seeing the best channel quality on s_i access the channel. That is, if $w_1, w_2, \ldots, w_j$ denote the j waiting CUs, $\gamma_{s_i}^{w_k}$ refers to the instantaneous SNR seen by CU w_k on channel s_i, and δ_i^j is defined as:

$$\delta_i^j \triangleq \max(\gamma_{s_i}^{w_1}, \gamma_{s_i}^{w_2}, \ldots, \gamma_{s_i}^{w_j}) \tag{4.2}$$

then the coordinator should let the waiting CU seeing SNR δ_i^j access channel s_i. The corresponding reward would be $c_i r(\delta_i^j) + V_{i+1}^{j-1}$. This must be true because the channels are assumed to be homogenous i.e., they have similar statistics, and hence if any other waiting CU w_k were given access, the reward would be $c_i r(\gamma_{s_i}^{w_k}) + V_{i+1}^{j-1}$ where $\gamma_{s_i}^{w_k} \neq \delta_i^j$, which by definition cannot exceed $c_i r(\delta_i^j) + V_{i+1}^{j-1}$. Note that the coordinator does not know anything about the instantaneous quality of the *future* channels in its sensing order that can aid in its decision making in the current sensing slot. Thus, we can write v_i^j ($j > 0$) as

$$v_i^j = \begin{cases} c_i r(\delta_i^j), & \text{if } i = N \\ c_i r(\delta_i^j) + V_{i+1}^{j-1}, & \text{if } i \neq N \text{ and } c_i r(\delta_i^j) + V_{i+1}^{j-1} > V_{i+1}^j \\ V_{i+1}^j & \text{otherwise} \end{cases} \tag{4.3}$$

Here, $c_i r(\delta_i^j) + V_{i+1}^{j-1} > V_{i+1}^j$ is the *access rule* for the coordinator, analogous to the traditional stopping rule in the single-user case (Jiang et al., 2009), and can be similarly shown to be optimal as in Fan & Jiang (2009).

Using equations 4.1 and 4.3 for V_{i+1}^j and v_i^j respectively, we can recursively compute the $N \times M$ matrix $\{V_i^j\}$ ($1 \leq i \leq N, 1 \leq j \leq M$), starting with $\{V_N^j\}_{j=1}^M$ and repeating the procedure for $i = N-1, N-2, \ldots, 1$ to eventually compute V_1^M. A little thought would reveal that V_1^M is the objective function we had set out to formulate, since V_1^M is the expected sum throughput of M waiting CUs in the CRN when the coordinator proceeds to sense the first channel in its sensing order which, in other words, is the cognitive throughput in the network.

Our aim is to determine the sensing-order s that maximizes V_1^M. Note that V_1^M is a function of s; this dependence, however, is not explicitly shown for notational simplicity.

Optimal Centralized Sensing Order without Rate Adaptation

If the CUs transmit at a fixed rate R whenever they are granted access to a channel by the coordinator i.e., $r(\delta_i^m) = R \; \forall i, m$, we show that the descending order of channels in terms of θ_i ($1 \leq i < N$) is optimal in terms of maximizing the cognitive throughput in the network. It also turns out that for optimal performance, the coordinator should never skip a primary-free channel but always assign it to some waiting CU. We present these results as the following lemma along with a proof.

Lemma 1: If the CUs transmit at a fixed rate whenever they access a channel, the coordinator should never skip a free channel unless there are no more waiting CUs, and the descending order of channels based on their primary-free probabilities is the optimal centralized sensing-order.

Proof: See Appendix A.

Optimal Centralized Sensing Order with Rate Adaptation

When the CUs in a multi-user CRN use rate adaptation to exploit the time-varying features of the wireless channel, the descending order of channels in terms of their primary-free probabilities may no longer be optimal. Similar to the single-user case, the general procedure to find the optimal sensing order is to express the cognitive throughput V_1^M as a function of the centralized sensing order s using the general procedure described earlier and then carry out a brute-force search over all possible sensing orders to find which sensing order results in the maximum V_1^M. An illustrative example is presented in Appendix B.

A Note on Complexity: In general, the computational complexity of finding the optimal centralized sensing order by a brute-force search is $\mathcal{O}(MN \cdot N!)$ if the complexity of computing one V_i^j is $\mathcal{O}(1)$. A dynamic programming algorithm similar to the one proposed in Jiang et al. (2009) can reduce the complexity of finding the optimal centralized sensing order to $\mathcal{O}(MN \cdot 2^{N-1})$. This algorithm is based on the observation that if $(s_1{}^*, s_2{}^*, \dots, s_N{}^*)$ is the optimal centralized sensing order, then $V_i^j(s_1{}^*, s_2{}^*, \dots, s_N{}^*)$ must be at least as large as $V_i^j\left(s_1{}^*, s_2{}^*, \dots, s_{i-1}{}^*, \Pi(s_i{}^*, \dots, s_N{}^*)\right)$, where $\Pi(s_i{}^*, \dots, s_N{}^*)$ is any permutation of $(s_i{}^*, \dots, s_N{}^*)$. In other words, the expected throughput at the i-th channel in the optimal sensing order is the maximal throughput that can arise from all possible permutations of the *later* channels. As a result, we can start with the final channel and compute the maximal expected throughput V_i^j at each preceding stage for each j until we reach the first channel. We summarize the steps of the algorithm below.

Round 1: Compute the maximal value of V_N^j associated with the final channel in the centralized sensing order for each $j = 1, 2, \dots, M$. There are ${}^N C_1 = N$ possible candidates for s_N and therefore the complexity of this round is $M \cdot {}^N C_1$.

Round $k+1$ ($2 \leq k+1 \leq N$): Compute the maximal value of V_{N-k}^j associated with the k-th-from-last channel in the centralized sensing order for each j. In this round, there are ${}^N C_k$ possible combinations for $(s_{N-k+1}, \dots, s_N)$, each of which can have k possible transitions from s_{N-k} to s_{N-k+1} at the k-th-from-last position in the sensing order. Therefore, the complexity of this round is $Mk \cdot {}^N C_k$.

After Round N, an optimal centralized sensing order can be retraced according to the optimal transition at each round. Thus, this algorithm has $\mathcal{O}\left(M \sum_{k=1}^{N} k \cdot {}^{N}C_{N-k} \right) = \mathcal{O}\left(MN \cdot 2^{N-1} \right)$ complexity which is significantly lower than that of a brute-force search and yet provides optimal performance.

DECENTRALIZED SENSING ORDERS

In this section, we consider different multi-user CRN scenarios in which spectrum access decisions are made in a decentralized manner by the individual CUs. The problem is to assign a sensing order s^j to each CU j in a way that maximizes the cognitive throughput in the network. Note that in such networks, multiple CUs may *simultaneously* end up requesting access to the same channel, so the network must pre-define a mechanism for resolving contentions. In order to distinguish from the centralized sensing order that we studied in the previous section, we refer to the set of sensing orders $\{s^j\}_{j=1}^{M}$ as *decentralized sensing orders*. The rest of this section is organized as follows:

1. In the subsection titled *Decentralized Sensing Orders with Preemptive Priority*, we consider the scenario where the network organizes the CUs into distinct preemptive priority levels as a mechanism for resolving contentions if and when it occurs. We present the procedures for determining the optimal decentralized sensing orders without and with rate adaptation.
2. In the subsection titled *Decentralized Sensing Orders with Contention Resolution*, we consider the scenario where all CUs have equal priority and the network uses a pre-defined contention resolution mechanism for resolving contentions if and when it occurs. We present the procedure for determining the decentralized sensing orders when the CUs do not use rate adaptation. The case with rate adaptation is still open; we discuss a potential solution approach.

Decentralized Sensing Orders with Preemptive Priority

Let us start off with a special network scenario in which the CUs are organized into distinct *preemptive priority* levels. That is, a higher priority CU in the network has the right to replace a lower priority user from the service channel, and the lower priority user finds another channel according to its sensing order to repeat its transmission in a given time slot. This is a practical scenario when the network carries a mixture of real-time (e.g. CBR voice packets) and non-real time (e.g. nrt-VBR data packets) traffic (Traffic Management Specification, 1999; Karthik, Misra & Sharma, 2011), and the CUs are assigned different priority levels depending on their service categories.

It is worth mentioning here that in the next subsection, we will study a weaker form of priority access where the CUs are assigned *contention-only* priority levels to be used only in the event of a contention, so if a CU with a lower contention-only priority is already communicating on a channel, it cannot be replaced by a higher contention-only priority CU. However, in the priority scheme that we study in this subsection, the CUs enjoy absolute priority for accessing the spectrum.

Optimal Decentralized Sensing Orders

In order to determine the optimal decentralized sensing orders in a CRN with preemptive priority among the CUs, we follow the following approach. First, we show how the cognitive throughput can be computed given a set of decentralized sensing orders for the CUs. We also show how the procedure simplifies when the CUs do not rate adaptation. In general, a brute-force search may then be employed to find the optimal sensing orders, although it requires a prohibitively high complexity. We conclude this subsection by discussing a procedure to determine the decentralized sensing orders that can achieve near-optimal performance at a much lower complexity.

In order to compute the cognitive throughput, let us label the CUs 1, 2, ... , M in descending order of their priorities i.e., CU j has preemptive resume priority over CU k $\forall 1 \leq j < k \leq M$. Given a sensing order $\mathbf{s}^j = (s_1^j, s_2^j, \ldots, s_N^j)$ for each CU j, let θ_i^j denote the *availability probability* of channel i for CU j, which is the probability that channel i is not used by the primary users as well as by CUs 1, 2, ... , $(j-1)$ in a given time slot. Let p_i^j denote the position of channel i in the sensing order of CU j i.e., $s_{p_i^j}^j = i$.

Let U_i^j denote the expected throughput if the CU proceeds to sense channel s_i^j from s_{i-1}^j (let s_0^j be defined as the initial state of CU j at the beginning of any time slot, just before it senses s_1^j, the first channel in its sensing order). With this notation, it can be seen that U_1^j represents the expected throughput of CU j given the set of sensing orders $\{\mathbf{s}^j\}_{j=1}^M$. The cognitive throughput in the network is therefore $U \triangleq \sum_{j=1}^M U_1^j$. Note that the dependence of U on $\{\mathbf{s}^j\}_{j=1}^M$ is not explicitly shown for simplicity of notation. The set $\{\mathbf{s}^j\}_{j=1}^M$ that maximizes U is the optimal set of sensing orders.

The next challenge is to determine the expected throughputs $\{U_1^j\}_{j=1}^M$ for each CU. We make a key observation - each CU in this CRN faces a single-CU *like* scenario because all CUs that are higher than it in priority appear *like* primary users while all CUs that are below it in priority do not affect its performance and hence are transparent to its operation. Therefore, the expected throughputs $\{U_1^j\}_{j=1}^M$ can be formulated one-by-one, starting with the CU having the highest priority i.e., CU 1, and successively moving down the priority levels. At each level j, the expected throughput U_1^j for CU j can be computed as if it were the only CU in the network but with the availability probability θ_i^j replacing the primary-free probability θ_i in the single-user solution. We present below the general algorithm for obtaining the cognitive throughput U in this network scenario when the CUs use rate adaptation. The case when the CUs do not use rate adaptation is described as a special case.

1. Choose one of the $(N!)^M$ possible sets of sensing-orders $\{\mathbf{s}^j\}_{j=1}^M$ for the CUs.

    For each CU j, starting with CU 1,

2. Compute the availability probability θ_i^j as follows for each channel i in its sensing order starting with the first channel s_1^j.

a. For CU 1, $\theta_i^1 = \theta_i$ $\forall i$ since it has the highest priority among all CUs.

b. For CU $j+1$ ($2 \le j+1 \le M$), let A_i^0 denote the event that i is primary-free and A_i^j denote the event that it remains CU j-free in any time slot. Compute θ_i^{j+1} using the following recursion.

$$\theta_i^{j+1} = P\{A_i^0 A_i^1 \ldots A_i^{j-1} A_i^j\} = P\{A_i^0 A_i^1 \ldots A_i^{j-1}\} \times P\{A_i^j \mid A_i^0 A_i^1 \ldots A_i^{j-1}\}$$
$$= \theta_i^j \times \big[1 - \underbrace{P\{(A_i^j)^C \mid A_i^0 A_i^1 \ldots A_i^{j-1}\}}_{\triangleq \rho_i^j}\big] = \theta_i^j (1 - \rho_i^j) \tag{5.1}$$

where $\rho_i^j = P\{\text{Ch } i \text{ is used by CU } j \mid \text{Ch } i \text{ is primary-free \& CU 1, \ldots, } (j-1)\text{-free}\}$. This equation simply states that the availability probability of channel i for CU $j+1$ is the availability probability of channel i for CU j multiplied by the probability $(1 - \rho_i^j)$ that it is not used by CU j in any time slot given that it was available. ρ_i^j can be written as

$$\rho_i^j = \begin{cases} P\{c_{p_i^j} r(\gamma_i^j) > U_{p_i^j+1}^j\} & p_i^j = 1 \\[2mm] \prod_{k=1}^{p_i^j-1}\big(1 - \theta_{s_k^j}^j P\{c_k r(\gamma_{s_k^j}^j) > U_{k+1}^j\}\big) & p_i^j = N \\[2mm] \underbrace{\big(\prod_{k=1}^{p_i^j-1} 1 - \theta_{s_k^j}^j P\{c_k r(\gamma_{s_k^j}^j) > U_{k+1}^j\}\big)}_{\text{Prob. that none of } s_1^j, s_2^j, \cdots, s_{p_i^j-1}^j \text{ is used}} \times \underbrace{P\{c_{p_i^j} r(\gamma_i^j) > U_{p_i^j+1}^j\}}_{\text{Cond. prob. that Ch. } i \text{ is used}} & \text{otherwise} \end{cases} \tag{5.2}$$

Note that $c_k r(\gamma_{s_k^j}^j) > U_{k+1}^j$ for any channel s_k^j is the traditional stopping rule, which says that the CU must access a channel whenever it is available as long as its instantaneous throughput from accessing the channel is greater than the expected throughput from skipping the channel. This stopping rule is optimal since each CU is a virtual single user in this scenario (Jiang et al., 2009). When the CUs do not use rate adaptation, the availability probabilities can be directly expressed as (let $k = p_i^j - 1$ and $\theta_{s_0^j} \triangleq 0$):

$$\theta_i^{j+1} = \theta_i^j \Big[\sum_{m=1}^{k} \theta_{s_m^j}^j \Big(\prod_{n=0}^{m-1}(1 - \theta_{s_n^j}^j)\Big)\Big] \tag{5.3}$$

3. Compute the expected throughput U_1^j as follows. Let u_i^j denote the instantaneous throughput of CU j at s_i^j given that the CU senses s_i^j to be available i.e., free of activity by primary users as well as CUs 1, 2, $\cdots$, $j-1$). We can express u_i^j as

$$u_i^j = \begin{cases} c_i r(\gamma_{s_i^j}^j) & \text{if } i = N \text{ (CU has to stop at } s_N^j) \\ c_i r(\gamma_{s_i^j}^j) & \text{if } i \neq N \text{ \& } c_i r(\gamma_{s_i^j}^j) > U_{i+1}^j \text{ (CU accesses } s_i^j) \\ U_{i+1}^j & \text{if } i \neq N \text{ \& } c_i r(\gamma_{s_i^j}^j) \leq U_{i+1}^j \text{ (CU skips } s_i^j) \end{cases} \tag{5.4}$$

where U_{i+1}^j is as defined earlier. As mentioned before, $c_i r(\gamma_{s_i^j}^j) > U_{i+1}^j$ is the traditional stopping rule, and is optimal for this scenario. These stopping rules provide a *stopping threshold* for each CU in terms of the instantaneous SNR $\gamma_{s_i^j}^j$ for every channel in its sensing order. We can express U_i^j as

$$U_i^j = \begin{cases} \theta_{s_i^j}^j E\{u_i^j\} + (1 - \theta_{s_i^j}^j) U_{i+1}^j, & \text{if } i < N \\ \theta_{s_i^j}^j E\{u_i^j\}, & \text{if } i = N \end{cases} \tag{5.5}$$

where $E\{\cdot\}$ denotes expectation. Note that when the CUs do not use rate adaptation, the optimal stopping rule is to stop at the first available channel (Jiang et al., 2009). Starting with U_N^j, we can use the recursive Equation (5.1) to compute U_1^j.

4.   Compute the cognitive throughput $U = \sum_{j=1}^M U_1^j$. Repeat the above procedure for all $(N!)^M$ possible

sets of sensing orders for the CUs. The set that results in the maximum value for U is the optimal set of decentralized sensing orders in this scenario.

When the CUs do not use rate adaptation, it is worthy to note that the first channel s_1^j in the sensing-order of CU j becomes unavailable for all lower priority CUs i.e., $\theta_{s_1^j}^k = 0 \; \forall k > j$. As a result, CU j does not need to sense any more than $N - (j-1)$ channels, because $(j-1)$ channels $\{s_1^k\}_{k=1}^{j-1}$ are always going to be unavailable for its use.

Heuristic Decentralized Sensing Orders

The complexity of brute-force search to determine the optimal decentralized sensing orders using the above procedure is $\mathcal{O}\big((N!)^M \cdot MN\big)$, where $\mathcal{O}(1)$ is the complexity of finding one U_i^j, which is enormous even for moderately large N and/or M. In this subsection, we describe a heuristic search algorithm for selecting the sensing orders of the CUs in a multi-user CRN that employs preemptive priority among its CUs. The resulting sensing orders would be referred to as the *heuristic sensing orders*.

The motivation behind this algorithm is the earlier observation that each CU in this scenario is a virtual single user. Therefore, we proceed user-by-user and use the single-user solution at each step to select the heuristic sensing order of the CU being considered in that step. This algorithm consists of M stages. In the j-th stage, select the heuristic sensing-order of CU j as follows. Compute the availability probabilities of the different channels for CU j using Equation 5.1 or 5.3 as the case may be. If the CUs do not use rate adaptation, the heuristic sensing-order of CU j is the descending order of channels in terms of θ_i^j. Otherwise, carry out a dynamic programming based search to find the heuristic sensing-order that maximizes U_1^j as obtained using Equations 5.4 and 5.5. The complexity of this heuristic search algorithm is $\mathcal{O}(2^{N-1} \cdot MN)$ which is significantly lower than the brute force search complexity, where $\mathcal{O}(1)$ is the complexity of finding one U_i^j.

In order to check how well this heuristic search algorithm works in comparison to the optimal algorithm, we simulate a 5-channel, 3-user CRN with the following *loading* scenarios for the channels (where a higher loaded channel is one that carries a higher volume of primary traffic and hence has a lower primary-free probability).

1. All channels are lightly loaded $(\theta_1 = 0.95, \ \theta_2 = 0.90, \ \theta_3 = 0.85, \ \theta_4 = 0.80, \ \theta_5 = 0.75)$.
2. All channels are heavily loaded $(\theta_1 = 0.50, \ \theta_2 = 0.45, \ \theta_3 = 0.40, \ \theta_4 = 0.35, \ \theta_5 = 0.30)$.

In addition to the optimal sensing orders (*Opt*) and the heuristic sensing orders (*Heu*), we also consider cyclic sensing orders (*Cyc*) consisting of the intuitive sensing order (descending order of channels based on their primary-free probabilities) and its cyclic shifts. Figure 4 plots the cognitive throughput as a function of the sensing overhead τ / T (as a percentage of the throughput achievable when $\tau = 0$).

When the channels are all lightly loaded, we see that not just the heuristic sensing orders but also the cyclic sensing orders result in near-optimal performance. This makes sense because when all the channels are lightly loaded, minor differences in their primary-free probabilities do not affect performance,

Figure 4. Cognitive throughput vs τ / T in a 5-channel, 3-user CRN with preemptive priority among CUs. The performance is plotted for optimal (Opt), heuristic (Heu) and cyclic (Cyc) sensing orders.

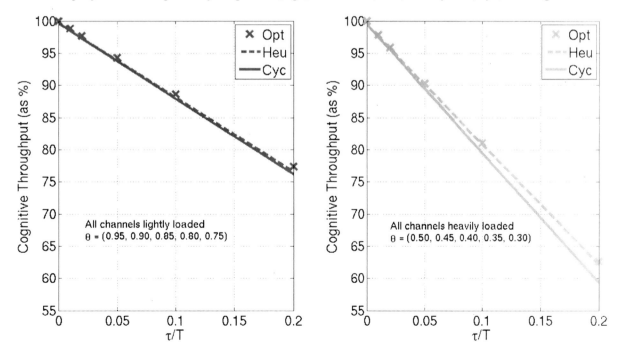

and any sensing order that assigns disjoint channels to the CUs in their first sensing slots i.e., $s_1^j \neq s_1^k \;\; \forall j \neq k$, will result in nearly similar performance. However, when all channels are heavily loaded, we see that the effect of minor differences in the primary-free probabilities is more pronounced. However, *Heu* performs near-optimally (throughput difference $< 1\%$) over all practical sensing overhead values. It is also interesting to note that the cyclic sensing orders, which are straightforward to obtain and involve no computational complexity, also perform reasonable close to *Opt* (throughput difference $< 5\%$ in all cases) and its performance is especially better at lower τ / T.

Decentralized Sensing Orders with Contention Resolution

Let us now turn to the scenario where all CUs in a multi-user CRN have equal priority for spectrum access and the network uses a pre-defined mechanism to resolve contentions whenever two or more CUs request access to the same channel simultaneously. We study the sensing-order problem for the following three contention-resolution mechanisms.

1. **BACKOFF:** Each CU uses a backoff mechanism to avoid a possible collision, similar to the ones used in IEEE 802.3 for Ethernet and IEEE 802.11 for Wireless LAN (Fruth, 2006). One CU wins and transmits in the channel for the remainder of the time slot, while the other CUs who fail in the contention continue to sense the further channels according to their respective sensing orders.
2. **PRIORITY:** Each CU is assigned a priority level just for the purpose of contention resolution. Whenever (and only when) there is a contention, the CU with the highest priority among the contending users gets to transmit until the end of the slot while the other CUs continue their search for another free channel. As before, we label the CUs in descending order of their contention-only priorities i.e., CU j has contention-only priority over CU $k \;\; \forall 1 \leq j < k \leq M$.
3. **COLLIDE:** Each contending CU transmits in the channel until the end of the slot. As a result, all of them collide and earn zero throughputs for that slot.

In the subsection titled *Optimal decentralized sensing orders without rate adaptation*, we present the optimal procedure to determine the sensing orders when the CUs do not use rate adaptation. In the subsection titled *Greedy decentralized sensing orders*, we discuss a greedy search algorithm to compute near-optimal sensing orders at a much lower complexity than brute-force search. Although the case with rate adaptation is still an open problem, we briefly discuss a potential solution approach for determining the optimal sensing orders in the subsection titled *Optimal decentralized sensing orders with rate adaptation*.

Optimal Decentralized Sensing Orders without Rate Adaptation

Let a_i denote the binary availability of channel i i.e., $a_i = 1$ if channel i is primary free and $a_i = 0$ if it is primary busy. Let $\mathbf{a} \triangleq (a_1, a_2, ..., a_N)$ denote a *channel availability combination* of the N channels. Figure 5 illustrates the concept of channel availability combination. Since the primary busy/free states of the different channels have been assumed to be independent of each other, we can write the probability of a given channel availability combination $\mathbf{a}$ as follows.

Figure 5. Illustration of channel availability combination **a**. *While primary-free probabilities are based on the statistics of usage of the channels over time, the availability combination represents the usage in a given time slot and is a binary vector indicating the primary ON/OFF (busy/free) status of the channels.*

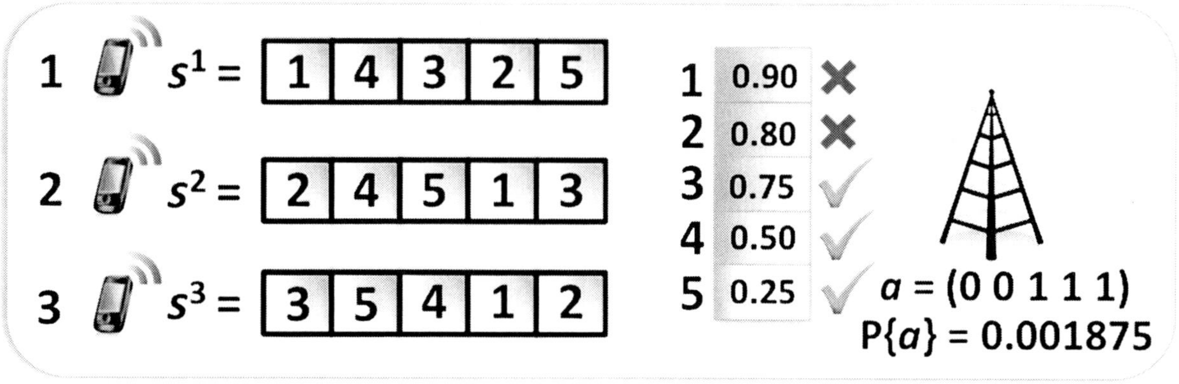

$$P\{\mathbf{a}\} = \prod_{i=1}^{N}\left[a_i\theta_i + (1-a_i)(1-\theta_i)\right] \tag{5.6}$$

When the CUs do not use rate adaptation, the cognitive throughput for a given set of sensing orders $\{\mathbf{s}^j\}_{j=1}^{M}$ can be computed by listing down all 2^N possible channel availability combinations, determining the sum throughput for each combination and computing an expectation over all possible channel availability combinations. We illustrate this by means of an example. Let us consider a 5-channel, 3-user CRN. Let the decentralized sensing orders of the 3 CUs be $\mathbf{s}^1 = (1,4,3,2,5)$, $\mathbf{s}^2 = (2,4,5,1,3)$ and $\mathbf{s}^3 = (3,5,4,1,2)$, and let $\mathbf{a} = (0,0,1,1,1)$. Note that $P\{\mathbf{a}\} = (1-\theta_1)(1-\theta_2)\theta_3\theta_4\theta_5$. In the first round of sensing, CUs 1 and 2 will sense channels 1 and 2 respectively to be busy, while CU 3 will sense channel 3 to be free and use it for transmission for the remainder of the slot to earn a throughput of $c_1 R$. In the second round of sensing, both CU 1 and CU 2 will sense channel 4 to be free.

1. If COLLIDE is being used, both CUs would transmit on channel 4 for the remainder of the slot, and hence collide, and earn zero reward. Therefore, the sum throughput of all CUs with COLLIDE is just $c_1 R$.

2. If PRIORITY is being used, CU 1 would get to transmit on channel 4 and earn a throughput of $c_2 R$, while CU 2 would continue sensing, sense channel 5 to be free in the third round of sensing and use it for its transmission to earn a throughput of $c_3 R$. Therefore, the sum throughput of all CUs with PRIORITY is $(c_1 + c_2 + c_3)R$.

3. If BACKOFF is used, both CUs 1 and 2 could win the contention with equal probabilities. If CU 1 wins the contention (probability = 0.5), CUs 1 and 2 would earn throughputs of $c_2 R$ and $c_3 R$ respectively as discussed above, whereas if CU 2 wins the contention (probability = 0.5), the corresponding rewards are $c_5 R$ and $c_2 R$. Therefore, the (expected) sum throughput for this case is

$$c_1R + \frac{1}{2}(c_2R + c_3R) + \frac{1}{2}(c_2R + c_5R) = (c_1 + c_2 + \frac{1}{2}c_3 + \frac{1}{2}c_5)R. \tag{5.7}$$

The cognitive throughput for a given $\{s^j\}_{j=1}^{3}$ can be obtained by repeating the above procedure for all 2^5 possible choices for **a** and then finding the expected value of sum throughput over all of them. See Appendix C for an illustrative example highlighting the entire exercise.

Greedy Decentralized Sensing Orders

The complexity of the brute-force search described above is $\mathcal{O}\big((N!)^M \cdot 2^N\big)$, if $\mathcal{O}(1)$ is the complexity of computing one u_X, which becomes too high to be practical even for moderately large N and/or M. In this subsection, we present a sub-optimal greedy search algorithm for selecting the decentralized sensing orders in a multi-user CRN that employs a contention resolution mechanism. The resulting sensing orders would be referred to as the *greedy sensing orders*.

The greedy search algorithm consists of N rounds. In the i-th round, we jointly select $\{s_i^j\}_{j=1}^{M}$ i.e., the i-th channel in the sensing orders of all CUs. We list below the steps of this algorithm.

Round 1: If all CUs were to be allowed to sense only one channel each, jointly select a set of M channels, one for each CU, that maximizes the cognitive throughput.

Round i ($2 \leq i \leq N$): If all CUs were to be allowed to sense up to a maximum of i channels each, and given that the first $(i-1)$ channels in their sensing orders have already been fixed in the previous $(i-1)$ rounds, jointly select a set of M channels, one for each CU, that maximizes the cognitive throughput.

In round i of the algorithm, the cognitive throughput can be obtained using the procedure described above with the constraint that each CU has only i ($1 \leq i \leq N$), and not N, channels in its sensing order. This set of channels would occupy the i-th position in the greedy sensing order of the respective CU. The complexity of round i is $\mathcal{O}\big((N-i+1)^M 2^i\big)$, if $\mathcal{O}(1)$ is the complexity of computing one u_X.

Therefore, the complexity of the greedy search algorithm is $\mathcal{O}\big(\sum_{k=2}^{N} k^M 2^{N-k+1}\big)$, which is significantly lower than that of a brute force search.

In order to see how well this greedy search algorithm performs in comparison to the optimal algorithm, we simulate the same 5-channel, 3-user CRN as described earlier. Figure 6 plots the cognitive throughput (in units of R) for different values of τ / T using BACKOFF (left) and COLLIDE (right).

It is interesting to see that the very simple cyclic sensing orders perform reasonably close to the optimal sensing orders and even outperform the more complex greedy sensing orders at very low τ / T. However, at higher τ / T, the greedy sensing orders perform near-optimally (throughput difference < 1%) and much better than *Cyc*.

Figure 6. Cognitive throughput vs τ / T in a 5-channel, 3-user CRN with BACKOFF (left) and COL-LIDE (right). The performance is plotted for optimal (Opt), greedy (Gre) and cyclic (Cyc) sensing orders.

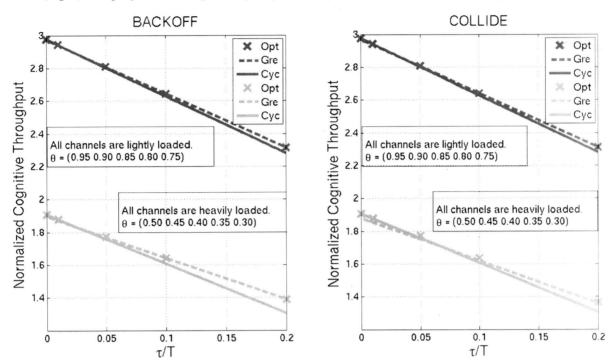

Optimal Decentralized Sensing Orders with Rate Adaptation

When the CUs use rate adaptation, not just the sensing orders but also the stopping thresholds need to be jointly solved for optimality, and designing an algorithm that solves this problem in its entire generality is still an open problem. In this subsection, we discuss a possible approach that can be used to develop such an algorithm.

In order to keep the problem tractable, let us assume that the CUs use discrete rate adaptation i.e., they choose from a finite set of transmission rates depending on the channel quality, which is in fact how rate adaptation is implemented in practice. We divide the range of possible SNRs $[0, \infty)$ into L regions by means of $L-1$ thresholds, and assume that whenever the instantaneous SNR lies in region i ($1 \leq i \leq L$), the transmission rate used is $\hat{R}_i$ (where the regions are numbered such that $\hat{R}_1 \geq \hat{R}_2 \geq \cdots \geq \hat{R}_L$).

In its i-th sensing, if CU j finds channel s_i^j to be free at the beginning of the i-th slot, then its instantaneous reward will be

$$u_i^j = \begin{cases} \beta_i^j c_i r(\gamma_i^j) & \text{if } \mathrm{E}\{\beta_i^j\} c_i r(\gamma_i^j) \geq U_{i+1}^j \text{ or if } i = N \ (\text{CU } j \text{ decides to access } s_i^j) \\ U_{i+1}^j & \text{otherwise} \end{cases} \tag{5.8}$$

where U_{i+1}^j is defined earlier, $r(\gamma_i^j)$ is one of $(\hat{R}_1, \hat{R}_2, \cdots, \hat{R}_L)$ and β_i^j is a binary indicator that indicates whether or not CU j's attempted transmission in the i-th slot is successful given that s_i^j was sensed to be free. It is useful to note that β_i^j is a function of the contention resolution mechanism, the decentralized sensing orders and the stopping rules of all CUs, and $E\{\beta_i^j\}$ can be computed and passed along to the CUs by the coordinator when it informs them of their respective sensing orders. The condition $E\{\beta_i^j\}c_i r(\gamma_i^j) \geq U_{i+1}^j$ is analogous to the traditional stopping rule and says that a CU must stop sensing and access a channel that it finds to be available iff its instantaneous *expected* reward is greater than its expected reward from skipping the channel.

A recursion for U_{i+1}^j can be written down similar to Equation 5.5. However, the challenge now is that unlike Equation 5.5, this recursion cannot be solved for sequentially starting with U_N^j. This is because each U_N^j now depends on $E\{\beta_N^j\}$ which in turn is a function of all $\{U_i^j\}_{N \times M}$, thereby leaving us with something like the chicken-and-egg problem.

A key observation that helps here is that with discrete rate-adaptation, the stopping rules depend not on the exact U_{i+1}^j but on the range that $U_{i+1}^j / (c_i \ E\{\beta_i^j\})$ belongs to. For e.g., if $\hat{R}_k < U_{i+1}^j / (c_i \ E\{\beta_i^j\}) < \hat{R}_{k+1}$, then as long as γ_i^j can support $\hat{R}_{k+1}$ or higher, CU j will decide to stop sensing and start transmitting irrespective of the exact value of U_{i+1}^j. This implies that we can consider all possible regions for each U_{i+1}^j and carry out a feasibility analysis to find which combination(s) of regions satisfy Equations 5.8 and 5.5. Solving this feasibility problem will tell us all feasible choices for every $E\{\beta_i^j\}$. We can then use Equations 5.8 and 5.5 again to solve for U_{i+1}^j recursively starting with U_N^j to find the maximum cognitive throughput among all feasible combinations of $E\{\beta_i^j\}$. A brute-force search over all possible sets of sensing orders can then give us the optimal decentralized sensing orders.

FUTURE RESEARCH DIRECTIONS

Although this chapter studied the sensing-order problem in multi-user multi-channel CRNs for several commonly-used spectrum access techniques, there were a few accompanying assumptions which could be relaxed in a more general study of the problem. Future research may proceed along the following directions.

- The problem of designing an optimal algorithm to find the decentralized sensing orders when the CUs adapt their data rates based on the channel fading conditions is still open. A discussion of this problem and a possible approach to finding the solution is presented in the subsection titled *Optimal decentralized sensing orders with rate adaptation*.
- One of the assumptions in this chapter is that the primary-free statistics are independent across slots and across the channels. However, several measurement studies (Kang, 2009; López-Benítez & Casadevall, 2011) have shown that the spectrum occupancy across time and frequency is correlated. Using a Markovian model to capture this time correlation of spectrum occupancy (Zhao,

Tong, Swami & Chen, 2007), a channel sensing procedure to maximize single-user CU throughput is derived by Umashankar and Kannu (2013). Extending this result to multi-user CRNs is a possible future research direction.

- In the model used in this chapter, we assumed that the channel sensing results are error-free. Future research can incorporate sensing errors in the derivation of optimal sensing orders for multi-user CRNs.
- In this work we focused on maximizing the expected sum throughput of cognitive users. One may be interested in delays experienced by packets in the CRN. Using queuing theoretic tools to study the sensing-order problem in multi-user CRNs is another possible area of research

CONCLUSION

This chapter followed a systematic approach to studying the sensing-order problem for multi-user CRNs and described how to compute the optimal sensing orders for a variety of commonly used spectrum access techniques. The focus was on maximizing the per-slot cognitive throughput in the network, and estimates of primary-free probabilities of the different channels were assumed to be known a priori to a central coordinator. The algorithms in this chapter however included a few assumptions, many of which could be relaxed in a more general study of this problem (see the section on *Future Research Directions*).

It is worth mentioning here that the study of the sensing-order problem in multi-user cognitive radio networks spans across several objective functions, several spectrum access techniques, several sensing techniques, several categories of primary users with different traffic models etc. Hence, it is not difficult to see why the problem is very broad and why it is practically impossible to devise a general algorithm that solves the problem in its entirety. It is also not surprising then that there exists a vast body of work on this topic, each trying to address a specific instance of the sensing-order problem (see *References* and *Additional Reading*). Although this chapter attempted to cover several instances of the problem that commonly arise in practice, the techniques and tools used in the chapter however are broadly applicable and can be suitably modified to solve the sensing-order problem arising in many different contexts. This chapter also highlighted the fact that very often, the performance of the optimal algorithm might only be imperceptibly higher than a much simpler heuristic algorithm and therefore the focus in practical deployments should be on exploiting the problem setting to design smarter algorithms with lower run-times and near-optimal performance.

REFERENCES

ATM Forum Technical Committee. (1999). Traffic management specification version 4.1. *af-tm-0121.000*.

Cheng, H. T., & Zhuang, W. (2011). Simple channel sensing order in cognitive radio networks. *IEEE Journal on* Selected Areas in Communications, *29*(4), 676–688.

Fan, R., & Jiang, H. (2009). Channel sensing-order setting in cognitive radio networks: A two-user case. *IEEE Transactions on* Vehicular Technology, *58*(9), 4997–5008.

Fruth, M. (2006, November). Probabilistic model checking of contention resolution in the IEEE 802.15.4 low-rate wireless personal area network protocol. In *Proceedings of Leveraging Applications of Formal Methods, Verification and Validation,* (pp. 290-297). IEEE.

Hamza, D., & Aissa, S. (2013). Wideband Spectrum Sensing Order for Cognitive Radios with Sensing Errors and Channel SNR Probing Uncertainty. *IEEE Wireless Communications Letters, 2*(2), 151–154. doi:10.1109/WCL.2012.121812.120809

Hossain, E., Niyato, D., & Han, Z. (2009). *Dynamic spectrum access and management in cognitive radio networks.* Cambridge University Press. doi:10.1017/CBO9780511609909

Jiang, H., Lai, L., Fan, R., & Poor, H. V. (2009). Optimal selection of channel sensing order in cognitive radio. *IEEE Transactions on* Wireless Communications, *8*(1), 297–307.

Kang, B.-J. (2009, September). Spectrum sensing issues in cognitive radio networks. In *Proceedings of Communications and Information Technology,* (pp. 824-828). IEEE.

Kartheek, M., Misra, R., & Sharma, V. (2011, January). Performance analysis of data and voice connections in a cognitive radio network. In *Proceedings of Communications (NCC),* (pp. 1-5). IEEE. doi:10.1109/NCC.2011.5734764

Khan, Z., Lehtomaki, J. J., Mustonen, M., & Matinmikko, M. (2011, June). Sensing order dispersion for autonomous cognitive radios. In *Proceedings of Cognitive Radio Oriented Wireless Networks and Communications (CROWNCOM),* (pp. 191-195). IEEE. doi:10.4108/icst.crowncom.2011.246247

Kim, H., & Shin, K. G. (2008, October). Fast discovery of spectrum opportunities in cognitive radio networks. In *Proceedings of New Frontiers in Dynamic Spectrum Access Networks,* (pp. 1-12). IEEE. doi:10.1109/DYSPAN.2008.30

Kim, H., & Shin, K. G. (2013). Optimal online sensing sequence in multichannel cognitive radio networks. *IEEE Transactions on* Mobile Computing, *12*(7), 1349–1362.

Lai, L., El Gamal, H., Jiang, H., & Poor, H. V. (2011). Cognitive medium access: Exploration, exploitation, and competition. *IEEE Transactions on* Mobile Computing, *10*(2), 239–253.

Lai, L., Jiang, H., & Poor, H. V. (2008, October). Medium access in cognitive radio networks: A competitive multi-armed bandit framework. In *Proceedings of Signals, Systems and Computers,* (pp. 98-102). IEEE. doi:10.1109/ACSSC.2008.5074370

López-Benítez, M., & Casadevall, F. (2011, June). Modeling and simulation of time-correlation properties of spectrum use in cognitive radio. In *Proceedings of Cognitive Radio Oriented Wireless Networks and Communications (CROWNCOM),* (pp. 326-330). IEEE. doi:10.4108/icst.crowncom.2011.246158

Misra, R., & Kannu, A. P. (2012a, June). Optimal sensing-order in cognitive radio networks with cooperative centralized sensing. In *Proceedings of Communications (ICC),* (pp. 1566-1570). IEEE. doi:10.1109/ICC.2012.6363863

Misra, R., & Kannu, A. P. (2012b, December). Optimal decentralized sensing-orders in multi-user cognitive radio networks. In *Proceedings of Global Communications Conference (GLOBECOM),* (pp. 1447-1452). IEEE. doi:10.1109/GLOCOM.2012.6503317

Shokri-Ghadikolaei, H., Sheikholeslami, F., & Nasiri-Kenari, M. (2013). Distributed Multiuser Sequential Channel Sensing Schemes in Multichannel Cognitive Radio Networks. *IEEE Transactions on* Wireless Communications, *12*(5), 2055–2067.

Umashankar, G., & Kannu, A. P. (2013). Throughput Optimal Multi-Slot Sensing Procedure for a Cognitive Radio. *IEEE Communications Letters, 17*(12), 2292–2295. doi:10.1109/LCOMM.2013.102613.131825

Xu, Y., Anpalagan, A., Wu, Q., Shen, L., Gao, Z., & Wang, J. (2013). Decision-Theoretic Distributed Channel Selection for Opportunistic Spectrum Access: Strategies, Challenges and Solutions. *IEEE Communications Surveys and Tutorials, 15*(4), 1689–1713. doi:10.1109/SURV.2013.030713.00189

Zhang, Z., Jiang, H., Tan, P., & Slevinsky, J. (2013). Channel exploration and exploitation with imperfect spectrum sensing in cognitive radio networks. *IEEE Journal on* Selected Areas in Communications, *31*(3), 429–441.

Zhao, J., & Wang, X. (2012, October). Channel sensing order in multi-user cognitive radio networks. In *Proceedings of Dynamic Spectrum Access Networks (DYSPAN),* (pp. 397-407). IEEE.

Zhao, Q., Tong, L., Swami, A., & Chen, Y. (2007). Decentralized cognitive MAC for opportunistic spectrum access in ad hoc networks: A POMDP framework. *IEEE Journal on* Selected Areas in Communications, *25*(3), 589–600.

ADDITIONAL READING

Akkarajitsakul, K., Hossain, E., Niyato, D., & Kim, D. I. (2011). Game theoretic approaches for multiple access in wireless networks: A survey. *IEEE Communications Surveys and Tutorials, 13*(3), 372–395. doi:10.1109/SURV.2011.122310.000119

Akyildiz, I. F., Lee, W. Y., Vuran, M. C., & Mohanty, S. (2008). A survey on spectrum management in cognitive radio networks. *Communications Magazine, IEEE, 46*(4), 40–48. doi:10.1109/MCOM.2008.4481339

Attar, A., Nakhai, M. R., & Aghvami, A. H. (2009). Cognitive radio game for secondary spectrum access problem. *Wireless Communications. IEEE Transactions on, 8*(4), 2121–2131.

Auer, P., Cesa-Bianchi, N., & Fischer, P. (2002). Finite-time analysis of the multiarmed bandit problem. *Machine Learning, 47*(2-3), 235–256. doi:10.1023/A:1013689704352

Bkassiny, M., Li, Y., & Jayaweera, S. K. (2013). A survey on machine-learning techniques in cognitive radios. *IEEE Communications Surveys and Tutorials, 15*(3), 1136–1159. doi:10.1109/SURV.2012.100412.00017

Chang, N. B., & Liu, M. (2007, September). Optimal channel probing and transmission scheduling for opportunistic spectrum access. In *Proceedings of the 13th annual ACM international conference on Mobile computing and networking* (pp. 27-38). ACM. doi:10.1145/1287853.1287858

Chaporkar, P., & Proutiere, A. (2008). Optimal joint probing and transmission strategy for maximizing throughput in wireless systems. *Selected Areas in Communications. IEEE Journal on, 26*(8), 1546–1555.

Chen, Y., Zhao, Q., & Swami, A. (2008). Joint design and separation principle for opportunistic spectrum access in the presence of sensing errors. *Information Theory. IEEE Transactions on, 54*(5), 2053–2071.

Chen, Y., Zhao, Q., & Swami, A. (2009). Distributed spectrum sensing and access in cognitive radio networks with energy constraint. *Signal Processing. IEEE Transactions on, 57*(2), 783–797. doi:10.1109/TSP.2008.2007928

Cormio, C., & Chowdhury, K. R. (2009). A survey on MAC protocols for cognitive radio networks. *Ad Hoc Networks, 7*(7), 1315–1329. doi:10.1016/j.adhoc.2009.01.002

Gai, Y., Krishnamachari, B., & Jain, R. (2010, April). Learning multiuser channel allocations in cognitive radio networks: A combinatorial multi-armed bandit formulation. In *New Frontiers in Dynamic Spectrum, 2010 IEEE Symposium on* (pp. 1-9). IEEE. doi:10.1109/DYSPAN.2010.5457857

Haddad, M., Elayoubi, S. E., Altman, E., & Altman, Z. (2011). A hybrid approach for radio resource management in heterogeneous cognitive networks. *Selected Areas in Communications. IEEE Journal on, 29*(4), 831–842.

Huang, J. W., & Krishnamurthy, V. (2010). Transmission control in cognitive radio as a Markovian dynamic game: Structural result on randomized threshold policies. *Communications. IEEE Transactions on, 58*(1), 301–310. doi:10.1109/TCOMM.2010.01.080157

Lee, W. Y., & Akyildiz, I. F. (2008). Optimal spectrum sensing framework for cognitive radio networks. *Wireless Communications. IEEE Transactions on, 7*(10), 3845–3857.

Li, B., Yang, P., Li, X. Y., Wang, J., & Wu, Q. (2011, June). Finding optimal action point for multistage spectrum access in cognitive radio networks. In *Communications (ICC), 2011 IEEE International Conference on* (pp. 1-5). IEEE. doi:10.1109/icc.2011.5962854

Li, B., Yang, P., Wang, J., Wu, Q., Tang, S. J., Li, X. Y., & Liu, Y. (2012). Optimal frequency-temporal opportunity exploitation for multichannel ad hoc networks. *Parallel and Distributed Systems. IEEE Transactions on, 23*(12), 2289–2302.

Liang, Y. C., Zeng, Y., Peh, E. C., & Hoang, A. T. (2008). Sensing-throughput tradeoff for cognitive radio networks. *Wireless Communications. IEEE Transactions on, 7*(4), 1326–1337.

Masonta, M. T., Mzyece, M., & Ntlatlapa, N. (2013). Spectrum decision in cognitive radio networks: A survey. *IEEE Communications Surveys and Tutorials, 15*(3), 1088–1107. doi:10.1109/SURV.2012.111412.00160

Niyato, D., Hossain, E., & Wang, P. (2011). Optimal channel access management with QoS support for cognitive vehicular networks. *Mobile Computing. IEEE Transactions on, 10*(4), 573–591.

Pei, Y., Liang, Y. C., Teh, K. C., & Li, K. H. (2011). Energy-efficient design of sequential channel sensing in cognitive radio networks: Optimal sensing strategy, power allocation, and sensing order. *Selected Areas in Communications. IEEE Journal on, 29*(8), 1648–1659.

Rashid, M. M., Hossain, M., Hossain, E., & Bhargava, V. K. (2009). Opportunistic spectrum scheduling for multiuser cognitive radio: A queueing analysis. *Wireless Communications. IEEE Transactions on, 8*(10), 5259–5269.

Sabharwal, A., Khoshnevis, A., & Knightly, E. (2007). Opportunistic spectral usage: Bounds and a multi-band CSMA/CA protocol. *Networking, IEEE/ACM Transactions on, 15*(3), 533-545.

Shin, K. G., Kim, H., Min, A. W., & Kumar, A. (2010). Cognitive radios for dynamic spectrum access: From concept to reality. *Wireless Communications, IEEE, 17*(6), 64–74. doi:10.1109/MWC.2010.5675780

Unnikrishnan, J., & Veeravalli, V. V. (2010). Algorithms for dynamic spectrum access with learning for cognitive radio. *Signal Processing. IEEE Transactions on, 58*(2), 750–760. doi:10.1109/TSP.2009.2028970

Wang, B., Wu, Y., & Liu, K. J. (2010). Game theory for cognitive radio networks: An overview. *Computer Networks, 54*(14), 2537–2561. doi:10.1016/j.comnet.2010.04.004

Xu, Y., Gao, Z., Wang, J., & Wu, Q. (2012). Multichannel opportunistic spectrum access in fading environment using optimal stopping rule. In *Wireless Communications and Applications* (pp. 275–286). Springer Berlin Heidelberg. doi:10.1007/978-3-642-29157-9_26

Xu, Y., Wang, J., Wu, Q., Anpalagan, A., & Yao, Y. D. (2012). Opportunistic spectrum access in cognitive radio networks: Global optimization using local interaction games. *Selected Topics in Signal Processing. IEEE Journal of, 6*(2), 180–194.

Xu, Y., Wu, Q., Wang, J., Anpalagan, A., & Xu, Y. (2012). Exploiting Multichannel Diversity in Spectrum Sharing Systems Using Optimal Stopping Rule. *ETRI Journal, 34*(2).

Yin, S., Chen, D., Zhang, Q., & Li, S. (2011). Prediction-based throughput optimization for dynamic spectrum access. *Vehicular Technology. IEEE Transactions on, 60*(3), 1284–1289.

Yucek, T., & Arslan, H. (2009). A survey of spectrum sensing algorithms for cognitive radio applications. *IEEE Communications Surveys and Tutorials, 11*(1), 116–130. doi:10.1109/SURV.2009.090109

Zhao, Q., Geirhofer, S., Tong, L., & Sadler, B. M. (2008). Opportunistic spectrum access via periodic channel sensing. *Signal Processing. IEEE Transactions on, 56*(2), 785–796. doi:10.1109/TSP.2007.907867

Zhao, Q., & Swami, A. (2007). A decision-theoretic framework for opportunistic spectrum access. *Wireless Communications, IEEE, 14*(4), 14–20. doi:10.1109/MWC.2007.4300978

Zhao, Y., Mao, S., Neel, J. O., & Reed, J. H. (2009). Performance evaluation of cognitive radios: Metrics, utility functions, and methodology. *Proceedings of the IEEE, 97*(4), 642–659. doi:10.1109/JPROC.2009.2013017

Zheng, D., Ge, W., & Zhang, J. (2009). Distributed opportunistic scheduling for ad hoc networks with random access: An optimal stopping approach. *Information Theory. IEEE Transactions on, 55*(1), 205–222.

KEY TERMS AND DEFINITIONS

Availability Combination: A binary vector indicating the ON/OFF state of each channel depending on whether or not it is in use by the primary users.

Availability Probability (of a channel for a cognitive user): The probability that the channel is *available* for use by the cognitive user in a given time slot or, in other words, the probability that the channel is not in use by the primary users of the channel as well as all higher preemptive-priority cognitive users, if any.

Centralized Sensing Order: A sensing order for a central node in a network that makes all spectrum access decisions in a centralized manner.

Cognitive Radio: A smart radio that can intelligently detect available wireless channels in its vicinity and dynamically configure its transmission or reception parameters to make the best use of these channels.

Cognitive Throughput: The expected sum throughput of all cognitive users in a network.

Decentralized Sensing Order: A sensing order for a local node in a network where spectrum access decisions are made in a decentralized manner by the individual nodes.

Primary-Free Probability (of a channel): The probability that the channel is free of activity by the primary users of the channel.

Rate Adaptation: A dynamic technique used by wireless transmitters to match their modulation, coding and other signal parameters that affect their effective transmission rates to the prevailing conditions on the wireless links.

Sensing Order: A sequential order of channels used for sensing by a cognitive radio that is capable of sensing only one channel at a time to find a free channel for its communication.

Stopping Rule: A rule associated with the sensing order that a cognitive radio uses whenever it senses a channel to be free to determine whether to stop sensing at the current channel (and start transmitting data) or to continue sensing the next channel in its sensing order.

APPENDIX A

Proof of Lemma 1

If at the i-th sensing with $j\,(>0)$ waiting CUs, the coordinator senses channel s_i to be primary-free, the instantaneous throughput if the coordinator stops here is $c_i R + V_{i+1}^{j-1}$ which can be shown to be greater than V_{i+1}^j, the instantaneous throughput if the coordinator proceeds to sense channel s_{i+1}. Therefore, using the optimal access rule, the coordinator should never skip a primary-free channel but always assign it to some waiting-CU for optimal performance.

The optimality of the sensing order mentioned in the lemma can be proved by contradiction. Let the optimal sensing-order be $(s_1, s_2, \ldots, s_N)$, and let there exist $k < N$ such that $\theta_{s_k} < \theta_{s_{k-1}}$. Then from equation 4.1, we have:

$$
\begin{aligned}
V_k^j &= \theta_{s_k}(c_k R + V_{k+1}^{j-1}) + (1 - \theta_{s_k})V_{k+1}^j \\
&= \theta_{s_k}[c_k R + \theta_{s_{k+1}}(c_{k+1}R + V_{k+2}^{j-2}) + (1 - \theta_{s_{k-1}})V_{k+2}^{j-1}] + (1 - \theta_{s_k})[\theta_{s_{k+1}}(c_{k+1}R + V_{k+2}^{j-1}) + (1 - \theta_{s_{k-1}})V_{k+2}^j] \\
&= \theta_{s_k}c_k R + \theta_{s_{k-1}}c_{k+1}R + \theta_{s_k}\theta_{s_{k+1}}V_{k+2}^{j-2} + (1 - \theta_{s_k})(1 - \theta_{s_{k-1}})V_{k+2}^j + [\theta_{s_k}(1 - \theta_{s_{k-1}}) + (1 - \theta_{s_k})\theta_{s_{k-1}}]V_{k+2}^{j-1}
\end{aligned}
$$

$$(1)$$

where $V_{N+1}^j \triangleq 0$. Let us now consider a new sensing order by interchanging channels s_k and s_{k+1} in the above sensing order, and preserving the rest of the order. Then, we have

$$
\hat{V}_k^j = \theta_{s_k}c_{k+1}R + \theta_{s_{k-1}}c_k R + \theta_{s_k}\theta_{s_{k+1}}V_{k+2}^{j-2} + (1 - \theta_{s_k})(1 - \theta_{s_{k+1}})V_{k+2}^j + [\theta_{s_k}(1 - \theta_{s_{k-1}}) + (1 - \theta_{s_k})\theta_{s_{k-1}}]V_{k+2}^{j-1}
$$

$$(2)$$

From Equations 1 and 2, we have the following.

$$
V_k^j - \hat{V}_k^j = (\theta_{s_k} - \theta_{s_{k-1}})(c_k - c_{k+1})R
$$

$$(3)$$

Since $\theta_{s_k} < \theta_{s_{k-1}}$ and $c_k > c_{k+1}$, it follows from Equation 3 that $V_k^j < \hat{V}_k^j$. Further, we have

$$
V_{k-1}^j = \theta_{s_{k-1}}(c_{k-1}R + V_k^{j-1}) + (1 - \theta_{s_{k-1}})V_k^j < \theta_{s_{k-1}}(c_{k-1}R + \hat{V}_k^{j-1}) + (1 - \theta_{s_{k-1}})\hat{V}_k^j = \hat{V}_{k-1}^j
$$

Similarly, we have $V_1^M < \hat{V}_1^M$ which contradicts the assumption that the sensing order $(s_1, s_2, \ldots, s_N)$ is optimal. This proves that the sensing order formed by arranging the channels in decreasing order of $\{\theta_i\}_{i=1}^N$ is the optimal centralized sensing order when the CUs do not use rate-adaptation.

APPENDIX B

Illustration of Computation of Optimal Centralized Sensing Order with Rate Adaptation

Let us consider homogenous Rayleigh-fading channels where Γ denotes the mean SNR seen on each channel by the CUs to their respective receivers. Hence, the instantaneous SNR γ_i^m seen on any channel i by any CU m follows the same exponential distribution whose pdf is given by

$$f_i(\gamma_i^m) = \frac{1}{\Gamma} e^{-\frac{\gamma_i^m}{\Gamma}}, \gamma \geq 0 \tag{4}$$

Let the transmission rate used on channel i by CU m be $r(\gamma_i^m) = \ln(1 + \gamma_i^m)$. The pdf of δ_i^j, where δ_i^j is defined by Equation 4.2 and is the maximum of j independent and identically distributed random variables, can be shown to be as follows.

$$f_\Delta(\delta_i^j) = j\left(1 - e^{-\frac{\delta_i^j}{\Gamma}}\right)^{j-1} \cdot \frac{1}{\Gamma} e^{-\frac{\delta_i^j}{\Gamma}}, \; \delta_i^j \geq 0 \; \forall i \tag{5}$$

For illustration, let us consider the special case of a 2-user CRN ($M = 2$). Our aim is to first find a closed-form expression for the cognitive throughput V_1^2 as a function of the sensing order $s = (s_1, s_2, \ldots, s_N)$ of the coordinator, and then determine the sensing order that maximizes V_1^2. We begin by using Equations 4.1, 4.3 and 5 for $i = N, j = 1$ to obtain V_N^1 as follows.

$$V_N^1 = \theta_{s_N} E[v_N^1] = \theta_{s_N} \int_0^\infty c_N \ln(1 + \delta_N^1) \cdot \frac{1}{\Gamma} e^{-\frac{\delta_N^1}{\Gamma}} d\delta_N^1 = \theta_{s_N} c_N e^{1/\Gamma} \psi\left(\frac{1}{\Gamma}\right) \tag{6}$$

where $\psi(.)$ is defined as $\psi(x) = \int_x^\infty \frac{e^{-t}}{t} dt$ and δ_N^1 is the variable of integration. We use Equations 4.1, 4.3 and 5 again, for $i = N, j = 2$ and integrate over δ_N^2 to obtain V_N^2 as follows.

$$V_N^2 = \theta_{s_N} E[v_N^2] = \theta_{s_N} \int_0^\infty c_N \ln(1 + \delta_N^2) \cdot \frac{2}{\Gamma}(1 - e^{-\frac{\delta_N^2}{\Gamma}}) \frac{1}{\Gamma} e^{-\frac{\delta_N^2}{\Gamma}} d\delta_N^2 = 2\theta_{s_N} c_N e^{1/\Gamma} \psi\left(\frac{1}{\Gamma}\right) - \theta_{s_N} c_N e^{2/\Gamma} \psi\left(\frac{2}{\Gamma}\right) \tag{7}$$

Similarly, for $i \neq N, j = 1$, we can obtain V_i^1 as follows.

$$V_i^1 = \theta_{s_i} E[v_i^1] + (1 - \theta_{s_i})V_{i+1}^1$$

$$= \theta_{s_i}\left[\int_{c_i \ln(1+\delta_i^1) > V_{i-1}^1} c_i \ln(1+\delta_i^1) \cdot \frac{1}{\Gamma} e^{\frac{-\delta_i^1}{\Gamma}} d\delta_i^1 + \int_{c_i \ln(1+\delta_i^1) \leq V_{i-1}^1} V_{i+1}^1 \cdot \frac{1}{\Gamma} e^{\frac{-\delta_i^1}{\Gamma}} d\delta_i^1\right] + (1-\theta_{s_i})V_{i+1}^1 \qquad (8)$$

$$= V_{i+1}^1 + \theta_{s_i} c_i e^{1/\Gamma} \psi\left(\frac{e^{V_{i-1}^1/c_i}}{\Gamma}\right)$$

To obtain V_i^2 $(i \neq N)$, we note that

$$v_i^2 = \begin{cases} c_i \ln(1+\delta_i^2) + V_{i+1}^1, & c_i \ln(1+\delta_i^2) + V_{i+1}^1 > V_{i+1}^2 \text{ (coordinator grants access to } s_i) \\ V_{i+1}^2, & \text{otherwise (coordinator skips } s_i) \end{cases} \qquad (9)$$

Thus, using equation 4.1, we can write V_i^2 as

$$V_i^2 = \theta_{s_i} E[v_i^2] + (1-\theta_{s_i})V_{i+1}^2$$

$$= \theta_{s_i}\left[\int_{c_i \ln(1+\delta_i^2)+V_{i+1}^1 > V_{i-1}^2} [c_i \ln(1+\delta_i^2) + V_{i+1}^1] \cdot f_\Delta(\delta_i^2)d\delta_i^2 + \int_{c_i \ln(1+\delta_i^2)+V_{i+1}^1 \leq V_{i-1}^2} V_{i+1}^2 \cdot f_\Delta(\delta_i^2)d\delta_i^2\right] + (1-\theta_{s_i})V_{i+1}^2$$

$$= V_{i+1}^2 + 2\theta_{s_i} c_i e^{\frac{1}{\Gamma}} \psi\left(\frac{e^{(V_{i-1}^2-V_{i+1}^1)/c_i}}{\Gamma}\right) - \theta_{s_i} c_i e^{\frac{2}{\Gamma}} \psi\left(\frac{2e^{(V_{i-1}^2-V_{i+1}^1)/c_i}}{\Gamma}\right)$$

$$\qquad (10)$$

Starting with Equations 6 and 7, and using the recursive Equations 8 and 10, we can derive a closed-form expression for the reward V_1^2 as a function of the sensing order s of the coordinator. As a specific example, these expressions for a 3-channel CRN ($N = 3$) would be the following.

$$V_3^1 = \theta_{s_3} c_3 e^{1/\Gamma} \psi\left(\frac{1}{\Gamma}\right), \quad V_2^1 = V_3^1 + \theta_{s_2} c_2 e^{1/\Gamma} \psi\left(\frac{e^{V_3^1/c_2}}{\Gamma}\right), \quad V_1^1 = V_2^1 + \theta_{s_1} c_1 e^{1/\Gamma} \psi\left(\frac{e^{V_2^1/c_1}}{\Gamma}\right)$$

$$V_3^2 = 2\theta_{s_3} c_3 e^{1/\Gamma} \psi\left(\frac{1}{\Gamma}\right) - \theta_{s_3} c_3 e^{2/\Gamma} \psi\left(\frac{2}{\Gamma}\right)$$

$$\qquad (11)$$

$$V_2^2 = V_3^2 + 2\theta_{s_2} c_2 e^{\frac{1}{\Gamma}} \psi\left(\frac{e^{(V_3^2-V_3^1)/c_2}}{\Gamma}\right) - \theta_{s_2} c_2 e^{\frac{2}{\Gamma}} \psi\left(\frac{2e^{(V_3^2-V_3^1)/c_2}}{\Gamma}\right)$$

$$V_1^2 = V_2^2 + 2\theta_{s_1} c_1 e^{\frac{1}{\Gamma}} \psi\left(\frac{e^{(V_2^2-V_2^1)/c_1}}{\Gamma}\right) - \theta_{s_1} c_1 e^{\frac{2}{\Gamma}} \psi\left(\frac{2e^{(V_2^2-V_2^1)/c_1}}{\Gamma}\right)$$

The sensing order s that maximizes V_1^2 is the optimal sensing order and the corresponding V_1^2 is the optimal cognitive throughput.

APPENDIX C

Illustration of Computation of Optimal Decentralized Sensing Order without Rate Adaptation

To see an illustration of how to compute the cognitive throughput for a given set of sensing orders, let us consider a 4-channel CRN, so that we need to list down only 16 possible values for $\mathbf{a}$. Let the sensing orders of the 3 CUs be $\mathbf{s}^1 = (1, 4, 2, 3)$, $\mathbf{s}^2 = (2, 4, 3, 1)$ and $\mathbf{s}^3 = (3, 2, 4, 1)$ respectively. We tabulate below the sum throughputs for each of the three contention resolution mechanisms, denoted by u_B, u_P and u_C for BACKOFF, PRIORITY and COLLIDE respectively, for each of the 16 possible channel availability combinations.

Once we construct a table similar to Table 1 for a given set of sensing orders, we can compute the cognitive throughput U_B, U_P or U_C for BACKOFF, PRIORITY and COLLIDE respectively, as the case may be, by computing the expected sum throughput as follows,

$$U_X = \sum_{\mathbf{a} \in \{0, 1\}^N} P\{\mathbf{a}\} u_X(\mathbf{a}) \tag{12}$$

where $X = B$ or P or C and $P\{\mathbf{a}\}$ is given by Equation 5.6. We then repeat this procedure for all possible sets of sensing-orders. The set $\{\mathbf{s}^j\}_{j=1}^{M}$ that maximizes the corresponding cognitive throughput U_X is the optimal set of decentralized sensing orders in this scenario.

Table 1. Sum throughputs (in units of R) for a 4-channel, 3-user CRN when the decentralized sensing orders are chosen to be $\mathbf{s}^1 = (1,4,2,3)$, $\mathbf{s}^2 = (2,4,3,1)$ *and* $\mathbf{s}^3 = (3,2,4,1)$

a_1	a_2	a_3	a_4	$u_B(\mathbf{a})$	$u_P(\mathbf{a})$	$u_C(\mathbf{a})$
0	0	0	0	0	0	0
0	0	0	1	c_2	c_2	0
0	0	1	0	c_1	c_1	c_1
0	0	1	1	$c_1 + c_2$	$c_1 + c_2$	$c_1 + c_2$
0	1	0	0	c_1	c_1	c_1
0	1	0	1	$c_1 + c_2$	$c_1 + c_2$	$c_1 + c_2$
0	1	1	0	$2c_1$	$2c_1$	$2c_1$
0	1	1	1	$2c_1 + c_2$	$2c_1 + c_2$	$2c_1 + c_2$
1	0	0	0	c_1	c_1	c_1
1	0	0	1	$c_1 + c_2$	$c_1 + c_2$	$c_1 + c_2$
1	0	1	0	$2c_1$	$2c_1$	$2c_1$
1	0	1	1	$2c_1 + c_2$	$2c_1 + c_2$	$2c_1 + c_2$
1	1	0	0	$2c_1$	$2c_1$	$2c_1$
1	1	0	1	$2c_1 + c_3$	$2c_1 + c_3$	$2c_1 + c_3$
1	1	1	0	$3c_1$	$3c_1$	$3c_1$
1	1	1	1	$3c_1$	$3c_1$	$3c_1$

Section 2
Radio Spectrum Management and Access

Chapter 10
On Fuzzy Logic-Based Channel Selection in Cognitive Radio Networks

Yong Yao
Blekinge Institute of Technology, Sweden

Alexandru Popescu
Blekinge Institute of Technology, Sweden

Adrian Popescu
Blekinge Institute of Technology, Sweden

ABSTRACT

Cognitive radio networks are a new technology based on which unlicensed users are allowed access to licensed spectrum under the condition that the interference perceived by licensed users is minimal. That means unlicensed users need to learn from environmental changes and to make appropriate decisions regarding the access to the radio channel. This is a process that can be done by unlicensed users in a cooperative or non-cooperative way. Whereas the non-cooperative algorithms are risky with regard to performance, the cooperative algorithms have the capability to provide better performance. This chapter shows a new fuzzy logic-based decision-making algorithm for channel selection. The underlying decision criterion considers statistics of licensed user channel occupancy as well as information about the competition level of unlicensed users. The theoretical studies indicate that the unlicensed users can obtain an efficient sharing of the available channels. Simulation results are reported to demonstrate the performance and effectiveness of the suggested algorithm.

INTRODUCTION

Today, one of the most active areas of research in Cognitive Radio (CR) is on Dynamic Spectrum Access (DSA), which refers to the method used to detect, to select and to access spectrum holes. Related to this, an important challenge for the research and industrial communities is to bridge the gap between the existent research results and the large-scale deployment of Cognitive Radio Networks (CRNs) (Akyildiz,

DOI: 10.4018/978-1-4666-6571-2.ch010

Lee, Vuran, & Mohanty, 2006). Sustained research efforts are needed to provide technological solutions able to take advantage of the great potential and commercial promises of CRNs. In a longer perspective, it is expected that DSA will go beyond the opportunistic spectrum access model and new technologies and policies will be developed for CRNs, to allow the access to a portfolio of different spectrum types like, e.g., licensed spectrum, unlicensed spectrum and leased spectrum. The radio devices are expected to be able to dynamically change the operating spectrum within the particular spectrum portfolio, and to do this on a "just-in-time" basis. Furthermore, the resources of the spectrum pool can be characterized in terms of context, location and technology. Parameters like price, QoS/QoE, energy saving and competition may be used in selecting the particular spectrum.

The focus of the paper is on the channel selection and access in CRNs for unlicensed users (also known as secondary users or SUs). We first provide an overview of the spectrum decision problem in CRNs. We also provide definition and description of the competition problem for SUs in CRNs. A new solution is advanced for channel access for SUs, which is based on their competition for the resources not used by licensed users (also known as primary users or PUs) as well as on the statistics of PUs behaviour. Simulation results are provided to demonstrate the performance and effectiveness of the suggested solution.

The rest of the paper is organized as follows. In Sections 2 and 3, we present the background, motivation and the adopted solution. In section 4, we describe the system model. Section 5 discusses the learning of idle time statistics. Section 6 describes the competition problem and the suggested method for alleviating the competition. Section 7 is about the hybrid decision making algorithm. The performance evaluation is presented in Section 8. Finally, we conclude the paper in Section 9

BACKGROUND AND MOTIVATION

In CR networks, the licensed channels are either exclusively reserved for PUs or temporarily used by SUs. Extensive research has been done to develop the concept of CR, based on which the SUs are allowed to access the available channels (also known as spectrum holes) not being occupied by PUs. Moreover, when the PU occupies a channel, the SU in the same channel must leave. Otherwise, the PU transmission would be impaired.

Since PUs do not need to notify SUs of their activities, a time-slotted transmission scheme is suggested for SUs to communicate in CR networks. In this scheme, the SU's transmission is divided into identical slots over time (Zhao, Tong, Swami, & Chen, 2007). During each slot, the SU first performs spectrum sensing to detect channel availability. The SU may then transmit data via an available channel (if it exists) within the remaining slot duration. Further, to alleviate the interruption from PUs, SUs need to learn from the statistical information about PUs' activity and, based on that, to select the most available channels to use. An existing solution along with this line is given by using the idle-time-based statistics. For a single channel, being idle indicates the PU absence and the idle time indicates how long this absence is. Yang et al. (2007) consider in their work that the longer an available channel remains idle in the near future, the higher the channel availability becomes. Further, by predicting the idle time, the most available channel is attributed to the characteristic of having the longest remaining idle time.

The problem however is regarding the limitation of the reported results, which is basically due to limited theoretical models considered. Most of these models do not consider the problem of competition existing among SUs in accessing available channels. This problem is addressed by the fact that the idle time statistics can be shared by all SUs within the same CRN. When multiple SUs simultaneously want to use available channels, the selection criterion based on the longest remaining idle time may lead to the same channels. In particular, the SUs that can perceive (by receiving radio signals) each other are likely to compete for the channel utilization over a single channel. We call them SUs competitors. As the channel capacity is limited, the single channel may not satisfy the requirements of all SU competitors. If a channel is overcrowded due to a large number of SU competitors, the QoS performance degrades (Akter, Natarajan, & Scoglio, 2009). The conclusion therefore is that there is strong need for new scientific results that consider the problem of internal competition among SUs in accessing the available channels as well. Furthermore, another important aspect that also needs to be considered is regarding the use of appropriated theoretical models to capture the competition among SUs in accessing the available spectrum.

SOLUTION APPROACH

The adopted solution is based on a joint consideration of the statistics of idle time of spectrum and the competition that may exist among SUs in accessing the available channels (Yao, Ngoga, Erman, & Popescu, 2012). To alleviate the interruption from PUs, the SUs first learn from the statistical properties of PUs activities. At the same time, the SUs learn from the own behaviour with the help of a Two-Step Information-Exchange (TSIE) mechanism in an ad-hoc environment. Based on this, and also on using a fuzzy logic based decision making algorithm, the SUs take decisions in accessing the spectrum. The goal is to provide channel selection with leverage for long remaining time of PUs being absent and low level SU competition. The reported performance results indicate the feasibility of this approach.

The majority of other solutions reported so far on solving the competition problem among SUs are based on using Markov chains or game theory to develop a forecasting strategy (Cheng et al., 2014; Deng et al., 2013). The limitations of these solutions are mainly due to the high computational complexity, with consequences like implementation difficulties and also the lack of the precision of reported results. Compared with these limitations, the solution advanced in this chapter is simple, it considers the problems related to cooperation among SUs and provides precise results as well. By learning from the statistics of both PUs and SUs activities and also by using a hybrid decision making algorithm, SUs are able to obtain better performance with regard to accessing the available channels. The positive consequences of this are especially in the form of better resource utilization. At the same time, the PUs are provided protection assurance in accessing the media.

Moreover, compared with a previous work (Yao, 2012), this paper presents new contributions in the form of a new simulator specially developed for the purpose of evaluating CRNs. The advantage of this simulator is the facility to jointly deal with the simulation of various CRN topologies and the performance evaluation of different spectrum decision strategies. In this paper, a brief description of the developed simulator is presented in Section 6.

System Model

We consider a CRN system with N licensed channels marked with indexes $1, 2, \ldots, N$. The PU's activity in these channels is assumed to be done in a synchronous time-slotted basis. Each PU's slot has an uniform length δ in time domain. In the system, there are M SUs having the labels $s_1, s_2, ..., s_M$. We define S as the set of M SUs, i.e., $S = \{s_1, s_2, ..., s_M\}$. These SUs are ready to transmit data to other SU receivers in a single-hop ad-hoc manner.

We assume that a central coordinator is used in the system. The coordinator can, e.g., be a secondary base station or a support node (Zhang & Su, 2011; Westerhagen, 2014). It is assumed that the collaborative spectrum sensing is done on both coordinator and SU sides, and thus the probabilities of missed detection and false alarm can be decreased (Zhang et al., 2011). We also assume that the sensing results are perfect. The central coordinator is also responsible for collecting PU's slot information, and thus periodically performs spectrum sensing with duration δ corresponding to PU's activity. Both PU's slot information and sensing results are broadcasted by the coordinator to SUs via a common control channel (Akyildiz et al., 2006; Westerhagen, 2014). Moreover, the coordinator helps every SU transmitter/ receiver pair in establishing a reliable communication. Zhao et al. (2007) suggest a *partially observable Markov decision process* based method to achieve the channel synchronization for a SU transmitter/ receiver pair. However, due to the dynamic nature of PU's and SU's activities, the design of precise and reliable channel synchronization is very challenging in ad-hoc CR networks.

The SUs use a time-slotted transmission scheme to opportunistically access available channels. By receiving broadcast messages from the central controller, the SUs can be synchronized with the PUs. To differentiate PU's signal from SU's signal, the SUs usually keep silence at the same time while doing sensing (Zhang et al., 2011). Further, the SU configures its transmission slot length as δ. More specifically, a SU's slot consists of three phases: i) the spectrum sensing and receiving broadcast are done in the first phase, and ii) the second and third phases are used by SUs to cooperatively exchange information and to transmit data, respectively.

The information exchange among SUs can be done by using either the common control channle or cooperative mechanisms in the form of signaling protocol (Chen, Teo, & Farhang-Boroujeny, 2011). For data transmission, Bae et al. (2009) suggest in their work that the third phase is divided into multiple identical sub-slots. By using a modified CSMA/CA protocol, several SU competitors for the same channel can use different sub-slots to transmit data with low-level collision.

Furthermore, we assume in our model that the above mentioned functions (i.e., perfect spectrum sensing, coordinator's broadcasting, information exchange, sub-slot and CSMA/CA based transmission) are applicable in the modeled system. These functions are not deeply studied in our paper. We instead focus on the joint consideration of idle time statistics and SUs' competition problem.

IDLE TIME STATISTICS

Unlike the work done by Yang et al. (2007), we do not assume that SUs have a priori knowledge about the distribution parameters characterizing the idle time like, e.g., the PU arrival and departure rates. Thus, the learning of idle time statistics requires a short-term historical information about PU channel occupancy.

Primary User Channel Occupancy

Given the current time t, the time interval $[t - H\delta, t]$ is identically divided into H time slots, within each of which PUs are either present or absent. The PU's activity may have a change at time points

$\{t - H\delta, t - (H-1)\delta, ..., t\}$. Further, we let h denote the slot $[t - (H-h)\delta, t - (H-h-1)\delta]$, where $h = 0, 1, ..., (H-1)$. We also let the random variable v_h^n denote the sensing result of detecting the PU's activity in channel $n \in \{1, 2, ..., N\}$ in slot h. Then, v_h^n is specified as:

$$v_h^n = \begin{cases} 1, & PU \ presence \\ 0, & PU \ absence \end{cases} \tag{1}$$

This gives a binary sequence indicating the PU channel occupancy. We schematically illustrate an example of this sequence in Figure 1.

At time t, if channel n is sensed to be idle, we have $v_H^n = 0$. This means that the channel n will be idle in the whole interval $[t, t + \delta]$ and it may remain idle in one or more consecutive slots after the time point $(t + \delta)$. As such, the capability of looking ahead the future trends of all channels is desirable for SUs. We are therefore faced with the task of knowing in advance the statistics of the remaining idle time on every channel.

Remaining Idle Time

To achieve the above mentioned task, we first compute the average idle time and ongoing PU absence on a channel n (available at time t) during the past interval $H\delta$.

To compute the average idle time, we are interested in the occurrence times of two events, namely, the event "$v_h^n = 0$", and the event "PU being absent". We observe in Figure 1 that the occurrence time of the first event is equal to:

$$\sum\nolimits_{h=0}^{H-1}(1 - v_h^n) \, .$$

The occurrence time of the second event is computed with:

$$\left(\left| \frac{1}{2} \sum\nolimits_{h=0}^{H-1} \Lambda(n,h) \right| + 1 \right), \text{ where } \Lambda(n,h)$$

means a change of PU activity and equals 1 if $v_h^n \neq v_{h+1}^n$, otherwise 0. We then have the average idle time on channel n in interval $[t - H\delta, t]$:

Figure 1. Example of binary sequence, which indicates the PU channel occupancy

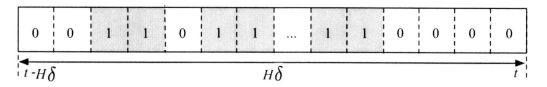

$$E_{idle}^n(t) = \frac{\left|\frac{1}{2}\sum_{h=0}^{H-1}\Lambda(n,h)\right|+1}{\sum_{h=0}^{H-1}(1-v_h^n)} \qquad (2)$$

Let $x^n(t)$ denote the time period of the ongoing PU absence on channel n until the time point t. For instance, in Figure 1, the last four slots before time t are associated with an ongoing PU absence. To compute $x^n(t)$, we find out the slot in which the latest event "$v_h^n = 1$" takes place. Let h' denote this slot, i.e., $h' = \arg\max\left\{h\left|v_h^n = 1 : h = 0, 1, ..., H-1\right.\right\}$. We then obtain:

$$x^n(t) = (H - h')\delta \qquad (3)$$

Actually, the parameter $E_{idle}^n(t)$ provides an insight into how long the duration of a PU absence is expected to be. In contrast to the remaining idle time, the larger $x^n(t)$ is, the lower the availability of channel n becomes.

COMPETITION AMONG SECONDARY USERS

For simplicity purposes, we assume that when the SUs do single-hop ad-hoc transmission, they have the same transmission range D. We let $d_{i,j}(t)$ denote at time t the distance between two different SUs, $s_i, s_j \in S$. Let $\zeta_{i,j}(t)$ denote a relation of whether or not s_i can perceive s_j. We have:

$$\zeta_{i,j}(t) = \begin{cases} 1\ (can) & , \quad d_{i,j}(t) \le D \\ 0\ (can\ not), & d_{i,j}(t) > D \end{cases} \qquad (4)$$

where $\zeta_{i,j}(t) = \zeta_{j,i}(t)$. When $\zeta_{i,j}(t) = 1$, s_i is said to be a neighbor of s_j. In this case, if both s_i and s_j switch to the same available channel for data transmission, they become then SU competitors against each other.

Competition Problem

To address the competition problem, we first introduce a measure called *sub-slot utilization*. Consider a slot h in which the transmission phase is identically divided into L sub-slots, each denoted by 1, 2, ..., L. Given a channel $n \in \{1, 2, ..., N\}$ available in slot h, a SU $s_m \in S$ attempts to access the channel n and to transmit data within the particular sub-slots. Clearly, when the activity of the SU s_i follows a CSMA/CA-like protocol model, its transmission may only take place in a subset of L sub-slots. Let $l_m^n(h)$

denote the number of used sub-slots in slot h, i.e., $0 \le l_m^n(h) \le L$. We thus define the sub-slot utilization $u_m^n(h)$ of s_m on channel n in slot h as being the ratio between the number of used sub-slots and the number of total sub-slots, i.e., $u_m^n(h) = l_m^n(h) / L$.

A reliable communication between a SU transmitter and a SU receiver is therefore constrained by the sub-slot utilization threshold, denoted by U. In other words, if $u_m^n(h)$ is less than U (due to other SU competitors), s_m may terminate the transmission since the QoS performance may not be satisfied by the receiver any more.

We illustrate in Figure 2 a competition example. In the figure, three SU transmitters, denoted by s_a, s_b and s_c, want to use the same available channel n within slot h. Assume a particular threshold $U=45\%$. For s_a and s_c, each of them can perceive another single SU transmitter, so that the number of competitors for them is *two*. Thus, the largest sub-slot utilization allowed for each of them to hold is 90%. However, for s_b, the number of competitors is *three*, since s_b can perceive two other SU transmitters. As a result, both s_a and s_c could successfully use channel n in slot h, but s_b may not do reliable transmission on channel n in slot h because of not enough existing sub-slots to use.

As another example, assume that the channel n was available in slot $(h-1)$ for SUs, and two SUs s_a and s_c have been using it. We also assume that the SU s_b newly starts using channel n in slot h. In this case, we call both s_a and s_c as ongoing SUs on channel n, and we call s_b as new SU on channel n. Further, if each of the three SUs has the same sub-slot utilization threshold $U=60\%$, the new SU s_b may interrupt the ongoing transmission of both s_a and s_c in slot h.

Two-Step Information-Exchange

To solve the above described competition problem, we suggest a simple method called Two-Step Information-Exchange (TSIE) for SUs. TSIE is accomplished by SUs during the second phase of every SU's slot. The process of doing TSIE is shown in Figure 3.

The first step is performed by ongoing SUs among themselves via accessed channels. The information in this case is about which available channels the ongoing SUs are using. Let $n_m(t)$ denote an available channel used by an ongoing SU $s_m \in S$ at time t. If no channel is used by s_m at time t, then

Figure 2. Example of three SUs competing for the use of the same channel

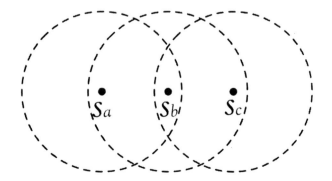

Figure 3. Two-Step Information-Exchange method for SU transmitters

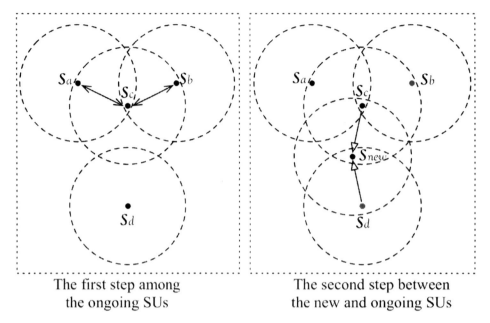

| The first step among the ongoing SUs | The second step between the new and ongoing SUs |

$n_m(t)$ equals zero. By exchanging information with neighboring ongoing SUs via channel $n_m(t)$, s_m can obtain the number, denoted by $\psi_m^n(t)$, of SU competitors on channel $n_m(t)$. This is given by:

$$\psi_m^n(t) = \sum_{i=1}^{M} \left[\zeta_{m,i}(t) \middle| n_i(t) = n_m(t) \right] \tag{5}$$

The second step is initiated by new SUs and it is conducted between new and ongoing SUs. The information in this case is about the competitor number perceived by different ongoing SUs. Consider that a newly arrived SU $s_{m'} \in S$ wants to use channel n in slot h. By communicating with a neighboring SU s_m at time t, SU $s_{m'}$ can get information of $\psi_m^n(t)$ from SU s_m. Similarly, after communicating with all neighboring ongoing SUs, SU $s_{m'}$ can learn the largest number of SU competitions on channel n as being:

$$y_{m'}^n(t) = \arg \max \left\{ \psi_i^n(t) \middle| n_i(t) = n, \zeta_{m',i}(t) = 1, i = 1, 2, ..., M \right\} \tag{6}$$

where $y_{m'}^n(t)$ is called competition level for $s_{m'}$ to access channel n at time t. If $(y_{m'}^n(t) + 1)U$ is not larger than one, then SU $s_{m'}$ can use channel n. Otherwise $s_{m'}$ has to look for other available channels in order to protect the ongoing SUs using channel n. We further let T_{com} denote the maximum of SU competitors accommodated by the same channel. We have $T_{com} = \left\lfloor \frac{1}{U} \right\rfloor$.

Clearly, for new SUs, the competition level on a channel indicates how heavily the channel is used by the ongoing SUs. The larger the competition level on a channel is, the lower the channel availability becomes.

CHANNEL SELECTION

So far, we have formulated two pairs of parameters (x'', E_{idle}^n) and (y_m^n, T_{com}). It is clear that these parameters vary in distinct metrics and measures. This gives rise to a two-constraint based decision problem of finding the most available channel for SUs. Fuzzy logic is suggested to solve this problem because of the capability of coping with various criteria for decision making purposes. We first introduce a parameter named Fuzzy Channel Availability (FCA).

Fuzzy Channel Availability

Proposition: FCA is a fuzzy logic based parameter used to represent three different levels of channel availability for SUs. The three levels are respectively formalized as three fuzzy sets, namely, *high-level*, *medium-level* and *low-level* channel availabilities.

The use of FCA is to map different types of parameter values to an uniform type, i.e., fuzzy membership degree. Let σ denote a parameter of either x'' or y_m^n, i.e., $\sigma \in \{x'', y_m^n\}$. The value of the parameter σ at time t is denoted by $\sigma(t)$. We use the notations α_σ, β_σ and γ_σ to denote three fuzzy sets *high-level*, *medium-level* and *low-level* under parameter σ, respectively. Their membership functions are denoted by g_σ^a, g_σ^i, and g_σ^-, respectively. $g_\sigma^a(\sigma(t))$, $g_\sigma^i(\sigma(t))$, $g_\sigma^-(\sigma(t)) \in [0.0, 1.0]$ are defined as *membership degrees* of $\sigma(t)$ to the fuzzy sets α_σ, β_σ and γ_σ, respectively. The three membership degrees form a vector:

$$V_\sigma(t) = \left(g_\sigma^a(\sigma(t)), \, g_\sigma^i(\sigma(t)), \, g_\sigma^-(\sigma(t)) \right) \tag{7}$$

We call $V_\sigma(t)$ the FCA-based *characterization* of parameter σ at time t. As $V_\sigma(t)$ is a three-coordinate vector, it is not convenient to carry out the numerical computing regarding decision making. This has prompted the development of methods to compound three coordinates into a joint value referred to the channel availability.

Fuzzy-Comparison

We adopt a fuzzy-comparison based algorithm developed by Saaty (1978). The algorithm is based on using a paired-comparison of three fuzzy sets' importances in deciding which channel is most available for SUs.

Let $\pi_\times$, π_+, and π_- denote the importances of *high-level*, *medium-level* and *low-level*, respectively. Since *high-level* has stronger importance over *low-level*, we assign $\pi_\times / \pi_-$ with 5. Since *high-level* (resp. *medium-level*) has weeker importance than *medium-level* (resp. *low-level*), we assign both $\pi_\times / \pi_+$

and π_+ / π_- with 3. Because *high-level*, *medium-level* or *low-level* has equal importance over itself, we have $\pi_\times / \pi_\times = \pi_+ / \pi_+ = \pi_- / \pi_- = 1$. We can therefore obtain a fuzzy-comparison matrix as:

$$\Pi = \begin{bmatrix} \pi_- / \pi_- & \pi_+ / \pi_- & \pi_\times / \pi_- \\ \pi_- / \pi_+ & \pi_+ / \pi_+ & \pi_\times / \pi_+ \\ \pi_- / \pi_\times & \pi_+ / \pi_\times & \pi_\times / \pi_\times \end{bmatrix} = \begin{bmatrix} 1 & 3 & 5 \\ 1/3 & 1 & 3 \\ 1/5 & 1/3 & 1 \end{bmatrix} \tag{8}$$

From fuzzy-logic viewpoint, the matrix Π is used to determine the relative weights of three importance parameters *high-level*, *medium-level* and *low-level*, when carrying out the fuzzy-comparison among them. The three weights are denoted by $\omega_\times$, ω_+ and ω_- for $\pi_\times$, π_+, and π_-, respectively. Given the eigen value λ and eigen vector W of matrix Π, they satisfy the eigen equation $\Pi W = \lambda W$ and characteristic equation $\det(\Pi - \lambda I) = 0$, where I is an unit matrix. The largest real eigen value corresponds to an eigen vector given by:

$$W^* = \left\{ \omega_\times^*, \omega_+^*, \omega_-^* \right\} = \left\{ 0.94, 0.31, 0.19 \right\} \tag{9}$$

The three coordinates of vector W^* refer to the solution of three weights $\omega_\times$, ω_+ and ω_-. Consequently, for the parameter σ, we compose three coordinates of $V_\sigma(t)$ in term of a linear combination:

$$\theta_\sigma(t) = \omega_\times g_\sigma^\alpha(\sigma(t)) + \omega_+ g_\sigma^\beta(\sigma(t)) + \omega_- g_\sigma^\gamma(\sigma(t)) \tag{10}$$

where $\theta_\sigma(t)$ is called the FCA-based *decision factor* of parameter σ at time t for channel selection. By using FCA-based decision factor, we develop the hybrid decision making (for channel selection) in the following subsection.

Hybrid Decision Making

Considering a channel n available at time t, we first map the idle time statistics, i.e., the parameter pair (x^n, E_{idle}^n), to fuzzy membership degree of x^n to $g_{x^n}^\alpha$, $g_{x^n}^\beta$ and $g_{x^n}^\gamma$, respectively. As an example, we consider three values 0, E_{idle}^n and $2E_{idle}^n$. Under these three values, the availability of channel n is defined to be exactly equivalent to *high-level*, *medium-level* and *low-level*, respectively, i.e.:

$$\begin{cases} g_{x^n}^\alpha(0) = 1 \\ g_{x^n}^\beta(E_{idle}^n) = 1 \\ g_{x^n}^\gamma(2E_{idle}^n) = 1 \end{cases} \tag{11}$$

Since the availability of channel n decreases with $x^n(t)$, this implies that: i) when $x^n(t)$ is increasing, the channel availability is far away from *high-level* and becomes closer to *low-level*, ii) when $x^n(t)$

is increasing and it is smaller than E_{idle}^n, the channel availability becomes closer to *medium-level*, and iii) when $x^n(t)$ is increasing and it is larger than E_{idle}^n, the channel availability is far away from the *medium-level*. We therefore have:

$$g_{x^n}^\alpha (x^n(t)) = \begin{cases} 1 - \dfrac{x^n(t)}{E_{idle}^n}, & 0 \le x^n(t) < E_{idle}^n \\ 0, & others \end{cases} \tag{12}$$

$$g_{x^n}^j (x^n(t)) = \begin{cases} \dfrac{x^n(t)}{E_{idle}^n}, & 0 \le x^n(t) < E_{idle}^n \\ 2 - \dfrac{x^n(t)}{E_{idle}^n} & E_{idle}^n \le x^n(t) < 2E_{idle}^n \\ 0 & others \end{cases} \tag{13}$$

$$g_{x^n}^\gamma (x^n(t)) = \begin{cases} \dfrac{x^n(t)}{E_{idle}^n} - 1, & E_{idle}^n \le x^n(t) < 2E_{idle}^n \\ 0, & others \end{cases} \tag{14}$$

For the parameter pair (y_m^n, T_{com}), we know that the larger $y_m^n(t)$ is, the lower the availability of channel n for SU s_m becomes. Therefore, we adopt similar membership functions with regard to Equations (12-14). The difference is that we set:

$$\begin{cases} g_{y_m^n}^\alpha (0) = 1 \\ g_{y_m^n}^j (T_{com}/2) = 1 \\ g_{y_m^n}^\gamma (T_{com}) = 1 \end{cases} \tag{15}$$

According to Equation (10), the FCA-based decision factors of x^n and y_m^n at time t, denoted by $\theta_{x^n}(t)$ and $\theta_{y_m^n}(t)$, can be computed.

Although, the values of $x^n(t)$ and $y_m^n(t)$ have the same type with respect to FCA, their respective weights for decision making still need to be configured. Therefore, we introduce a variable $p_r \in [0.0, 1.0]$, in which the decision maker configures $\theta_{x^n}(t)$ with weight $(1 - p_r)$ and $\theta_{y_m^n}(t)$ with weight p_r. For a given SU s_m, the numerical channel availability of channel n at time t is finally given by:

$$\eta_m^n(t) = (1 - p_r)\theta_{x^n}(t) + p_r\theta_{y_m^n}(t) \tag{16}$$

In the equation, p_r is called *hybrid coefficient* of integrating both $\theta_{x^n}(t)$ and $\theta_{y_m^n}(t)$ when doing decision making. For instance, at $p_r = 0$, the pure idle time based selection is performed. By computing the numerical channel availabilities of the channels of interest, the most available channel in this particular case is determined by the largest value of $\eta_m^n(t)$.

PERFORMANCE EVALUATION

In this section we report the simulation results for performance evaluation of the suggested hybrid decision making algorithm.

Cognitive Radio Network Simulator

To evaluate the effectiveness of the suggested hybrid decision making algorithm, we have developed a CRN simulator (in C/C++ programming language) to conduct experimental simulations (Yao, 2014).

The simulator provides two main functionalities. The first functionality is regarding the definition and configuration of a particular CRN topology, together with relevant parameters. For instance, a particular CRN topology can be configured as an ad-hoc network, where the channel availability is spatially invariant for all SUs. Further, for time-slotted transmission scheme, relevant parameters are, e.g., the mean durations of PU being present and being absent on a particular channel. The second functionality is to simulate the dynamic behaviours of both PUs and SUs, and to deal with the interactions among them. The corresponding interactions can be, e.g., a PU accesses or releases a channel, a SU vacates a channel due to the channel occupancy by PUs.

Figure 4 shows the brief structure of the above mentioned simulator. The features of this simulation are as follows:

- **Setting up an Uniform Decision Criterion:** This is done by using Computational Intelligence techniques, such as Fuzzy Logic and Neural Network.

Figure 4. Brief structure of the developed CRN simulator

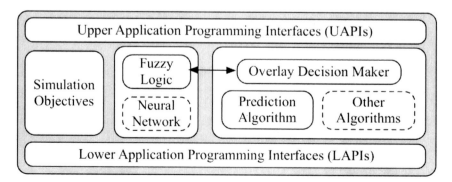

- **Setting up an Overlay Manager:** The overly manager provides the simulator users (e.g., academic researchers, industrial developers) with several alternative CRN models, based on which the theoretical analysis, simulation evaluation and practical experiments can be done. For a particular model, e.g., CR ad-hoc network, the overlay manager intelligently provides spectrum decision maker with different channel characterization parameters and suitable spectrum decision making algorithms.

- **Provision of the Upper Application Programming Interface (UAPI):** The UAPIs are platform-independent interfaces to the simulation users. In other words, the simulator users can conveniently embed the simulator into their own research tasks called *upper objectives*. Examples of upper objectives are like, e.g., simulation based study of CRNs, testbed setting up, performance evaluation, algorithm development.

- **Provision of the Lower Application Programming Interface (LAPI):** The LAPIs help the simulator users in updating old decision making algorithms existing in the simulation and adding newly developed algorithms into the simulation. In particular, the updated and added algorithms can be evaluated by experiments or simulations on the basis of specific upper objectives. Therefore, the functionalities of our developed simulator can be improved and extended.

Simulation Scenarios and Performance Metrics

By using the developed CRN simulator, we simulated a CRN environment, where 100 SU transmitters are uniformly distributed over a $500\,m^2$ area. The time periods of *PU presence* and *PU absence* are exponentially distributed with mean values T_{p1} and T_{p2}, respectively. Every SU holds a time period T_{s1} before performing an access. The expected time period of a SU transmission is equal to a constant value T_{s2}. Once an ongoing SU transmission is interrupted by PU's channel occupancy, the SU will restart transmission after a time interval equal to T_{s1} plus the remaining duration of T_{s1}. The simulation parameter settings are presented in Table 1. These paramter settings have been selected to reflect usual radio transmission scenarios of PUs and SU in CRNs.

To compare the performance of the hybrid spectrum decision with other solutions, we consider three different channel selection algorithms: random based selection, pure idle time based selection and fuzzy logic based hybrid decision making based selection. Since the value range of hybrid coeffeicent is [0.0, 1.0], we choose four typical values $p_r \in \{0.1, 0.2, 0.4, 0.8\}$ to study how well the hybrid coefficient affects. In other words, we consider six different simulation scenarios, which are denoted by random, idle time, $p_r - 0.1$, $p_r - 0.2$, $p_r - 0.4$ and $p_r - 0.8$. Further, the TSIE method is used in all six scenarios. For each simulation scenario, the simulator runs in looping manner, and each loop indicates $10ms$. Furthermore, we run each scenario 40 times, and the simulation time of each run is $10000s$.

Three metrics are used for performance evaluation, namely, the average dropping probability, the average blocking probability and the average success probability. They are denoted by P_d, P_b and P_s, respectively. Considering the ith simulation run of a scenario, assume that the mth SU performed $z_{i,m}^{(1)}$ time of attempting channel access, while this SU got $z_{i,m}^{(2)}$ time of being dropped and $z_{i,m}^{(3)}$ time of being blocked. For 40 runs, we have:

Table 1. Parameter settings

Parameters	Values
Sensing setting	$\delta = 10ms$; $H = 20000$
Radio setting	$D = 200m$; $T_{com} = 6$
Number of channels	$N = 10, 15$
Absence duration of SU	T_{s1}: uniform in $[1.0s, 10.0s]$
SU transmission duration	$T_{s2} = 1.0s$
Absence duration of PU	Exponential distribution with mean value T_{p1}, which is uniform in $[1.0s, 10.0s]$; The value is also constrained by interval $[2.0s, 20.0s]$
Presence duration of PU	Exponential distribution with mean value T_{p2}, which is uniform in $[1.0s, 10.0s]$

$$P_d = \frac{1}{40} \sum_{i=1}^{40} \left[\frac{1}{M} \cdot \frac{\sum_{m=1}^{M} z_{i,m}^{(2)}}{\sum_{m=1}^{M} z_{i,m}^{(1)}} \right] \tag{17}$$

$$P_b = \frac{1}{40} \sum_{i=1}^{40} \left[\frac{1}{M} \cdot \frac{\sum_{m=1}^{M} z_{i,m}^{(3)}}{\sum_{m=1}^{M} z_{i,m}^{(1)}} \right] \tag{18}$$

$$P_s = 1 - P_d - P_b \tag{19}$$

Results and Discussion

The simulation results regarding the SUs' average dropping and blocking probabilities are shown in Figures 5 and 6, respectively. The 95% confidence interval is shown as well. The SUs' average success probability is shown in Figure 6.

As observed in Figure 5, for a fixed channel number N (10 or 15), the pure idle time scenario leads to the smallest dropping probability P_d, while P_d in the random scenario is largest. The reason for this is that in pure idle time scenario the SUs have learned in advance the channel availability based on idle time statistics. Thus, SUs may have more concentration on good channels, and the possibility of being

Figure 5. Average dropping probability of SUs

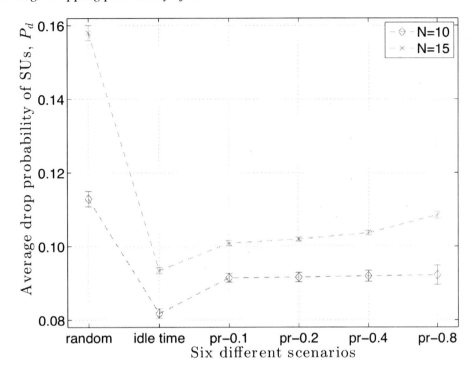

dropped due to PUs is reduced. Since in hybrid scenarios the SUs use statistical information in part, the corresponding values of P_d under fixed N are between the values used in random and pure idle time scenarios.

In Figure 6, we observe that, under fixed N, the pure idle time scenario stands out as showing the worst performance in a form of the largest blocking probability P_b. This is because the limited channel capacity may have the consequence that parts of SUs access the best channels, whereas other SUs are left with the worse channels. Figure 6 further shows that, for the same scenario, P_b under $N =15$ is smaller than the one under $N =10$. This is because of the larger number of channels provided for SUs under $N =15$ than under $N =10$. However, the possibility of SUs being dropped may increase with N (i.e., from 10 to 15). Hence, for the same scenario, P_d under $N=15$ is larger than the one under $N=10$, as shown in Figure 5.

To investigate the overall performance of different channel selection algorithms, we study the SU's average success probability P_s. In Figure 7, we observe that: i) under $N=15$, the value of P_s in pure idle time scenario is smaller than the ones in hybrid scenarios, and ii) under $N=10$, P_s in pure idle time scenario is smallest. This means that: i) by learning only from idle time statistics, SUs are good at looking for the best channels, yet ii) this may increase the number of SUs with the starvation of available channels. By using the hybrid decision making, every SU can learn how heavily the interested channels are used by other SUs (on average). Thus, SUs are able to make a trade-off between the long remaining idle time and the low SU competition level when doing channel selection. In Figure 7, the gain from trade-off is such as, under $N=15$, P_s in scenario $p_r = 0.2$ is about 6.6% and 4.2% larger than the ones in random and pure idle time scenarios, respectively.

Figure 6. Average blocking probability of SUs

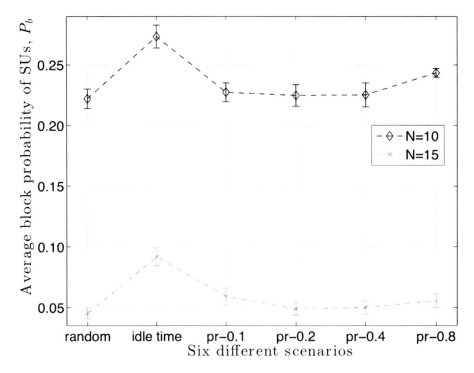

Figure 7. Average success probability of SUs

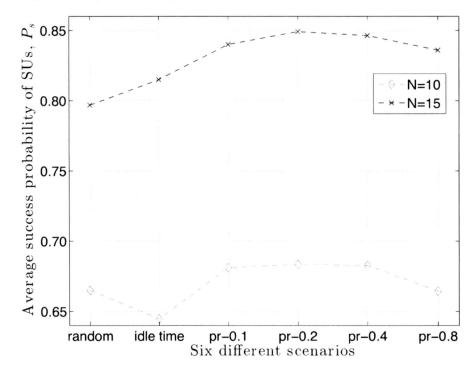

CONCLUSION

A new algorithm of fuzzy logic-based channel selection for secondary users in cognitive radio networks has been developed and reported. Based on the historical information about the channel occupancy by primary users, idle time statistics are derived. Furthermore, an information exchange-based method is used to learn the competition level of secondary users on the available radio channels. The idle time statistics and secondary users competition level are then integrated into a hybrid decision criterion, according to fuzzy logic concepts. Finally, the channel selection is optimized based on that the longest value referred to the hybrid decision criterion indicates the most available channel. Simulation studies show that the overall performance of the suggested algorithm outperforms both random and pure idle time based channel selection algorithms.

Our future work is to study practical ad-hoc cognitive radio networks, where the radio transmission between two secondary users may need other secondary users serving as relay nodes. Accordingly, the practice implementation of such cognitive radio networks and the performance evaluation of hybrid spectrum decision will be done.

REFERENCES

Akter, L., Natarajan, B., & Scoglio, C. (2009). Spectrum usage modeling and forecasting in cognitive radio networks. In Y. Xiao & F. Hu (Eds.), *Cognitive Radio Networks* (pp. 37–60). Boca Raton, FL: CRC Press Taylor & Francis Group. doi:10.1201/9781420064216.ch2

Akyildiz, I. F., Lee, W. Y., Vuran, M. C., & Mohanty, S. (2006). Next generation/dynamic spectrum access/cognitive radio wireless networks: A survey. *Computer Networks Journal (Elsevier)*, *50*(13), 2127–2159. doi:10.1016/j.comnet.2006.05.001

Bae, Y. H., Alfa, A. S., & Choi, B. D. (2009). Performance analysis of modified IEEE 802.11-based cognitive radio networks. *IEEE Communications Letters*, *14*(10), 975–977. doi:10.1109/LCOMM.2010.082310.100322

Chen, R. R., Teo, K. H., & Farhang-Boroujeny, B. (2011). Random access protocols for collaborative spectrum sensing in multi-band cognitive radio networks. *IEEE Journal on Selected Areas in Communications*, *5*(1), 124–136.

Cheng, N., Zhang, N., Lu, N., Shen, X. W., Mark, J., & Liu, F. (2014). Opportunistic spectrum access for CR-VANETs: A game-theoretic approach. *IEEE Transactions on Vehicular Technology*, *63*(1), 237–251. doi:10.1109/TVT.2013.2274201

Deng, S., Du, L., Wang, B., Wang, W., Zheng, Y., & Chai, K. K. (2013). Dynamic channel reservation based on forecast in cognitive radio. In *Proceedings of IEEE Vehicular Technology Conference*. Dresden, Germany: IEEE. doi:10.1109/VTCSpring.2013.6692522

Saaty, T. L. (1978). Exploring the interface between hierarchies, multiplied objectives and fuzzy sets. *Fuzzy Sets and Systems Journal (Elsevier)*, *1*(1), 57–68. doi:10.1016/0165-0114(78)90032-5

Westerhagen, A. P. (2014). *Cognitive radio networks: Elements and architectures.* (Doctoral dissertation). Blekinge Institute of Technology, Sweden.

Yang, L., Cao, L., & Zheng, H. (2007). Proactive channel access in dynamic spectrum networks. In *Proceedings of IEEE International Conference on Cognitive Radio Oriented Wireless Networks and Communications.* Orlando, FL: IEEE.

Yao, Y. (2014). *A software framework for prioritized spectrum access in heterogeneous cognitive radio networks.* (Doctoral dissertation). Blekinge Institute of Technology, Sweden.

Yao, Y., Ngoga, S. R., Erman, D., & Popescu, A. P. (2012). Competition-based channel selection for cognitive radio networks, In *Proceedings of IEEE Wireless Communications and Networking Conference.* Paris, France: IEEE. doi:10.1109/WCNC.2012.6214006

Zhang, X., & Su, H. (2011). Opportunistic spectrum sharing schemes for CDMA-based uplink MAC in cognitive radio networks. *IEEE Journal on Selected Areas in Communications, 29*(4), 716–730. doi:10.1109/JSAC.2011.110405

Zhao, Q., Tong, L., Swami, A., & Chen, Y. (2007). Decentralized cognitive MAC for opportunistic spectrum access in ad hoc networks: A POMDP framework. *IEEE Journal on Selected Areas in Communications, 25*(3), 589–600. doi:10.1109/JSAC.2007.070409

ADDITIONAL READING

Akyildiz, I. F., Lee, W. Y., Vuran, M. C., & Mohanty, S. (2008). A survey on spectrum management in cognitive radio networks. *IEEE Communications Magazine, 46*(4), 40–48.

Blin, J. M. (1974). Fuzzy relations in group decision theory. *Journal of Cybernetics, 4*(2), 17–22.

Chen, R. R., Teo, K. H., & Farhang-Boroujeny, B. (2011). Random access protocols for collaborative spectrum sensing in multi-band cognitive radio networks. *IEEE Journal on Selected Areas in Communications, 5*(1), 124–136.

Gelabert, X., Akyildiz, I.F., & Sallent, O., & Agusti. Operating point selection for primary and secondary users in cognitive radio networks, *Computer Networks Journal (Elsevier), 53*(8), 1158–1170, 2009.

Haykin, S. (2005). Cognitive radio: Brain-empowered wireless communications. *IEEE Journal on Selected Areas in Communications, 23*(2), 201–220.

Hoang, A. T., Wong, D. T. C., & Liang, Y. C. (2009). Design and analysis for an 802.11-based cognitive radio network, In *Proceedings of IEEE Wireless Communications and Networking Conference,* Budapest, Hungary.

Katsaros, D., & Manolopoulos, Y. (2009). Prediction in wireless networks by Markov Chains. *IEEE Wireless Communications, 16*(2), 56–63.

Lai, L. F., Gamal, H. E., Jiang, H., & Poor, H. V. (2011). Cognitive medium access: Exploration, exploitation, and competition. *IEEE Transactions on Mobile Computing, 10*(2), 239–253.

Mitola, J. (2000). *Cognitive radio: An integrated agent architecture for software defined radio*, (Doctoral dissertation), Royal Institute of Technology, Sweden. (ISRN KTH/IT/AVH-00/01-SE)

Ngoga, S. R. (2014). *On dynamic spectrum access in cognitive radio networks* (Doctoral dissertation), Blekinge Institute of Technology, Sweden. (ISBN: 978-91-7295-267-6)

Niyato, D., Hossain, E., & Han, Z. (2009). Dynamic spectrum access in IEEE 802.22-based cognitive wireless networks: A game theoretic model for competitive spectrum bidding and pricing. *IEEE Wireless Communications*, *16*(2), 16–23.

Rakus-Andersson, E. (1999). A fuzzy group-decision making model applied to the choice of the optimal medicine in the case of symptoms not disappearing after the treatment, In *Proceedings of The 4th Meeting of the EURO Working Group on Fuzzy Sets*, Budapest, Hungary.

Rakus-Andersson, E. (2007). The Choice of Optimal Medicines. In E. Rakus-Andersson (Ed.), *Fuzzy and rough techniques in medical diagnosis and medication* (pp. 127–154). Berlin, Heidelberg: Springer.

Staple, G., & Werbach, K. (2004). The end of spectrum scarcity. *IEEE Spectrum*, *41*(3), 48–52.

Tang, P. K., Chew, Y. H., Ong, L. C., & Halder, M. K. (2006). Performance of secondary radios in spectrum sharing with prioritized primary access, In *Proceedings of IEEE Military Communications Conference*, Washington, D.C., USA.

Wang, B., Ji, Z., & Liu, K. J. R. (2009). Primary-prioritized Markov approach for dynamic spectrum allocation. *IEEE Transactions on Wireless Communications*, *8*(4), 1854–1865.

Wang, B., Wu, Y., & Liu, K. J. R. (2010). Game theory for cognitive radio networks: An overview. *Computer Networks Journal (Elsevier)*, *54*(14), 2537–2561.

Wang, W., Zhou, Z. H., Ge, M., & Wang, C. (2013). Resource allocation for heterogeneous cognitive radio networks with imperfect spectrum sensing. *IEEE Journal on Selected Areas in Communications*, *31*(3), 464–475.

Wang, X. Y., Wong, A., & Ho, P. H. (2011). Stochastic medium access for cognitive radio ad hoc networks. *IEEE Journal on Selected Areas in Communications*, *29*(4), 770–783.

Westerhagen, A. P. (2014). *Cognitive radio networks: Elements and architectures* (Doctoral dissertation), Blekinge Institute of Technology, Sweden. (ISBN: 78-91-7295-272-0)

Xie, R., Ji, H., & Si, P. (2008). Optimal joint transmission time and power allocation for heterogeneous cognitive radio networks, In *Proceedings of IEEE International Conference on Communications*, Kyoto, Japan.

Xing, Y., Chandramouli, R., Mangold, S., & Nandagopalan, S. S. (2006). Dynamic spectrum access in open Spectrum in wireless networks. *IEEE Journal on Selected Areas in Communications*, *24*(3), 626–637.

Yao, Y. (2012). *A spectrum decision support system for cognitive radio networks* (Licentiate dissertation), Blekinge Institute of Technology, Sweden.

Zhao, Q., & Sadler, B. (2007). A survey of dynamic spectrum access: Signal processing, networking, and regulatory policy. *IEEE Signal Processing Magazine*, *24*(3), 79–89.

Zhu, X., Shen, L., & Yum, T. S. P. (2007). Analysis of cognitive radio spectrum access with optimal band reservation. *IEEE Communications Letters*, *11*(4), 304–306.

KEY TERMS AND DEFINITIONS

Cognitive Radio (CR): It is envisioned to act as a highly intelligent radio unit where transmission parameters like frequency range, transmit power and modulation type are altered by learning the radio environment.

Cognitive Radio Networks (CRNs): The novel communication paradigm of Cognitive Radio leads to an enabling framework "CR Networks", which is first suggested by the Federal Communication Commission (FCC).

Competition Problem: The competition problem is created by that multiple SUs select and use the same spectrum opportunity during the same time period.

Primary User (PU): The licensed mobile user who is authorized to exclusively use the licensed spectrum.

Secondary User (SU): The unlicensed mobile user who is allowed to use spectrum opportunities as long as not harmfully interfering with PUs.

Spectrum Decision: Since multiple spectrum opportunities may be obtained by a particular SU at a time, this SU needs to decide which channel should be selected for the use in the near future. The selection process is often referred to as spectrum decision.

Spectrum Opportunity: If and when a spectrum band is not used by PUs, it becomes available for the use by SUs, so-called a spectrum opportunity for SUs.

Chapter 11
Routing through Efficient Channel Assignment in Cognitive Radio Networks

Yasir Saleem
Sunway University, Malaysia

Farrukh Salim
Technische Universität Ilmenau, Germany & NED University of Engineering and Technology, Pakistan

Mubashir Husain Rehmani
COMSATS Institute of Information Technology, Wah Cantt, Pakistan

Bushra Rashid
COMSATS Institute of Information Technology, Wah Cantt, Pakistan

ABSTRACT

In Cognitive Radio Networks (CRNs) there is much dynamicity due to the activities of primary users which results in instability of routes. Therefore, an efficient routing protocol based on good channel assignment strategy is required in CRNs. A good channel selection strategy makes route stable by selecting channels having larger capacity and greater availability time. Therefore, the focus of this chapter is joint channel assignment and routing in CRNs, which provides a comprehensive survey on routing and channel assignment in CRNs. First, the importance of joint channel assignment and routing for successful communication in cognitive radio networks is discussed. Then classification and challenges related to channel assignment and routing are discussed in detail. In order to establish reliable routes in CRNs, some factors are discussed that further enhance the communication in CRNs. Finally, guidelines for the development of efficient routing protocols are discussed.

1. INTRODUCTION

Federal Communications Commission (FCC, 2004), USA and Ofcom, UK have found that most of the radio spectrum is under-utilized due to the fact that in traditional network, e.g., paging and military bands, spectrum allocation is fixed and spectrum cannot be re-used even when it is not occupied by licensed user (Akyildiz et al., 2009) . Cognitive Radio Networks (CRNs) aim to intelligently utilize the licensed band when it is unoccupied without causing interference to licensed users. Federal Communications Commission (FCC) (FCC, 2004) has approved the usage of licensed spectrum band by unlicensed de-

DOI: 10.4018/978-1-4666-6571-2.ch011

vices subject to the condition that communication of licensed users should not be interfered/interrupted. There are two types of users in CRNs, one is Primary User (PU or licensed user) which operates on a licensed spectrum band and other is Secondary User (SU or unlicensed user), which operates either on unlicensed spectrum band or on licensed spectrum band whenever PU is not utilizing its spectrum band. When licensed spectrum band is in use of SU, SU has to vacate this spectrum band whenever PU needs to occupy the said spectrum band at any time and switch to another one by selecting spectrum from spectrum pool (Weiss et al., 2004) so that PU should not be interfered. This is known as spectrum handoff (Akyildiz et al., 2009).

Packet is a basic unit of information in computer networks, and routing is the fundamental function that allows successful and timely reception of data packets for any wireless network. Routing is used for information exchange by transferring packets from one host to another across the network. Significant amount of work has been carried out for routing in different wireless networks. These networks include Wireless Sensor Networks (WSNs), Wireless Mesh Networks (WMNs), Mobile Ad-Hoc Networks (MANETs), Vehicular Ad-Hoc Networks (VANETs) etc. But in CRNs, routing is much more complex due to the dynamic nature of the available frequency bands. This dynamicity is due to various factors such as PU activity, diversity in available channels, heterogeneous channels etc.

Channel assignment plays an important role in the performance and stability of routing protocols. In CRNs, if channel assignment strategy selects channels with low PU activity, high channel availability and high connectivity with neighbors, then routing will be very efficient, stable and CR user can stay on the same channel for longer time interval. Therefore, channel assignment should be properly investigated with reference to its effect on routing in CRNs so that routing can be performed effectively and efficiently.

The importance of CRNs can be seen by examining its various applications in the world. These applications include military applications (Leschhorn & Buchin, 2004), vehicular networks (Maldonado et al., 2005), Delay Tolerant Networks (DTNs) (Van der Schaar & Fu, 2009), etc. CRNs enable communication and restore network connectivity in situations when existing infrastructure is destroyed or disabled by natural calamities in disaster response (Rehmani et al., 2010), emergency and public safety networks by using existing spectrum without needing any infrastructure. The most crucial point in all these applications is timely and reliable data transfer which can only be accomplished through a robust routing protocol. Robustness of routing protocol, in turn, hinges on the selection of channels that are available for transmission for long time periods and cause very little transmission losses commonly known as channel assignment strategy.

Therefore in this chapter, joint channel assignment and routing in CRNs and role of channel assignment in improving network performance is extensively discussed. A lot of work in the form of surveys has been done on channel assignment and routing in CRNs individually (Abdelaziza et al., 2012; Al-Rawi et al., 2012; Tragos et al., 2013; Masonta et al., 2013; Youssef et al., 2013; Xu et al., 2013). But none of them focused on joint channel assignment and routing in CRNs. Therefore, the focus of this chapter is to present a survey on joint effects of routing and channel assignment in CRNs.

Main Contribution

The main points discussed in this chapter are summarized as follows:

- A comprehensive survey on routing; its classification and challenges.

- A detailed study on channel assignment, its classification with major goals that are unique to each channel assignment strategy, nature of behavior of channel assignment strategy in response to PU activity, types and its challenges.
- Extensive discussion on routing with efficient channel assignment in CRNs.
- Guidelines for the development of efficient routing protocols in CRNs

Moreover, channel aggregation technique combines several channels together into a single channel and improves network performance by providing greater bandwidth. Therefore, if channel aggregation is performed during channel assignment in CRNs, then applications' desired bandwidth can be increased and routes will be more stable. Channel aggregation in CRNs is performed in (Jiao et al., 2011) and (Jiao et al., 2012) which provide channel adaptation by dynamically adjusting channel occupancy through channel aggregation and therefore improving network performance.

The remainder of this chapter is organized as follows. Section 2 describes classification of routing protocols in CRNs. In section 3, challenges of routing protocols in CRNs are discussed. Subsequently, classification of channel assignment strategies with major goals that are unique to each channel assignment strategy, nature of behavior of channel assignment strategy in response to PU activity, types and its challenges are explained in section 4. Afterwards, challenges of channel assignment are discussed in section 5. Then routing with efficient channel assignment is discussed in section 6 along with routing protocols' nature. Next, some guidelines for the development of efficient routing protocols for CRNs are discussed in section 7. Lastly, the chapter is concluded in section 9.

Next we discuss classification of routing protocols in CRNs on the basis of some specific metrics or parameters which are set as criteria for optimization e.g. best route is chosen keeping link stability as criteria so that channel with greatest link stability is chosen from the channel pool for communication.

2. CLASSIFICATION OF ROUTING PROTOCOLS IN CRNs

As described above, routing protocols are classified on the basis of performance metric to be optimized. Routing protocols in CRNs can be classified into four main categories (Samar & ElNainay) which are illustrated in Figure 1. These categories are: (1) Delay based routing protocols (2) Link Stability based routing protocols (3) Throughput based routing protocols and (4) Location based routing protocols. These four categories are explained in detail as below:

A. Delay-Based Routing Protocols

Delay-based routing protocols consider delay in order to measure the quality of routing protocols. Generally there are three components of delay:

- **Switching Delay:** This delay occurs when a node switches from its current channel to another channel.
- **Backoff Delay:** While working on identical frequency band, when MAC protocol tries to solve exposed-terminal or hidden-terminal problems, the delay occurs commonly known as backoff delay.
- **Queuing Delay:** When a node transmits data on a given channel, the incurred delay caused due to queuing of data at each step is known as queuing delay.

Figure 1. Routing protocol classification in CRNs

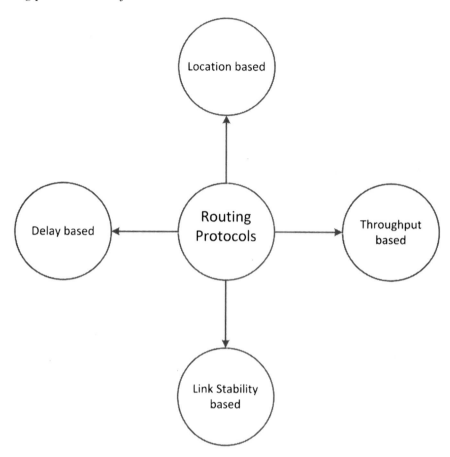

B. Link Stability Based Routing Protocol

In CRNs, CR users use channels of PUs when they are idle and whenever PUs arrive on their licensed channels then CR users have to vacate these channels and occupy the next available ones. Therefore, as CR users have to frequently switch to different channels on appearance of PUs, therefore the communication link between CR nodes can become highly unstable. Therefore, link-stability based routing protocols work on link stability by modeling PU activity and considering spectrum fluctuations.

C. Throughput Based Routing Protocol

Throughput is defined as average rate of successful delivery of packets per second. This is the basic metric of routing protocols and is used extensively. Throughput based routing protocols work on maximizing network throughput and measure network performance on the basis of successful delivery of packets.

D. Location Based Routing Protocol

Nowadays, most of the wireless devices are location enabled and as businesses have bloomed resulting in creation of franchise chains where customers can use single telephone number for all outlets in the country thus many routing protocols have been designed where performance metric to be optimized is location. The location of CR nodes can be obtained through FCC Geolocation-Databases(FCC, 2004). Therefore, location based routing protocols can work in CRNs but there will be new challenges in their incorporation due to dynamic conditions of the network and PU activity.

The next section is devoted to the discussion of challenges of routing in cognitive radio networks. The biggest challenge in realizing most efficient routing protocol is dynamic nature of the entire network as it is harder to predict the channel versus PU response at any instant of time when PU can occupy any licensed band at any instant of time. Other challenges are discussed in detail in the coming paragraphs.

3. CHALLENGES OF ROUTING IN CRNs

This section describes routing challenges in CRNs. Normally routing challenges in CRNs are divided into three categories (Al-Rawi&Yau, 2012) which are channel-based, host-based and network-based. These categories along with challenges in each category are illustrated in Figure 2. Routing in CRNs is especially challenging due to the dynamic and unconventional nature of network.

A. Channel-Based Challenges

Channel's behavior to any signal is usually unpredictable due to random environment conditions at different time intervals. Therefore, channel-based challenges are associated with nature of channel and

Figure 2. Routing challenges in CRNs

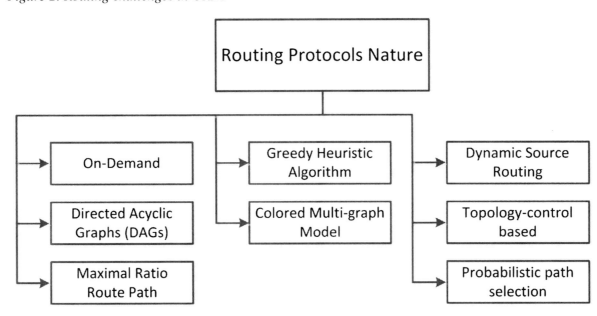

operating environment. The four challenges mentioned in Figure 2 are; dynamicity in channels' availability, operating channels diversity, lack of common control channel and integration of route discovery with channel selection in CRNs.

The basic motivation behind cognitive radio networks is that CR users should not interfere with the PUs' transmissions. Therefore, whenever PU needs to occupy a channel, CR user has to vacate it and search for other channel in order to continue its own transmission which causes link instability. Hence, dynamicity in cognitive radio networks poses challenges in routing protocols' efficiency.

Communication between two nodes, in general occurs when both of them are tuned to same channel and have similar data rates for successful wireless transmission. Similarly, for successful communication between CR nodes in Cognitive Radio Networks, it is necessary that both have common channel on which they can communicate. But due to dynamicity in channels availability, in case PU becomes active during CR communication, then CR user has to vacate the said channel and search for a new and reliable channel in order to continue communication. But the problem that arises after changing channel is that now both CR users are on different frequencies. Therefore routing protocol has to incorporate some solution for this as well. This problem is commonly known as operating channel diversity. Two types of channels exist for communication. One is common control channel (CCC) and other is data channel. CCC is available to all nodes in the network and is used for exchange of control packets such as Route Request (RREQ) and Route Reply (RREP) while data channel is used solely for transmission of actual data. But due to dynamicity and operating channel diversity in CRNs, it is not possible to dedicate one channel as CCC for exchange of control information. Therefore exchange of control packets between CR nodes is another major challenge in CRNs.

In routing protocols, normally a route selection is carried out with inherent information of channels for transmission within the control packets. But in Cognitive Radio Networks, CR users continuously keep exchanging their channel whenever PU becomes active on the CR occupied channel, so it is not feasible to send channel information along with route discovery. Therefore, it is another challenge to integrate route discovery with channel decision.

B. Host-Based Challenges

Host-based challenges are associated with CR nodes. CR nodes are quite different from conventional host nodes e.g. CR nodes have to intelligently configure themselves according to available channels and perform continuous updating of their channels availability and PU activity in CRNs. Therefore challenges associated with CR nodes are of three types: channel switching and back-off delay, broadcast/multicast multiple transmissions and CR nodes heterogeneity and mobility.

Whenever PU becomes active on a channel which is currently being used by CR nodes, CR users have to choose between two actions. Either they can switch to another available channel which will incur channel switching delay or they can wait for PU to complete its activity and then resume their communication on the same channel if they do not want to switch channel. If the CR node chooses to wait for PU to complete its communication, it will result in backoff delay. Routing protocols should also consider these challenges and some work should be done in order to minimize both switching and backoff delay.

Generally in routing, broadcasting/multicasting is done for exchanging control messages between nodes. In CRNs, since there is diversity in number of available channels and common control channel is not present, therefore each CR node can have different number of available channels. Hence, a broadcast/multicast message sent on a single channel, might not be received by all CR nodes, therefore, it is compulsory to broadcast on multiple channels, i.e., multicast in order to exchange control messages . This multiple broadcast/multicast could have adverse impact on network throughput. Thus a tradeoff between network throughput and network control has to be done. This is a complex challenge for routing in CRNs.

In CRNs, there are multiple CR nodes which have different capabilities like processing speed, transmission power etc. Thus if a node with limited capabilities is selected as relay node then it will degrade the performance of the entire network. Thus in routing, a relay node should always be chosen in such a way that it does not deteriorate the network performance. Another issue is that CR nodes can be mobile and in each region, channel availability can be different. Higher mobility of CR nodes will demand faster channel switching and reduced channel access time, which is challenging in fulfillment of QoS requirements and PU interference minimization.

C. Network-Based Challenges

Network-based challenges are associated with network-wide CR nodes. These are related to challenges that affect the network performance and each node is affected equally by it. Normally there are three types of challenges i.e. number of hops and network-wide performance tradeoff, network-wide energy consumption and, fast and spectrum-aware route recovery.

If a route has lower number of hops or transmission range among CR nodes is longer then it will cause more interference to PU due to long propagation delay and there will be more link failures which will result in higher cost for route maintenance. Also due to long transmission range, there will be higher energy consumption. Therefore number of hops should be carefully considered while designing a routing protocol.

Due to PUs' activities, CR nodes have to switch to another available channel for communication. This channel switching will nullify the previous route. Therefore, a new optimal route has to be chosen so that there will be fast and spectrum-aware route recovery which enables CR nodes to continue their transmission efficiently.

Based on the routing challenges discussed above, new solutions should be proposed to solve these challenges. Furthermore, channel assignment strategies can help in reducing the problems in routing because intelligent channel allocation not only reduces back off and channel switching delay but also enhances network performance on the whole. Next, classification of channel assignment strategies in CRNs are illustrated in detail.

4. CLASSIFICATION OF CHANNEL ASSIGNMENT STRATEGIES IN CRNs

Channel assignment in Cognitive Radio Networks is necessary component because channels are selected strategically in such a way that channel occupancy is ensured for maximum time period without causing interference to PUs. Channel assignment is essentially influential in successful operation of routing protocols because dynamic nature of Cognitive Radio Networks can be handled due to intelligent allocation of channels to CR users. Thus, in this section, our focus will be on detailed discussion regarding channel assignment strategies proposed for CRNs. There are different goals that must be fulfilled in channel assignment strategies, i.e., optimization of throughput, delay, channel switching etc. Every channel assignment strategy in CRNs has a nature, based on how CR nodes react when PUs become active. Normally, on the basis of nature, channel assignment strategies are divided into three major categories:

(1) Proactive (2) Threshold-based and (3) Reactive. From the algorithmic perspective, generally channel assignment strategies are divided into two main categories which are (1) Centralized and (2) Distributed (Rehmani & Viana, 2011).

A. Goals of Channel Assignment Strategies

There are many channel assignment strategies each with its own goals for optimization. One of the major goals of channel assignment strategies is throughput maximization. Throughput in general is defined as *average rate of successful delivery of messages per second over a communication link.* In CRNs, Nguyen, Kompella, Wieselthier & Ephremides (2010) and Yang, Cao & Zheng (2008) worked on maximizing throughput as a goal for channel assignment strategy.

Delay minimization is another important goal of channel assignment strategy. In channel assignment, delay minimization results in lower channel switching delay. Channel switching delay occurs when a node switches from one channel to another channel on its radio interface. Another important goal is efficient routing in which the goal of channel assignment strategy is to assign channels in-order to fulfill timely routing requirements. In CRNs, routing requirements for channel assignment include channels with low PUs' activities, high bandwidth, less interference, maximum connectivity etc. Therefore, channel assignment in CRNs is quite complex as compared to conventional methods of channel allocation.

B. Nature of Channel Assignment Strategies

In CRNs, when PU become active on CR communication channel, every channel assignment strategy has a nature with which it reacts to this change. Thus channel assignment strategies in CRNs can be classified into three types which are proactive (predictive), threshold-based and reactive (Rehmani & Viana, 2011).

- **Proactive (Predictive):** Proactive (predictive) channel assignment schemes predict the activities of PUs and move CR nodes to appropriate channels based on the prediction. This prediction is used to find channels with longest idle time so that CR nodes can use these channels. This in-turn helps in reducing number of channel switching and delay.
- **Threshold-Based:** In threshold-based channel assignment schemes, PUs use their channels all the time and CR nodes have no available idle channel. Thus for such channels, a threshold is defined below which there is no harmful interference to PUs. Thus, CR nodes utilize these channels as long as interference is below the defined threshold level.
- **Reactive:** Reactive channel assignment strategies are those which react on occurrence of PUs activities by channel switching. In reactive schemes, CR nodes monitor the spectrum for PUs and whenever PU activity is detected, this information is passed on to other CR nodes and then CR nodes switch to another available channel.

C. Types of Channel Assignment Strategies

There are two types of channel assignment strategies. One is centralized and other is distributed. Both have their own advantages and disadvantages. These are described below in some detail:

- **Centralized Channel Assignment:** In centralized channel assignment, a central entity known as 'spectrum administrator' is responsible for gathering information about channels in the network and then assigning these channels to CR nodes. But this approach is not feasible for multi-hop networks because an attack like Denial of Service (DoS) on spectrum administrator will jam the whole network. This approach is easy to implement but is not scalable for large networks because when the network expands, then spectrum administrator becomes overloaded.
- **Distributed Channel Assignment:** Distributed channel assignment is another type of channel assignment strategy in which there is no centralized entity, rather all the nodes sense channels themselves individually and share this information with their neighbors. Then based on sensed and shared information, each node decides itself which channel to assign on its radio interface(s).

The next section is devoted to the discussion of challenges in channel assignment in Cognitive Radio Networks.

5. CHALLENGES OF CHANNEL ASSIGNMENT IN CRNs

Cognitive Radio Networks have a continuously changing environment e.g. channel diversity; back off delay etc., due to which channel assignment is very complex. The factors which contribute to the dynamicity of environment are as follows:

- **Primary User (PU) Activity:** In CRNs, channels are licensed to PUs and they cannot be occupied by CR nodes as long as PUs are using them. Whenever PUs vacate their channels, these channels become idle and are available to CR nodes. Therefore, number of available channels changes with time as well as location for CR nodes and this in-turn causes diversity in available channel pool.
- **Availability of Multiple Channels:** Multiple channels are available to CR nodes in their channel set. These channels might have different PU activity pattern and bandwidth. Thus utilizing the channels according to application scenario is another challenge.
- **Diversity in Number of Available Channels:** Due to dynamic conditions of CRNs, there is much diversity in number of available channels for CR nodes. Thus, channel diversity results in complexity in channel assignment strategies so it should be considered carefully.

By considering these challenges, here we provide the key features which will improve channel assignment in CRNs.

- **Primary User Constraints:** During transmission, channel assignment strategy must guarantee that there will be no interference to PUs' communication.
- **Autonomous decision by CR nodes:** In distributed multi-hop CRNs, CR nodes should sense and share the channel availability information and then take decision themselves about which channels to assign by considering PUs activities and varying channel conditions.
- **Sender/Receiver Tuning:** In order to enable communication, channel assignment strategy should ensure that sender and receiver are tuned to same channel so that they can communicate with each other.

Next the solution to above mentioned challenges is discussed in the form of routing with efficient channel assignment in CRNs. Efficient routing protocol has to be amalgamated with competent channel assignment strategies in order to improve network performance.

6. ROUTING WITH EFFICIENT CHANNEL ASSIGNMENT IN CRNs

We have discussed classification and challenges of routing in CRNs which should be considered for good routing protocol. Since channel assignment plays a vital role in the performance of routing protocol, therefore we also have discussed classification of channel assignment strategies which includes goals, nature and types of channel assignment. Then, we have illustrated challenges for channel assignment in CRNs. Thus for routing protocol to perform better, an efficient channel assignment strategy is required which should consider PU activity, diversity in number of available channels and other dynamic conditions of CRNs so that routing protocol is stable and can exist for longer time period so that communication can be performed smoothly without noticeable distortion.

Here we provide a list of routing protocols that have been designed for CRNs. Next we describe those routing strategies which use channel selection schemes in their routing decisions.

Table I illustrates existing routing protocols in CRNs extensively with their properties. In (Lin et al., 2007), authors proposed a distributed joint channel assignment, scheduling and routing algorithm for CRNs. This algorithm is on-demand or reactive in nature and considers queue-length as routing metric. It is not spectrum-aware and has no measures to model PU activity which is very important for getting realistic results in CRNs. This strategy uses 'Mean Total Backlog' (unfulfilled data transmissions on the network) as parameter for its performance evaluation.

Chehata et al. (2011) proposed an on-demand or reactive routing protocol for multi-hop multi-radio and multi-channel CRNs known as 'CR-AODV'. This protocol is based on AODV and uses Weighted Cumulative Expected Transmission Time (WCETT) as routing metric. Its performance is evaluated with 'end-to-end throughput' and 'route disconnectivity ratio' parameters. This protocol considers varying spectrum conditions and PU activity. Due to these factors, its results are much more realistic as compared to those which do not consider these factors.

In (Nandiraju et al., 2009), a routing protocol 'Adaptive State-based Multi-path Routing Protocol' (ASMRP) for multi-radio, multi-channel and multi-path networks is presented. It uses Directed Acyclic Graphs (DAGs) for the discovery of optimal paths between Mesh Routers and Internet Gateways. But it is not for CRNs rather is for wireless mesh networks. Therefore it is not spectrum-aware and it does not incorporate PU activity awareness metric into the routing protocol. It uses 'aggregate throughput' and 'delay' as performance parameters.

Zeeshan et al. (2010) introduced another on-demand joint cross-layer routing or channel assignment protocol 'Backup Channel and Cooperative Channel Switching' (BCCCS). It is based on AODV and can work without central control channel. It considers resource consumption and route stability as routing metrics and considers connectivity against number of channels as performance parameter. It considers varying conditions of the spectrum but does not model PU activity.

In (Aduwo et al., 2004), authors proposed a channel-aware inter-cluster routing protocol known as Maximal Ratio Route Path (MRRP). For routing metric, it considers fading and uses 'outage probability' and 'end-to-end average bit error rate' as a standard for evaluating its' performance. Since it is for wireless ad-hoc networks therefore it is neither spectrum-aware nor model PU activity.

In (Liu et al., 2005), a greedy heuristic algorithm for channel-aware routing is proposed for wireless ad-hoc networks known as MRPS-T. This routing protocol is a Multiple Path Route Selection (MRPS) which aims at selecting best T paths for forwarding a packet to its next hop. It uses 'channel condition' as a routing metric and considers 'end-to-end outage probability' as a performance evaluation parameter. It is spectrum-aware but it does not consider PU activity.

In (Deng et al., 2007), a collaborative strategy known as 'Maximize the Minimum Interference Margin Principle' (MMIM) is introduced for route and spectrum selection in CRNs. Spectrum stability and path stability are considered as routing metrics in this strategy and 'probability related to stability' is used as parameter for performance evaluation. It is spectrum-aware but it does not consider PU activity.

(Krishnamurthy et al., 2005) present a strategy for efficient routing of packets by MAC layer configuration in CRNs. This strategy is spectrum and situation aware where situation awareness is achieved by information sharing about physical location among the nodes. This strategy does not model PU activity. For routing metrics, it considers 'Number of channel switches along a route' and 'frequency of channel switches over a link' as routing metrics. It performs mathematical formulation and claims that its algorithm determines global network topology in $O(N^2)$ timeslots.

In (Pan et al., 2008), a cost design to select candidate forwarder list for opportunistic routing in CRNs known as CROR is proposed. It uses link-cost as routing metric and 'number of transmissions' and 'end-to-end delay' as performance evaluation parameters. It is spectrum-aware and it also models PU activity.

A cross-layer routing and dynamic spectrum assignment strategy for CRNs is proposed in (Ding et al., 2010) known as 'Routing and Spectrum Allocation Algorithm (ROSA)'. For routing metrics, it considers spectrum utility and spectrum holes. It also performs spectrum assignment which is based on spectrum utility. Throughput, fairness index, network spectrum utility and average delay are used as performance parameters. This strategy also models PU activity.

A cross-layer routing design using colored multigraph model for CRNs is proposed in (Zhou et al., 2009). This algorithm uses $O(n^2)$ complexity polynomial time for interface assignment and routing. It is not spectrum-aware and does not incorporate PU activity.

A routing protocol for delay-sensitive applications in CRNs is proposed by Jashni et al., (2010). Queuing and transmission delay are used as routing metrics for this protocol. This protocol uses end-to-end delay and packet loss rate as performance parameters. It is not aware of varying spectrum conditions but it considers PU activity.

Kamruzzaman et al., (2012) proposes spectrum and energy aware routing protocol for CRNs which is based on Dynamic Source Routing (DSR). It also performs channel selection based on spectrum and energy awareness. It uses system throughput, mean message delay, ratio of survival note and consumed energy per packet for performance evaluation parameters. It also incorporates PU activity in routing and channel assignment.

Connectivity based routing scheme known as Gymkhana is presented for CRNs by Abbagnale et al., (2010). Routing metric for this strategy is based on path connectivity. Its performance is evaluated by considering connectivity of different paths. But it does not consider varying spectrum conditions and does not incorporate PU activity.

'Spectrum aware routing protocol' (SPEAR) is proposed by Sampath et al., (2008) with considerations for channel assignment in CRNs. It considers system integration and route discovery as routing metrics. Channel assignment is modeled by graph coloring problem in this protocol. It uses system throughput and end-to-end packet delivery latency as parameters for performance evaluation. This protocol is spectrum-aware, i.e., considers varying spectrum condition and also models PU activity.

Table 1. Existing routing protocols in cognitive radio networks

Strategy	Routing Protocol Nature	Routing Metric	Performance Parameters	Spectrum Awareness	Routing Protocol Based On	Simulators Used
(Lin & Rasool, 2007)	On demand	Queue-length	Mean total backlog	No		Self-made
CR-AODV (Chehata, Ajib & Elbiaze, 2011)	on demand	Weighted Cumulative Expected Transmission Time (WCETT)	End-to-end throughput, route disconnectivity ratio	Yes	AODV	NS-2
ASMRP (Nandiraju, Nagesh & Dharma, 2009)	Directed Acyclic Graphs (DAGs)		Aggregate throughput, delay	No		NS-2
BCCCS (Zeeshan, Manzoor & Qadir, 2010)		resource consumption and route stability	Connectivity against number of channels	Yes	AODV	MATLAB
MRRP(Aduwo & Annamalai, 2004)	Maximal Ratio Route Path	Fading	Outage probability, end-to-end average bit error rate,	No		Not mentioned
MRPS-T (Liu & Annamalai, 2005)	greedy heuristic algorithm	Channel condition	End-to-end outage probability	Yes		Not mentioned
MMIM (Deng, Chen, He & Tang, 2007)		Spectrum stability, path stability	probability of stability	Yes		Self-made
(Krishnamurthy, Thoppian, Venkatesan & Prakash, 2005)		Number of channel switches along a route and frequency of channel switches over a link		Yes	MAC Layer configuration	Not mentioned
CROR (Pan, Rongsheng & Yuguang, 2008)		Link-Cost based (for prioritizing forwarder candidates)	Number of transmissions, end-to-end delay	Yes		Not mentioned
(Zhou, Lin, Wang & Zhang, 2009)	Colored multigraph model		Algorithm with complexity of $O(n^2)$ is proposed where n is number of nodes	No		Mathematical modeling
(Jashni, Tadaion & Ashtiani, 2010)		Delay-based (Queuing and Transmission delay)	End-to-end delay, packet loss rate,	No		Not mentioned
(Kamruzzaman, Kim, Jeong & Jeon, 2012)	Dynamic source routing protocol	Spectrum and energy aware	System throughput, mean message delay, ratio of survival node, consumed energy per packet	Yes	DSR	Not mentioned
Gymkhana (Abbagnale & Cuomo, 2010)		Path connectivity	Evaluates connectivity of different paths	No		Mathematical modeling
SPEAR (Sampath, Yang & Zheng, 2008)		Spectrum integration and route discovery	System throughput, end-to-end packet delivery latency	Yes		QualNet

continued on following page

Table 1. Continued

Strategy	Routing Protocol Nature	Routing Metric	Performance Parameters	Spectrum Awareness	Routing Protocol Based On	Simulators Used
J-CAR (Chiu, Yeung & King-Shan, 2009)		Interference minimization	Aggregate goodput and average end-to-end delay	No		NS-2
(Zin, Weng, Cao & Feng, 2011)	greedy heuristic approximation, exact solution, poly-logarithmic approximation,	interference constraints, QoS requirements	maximize the coverage of BS for CRs	No		Mathematical modeling
(Alicherry, Bhatia, Randeep & Li, 2005)		Interference constraint	Throughput	No		Mathematical modeling
DORP (Cheng, Wei, Yunzhao & Wenqing, 2007)		Delay	Cumulative delay, number of switching	Yes	AODV	Glomosim
MILP (Ma & Tsang, 2008)		Interference constraints	Results confirmed that spectrum sharing among CRs is interference free and fair routing is guaranteed	No		Not mentioned
(Yang, Cheng, Liu, Yuan & Cheng, 2008)		End-to-end delay	Queuing delay, cumulative delay, end-to-end delay	Yes	AODV	MATLAB
(Filippini, Ekici, & Cesana, 2009)		Route maintenance cost	Route Maintenance Cost	Yes		Mathematical modeling
(Fujii & Yasushi, 2006)		Multi-bands (packet loss)	Packet loss rate	Yes	AODV	Mathematical modeling
(Chen et al., 2011)		End-to-end delay with QoS provisioning in underlay CRN	Minimized mean end-to-delay, Average buffer occupancy of a relaying SU	Yes		Mathematical modeling
(Ghahremani, Khokhar, Noor, Naebi & Kheyrihassankandi, 2012)		Throughput	Packet delivery ratio, average end-to-end delay, throughput	No		MATLAB
JRCS (Mumey, Judson & Stevens, 2012)		End-to-end throughput	Average throughput	Yes		Mathematical modeling
(Badarneh & Salameh, 2011)		Spectrum availability time & CR required transmission time	Throughput	Yes		MATLAB
PCTC (Guan, Yu, Jiang & Wei, 2010)	Topology-control based	Primary users awareness and link duration (get from topology control)	Node degree, link duration, end-to-end delay, throughput	Yes		Mathematical modeling
MPP (Khalife, Ahuja, Malouch & Krunz, 2008)	Probabilistic path selection	Probabilistic capacity	Number of accepted flows	Yes		NS-2

continued on following page

Table 1. Continued

Strategy	Routing Protocol Nature	Routing Metric	Performance Parameters	Spectrum Awareness	Routing Protocol Based On	Simulators Used
SURF (Rehmani, Viana, Khalife & Fdida, 2013)			% of messages received, PU harmful interference ratio, ratio of accumulate receivers, average delivery ratio, packet ratio	Yes		NS-2
RL (Xia, Wahab, Yang, Fan & Sooriyabandara, 2009)		Routes with more available channels (reinforcement learning)	Average packet delivery time	Yes		OMNET++
ROPCORN (Cagatay Talay, & Altilar, 2009)		Spectrum availability cost and link estimation	Aggregate throughput, end-to-end delay	Yes		Not mentioned
(Wang & Zheng, 2006)		Conflict free channel	Aggregated throughput	No		NS-2

In (Zhu et al., 2012), interference-aware routing protocol is introduced for CRNs. It uses interference minimization for PUs as routing metric. Normalized interference and normalized delay are considered for performance evaluation parameters. This protocol does not take into account varying spectrum conditions but it models PU activity.

IPSAG, an IP spectrum aware geographic routing protocol is presented in (Badoi et al., 2012) for multi-hop CRNs. It is based on IP hop-by-hop routing, spectrum-aware routing and geographic routing. For routing metric, it integrates spectrum opportunity, channel quality and current nodes' position towards destination. It considers varying spectrum condition but does not model PU activity. Also, it presents an algorithm but does not evaluate its performance.

A joint channel assignment and routing protocol is proposed in (Hon et al., 2009) for IEEE 802.11 based multi-channel multi-interface mobile ad-hoc networks. This protocol takes into account interference minimization as a routing metric. In this protocol, a channel with minimized interference is selected for allocation to CR user. It uses aggregate goodput and average end-to-end delay as performance parameters. It is not spectrum-aware and does not model PU activity.

Xin et al., (2011) uses greedy heuristic approximation and proposes joint channel assignment and QoS provision routing for coverage optimization in cognitive radio cellular networks. It considers interference constraints and QoS requirements for routing metrics. Maximization of base station coverage for CR nodes is considered as performance parameter. This protocol does not model PU activity and is not spectrum-aware.

Alicherry et al., (2005) introduces a joint channel assignment and routing in multi-radio wireless mesh networks for optimization of throughput. It considers interference constraint as a routing metric and for parameter of performance evaluation, it considers throughput. It does not incorporate PU activity and is not aware of varying spectrum conditions.

A Delay motivated On-demand Routing Protocol (DORP) is proposed in (Cheng et al., 2007) for joint routing and channel assignment in CRNs. For routing metric, this protocol considers delay. It uses

Node Analytical Model (NAM) for channel assignment which is a scheduling based channel assignment scheme for minimizing inter-flow interference and frequent switching delay. This protocol is spectrum-aware and is based on AODV. It uses cumulative delay and number of channel switching as performance parameters. It does not model PU activity.

Mixed Integer Linear Programming (MILP) is another strategy introduced in (Ma et al., 2008) for joint spectrum sharing and fair routing in CRNs. This strategy takes into account interference constraint as a routing metric. Results of this strategy confirm that spectrum sharing among secondary users is interference free and fair routing is guaranteed. PU activity is not modeled in this strategy and it does not consider spectrum varying conditions.

In (Yang et al., 2008), a local coordination based spectrum assignment and routing is proposed for multi-hop CRNs. Routing metric for this strategy is end-to-end delay. This strategy is spectrum-aware and is based on AODV. Channel assignment in this strategy is switch-aware and k-hop distinct. Performance parameters for this strategy are Queuing delay, cumulative delay and end-to-end delay.

Minimum maintenance cost routing is introduced in (Filippini et al., 2009) for CRNs. Route maintenance is considered as a routing metric in this protocol. This protocol incorporates and models PU activity and is spectrum-aware. It takes route maintenance cost parameter for performance evaluation.

(Fuji et al., 2006) discusses multi-band routing for CRNs in-order to minimize interference to primary users. This protocol is based on AODV and it uses packet loss as a routing metric. It does not model PU activity but is spectrum-aware. It considers packet loss rate parameter for the performance evaluation of this protocol.

In (Chen et al., 2011), a multi-path routing protocol for underlay CRNs is proposed. For routing metric, this protocol considers end-to-end delay with QoS provisioning. It does not incorporate PU activity. It uses minimized mean end-to-delay and average buffer occupancy of a relaying CR user as performance parameters and it is based on AODV routing protocol. It also incorporates varying conditions of spectrum.

(Ghahremani et al., 2012) presents a QoS aware routing protocol for mobile WiMAX CRNs which considers throughput as a routing metric. It considered packet delivery ratio, average end-to-end delay and throughput as performance parameters. It does not model PU activity and also does not incorporate varying spectrum conditions.

A joint routing and channel selection algorithm (JRCS) is presented in (Mumey et al., 2012) for cognitive radio mesh networks. It used dynamic programming approach for channel selection. In dynamic programming approach, solution to full problem is solved by assembling solution to partial problems. It uses end-to-end throughput as routing metric and average throughput as performance parameter. It does not incorporate PU activity but considers varying spectrum conditions.

In (Badarneh et al., 2011), opportunistic routing is introduced in CRNs which takes into account spectrum availability time and CR required transmission time as routing metrics. It is spectrum-aware and also models PU activity. For performance evaluation of this protocol, throughput has been considered.

Prediction-based Cognitive Topology Control (PCTC) is a topology control based routing protocol which is discussed in (Guan et al., 2010) for cognitive radio mobile ad-hoc networks. It uses primary user awareness and link duration as routing metrics. Link duration is extracted from topology control. It is spectrum-aware and it also models PU activity. Parameters for the performance evaluation of this protocol are predicted link duration, node degree, link duration, end-to-end delay and throughput.

Another probabilistic path selection strategy known as MPP (Most Probable Path) is explained in (Khalife et al., 2008) for multi-channel CRNs. For every channel, this strategy estimates the available capacity probabilistically, over every CR-to-CR link while considering PU activity model. In this strategy, probabilistic capacity is used as a routing metric. It considers number of accepted flows as performance parameter. It models PU activity and is also aware of varying conditions of spectrum.

In (Parvin et al., 2012), radio environment aware stable routing (RASP) strategy is described for multi-hop CRNs. This strategy considers route stability as a routing metric. The route stability is mostly affected by two radio environmental issues, i.e., Primary Exposed Node (PEN) and Primary Hidden Node (PHN). This strategy minimizes these problems by keeping a channel priority list for each CR user. For performance evaluation, it considers throughput, end-to-delay, packet loss ratio and route repair success ratio. It takes into account varying spectrum conditions and also incorporates PU activity in its results.

SURF, proposed in (Rehmani et al., 2013) is a distributed channel selection scheme for data dissemination in CRNs. It classifies available channels based on PU un-occupancy and number of CR neighbors using the channel. SURF uses percentage of messages received, PU harmful interference ratio, ratio of accumulated receivers, average delivery ratio and packet ratios as performance evaluation parameters. It also models PU activity and is aware of varying spectrum conditions.

Another spectrum-aware routing protocol for CRNs is proposed in (Xia et al., 2009). This protocol is based on reinforcement learning (RL), therefore from local information, it can learn about good routes having more available channels. It considers average packet delivery time as a performance metric and it also incorporates PU activity. It is aware about changing conditions of spectrum.

ROPCORN, a routing protocol for Cognitive radio ad-hoc networks is discussed in (Cagatay Talay et al., 2009). Its main objectives are to minimize delay and maximize data rates. Two routing metrics are used in this protocol which are spectrum availability cost and link estimation. It considers aggregate throughput and end-to-end delay for performance parameters. It is spectrum-aware but does not model PU activity.

In (Wang et al., 2006), a collaborative design for route and spectrum selection in dynamic networks is described. This strategy integrates routing and spectrum selection into a single task. In this strategy, only conflict free channels are used for routing and channel selection. This strategy uses aggregate throughput as performance evaluation parameter. It does not consider varying spectrum conditions and does not model PU activity.

A. Routing Protocol with Channel Selection Schemes

Table II contains routing strategies which consider channel selection schemes in their routing decisions. Channel selection scheme used in ROSA (Ding et al., 2010) is based on spectrum utility. It considers availability of spectrum holes at a particular geographic location and their variability with time. In (Kamruzzaman et al., 2012), Dynamic Source Routing (DSR) protocol is used in which channel selection scheme selects channels based on spectrum awareness and energy awareness.

In SPEAR (Sampath et al., 2008), a joint routing and channel assignment scheme is proposed. The purpose of channel selection assignment strategy is to minimize inter-flow and intra-flow interference and integrate channel assignment and route discovery for robust path formation. It uses graph coloring problem for minimizing inter-flow and intra-flow interference.

J-CAR (Chiu et al., 2009) is another multi-channel multi-interface joint routing protocol and channel assignment strategy for mobile ad-hoc networks (MANETs). It uses multiple non-overlapping channels and multiple interfaces in-order to reduce collision and channel interference. In its channel assignment strategy, data interface switches between send and receive mode dynamically which optimizes inter-

Table 2. Routing protocol with channel assignment strategies

Strategy	Routing Metric	Performance Parameters	Channel Assignment Scheme	Spectrum Awareness	Routing Protocol Based On
ROSA (Ding et al., 2010)	Spectrum utility and spectrum holes	Throughput, fairness index, network spectrum utility, average delay	Based on spectrum utility	Yes	
(Kamruzzaman et al., 2012)	Spectrum and energy aware	System throughput, mean message delay, ratio of survival note, consumed energy per packet	Spectrum and energy aware	Yes	DSR
SPEAR (Sampath et al., 2008)	Spectrum integration and route discovery	System throughput, end-to-end packet delivery latency	Graph coloring problem	Yes	
J-CAR (Chiu et al., 2009)	Interference minimization	Aggregate goodput and average end-to-end delay	Minimized interfered channel is selected	No	
(Yang et al., 2008)	End-to-end delay	Queuing delay, cumulative delay, end-to-end delay	Switch-aware and k-hop distinct	Yes	AODV
JRCS (Mumey et al., 2012)	End-to-end throughput	Average throughput	Dynamic programming approach (in which solutions to partial problems are used to assemble a solution to the full problem).	Yes	
SURF (Rehmani et al., 2013)		% of messages received, PU harmful interference ratio, ratio of accumulate receivers, average delivery ratio, packet ratio	SURF	Yes	
(Bayhan et al., 2014)	Nil	Error in steady state probability, Average number of transmitting SUs, Probability of success and Spectrum opportunity utilization	Best-fit Channel (BFC)	Yes	
(Zhang et al., 2012)	Interference impact	Throughput, Path Longevity	Minimized interference	Yes	
(Xie et al., 2012)	Nil	End-to-end delay	Formulated as optimization problem and solved using heuristic method	Yes	

face and channel utilization. J-CAR uses channel interference index by considering both protocol and physical interference models and selects the least interfered channel at each hop by considering channel interference model.

(Yang et al., 2008) proposed an on-demand joint routing and channel assignment protocol for CRNs. Its channel assignment scheme is switch-aware and k-hop distinct. Its purpose is to minimize per-node delay. It is based on AODV and it considers varying spectrum conditions. But it does not model PU activity.

In JRCS (Mumey et al., 2012), a routing and channel selection scheme for CRNs is proposed. It considers the problem of selecting channels based on maximizing end-to-end throughput. For this purpose, it uses dynamic programming approach for channel selection in which solutions to partial problems are used to assemble a solution to full problem.

SURF (Rehmani et al., 2013) is also a channel selection strategy proposed for data dissemination in multi-hop CRNs. Its goal is to increase reliability and minimize harmful interference to PU. It selects channels on the basis of high un-occupancy of PU and high number of CR neighbors.

B. Routing Protocol Nature

Routing protocols have different types of nature, signifying the way in which they respond to PU occupancy of a channel. Some well-known natures of routing protocols are illustrated in Figure 3. Nature of routing protocols discussed in (Lin et al., 2007) and (Chehata et al., 2011) are on-demand. Directed Acyclic Graphs (DAGs) are used in (Nandiraju et al., 2009). Nature of (Aduwo et al., 2004) is Maximal Ratio Route Path. (Liu et al., 2005) and (Xin et al., 2011) belong to greedy heuristic algorithm. In (Zhou et al., 2009), routing protocol uses Colored multi-graph model. Nature of (Kamruzzaman et al., 2012) is dynamic source routing. PCTC (Guan et al., 2010) is topology-control based routing protocol and MPP (Khalife et al., 2008) is probabilistic path selection based routing protocol.

7. GUIDELINES FOR EFFICIENT ROUTING IN CRNs

There are many things to consider while designing an efficient routing protocol in CRNs. Some of the guidelines for efficient routing are summarized as follows:

Figure 3. Routing protocols nature

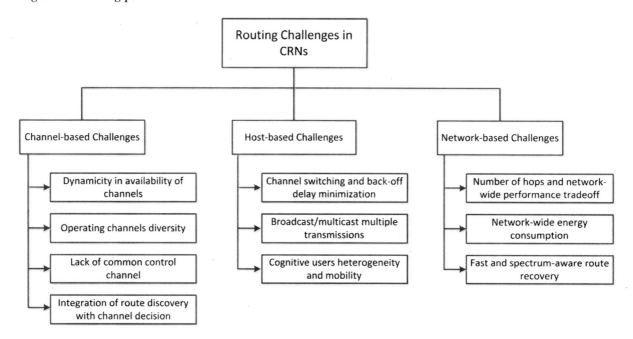

- Channel selection should be considered in conjunction with routing, because in CRNs, there is dynamicity in availability of channels. Therefore, if joint channel selection and routing are considered, then routes will be more stable, which in turn will improve network performance.

- PU activity modeling is very important in CRNs due to dynamic nature of the network. Therefore it should be modeled appropriately by selecting a PU activity model in order to get realistic results and to analyze effects of PUs on the networks. Many PUs activities models have been proposed in the literature. For instance, (Saleem et al., 2014) classifies existing PUs activities models by presenting a survey.

- Simulators play a vital role for the development of routing protocols in CRNs. They are gateways for reaching realistic results for any scenario without performing actual hardware simulations. Therefore, a good network simulator must be considered while developing a routing protocol for CRNs.

- New routing metrics should be developed for CRNs, because existing routing metrics for traditional wireless networks may not work well in CRNs due to the dynamic nature of the network. Therefore, novel routing metrics having dynamic nature should be developed for efficient routing protocol in CRNs.

- Appropriate performance parameters should be considered for measuring the performance of routing protocols in CRNs. For example, some of the performance parameters for CRNs can be interference ratio of CR users to PUs, number of channels switching etc.

CONCLUSION

Routing is the backbone of communication network and a good routing protocol is highly dependent upon good channel assignment scheme. Because if channel assignment scheme selects channels with longer availability time, low interference to primary users and minimum contention among secondary users, then routing links will be more stable and can exist for longer time period. Therefore, this chapter is focused on joint channel assignment and routing with reference to cognitive radio networks. Firstly, an introduction to routing, CRNs and applications of CRNs are elaborated. Then, classification of routing protocols in CRNs is discussed. Next we highlighted the challenges of routing in CRNs. Afterwards, classification of channel assignment strategies in CRNs and subsequently, challenges of channel assignment in CRNs are highlighted. Next, we explained routing with efficient channel assignment in CRNs in which we described routing proposals in CRNs, routing protocols with channel selection schemes, routing protocols nature, routing metrics, performance parameters, PR activity modeling, spectrum awareness and simulators used for performance evaluation of routing protocols. In the end, guidelines for the development of efficient routing protocols for cognitive radio networks are discussed.

REFERENCES

Abbagnale, A., & Cuomo, F. (2010). Gymkhana: A connectivity-based routing scheme for cognitive radio ad hoc networks. In *Proceedings of IEEE International Conference on Computer Communications Workshops (INFOCOM)*, (pp. 1-5). IEEE. doi:10.1109/INFCOMW.2010.5466618

Abdelaziza, S., & ElNainay, M. (2012). *Survey of routing protocols in cognitive radio networks.* Academic Press.

Aduwo, A., & Annamalai, A. (2004). Channel-aware inter-cluster routing protocol for wireless ad-hoc networks exploiting network diversity. In *Proceedings of IEEE 60th Vehicular Technology Conference (VTC)*, (vol. 4, pp. 2858-2862). IEEE. doi:10.1109/VETECF.2004.1400581

Akyildiz, I. F., Lee, W.-Y., & Chowdhury, K. R. (2009). CRAHNs: Cognitive radio ad hoc networks. *Ad Hoc Networks*, *7*(5), 810–836. doi:10.1016/j.adhoc.2009.01.001

Al-Rawi, H. A., & Yau, K.-L. A. (2013). Routing in distributed cognitive radio networks: A survey. *Wireless Personal Communications*, *69*(4), 1983–2020. doi:10.1007/s11277-012-0674-7

Alicherry, M., Bhatia, R., & Li, L. E. (2005). Joint channel assignment and routing for throughput optimization in multi-radio wireless mesh networks. In *Proceedings of the 11th Annual International Conference on Mobile Computing and Networking*, (pp. 58-72). Academic Press. doi:10.1145/1080829.1080836

Badarneh, O. S., & Salameh, H. B. (2011). Opportunistic routing in cognitive radio networks: exploiting spectrum availability and rich channel diversity. In *Proceedings of IEEE Global Telecommunications Conference (GLOBECOM)*, (pp. 1-5). IEEE. doi:10.1109/GLOCOM.2011.6134241

Badoi, C.-I., Croitoru, V., & Prasad, R. (2010). IPSAG: an IP spectrum aware geographic routing algorithm proposal for multi-hop cognitive radio networks. In *Proceedings of IEEE 8th International Conference on Communications (COMM)*, (pp. 491-496). IEEE. doi:10.1109/ICCOMM.2010.5509020

Bayhan, S., & Alagöz, F. (2014). A Markovian approach for best-fit channel selection in cognitive radio networks. *Ad Hoc Networks*, *12*, 165–177. doi:10.1016/j.adhoc.2011.08.007

Cagatay Talay, A., & Altilar, D. T. (2009). ROPCORN: Routing protocol for cognitive radio ad hoc networks. In *Proceedings of IEEE International Conference on Ultra Modern Telecommunications & Workshops (ICUMT)*, (pp. 1-6). IEEE.

Chehata, A., Ajib, W., & Elbiaze, H. (2011). An on-demand routing protocol for multi-hop multi-radio multi-channel cognitive radio networks. In Proceedings of IEEE Wireless Days (WD) (pp. 1–5). IEEE. doi:10.1109/WD.2011.6098215

Chen, P.-Y., Cheng, S.-M., & Ao, W. C. (2011). Multi-path routing with end-to-end statistical QoS provisioning in underlay cognitive radio networks. In *Proceedings of IEEE International Conference on Computer Communications Workshops (INFOCOM WKSHPS)*, (pp. 7-12). IEEE.

Cheng, G., Liu, W., Li, Y., & Cheng, W. (2007). Joint on-demand routing and spectrum assignment in cognitive radio networks. In *Proceedings of IEEE International Conference on Communications (ICC)*, (pp. 6499-6503). IEEE. doi:10.1109/ICC.2007.1075

Chiu, H. S., Yeung, K. L., & Lui, K.-S. (2009). J-CAR: An efficient joint channel assignment and routing protocol for IEEE 802.11-based multi-channel multi-interface mobile ad hoc networks. *IEEE Transactions on Wireless Communications*, *8*(4), 1706–1715. doi:10.1109/TWC.2009.080174

Deng, S., Chen, J., He, H., & Tang, W. (2007). Collaborative strategy for route and spectrum selection in cognitive radio networks. *IEEE Future Generation Communication and Networking, 2*, 168–172. doi:10.1109/FGCN.2007.88

Ding, L., Melodia, T., Batalama, S. N., Matyjas, J. D., & Medley, M. J. (2010). Cross-layer routing and dynamic spectrum allocation in cognitive radio ad hoc networks. *IEEE Transactions on Vehicular Technology, 59*(4), 1969–1979. doi:10.1109/TVT.2010.2045403

FCC. (2004). *FCC adopted rules for unlicensed use of television white spaces. FCC adopted rules for unlicensed use of television white spaces* (Docket no. 04-186, Second Report and Order and Memorandum Opinion and Order). FCC.

Filippini, I., Ekici, E., & Cesana, M. (2009). Minimum maintenance cost routing in cognitive radio networks. In *Proceedings of IEEE 6th International Conference on Mobile Ad hoc and Sensor Systems (MASS),* (pp. 284-293). IEEE. doi:10.1109/MOBHOC.2009.5336987

Fujii, T., & Yamao, Y. (2006). *Multi-band routing for ad-hoc cognitive radio networks.* IEEE SDR.

Ghahremani, S., Khokhar, R. H., Noor, R. M., Naebi, A., & Kheyrihassankandi, J. (2012). On QoS routing in Mobile WiMAX cognitive radio networks. In *Proceedings of IEEE International Conference on Computer and Communication Engineering (ICCCE),* (pp. 467-471). IEEE. doi:10.1109/ICCCE.2012.6271231

Guan, Q., Yu, F. R., Jiang, S., & Wei, G. (2010). Prediction-based topology control and routing in cognitive radio mobile ad hoc networks. *IEEE Transactions on Vehicular Technology, 59*(9), 4443–4452. doi:10.1109/TVT.2010.2069105

Jashni, B., Tadaion, A. A., & Ashtiani, F. (2010). Dynamic link/frequency selection in multi-hop cognitive radio networks for delay sensitive applications. In *Proceedings of IEEE 17th International Conference on Telecommunications (ICT),* (pp. 128-132). IEEE.

Jiao, L., Li, F. Y., & Pla, V. (2011). Dynamic channel aggregation strategies in cognitive radio networks with spectrum adaptation. In *Proceedings of IEEE Global Telecommunications Conference* (GLOBE-COM), (pp. 1-6). IEEE.

Jiao, L., Li, F. Y., & Pla, V. (2012). Modeling and performance analysis of channel assembling in multichannel cognitive radio networks with spectrum adaptation. *IEEE Transactions on Vehicular Technology, 61*(6), 2686–2697.

Kamruzzaman, S., Kim, E., Jeong, D. G., & Jeon, W. S. (2012). Energy-aware routing protocol for cognitive radio ad hoc networks. *IET Communications, 6*(14), 2159–2168. doi:10.1049/iet-com.2011.0698

Khalife, H., Ahuja, S., Malouch, N., & Krunz, M. (2008). Probabilistic path selection in opportunistic cognitive radio networks. In *Proceedings of IEEE Global Telecommunications Conference (GlobeCom),* (pp. 1-5). IEEE. doi:10.1109/GLOCOM.2008.ECP.931

Kim, H., & Shin, K. G. (2008). Efficient discovery of spectrum opportunities with MAC-layer sensing in cognitive radio networks. *IEEE Transactions on Mobile Computing, 7*(5), 533–545. doi:10.1109/TMC.2007.70751

Kondareddy, Y. R., & Agrawal, P. (2008). Selective broadcasting in multi-hop cognitive radio networks. In *Proceedings of IEEE Sarnoff Symposium*, (pp. 1-5). IEEE. doi:10.1109/SARNOF.2008.4520042

Krishnamurthy, S., Thoppian, M., Venkatesan, S., & Prakash, R. (2005). Control channel based MAC-layer configuration, routing and situation awareness for cognitive radio networks. In *Proceedings of IEEE Military Communications Conference (MILCOM)*, (pp. 455-460). IEEE. doi:10.1109/MILCOM.2005.1605725

Leschhorn, R., & Buchin, B. (2004). Military software defined radios-rohde and schwarz status and perspective. In *Proceedings of the SDR 04 Technical Conference and Product Exposition*. Academic Press.

Lin, X., & Rasool, S. (2007). A distributed joint channel-assignment, scheduling and routing algorithm for multi-channel ad-hoc wireless networks. In *Proceedings of IEEE 26th International Conference on Computer Communications (INFOCOM)*, (pp. 1118-1126). IEEE. doi:10.1109/INFCOM.2007.134

Liu, J., & Annamalai, A. (2005). Channel-Aware Routing Protocol for Wireless Ad-Hoc Networks: Generalized Multiple-Route Path Selection Diversity. *IEEE Vehicular Technology Conference*, 62, 2258.

Ma, M., & Tsang, D. H. (2008). Joint spectrum sharing and fair routing in cognitive radio networks. In *Proceedings of IEEE 5th Consumer Communications and Networking Conference (CCNC)*, (pp. 978-982). IEEE. doi:10.1109/ccnc08.2007.225

Maldonado, D., Le, B., Hugine, A., Rondeau, T. W., & Bostian, C. W. (2005). Cognitive radio applications to dynamic spectrum allocation: a discussion and an illustrative example. In *Proceedings of IEEE International Symposium on New Frontiers in Dynamic Spectrum Access Networks (DySPAN)*, (pp. 597-600). IEEE. doi:10.1109/DYSPAN.2005.1542677

Masonta, M. T., Mzyece, M., & Ntlatlapa, N. (2013). Spectrum decision in cognitive radio networks: A survey. *IEEE Communications Surveys and Tutorials*, 15(3), 1088–1107. doi:10.1109/SURV.2012.111412.00160

Min, A. W., & Shin, K. G. (2008). Exploiting multi-channel diversity in spectrum-agile networks. In *Proceedings of IEEE 27th International Conference on Computer Communications (INFOCOM)*. IEEE. doi:10.1109/INFOCOM.2008.256

Mumey, B., Tang, J., Judson, I. R., & Stevens, D. (2012). On routing and channel selection in cognitive radio mesh networks. *IEEE Transactions on Vehicular Technology*, 61(9), 4118–4128. doi:10.1109/TVT.2012.2213310

Nandiraju, D. S., Nandiraju, N. S., & Agrawal, D. P. (2009). Adaptive state-based multi-radio multi-channel multi-path routing in Wireless Mesh Networks. *Pervasive and Mobile Computing*, 5(1), 93–109. doi:10.1016/j.pmcj.2008.11.003

Nguyen, G. D., Kompella, S., Wieselthier, J. E., & Ephremides, A. (2010). Channel sharing in cognitive radio networks. In *Proceedings of IEEE Military Communications Conference (MILCOM)*, (pp. 2268-2273). IEEE.

Pan, M., Huang, R., & Fang, Y. (2008). Cost design for opportunistic multi-hop routing in cognitive radio networks. In *Proceedings of IEEE Military Communications Conference (MILCOM)*, (pp. 1-7). IEEE.

Parvin, S., & Fujii, T. (2012). Radio environment aware stable routing scheme for multi-hop cognitive radio network. *IEEE IET Networks*, *1*(4), 207–216. doi:10.1049/iet-net.2012.0103

Rehmani, M. H. (2011). *Opportunistic Data Dissemination in Ad-Hoc Cognitive Radio Networks.* (Ph.D. Dissertation). Universit'e Pierre et Marie Curie-Paris VI.

Rehmani, M. H., Viana, A. C., Khalife, H., & Fdida, S. (2010). A cognitive radio based internet access framework for disaster response network deployment. In *Proceedings of International Symposium on Applied Sciences in Biomedical and Communication Technologies (ISABEL)*. Academic Press. doi:10.1109/ISABEL.2010.5702851

Rehmani, M. H., Viana, A. C., Khalife, H., & Fdida, S. (2013). Surf: A distributed channel selection strategy for data dissemination in multi-hop cognitive radio networks. *Computer Communications*, *36*(10), 1172–1185. doi:10.1016/j.comcom.2013.03.005

Saleem, Y., Bashir, A., Ahmed, E., Qadir, J., & Baig, A. (2012). Spectrum-aware dynamic channel assignment in cognitive radio networks. In *Proceedings of IEEE International Conference on Emerging Technologies (ICET)*, (pp. 1-6). IEEE. doi:10.1109/ICET.2012.6375468

Saleem, Y., & Husain Rehmani, M. (2014). Primary radio user activity models for cognitive radio networks: A survey. *Journal of Network and Computer Applications*, *43*, 1–16. doi:10.1016/j.jnca.2014.04.001

Sampath, A., Yang, L., Cao, L., Zheng, H., & Zhao, B. Y. (2008). High throughput spectrum-aware routing for cognitive radio networks. In *Proceedings of IEEE International Conference on Cognitive Radio Oriented Wireless Networks (CrownCom)*. IEEE.

Tragos, E., Zeadally, S., Fragkiadakis, A., & Siris, V. (2013). Spectrum assignment in cognitive radio networks: A comprehensive survey. *IEEE Communications Surveys and Tutorials*, *15*(3), 1108–1135. doi:10.1109/SURV.2012.121112.00047

Van der Schaar, M., & Fu, F. (2009). Spectrum access games and strategic learning in cognitive radio networks for delay-critical applications. *Proceedings of the IEEE*, *97*(4), 720–740.

Wang, Q., & Zheng, H. (2006). Route and spectrum selection in dynamic spectrum networks. In *Proceedings of IEEE Consumer Communications and Networking Conference (CNCC)*, (pp. 342-346). IEEE.

Weiss, T. A., & Jondral, F. K. (2004). Spectrum pooling: An innovative strategy for the enhancement of spectrum efficiency. *IEEE Communications Magazine*, *42*(3), S8–S14. doi:10.1109/MCOM.2004.1273768

Xia, B., Wahab, M. H., Yang, Y., Fan, Z., & Sooriyabandara, M. (2009). Reinforcement learning based spectrum-aware routing in multi-hop cognitive radio networks. In *Proceedings of IEEE 4th International Conference on Cognitive Radio Oriented Wireless Networks and Communications (CrownCom)*, (pp. 1-5). IEEE. doi:10.1109/CROWNCOM.2009.5189189

Xie, L., Heegaard, P. E., Zhang, Y., & Xiang, J. (2012). Reliable Channel Selection and Routing for Real-Time Services over Cognitive Radio Mesh Networks. In Quality, Reliability, Security and Robustness in Heterogeneous Networks (pp. 41-57). Springer. doi:10.1007/978-3-642-29222-4_4

Xin, Q., Wang, X., Cao, J., & Feng, W. (2011). Joint Admission Control, Channel Assignment and QoS Routing for Coverage Optimization in Multi-hop Cognitive Radio Cellular Networks. In *Proceedings of IEEE 8th International Conference on Mobile Adhoc and Sensor Systems (MASS)*, (pp. 55-62). IEEE.

Xu, Y., Anpalagan, A., Wu, Q., Shen, L., Gao, Z., & Wang, J. (2013). Decision-theoretic distributed channel selection for opportunistic spectrum access: Strategies, challenges and solutions. *IEEE Communications Surveys and Tutorials, 15*(4), 1689–1713. doi:10.1109/SURV.2013.030713.00189

Yang, L., Cao, L., & Zheng, H. (2008). Proactive channel access in dynamic spectrum networks. *Physical Communication, 1*(2), 103–111. doi:10.1016/j.phycom.2008.05.001

Yang, Z., Cheng, G., Liu, W., Yuan, W., & Cheng, W. (2008). Local coordination based routing and spectrum assignment in multi-hop cognitive radio networks. *Mobile Networks and Applications, 13*(1-2), 67–81. doi:10.1007/s11036-008-0025-9

Youssef, M., Ibrahim, M., Abdelatif, M., Chen, L., & Vasilakos, A. (2014). Routing Metrics of Cognitive Radio Networks: A Survey. *IEEE Communications Surveys and Tutorials, 16*(1), 92–109. doi:10.1109/SURV.2013.082713.00184

Zeeshan, M., Manzoor, M. F., & Qadir, J. (2010). Backup channel and cooperative channel switching on-demand routing protocol for multi-hop cognitive radio ad hoc networks (BCCCS). In *Proceedings of IEEE 6th International Conference on Emerging Technologies (ICET)*, (pp. 394-399). IEEE.

Zhang, J., Yao, F., Liu, Y., & Cao, L. (2012). Robust route and channel selection in cognitive radio networks. In *Proceedings of IEEE 14th International Conference on Communication Technology (ICCT)*, (pp. 202-208). IEEE.

Zhou, X., Lin, L., Wang, J., & Zhang, X. (2009). Cross-layer routing design in cognitive radio networks by colored multigraph model. *Wireless Personal Communications, 49*(1), 123–131. doi:10.1007/s11277-008-9561-7

Zhu, Q., Yuan, Z., Song, J. B., Han, Z., & Basar, T. (2012). Interference aware routing game for cognitive radio multi-hop networks. *IEEE Journal on Selected Areas in Communications, 30*(10), 2006–2015. doi:10.1109/JSAC.2012.121115

KEY TERMS AND DEFINITIONS

Channel Assignment: Channel assignment means assigning channels or spectrum bands to radio interfaces for communication.

Cognitive Radio Networks: Cognitive radio networks are composed of cognitive radio devices which exploit Cognitive Radio Technology in-order to utilize the spectrum opportunistically.

Cognitive Radio Technology: Cognitive Radio technology enables Cognitive Radio Networks to exploit the spectrum dynamically by using Cognitive Radio devices.

Cognitive Radio: Cognitive radio is a radio which can adapt its parameters according to the operating environment.

Primary User: Primary user is a licensed user who has assigned a fixed spectrum band and it can use its spectrum band without any interference or disruption from other users in the network.

Routing: Routing is the process of selecting paths in order to send data from one host to another in the network

Secondary User: Secondary user is an unlicensed user who either utilizes unlicensed spectrum band or licensed spectrum band of primary user when it is idle, i.e., not utilizing by the primary user subject to the condition that primary user should not be interfered.

Chapter 12
Dynamic Spectrum Management Algorithms for Multiuser Communication Systems

Sean Huberman
McGill University, Canada

Tho Le-Ngoc
McGill University, Canada

ABSTRACT

Dynamic Spectrum Management (DSM) is an effective method for reducing the effect of interference in both wireless and wireline communication systems. This chapter discusses various DSM algorithms, including Optimal Spectrum Balancing (OSB), Iterative Spectrum Balancing (ISB), Iterative Water-Filling (IWF), Selective Iterative Water-filling (SIW), Successive Convex Approximation for Low complExity (SCALE), the Difference of Convex functions Algorithm (DCA), Distributed Spectrum Balancing (DSB), Autonomous Spectrum Balancing (ASB), and Constant Offset ASB using Multiple Reference Users (ASB-MRU). They are compared in terms of their performance (achievable data-rate) by extensive simulation results and their computational complexity.

INTRODUCTION

In order to meet the increasing demand of data-intensive services in a multiuser multi-sub-carrier environment for both wireless and wireline communication systems, effective spectrum management techniques are required. The most basic form of spectrum management is Static Spectrum Management (SSM). Typically, SSM performs spectrum management based on a worst-case scenario assumption for all users. Clearly, this leads to an inefficient spectrum utilization whenever the scenario is not the worst-case and leads to highly sub-optimal performance.

The poor performance of SSM approaches led to the introduction of Dynamic Spectrum Management (DSM) (Starr, Sorbara, Cioffi, & Silverman, 2003). DSM is a wide field which looks to adaptively apply different spectrum allocations for each user with some optimization criteria in mind (e.g., maximizing

DOI: 10.4018/978-1-4666-6571-2.ch012

throughput, minimizing power). DSM allows for a far more efficient use of the frequency spectrum than SSM does. As a result, many different DSM algorithms have been proposed. This chapter will present some of the most significant DSM algorithms and compare them.

The main criteria in comparing different DSM algorithms are their performance and their complexity. The performance relates how well an approach succeeds achieving its objective (e.g., maximizing the achievable data-rate, minimizing transmission power) as compared to the theoretical optimum. The complexity of the algorithm is related to the amount of time required to derive the power allocation as the number of users and frequency sub-carriers increase.

There are two main types of DSM algorithms: Centralized and distributed. Centralized systems require a central hub with full knowledge of the network. In general, this system allows for better performance at a cost of increasing the complexity and computational time. On the other hand, distributed systems allow for every user to self-optimize fully autonomously without the need of explicit message passing. In general, distributed systems reduce the complexity and computational time but often sacrifice some optimality in terms of performance.

In particular, nine DSM algorithms will be covered in this chapter. Three of the DSM algorithms are centralized algorithms: Optimal Spectrum Balancing (OSB), Iterative Spectrum Balancing (ISB), and the Difference of Convex functions Algorithm (DCA). Three of the DSM algorithms are semi-centralized DSM algorithms where the users self-optimize but require some per-iteration messaging passing with a central hub. The three semi-centralized DSM algorithms are: Selective Iterative Water-filling (SIW), Successive Convex Approximation for Low complExity (SCALE), and Distributed Spectrum Balancing (DSB). Finally, the three distributed DSM algorithms are Iterative Water-Filling (IWF), Autonomous Spectrum Balancing (ASB), and Constant Offset ASB using Multiple Reference Users (ASB-MRU).

BACKGROUND

This section provides some background material on the topic of dynamic spectrum management. For more detailed background material, interested readers are referred to Leaves et al. (2004), Akyildiz, Lee, Vuran, & Mohanty (2006), Zhao & Sadler (2007), and the references therein. A Venn diagram illustrating the similarity and differences between the algorithms discussed in the background material is shown in Figure 1.

Yu, Ginis, & Cioffi (2002) introduced Iterative Water Filling (IWF), one of the first distributed DSM algorithms. In IWF, each user selfishly performs their own power allocation without any regard for their effect on the other users until a point where no user benefits from changing its current power allocation is reached. While IWF gives significant performance improvements over SSM techniques, in many situations, it leads to sub-optimal performance.

The sub-optimality of IWF is caused by inefficient use of the frequency spectrum. In an attempt to increase the efficiency of the frequency spectrum, many other heuristic variations were proposed. For example, shifting the user's spectrums away from one another (Liu & Su, 2007), accelerating convergence (Bagheri, Pakravan, & Khalaj, 2004), and incorporating user priorities in a centralized manner (Statovci, Nordstrom, & Nilsson, 2006).

Two similar algorithms which improve on the performance of IWF were also introduced. More specifically, Lee, Kim, Brady, & Cioffi (2006) introduced Band Preference Spectrum Management (BPSM), and Noam & Leshem (2009) introduced Iterative Power Pricing (IPP). Both apply the general

Figure 1. Venn diagram illustrating the similarity and differences between the algorithms discussed in the background material

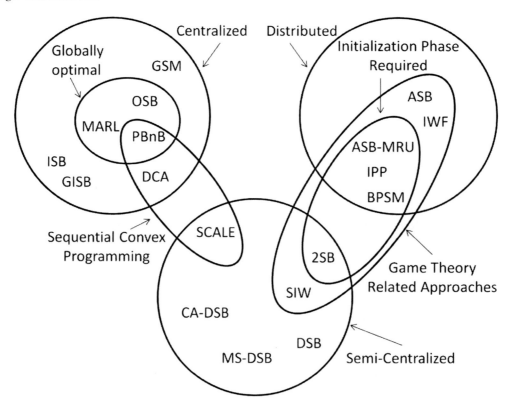

IWF algorithm but BPSM gives preferences to different users in each frequency band, while IPP uses power pricing for each frequency sub-carrier to allow weaker users to compete more fairly. The BPSM and IPP algorithms are intuitively similar, except they solve different optimization problems. BPSM is used to maximize the achievable rate, while IPP is used to minimize the required power.

Xu, Panigrahi, & Le-Ngoc (2005) introduced an algorithm known as Selective Iterative Water-filling (SIW). SIW selectively applies IWF to users in the under-utilized sections of the frequency spectrum until all users use up their total power. While SIW shows significant performance gains over IWF, the performance is still sub-optimal.

Cendrillon, Yu, Moonen, Verlinden, and Bostoen (2006) introduced a centralized algorithm called Optimal Spectrum Balancing (OSB), which maximizes the weighted sum-rate across the users. OSB uses dual decomposition to solve for the optimal transmit powers for each user separately on each frequency sub-carrier by exhaustive search. While OSB is not computationally tractable for many users, it serves as an upper bound on the performance of other DSM algorithms for cases with few users.

Wei, Youming, & Miaoliang (2009) introduced a similar centralized algorithm called Grouping Spectrum Management (GSM). GSM groups users into clusters and calculates a reference user for each cluster. GSM then applies OSB on the reference users for each cluster in order to reduce the computational complexity.

Qian & Jun (2009) introduced another centralized algorithm, based on Monotomic Optimization (MO), called MO-bAsed poweR controL (MARL). MARL finds globally optimal points by constructing a series of poly-blocks that approximate the feasible region with increasing precision; however, MARL is currently not feasible in the case of many users or frequency sub-carriers.

By representing the objective function of the non-convex optimization problem in OSB explicitly as the Difference of two Convex functions (DC), (i.e., $f = g - h$), a modified prismatic branch-and-bound algorithm was proposed by Xu, Le-Ngoc, & Panigrahi (2008). The constraint set is convex since it consists of a system of linear equations and one nonlinear constraint. The Prismatic Branch and Bound (PBnB) algorithm operates by successively approximating the nonlinear constraint by a piecewise linear function, and finds a globally optimal solution by solving a sequence of Linear Programming (LP) sub-problems (Horst, Phong, Thoai, & Vries, 1991). Although its computational complexity is still high, the algorithm proposed by Xu et al. (2008) can substantially reduce the complexity of OSB in finding a globally optimal solution, especially for a larger number of users. It can provide an upper-bound on the performance in many situations where OSB is computationally intractable, and hence, can serve as a comparison measure for other more practical DSM algorithms.

Since the previously mentioned global optimizing algorithms are intractable for many users, a centralized algorithm called Iterative Spectrum Balancing (ISB) was introduced (Cendrillon & Moonen, 2005; Lui & Yu, 2005). ISB formulates the optimization problem similar to OSB, using the weighted sum-rate and dual decomposition, but solves for the power allocations in an iterative fashion using one-dimensional instead of K-dimensional exhaustive searches (where K is the number of users). This allows for ISB to be computationally tractable for many users.

Forouzan & Garth (2005) proposed a similar algorithm called Generalized Iterative Spectrum Balancing (GISB). GISB operates similarly to ISB and OSB, where the difference lies in terms of the number of users over which the exhaustive search is performed. For GISB, the exhaustive search is performed over K_{GISB} users where $1 \le K_{GISB} \le K$. Note that ISB and OSB are the special cases where $K_{GISB} = 1$ and $K_{GISB} = K$, respectively. GISB allows for a trade-off between performance and computational complexity. The closer the value of K_{GISB} is to K, the better the performance will be at the expense of computational complexity.

Centralized systems are generally harder to implement in practice. It is for this reason that many other algorithms were introduced. Cendrillon, Huang, Chiang, & Moonen (2007) introduced one such algorithm, known as Autonomous Spectrum Balancing (ASB). ASB uses the same problem formulation as OSB and ISB but operates in a distributed fashion without the need for any explicit message passing. ASB uses the concept of a virtual user (referred to as a reference user), which represents the typical victim in the network. Each user self-optimizes to protect the reference user and hence attempts to better the overall network. Generally ASB cannot find a globally optimal solution; however, its performance has been shown to be near-optimal in some situations while maintaining a relatively low complexity.

One issue regarding the ASB algorithm is that the spectrum update formula can be relatively time consuming. It is for this reason that Tsiaflakis, Diehl, & Moonen (2008) proposed the ASB-2 algorithm. The ASB-2 algorithm works exactly like the ASB algorithm but uses a slightly different spectrum update formula. This spectrum update formula has a significantly lower complexity. ASB and ASB-2 do not necessarily converge to the same solution; however, the ASB-2 algorithm converges significantly faster than ASB, especially when the number of users and/or frequency sub-carriers is very large.

The ASB and ASB-2 algorithms presented by Cendrillon et al. (2007) and Tsiaflakis et al. (2008) only use one reference user; however, Leung, Huberman, & Le-Ngoc (2010) showed that using Multiple

Reference Users (MRU) can significantly improve performance while still maintaining a low complexity. While guidelines for choosing reference users to ensure consistently strong performance were also presented by Leung et al. (2010), a systematic approach for the reference user selection for both single and multiple reference users still remained an open question. Huberman, Leung, & Le-Ngoc (2012) introduced a systematic method for selecting the reference users and their respective parameters that can be blindly applied to practical networks. Huberman et al. (2012) introduced an algorithm referred to as constant offset ASB-MRU using the concept of constant Lagrange multiplier offsets which are calculated based on a virtual network of reference users. The offsets act as per-sub-carrier quotas for each user and are computed during a semi-centralized initialization phase prior to the fully distributed optimization phase using only local channel information. Huberman et al. (2012) showed that the constant offset ASB-MRU algorithm can provide near-optimal performance while operating at a low computational complexity and requiring significantly fewer iterations to converge.

Moraes, Dortschy, Klautau, & Rius i Riuy (2010) introduced an algorithm called Semi-Blind Spectrum Balancing (2SB) which builds on the concepts introduced in ASB. 2SB operates in the same fashion as ASB but also dynamically updates the virtual reference user parameters for each user separately, to more accurately represent the network. The virtual users are updated at a central hub based on message passing to and from the users. This removes the autonomous aspect of ASB, resulting in a semi-centralized algorithm, but generally leads to an improved data-rate.

Several Sequential Convex Programming (SCP) algorithms (iterative algorithms which solve a sequence of convex sub-problems to find a locally optimal solution) will be discussed in this chapter. One SCP algorithm, proposed by Papandriopoulos & Evans (2006, 2009) called Successive Convex Approximation for Low-complExity (SCALE) is a semi-centralized algorithm that applies a series of concave lower-bounds to the maximization problem. This enables SCALE to make use of the well-researched area of convex optimization to maximize the concave lower-bound. Each successive iteration tightens the lower-bound towards a locally optimal solution.

The PBnB algorithm (discussed above) makes use of the DC property of the objective function to solve for globally optimal solutions. Finding globally optimal solutions has a large complexity; hence, several SCP algorithms (discussed below) which make use of the DC property of the objective function to solve for locally optimal solutions are more computationally attractive.

One such SCP algorithm discussed in this chapter makes use of the Difference of Convex functions Algorithm (DCA), originally developed by Tao & Hoai An (1997). DCA is a centralized algorithm that begins by re-writing the non-convex objective function in terms of the difference of two convex functions (i.e., $f = g - h$); however, DCA iteratively creates an affine minorization (multivariate first-order approximation) of h, denoted by h', which is used to make the objective function, $f' = g - h'$, convex. Each successive iteration more closely approximates the locally optimal solution. For any function f, many DC decompositions exist (e.g., $g - h = (g + \varphi) - (h + \varphi)$). The choice of decomposition has a crucial impact on the convergence speed as well as the performance. There are still a lot of heuristics regarding the DCA implementation which have yet to be explored in great detail.

A similar algorithm called Convex Approximation Distributed Spectrum Balancing (CA-DSB) was proposed by Tsiaflakis et al. (2008). CA-DSB applies the DC property of the objective function and follows a similar algorithm to DCA; however, CA-DSB uses a different DC decomposition than the DCA proposed by Tao & Hoai An (1997) and therefore can operate in a semi-centralized manner. This makes use of parallel processing to reduce the complexity while still maintaining strong performance through the use of message passing.

Tsiaflakis et al. (2008) also proposed another algorithm called Distributed Spectrum Balancing (DSB). DSB also writes the objective function as a DC function but it applies the Karush-Kuhn-Tucker (KKT) conditions directly. With the use of a semi-centralized message passing system, the DSB algorithm solves for a locally optimal solution in an iterative fashion. The DSB algorithm and CA-DSB algorithms are nearly identical in terms of performance; however, in general the runtime of DSB is slightly faster.

Tsiaflakis et al. (2008) also introduced the Multiple Starting Point DSB (MS-DSB) algorithm, as an extension of the DSB algorithm which makes use of multiple starting points. This allows for the algorithm to find solutions that are at least as good as the DSB algorithm without significantly increasing the complexity.

DSM is a wide-field with many applications to both wireline and wireless systems. For example, Subramanian, Al-Ayyoub, Gupta, Das, & Buddhikot (2008) proposed a near-optimal DSM scheme using a dynamic action based approach for cellular networks. Im et al. (2011) makes use of a spectrum broker to coordinate the spectrum allocation of secondary users within the licensed spectrum. Alnwaimi, Arshad, & Moessner (2011) proposed a centralized DSM scheme for multi-cell multi-operator interference management. Feng, Li, Li, Le, & Gulliver (2011) propose a DSM scheme for Wideband Code Division Multiple Access (WCDMA) / Digital Video Broadcasting (DVB) heterogeneous systems. Shi, Li, Ma, & Li (2013) designed a DSM scheme for heterogeneous cellular wireless networks which guarantees a certain cell coverage probability using a graph coloring based approach.

DYNAMIC SPECTRUM MANAGEMENT PROBLEM FORMULATION

This chapter assumes an Orthogonal Frequency-Division Multiplexing (OFDM) or Discrete Multi-Tone (DMT) transmission scheme. OFDM and DMT are multi-channel transmission techniques that divide the available spectrum into smaller sub-channels or frequency sub-carriers. DMT differs from OFDM in that it is also capable of optimizing the bit and energy distribution over the sub-channels (e.g., channel partitioning or bit-loading) (Starr et al., 2003). The basic idea is to transmit data in parallel over each frequency sub-carrier where some sub-carriers might transmit no data, while others can transmit a lot of data.

Figure 2 shows the OFDM/DMT transmission block diagram. The data is put through a serial to parallel converter. This formats the word size required for the parallel transmission. As well, this transforms the wide-band frequency selective channel into many parallel narrow-band frequency flat tones. Each frequency tone is modulated independently using an M-ary modulation scheme (e.g., Quadrature Amplitude Modulation (QAM), Phase-Shift Keying (PSK)). The Inverse Fast Fourier Transform (IFFT) is used to convert the waveform from the frequency domain to the time domain. A Cyclic Prefix (CP) is then added in order to avoid Inter-Symbol Interference (ISI) and Inter-Channel Interference (ICI). The Digital-to-Analog Converter (DAC) converts the data from the digital domain to the analog domain. After the data passes through the wireless or wireline channel, the receiver performs the reverse operations in order to recover the transmitted data.

Consider a wireless or wireline communication system with K users and F frequency sub-carriers. Using synchronous modulation, there is no ICI and transmission can be modeled independently on each sub-carrier, f, as follows:

Figure 2. OFDM/DMT block diagram

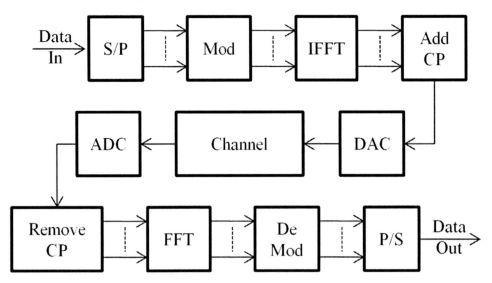

$$\mathbf{y}_f = \mathbf{H}_f \mathbf{x}_f + \mathbf{z}_f,$$

The vector $\mathbf{x}_f = (x_f^1, \ldots, x_f^K)^T$ contains the transmitted signals for all users on the f-th frequency sub-carrier, where x_f^k is the signal transmitted by user k on frequency sub-carrier f. Similarly, $\mathbf{y}_f = (y_f^1, \ldots, y_f^K)^T$ and $\mathbf{z}_f = (z_f^1, \ldots, z_f^K)^T$, where y_f^k is the received signal for user k on frequency sub-carrier f. Likewise, z_f^k is the additive noise for user k on frequency sub-carrier f which contains thermal noise and radio frequency interference. $\mathbf{H}_f$ is a K x K matrix such that $[\mathbf{H}_f]_{k,l}$ is the channel gain from transmitter l to receiver k and is defined as $h_f^{k,l}$. The transmit power of user k on frequency sub-carrier f is defined as $p_f^k = E[\,|x_f^k|^2\,]$, where $E[\bullet]$ denotes expected value. The vector containing the transmit power of user k on all frequency sub-carriers is defined as $\mathbf{p}^k = (p_1^k, \ldots, p_F^k)^T$.

When the number of users is large enough, the interference is well approximated by a Gaussian distributed random variable, and hence the bit-loading (bits/s/Hz) of user k on frequency sub-carrier f is defined as:

$$b_f^k = \log_2\left(1 + \frac{1}{\Gamma}\frac{\left|h_f^{k,k}\right|^2 p_f^k}{\sum_{l \ne k}\left|h_f^{k,l}\right|^2 p_f^l + \sigma_f^k}\right),$$

Where Γ is the Signal to Noise Ratio (SNR) gap which is a function of the desired Bit Error Rate (BER), coding gain, and noise margin and $\sigma_f^k = E[\,|z_f^k|^2]$ is the noise power of user k on frequency sub-carrier f. The achievable data-rate for user k is given by:

$$R_k = f_s \sum_{f=1}^{F} b_f^k \tag{1}$$

DSM dynamically allocates transmit powers for all users (i.e., $\mathbf{p}^k$, $k = 1, \ldots, K$) in response to changes in channel conditions (Starr et al., 2003). There are several physical constraints imposed on the transmit powers for each user. One such constraint is the maximum power which each user is allowed to allocate over all of its frequency sub-carriers. The maximum power constraint for user k is denoted by $P_{k,\max}$. Another constraint is the maximum power which each user is allowed to allocate on any particular frequency sub-carrier referred to as a spectral mask. The spectral mask for user k on frequency tone f is denoted by $p_f^{k,\text{mask}}$. Therefore the power constraints can be summarized as:

$$\sum_{f=1}^{F} p_f^k = P_{k,\max}, \; \forall \, k$$
$$0 \leq p_f^k \leq p_f^{k,\text{mask}}, \; \forall \, k, f \tag{2}$$

There are many different possible spectrum management problems depending on the specific goal of the network. For example, maximizing the total achievable data rate (i.e., Rate Adaptive (RA)) or minimizing the total power allocated (i.e., Fixed Margin (FM)) while ensuring some Quality of Service (QoS) minimum data-rate requirement for each user (Sadr, Anpalagan, & Raahemifar, 2009). Regardless of the specific goal of the network, the physical constraints shown in (2) are always present.

One approach to DSM focuses on maximizing the achievable data-rate of one particular user (e.g., user 1), given that the rest of the users satisfy some QoS data-rate requirement. The QoS data-rate requirement for user k is denoted by $R_{k,\text{target}}$. This approach to DSM can be summarized in the following RA optimization problem:

$$\max_{p_f^k, \forall k, f} \quad R_1$$
$$\text{s.t.} \quad R_k \geq R_{k,\text{target}}, \; k = 2, \ldots, K$$
$$\sum_{f=1}^{F} p_f^k = P_{k,\max}, \; \forall \, k \tag{3}$$
$$0 \leq p_f^k \leq p_f^{k,\text{mask}}, \; \forall \, k, f.$$

Cendrillon et al. (2006) showed that the optimal solution to the above RA optimization problem is equivalent to the optimal solution of the RA optimization problem shown below for some set of weighting factors $w_1, \ldots, w_K$ where w_k represents the importance of user k.

$$\max_{p_f^k, \forall k, f} \quad \sum_{k=1}^{K} w_k R_k$$
$$\text{s.t.} \quad \sum_{f=1}^{F} p_f^k = P_{k,\max}, \; \forall \, k \tag{4}$$
$$0 \leq p_f^k \leq p_f^{k,\text{mask}}, \; \forall \, k, f.$$

In the most general form, one can incorporate both RA and FM into the spectrum management problem (Papandriopoulos & Evans, 2009). Let *RA* and *FM* denote the index sets of the RA and FM users, respectively. Assuming there are some users which are trying to maximize their rate, the joint RA and FM spectrum management problem is shown below. Note that the rate for all of the users in *RA* are being maximized while the rate for all of the users in *FM* are being fixed at their respective QoS data-rate requirement in order to minimize their power consumption.

$$\max_{p_f^k, \forall k, f} \quad \sum_{k \in RA} w_k R_k$$
$$\text{s.t.} \quad R_k = R_{k,\text{target}}, \, k \in FM$$
$$\sum_{f=1}^{F} p_f^k = P_{k,\max}, \, \forall \, k$$
$$0 \le p_f^k \le p_f^{k,\text{mask}}, \, \forall \, k, f.$$

If there are no users trying to maximize their rate (i.e., *RA* is the empty set), the spectrum management problem reduces to the FM case. The FM optimization problem is shown below. Here the total power consumed by all users on all frequency sub-carriers is being minimized while ensuring each user still meets their respective QoS data-rate requirements.

$$\min_{p_f^k, \forall k, f} \quad \sum_{k=1}^{K} p_f^k$$
$$\text{s.t.} \quad R_k \ge R_{k,\text{target}}, \, \forall \, k$$
$$\sum_{f=1}^{F} p_f^k = P_{k,\max}, \, \forall \, k$$
$$0 \le p_f^k \le p_f^{k,\text{mask}}, \, \forall \, k, f.$$

The DSM algorithms discussed in this book chapter can be adapted for any interference control model since they take the interference control or management into account as part of their objective functions and/or constraints.

DSM ALGORITHMS UNDER CONSIDERATION

Optimal Spectrum Balancing

Solving for the optimal power allocation of all K users on each of the F frequency sub-carriers using a brute force exhaustive search has a computational complexity which is exponential in both the number of users and the number of frequency sub-carriers. OSB begins by re-writing the RA optimization problem, shown in (3), as a weighted sum-rate optimization problem shown in (4) (Cendrillon et al., 2006). The OSB algorithm then uses the concept of dual decomposition to eliminate the exponential complexity in terms of the number of frequency subcarriers, F, and converts it into a linear complexity in terms of the number of frequency sub-carriers, F. OSB converts the constrained optimization problem, (4), into an unconstrained optimization problem using duality theory. Interested readers are referred to Chapter 5 of Boyd & Vandenberghe (2004) for more details on duality theory. Therefore, the primal problem, (4), is replaced by the dual problem:

$$\max_{p_f^k, \forall k, f} \quad L\left(\mathbf{p}^1, ..., \mathbf{p}^K\right) \tag{5}$$

where $L(\mathbf{p}^1, ..., \mathbf{p}^K)$ is given by:

$$L\left(\mathbf{p}^1, ..., \mathbf{p}^K\right) = \sum_{k=1}^{K} w_k R_k - \sum_{k=1}^{K} \sum_{f=1}^{F} \lambda_k p_f^k .$$

The Lagrange multipliers, $w_1, ..., w_K, \lambda_1, ..., \lambda_K$, are chosen such that the KKT conditions are satisfied:

$$\lambda_k \left(P_{k,\max} - \sum_{f=1}^{F} p_f^k \right) = 0, \quad \forall\, k$$

$$w_k \left(R_{k,\text{target}} - \sum_{f=1}^{F} b_f^k \right) = 0, \quad \forall\, k \tag{6}$$

provided that the KKT conditions are satisfied, the dual problem, (5), is equivalent to the primal optimization problem, (4). Since synchronous transmission is assumed, there is no ICI and the Lagrangian function can be decomposed over the frequency sub-carriers. This reduces the computational complexity of the algorithm from exponential to linear in the number of frequency sub-carriers, since the exhaustive search can be performed independently on each sub-carrier. Therefore, OSB reduces a *KF*-dimensional exhaustive search into *F* different *K*-dimensional exhaustive searches,

$$\max_{p_f^1, ..., p_f^K} \quad L_f\left(p_f^1, ..., p_f^K\right), \tag{7}$$

where $L_f(p_f^1, ..., p_f^K)$ is given by:

$$L_f\left(p_f^1, ..., p_f^K\right) = f_s \sum_{k=1}^{K} w_k b_f^k - \sum_{k=1}^{K} \lambda_k p_f^k .$$

Solving (7) over each frequency sub-carrier is equivalent to solving (5) since:

$$L\left(\mathbf{p}^1, ..., \mathbf{p}^K\right) = \sum_{f=1}^{F} L_f\left(p_f^1, ..., p_f^K\right).$$

The process of writing out the Lagragian dual problem and decomposing the Lagrangian across all frequency sub-carriers is known as dual decomposition. The full OSB algorithm is outlined in Algorithm 1.

One method for specifying the choice of ε is given by Tsiaflakis, Vangorp, Moonen, Verlinden, & Van Acker (2006). The spectral masks can be easily incorporated into the OSB algorithm by setting the value of L_k to a very large negative number if $p_f^k > p_f^{k,\text{mask}}$ for each user k.

OSB reduces the complexity to linear in terms of the number of sub-carriers. As such, OSB becomes computationally tractable for large number of sub-carriers; however, OSB is still computationally intractable for large numbers of users. As such, OSB serves as an upper-bound with which to compare the performance of other DSM algorithms for few users.

Iterative Spectrum Balancing

Since OSB is computationally intractable for large number of users, Cendrillon & Moonen (2005) and Lui & Yu (2005) proposed an iterative algorithm which is tractable for large numbers of users. Iterative Spectrum Balancing (ISB) operates in a similar fashion to OSB by re-writing optimization problem (3) as optimization problem (4). ISB also uses the concept of dual decomposition, as discussed in the OSB section, to re-write the optimization problem as (5).

The difference lies in the fact that in ISB, the transmit power of each user is searched for in an iterative fashion by updating one user's power at a time while keeping the other user's powers fixed,

$$\max_{p_f^k} \quad L_f\left(p_f^1,...,p_f^K\right), \tag{8}$$

where $L_f(p_f^1,...,p_f^K)$ is given by:

$$L_f\left(p_f^1,...,p_f^K\right) = f_s \sum_{k=1}^{K} w_k b_f^k - \sum_{k=1}^{K} \lambda_k p_f^k .$$

Algorithm 1. OSB algorithm

```
Let ε > 0 be given ;
Repeat
        For each f = 1, ..., F
                Solve (7) by K-dimensional exhaustive search ;
        End
        For each k = 1, ..., K
```

$$\lambda_k = \max\left\{\lambda_k + \varepsilon\left(P_{k,\max} - \sum_{f=1}^{F} p_f^k\right), 0\right\} ;$$

$$w_k = \max\left\{w_k + \varepsilon\left(R_{k,\text{target}} - f_s \sum_{f=1}^{F} b_f^k\right), 0\right\} ;$$

```
        End
Until p_f^k converges for all f and k ;
```

Note that in (8), since the maximization is over a single user's power, each user is performing one-dimensional searches for their own power while keeping the power of other users fixed. As with OSB, the KKT conditions outlined in (6) must still be satisfied to ensure the dual and primal problems are equivalent. The full ISB algorithm is outlined in Algorithm 2.

As with OSB, the spectral masks can be easily incorporated into the algorithm by setting the value of L_k to a very large negative number if $p_f^k > p_f^{k,\text{mask}}$ for each user k. While ISB is not guaranteed to find optimal power allocations, it has been shown to lead to optimal power allocations in many test cases, especially with few users. Since the complexity of ISB is quadratic in the number of users and linear in the number of sub-carriers, it has significant computational complexity advantages over OSB and as such is computationally tractable for scenarios involving more users. As such, it can be used as a comparative measure since its performance has been shown to be close to optimal.

Iterative Water-Filling

Water-Filling (WF) is the process by which one user attempts to optimize its power allocation regardless of the effect on other users. More specifically, water-filling attempts to solve the following maximization problem:

$$\max_{p_1^k,\dots,p_F^k} \quad R_k$$

$$\text{s.t.} \quad \sum_{f=1}^{F} p_f^k = P_{k,\max},$$

$$0 \le p_f^k \le p_f^{k,\text{mask}}.$$

Algorithm 2. ISB algorithm

```
Let ε > 0 be given ;
Repeat
      For each k = 1, …, K
            Repeat
                  For each f = 1, …, F
                        Fix pfˡ for l ≠ k ;
                        Solve (8) by one-dimensional exhaustive search ;
                  End
                  For each k = 1, …, K
```

$$\lambda_k = \max\left\{\lambda_k + \varepsilon\left(P_{k,\max} - \sum_{f=1}^{F} p_f^k\right), 0\right\} ;$$

$$w_k = \max\left\{w_k + \varepsilon\left(R_{k,\text{target}} - f_s \sum_{f=1}^{F} b_f^k\right), 0\right\} ;$$

```
                  End
            Until w_k and λ_k converge ;
      End
```

Since the maximization problem is concave, it can be solved by setting the derivative of the Lagrangian to zero, resulting in the following power update formula:

$$p_f^k = \left[\frac{1}{\lambda} - \frac{\sum_{l \neq k} \left| h_f^{k,l} \right|^2 p_k^l + \sigma_f^k}{\left| h_f^{k,k} \right|^2 / \Gamma} \right]_0^{p_f^{k,mask}}, \tag{9}$$

where $[a]_b^c = \min\{\max\{a, b\}, c\}$. The optimal value of λ can be easily found using a bisection search, as shown in Algorithm 3.

The Iterative Water-Filling (IWF) algorithm iteratively performs WF one user at a time. The users continuously perform WF in turn until a Nash Equilibrium (NE) point is reached. A NE point is a point where no user can benefit from changing its power allocation.

When applying rate constraints, each user adjusts their WF process to ensure that they achieve their target rate. Hence, if a user achieves a rate higher than its target rate, that user's allowable total power is reduced. Similarly, if a user achieves a rate lower than its target rate, that user's allowable total power is increased. Note that the allowable total power can never exceed the total power constraint. Therefore, the target rates for the users must be feasible in order for the algorithm to converge. The IWF algorithm is shown in Algorithm 4.

Algorithm 3. Single-user WF algorithm (for user k)

```
Let  ε > 0 be given ;
Initialize:
p_f^k = 0, for all f ;
λ_min = 0 ;
λ_max = 1 ;
Update power using (9) with λ = λ_max ;
While Σ_f p_f^k > P_k,max do
      λ_max = 2 λ_max ;
      Update power using (9) with λ = λ_max ;
End
Repeat
      λ = (λ_max + λ_min) / 2 ;
      Update power using (9) with λ ;
      If Σ_f p_f^k > P_k,max
            λ_min = λ ;
      Else
            λ_max = λ ;
      End
Until P_k,max - Σ_f p_f^k ≤ ε ;
```

Algorithm 4. IWF algorithm

```
Let ΔP > 0 be given ;
Initialize:
pᶠᵏ = 0, for all f, k ;
Ψ_{k,max} = P_{k,max}, for all k ;
Repeat
      For each k = 1, … , K
              Perform single-user WF on user k ;
              If R_k > R_{k,target}
                      Ψ_{k,max} = Ψ_{k,max} - ΔP ;
              Elseif R_k < R_{k,target}
                      Ψ_{k,max} = Ψ_{k,max} + ΔP ;
              End
              If Ψ_{k,max} > P_{k,max}
                      Ψ_{k,max} = P_{k,max}
              End
      End
Until pᶠᵏ converges for all k, f ;
```

Selective Iterative Water-Filling

Selective Iterative Water-filling (SIW) takes advantage of the fact that IWF with rate constraints forces users with better channels to restrict their total power consumption. As well, IWF is limited by the fact that each user has only has a single WF level on each frequency sub-carrier. SIW attempts to remove these restrictions by allowing some users to perform another IWF process. More specifically, at the end of every IWF process, SIW allows users who did not use up all their transmit power to perform another IWF process without affecting the users that already used up their transmit power. This allows some users to increase their WF level on some frequency sub-carriers without affecting the other users, which gives performance benefits over IWF. As such, SIW allows for users to make use of their excess allowable total power that is wasted using IWF.

The SIW algorithm is shown in Algorithm 5. The modified IWF algorithm used by SIW must guarantee that at least one user is using all of its allowable power. This ensures that at least one user is removed from the process after each outer iteration. One method for accomplishing this is by maximizing one user's rate while forcing the remaining users to meet their target rates.

At the end of each modified IWF process, the SIW algorithm removes users that have used up all of their allowable power for the next round. As well, the frequency sub-carriers used by those users are also removed. As such, future rounds do not affect the achievable rate of the users removed from the process. The process is repeated until either all frequency sub-carriers or all users have been removed.

Algorithm 5. SIW algorithm

```
Initialize:
Ψ_{k,max} = P_{k,max}, for all k ;
K = {1, … , K} be the set of all users ;
F = {1, … , F} be the set of all sub-carriers;

While K ≠ {} and F ≠ {}
        Perform modified IWF with K, F, and power constraints Ψ_{k,max} for all k ;
        For each k ∈ K
                If Σ_f p_f^k = P_{k,max}
                        K = K \ {k} ;
                        For each f ∈ F
                                If p_f^k ≠ 0
                                        F = F \ {f} ;
                                End
                        End
                End
        End
        For each k ∈ K
                P_{k,used} = Σ_{f ∈{1,…,F}\ F} p_f^k ;
                Ψ_{k,max} = P_{k,max} - P_{k,used} ;
        End
End
```

The SIW algorithm is guaranteed to increase the rate of IWF; however, it requires a message passing system to identify the removed users and sub-carriers. As such, SIW must operate in a semi-centralized manner, while IWF can operate in a distributed manner.

Successive Convex Approximation for Low Complexity

The Successive Convex Approximation for Low complExity (SCALE) algorithm, introduced by Papandriopoulos & Evans (2006, 2009), solves the weighted sum-rate optimization problem, (4), by approximating the non-convex objective function with a concave lower-bound, maximizing the approximation and repeating the process with another approximation. SCALE introduces a method of distributing the processing over multiple users, which allows for the algorithm to be run in a semi-centralized manner.

SCALE uses the following approximation to $\log(1 + z)$ around the point z_0: $\log(1 + z) \geq \alpha \log(z) + \beta$, where $\alpha = z_0 / (1 + z_0)$ and $\beta = \log(1 + z_0) - z_0 \log(z_0) / (1 + z_0)$. The inequality is tight when $z = z_0$. By applying the inequality to the sum-rate equation and replacing the variables p_f^k with $\exp(\rho_f^k)$, a concave maximization problem is obtained:

$$\max_{\rho^k_f, \forall k,f} \quad \sum_{k=1}^{K}\sum_{f=1}^{F} w_k \alpha^k_f \log_2 \left(\frac{\left|h^{k,k}_f\right|^2 \exp\left(\rho^k_f\right)/\Gamma}{\sum_{l\neq k}\left|h^{k,l}_f\right|^2 \exp\left(\rho^l_f\right) + \sigma^k_f} \right) + \beta^k_f$$

$$\text{s.t.} \quad \sum_{f=1}^{F} \exp\left(\rho^k_f\right) = P_{k,\max}, \ \forall\, k$$

$$0 \leq \exp\left(\rho^k_f\right) \leq p^{k,\text{mask}}_f, \ \forall\, k,f,$$

which can be re-written as:

$$\max_{\rho^k_f, \forall k,f} \quad \sum_{k=1}^{K}\sum_{f=1}^{F} w_k \alpha^k_f \left[\log_2\left(\left|h^{k,k}_f\right|^2/\Gamma\right) + \rho^k_f - \log_2\left(\sum_{l\neq k}\left|h^{k,l}_f\right|^2 \exp\left(\rho^l_f\right) + \sigma^k_f\right)\right] + \beta^k_f$$

$$\text{s.t.} \quad \sum_{f=1}^{F} \exp\left(\rho^k_f\right) = P_{k,\max}, \ \forall\, k \tag{10}$$

$$0 \leq \exp\left(\rho^k_f\right) \leq p^{k,\text{mask}}_f, \ \forall\, k,f.$$

Since (10) is concave, any convex optimization tool can be used to solve the optimization problem; however, Papandriopoulos & Evans (2006, 2009) proposed a gradient approach that allows for the computations to be distributed among each user. The gradient approach is applied to the Lagrangian function shown below:

$$L = \sum_{k=1}^{K}\sum_{f=1}^{F} w_k \alpha^k_f \left[\log_2\left(\left|h^{k,k}_f\right|^2/\Gamma\right) + \rho^k_f - \log_2\left(\sum_{l\neq k}\left|h^{k,l}_f\right|^2 \exp\left(\rho^l_f\right) + \sigma^k_f\right)\right] + \beta^k_f - \sum_{k=1}^{K}\lambda_k\left(\sum_{f=1}^{F}\exp\left(\rho^k_f\right) - P_{k,\max}\right)$$

By setting the gradient of the Lagrangian to zero, the following power update formula for the *k*-th user on the *f*-th frequency sub-carrier is obtained:

$$p^k_f = \frac{w_k \alpha^k_f}{\lambda_k + \sum_{l\neq k}\dfrac{\left|h^{k,l}_f\right|^2 w^l \alpha^l_f}{\sum_{q\neq l}\left|h^{l,q}_f\right|^2 p^q_f + \sigma^l_f}} \tag{11}$$

Since the denominator in (11) requires information from other users, the algorithm is capable of converging to a locally optimal point; however, the SCALE algorithm must operate in a semi-centralized manner, making use of a message passing system so that each user can acquire information from other users. Papandriopoulos & Evans (2006, 2009) proposed the following message passing system: during the first message passing phase, every user measures their total interference plus noise on every sub-carrier and transmits the information to a Spectrum Management Center (SMC). The SMC gathers the data from all users and sends each user their respective new message, which is computed using the data acquired by the SMC through the first message passing phase.

As such each user updates their power using the following update formula:

$$p_f^k = \frac{w_k \alpha_f^k}{\lambda_k + M_f^k},$$
(12)

where M_f^k is the message sent by the SMC to the k-th user regarding the f-th sub-carrier (during the second message passing phase) and is defined as:

$$M_f^k = \sum_{l \neq k} \left| h_f^{k,l} \right|^2 N_f^l,$$
(13)

where N_f^l is the message sent by the k-th user regarding the f-th sub-carrier (during the first message passing phase) and is defined as:

$$N_f^l = \frac{w^l \alpha_f^l}{\sum_{q \neq l} \left| h_f^{l,q} \right|^2 p_f^q + \sigma_f^l}.$$
(14)

The SCALE message passing system can be summarized as follows. For each iteration, every user k, computes their respective N_f^k on each sub-carrier, f, using (14) and sends it to the SMC. The SMC produces the M_f^k values using (13) and distributes them to each user k. The full SCALE algorithm is summarized in Algorithm 6.

Algorithm 6. SCALE algorithm

```
Let ζ be a given integer ;
At each user k:
Initialize:
p_f^k = 0, for all f ;
α_f^k = 1, for all f ;
Repeat
        Receive M_f^k from SMC ;
        Update power using (12) ;
        Every ζ iterations, update α_f^k ;
        Generate N_f^k using (14) and send to SMC ;
Indefinitely ;

At the SMC:
Repeat
        Receive N_f^k from each user ;
        Generate M_f^k using (13) and send to each user k ;
Indefinitely ;
```

Difference of Convex Functions Algorithm

The Difference of Convex Functions Algorithm (DCA) was originally developed by Pham & Hoai An (1997). DCA is based on the local optimality conditions of a minimization problem involving convex constraints and an objective function which can written in the form of a Difference of Convex functions (DC), $f = g - h$, where g and h are both convex. The general DCA approach makes use of the concept of sub-differentials, which generalizes the derivative at points where the objective function is non-differentiable. However, since for DSM optimization problems the objective functions are differentiable, when applying DCA to the DSM problem, this chapter uses derivates instead of sub-differentials.

Starting from an initial point, $\mathbf{p}^{(0)} = \{p_f^k,$ for $f = 1, \ldots, F,$ and $k = 1, \ldots, K\}$, two sequences $\mathbf{p}^{(i)}$ and $\mathbf{q}^{(i)}$ are constructed such that:

$$\mathbf{q}^{(i)} \in \nabla h\left(\mathbf{p}^{(i)}\right),$$
$$\mathbf{p}^{(i+1)} \in \nabla g^*\left(\mathbf{q}^{(i)}\right),$$

where g^* is the conjugate of g. Recall that g and h come from the DC decomposition of the objective function, f.

The first step of applying DCA to the DSM problem is to re-write optimization problem (4) as a minimization problem:

$$\min_{p_f^k, \forall k, f} \quad -\sum_{k=1}^{K} w_k R_k$$
$$\text{s.t.} \quad \sum_{f=1}^{F} p_f^k = P_{k,\max}, \ \forall \ k \tag{15}$$
$$0 \le p_f^k \le p_f^{k,\text{mask}}, \ \forall \ k, f.$$

Next, the objective function, $f = -\sum_k w_k R_k$, must be re-written as a DC. For simplicity, the following variables are defined:

$$\left|\tilde{h}_f^{k,l}\right|^2 = \begin{cases} \left|h_f^{k,k}\right|^2 \Big/ \Gamma, & \text{if } l = k \\ \left|h_f^{k,l}\right|^2, & \text{if } l \ne k \end{cases}$$
$$A_f^k\left(\mathbf{p}\right) = \sum_{l \ne k} \left|\tilde{h}_f^{k,l}\right|^2 p_f^l + \sigma_f^k,$$
$$B_f^k\left(\mathbf{p}\right) = A_f^k\left(\mathbf{p}\right) + \left|\tilde{h}_f^{k,k}\right|^2 p_f^k.$$

Then, the objective function of the DCA optimization problem can be re-written as: $f(\mathbf{p}) = g(\mathbf{p}) - h(\mathbf{p})$ where $g(\mathbf{p})$ and $h(\mathbf{p})$ are given by:

$$g\left(\mathbf{p}\right) = -f_s\sum_{f=1}^{F}\sum_{k=1}^{K} w_k \log_2\left(B_f^k\left(\mathbf{p}\right)\right)$$

$$h\left(\mathbf{p}\right) = -f_s\sum_{f=1}^{F}\sum_{k=1}^{K} w_k \log_2\left(A_f^k\left(\mathbf{p}\right)\right)$$

Since the summation of convex functions is a convex function and due to the fact that the negative log-sum is convex, and that w_k and f_s are positive constants, it follows that both $g(\mathbf{p})$ and $h(\mathbf{p})$ are convex functions.

The partial derivatives of $g(\mathbf{p})$ and $h(\mathbf{p})$ are given below:

$$\frac{\partial g\left(\mathbf{p}\right)}{\partial p_{\tilde{f}}^{\tilde{k}}} = -f_s\sum_{k=1}^{K} \frac{w_k}{\ln\left(2\right)} \frac{\left|\tilde{h}_{\tilde{f}}^{k,\tilde{k}}\right|^2}{A_f^k\left(\mathbf{p}\right)}, \tag{16}$$

$$\frac{\partial h\left(\mathbf{p}\right)}{\partial p_{\tilde{f}}^{\tilde{k}}} = -f_s\sum_{k\neq\tilde{k}} \frac{w_k}{\ln\left(2\right)} \frac{\left|\tilde{h}_{\tilde{f}}^{k,\tilde{k}}\right|^2}{B_f^k\left(\mathbf{p}\right)}. \tag{17}$$

Note that the denominators of (16) and (17) require evaluating $(K-1)$- and K-dimensional summations, respectively.

While there are many different DCA approaches that can be taken to solve this optimization problem, the approach presented in this chapter was presented by Pham & Hoai An (1997) and Blanquero & Carrizosa (2000). The approach involves creating a new DC decomposition to simplify the sub-problem calculation by defining a new objective function as:

$$f_2\left(\mathbf{p}\right) = \tfrac{1}{2}\,\xi\left\|\mathbf{p}\right\|^2 - \left(\tfrac{1}{2}\,\xi\left\|\mathbf{p}\right\|^2 - f\left(\mathbf{p}\right)\right),$$

where $\xi > 0$ to enforce convexity on $g_2(\mathbf{p}) = \tfrac{1}{2}\xi\|\mathbf{p}\|^2$. If $\xi > \|\nabla^2 g(\mathbf{p})\|_\infty$, then $h_2(\mathbf{p}) = \tfrac{1}{2}\xi\|\mathbf{p}\|^2 - f(\mathbf{p})$ is convex over the region of interest (Huberman, Leung, & Le-Ngoc, 2012b). The DCA algorithm is applied to the new DC decomposition, $f_2(\mathbf{p})$, which converts a difficult non-linear optimization problem into a simple quadratic optimization problem, which is a well-researched field containing many algorithms to find solutions.

As such, on the i-th iteration, the next term in the sequence $\mathbf{q}^{(i)}$ can be solved for using:

$$\mathbf{q}^{(i)} = \nabla\tilde{h}\left(\mathbf{p}^{(i)}\right) = \xi\,\mathbf{p}^{(i)} - \nabla f_2\left(\mathbf{p}^{(i)}\right). \tag{18}$$

As well, on the i-th iteration, the next term in the sequence can be found by solving the following convex optimization problem:

$$\min_{\mathbf{p}^{(i-1)}} \quad \tfrac{1}{2}\xi\left\|\mathbf{p}^{(i+1)}\right\|^2 - \left\langle \mathbf{p}^{(i+1)}, \mathbf{q}^{(i)}\right\rangle$$

$$\text{s.t.} \quad \sum_{f=1}^{F}\left(p^{(i+1)}\right)_f^k = P_{k,\max}, \ \forall \ k \tag{19}$$

$$0 \le \left(p^{(i+1)}\right)_f^k \le \left(p^{(i+1)}\right)_f^{k,\mathrm{mask}}, \ \forall \ k, f,$$

where $(p^{(i+1)})_f^k$ refers to the transmit power in the vector $\mathbf{p}^{(i+1)}$ associated with the k-th user on the f-th sub-carrier. Hence, the DCA approach is summarized in Algorithm 7.

Distributed Spectrum Balancing

Tsiaflakis et al. (2008) proposed the Distributed Spectrum Balancing (DSB) algorithm by applying the KKT conditions of the DC problem directly. In contrast to the DCA approach, the DSB algorithm can operate in a semi-centralized manner making use of message passing, while the DCA approach requires centralized processing. Optimization problem (4) can be re-written as follows:

$$\max_{p_f^k, \forall k, f} \left\{ \sum_{f=1}^{F}\sum_{k=1}^{K} w_k \log_2\left(\Gamma\sum_{l\ne k}\left|h_f^{k,l}\right|p_f^l + \left|h_f^{k,k}\right|p_f^k + \Gamma\sigma_f^k\right) \right.$$

$$\left. -\sum_{f=1}^{F}\sum_{k=1}^{K} w_k \log_2\left(\Gamma\sum_{l\ne k}\left|h_f^{k,l}\right|p_f^l + \Gamma\sigma_f^k\right)\right\} \tag{20}$$

$$\text{s.t.} \quad \sum_{f=1}^{F} p_f^k = P_{k,\max}, \ \forall \ k$$

$$0 \le p_f^k \le p_f^{k,\mathrm{mask}}, \ \forall \ k, f.$$

By setting the derivative of the Lagrangian function to zero, the following equation can be derived in order to satisfy the KKT conditions for all k and f.

Algorithm 7. DC algorithm

```
Initialize p⁽⁰⁾ to a best guess ;
Initialize i = 0 ;
Repeat
      Calculate q⁽ⁱ⁾ using (18) ;
      Calculate p⁽ⁱ ⁺ ¹⁾ using (19) ;
Until p⁽ⁱ⁾ converges ;
```

$$0 = \lambda_k + \frac{w_k f_s \left|h_f^{k,k}\right|^2 / \ln(2)}{\left|h_f^{k,k}\right|^2 \underbrace{p_f^k}_{*} + \sum_{l \neq k} \Gamma \left|h_f^{k,l}\right|^2 p_f^l + \Gamma \sigma_f^k}$$

$$+ \sum_{l \neq k} \frac{w_l f_s \Gamma \left|h_f^{l,k}\right|^2 / \ln(2)}{\sum_{q \neq l} \Gamma \left|h_f^{l,q}\right|^2 p_f^q + \left|h_f^{l,l}\right|^2 p_f^l + \Gamma \sigma_f^l} - \sum_{l \neq k} \frac{w_l f_s \left|h_f^{l,k}\right|^2 / \ln(2)}{\sum_{q \neq l} \left|h_f^{l,q}\right|^2 p_f^q + \sigma_f^l} \qquad (21)$$

The DSB power update formula can be found by isolating the p_f^k term indicated by the (*) in (21), resulting in:

$$p_f^k = \left[\frac{w_k f_s / \ln(2)}{\lambda_k + \Psi_f^{\text{DSB},k}} - \frac{\text{int}_f^k}{\left|h_f^{k,k}\right|^2} \right]_0^{p_f^{k,\text{mask}}}, \qquad (22)$$

where $\Psi_f^{\text{DSB},k}$ and int_f^k are given by:

$$\Psi_f^{\text{DSB},k} = \sum_{l \neq k} \frac{w_l f_s \Gamma \left|h_f^{l,k}\right|^2}{\ln(2)} \left(\frac{1}{\text{int}_f^k} - \frac{1}{\text{rec}_f^k} \right),$$

$$\text{int}_f^k = \sum_{l \neq k} \Gamma \left|h_f^{k,l}\right|^2 p_f^l + \Gamma \sigma_f^k,$$

$$\text{rec}_f^k = \text{int}_f^k + \left|h_f^{k,k}\right|^2 p_f^k.$$

The DSB algorithm applies the following message passing protocol. First, each user k measures their respective int_f^k for $f = 1, \ldots, F$ and sends the message $((\text{int}_f^k)^{-1} - (\text{rec}_f^k)^{-1})$ to the SMC. The SMC computes the corresponding $\Psi_f^{\text{DSB},k}$ for each k and f and sends the values to each user. Finally, each user updates their transmit power using (22). The DSB algorithm is summarized in Algorithm 8.

While the DSB algorithm has been shown to provide significant performance improvements over some of the existing techniques, its performance can depend on the choice of initial point. As such, Tsiaflakis et al. (2008) extended the DSB algorithm to the Multiple Starting point DSB (MS-DSB). For each sub-carrier, the MS-DSB algorithm tries multiple starting points, runs a few iterations with each one and then continues with the one corresponding to the best performance based on the those few iterations. This approach increases the probability of improving the locally optimal solution on a per-sub-carrier basis. Tsiaflakis et al. (2008) proposed testing $K + 2$ different initial transmit powers for each sub-carrier, corresponding to three specific cases. The first case being the all-zeros transmit power, the second being all transmit powers set to $p_f^{k,\text{mask}}/2$ (as in DSB), and the third being where each user has a zero transmit power except one user has their power initialized to 1 (i.e., the k-th unit vector). Intuitively these cases correspond to the cases of no users transmitting, all users transmitting equally, and each user transmitting while the other users remain silent. After the initial point selection phase, the MS-DSB algorithm operates identically to the DSB algorithm.

Algorithm 8. DSB algorithm

```
At each user k:
Initialize:
p_f^k = p_f^{k,mask}/2, for all f ;
Repeat
      Receive Ψ_f^{DSB,k} from SMC ;
      Update power using (22)[REMOVED REF FIELD][REMOVED REF FIELD][REMOVED
REF FIELD][REMOVED REF FIELD][REMOVED REF FIELD] ;
```

$$\text{Generate } \left(\frac{1}{\text{int}_f^k} - \frac{1}{\text{rec}_f^k} \right) \text{ and send to SMC ;}$$

```
Indefinitely ;


At the SMC:
Repeat
```

$$\text{Receive } \left(\frac{1}{\text{int}_f^k} - \frac{1}{\text{rec}_f^k} \right) \text{ from each user ;}$$

```
      Generate Ψ_f^{DSB,k} and send to each user k ;
Indefinitely ;
```

Autonomous Spectrum Balancing

Cendrillon et al. (2007) proposed an algorithm referred to as Autonomous Spectrum Balancing (ASB) which operates in a distributed manner and requires no per-iteration message passing. The key concept behind ASB is that each user attempts to maximize the throughput of the system while ensuring their performance is above some threshold value. The result is a more selfless system and hence, the overall performance is expected to improve since each user is attempting to act more selflessly.

The ASB algorithm uses the concept of a virtual reference user which attempts to mimic a typical victim user in the network. Each user ensures their respective target rate constraint is met while trying to minimize the damage done to the reference user. The reference user must be predetermined but makes use of the network statistics to determine the reference user parameters autonomously. The ASB algorithm does not specify how one should choose the reference user; hence, the reference user selection process must be done in a heuristic manner. Some choices of reference users could include the user which is expected to have the weakest channel gain or is expected to be most sensitive to interference from other users. Once the reference user is selected, each user can self-optimize using locally available information allowing ASB to operate in a distributed manner while optimizing the network.

From the k-th user's perspective, the reference user's rate is defined as:

$$R_{k,\text{ref}} = f_s \sum_{f=1}^{F} b_f^{k,\text{ref}} ,$$

where

$$b_f^{k.\text{ref}} = \log_2 \left(1 + \frac{1}{\Gamma} \frac{\left| h_f^{\text{ref}} \right|^2 p_f^{\text{ref}}}{\left| h_f^{\text{ref}.k} \right|^2 p_f^k + \sigma_f^{\text{ref}}} \right),$$

and where p_f^{ref} is the reference user's *virtual* transmit power, $|h_f^{\text{ref}}|^2$ is the reference user's *virtual* direct channel gain, $|h_f^{\text{ref},\,k}|^2$ is the *virtual* channel gain from the k-th user to the reference user, and σ_f^{ref} is the *virtual* noise power of the reference user.

The ASB per-user optimization problem is formulated as follows for the k-th user:

$$\begin{aligned} \max_{p_f^k,\,\forall f} \quad & R_{k.\text{ref}} \\ \text{s.t.} \quad & R_k \geq R_{k.\text{target}}, \\ & \sum_{f=1}^{F} p_f^k = P_{k.\max}, \\ & 0 \leq p_f^k \leq p_f^{k.\text{mask}}, \; \forall f. \end{aligned} \tag{23}$$

Each user k solves a different version of optimization problem (23). Cendrillon et al. (2007) solved (23) by applying a two-user dual decomposition approach. The Lagrange function was written out as:

$$L_f^k = w_k b_f^k + w_{k.\text{ref}} b_f^{k.\text{ref}} - \lambda_k p_f^k,$$

where $w_{k.\text{ref}}$ is the weight of the reference user, which is set to $1 - w_k$ when applying rate constraints and set to the weight of the virtual user it is attempting to emulate if rate constraints are not used. As such, (23) can be solved by taking the partial derivative of L_f^k with respect to p_f^k and setting it equal to zero.

$$\lambda_k + \frac{w_k f_s \left| h_f^{k,k} \right|^2 / \ln(2)}{\left| h_f^{k,k} \right|^2 \underbrace{p_f^k}_{*} + \sum_{l \neq k} \Gamma \left| h_f^{k,l} \right|^2 p_f^l + \Gamma \sigma_f^k} - \frac{w_{k.\text{ref}} f_s \left| h_f^{\text{ref}} \right|^2 p_f^{\text{ref}} \Gamma \left| h_f^{\text{ref},k} \right|^2 / \ln(2)}{\left(\Gamma \left| h_f^{\text{ref},k} \right|^2 p_f^k + \Gamma \sigma_f^{\text{ref}} \right) \left(\left| h_f^{\text{ref}} \right|^2 p_f^{\text{ref}} + \Gamma \left| h_f^{\text{ref},k} \right|^2 p_f^k + \Gamma \sigma_f^{\text{ref}} \right)} = 0$$

This results in a cubic equation in terms of p_f^k and hence has three roots that can be solved for in closed form. Each of these roots and the boundary conditions must be tested (i.e., $p_f^k = 0$ and $p_f^k = p_f^{k.\text{mask}}$ or $p_f^k = P_{k.\max}$ if no spectral mask is used). The optimal p_f^k is selected as a feasible power corresponding to the maximum value of L_f^k.

In order to reduce the complexity and computational time associated with the ASB power update, Tsiaflakis et al. (2008) proposed an algorithm they refer to as ASB-2, which operates identically to the ASB algorithm except that it uses an simplified transmit power update formula. In particular, the ASB-

2 update formula is based on re-arranging the partial derivative of L_f^k with respect to p_f^k to the p_f^k term indicated by (*). The p_f^k value from the previous iteration are used for the other p_f^k terms and are denoted by $(p_{\text{prev}})_f^k$. Hence, the ASB-2 spectrum update formula is given by:

$$
p_f^k = \left[\frac{w_k \, f_s / \ln(2)}{\lambda_k + \Psi_f^{ASB-2.k}} - \frac{\sum_{l \neq k} \Gamma \left| h_f^{k,l} \right|^2 p_f^l + \Gamma \sigma_f^k}{\left| h_f^{k,k} \right|^2} \right]_0^{p_f^{k,\text{mask}}}, \tag{24}
$$

where $\Psi_f^{\text{ASB-2},k}$ is given by:

$$
\Psi_f^{\text{ASB-2.k}} = \frac{w_{k.\text{ref}} f_s \left| h_f^{\text{ref}} \right|^2 p_f^{\text{ref}} \Gamma \left| h_f^{\text{ref}.k} \right|^2 / \ln(2)}{\left(\Gamma \left| h_f^{\text{ref}.k} \right|^2 \left(p_{\text{prev}} \right)_f^k + \Gamma \sigma_f^{\text{ref}} \right) \left(\left| h_f^{\text{ref}} \right|^2 p_f^{\text{ref}} + \Gamma \left| h_f^{\text{ref}.k} \right|^2 \left(p_{\text{prev}} \right)_f^k + \Gamma \sigma_f^{\text{ref}} \right)}.
$$

The ASB and ASB-2 solutions do not necessarily converge to the same solution; however, the ASB-2 approach requires significantly less computational time per iteration. After solving (23) using either the ASB or ASB-2 approach, the interference values in the network change, as such each user repeatedly updates their transmit power by solving (23) until convergence. The ASB and ASB-2 algorithms are summarized in Algorithm 9 using a simple bisection search approach.

Note that the parameter $\Lambda_{k,\text{max}}$ must be found prior to performing the optimization. One method for solving for $\Lambda_{k,\text{max}}$ is to initialize it to 1 and then continuously increase it (e.g., doubling it) until the allocated power is less than or equal to the total power constraint. This will ensure that λ_k has a feasible solution.

Constant Offset ASB Using Multiple Reference Users

The constant offset Autonomous Spectrum Balancing using Multiple Reference Users (ASB-MRU) algorithm was proposed by Huberman et al. (2012a). There are several key components to the constant offset ASB-MRU approach.

First, the algorithm makes use of a constant Lagrange multiplier offset. Earlier work showed how the use of a Lagrange multiplier offset (e.g., DSB, SCALE, ASB-2) could significantly improve the performance of the iterative water-filling approach. The Lagrange multiplier offsets represent per-frequency sub-carrier penalties or quotas and hence, by effectively selecting or tuning the offsets, the water-filling power allocation process can be influenced resulting in a more effective use of the frequency spectrum and leading to better performance.

One limitation of the existing approaches was that they required that the Lagrange multiplier offset be tuned on a per-iteration basis. As such, often the existing approaches exhibited slow convergence (i.e., they require many iterations). The use of constant Lagrange multiplier offsets avoids the tuning process and hence, allows for significantly faster convergence (i.e., it requires fewer iterations).

Next, since the Lagrange multiplier offsets remain constant throughout the duration of the optimization process, an approach for selecting each user's constant offsets was required. The constant offset ASB-MRU algorithm generates the constant offsets using the concept of a *virtual network* of reference

Algorithm 9. ASB and ASB-2 algorithm

```
Let ε_R > 0, ε_P > 0, and Λ_{k,max} > 0 be given ;
Initialize: p_f^k = P_{k,max}/F, for all k, f ;
Repeat
        For each k do
                Initialize w_{k,min} = 0, w_{k,max} = 1 ;
                While | Σ_f b_f^k - R_{k,target} | > ε_R
                      w_k = (w_{k,min} + w_{k,max})/2 ;
                      Initialize λ_{k,min} = 0, λ_{k,max} = Λ_{k,max} ;
                      While | Σ_f p_f^k - P_{k,max} | > ε_P and Σ_f p_f^k ≤ P_{k,max}
                            λ_k = (λ_{k,min} + λ_{k,max})/2 ;
                            Solve (23) for p_f^k for all f using either the ASB or
ASB-2 approach ;
                            If Σ_f p_f^k > P_{k,max}
                                  λ_{k,min} = λ_k ;
                            Else
                                  λ_{k,max} = λ_k ;
                            end
                      end
                      If Σ_f b_f^k > R_{k,target}
                            w_{k,max} = w_k ;
                      Else
                            w_{k,min} = w_k ;
                      end
                end
        end
Until  p_f^k converges for all k, f ;
```

users. The virtual network of reference users makes use of MRU which should be representative of the true network. Intuitively, users self-optimize with approximate global knowledge introduced by the virtual network, representing both the disturbers and the victims.

Finally, the constant offset ASB-MRU algorithm introduces a systematic approach to the selection of the virtual network of reference users and their respective parameters that can be blindly applied to any network. The constant offset ASB-MRU removes heuristic reference user selection process required by ASB and ASB-2 by making use of a clustering theory approach which builds on the approach presented by Huberman, Leung, & Le-Ngoc (2011).

The constant offset ASB-MRU algorithm uses the following spectrum update formula:

$$p_f^k = \left[\frac{w_k}{\lambda_k + \Delta\lambda_f^k} - \frac{\Gamma \operatorname{int}_f^k}{\left|h_f^{k,k}\right|^2} \right]_0^{p_f^{k,\text{mask}}} , \tag{25}$$

where $\mathrm{int}_f^k = \Sigma_{l \neq k} |h_f^{k,l}|^2 p_f^l + \sigma_f^k$. The constant offset ASB-MRU power update formula shown in (25) generalizes the power updates of IWF, DSB, and ASB-2; however, for DSB and ASB-2, the value of $\Delta \lambda_f^k$ is updated each iteration. The SCALE power update has a slightly different form but also requires the tuning of an offset on a per-iteration basis. Moreover, DSB and SCALE tune the offsets using per-iteration message passing and central computations. The constant offset ASB-MRU approach removes the offset tuning process by approximately pre-computing the converged Lagrange multiplier offsets. In particular, the virtual network of reference users is used to generate offsets that well-approximate the offsets obtained using other approaches (e.g., DSB).

The constant offset ASB-MRU algorithm is separated into four steps: setting up the virtual network of reference users, selecting each reference user's parameters, computing the constant Lagrange multiplier offsets, and finally, the optimization phase. The first three steps make up the initialization phase.

Clustering is used to determine how many and which reference users are selected in a systematic manner which can be easily applied to any network. Each user cluster is replaced by the mean representative and the weight of the reference user is selected as the mean of the users in the cluster. When the clustering algorithm returns a cluster of a single user, that reference user is doubled (i.e., two identical reference users) since each reference user requires other representative reference users to accurately estimate the corresponding constant Lagrange multiplier offsets. Let K_R be the number of virtual reference users in the virtual network of reference users, where $K_R < K$.

The clustering is applied based on a numerical value computed using a payoff function to represent how "weak" the users are (i.e., the larger the payoff function value, the weaker the user). The payoff function provides a method to systematically compare each user's relative strength/weakness based on their respective channel conditions.

The constant offset ASB-MRU payoff function was derived to be:

$$\mathrm{INTsum}(k) = \sum_{f=1}^{F} \log_2 \left(1 + \frac{\sigma_f^k + \sum_{l=k} \left| h_f^{k,l} \right|^2 p_f^{\mathrm{fixed}}}{\left| h_f^{k,k} \right|^2 p_f^{\mathrm{fixed}}} \right) = \sum_{f=1}^{F} \log_2 \left(\frac{\left(\mathrm{rec}_{\mathrm{INT}} \right)_f^k}{\left| h_f^{k,k} \right|^2 p_f^{\mathrm{fixed}}} \right), \tag{26}$$

where $(\mathrm{rec}_{\mathrm{INT}})_f^k = |h_f^{k,k}|^2 p_f^{\mathrm{fixed}} + \sigma_f^k + \Sigma_{l \neq k} |h_f^{k,l}|^2 p_f^{\mathrm{fixed}}$. $(\mathrm{rec}_{\mathrm{INT}})_f^k$ represents the full received signal for user k on sub-carrier f when p_f^{fixed} is transmitted by each user. The payoff function can be computed using locally available channel knowledge by making use of already available channel measurements and hence, does not add any significant additional complexity.

The virtual network of reference users is constructed as follows. First, the payoff function is computed for each user and sent to a SMC. The SMC clusters the users into groups and selects the corresponding virtual network.

Once the virtual network is constructed, the SMC uses standard channel models to construct the channel gains of the reference users. The reference users' virtual transmit power is determined by reusing the $\mathrm{INT}_{\mathrm{sum}}$ payoff function; however, here, the $\mathrm{INT}_{\mathrm{sum}}$ payoff function is evaluated for each *reference user*, r. Hence, $\mathrm{INT}_{\mathrm{sum}}(r)$ is defined as in (26) except that it uses only the virtual network of reference users. The resulting virtual reference user transmit power is given by single-user water-filling (i.e., ignoring the effects of interference) but scaled down by the ratio $\mathrm{INT}_{\mathrm{sum}}(r) / \max_r \{ \mathrm{INT}_{\mathrm{sum}}(r) \}$. The scaling causes weaker reference users to use a higher total power than stronger reference users proportional to each reference user's relative strength as determined by the payoff function, which provides a more realistic approximation of the power consumption that actual users would use.

The reference users' virtual transmit powers do not need to be very accurate, they just need to be "proportional enough" so that when the constant offsets are computed that they can more accurately estimate the per-sub-carrier quotas for each user.

Next, the constant Lagrange multiplier offset for the k-th user on the f-th sub-carrier is selected as follows:

$$\Delta\lambda_f^k = \sum_{r \in \text{Ref}} w_{ref.r} \left|h_f^{r,k}\right|^2 \left(\frac{1}{\text{int}_f^r} - \frac{1}{\text{rec}_f^r}\right), \tag{27}$$

where Ref represents the set of K_R virtual reference users and

$$\text{int}_f^r = \sigma_f^r + \sum_{u \in \text{Ref}\backslash\{r\}} \left|h_f^{r,u}\right|^2 p_f^u,$$

$$\text{rec}_f^r = \text{int}_f^r + \tfrac{1}{\Gamma}\left|h_f^{r,r}\right|^2 p_f^r.$$

Finally, the SMC sends each user their respective constant offset and each user self-optimizes using their respective constant offsets. The constant offset ASB-MRU algorithm is summarized in Algorithm 10.

For the ASB-MRU algorithm, there is a relationship between the performance and the complexity of the algorithm. In particular, as the number of reference users are increased, the accuracy of the virtual network also increases; however, increasing the number of reference users also increases the computational complexity of the algorithm as will be shown in the computational complexity section of this chapter.

Algorithm 10. Constant offset ASB-MRU algorithm

```
Initialization phase:
Each user computes their respective payoff function using (26) and sends it to
the SMC ;
SMC constructs virtual network and computes each reference user's parameters ;
SMC computes Δλ_f^k for all f, k using (27) and sends them to each user ;


Optimization phase:
Repeat
        For each user k and frequency sub-carrier f
                Apply (25) using Δλ_f^k ;
        end
Until p_f^k converges for all f, k ;
```

As such, the systematic approach referred to as the constant offset ASB-MRU algorithm automatically determines how many and which reference users should be selected in order to consistently find an effective balance between the performance and complexity.

COMPUTATIONAL COMPLEXITY COMPARISON

The computational complexity represents the number of operations required for the execution of an algorithm. In particular, it demonstrates the behaviour of different algorithms as the parameters (e.g., number of users, sub-carriers) increase. This is represented using big-O notation, which shows the limiting behaviour of a function when the arguments tend towards infinity. Mathematically, the big-O notation is described as follows:

$$f(x) = O(g(x)) \Leftrightarrow |f(x)| \leq c|g(x)| \quad \forall \ x \text{ as } x \to \infty,$$

where c is some constant. Suppose two algorithms are dependent on a variable X, where one algorithm has a complexity of $O(X)$ while the other has a complexity of $O(X^2)$. Then, the big-O notation implies that the $O(X)$ algorithm will converge faster than the $O(X^2)$ algorithm for sufficiently large X. The convergence time of an algorithm is said to be exponential if it is $O(e^X)$ and is said to be polynomial if it is $O(X^a)$ for a > 1. In algorithmic terms, exponential time is generally significantly slower than polynomial time. One key factor which the big-O notation fails to capture is the value of the constants in front of the key terms (e.g., number of iterations required for the algorithm to be run) since it only looks at the behaviour of the algorithms as the parameters tend towards infinity. Hence, for a fixed value of X, it is possible for a $O(X)$ algorithm to take longer to converge than a $O(X^2)$ algorithm. More specifically, for a fixed value of X, suppose the number of iterations required to run the $O(X)$ algorithm is denoted by v_1 and the number of iterations required to run the $O(X^2)$ algorithm is v_2. It is possible that $v_1 X > v_2 X^2$, for some values of X; however, for sufficiently large X, the $O(X)$ algorithm will run faster than the $O(X^2)$.

For DSM purposes, there are two main parameters: the number of users K, and the number of frequency sub-carriers, F. Hence, almost all the algorithms will be of the form $O(f(K,F))$ for some function $f(K,F)$. The only exception is the constant offset ASB-MRU algorithm, which also has the parameter for the number of reference users in the virtual network, denoted as K_R, hence it will be of the form $O(f(K,F,K_R))$.

Due to the total power constraint, the coupling between sub-carriers requires the transmit powers be searched for jointly over all frequency sub-carriers. Solving an X-dimensional exhaustive search has a computational complexity of $O(e^X)$. Therefore, solving the DSM optimal resource allocation using an exhaustive search over both the number of users and the number of sub-carriers has a complexity of $O(e^{KF})$, which is computationally intractable for large K and/or F.

Another source of algorithm complexity is the message passing requirements, depending on structure of the algorithm (i.e., centralized, semi-centralized, or distributed). In general, centralized algorithms require some initial message passing to gain global channel knowledge from all users, semi-centralized algorithms require per-iteration message passing, and distributed algorithms require no message passing. After each iteration, centralized algorithms must compute each user's updated interference plus noise term. For the k-th user, this computation requires cycling through all users $l \neq k$ and multiply their channel gain by the l-th user's transmit power for the specified f-th sub-carrier and add them together. This process must be repeated for each user on each sub-carrier, resulting in a computational complexity of:

$$F \times K \times (K - 1) = FK^2 - FK$$

Therefore, the computational complexity of calculating the interference plus noise for all K users on all F sub-carriers will be $O(FK^2)$.

Generally, semi-centralized and distributed algorithms can avoid manually computing the total interference plus noise by making use of per-iteration channel measurements and message passing, in the semi-centralized case.

Optimal Spectrum Balancing

Assuming the discrete search space for each user's transmit power is X, then the search space for all users' transmit powers will contain X^K elements. The OSB approach performs an exhaustive search over each of the X^K combinations. For each combination, the sum-rate must be calculated, which adds a multiplicative factor of K^2.

The exhaustive searches are applied independently on each of the F sub-carriers, and the process is repeated until convergence. Assuming the number of iterations until convergence is v, the complexity of the OSB algorithm is $v \times F \times K^2 \times X^K$, which implies:

$$C_{\text{OSB}} = O\left(FK^2 X^K\right)$$

Iterative Spectrum Balancing

The ISB algorithm operates similarly to the OSB algorithm, except that the exhaustive searches are one-dimensional instead of K-dimensional. Again, assuming the discrete search space for each user's transmit power is X, then each of the ISB exhaustive search spaces will contain X elements. The ISB algorithm contains two loops. The outer loop iterates amongst the K users, while the inner loop performs K different one-dimensional exhaustive searches over the X elements. For each combination, the total rate must be calculated, which adds a multiplicative factor of K^2. Let the number of outer iterations before convergence be v_1 and the number of inner iterations before convergence be v_2. Then, the complexity of the ISB algorithm is $v_1 \times v_2 \times F \times K^2 \times K \times X$, which implies:

$$C_{\text{ISB}} = O\left(FK^3\right)$$

Iterative Water-Filling

The IWF algorithm is a distributed algorithm which consists of two loops. The outer loop cycles through all the K users, while the inner loop performs water-filling for the k-th user over the F sub-carriers.

The water-filling process involves a simple bisection search. Assuming the number of outer iterations required for convergence is v and assuming the bisection loop is run until an accuracy of ϵ is achieved, the total number of iterations required will be $v \times \log_2(1/\epsilon)$. Therefore, the overall complexity of the IWF algorithm is $v \times \log_2(1/\epsilon) \times K \times F$, which implies:

$$C_{\mathrm{IWF}} = O\left(FK\right)$$

Selective Iterative Water-Filling

The SIW algorithm performs two loops, where the outer loop determines which users are still participating while the inner loop performs IWF for all participating users. At least one user and sub-carrier is removed during each outer loop iteration. This process is repeated until all users (a maximum of K iterations) or all sub-carriers have been removed (possibly in less than K iterations). Therefore, the maximum number of outer loop iterations is K, which leads to a complexity of up to K times that of IWF which implies:

$$C_{\mathrm{SIW}} = O\left(FK^2\right)$$

Successive Convex Approximations for Low Complexity

The SCALE algorithm complexity is broken down into the complexity at each user and the complexity at the SMC. Each user optimizes their transmit power, measures their total interference-plus-noise and sends the corresponding values to the SMC. Each user then waits for a message from the SMC before re-updating their transmit power. Assuming that w is the number of operations required to update each user's transmit power, the complexity at each user is given by wF.

The SMC receives a message from each user and generates the messages required for each user. Generating the message for each user requires $F \times K$ additions. This process must be repeated for each of the K users. Hence, the complexity at the SMC is given by FK^2. Therefore, assuming the number of iterations for the transmit powers of each user to converge is v, the overall complexity of the SCALE algorithm is $v(FK^2 + wK)$, which implies:

$$C_{\mathrm{SCALE}} = O\left(FK^2\right)$$

Difference of Convex Functions Algorithm

The complexity of DCA is dominated by the calculation of the gradient of the objective function and applying convex optimization techniques. Since there are a total of $F \times K$ variables (i.e., the transmit power for each of the K user on each of the F sub-carrier), calculating the gradient requires computing $F \times K$ derivatives. Each derivative involves a nested summation over K and $K-1$ elements, respectively. This requires $F \times K \times K \times (K-1) = FK^3 - FK^2$ computations at each iteration.

As well, for each iteration, a convex quadratic program must be solved. Goldfarb & Liu (1991) showed that a convex quadratic program can be solved in $O(n^3L)$ arithmetic operations, where n is the number of variables and L is the encoding length (a constant). Therefore, solving the convex quadratic program for each user on each subcarrier has a complexity of $O(FK^3)$. Therefore, assuming the number of iterations required until convergence is v, the complexity of DCA is v x $(FK^3 - FK^2 + FK^3)$, which implies:

$$C_{\mathrm{DCA}} = O\left(FK^3\right)$$

Distributed Spectrum Balancing

The DSB algorithm consists of each user measuring its channel and sending a message to the SMC (one per sub-carrier). The SMC combines the received messages from all K users and computes each user's Lagrange multiplier offset for each sub-carrier, $\Psi_f^{\mathrm{DSB},k}$. Each user then updates their transmit power on each of the F sub-carriers using the received offset values. The process is repeated until the offset values for each user on each sub-carrier converge. Assuming the w is the number of operations required to update each user's transmit power, the complexity at each user is given by wF.

Calculating the message that the SMC sends to each user requires a summation over K-1 users of the product between the corresponding interference channel gain and the received message from that particular user. Hence, the generation of the message for each user requires F x $(K-1)$ additions. The process is repeated for each of the K users, leading to a complexity of $FK^2 - FK$ at the SMC. Therefore, assuming the number of iterations required until convergence is achieved is v, the complexity of the DSB algorithm is v x $(FK^2 - FK + wF)$, which implies:

$$C_{\mathrm{DSB}} = O\left(FK^2\right).$$

Autonomous Spectrum Balancing

The ASB algorithm consists of three nested loops. The outmost loop cycles through all K users, the next loop updates adjusts the weighting factors until the target rate is achieved, and the innermost loop adjusts the water-filling level until the total power constraint is satisfied. Assuming each bisection loop is run until an accuracy of ϵ_R and ϵ_p are achieved, respectively, the total number of iterations required will be $\log_2(1/\epsilon_R)$ x $\log_2(1/\epsilon_p)$.

The complexity of each iteration is dominated by finding the roots of the cubic polynomial. Nickalls (1993) showed that finding the roots of a cubic equation requires 44 operations in total. A cubic equation must be solved for each of the F sub-carriers, leading to a total complexity of $44F$ for solving the cubic equations. Assuming the number of iterations required until convergence is v, the complexity of the ASB algorithm is v x K x $\log_2(1/\epsilon_R)$ x $\log_2(1/\epsilon_p)$ x $44F$, which implies:

$$C_{\mathrm{ASB}} = O\left(FK\right)$$

Note that the computational complexity of ASB-2 is also $O(FK)$ even though the transmit power update complexity is reduced from solving a cubic equation into solving a simple water-filling-like update. In particular, while their computational complexities are the same, the ASB-2 algorithm has a significantly faster runtime.

Constant Offset ASB Using Multiple Reference Users

The constant offset ASB-MRU algorithm consists of two phases: the initialization phase and the optimization phase. It is important to point out that the initialization complexity and message passing requirements are one-time costs, whereas the optimization complexity and message passing are per-iteration costs.

Computing the payoff function requires $O(F)$ computations per user. The clustering algorithm complexity is dominated by the computation of the pair-wise Euclidean distances of the payoff functions, requiring $O(K^2)$ computations. Generating the virtual network of reference users requires a complexity of $O(FKK_R)$, where K_R is the number of reference users in the virtual network. Computing the pair-wise interference for the virtual network requires $O(FK_R^2)$ computations. Hence, the initialization phase requires a complexity of $O(FK + K^2 + FKK_R + FK_R^2) = O(FKK_R)$, since $K > K_R$ and typically, $F > K$. In terms of message passing, during the initialization phase, each user sends their K-dimensional payoff function to the SMC and the SMC sends the K-dimensional offset to each user. Therefore, the initialization phase requires $2K$ messages to be sent.

The optimization phase is implemented in a fully distributed manner and runs at the same complexity of IWF, where the only difference lies in the constant offset value. Hence, the per user optimization phase complexity is $O(F)$. Due to the fact that the offsets are constant, the constant offset ASB-MRU optimization phase can be run without any message passing or central computations. Furthermore, the constant offset re-initialization phase does not need to be run every time a new user becomes active or inactive. Instead, the virtual network should only be updated periodically. Once the virtual network is established, the initialization phase only consists of computing the offset values for the new users. Therefore, from a practical implementation perspective, the initialization phase does not create a significant source of algorithmic complexity.

Practical Implementation Aspects

The key practical implementation aspects for DSM algorithms include the computational complexity and amount of signaling (message passing) required. These are summarized in Table 1. The initialization phase performs computations required to setup the variables for the optimization phase. The optimization phase consists of computations performed by the users and computations performed by the SMC. Note that the initialization complexity and signaling are *one-time* costs, whereas the optimization complexity and signaling are *per-iteration* costs.

As expected, Table 1 shows that the centralized algorithms (e.g., OSB, ISB, DCA) lead to a higher computational complexity than the semi-centralized algorithms (e.g., SIW, SCALE, DSB) and the distributed algorithms (e.g., IWF, ASB, constant offset ASB-MRU). Similarly, while the overhead associated with the distributed algorithms is significantly reduced compared to the semi-centralized algorithms, the semi-centralized and centralized algorithms allow for more service provider control.

Table 1. Summary of DSM algorithm computational complexities

DSM Algorithms	Computational Complexity				Messages / User	
	Initialization Phase	Optimization Phase			Initialization Phase	Optimization Phase
		Overall	SMC / Iteration	At User		Per Iteration
OSB	–	$O(FK^2X^K)$	$O(FK^2X^K)$	–	F	–
ISB	–	$O(FK^3)$	$O(FK^3)$	–	F	–
IWF	–	$O(FK)$	–	$O(F)$	–	–
SIW	–	$O(FK^2)$	$O(K)$	$O(F)$	–	$\leq F$
SCALE	–	$O(FK^2)$	$O(FK^2)$	$O(F)$	–	$2F$
DCA	–	$O(FK^3)$	$O(FK^3)$	–	F	–
DSB	–	$O(FK^2)$	$O(FK^2)$	$O(F)$	–	$2F$
ASB	–	$O(FK)$	–	$O(F)$	–	–
Constant Offset ASB-MRU	$O(FKK_R)$	$O(FK)$	–	$O(F)$	$2F$	–

In particular, for centralized and semi-centralized algorithms, the service provider has more opportunity to influence the optimization process. This allows for more control over user preferences, constraints, and application-based priorities. For centralized algorithms, the service provider has full control throughout the process. For semi-centralized algorithms, the service provider can influence the network on a per-iteration basis. Typically, distributed algorithms do not provide the service provider with much opportunity to influence the network; however, the constant offset ASB-MRU is an exception since its initialization phase requires some central coordination, and hence, service providers can prioritize customers by modifying the offsets accordingly.

ILLUSTRATIVE EXAMPLES

For comparison, the DSM algorithms were implemented in a wireline VDSL environment. While the DSM algorithms are implemented in a wireline VDSL environment, they are also applicable to low-mobility wireless environments. Three sets of illustrative examples will be provided: a two-user uplink and two 48-user downlink and uplink test cases.

All test cases assume 26-gauge (0.4 mm) lines are used. The target symbol error probability is 10^{-7}. The coding gain and noise margin are 3 dB and 6 dB, respectively. The frequency tone spacing is 4.3125 kHz and the symbol rate is 4 kHz. The frequency division duplexing band plan 998 is used, which consists of two separate uplink transmission bands: 3.75 – 5.2 MHz and 8.5 – 12 MHz and three separate downlink transmission bands: 0.276 – 3.75 MHz, 5.2 – 8.5 MHz, and 12 – 17.664 MHz. A maximum transmit power of 11.5 dBm was applied to each user.

Two-User Scenario

One way to visualize the performance of different DSM algorithms is to plot and compare the rate regions for each algorithm. The rate region is a plot where each axis represents the rate achieved for a different user in the system. In particular, the rate region shows all possibly combinations of achievable data-rates for the users. Hence, in order to more easily visualize the rate region, this scenario focuses on the two-user case. The further the rate region curve is from the origin, the higher the algorithm's achievable data-rate is, and the better the performance. The goal of DSM algorithms are to generate points which are close to the optimal values (i.e., the boundary of the rate region), while operating at a low complexity. This concept extends naturally to the multi-user case which will be discussed later.

This test case represents a non-symmetric frequency-selective network that considers uplink transmission between two users. The first user is located 457 m (1500 ft) from the Central Office (CO), and the second user is located 914 m (3000 ft) from the CO.

The direct (H11, H22) and interference (H12 and H21) channel transfer functions for the two-user scenario are shown in Figure 3. Figure 3 shows the channel transfer function over all 4096 frequency sub-carrier, but the bolded parts of the curves represent the sections dedicated to uplink transmission and used in the simulations to follow. Clearly, user one generates a significant amount of interference to user two. This is an example of the classic near-far problem.

Figure 3. Two-user scenario

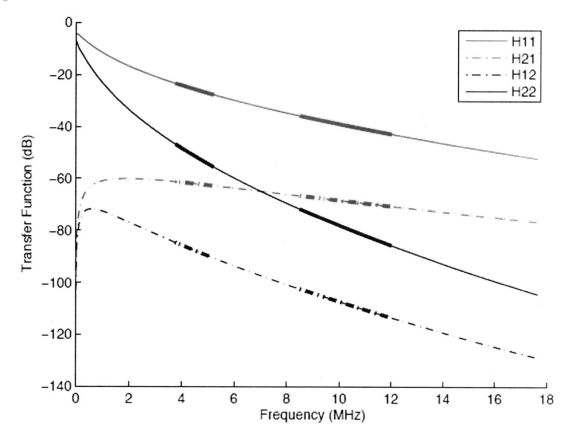

Simulations were run for each of the DSM algorithms, as well as the flat Power Back-Off (PBO) algorithm, which is a SSM algorithm. The results are summarized in Figure 4. The flat PBO curve demonstrates the significant performance improvements of DSM over SSM approaches. The benefits of DSM are even more evident when the number of users increases.

The two-user scenario is an overly simplistic scenario due to the limited number of users (i.e., interference) in the system, and hence, the performances of the DSM algorithms are more optimistic than usual. In particular, due to the limited number of users in the system, this scenario exhibits fewer locally optimal points and therefore, it is easier for the locally optimal algorithms to solve for globally optimal points.

The two-user scenario does provide some insight into what the expected performance of the DSM algorithms should be in a multi-user setting; however, in order to more appropriately differentiate the higher-caliber algorithms, a multiuser scenario is necessary.

Figure 4. Two-user scenario rate region

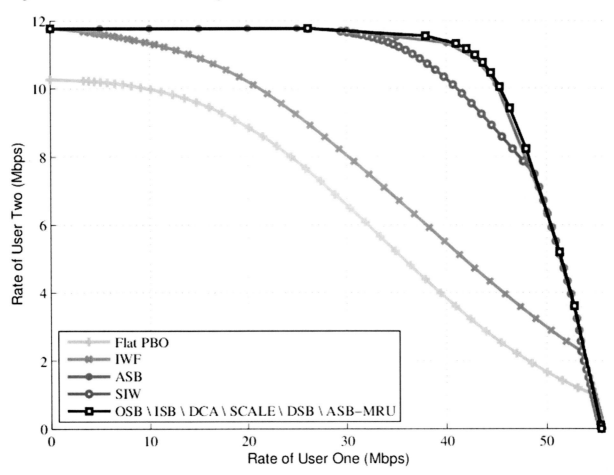

48-User Scenarios

This section discusses two 48-user Monte-Carlo test cases. The first test-case represents a typical Fiber-To-The-Curb (FTTC) scenario and the second test-case represents a typical Fiber-To-The-Node (FTTN) scenario.

Each Monte Carlo test-case consisted of 1000 downlink and 1000 uplink realizations. The FTTC test case consisted of 24 users whose line lengths were uniformly distributed from 100 to 500 m and 24 users who were offset by 250 m and whose line lengths were uniformly distributed between 100 to 500 m. The FTTN test-case consisted of 48 users uniformly distributed between 100 to 1000 m.

Due to the large complexity of the centralized algorithms, it was not practical to apply them to the 48-user scenarios. In particular, in order to distinguish between the higher-caliber algorithms, the DSB, SCALE, IWF, and ASB-MRU algorithms were compared.

The performance of the DSM algorithms were compared in terms of their Percentage of Maximum (PoM) weighted sum-rate, on a per-realization basis. For a fair comparison, the maximum and minimum PoM per-realization was also computed, representing the variation in results from one realization to the next. The number of iterations and number of iterations required to reach 98% of the algorithm's final converged rate are also given. Often, the limiting factor in terms of the practical runtime of an algorithm is the number of iterations required, since each user must take new interference measurements after each iteration. As well, the semi-centralized algorithms require per-iteration message passing and central computations in addition to the interference measurements. The number of iterations required to reach 98% of the final converged rate is also extremely relevant from a practical perspective since it shows how easily each of the algorithms can trade-off a slight (2%) loss in performance for an improved computational time. In practice, it is often less important to gain this additional 2% performance improvements than it is to be able to compute the appropriate power allocation scheme in a timely manner.

Table 2 summarizes the results of the 48-user FTTC test case. The results show that typically, the SCALE approach slightly outperforms the DSB and constant offset ASB-MRU approaches; however, it requires significantly more iterations to converge than the constant offset ASB-MRU approach. On average, the constant offset ASB-MRU approach achieved near-DSB and -SCALE performance for both uplink and downlink FTTC transmission while requiring significantly less iteration. Moreover, beyond what is shown in Table 2, the constant offset ASB-MRU algorithm outperformed DSB in 17.1% and 63.3% of realizations for uplink and downlink transmission, respectively. As well, for uplink transmission, the minimum percentage difference between the constant offset ASB-MRU algorithm and IWF was 11.4%, the maximum percent difference was 26.0%, and the average percent difference was 18.5%. For downlink transmission, the minimum percent difference between the constant offset ASB-MRU and IWF was 2.1%, the maximum percent difference was 9.0%, and the average percent difference was 5.7%.

Figure 5 shows the rate of convergence for a typical uplink FTTC realization. The convergence plot shows that only one iteration of the constant offset ASB-MRU algorithm is required to obtain strong performance. The ability of the constant offset ASB-MRU algorithm to achieve near-DSB and -SCALE performance after just one iteration allows for the algorithm to very quickly adapt to changes in the network (i.e., new users entering the system or users exiting the system).

The results also show that while the DSB and SCALE algorithms tend to outperform the constant offset ASB-MRU algorithm, it takes many more iterations (i.e., interference measurements, message passing, and central computations) to do so.

Table 2. Summary of 48-user FTTC test case results

FTTC Uplink				
	DSB	SCALE	IWF	Constant Offset ASB-MRU
Avg PoM	99.2	100.0	79.4	97.9
Avg # Iters	303.3	351.6	6.7	8.8
Avg # iters (98%)	36.4	37.8	1.9	1.0
Max PoM	100.0	100.0	86.0	100.0
Min PoM	93.2	97.8	72.1	91.1
Avg # Refs	–	–	–	15.8
FTTC Downlink				
	DSB	SCALE	IWF	Constant Offset ASB-MRU
Avg PoM	99.5	99.8	94.0	99.7
Avg # Iters	323.4	490.6	74.3	46.8
Avg # iters (98%)	6.9	47.5	1.7	1.0
Max PoM	100.0	100.0	96.8	100.0
Min PoM	97.1	98.6	90.8	98.4
Avg # Refs	–	–	–	15.1

Table 3 summarizes the results of the 48-user FTTN test case. The results are similar to the FTTC test case in that DSB and SCALE tend to provide the best overall performance, but require significantly more iterations to converge. As well, the constant offset ASB-MRU algorithm achieved near-DSB and -SCALE performance for both uplink and downlink transmission in significantly fewer iterations. Again, only one iteration was necessary for the constant offset ASB-MRU algorithm to achieve over 98% of its final converged value. Moreover, for uplink transmission, the minimum percent difference between the constant offset ASB-MRU and IWF was 23.8%, the maximum percent difference was 42.3%, and the average percent difference was 32.5%.

CONCLUDING REMARKS

This chapter presented an overview of algorithms that adaptively allocate transmit power in a multiuser wireline or wireless system. The concept of DSM was introduced and its advantages over SSM was demonstrated using a simple two-user illustrative example. Nine DSM algorithms were discussed in detail. More specifically, the OSB, ISB, IWF, SIW, SCALE, DCA, DSB, ASB, and constant offset ASB-MRU algorithms were presented in detail. The nine DSM algorithms were compared using a computational complexity analysis and through extensive simulation results.

Figure 5. FTTC uplink convergence rate comparison

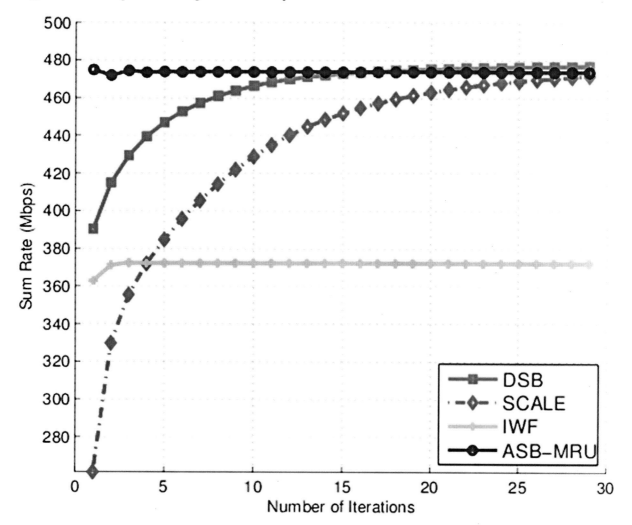

The illustrative examples showed that, in general, the centralized approaches were infeasible when there were many users present. As such, semi-centralized approaches (e.g., DSB and SCALE) were shown to provide the largest data-rate improvements but required many iterations to do so. Moreover, it was shown that SCALE often provided marginal performance improvements over DSB, but required more iterations to converge.

The computational complexity analysis showed that the distributed algorithms (e.g., IWF, ASB, constant offset ASB-MRU) provided very significant computational advantages over the semi-centralized algorithms (e.g., DSB, SCALE, SIW). Due to the relatively fast per-user transmit power update formulas for the non-centralized algorithms, the main time-consuming limitations are caused by the number of iterations and the corresponding centralized per-iteration computations.

Table 3. Summary of 48-user FTTN test case results

FTTN Uplink				
	DSB	**SCALE**	**IWF**	**Constant Offset ASB-MRU**
Avg PoM	100.0	100.0	66.0	98.4
Avg # Iters	267.7	225.5	5.1	10.8
Avg # iters (98%)	35.9	10.9	2.0	1.0
Max PoM	100.0	100.0	75.9	99.8
Min PoM	100.0	100.0	56.1	93.8
Avg # Refs	–	–	–	15.9
FTTN Downlink				
	DSB	**SCALE**	**IWF**	**Constant Offset ASB-MRU**
Avg PoM	100.0	99.9	96.0	99.2
Avg # Iters	158.3	500.0	12.6	12.0
Avg # iters (98%)	2.0	74.4	1.0	1.0
Max PoM	100.0	100.0	97.6	99.7
Min PoM	100.0	99.7	95.0	97.5
Avg # Refs	–	–	–	14.6

To this end, the constant offset ASB-MRU algorithm was shown to provide near-DSB and -SCALE performance after only one iteration. Therefore, while the DSB and SCALE algorithms provide the benchmark for performance for systems involving many users, the constant offset ASB-MRU algorithm provides a competitive balance between performance and complexity.

FURTHER RESEARCH DIRECTIONS

Future issues in the area of dynamic spectrum management include applications to small-cells (e.g., femtocells, picocells). For example, Ngo, Khakurel, & Le-Ngoc (2014) investigate the issue of joint sub-channel and power allocation for OFDMA femtocell networks. More generally, Saad, Ismail, & Nordin (2013) provides a survey on some preliminary power control techniques for femtocell networks. Another potential application of dynamic spectrum management is Cooperative Multi-Point (CoMP) systems. In CoMP systems, multiple base stations coordinate when servicing users. For example, Zhang et al. (2013) proposed a distributed power control algorithm for CoMP systems.

REFERENCES

Akyildiz, I. F., Lee, W.-Y., Vuran, M. C., & Mohanty, S. (2006). Next generation/dynamic spectrum access/cognitive radio wireless networks: A survey. *Computer Networks*, *50*(13), 2127–2159. doi:10.1016/j.comnet.2006.05.001

Alnwaimi, G., Arshad, K., & Moessner, K. (2011). Dynamic spectrum allocation algorithm with interference management in co-existing networks. *IEEE Communications Letters, 15*(9), 932–934. doi:10.1109/LCOMM.2011.062911.110248

Bagheri, H., Pakravan, M. R., & Khalaj, B. H. (2004). Iterative multi-user power allocation: performance evaluation & modification. In Proceedings of IEEE Region 10 (pp. 216–219). TENCON. doi:10.1109/TENCON.2004.1414746

Blanquero, R., & Carrizosa, E. (2000). On covering methods for d.c. optimization. *Journal of Global Optimization, 18*(3), 265–274. doi:10.1023/A:1008366808825

Boyd, S., & Vandenberghe, L. (2004). *Convex Optimization*. Academic Press.

Cendrillon, R., Huang, J., Chiang, M., & Moonen, M. (2007). Autonomous spectrum balancing for digital subscriber lines. *IEEE Transactions on Signal Processing, 55*(8), 4241–4257. doi:10.1109/TSP.2007.895989

Cendrillon, R., & Moonen, M. (2005). Iterative spectrum balancing for digital subscriber lines. In *Proceedings of IEEE International Conference on Communications* (pp. 1937–1941). IEEE. doi:10.1109/ICC.2005.1494677

Cendrillon, R., Yu, W., Moonen, M., Verlinden, J., & Bostoen, T. (2006). Optimal multiuser spectrum balancing for digital subscriber lines. *IEEE Transactions on Communications, 54*(5), 922–933. doi:10.1109/TCOMM.2006.873096

Feng, Z., Li, W., Li, Q., Le, V., & Gulliver, T. A. (2011). Dynamic spectrum management for wcdma/dvb heterogeneous systems. *IEEE Transactions on Wireless Communications, 10*(5), 1582–1593. doi:10.1109/TWC.2011.030911.100915

Forouzan, A. R., & Garth, L. M. (2005). Generalized iterative spectrum balancing and grouped vectoring for maximal throughput of digital subscriber lines. In *Proceedings of IEEE Global Communications Conference* (pp. 2359–2363). IEEE. doi:10.1109/GLOCOM.2005.1578085

Goldfarb, D., & Liu, S. (1991). An o(n3l) primal interior point algorithm for convex quadratic programming. *Mathematical Programming, 49*(3), 325–340.

Horst, R., Phong, T. Q., Thoai, N. V., & Vries, J. (1991). On solving a d.c. programming problem by a sequence of linear programs. *Journal of Global Optimization, 1*(2), 183–293. doi:10.1007/BF00119991

Huberman, S., Leung, C., & Le-Ngoc, T. (2011). A clustering approach to autonomous spectrum balancing using multiple reference lines for dsl. In *Proceedings of IEEE Global Telecommunications Conference* (pp. 1–5). IEEE. doi:10.1109/GLOCOM.2011.6133937

Huberman, S., Leung, C., & Le-Ngoc, T. (2012a). Constant offset autonomous spectrum balancing using multiple reference lines for vdsl. *IEEE Transactions on Signal Processing, 60*(12), 6719–6723. doi:10.1109/TSP.2012.2215606

Huberman, S., Leung, C., & Le-Ngoc, T. (2012b). Dynamic spectrum management (dsm) algorithms for multi-user xdsl. *IEEE Communications Surveys and Tutorials, 14*(1), 109–130. doi:10.1109/SURV.2011.092110.00090

Im, S., Kang, Y., Kim, W., Kim, S., Kim, J., & Lee, H. (2011). Dynamic spectrum allocation with efficient sinr-based interference management. In *Proceedings of IEEE Vehicular Technology Conference* (pp. 1–5). IEEE. doi:10.1109/VETECF.2011.6092824

Leaves, P., Moessner, K., Tafazolli, R., Grandblaise, D., Bourse, D., Tonjes, R., & Breveglieri, M. (2004). Dynamic spectrum allocation in composite reconfigurable wireless networks. *IEEE Communications Magazine*, *42*(5), 72–81. doi:10.1109/MCOM.2004.1299346

Lee, W., Kim, Y., Brady, M. H., & Cioffi, J. M. (2006). Band preference in dynamic spectrum management. In *Proceedings of IEEE Global Communications Conference* (*Vol. 2*, pp. 1–5). IEEE.

Leung, C., Huberman, S., & Le-Ngoc, T. (2010). Autonomous spectrum balancing using multiple reference lines for digital subscriber lines. In *Proceedings of IEEE Global Communications Conference* (pp. 1–6). IEEE. doi:10.1109/GLOCOM.2010.5683960

Liu, Y.-S., & Su, Z.-J. (2007). Distributed dynamic spectrum management for digital subscriber lines. *IEICE Transactions on Communications*, *E90-B*(3), 491–498. doi:10.1093/ietcom/e90-b.3.491

Lui, R., & Yu, W. (2005). Low-complexity near-optimal spectrum balancing for digital subscriber lines. In *Proceedings of IEEE International Conference on Communications* (pp. 1947–1951). IEEE. doi:10.1109/ICC.2005.1494679

Moraes, R. B., Dortschy, B., Klautau, A., & Rius i Riuy, J. (2010). Semiblind spectrum balancing for dsl. *IEEE Transactions on Signal Processing*, *58*(7), 3717–3727. doi:10.1109/TSP.2010.2045553

Ngo, D. T., Khakurel, S., & Le-Ngoc, T. (2014). Joint subchannel assignment and power allocation for ofdma femtocell networks. *IEEE Transactions on Wireless Communications*, *13*(1), 342–355. doi:10.1109/TWC.2013.111313.130645

Nickalls, R. (1993). A new approach to solving the cubic: Cardan's solution revealed. *Mathematical Gazette*, *77*(480), 354–359. doi:10.2307/3619777

Noam, Y., & Leshem, A. (2009). Iterative power pricing for distributed spectrum coordination in dsl. *IEEE Transactions on Communications*, *57*(4), 948–953. doi:10.1109/TCOMM.2009.4814362

Papandriopoulos, J., & Evans, J. S. (2006). Low-complexity distributed algorithms for spectrum balancing in multi-user dsl networks. In *Proceedings of International Conference on Communications* (pp. 3270–3275). Academic Press. doi:10.1109/ICC.2006.255311

Papandriopoulos, J., & Evans, J. S. (2009). Scale: A low-complexity distributed protocol for spectrum balancing in multiuser dsl networks. *IEEE Transactions on Information Theory*, *55*(8), 3711–3724. doi:10.1109/TIT.2009.2023751

Pham, D. T., & Hoai An, L. T. (1997). Convex analysis approach to d.c. programming: Theory, algorithms, and applications. *Acta Mathematica Vietnamica*, *22*(1), 289–355.

Qian, L. P., & Jun, Y. (2009). Monotonic optimization for non-concave power control in multiuser multicarrier network systems. In *Proceedings of IEEE International Conference on Computer Communications* (pp. 172–180). IEEE. doi:10.1109/INFCOM.2009.5061919

Saad, S., Ismail, M., & Nordin, R. (2013). A survey on power control techniques in femtocell networks. *Journal of Communication, 8*(12), 845–854. doi:10.12720/jcm.8.12.845-854

Sadr, S., Anpalagan, A., & Raahemifar, K. (2009). Radio resource allocation algorithms for the downlink of multiuser ofdm communication systems. *IEEE Communications Surveys and Tutorials, 11*(3), 92–106. doi:10.1109/SURV.2009.090307

Shi, H., Li, J., Ma, Y., & Li, Z. (2013). Coverage probability driven dynamic spectrum allocation in heterogeneous wireless networks. *Science China Information Sciences, 56*(8), 1–7. doi:10.1007/s11432-013-4920-8

Starr, T., Sorbara, M., Cioffi, J. M., & Silverman, P. J. (2003). *DSL Advances*. Prentice-Hall.

Statovci, D., Nordstrom, T., & Nilsson, R. (2006). The normalized-rate iterative algorithm: A practical dynamic spectrum management method for dsl. *EURASIP Journal on Applied Signal Processing, 2006*, 1–17. doi:10.1155/ASP/2006/95175 PMID:16758000

Subramanian, A. P., Al-Ayyoub, M., Gupta, H., Das, S. R., & Buddhikot, M. M. (2008). Near-optimal dynamic spectrum allocation in cellular networks. In *Proceedings of IEEE Symposium on New Frontiers in Dynamic Spectrum Access Networks*, (pp. 1–11). IEEE. doi:10.1109/DYSPAN.2008.41

Tsiaflakis, P., Diehl, M., & Moonen, M. (2008). Distributed spectrum management algorithms for multiuser dsl networks. *IEEE Transactions on Signal Processing, 56*(10), 4825–4843. doi:10.1109/TSP.2008.927460

Tsiaflakis, P., Vangorp, J., Moonen, M., Verlinden, J., & Van Acker, K. (2006). An efficient search algorithm for the lagrange multipliers of optimal spectrum balancing in multi-user xdsl systems. In *Proceedings of IEEE International Conference on Acoustics, Speech and Signal Processing* (pp. 101–104). IEEE. doi:10.1109/ICASSP.2006.1660915

Wei, S., Youming, L., & Miaoliang, Y. (2009). Low-complexity grouping spectrum management in multi-user dsl networks. In *Proceedings of International Conference on Communications and Mobile Computing* (pp. 381–385). Academic Press. doi:10.1109/CMC.2009.125

Xu, Y., Le-Ngoc, T., & Panigrahi, S. (2008). Global concave minimization for optimal spectrum balancing in multi-user dsl networks. *IEEE Transactions on Signal Processing, 56*(7), 2875–2885. doi:10.1109/TSP.2008.917378

Xu, Y., Panigrahi, S., & Le-Ngoc, T. (2005). Selective iterative water-filling for digital subscriber loops. In *Proceedings of International Conference on Information, Communications and Signal Processing* (pp. 101–105). Academic Press.

Yu, W., Ginis, G., & Cioffi, J. M. (2002). Distributed multiuser power control for digital subscriber lines. *IEEE Journal on Selected Areas in Communications, 20*(5), 1105–1115. doi:10.1109/JSAC.2002.1007390

Zhang, X., Sun, Y., Chen, X., Zhou, S., Wang, J., & Shroff, N. B. (2013). Distributed power allocation for coordinated multipoint transmissions in distributed antenna systems. *IEEE Transactions on Wireless Communications*, *12*(5), 2281–2291. doi:10.1109/TWC.2013.040213.120863

Zhao, Q., & Sadler, B. M. (2007). A survey of dynamic spectrum access. *IEEE Signal Processing Magazine*, *24*(3), 79–89. doi:10.1109/MSP.2007.361604

ADDITIONAL READING

Akbar, I. A., & Tranter, W. H. (2007). Dynamic spectrum allocation in cognitive radio using hidden Markov models: poisson distributed case. In IEEE SoutheastCon (pp. 196–201). doi:10.1109/SECON.2007.342884

Akyildiz, I. F., Lee, W., Vuran, M. C., & Mohanty, S. (2008). A survey on spectrum management in cognitive radio networks. *IEEE Communications Magazine*, *46*(4), 40–48. doi:10.1109/MCOM.2008.4481339

Al-Khasib, T., Shenouda, M. B., & Lampe, L. (2011). Dynamic spectrum management for multiple-antenna cognitive radio systems: Designs with imperfect csi. *IEEE Transactions on Wireless Communications*, *10*(9), 2850–2859. doi:10.1109/TWC.2011.070511.091419

Buddhikot, M. M., & Ryan, K. (2005). Spectrum management in coordinated dynamic spectrum access based cellular networks. In *IEEE International Symposium on New Frontiers in Dynamic Spectrum Access Networks* (pp. 299–307). Ieee. doi:10.1109/DYSPAN.2005.1542646

Ding, L., Melodia, T., Batalama, S. N., Matyjas, J. D., & Medley, M. J. (2010). Cross-layer routing and dynamic spectrum allocation in cognitive radio ad hoc networks. *IEEE Transactions on Vehicular Technology*, *59*(4), 1969–1979. doi:10.1109/TVT.2010.2045403

Gür, G., Bayhan, S., & Alagöz, F. (2010). Cognitive femtocell networks: An overlay architecture for localized dynamic spectrum access. *IEEE Wireless Communications*, *17*(4), 62–70. doi:10.1109/MWC.2010.5547923

Hossain, E., Niyato, D., & Han, Z. (2009). *Dynamic spectrum access and management in cognitive radio networks*. Cambridge University Press. doi:10.1017/CBO9780511609909

Hwang, J., & Yoon, H. (2008). Dynamic spectrum management policy for cognitive radio: an analysis of implementation feasibility issues. In *IEEE Symposium on New Frontiers in Dynamic Spectrum Access Networks* (pp. 1–9). doi:10.1109/DYSPAN.2008.64

Jabbari, B., Pickholtz, R., & Norton, M. (2010). Dynamic spectrum access and management. *IEEE Wireless Communications*, *17*(4), 6–15. doi:10.1109/MWC.2010.5547916

Ji, Z., & Ray Lui, K. J. (2007). Dynamic spectrum sharing: A game theoretical overview. *IEEE Communications Magazine*, *45*(5), 88–94. doi:10.1109/MCOM.2007.358854

Jia, J., Zhang, Q., Zhang, Q., & Liu, M. (2009). Revenue generation for truthful spectrum auction in dynamic spectrum access. In *International Symposium on Mobile ad hoc networking and computing* (pp. 3–12). New York, New York, USA: ACM Press. doi:10.1145/1530748.1530751

Khozeimeh, F., & Haykin, S. (2010). Self-organizing dynamic spectrum management for cognitive radio networks. In *Communication Networks and Services Research Conference* (pp. 1–7). Ieee. doi:10.1109/CNSR.2010.51

Khozeimeh, F., & Haykin, S. (2009). Dynamic spectrum management for cognitive radio: An overview. *Wireless Communications and Mobile Computing*, 9(11), 1447–1459. doi:10.1002/wcm.732

Le Nir, V., & Scheers, B. (2010). Autonomous dynamic spectrum management for coexistence of multiple cognitive tactical radio networks. In *International Conference on Cognitive Radio Oriented Wireless Networks & Communications* (Vol. 2, pp. 1–5). doi:10.4108/ICST.CROWNCOM2010.9228

Luo, Z.-Q., & Zhang, S. (2008). Dynamic spectrum management: Complexity and duality. *IEEE Journal of Selected Topics in Signal Processingin Signal Processing*, 2(1), 57–73. doi:10.1109/JSTSP.2007.914876

Luo, Z.-Q., & Zhang, S. (2009). Duality gap estimation and polynomial time approximation for optimal spectrum management. *IEEE Transactions on Signal Processing*, 57(7), 2675–2689. doi:10.1109/TSP.2009.2016871

Niyato, D., & Hossain, E. (2008). Spectrum trading in cognitive radio networks: A market-equilibrium-based approach. *IEEE Wireless Communications*, 15(6), 71–80. doi:10.1109/MWC.2008.4749750

Pan, M., Liang, S., Xiong, H., Chen, J., & Liu, G. (2006). A novel bargaining based dynamic spectrum management scheme in reconfigurable systems. In *International Conference on Systems and Networks Communications* (pp. 1–5). doi:10.1109/ICSNC.2006.10

Salami, G., Member, S., Durowoju, O., Attar, A., Holland, O., Tafazolli, R., & Aghvami, H. (2011). A comparison between the centralized and distributed approaches for spectrum management, 13(2), 274–290.

Song, K. B., & Chung, S. T. (2002). Dynamic spectrum management for next-generation dsl systems. *IEEE Communications Magazine*, 40(10), 101–109. doi:10.1109/MCOM.2002.1039864

Subramanian, A. P., Gupta, H., Das, S. R., & Buddhikot, M. M. (2007). Fast spectrum allocation in coordinated dynamic spectrum access based cellular networks. In *IEEE International Symposium on New Frontiers in Dynamic Spectrum Access Networks* (pp. 320–330). Ieee. doi:10.1109/DYSPAN.2007.50

Tsiaflakis, P., Necoara, I., Suykens, J. A. K., & Moonen, M. (2010). Improved dual decomposition based optimization for dsl dynamic spectrum management. *IEEE Transactions on Signal Processing*, 58(4), 2230–2245. doi:10.1109/TSP.2009.2039825

Tsiaflakis, P., Yi, Y., Chiang, M., & Moonen, M. (2012). Throughput and delay performance of a dsl broadband access with cross-layer dynamic spectrum management. *IEEE Transactions on Communications*, 60(9), 2700–2711. doi:10.1109/TCOMM.2012.062512.110385

Wolkerstorfer, M., Statovci, D., & Nordstr, T. (2008). Dynamic spectrum management for energy-efficient transmission in dsl. In *IEEE Singapore International Conference on Communication Systems* (pp. 1015–1020). doi:10.1109/ICCS.2008.4737336

Xie, Y., Armbruster, B., & Ye, Y. (2010). Dynamic spectrum management with the compeititve market model. *IEEE Transactions on Signal Processing, 58*(4), 2442–2446. doi:10.1109/TSP.2009.2039820

Xing, Y., Chandramouli, R., Mangold, S., & Shankar, N. S. (2006). Dynamic spectrum access in open spectrum wireless networks. *IEEE Journal on Selected Areas in Communications, 24*(3), 626–637. doi:10.1109/JSAC.2005.862415

Yang, L., Cao, L., & Zheng, H. (2008). Physical interference driven dynamic spectrum management. In *IEEE Symposium on New Frontiers in Dynamic Spectrum Access Networks* (pp. 1–12). Ieee. doi:10.1109/DYSPAN.2008.47

Yu, W., & Lui, R. (2006). Dual methods for nonconvex spectrum optimization of multicarrier systems. *IEEE Transactions on Communications, 54*(7), 1310–1322. doi:10.1109/TCOMM.2006.877962

Zhao, J., Zheng, H., & Yang, G.-H. (2005). Distributed coordination in dynamic spectrum allocation networks. In *IEEE International Symposium on New Frontiers in Dynamic Spectrum Access Networks* (pp. 259–268). Ieee. doi:10.1109/DYSPAN.2005.1542642

KEY TERMS AND DEFINITIONS

Centralized Algorithms: Algorithms which make use of a central hub with full knowledge of the network to perform the power allocations.

Distributed Algorithms: Algorithms in which users self-optimize fully autonomously without the need of explicit message passing.

Dynamic Spectrum Management: Adaptively applies different spectrum allocations for each user with some optimization criteria in mind (e.g., maximizing throughput, minimizing power).

Global Optimality: Refers to an operating point which is the best possible over the entire domain with respect to some objective (e.g., maximum sum-rate, minimum transmit power).

Locally Optimality: Refers to an operating point which is the best within some neighbourhood (i.e., subset of the domain) with respect to some objective (e.g., maximum sum-rate, minimum transmit power).

Nash Equilibrium: A non-cooperative game theoretic solution where no user can gain by changing their current power allocation strategy.

Semi-Centralized Algorithms: Algorithms which make use of both a central hub and distributed processing to perform the power allocations using a per-iteration message passing system.

Sequential Convex Programming: Iterative algorithms which solve non-convex optimization problems by solving a sequence of convex sub-problems to find a locally optimal solution.

Chapter 13
Performance Studies for Spectrum-Sharing Cognitive Radios under Outage Probability Constraint

Abdallah K. Farraj
University of Toronto, USA

Eman M. Hammad
University of Toronto, USA

Scott L. Miller
Texas A&M University, USA

ABSTRACT

This chapter investigates the performance of primary and secondary users in a spectrum-sharing cognitive environment. In this setup, multiple secondary users compete to share a channel dedicated to a primary user in order to transmit their data to a receiver unit. One secondary user is scheduled to share the channel, and to do so, its transmission power should satisfy the outage probability requirement of the primary user. Secondary users are ranked according to their channel strength, and performance measures are derived as a function of a generic channel rank. The performance of different scheduling schemes is also investigated. Further, the performance of the primary user is investigated in this environment. Numerical results are presented to verify the theoretical analysis and investigate the relation between the parameters of the communication environment and the performance measures of the users of the system.

INTRODUCTION

Extensive use of wireless technologies made the frequency spectrum a very limited resource for modern wireless communication systems; although limited, recent studies indicate that the spectrum is actually under-utilized (Federal Communications Commission, 2002). Cognitive communications technology represents a promising solution to achieve more efficient spectrum utilization (Mitola, 2000). In a cog-

DOI: 10.4018/978-1-4666-6571-2.ch013

nitive communication environment, the primary user (PU) refers to a transceiver unit that is licensed to use a specific wireless channel. Further, the cognitive radio, also referred to as the secondary user (SU), refers to a radio unit that adapts its communication settings in order to transmit its data over the primary channel even though the secondary user is not licensed to operate over that channel.

Cognitive communications can be carried out using spectrum-sensing or spectrum-sharing methods. In a spectrum-sensing mode, the secondary user can only communicate over the channel when the channel is detected as idle. In a spectrum-sharing technique, the secondary user can transmit data simultaneously along with the primary user over the same channel as long as the secondary user's transmission does not deteriorate the quality of service (QoS) of the primary user below a certain requirement. In order to make a successful cognitive communication system, the channel under consideration has to be detected, the channel parameters have to be estimated, and the cognitive communication settings need to be chosen in order to meet the quality of service requirements of the primary user of the channel. Quality of service requirements might include limiting the maximum or average interference caused by the secondary user, having a minimum value of the signal-to-interference plus noise ratio (SINR) of the primary user's signal, or limiting the primary user's outage probability to some threshold. In a multiuser cognitive environment, multiple secondary users compete to transmit their data over the wireless cognitive channel. Some authority decides which secondary user is assigned to use the channel according to a defined scheduling criterion. In addition, because the channels of the secondary users can change over time, a dynamic channel scheduling is anticipated.

Cognitive communication systems have seen a surge in research activity in the past few years. Some recent work on resource allocation and performance in cognitive systems includes Le & Liang (2007), Hamdi et al. (2007), Kang et al. (2009), Zhang et al. (2008), Ban et al. (2009), Li (2011), Ekin et al. (2009), and Li (2010). A fuzzy logic system was proposed in Le & Liang (2007) to opportunistically control the transmission power of the secondary users in order to provide them the ability to coexist with primary users in the same frequency band. In order to limit the interference caused by a secondary user over the primary user, a power control approach was designed in Hamdi et al. (2007). This approach adjusts the transmission power of the secondary user, depending on side information, while maintaining a quality of service requirement for the primary user's signal. The power allocation problem of a single secondary user under the primary user's outage probability loss constraint was considered in Kang et al. (2009), and both the average and peak cognitive power constraints were considered. The fading cognitive multiple-access channel was studied in Zhang et al. (2008), and the maximum sum-rate capacity of secondary users that share the wireless spectrum with an existing primary network was found. The effect of multiuser diversity in a spectrum-sharing system was investigated in Ban et al. (2009), and both single and multiple primary user cases were studied. The power allocation problem of multiple secondary users was considered in Li (2011). An optimization problem that minimizes the total transmitted power, under the constraint of minimum required secondary user's signal-to-noise ratio and maximum acceptable interference on the primary user, was developed and numerically solved. The hyper-fading channels case was considered in Ekin et al. (2009), and the capacity limits for such channels in spectrum-sharing systems under the interference-temperature constraint were studied.

The primary user's outage probability was recently investigated as a quality of service constraint in spectrum-sharing cognitive systems. For example, the queueing model of the secondary user was investigated in Farraj et al. (2011) and Farraj (2013b). The performance of the primary user in a cognitive environment was studied in Farraj & Hammad (2013b) and Farraj (2013a), and the queueing system of the primary user was investigated in Farraj (2014a). Moreover, the quality of service impact on the secondary users was investigated in Farraj & Hammad (2013a). The work of Ekin et al. (2009) was extended in Farraj & Ekin (2013) by considering the case of outage probability constraint. In addition, scheduling of secondary users in a spectrum-sharing cognitive environment was investigated in Farraj & Miller (2013) and was revisited in Farraj (2014b).

The case of multiple secondary users is considered in this chapter. In order to transmit their data to a receiver unit, the secondary users compete to share the wireless channel with the primary user using a spectrum-sharing method. Only one secondary user is scheduled to communicate over the channel at each time interval, and the usage of the channel by the scheduled secondary user should satisfy, on average, the outage probability requirement of the primary user. The work of Li (2010) proposed to schedule the secondary user that has the least channel gain. This problem is generalized in this work. The performance measures when choosing a specific secondary user, depending on its channel gain rank, are investigated. In this multiuser communication environment, the secondary users are ranked according to their channel strength, and the channel is assigned to a secondary user with a generic channel rank. This approach includes the work in Li (2010) as a special case.

The performance of the secondary users in terms of the average bit error rate and channel capacity is evaluated. The impact of choosing a specific channel rank on the resultant secondary transmission power is also considered. In addition, the effect of the number of secondary users on the secondary transmission power is considered. Three scheduling algorithms are investigated in this work as well. The average performance measures of the random-selection, round-robin, and proportional-fair scheduling algorithms are found. The performance of the primary user in a cognitive environment, although an important issue, was overlooked in recent works. Consequently, the performance measures of the primary user are investigated under different cognitive scheduling schemes. Furthermore, the communication setup is then extended to a more general case in which the secondary and primary receiver units are distinct. Performance measures of the secondary users are developed in this setup as a function of the users' rank in both the intended and the interference channels.

The rest of this chapter is organized as follows. The system model and the parameters of the communication environment are introduced in Section 2. The performance measures for the secondary users are developed in Section 3. The performance measures of the primary user under different scheduling strategies are developed in Section 4. The model is extended into distinct-receivers setup in Section 5 and a brief performance analysis is shown. Finally, conclusions are discussed in Section 6.

SYSTEM MODEL

In this environment, there is one PU that uses the wireless communication channel to transmit its data to the receiver unit (RU). At the same time, there are N SUs that plan to share this channel with the PU. The SUs intend to transmit their data to the RU as well. It is assumed that only one SU will be assigned the channel according to a specific scheduling mechanism, and the cognitive communication of that SU will take place while the PU is using the channel. As a consequence of the concurrent usage of the channel, the PU's signal will experience an extra interference due to the SU's signal, and vice versa. The described system setup is shown in Fig. 1. The RU acts as a receiver for both the PU and the SUs in this model. The RU is also assumed to have the capability to decode both signals of the PU and the scheduled SU. As an example, the RU works similar to a base station that is dedicated to support licensed users, but at the same time it has the capability to act as a receiver for lower-priority unlicensed users.

Figure 1. Cognitive system model

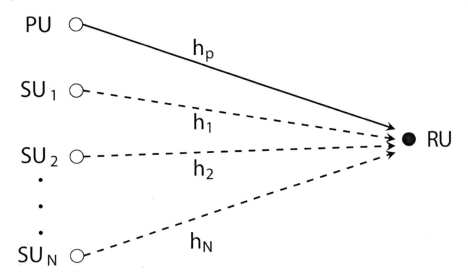

The primary outage probability is considered as the quality of service measure. If a certain SU is selected to share the channel with the PU, its usage of the channel (i.e., the transmission of its data concurrently with the PU) should not increase the PU's outage probability above a certain specified limit. Outage probability is an indication of the quality of the communication channel, and it is measured by finding the probability that a specific transmission rate is not supported because of the varying nature of the communication channel. When a channel outage occurs, it is more probable that the receiver unit will decode the received data incorrectly.

The transmission power of the PU is assumed to be constant and is equal to P_p, its bandwidth-normalized data rate is R_p, and its outage probability requirement is bounded by ξ. The outage probability of the PU is defined as $P\left\{\log_2\left(1+\gamma_p\right)\right\} \leq \xi$, where $P\left\{.\right\}$ is the probability operator and γ_p is the SINR of the PU (Goldsmith, 2005). The noise at the RU is assumed to be additive white Gaussian with zero mean and a variance of σ^2. The channel gain between the PU and the RU is denoted as h_p, and the channel gain between the nth SU and the RU is denoted as h_n, where $1 \leq n \leq N$. All channels are assumed to experience independent and identically distributed (i.i.d.) block fading random processes, with each having a Rayleigh distribution. The channel power gain, denoted as G, is defined as $G = |h|^2$. Because all channels are assumed to have an i.i.d. Rayleigh distribution, G_p and G_n will have an i.i.d. exponential distribution with mean equal to one.

PERFORMANCE OF THE SECONDARY USERS

Conducting cognitive communication will generate an extra interference on the already-established PU wireless communication link, and it is important to characterize the nature of this interference and study its impact on the performance of the scheduled SU. The generated interference is dependent on the secondary channel nature and the transmission power. In the case of multiple secondary users, as the generated secondary interference is dependent on the scheduling scheme, it is necessary to study the impact of choosing a specific SU to share the channel with the PU. In order to characterize the performance of the secondary users, a generic analysis approach, that is a function of the rank of the SU's channel power gain, is devised. Because the developed performance measures are functions of the user's channel gain rank, it is easier to see the performance difference between scheduling a secondary user that has a lower channel rank (i.e., a user with a weaker channel) over another user that has a higher channel gain rank (i.e., a user that has a stronger channel).

Analysis Framework

Different secondary users experience, in general, different channel conditions. Accordingly, it is expected that, when trying to communicate with the RU, they will have different channel gains. Because the channels are assumed to have a block fading nature, the channel gains can be considered constant at any specific time interval. The channel gains for the different secondary users can be estimated frequently, and so the channel gain rank of each user can be updated in a regular manner. Without loss of generality, the secondary users are assumed to be sorted according to the strength of their communication channels; that is to say, $G_1 \leq G_2 \cdots \leq G_N$. Accordingly, SU_1 is assumed to have the lowest channel power gain, and SU_N has the strongest channel. The transmission power of the scheduled SU is also denoted as P_s, and the channel power gain between the scheduled SU and the RU is termed as G_s.

In order to find the distribution of G_s of the scheduled user, the concept of order statistics is used. Let $X_1, X_2, \cdots X_N$ be a set of i.i.d. random variables, each having a probability density function (PDF) of $f_X(x)$ and a cumulative distribution function (CDF) of $F_X(x)$. To create the order statistics of the X s, a new set of random variables, namely $Y_1, Y_2, \cdots Y_N$, is created. In this case, Y_1 is equal to the smallest of the X s, Y_2 is equal to the second smallest, $\cdots$ and Y_N is equal to the largest of the X s. The PDF of Y_m (i.e., the mth smallest of the X s) is found as (Miller & Childers, 2004)

$$f_{Y_m}(z) = m \binom{N}{m} f_X(z) \left(F_X(z) \right)^{m-1} \left(1 - F_X(z) \right)^{N-m}. \tag{1}$$

To keep the analysis and derivation of the performance measures as general as possible, let the secondary user which has the mth weakest channel (i.e., SU_m) be scheduled to share the channel with the primary user. In this case, $G_s = G_m$, where $1 \leq m \leq N$. With the mentioned P_s and G_s, the SINR of the PU is expressed as $\gamma_p = \dfrac{P_p G_p}{P_s G_s + \sigma^2}$. In an interference-limited communication environment, the noise term in γ_p can be ignored, and so $\gamma_p \approx \dfrac{P_p G_p}{P_s G_s}$. To find the PDF of γ_p, denoted as f_{γ_p}, let $Z = X/Y$, where $X = P_p G_p$ and $Y = P_s G_s$, then $f_Z(z) = \int_0^x y f_X(yz) f_Y(y) dy$ (Papoulis and Pillai, 2002). Because the primary channel's power gain (i.e., G_p) has an exponential distribution with mean equal to one, then

$f_X(x) = \dfrac{1}{P_p} \exp\left(\dfrac{-1}{P_p} x\right)$. To find the PDF of Y, Eq. (1) is used to get

$$f_Y(y) = \binom{N}{m} \frac{m}{P_s} \exp\left(-\frac{N+1-m}{P_s} y\right)\left[1 - \exp\left(-\frac{1}{P_s} y\right)\right]^{m-1}. \tag{2}$$

Accordingly, the PDF of γ_p is given by

$$f_{\gamma_p}(z) = \frac{\beta_m}{P_p P_s} \int_0^\infty y \exp\left(-\left(\frac{z}{P_p} + \frac{N+1-m}{P_s}\right)y\right)\left[1 - \exp\left(-\frac{1}{P_s} y\right)\right]^{m-1} dy, \tag{3}$$

where $\beta_m = m\binom{N}{m}$. Using the Binomial expansion of $\left[1 - \exp\left(-\frac{1}{P_s} y\right)\right]^{m-1}$, the PDF of γ_p is expressed as (Farraj & Miller, 2013)

$$f_{\gamma_p}(z) = P_r \beta_m \sum_{k=0}^{m-1} \alpha_{mk} \frac{1}{\left(z + P_r \delta_{mk}\right)^2}, \tag{4}$$

where $\delta_{mk} = N + k + 1 - m$, $\alpha_{mk} = \binom{m-1}{k}(-1)^k$, and $P_r = \dfrac{P_p}{P_s}$. Further, the CDF of γ_p, denoted as F_{γ_p}, is found as (Farraj & Miller, 2013)

$$F_{\gamma_p}(z) = \beta_m \sum_{k=0}^{m-1} \frac{\alpha_{mk}}{\delta_{mk}} \frac{z}{z + P_r \delta_{mk}}. \tag{5}$$

The value of P_s that satisfies, on average, the outage probability requirement of the PU can be found from $P\left\{\log_2(1+\gamma_p) \le R_p\right\} \le \xi$ or $F_{\gamma_p}(2^{R_p} - 1) \le \xi$. In this case,

$$\sum_{k=0}^{m-1} \frac{\alpha_{mk}}{\delta_{mk}} \frac{\beta_m}{2^{R_p} - 1 + P_r \delta_{mk}} \le \frac{\xi}{\left(2^{R_p} - 1\right)}, \tag{6}$$

and from this relation the value of P_r can be obtained. The transmission power of the scheduled secondary user that is needed to satisfy the outage probability requirement of the PU is found as $P_s = \dfrac{1}{P_r} P_p$.

An interesting observation from Eq. (6) is that the value of P_r depends only on R_p, ξ, N, and m. Because the value of P_r is independent of anything else (including P_p), then the value of P_s is linear with P_p.

A similar approach can be used to find the secondary user's SINR, termed as γ_s, in an interference-limited environment. In this case, $\gamma_s \approx \dfrac{P_s G_s}{P_p G_p}$. To find the PDF of γ_s, let $Z = X/Y$, where

$X = P_s G_s$ and $Y = P_p G_p$. Consequently, $f_Y(y) = \dfrac{1}{P_p} \exp\left(\dfrac{-1}{P_p} y\right)$ and

$f_X(x) = \dfrac{\beta_m}{P_s} \exp\left(-\dfrac{N+1-m}{P_s} x\right)\left(1 - \exp\left(-\dfrac{1}{P_s} x\right)\right)^{m-1}$. The PDF of γ_s is consequently expressed as (Farraj & Miller, 2013)

$$f_{\gamma_s}(z) = \frac{\beta_m}{P_r} \sum_{k=0}^{m-1} \frac{\alpha_{mk}}{\delta_{mk}^2} \frac{1}{\left(z + \dfrac{1}{P_r \delta_{mk}}\right)^2} . \tag{7}$$

Next, based on the developed probability density of γ_s, the performance measures of the SUs are developed.

Bit Error Rate: Assuming that the scheduled SU (i.e., SU$_m$) uses a Binary Phase Shift Keying (BPSK) modulation scheme, and for an SINR of γ_s, the instantaneous bit error rate is $Q\left(\sqrt{2\gamma_s}\right)$ (Sklar, 1988), where $Q(.)$ is the Gaussian Q-function. The average bit error rate of the scheduled SU, denoted as BER_m, is found as $BER_m = E\left[Q\left(\sqrt{2\gamma_s}\right)\right]$, where $E[.]$ is the expectation operator. Accordingly,

$$BER_m = \frac{\beta_m}{P_r} \int_0^\infty Q\left(\sqrt{2z}\right) \sum_{k=0}^{m-1} \frac{\alpha_{mk}}{\delta_{mk}^2} \frac{1}{\left(z + \dfrac{1}{P_r \delta_{mk}}\right)^2} dz . \tag{8}$$

It can be shown that the average bit error rate of SU$_m$ will be equal to

$$BER_m = \beta_m \sum_{k=0}^{m-1} \frac{\alpha_{mk}}{\delta_{mk}} \left(\frac{1}{2} - \sqrt{\frac{\pi}{P_r \delta_{mk}}} \exp\left(\frac{1}{P_r \delta_{mk}}\right) Q\left(\sqrt{\frac{2}{P_r \delta_{mk}}}\right)\right) . \tag{9}$$

It is noted here that BER_m is function of m, N, R_p, and ξ. Because P_r is independent of P_p and P_s, the values of P_p and P_s do not affect the average bit error rate of the secondary user.

Channel Capacity: For an SINR of γ_s, the SU's instantaneous bandwidth-normalized channel capacity is $\log_2\left(1+\gamma_s\right)$ (Cover & Thomas, 1991). Accordingly, the average capacity of the secondary channel, termed as C_m, is found as $C_m = E\left[\log_2\left(1+\gamma_s\right)\right]$. In this case, the average channel capacity is expressed as

$$C_m = \frac{\beta_m}{P_r} \int_0^\infty \log_2\left(1+z\right) \sum_{k=0}^{m-1} \frac{\alpha_{mk}}{\delta_{mk}^2} \frac{1}{\left(z+\frac{1}{P_r\delta_{mk}}\right)^2} dz, \qquad (10)$$

and it can be shown that this will result in (Farraj & Miller, 2013)

$$C_m = \frac{\beta_m}{P_r} \sum_{k=0}^{m-1} \frac{\alpha_{mk}}{\delta_{mk}^2} \begin{cases} \log_2\left(e\right) & P_r\delta_{mk} = 1 \\ \frac{P_r\delta_{mk}}{P_r\delta_{mk}-1}\log_2\left(P_r\delta_{mk}\right) & P_r\delta_{mk} \neq 1 \end{cases}. \qquad (11)$$

This result is consistent with the findings of Eq. (9) in that the performance measure is function of the parameters of the communication environment, number of secondary users, and the rank of the scheduled user.

Numerical Results

In this subsection, the performance of the secondary users is numerically investigated. The effect of choosing the secondary user that has the mth weakest channel gain as a candidate to share the channel with the PU is considered. The impact on power, bit error rate, and channel capacity is studied. For the following results, P_p is set to 10 dBm and N is 7.

Transmission Power: First, the transmission power that is needed to satisfy the primary user's outage probability requirement is considered. Fig. 2 presents P_s versus the order of channel gain, m, for different values of R_p and ξ. It is not surprising to see that with increasing channel gain rank, less transmission power is required to satisfy the outage probability requirement. In other words, when an SU with a stronger channel is assigned to share the channel with the PU, this SU will need less transmission power. It is interesting to see that although there is a handful number of secondary users, the transmission power difference is substantial. A saving of about 13 dB is possible if the system schedules the secondary user with the strongest channel compared to the case when the user that has the weakest channel is scheduled to share the channel. The figure also shows that the SU can transmit at a higher power when the primary outage probability increases. At the same time, with an increase of the primary user's data rate, the secondary user has to reduce its transmission power.

Bit Error Rate: The average bit error rate results are shown in Fig. 3. It is seen that assigning the channel to the user with the weakest channel (i.e., $m=1$) will result in the highest bit error rate. In this case, a secondary user with a stronger channel will get better results although it transmits at a lower

Figure 2. Cognitive power (dBm)

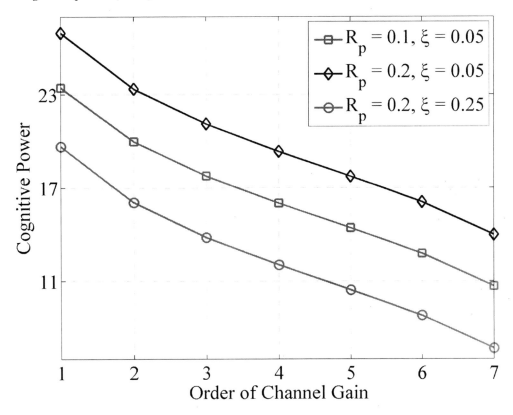

transmission power. To explain this behavior, the average bit error rate when $m = 1$ was shown to be identical to that of the random-selection scheme. Accordingly, any scheme that schedules a user that has a better channel than the weakest channel will achieve better performance. It is also noted that as the outage probability requirement gets higher, the SU experiences lower bit error rates. On the other hand, with increasing the data rate of the primary user, the bit error rate becomes higher. As Fig. 2 shows, P_s increases with increasing ξ and decreasing R_p. When P_s increases, γ_s increases, and so $Q\left(\sqrt{2\gamma_s}\right)$ decreases; accordingly, a lower bit error rate can be achieved.

Channel Capacity: Fig. 4 displays the channel capacity results. It can be seen that the average channel capacity increases with increasing the channel gain rank. Similar to the bit error rate results, the user with the lowest channel gain will achieve the lowest capacity. Because the average channel capacity when $m = 1$ is identical to that of the random-selection scheduling scheme, any scheme that schedules an SU with a higher channel gain rank will achieve better results. As the outage probability requirement becomes higher, the SU achieves more channel capacity. Further, with increasing primary user's transmission rate, the channel capacity that can be achieved by the secondary user becomes lower. This behavior has a similar explanation to that given for the results in Fig. 3.

Figure 3. Cognitive bit error rate

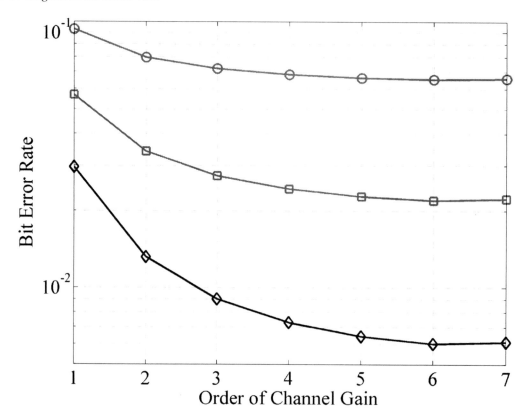

Special Cases

Three special cases are considered in this subsection. Specifically, the performance measures are found when there is only one secondary user and when the scheduled user has the strongest or weakest channel.

Single Secondary User Case: There is only one SU in the communication environment in this case; consequently, it can be treated as a special case of the multiple-users scenario where $N = 1$. By setting the value of N to 1 in Eqs. (4), (5), and (7), the following results are found

$$f_{\gamma_p}(z) = \frac{P_r}{\left(z + P_r\right)^2}$$

$$F_{\gamma_p}(z) = \frac{z}{z + P_r}$$

Figure 4. Cognitive channel capacity (bit/sec/Hz)

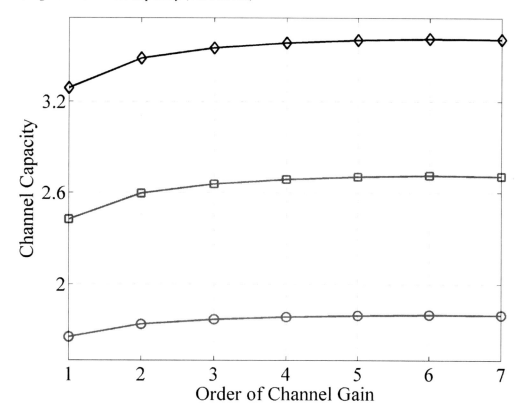

$$f_{\cdot_s}(z) = \frac{1}{P_r} \frac{1}{\left(z + \dfrac{1}{P_r}\right)^2}.$$ (12)

It can be shown that $P_s = \dfrac{\xi}{1-\xi} \dfrac{P_p}{2^{R_p}-1}$ satisfies the outage probability requirement of the PU, and

in this case $P_r = \dfrac{1-\xi}{\xi}\left(2^{R_p}-1\right)$. In addition, the average bit error rate can be expressed as

$$BER_m = \frac{1}{2} - \sqrt{\frac{\pi}{P_r}} \, \exp\left(\frac{1}{P_r}\right) Q\left(\sqrt{\frac{2}{P_r}}\right).$$

Further, the average channel capacity is also given by

$$C_m = \begin{cases} \log_2(e) & P_r = 1 \\ \dfrac{\log_2(P_r)}{P_r - 1} & P_r \neq 1 \end{cases}.$$

Strongest Channel Case: The secondary user that has the strongest channel is scheduled to share the channel with the PU in this case. The performance measures of this case can be found by setting $m = N$. Accordingly, the density and distribution functions are found as

$$f_{\gamma_p}(z) = P_r N \sum_{k=0}^{N-1} \alpha_{Nk} \frac{1}{\left(z + P_r(k+1)\right)^2}$$

$$F_{\gamma_p}(z) = N \sum_{k=0}^{N-1} \frac{\alpha_{Nk}}{k+1} \frac{z}{z + P_r(k+1)}$$

$$f_{\gamma_s}(z) = \frac{N}{P_r} \sum_{k=0}^{N-1} \frac{\alpha_{Nk}}{(k+1)^2} \frac{1}{\left(z + \dfrac{1}{P_r(k+1)}\right)^2}. \tag{13}$$

Going back to the outage probability requirement of the PU (i.e., $F_{\gamma_p}\left(2^{R_p} - 1\right) \leq \xi$), and using the results from Eq. (13), then

$$\sum_{k=0}^{N-1} \frac{\alpha_k}{k+1} \frac{1}{2^{R_p} - 1 + (k+1)\dfrac{P_p}{P_s}} \leq \frac{\xi}{N\left(2^{R_p} - 1\right)},$$

and the value of P_s that satisfies, on average, the outage probability requirement of the PU can be found. In addition, the average bit error rate in this case is found as

$$BER_m = N \sum_{k=0}^{N-1} \frac{\alpha_{Nk}}{2(k+1)} - \sqrt{\pi P_r(k+1)} \frac{\alpha_{Nk}}{P_r(k+1)^2} \exp\left(\frac{1}{P_r(k+1)}\right) Q\left(\sqrt{\frac{2}{P_r(k+1)}}\right),$$

and the average channel capacity is expressed as

$$C_m = \frac{N}{P_r} \sum_{k=0}^{N-1} \frac{\alpha_{Nk}}{(k+1)^2} \begin{cases} \log_2(e) & P_r(k+1) = 1 \\ \dfrac{P_r(k+1)}{P_r(k+1)-1} \log_2\left(P_r(k+1)\right) & P_r(k+1) \neq 1 \end{cases}.$$

Weakest Channel Case: In this case, the secondary user that has the weakest channel to the RU is scheduled to share the channel with the PU. For the case of $m = 1$, the PDF and CDF of γ_p and the PDF of γ_s are derived as

$$f_{\gamma_p}(z) = \frac{NP_r}{\left(z + NP_r\right)^2}$$

$$F_{\gamma_p}(z) = \frac{z}{z + NP_r}$$

$$f_{\gamma_s}(z) = \frac{1}{NP_r} \frac{1}{\left(z + \dfrac{1}{NP_r}\right)^2}. \tag{14}$$

The value of P_s that satisfies, on average, the outage probability requirements of the PU needs to be found. It is required that $P\left\{\log_2\left(1+\gamma_p\right) \leq R_p\right\} \leq \xi$ or $F_{\gamma_p}\left(2^{R_p}-1\right) \leq \xi$. Using the results from Eq.

(14), $P_s = \dfrac{\xi}{1 - \xi} \dfrac{NP_p}{2^{R_p}-1}$, and so $P_r = \dfrac{1-\xi}{\xi} \dfrac{2^{R_p}-1}{N}$ which is consistent with the findings in Li (2010). It can also be shown that when the scheduler picks the secondary user that has the lowest channel gain, the average bit error rate is expressed as

$$BER_m = \frac{1}{2} - \sqrt{\frac{\pi}{NP_r}} \exp\left(\frac{1}{NP_r}\right) Q\left(\sqrt{\frac{2}{NP_r}}\right).$$

Further, the average channel capacity can be found as

$$C_m = \begin{cases} \log_2(e) & NP_r = 1 \\ \dfrac{\log_2\left(NP_r\right)}{NP_r - 1} & NP_r \neq 1 \end{cases}.$$

Both values of BER_m and C_m are consistent with the findings of Li (2010).

Scheduling Schemes

In this subsection, three scheduling algorithms, namely, random selection, round-robin, and proportional fair, are considered. The channel is assigned to a random secondary user, regardless of its channel strength, in the random-selection scheduling algorithm. The round-robin algorithm schedules all secondary users over the wireless channel in a periodic manner. Finally, the proportional-fair algorithm schedules a secondary user over the channel when its instantaneous channel quality is high relative to its own average channel condition. Making use of the generic performance approach developed above, the performance measures of these scheduling algorithms are easily developed.

Random Selection: For the random-selection scheduling algorithm, there are multiple secondary users that want to share the channel with the PU, and the system assigns the channel to a certain user randomly regardless of its channel status. When selected, the secondary user has to satisfy the outage probability requirement of the primary user in order to transmit over the channel. It is noted that

$f_X(x) = \dfrac{1}{P_p} \exp\left(\dfrac{-1}{P_p} x\right)$ and $f_Y(y) = \dfrac{1}{P_s} \exp\left(\dfrac{-1}{P_s} y\right)$ in this case. This makes the random-selection

scheduling algorithm, performance wise, equivalent to the single secondary user case. Consequently,

the value of f_{z_s} is found in Eq. (12). Similarly, $P_s = \dfrac{\xi}{1 - \xi} \dfrac{P_p}{2^{R_p} - 1}$ satisfies the outage probability re-

quirement of the PU. In addition, the average bit error rate of this scheduling algorithm, termed as BER_{rs}, can be also expressed as

$$BER_{rs} = \frac{1}{2} - \sqrt{\frac{\pi}{P_r}} \exp\left(\frac{1}{P_r}\right) Q\left(\sqrt{\frac{2}{P_r}}\right). \tag{15}$$

Moreover, the average channel capacity of this scheme, denoted as C_{rs}, is given by

$$C_{rs} = \begin{cases} \log_2(e) & P_r = 1 \\ \dfrac{\log_2(P_r)}{P_r - 1} & P_r \neq 1 \end{cases}. \tag{16}$$

The performance measures developed for the random scheduling are identical to those of the $\left(m = 1\right)$ case. On the other hand, the average transmission power of the $\left(m = 1\right)$ case is N times that of the random-selection scheduling algorithm.

Round-Robin: The channel is shared between the secondary users on a time division basis in the round-robin scheduling algorithm. Equal time slots are assigned in circular order to each secondary user. Channel assignment is implemented without priority, and secondary users are assigned the channel in a timely manner regardless of their channel gain.

Let the scenario where N repetitive time slots are assigned for the N secondary users be considered. The first secondary user (i.e., SU_1) is assigned the first time slot in the round, SU_2 is assigned the second time slot, and so on till the round repeats after SU_N is assigned the Nth time slot. In time slot m, where $1 \leq m \leq N$, SU_m is assigned the channel, and this user has to satisfy the QoS requirement in order to communicate over the channel. Because SU_m has the mth weakest channel, the PDFs of γ_p and γ_s are identical to those developed in Eqs. (4) and (7), respectively. Accordingly, its bit error rate and channel capacity are found in Eqs. (9) and (11), respectively. Because this algorithm schedules all the secondary users in a given round, the average bit error rate of this scheduling algorithm, denoted as BER_{rr}, is found by averaging the bit error results for the N secondary users. In this case, the average bit error rate is expressed as

$$BER_{rr} = \frac{1}{N} \sum_{m=1}^{N} \beta_m \sum_{k=0}^{m-1} \frac{\alpha_{mk}}{\delta_{mk}} \left(\frac{1}{2} - \sqrt{\frac{\pi}{P_r \delta_{mk}}} \exp\left(\frac{1}{P_r \delta_{mk}} \right) Q\left(\sqrt{\frac{2}{P_r \delta_{mk}}} \right) \right). \tag{17}$$

In a similar manner, the average channel capacity of this algorithm, termed as C_{rr}, is found as

$$C_{rr} = \frac{1}{N} \sum_{m=1}^{N} \frac{\beta_m}{P_r} \sum_{k=0}^{m-1} \frac{\alpha_{mk}}{\delta_{mk}^2} \begin{cases} \log_2(e) & P_r \delta_{mk} = 1 \\ \frac{P_r \delta_{mk}}{P_r \delta_{mk} - 1} \log_2(P_r \delta_{mk}) & P_r \delta_{mk} \neq 1 \end{cases}. \tag{18}$$

Proportional Fair: In this algorithm, the scheduler decides which secondary user to communicate over the channel at each time slot based on the requested rates from the secondary users. The scheduler keeps track of the average throughput of each secondary user in an exponentially weighted window of length t_c. In time slot k, the scheduler assigns the channel to user SU_m* that has the largest $\frac{R_m[k]}{T_m[k]}$, where $R_m[k]$ is the rate of SU_m at time slot k and $T_m[k]$ is the average throughput of SU_m at time slot k (Tse & Viswanath, 2005). For the case of channels with identical fading statistics, if t_c is much larger than the coherence time of the channels, then the throughput of each user $T_m[k]$ converges to the same quantity. The proportional-fair scheduling algorithm reduces to always selecting the secondary user with the highest rate. Thus, a secondary user is scheduled to transmit over the channel when its channel is the best, and at the same time the scheduling algorithm is perfectly fair in the long-term. More details about this algorithm can be found in Tse & Viswanath (2005).

Because this algorithm assigns the channel, at any given time slot, to the secondary user with the best channel (i.e., the user with the highest channel rank), the performance measures of this algorithm can be found from Eqs. (9) and (11) by setting $m = N$ (i.e., similar to the strongest channel case discussed above). Accordingly, the average bit error rate of this algorithm, termed as BER_{pf}, is expressed as

$$BER_{pf} = \beta_N \sum_{k=0}^{N-1} \frac{\alpha_{Nk}}{\delta_{Nk}} \left(\frac{1}{2} - \sqrt{\frac{\pi}{P_r \delta_{Nk}}} \exp\left(\frac{1}{P_r \delta_{Nk}}\right) Q\left(\sqrt{\frac{2}{P_r \delta_{Nk}}}\right) \right), \tag{19}$$

and the average channel capacity, denoted as C_{pf}, is found to be

$$C_{pf} = \frac{\beta_N}{P_r} \sum_{k=0}^{N-1} \frac{\alpha_{Nk}}{\delta_{Nk}^2} \begin{cases} \log_2(e) & P_r \delta_{Nk} = 1 \\ \dfrac{P_r \delta_{Nk}}{P_r \delta_{Nk} - 1} \log_2(P_r \delta_{Nk}) & P_r \delta_{Nk} \neq 1 \end{cases}. \tag{20}$$

PERFORMANCE OF THE PRIMARY USER

The performance measures of the primary user in a multiuser cognitive environment are found in this section. As shown in Fig. 1, there is a single primary user that uses the channel to transmit its data to the RU, and there are N secondary users that compete to share this channel in order to transmit their data to the RU as well. The cognitive communication will occur at the same time while the PU is using the channel. Only one SU will be scheduled to communicate over the channel according to some scheduling criterion. At first, the performance of the primary user is found when there is only one secondary user in the setup. Then, the performance measures of the primary user are found for a scheme that schedules the secondary user that has the strongest channel gain. Further, the performance of the primary user when the scheduled secondary user has the least channel gain, as was proposed in Li (2010), is investigated.

Single Secondary User Case

There is only one SU with parameters P_s and G_s in this case and $N = 1$. The PDF and CDF of γ_p were previously found in Eq. (12). The average bit error rate, denoted as BER_p, when the PU is using BPSK modulation scheme is found as

$$BER_p = \frac{1}{2} - \sqrt{\pi P_r} \exp(P_r) Q\left(\sqrt{2P_r}\right). \tag{21}$$

It is interesting to see that this result is independent of the transmission powers of the PU and the SU. The average bit error rate is only function of the parameters of the communication environment. The primary user's average channel capacity, termed as C_p, is also developed as

$$C_p = \begin{cases} \log_2(e) & P_r = 1 \\ \dfrac{P_r}{P_r - 1} \log_2(P_r) & P_r \neq 1 \end{cases}. \tag{22}$$

Strongest Channel Case

The secondary user that has the strongest channel is scheduled to share the channel with the PU (i.e., $m=N$ or $G_s = \max(G_n)$) in this case. The density function of γ_p is found in Eq. (13). Further, the average bit error rate of the PU can be found as (Farraj & Hammad, 2013b)

$$
BER_p = N \sum_{k=0}^{N-1} \frac{\alpha_k}{2(k+1)} - \alpha_k \sqrt{\frac{\pi P_r}{k+1}} \, \exp\left(P_r\left(k+1\right)\right) Q\left(\sqrt{2P_r\left(k+1\right)}\right). \tag{23}
$$

Similarly, the average channel capacity of the PU, when the SU with the highest channel gain is scheduled to use the channel, is expressed as (Farraj & Hammad, 2013b)

$$
C_p = NP_r \sum_{k=0}^{N-1} \alpha_k \begin{cases} \log_2\left(e\right) & P_r\left(k+1\right)=1 \\ \dfrac{1}{P_r\left(k+1\right)-1} \log_2\left(P_r\left(k+1\right)\right) & P_r\left(k+1\right) \neq 1 \end{cases}. \tag{24}
$$

Weakest Channel Case

In this case, the secondary user that has the weakest channel to the RU is scheduled to share the channel with the PU. The channel power gain between that SU and the RU as G_s, where $G_s = \min(G_n)$, where $1 \leq n \leq N$ (i.e., $m=1$). The performance measures of the PU are derived using the findings of Eq. (14). The average bit error rate of the PU is expressed as

$$
BER_p = \frac{1}{2} - \sqrt{\pi NP_r} \, \exp\left(NP_r\right) Q\left(\sqrt{2NP_r}\right), \tag{25}
$$

and this result is identical to that of Eq. (21). In other words, the average bit error rate for the PU in a single-secondary user case is identical to that of a multiple-secondary user case when the SU with the weakest channel is scheduled to use the channel. Moreover, the average capacity of the primary channel is found as

$$
C_p = \begin{cases} \log_2\left(e\right) & NP_r=1 \\ \dfrac{NP_r}{NP_r-1} \log_2\left(NP_r\right) & NP_r \neq 1 \end{cases}. \tag{26}
$$

Again, it is seen here that the average channel capacity is only dependent on the variables of the communication environment.

GENERAL SYSTEM MODEL

In this section, the model described in Fig. 1 is extended into a more general setup where the secondary receiver is distinct from the primary receiver. The model is illustrated in Fig. 5. In this case, the PU sends its data to the primary receiver unit (PR), and the secondary users intend to transmit their data to the secondary receiver unit (SR). Secondary interference can occur over the PR, and it is a function of the secondary power and the channel characteristics between the scheduled SU and the PR. The effect of selecting a specific SU on the performance measures is to be investigated. In this case, the secondary interference channel from the SU to the PR (i.e., P-S channel) is different from the intended channel from the SU to the SR (i.e., S-S channel). Consequently, an SU that has a very strong P-S channel might have a weak S-S channel, and so on.

Performance Analysis

As noted before, it is in the best interest of the scheduled SU to limit its interference on the PR and at the same time have the maximum received power at the SR. Let the case of an SU that has the weakest P-S channel and the strongest S-S channel be first considered. In order to satisfy the outage probability requirement, and because of having the weakest interference channel, the transmission power of that SU will be the highest as was explained in Fig. 2. At the same time, with the maximum transmission power and the best S-S channel, the received power at the intended receiver (i.e., SR) will be the highest. Consequently, it is anticipated that the best candidate to share the channel with the PU is an SU that has the strongest secondary channel and weakest interference channel. Because the different secondary users experience different channel conditions, it is uncommon to have an SU that has both the strongest S-S channel and weakest P-S channel at the same time. Accordingly, out of N SUs, to decide which SU will achieve the best performance, and so be scheduled to use the secondary channel, the SUs are ranked into

Figure 5. General cognitive system model

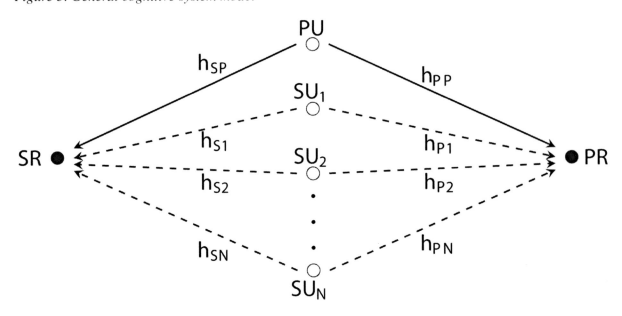

two lists; one list according to their P-S channel power gain rank, and another list according to their S-S channel power gain rank. A similar analysis to the one that was presented in Section 3 is conducted to derive the performance measures of the secondary users.

Each SU can be represented by an (m, n) pair, where m represents its P-S channel's power gain rank, and n represents the S-S channel's power gain rank, where $1 \leq m, n \leq N$. The SU that has the mth weakest P-S channel and the nth weakest S-S channel is assumed to be scheduled to share the channel with the PU. For this case, let $G_{PS} = G_{Pm}$ and $G_{SS} = G_{Sn}$. The SINR of the PU's signal at the PR is approximated in an interference-limited environment as $\gamma_p \approx \dfrac{P_p G_{PP}}{P_s G_{PS}}$. Following the same logic presented in Section 3, and assuming that all channels experience i.i.d. Rayleigh block fading, the secondary transmitted power that satisfies the outage probability requirement is found according to Eq. (6). On the other hand, the SINR of the secondary user's signal at the SR is found as $\gamma_s \approx \dfrac{P_s G_{SS}}{P_p G_{SP}}$. Similar to Eq. (7), the PDF of γ_s is expressed as

$$f_{\gamma_s}(z) = \frac{\beta_n}{P_r} \sum_{k=0}^{n-1} \frac{\alpha_{nk}}{\delta_{nk}^2} \frac{1}{\left(z + \dfrac{1}{P_r \delta_{nk}}\right)^2}, \tag{27}$$

where $\beta_n = n \binom{N}{n}$, $\delta_{nk} = N + k + 1 - n$, $\alpha_{nk} = \binom{n-1}{k}(-1)^k$. Moreover, for a BPSK modulation scheme, the average bit error rate can be found as

$$BER_{m,n} = \beta_n \sum_{k=0}^{n-1} \frac{\alpha_{nk}}{\delta_{nk}} \left(\frac{1}{2} - \sqrt{\frac{\pi}{P_r \delta_{nk}}} \exp\left(\frac{1}{P_r \delta_{nk}}\right) Q\left(\sqrt{\frac{2}{P_r \delta_{nk}}}\right)\right). \tag{28}$$

The average secondary channel capacity can also be expressed as (Farraj & Miller, 2013)

$$C_{m,n} = \frac{\beta_n}{P_r} \sum_{k=0}^{n-1} \frac{\alpha_{nk}}{\delta_{nk}^2} \begin{cases} \log_2(e) & P_r \delta_{nk} = 1 \\ \dfrac{P_r \delta_{nk}}{P_r \delta_{nk} - 1} \log_2(P_r \delta_{nk}) & P_r \delta_{nk} \neq 1 \end{cases}. \tag{29}$$

As seen from the above equations, the rank of the SU's interference channel (i.e., P-S channel) will determine its transmission power. On the other hand, its rank in the intended secondary channel (i.e., S-S channel) will determine the performance measures. As obvious, the resultant performance for a specific SU will be a function of its (m, n) channel rank pair.

Numerical Example

The following is an example on how to use the results in Eq. (29) to decide which SU should be scheduled to share the channel with the PU. As explained before, the performance of an SU is dependent on its (m, n) rank pair. In a given situation, each secondary user is ranked in the m-list and n-list according to its corresponding channel gains, and the corresponding performance measures are found for each rank pair. Finally, the channel is assigned to the secondary user that will achieve the best performance.

It is assumed that there are 4 SUs that want to share the channel with the PU, let them be called SU_a, SU_b, SU_c, and SU_d. It is also assumed that their (m, n) rank pairs are as follows: $(2, 3)$, $(1, 1)$, $(3, 2)$, and $(4, 4)$, respectively. Apparently, SU_d has the strongest P-S and S-S channels; on the other hand, SU_b has the weakest channels, and so on. Table 1 displays the average channel capacity for a cognitive system with $N = 4$, $R_p = 0.05$ bit/sec/Hz, $\xi = 0.2$, and $P_p = 10$ dBm. According to Table 1, the average channel capacity for SU_a will be 4.35 bit/sec/Hz, that of SU_b is 3.29 bit/sec/Hz, SU_c will have an average capacity of 2.75 bit/sec/Hz, and the capacity of SU_d is 3.57 bit/sec/Hz. Then, the scheduler decision will be to select SU_a to communicate over the secondary channel. Although SU_a does not have the strongest channel, this user will achieve the highest secondary channel capacity.

CONCLUSION

This work explores some performance measures of the primary and secondary users in a spectrum-sharing cognitive environment. Multiple secondary users want to share a channel dedicated to a primary user in order to transmit their data to a common receiver unit using spectrum sharing. In this case, only one secondary user is allowed to share the channel with the licensed user, and to do so, it should not increase the primary user's outage probability above a certain limit.

Secondary users are ranked according to their channel strength, and the impact of choosing a secondary

Table 1. Secondary channel capacity (bit/sec/Hz)

n/m	1	2	3	4
1	3.29	2.34	1.77	1.28
2	4.63	3.48	2.75	2.08
3	5.57	4.35	3.55	2.78
4	6.52	5.26	4.41	3.57

user with a generic channel gain rank on the resultant performance is studied. The effects of changing the primary outage probability requirement and the primary user's transmission rate are considered as well. It is found that as the channel gain rank increases (i.e., the secondary channel becomes stronger), the required transmission power becomes lower, and the performance measures of the secondary user become better.

The performance of the primary user of the channel is also investigated under different secondary user scheduling schemes. For each case, the primary user's average bit error rate and channel capacity are derived. The performance measures of the primary user are found to depend on the secondary user's scheduling criterion and on the parameters of the communication environment. This work also considers the case where the secondary receiver is different from the primary receiver. It is noted that, for the same interference channel rank, better performance can be achieved by scheduling a user with a better secondary channel rank. The performance of a specific secondary user is found to be dependent on both its interference channel rank and its secondary channel rank.

REFERENCES

Ban, T. W., Choi, W., Jung, B. C., & Sung, D. K. (2009). Multi-user diversity in a spectrum sharing system. *IEEE Transactions on Wireless Communications*, 8(1), 102–106. doi:10.1109/T-WC.2009.080326

Cover, T. M., & Thomas, J. A. (1991). *Elements of Information Theory*. John Wiley & Sons, Inc. doi:10.1002/0471200611

Ekin, S., Yilmaz, F., Celebi, H., Qaraqe, K., Alouini, M.-S., & Serpedin, E. (2009). Achievable capacity of a spectrum sharing system over hyper fading channels. In *Proceedings of IEEE Global Communications Conference (GLOBECOM)*. IEEE. doi:10.1109/GLOCOM.2009.5425715

Farraj, A. K. (2013a). Impact of cognitive communications on the performance of the primary users. *Wireless Personal Communications*, 71(2), 975–985. doi:10.1007/s11277-012-0855-4

Farraj, A. K. (2013b). Queue model analysis for spectrum-sharing cognitive systems under outage probability constraint. *Wireless Personal Communications*, 73(3), 1021–1035. doi:10.1007/s11277-013-1245-2

Farraj, A. K. (2014a). Analysis of primary users' queueing behavior in a spectrum-sharing cognitive environment. *Wireless Personal Communications*, 75(2), 1283–1293. doi:10.1007/s11277-013-1423-2

Farraj, A. K. (2014b). Switched-diversity approach for cognitive scheduling. *Wireless Personal Communications*, 74(2), 933–952. doi:10.1007/s11277-013-1331-5

Farraj, A. K., & Ekin, S. (2013). Performance of cognitive radios in dynamic fading channels under primary outage constraint. *Wireless Personal Communications*, 73(3), 637–649. doi:10.1007/s11277-013-1207-8

Farraj, A. K., & Hammad, E. M. (2013a). Impact of quality of service constraints on the performance of spectrum sharing cognitive users. *Wireless Personal Communications*, 69(2), 673–688. doi:10.1007/s11277-012-0606-6

Farraj, A. K., & Hammad, E. M. (2013b). Performance of primary users in spectrum sharing cognitive radio environment. *Wireless Personal Communications*, 68(3), 575–585. doi:10.1007/s11277-011-0469-2

Farraj, A. K., & Miller, S. L. (2013). Scheduling in a spectrum-sharing cognitive environment under outage probability constraint. *Wireless Personal Communications, 70*(2), 785–805. doi:10.1007/s11277-012-0722-3

Farraj, A. K., Miller, S. L., & Qaraqe, K. A. (2011). Queue performance measures for cognitive radios in spectrum sharing systems. In *Proceedings of IEEE International Workshop on Recent Advances in Cognitive Communications and Networking (RACCN) – Global Telecommunications Conference (GLOBECOM) Workshop*. IEEE. doi:10.1109/GLOCOMW.2011.6162606

Federal Communications Commission. (2002). *Spectrum Policy Task Force Report* (ET Docket No. 02-135). Washington, DC: FCC.

Goldsmith, A. (2005). *Wireless Communications*. Cambridge University Press. doi:10.1017/CBO9780511841224

Hamdi, K., Zhang, W., & Letaief, K. B. (2007). Power control in cognitive radio systems based on spectrum sensing side information. In *Proceedings of IEEE International Conference on Communications (ICC)*. IEEE. doi:10.1109/ICC.2007.853

Kang, X., Zhang, R., Liang, Y.-C., & Garg, H. K. (2009). Optimal power allocation for cognitive radio under primary user's outage loss constraint. In *Proceedings of IEEE International Conference on Communications (ICC)*. IEEE.

Le, H. S. T., & Liang, Q. (2007). An efficient power control scheme for cognitive radios. In *Proceedings of IEEE Wireless Communications and Networking Conference (WCNC)*. IEEE. doi:10.1109/WCNC.2007.476

Li, D. (2010). Performance analysis of uplink cognitive cellular networks with opportunistic scheduling. *IEEE Communications Letters, 14*(9), 827–829. doi:10.1109/LCOMM.2010.072910.100962

Li, D. (2011). Efficient power allocation for multiuser cognitive radio networks. *Wireless Personal Communications, 59*(4), 589–597. doi:10.1007/s11277-010-9926-6

Miller, S. L., & Childers, D. G. (2004). *Probability and Random Processes: With Applications to Signal Processing and Communications*. Elsevier Academic Press.

Mitola, J. (2000). *Cognitive radio: An integrated agent architecture for software defined radio.* (PhD Thesis). Royal Institute of Technology (KTH), Stockholm, Sweden.

Papoulis, A., & Pillai, S. U. (2002). *Probability, Random Variables, and Stochastic Processes* (4th ed.). McGraw-Hill.

Sklar, B. (1988). *Digital Communications: Fundamentals and Applications*. Prentice Hall.

Tse, D., & Viswanath, P. (2005). *Fundamentals of Wireless Communication*. Cambridge University Press. doi:10.1017/CBO9780511807213

Zhang, R., Cui, S., & Liang, Y.-C. (2008). On ergodic sum capacity of fading cognitive multiple-access channel. In *Proceedings of Forty-Sixth Annual Allerton Conference*. Academic Press. doi:10.1109/ALLERTON.2008.4797650

KEY TERMS AND DEFINITIONS

Bit Error Rate: A measure of the quality of the received digital signal. It is calculated by dividing the error bits by the total number of transmitted bits.

Channel Capacity: An upper bound on the rate of data that can be reliably transmitted over the communications channel.

Outage Probability: An indication of the quality of the communication channel. It is measured by finding the probability that a specific transmission rate is not supported.

Primary User: A transceiver unit that is licensed to use a specific wireless channel in a cognitive communication system.

Secondary User: A radio unit that adapts its communication settings in order to transmit its data over the primary channel even though it is not licensed to operate over that channel.

Signal-To-Interference Plus Noise Ratio: A measure that describes the quality of the received signal of the desired user. It is calculates by dividing the power of the desired user's signal over the interference power and the noise power.

Spectrum-Sharing Cognitive Communications: A mode of cognitive communications in which the secondary user transmits data simultaneously along with the primary user over the same channel.

Chapter 14
Distributed Mechanisms for Multiple Channel Acquisition in a System of Uncoordinated Cognitive Radio Networks

Kenneth Ezirim
City University of New York, USA

Shamik Sengupta
University of Nevada – Reno, USA

Ping Ji
City University of New York, USA

ABSTRACT

Due to the constraint imposed by the Dynamic Spectrum Access paradigm, Cognitive Radio (CR) networks are entangled in persistent competition for opportunistic access to underutilized spectrum resources. In order to maintain quality of service, each network faces the challenge of acquiring dynamic enough channels to meet channel size requirement. The main goal of every CR network is to minimize the amount of contention experienced during channel acquisition and to maximize the utility derived from acquired channels. This is a major challenge, especially without a global communication protocol that can facilitate communication between the networks. This chapter discusses self-coexistence of CR networks in a decentralized system with no support for coordinated radio transmission activities. Channel acquisition mechanisms that can help networks minimize contention and maximize utility are also discussed. The mechanisms guarantee fast convergence of the system leading to an equilibrium state whereby networks are able to operate on acquired channel with minimal or zero contention.

DOI: 10.4018/978-1-4666-6571-2.ch014

INTRODUCTION

Federal regulators have allocated most of the radio spectrum relevant for wireless communication. The static allocation makes it difficult for emerging wireless operators to meet up with bandwidth requirements. Studies have shown that the licensed owners underutilize allocated spectrum bands. Only about five percent or less of the allocated spectrum bands is actually utilized (First IEEE Symposium, November 2005). Dynamic Spectrum Access (DSA) was proffered as possible solution to these problems of spectrum scarcity and underutilization. Different approaches have been suggested that allow secondary users to operate alongside the licensed owners while strictly adhering to interference constraints (Zhao & Sadler, 2007).

A cognitive radio (CR) is a radio device equipped with the capability to dynamically discover and access fallow spectrum bands. Formerly, cognitive radio is defined as radio that can change its transmitter properties based on interaction with its environment (Federal Communication Commuication (FCC), 2003). Cognitive radios are characterized by their cognitive ability and reconfigurability (Haykin, 2005). The cognitive ability allows them to identify temporal unused spectrum at any specific location (Akyildiz, Lee, Vuran, & Mohanty, 2008). Cognitive radio can also be dynamically reconfigured to transmit and receive on a variety of frequencies and use different access technologies supported by its hardware design (Jondral, 2005). Cognitive radios are required to adhere to the DSA paradigm, which allows them to share spectrum bands with the licensed owners in an opportunistic manner. The most important regulatory aspect of the DSA is that cognitive radios must not interfere with the operation in the licensed bands and must identify and avoid such bands in timely manner (Federal Communication Commission (FCC), 2004) (Defence Advanced Research Project Agency (DARPA)). If any of the spectrum bands being used by the cognitive radios are accessed by the licensed incumbents, they are required to immediately vacate the spectrum band within the channel move time and switch to another channel (Cordeiro, Challapali, & Sai Shankar, 2005). A CR network is formed by a group of CR users that do not have the license to operate in any desired band. CR networks are sometimes referred to as secondary networks or DSA networks. One of the major challenges facing the existence of CR networks is the problem of self-coexisting with other CR networks operating in the same spectral space. Self-coexistence by definition simply means the ability of CR networks to independently exploit the spectrum to maintain quality of service with minimal conflict with neighboring networks.

Some suggestions have been made that permit distributed coordination among CR networks via the use of common control channels (Wu, Lin, Tseng, & Sheu, 2000) (So & Vaidya, 2004) (Zhao, Zheng, & Yang, 2005). But the availability of control channels is not always guaranteed. The common control channels are likely to be reclaimed by the licensed owners at any time. Their return to the control channels entails they can no longer be used for coordination between networks. In the case of a simple jamming attack the control channels will be rendered useless, disrupting the activities in the network. Similarly, in a situation where spectrum bands are scarce, using some of the channels as control channels would be inefficient and can lead to spectrum starvation. Those channels could be used instead for critical data transmission purposes. Thus, in our study we consider such situations where communication between CR networks is a major challenge, making it difficult for them to coordinate. Under such a condition, it becomes inconvenient for CR networks to operate in the vacant spectral space without getting into conflict with one another. For CR networks to remain functional amidst scarce spectrum resources, they need to implement mechanisms that maximize spectrum utilization. Such mechanisms should include some conflict avoidance mechanisms to minimize contention that could arise.

Generally, this chapter discusses new approaches that can be implemented in a system of CR networks to reduce the amount of contention experienced in the system as a result of multiple channel exploitation. In an attempt to enhance performance and improve quality of service (QoS), CR networks usually lean towards acquiring more channels that can serve as back-up channels. While acquiring new channels, CR networks target channels with better spectral characteristics that can enhance performance. However, the acquisition of multiple channels exposes the entire system to the risk of poor performance resulting from the competition for spectrum resources. Existing spectrum management approaches are considered not suitable to handle these self-coexistence challenges in the absence of coordination between the CR networks. In the absence of efficient techniques for channel acquisition and contention handling, spectrum utilization among CR networks might become unfair. Some networks benefit from the competitive scenario while others suffer from spectrum starvation. The mechanisms presented in this chapter ensure that CR networks share, in a fair manner, the available spectrum resources and stabilize over time. An earlier version of this work can be seen in (Ezirim, Sengupta, & Troja, 2013).

BACKGROUND

The acquisition of channels to satisfy spectrum demand undoubtedly increases network's utility and ability to handle unprecedented service requests from clients. Even though this process of channel acquisition can be beneficial to the CR networks, it can create contentious scenario because of multiple access by different networks to the available spectrum bands. In the case of a decentralized system of multiple overlapping cognitive radio networks, the system performance degrades as a result of increased contention for multiple spectrum bands. Usually in a decentralized system of CR networks, no central authority or coordinator is involved. The knowledge of the distribution of intended spectrum access to the available spectrum bands is very much limited, thereby creating a self-coexistence problem. In an earlier work by Sengupta et al. (Sengupta, Chandramouli, & Chatterjee, 2008), the problem of self-coexistence between CR networks was approached from a Game theory point of view. The problem was addressed as a Modified Minority Game model (MMG), where CR networks are allowed to acquire only one of the available spectrum bands. CR networks are expected to select a channel without the knowledge of the channels selected by other networks. Tosh et al. (Tosh & Sengupta, 2013) discussed the incorporation of perception learning in addressing self-coexistence problem involving heterogeneous spectrum bands. The perception learning mechanism involves networks strategizing their actions in order to maximize utility. In this chapter, we address the case where CR networks can acquire more than one channel. The main goal is to find means of making multiple channel acquisition decision to improve self-coexistence and reduce contention to a minimal. In contrast to the MMG model, where decisions are made to "stay" on a channel or "switch" to a different one, networks will have to make similar decisions but for multiple channels. This raises the following questions: What type of mechanism can allow cognitive radio networks to operate on multiple channels and yet be able to coexist with each other with minimal contention and maximum utilization of available spectrum resources? How quick does the system of CR networks converge to an equilibrium state with tolerable level of channel contention?

In some of the recent literatures, the problem of spectrum assignment in DSA systems has been addressed as a graph-coloring problem (Cao, Li, & Ye, 2011) (Anand, Sengupta, & Chandramouli, 2012) (Cao & Zheng, 2007) using coordinated approach. However, this approach is not adaptive to a scenario where multiple channel constraints are accompanied by the dynamic changes in the systems such as the number of networks present and the number of idle channels available. Sengupta et al. (Sengupta, Brahma, Chatterjee, & Sai Shankar, 2013) introduced the Utility Graph Coloring (UGC) to increase spectral efficiency and spectrum utilization in a system of cognitive radio-based Wireless Regional Area Networks (WRANs). UGC is adaptable only to centralized system where dynamic multiple broadcast messaging, aggressive contention resolution amongst other techniques are implemented. The learning-BEB algorithm (Barcelo, Bellalta, & Oliver, 2011) is an algorithm devised for achieving collision-free scheduling in 802.11 networks. The algorithm is a modified version of the CSMA/CA mechanism with truncated exponential back-off. The algorithm is known to suffer from slow convergence rates (Fang, Malone, Duffy, & Leith, 2013) but very simple to implement. The Communication Free Learning (CFL) algorithm, proposed by Duffy et al., uses stochastic learning mechanism to update the probability of choosing color (channel) based on local information. The CFL algorithm is quite adaptive to changes in topology and quite fast in convergence. All the afore-mentioned decentralized coloring algorithms are only applicable to only scenarios where there is a single channel requirement by the networks. A modified version of the CFL algorithm, known as the simplified CFL (SCFL) (Checco & Leith, 201a), gives similar convergence rate performance like the CFL algorithm but with less memory and processing power.

As Duffy et al. (Duffy, 2013) correctly identified in their paper on decentralized constraint satisfaction, it is possible to have limited communication between entities in system, especially in wireless network settings. Most of the algorithms for channel selection assume the existence of an end-to-end communication for centralized solutions, for control messaging solutions or some sort of global coordination in simulated annealing proposals (Wu, Yang, Tan, Chen, Zhang, & Zhang, 2006) (Kauffmann, Baccelli, Chaintreau, Papagiannaki, & Diot, 2005) (Jorge, Wu, & Shu, 2008) (Subramanian, Gupta, Das, & Cao, 2008).

Some distributed algorithms for wireless access in distributed systems, proposed in (Mishra, Banerjee, & Arbaugh, 2005) (Zhao, Zheng, & Yang, 2005) (Nie & Comaniciu, 2006), rely on control message exchange to communicate and learn their environment, which actually introduces an overhead cost and an additional cost of keeping track of a network system's topology. A number of protocol designs have been proposed for self-coexistence in both centralized and distributed networks (Raychaudhuri, Mandayam, Evans, Ewy, Seshan, & Steenkiste, 2006) (Huang, Jing, & Raychaudhuri, 2009) (Cormio & Chowdhury, 2009) (Zhao & Sadler, 2007). In an uncoordinated system of CR networks, the performance of these protocols becomes critically important. There is no incentive for the CR networks to implement these complex protocols because of time constraints and stringent demand to transmit. Under this condition, every network's priority is to discover and transmit over multiple spectrum bands as soon as possible. Thus there is need for a mechanism that a CR network can use to access multiple channels dynamically and minimize unnecessary overhead cost and cost due to spectrum contention. Such mechanism should reduce contention to the minimum and maximize spectrum utilization by the CR networks. An efficient channel acquisition mechanism would ensure that CR networks accomplish their goal of satisfying channel size requirements within a short period of time.

CHANNEL ACQUISITION MECHANISMS

In this section, we shall discuss the mechanisms that could be implemented to promote self-coexistence in a system of CR networks. Our primary focus is to understand how CR networks can access the spectrum in a distributed manner without actually exchanging any control information. Considering the volatility of the spectrum environment it rather difficult for CR networks to operate without minimal contention. The problem of spectrum access is further exacerbated by the nature of wireless medium. The open nature of the wireless medium makes it rather difficult to implement efficient MAC protocols. So far, there are few protocols that addresses the problem of contention among cognitive radio networks without significant message overhead and time synchronization (Cordeiro, Challapali, & Sai Shankar, 2005) (Raychaudhuri, Mandayam, Evans, Ewy, Seshan, & Steenkiste, 2006) (Huang, Jing, & Raychaudhuri, 2009) (Cormio & Chowdhury, 2009) (Zhao & Sadler, 2007). In particular, these protocols focus more on avoiding collision with the license owners of the spectrum bands. Little is discussed about conflicts that could arise when CR networks access the same set of spectrum bands at the same time. The channel acquisition mechanisms discussed in this section address such scenarios. The mechanisms do not include practices that could lead to unnecessary overhead due messaging and network synchronization.

In a decentralized system of CR networks, operating in the radio spectrum, it is difficult to coordinate radio transmission activities for several reasons. The networks practically have to sense and discover fallow spectrum bands and carry out transmission on the discovered bands. As we know, the spectrum bands are scarce and their availability usually depends on the primary users activities on those bands. The activities of secondary users are short lived once the primary users reappear to regain control of their channels. At that stage, secondary users are required to vacate within the designated channel move time to avoid conflict with primary users. The volatility and unpredictability of spectrum bands pose a serious challenge to secondary users in conducting efficient data transmission during the period when primary users are inactive. Under the same constraint, it is not beneficial to the CR networks to convert hard-to-find spectrum bands into common control channels to exchange control information. The selection of the control channels, also known as rendezvous process, involves a rigorous process of message exchange and negotiation between networks. This process carries a significant overhead cost due to messaging and time spent on coordination, especially in a case where the negotiation process repeatedly fails. It is important to note that the control channels are part of the spectrum resources that belong to the primary users. Therefore, their owners can reclaim the channels at any time, which subsequently leads to a repeated search for a new set of common control channels. It is therefore inefficient for CR networks to operate in this manner because of the tremendous amount of time and resources that could be wasted in negotiation process and selection of common control channels. Sustaining cooperation in a decentralized system by means of control channel is not only ineffective but also difficult to implement. Even though coordination between networks might be possible, relying on such possibility in a decentralized system setting might prove futile, especially in emergency situations.

The basic idea of the channel acquisition and contention handling (CACH) mechanisms stems from the basic TCP congestion control mechanisms. In TCP, the end systems do not rely on the network layer for explicit support on avoiding congestion. Rather, when an end system detects transmission failure, TCP modifies its congestion window, which directly places a constraint on the number of bytes that can be sent by the sender during a transmission cycle. Our approach simply suggests that CR networks reduce the number of channels used for transmission whenever contention is experienced in the system. If CR networks greedily contend the available channels, they might end up conflicting with one another and wasting the spectrum opportunities. A greedy approach can actually benefit some CR networks but to the disadvantage of other networks. It is also not best suited for a dynamic environment characterized by spontaneous changes in spectrum availability caused by primary users.

The channel acquisition mechanisms discussed in this section are based on some basic assumptions. One of the main assumptions is that networks are capable of sensing and discovering multiple vacant spectrum bands. The time frame for spectrum usage consists of sensing phase and data transmission phase. Before embarking on data transmission, CR networks sense the spectrum bands for the presence of primary users. The sensing and data transmission phases are synchronous across the system. Once a network completes spectrum sensing, it commences transmission provided no primary user was detected. Suppose there are N CR networks operated by N separate wireless service providers in a region are competing for M idle frequency channels. The set of CR networks is denoted as $\{n_1, n_2, \ldots, n_N\}$. The set of channel size requirements of the CR networks is denoted $\{m_1, m_2, \ldots, m_N\}$ where $m_1 + m_2 + \ldots + m_N \leq M$. For any CR network n_i, the channel size requirement satisfies the condition $m_i \geq 1$. If $m = 1$ for all CR networks, then the self-coexistence problem reduces to single channel acquisition which can be addressed using any of the decentralized approaches mentioned earlier. We must recall that the CR networks partially or completely overlap geographically. So the interference or conflict graph generated for such a system of CR networks is a complete graph. In such scenarios, when a channel is shared by more than one CR network, the QoS of the channel degrades.

Equipped with sensing capability, CR networks are able to discover idle spectrum bands and use them for transmission. A CR network uses a channel selection scheme to select the channels that it deems suitable for its transmission purposes. The channel selection scheme ensures that a network selects the best channels to operate with in each transmission stage. Later in this chapter, we shall discuss in details the specifics of the channel selection scheme. Since the CR networks are operating in an uncoordinated system, they are very susceptible to conflicts with their neighbors. We propose channel acquisition and contention handling mechanisms that allow CR networks acquire channels using two different approaches, namely exponential and linear channel acquisition approaches. A CR Network commences the channel acquisition process in an exponential order and then switches to a linear order of acquisition once the specified channel size threshold has been attained. A network sets it channel size threshold, which specifies the minimal number of channels required by the network to commence normal operations. Another approach to channel acquisition is for networks to attempt transmission on a number of channels slightly more than their channel size requirements. This approach is known to be prone to high contention when implemented because the possibility of contention increases with increase in average channel size requirement. On the other hand, it allows the networks to build profiles for the channels faster compared to the stepwise approach of channel acquisition. With more information on the available channels and the channel selection scheme, a network would eventually reach it channel size requirement.

When a network experiences contention, a network needs to do some adjustment in the number of channels used for transmission, also referred to as channel size, to avoid further contention with neighbors. CR networks implement a back-off mechanism targeted towards reducing contention in the system. The back-off mechanism is implemented by reducing the channel size whenever a network experiences contention. By reducing the channel size, the chance of contention between contending networks decreases. Assuming the number of channels used by a network before contention was experienced is m and it decides to drop k channels. By dropping k channels, the network is not only avoiding contention with neighbors on the k dropped channels but also consolidating its hold of $m - k$ channels. The channel selection scheme ensures that the best $m - k$ channels are selected in the next transmission. This approach to channel acquisition guarantees that the number of channels required to meet the channel size requirement is reduced in a stepwise fashion. Once each network is able to consolidate their presence on their set of channels with best channel profile, the contention for those channels by their neighbors will significantly reduce.

We propose the following channel acquisition mechanisms: RENO, TAHOE, linear increment and linear decrement, and Probabilistic Dynamic Incremental Mechanisms. The other channel acquisition mechanisms, RENOFT and TAHOEFT, were derived from RENO and TAHOE respectively, by slight modification of their heuristics.

RENO Mechanism

The basic idea of RENO mechanism was derived from the TCP RENO congestion control and avoidance mechanism. An illustration of RENO mechanism can be seen in Figure 1. With RENO mechanism, a CR network acquires channels in an exponential fashion until a predefined channel size threshold τ is reached. Once the network channel size exceeds the threshold, subsequent acquisition of channels is carried out in a linear order. When a CR network experience contention on one or more channels, it reduces the size of the channel set used for transmission. The channel set size is halved and the threshold assumes a value equal to the new channel set size. The best channels are incorporated into the channel set used by the network in the next transmission phase. Once the channel size threshold is reached, channel acquisition follows a linear pattern of increase in number of channel acquired. Let us introduce a predicate $q(\sigma_i)$ that determines when the channel size threshold is exceeded, where σ_i represents the number of channels that a network i currently has. The predicate $q(\sigma_i)$ is defined as:

$$q(\sigma_i) = \begin{cases} 1 & if\ \tau > \sigma_i \\ 0 & otherwise \end{cases} \tag{1.1}$$

Figure 1. RENO Channel Acquisition and Contention Handling Mechanism

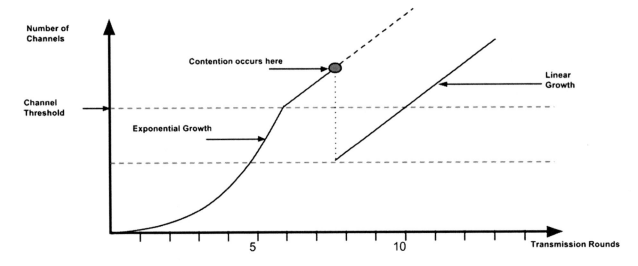

The actual σ_i' number of channels that a network solicits for in the next time frame is given as:

$$\sigma_i' = 2 \cdot \sigma_i \cdot q(\sigma_i) + (\sigma_i + 1) \cdot (1 - q(\sigma_i)) \tag{1.2}$$

Every network is expected to have a certain spectrum channel size requirement that it needs to maintain its quality of service (QoS). Let $\phi(.)$ be a random function that returns the actual outcome of a transmission phase, then the *i*-th network would satisfy its channel size requirement only when the indicator function $I(n_i) = 1$ (true) where,

$$I(n_i) = \begin{cases} 1 & if\ \phi(\sigma_i') > \sigma_i^{max} \\ 0 & otherwise \end{cases} \tag{1.3}$$

σ_i^{max} is the channel size requirement of the network n_i. Convergence is attained in the system only when all networks must have satisfied the spectrum requirement. The satisfaction stems from the fact that networks have settled on preferred bands and have no more incentive to contend on other bands. Thus, a state of convergence for a system consisting of N networks is achieved when $\prod_{i=1}^{N} I(n_i) = 1$.

Contention handling for this mechanism is carried out as follows. After a network experiences contention, which could be in one or more channels, it reduces the number of channels by half, that is, $\sigma_i' = \sigma_i / 2$. The threshold of the network is modified the same way such that $\tau = \sigma_i / 2$. This mechanism is implemented such that the threshold is always updated on any incident of contention. The interesting feature about RENO is that after the first contention, the network would no longer embark on exponential increment but rather a linear increment in the number of channels being acquired. Consequently, after the update, τ becomes equivalent to σ_i, and the CR network will no longer implements exponential increment.

TAHOE Mechanism

TAHOE mechanism and RENO mechanism share almost the same channel acquisition heuristics. An illustration of TAHOE mechanism can be seen in Figure 2. The two mechanisms differ in the way they handle contention. TAHOE, in contrast to RENO, reduces the number of channels that it currently has to one, that is, $\sigma_i' = 1$, when contention is detected. This implies that any network implementing the mechanism restarts channel acquisition upon encountering contention. Also, the channel size threshold is reduced to half the number of channels that the network had acquired prior to contention.

RENO Mechanism with Fixed Channel Size Threshold

RENOFT mechanism is a modification of the RENO channel acquisition and contention handling mechanism with the channel size threshold τ set to be fixed. The fixed threshold allows the CR network to continue to acquire channels exponentially provided that the number of channels acquired is below

Figure 2. TAHOE Channel Acquisition and Contention Handling Mechanisms

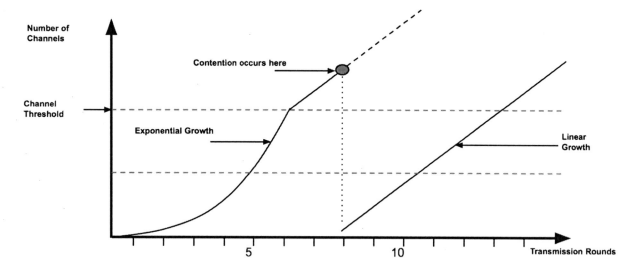

the threshold. This is clearly different from the earlier implementation we saw in RENO mechanism where a linear pattern of channel acquisition is followed right after the first contention is experienced. As we shall see later, the performance of this mechanism depends on the channel threshold τ which remains fixed most of the time. Depending on the value of the threshold channel acquisition using this mechanism can be greedy thereby having significant impact on the convergence time of the system.

TAHOE Mechanism with Fixed Channel Size Threshold

Just like RENOFT mechanism, TAHOEFT is simply a modification of TAHOE mechanism with fixed channel threshold, τ. The modification allows a network reset the number of channels it acquire to one upon contention and maintain the channel threshold to the next transmission round. Channel acquisition commences in an exponential manner after the contention, which might actually lead to a quicker recovery from contention, especially when $\tau > 1$.

Probabilistic Dynamic Incremental Mechanism (INCR)

Unlike the previously discussed mechanisms where channel acquisition depends on channel size threshold, the Probabilistic and Dynamic Incremental Mechanism ensures that channel acquisition is dynamic. The mechanism relies on a few parameters to determine the number of channels to acquire. The parameters include the number of channels that the network currently has in its possession, the number of channels remaining to reach its channel size requirement and the severity of the contention experienced in the system. The channel size requirement is the maximum number of channels, above which the network does not seek to acquire any extra channel.

Suppose a cognitive radio network n_i had initially σ_i number of channels and its channel size requirement for satisfactory performance is σ_i^{max}. We define $\phi(.)$ as a function that returns the number of channels that the network recorded having successful transmission in the previous transmission phase.

If the channel size requirement is not satisfied, that is $\sigma_i^{max} > \sigma_i$, then the network would need $\delta_i = \sigma_i^{max} - \sigma_i$ number of channels to reach its goal of channel size requirement. The INCR mechanism stipulates that the number of channels ρ_i that a network would seek to acquire in the next phase is given by the following expression,

$$\rho_i = \delta_i + \lambda\left(\delta_i\right) \tag{1.4}$$

where $\lambda\left(.\right)$ is a uniform and randomly distributed function such that $\lambda\left(x\right)\epsilon\left\{0,...,x-1\right\}$. The above equation shows a contending CR network attempts to acquire twice the number of channels that it requires to attain the channel size requirement. The network starts off in a greedy way to acquire channels but as it approaches its channel size requirement it contends for much less channels. We have that ρ_i always satisfies the inequality $\delta_i \leq \rho_i \leq 2.\delta_i$ and in the case where $\sigma_i = 0$, we have $\sigma_i^{max} \leq \rho_i \leq 2.\sigma_i^{max}$. The tendency towards convergence would strictly depend on the outcome of subsequent transmission phases and the amount of spectrum resources available. Given that $\sigma_i' = \rho_i + \sigma_i$ is the actual number of channels that a network senses in the next phase, and $\phi\left(\sigma_i'\right)$ is a random function that returns the actual number of channels as the outcome of the transmission phase. The condition for satisfaction of channel size requirement is given by

$$\phi\left(\sigma_i'\right) \geq \sigma_i^{max} \tag{1.5}$$

The function $\phi\left(x\right)$ is a uniform and randomly distributed function such that $\phi\left(x\right)\epsilon\left\{0,...,x\right\}$. If the condition for convergence does not hold, then the number of channels with successful sensing and transmission is set as $\sigma_i = \phi\left(\sigma_i'\right)$ and the process is repeated again until convergence is attained. Rewriting the relation described by the inequality (1.5), we have that

$$\phi\left(\rho_i + \sigma_i\right) \geq \sigma_i^{max} \tag{1.6}$$

After substituting ρ_i, the inequality becomes:

$$\phi\left(\delta_i + \lambda\left(\delta_i\right) + \sigma_i\right) \geq \sigma_i^{max} \tag{1.7}$$

We can replace δ_i with $\sigma_i^{max} - \sigma_i$ to get

$$\phi\left(\lambda\left(\sigma_i^{max} - \sigma_i\right) + \sigma_i^{max}\right) \geq \sigma_i^{max} \tag{1.8}$$

As previously stated, a network n_i satisfies its channel size requirement when the indicator function $I\left(n_i\right) = 1$. The indicator function is defined as follows:

$$I(n_i) = \begin{cases} 1 & if\ \phi\left(\rho_i + \sigma_i\right) \geq \sigma_i^{max} \\ 0 & otherwise \end{cases} \tag{1.9}$$

Similarly, the entire system of CR networks converges when $\prod_{i=1}^{N} I(n_i) = 1$ must hold. That is the condition under which no CR network is in contention with neighboring networks. The INCR mechanism handles contention by implementing the Channel Selection Scheme. The Channel Selection Scheme ensures that a CR network always selects channels with high success rate for transmission purposes. Due to the persistent nature of the INCR mechanism, coupled with an efficient Channel Selection Scheme, CR networks are able to retain good channels with minimal contention. The persistence of the networks to retain channels with high quality makes it possible for the system to converge fast.

Linear Increment and Linear Decrement Mechanism

The linear increment and linear decrement mechanism involve increasing the number of channels that a network in the next phase by one given that there was no record of contention in the previous phase. The CR network implementing this mechanism also reduces the number of channels it acquires by one if there was contention in the previous transmission phase. The mechanism assures a steady but seemingly slow pace of channel acquisition. This plays to the advantage of the network as there lesser contention experienced with linear pattern that is involved in the increment and decrement of channels marked out for acquisition. It also guarantees quicker convergence in part for the system.

CHANNEL SELECTION SCHEME

The channel acquisition mechanisms use a channel selection scheme to determine the channel to acquire. The basic idea is that a CR network chooses channels with high success rate. The success rate is measured in terms of the number of successful transmissions carried out by the network after discovering the channel to be idle. But since the network is ``blind'' to the intent of its neighbors, it is possible that they can interfere with each other's transmission. We assume that the channels are homogenous in capacity, so no one channel is better than the other. By probing the spectrum environment, the networks gather profile information about the available set of channels. The profile information contains the following: channel identity c_j, number of successful transmission s, and number of failed transmissions f. The failed transmission count f includes both the frequency of PU activity on the channel and the frequency of interference received when PU is absent.

The channel selection scheme works as follows. A CR networks sorts the channels according to their success rate such that the first channel has the highest success rate during transmission. If two or more channels share the same success rate, then they are randomly ordered. Let M be the total number of peers and μ_j be the success rate of channel c_j. Initially, $s = 1$ for all channels to ensure that the channels are selected uniformly at random from the set of available channels. At any stage of channel selection, available channels are selected according to the following rules:

1. If the T rounds has not elapsed, then for any channel c, the success rate is given as $\mu = \dfrac{s}{\sum_j s_j}$, where the denominator of the expression is the total number of successful transmissions.

2. If T rounds has elapsed, then for any channel c, the success rate is measured as $\mu = \dfrac{s}{s+f}$.

3. Given two channels c_j and c_k channel c_j is prioritize over channel c_k, if $\mu_j > \mu_k$.

4. If $\mu_j = \mu_k$ then a channel is chosen uniformly at random from the channel set $\{c_j, c_k\}$.

The parameter T indicates the length of the probing rounds during which a network probes the available channel set. Thus, the parameter is strictly network-defined. Using the first rule μ is taken as parameter that measures the relative performance or availability of a given channel. The second rule, on the other hand, treats μ as a measure strictly based on history of actual experience on a channel. Both formulas can be combined together to generate global expression for μ. That is,

$$\mu = \alpha . \frac{s}{\sum_j s_j} + \left(1 - \alpha\right) . \frac{s}{s+f}$$

where $\alpha \epsilon \left[0,1\right]$ is a coefficient that the relative importance of both rules. As a matter fact, when there are not enough history about the available channel set, the relative performance takes credence, so $\alpha = 1$. A network can decide to keep $\alpha = 1$ till after T rounds has elapsed and then $\alpha = 0$. Instead of keeping α constant until T-th round, the value of α can be gradually decreased by a specified quantum of change. By so doing the significance of the relative success rate measure is diminished gradually until enough channel experience is gathered. The channel selection scheme ensures that channels with high μ values are always selected. Because the occupation of a channel is also counted as failure to transmit, the scheme helps the networks to avoid channels with high PU activity.

PERFORMANCE COMPARISON OF CHANNEL ACQUISITION MECHANISMS

In order to compare the performance of the mechanisms discussed above, we shall use the following metrics: average utility and convergence time. Average utility derived by the CR networks is closely related to the convergence time of the system. An early convergence of the system, offers CR networks an opportunity to commence operation on time in a contention-free spectrum environment. The CR networks are thus able achieve higher utility, spurred by the early start of spectrum activities. The unit of system utility is measured in terms of one complete transmission slot devoid of contention. This implies that a contention-free transmission carried out on a single channel is considered as a unit of utility. Thus a CR network operating on m spectrum bands, during a particular transmission slot, derives m units of utility, provided there is no record of contention on any of the channel. Convergence time is measured in terms of the number of transmission slots that is required for the system to stabilize. Convergence is achieved when all CR networks in the system have settled on their preferred spectrum bands and are operating with little or no contention. We must emphasize that convergence is possible while the primary users are away. The return of primary users to their bands upsets the stability of the system. Displaced CR networks will have to resume new search for unoccupied spectrum bands and in the process might interfere with the operations of other stable CR networks. An adaptive mechanism that converges quickly will ensure that the system's stability is restored within a short period of time.

The simulation results were obtained with the following configuration. The simulations were conducted for 1000 timeslots. The networks shared a total of 225 channels with an average channel size requirement for each network being 5 channels per network. For simulation involving primary users, the probability of a channel going from idle to busy state was less than 0.1 with average channel occupancy (duration of busy state) was kept at 10 timeslots. The simulation results show the dependency of the following system parameters – utility, convergence and contention – on number of CR networks in the system. All system parameters are measured in timeslots.

System Utility

The performance of the various mechanisms can be compared based on system utility. Two scenarios are considered: the performance of the system with primary users present and the performance when the primary users are absent. By absence of the primary users, we mean that CR networks can operate on the available spectrum bands for a considerably long period of time without experiencing any interruption due to primary users return. In each CR network there is a channel size requirement required to maintain quality of service. We assume that all participating CR networks share the same channel size requirement. However, the number of available channels in the system is limited. The appearance and disappearance of primary users on spectrum varies the number of channels available to the CR networks. The channel size threshold, which decides whether exponential or linear acquisition of channels is implemented, is kept below the channel size requirement of the networks for RENO and TAHOE mechanisms. In contrast, the LINEAR and INCR mechanisms do not use channel size threshold and thus the condition is not enforced during implementation.

Figures 3 and 4 illustrate the performances of the channel acquisition mechanisms in terms of the average utility of the system during primary users' presence and absence respectively. The channel acquisition mechanisms show different system performance trend with increasing number of participating CR networks. The system performance achieved in the absence of primary users is notably better than the system performance achieved in their presence. This is anticipated because of the impact that the sudden appearance of primary users on the spectrum bands has on the stability of the system. Generally, an increase in the number of participating networks decreases the number of channels available for use and thus causes a decrease in average utility. The INCR mechanism gives the best performance compared to other mechanisms. The mechanism uses a slightly greedy approach that is strictly controlled by the channel size requirement of the CR network. The persistent attempt to attain its channel size requirement goal explains the mechanism's better performance over the other channel acquisition mechanisms. With an increasing number of networks, LINEAR gives a better performance than RENO and TAHOE mechanisms because the mechanism adopts a "slow and steady" approach in acquiring channels. The poorer performance of RENO and TAHOE can be attributed to the contention avoidance measure of halving channel size whenever there is contention. TAHOE mechanism's poor performance can be attributed to the drastic reduction in the number of acquired channel after contention. Even though subsequent successful transmissions are followed by an exponential increase in channel size (up until channel size threshold is reach), the mechanism may not be as fast as RENO in helping the network attain its channel size requirement.

Figure 3. Simulation results showing the dependence of system utility on the number of CR networks using different mechanisms. System utility measured in timeslots. Primary users were absent.

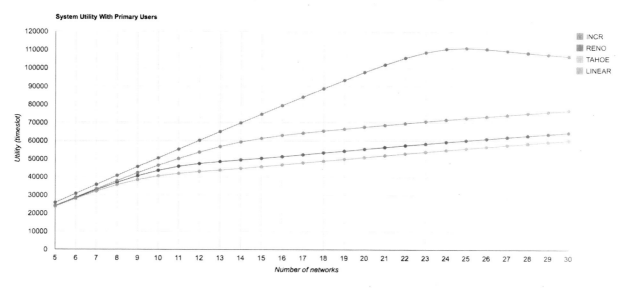

Figure 4. Simulation results showing the dependence of system utility on the number of CR networks using different mechanisms. System utility measured in timeslots. Primary users were present.

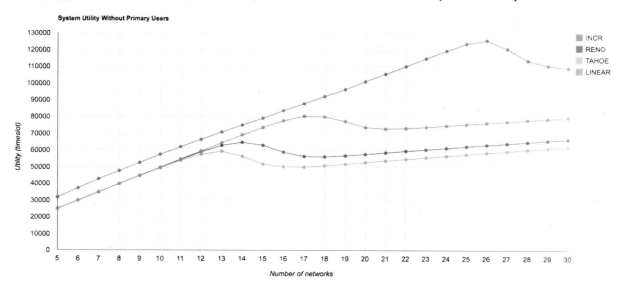

Convergence Time

Convergence time is the time it takes for the system of CR networks to finally stabilize to a contention-free state. Convergence time is dependent on those factors that affect the availability of spectrum bands. Some of those factors include the number of CR networks in the system, the amount of spectrum resources available and the primary user presence. Figures 5 and 6 show the dependency of convergence time on the number of CR networks in the system, given that the number of available channels is limited. We

Figure 5. Simulation results showing the dependence of convergence time, measured in timeslots, on the number of CR networks. Primary users were absent.

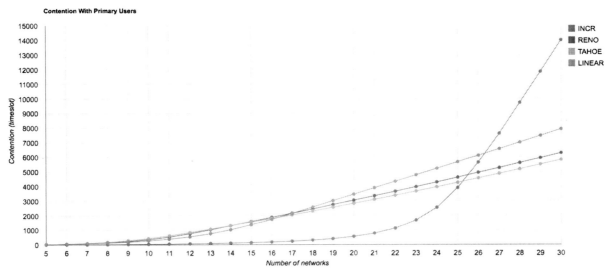

Figure 6. Simulation results showing the dependence of convergence time, measured in timeslots, on the number of CR networks in the system. Primary users were present.

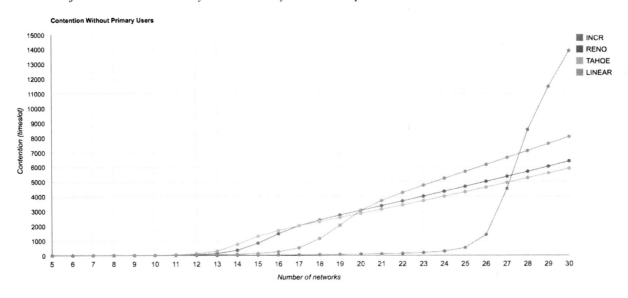

can see that convergence time increases with increasing number of CR networks. The increasing number of networks increases contention in the system as a result of reduction in the average number of channels available per network. Therefore it takes the system more time to converge. We can observe that the convergence time increases exponentially as the number of networks increases. But the each of the mechanisms exhibits a different trend. INCR mechanism gives the best convergence time. Compared to other mechanisms, convergence time remained significantly low for most values of N. By observing Figures 3 and 4, we can see that the fast convergence correlates with higher average system utility. The faster a system converges with a mechanism, the higher the average utility it derives. This is independent of the channel size requirement. Another important observation to note is convergence time trend with respect to primary user presence. Even though it appears that the trend invariant to primary users, on close observation we can notice that the system converges faster when primary users are absent in the system.

System Contention

Simulation results show that contention varied with the number of CR networks. In our simulation settings, contention parameter is a measure of the number of timeslots wasted due to contention between networks. System contention zeroes out once convergence is attained. Due to the different techniques used by the CACH mechanism in handling contention, their performances in terms of system contention are expected to vary. The performance of mechanisms in terms of system contention is presented in Figures 7 and 8. In both figures, we can observe interesting trends of the CACH mechanisms. The LINEAR mechanism shows an increase in the rate of contention when the number of networks exceeds 20. The persistent nature of the INCR mechanism makes it vulnerable to contention as the number of networks in the system increases. On the contrary, the RENO and TAHOE mechanisms show significant improvement in the number of contentions in denser system. Their system contention rate, compared against LINEAR and INCR, is lower for *N>28* irrespective of primary users presence. However, when the mechanisms are compared based on net system utility, the INCR and LINEAR still give better system performance.

Figure 7. Simulation results showing the dependence of system contention on the number of CR networks in the system in the absence of primary users.

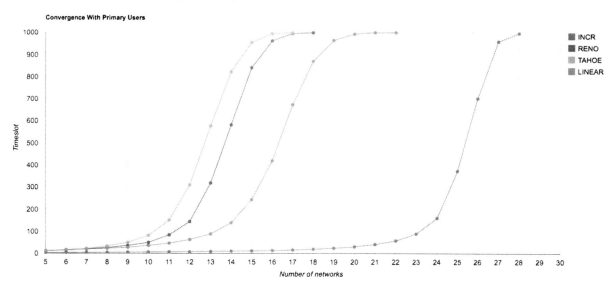

Figure 8. Simulation results showing the dependence of system contention on the number of CR networks in the system in the presence of primary users.

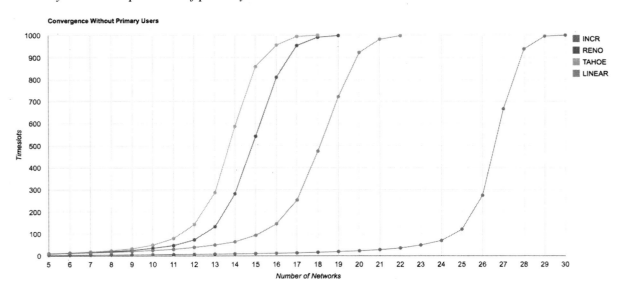

CONCLUSION

In this chapter, we discussed mechanisms that can be used by networks to operate in a spectral environment where self-coexistence is necessary to maintain quality of services to subscribers and clients. We argued that control messaging via common control channels does not completely address the self-coexistence problem in a system of CR networks. The CR networks cannot sufficiently know the channel selection of their neighbors. Given the dynamic nature of the spectrum and the cost of setting up common control channel in the midst of rather scarce spectrum resources, self-coexistence in an uncoordinated system requires a decentralized approach. The channel acquisition and contention handling mechanisms are decentralized mechanisms that require no coordination to implement on each of the CR network. The Channel Selection Scheme assures that the best channels are selected in transmission round. The mechanisms together with the Channel Selection Scheme guarantees quick convergence. We compared the performance of the CACH mechanisms in terms of average system utility and convergence time. Among the CACH mechanisms discussed, the INCR mechanism was found be the most effective mechanism. We attribute the effectiveness of the INCR mechanism to the persistent channel acquisition heuristics implemented by the mechanism. The performance of INCR is also shown to be consistent and fairly robust to the occasional presence of primary users in the system.

In conclusion, we emphasize that CACH mechanisms are well suited for such wireless network scenarios where there is no prior knowledge of channel availability as well as no history profile about both primary and secondary users activities. The mechanisms are also adaptable to scenarios where CR networks cannot exchange control messages to avoid contention with one another. The CACH mechanisms are quite simple to implement, efficient and inexpensive. There is no overhead cost due to messaging, and the memory cost is quite insignificant.

REFERENCES

Akyildiz, I. F., Lee, W.-Y., Vuran, M. C., & Mohanty, S. (2008). A Survey on Spectrum Management in Cognitive Radio Networks. *IEEE Communication Magazine, 46*(4), 40–48. doi:10.1109/MCOM.2008.4481339

Anand, S., Sengupta, S., & Chandramouli, R. (2012). MAximum SPECTrum packing: A distributed opportunistic channel acquisition mechanism in dynamic spectrum access networks. *IET Communications, 6*(8), 872–882. doi:10.1049/iet-com.2010.0607

Barcelo, J., Bellalta, B., & Oliver, M. (2011). Learning-BEB: Avoiding collisions in WLAN. *Other IFIP Publications,* (1).

Cao, L., & Zheng, H. (2007). On the Efficiency and Complexity of Distributed Spectrum Allocation. In *Proceedings of Cognitive Radio Oriented Wireless Networks and Communications,* (pp. 357-366). CROWNCOM.

Cao, Y., Li, Y., & Ye, F. (2011, December). Improved Fair Spectrum Allocation Algorithms Based on Graph Coloring Theory in Cognitive Radio Networks. *Journal of Computer Information Systems, 7*(13), 4694–4701.

Cordeiro, C., Challapali, K., & Sai Shankar, N. (2005, November). IEEE 802.22: The First Worldwide Wireless Standard Based On Cognitive Radios. *New Frontier in Dynamic Spectrum Access Networks, 2005,* 328–337.

Cormio, C., & Chowdhury, K. R. (2009). A survey on MAC protocols for cognitive radio networks. *Ad Hoc Networks, 7*(7), 1315–1329. doi:10.1016/j.adhoc.2009.01.002

Defence Advanced Research Project Agency (DARPA). (n.d.). DARPA Next Generation (XG) Communication Program.

Duffy, K. R. (2013). Article. *Decentralized Constraint Satisfaction., 21*(4), 1298–1308.

Ezirim, K., Sengupta, S., & Troja, E. (2013, January). (Multiple) Channel acquisition and contention handling mechanisms for dynamic spectrum access in a distributed system of cognitive radio networks. In *Proceedings of the 2013 International Conference on Computing, Networking and Communications (ICNC) (ICNC '13),* (pp. 252-256). ICNC.

Fang, M., Malone, D., Duffy, K. R., & Leith, D. J. (2013). Decentralised learning MACs for collision-free access in WLANs. *Wireless Networks, 19*(1), 83–98. doi:10.1007/s11276-012-0452-1

Federal Communication Commission (FCC). (2004, May). *Notice of Proposed Rule Making* (ET Docket no. 04-113). Washington, DC: FCC.

Federal Communication Commuication (FCC). (2003, December). *Notice of Proposed Rule Making and Order* (ET Docket no. 03-322). Washington, DC: FCC.

First IEEE Symposium. (November 2005). *Proceedings of the First IEEE Symposium on New Frontiers in Dynamic Spectrum Access Network.* IEEE.

Haykin, S. (2005). Cognitive Radio: Brain-empowered Wireless Communications. *IEEE Journal on Selected Areas in Communications, 23*(2), 201–220.

Huang, K.-C., Jing, X., & Raychaudhuri, D. (2009). MAC protocol adaptation in cognitive radio networks: An experimental study. In *Proceedings of Computer Communications and Networks,* (pp. 1-6). IEEE.

Jondral, F. K. (2005). Software-Defined Radio: Basics and Evolution to Cognitive Radio. *EURASIP Journal on Wireless Communications and Networking,* (3): 275–283.

Jorge, C., Wu, M.-Y., & Shu, W. (2008). Protocols and architectures for channel assignment in wireless mesh networks. *Ad Hoc Networks, 6*(7), 1051–1077. doi:10.1016/j.adhoc.2007.10.002

Kauffmann, B., Baccelli, F., Chaintreau, A., Papagiannaki, K., & Diot, C. (2005). *Self organization of interfering 802.11 wireless access networks.* Academic Press.

Maskery, M., Krishnamurthy, V., & Zhao, Q. (2009). Decentralized dynamic spectrum access for cognitive radios: Cooperative design of a non-cooperative game. *IEEE Transactions on* Communications, *57*(2), 459–469. doi:10.1109/TCOMM.2009.02.070158

Mishra, A., Banerjee, S., & Arbaugh, W. (2005). Weighted coloring based channel assignment for WLANs. *SIGMOBILE Mob. Comput. Commun. Rev., 9*(3), 19–31. doi:10.1145/1094549.1094554

Mitola Iii, J. (2000). *An Integrated Agent Architecture for Software Defined Radio.* Academic Press.

Nie, N., & Comaniciu, C. (2006). Adaptive Channel Allocation Spectrum Etiquette for Cognitive Radio Networks. *Mobile Network Application,* 779-797.

Raychaudhuri, D., Mandayam, N. B., Evans, J. B., Ewy, B. J., Seshan, S., & Steenkiste, P. (2006). CogNet: an Architectural Foundation for Experimental Cognitive Radio Networks within the Future Internet. In *Proceedings of First ACM/IEEE International Workshop on Mobility in the Evolving Internet Architecture* (pp. 11-16). ACM. doi:10.1145/1186699.1186707

Sengupta, S., Brahma, S., Chatterjee, M., & Sai Shankar, N. (2013). Self-coexistence among interference-aware IEEE 802.22 networks with enhanced air-interface. *Pervasive and Mobile Computing, 9*(4), 454–471. doi:10.1016/j.pmcj.2011.08.003

Sengupta, S., Chandramouli, R. a., & Chatterjee, M. (2008). A Game Theoretic Framework for Distributed Self-Coexistence Among IEEE 802.22 Networks. In *Proceedings of Global Telecommunications Conference,* (pp. 1-6). IEEE. doi:10.1109/GLOCOM.2008.ECP.598

So, J., & Vaidya, N. H. (2004). Multi-channel mac for ad hoc networks: handling multi-channel hidden terminals using a single transceiver. In *Proceedings of the 5th ACM International Symposium on Mobile Ad Hoc Networking and Computing* (pp. 222-233). ACM. doi:10.1145/989459.989487

Subramanian, A. P., Gupta, H., Das, S. R., & Cao, J. (2008). Minimum interference channel assignment in multiradio wireless mesh networks. *IEEE Transactions on* Mobile Computing, *7*(12), 1459–1473.

Tosh, D., & Sengupta, S. (2013, September). Self-Coexistence in Cognitive Radio Networks using Multi-Stage Perception Learning. In *Proceedings of Vehicular Technology Conference,* (pp. 1-5). Academic Press. doi:10.1109/VTCFall.2013.6692404

Wu, H., Yang, F., Tan, K., Chen, J., Zhang, Q., & Zhang, Z. (2006). Distributed Channel Assignment and Routing in MultiRadio Multichannel Multihop Wireless Network. *Selected Areas in Communications, 24*(11), 1972–1983. doi:10.1109/JSAC.2006.881638

Wu, S.-L., Lin, C.-Y., Tseng, Y.-C., & Sheu, J.-P. (2000). A new multi-channel MAC protocol with on-demand channel assignment for multi-hop mobile ad hoc networks. In *Proceedings of Parallel Architectures, Algorithms and Networks,* (pp. 232-237). IEEE.

Zhang, Q., Jia, J., & Zhang, J. (2009). Cooperative relay to improve diversity in cognitive radio networks. *IEEE Communications Magazine, 47*(2), 111–117. doi:10.1109/MCOM.2009.4785388

Zhao, J., Zheng, H., & Yang, G.-H. (2005). Distributed coordination in dynamic spectrum allocation networks. In *Proceedings of New Frontiers in Dynamic Spectrum Access Networks,* (pp. 259-268). IEEE.

Zhao, Q., & Sadler, B. M. (2007). A Survey of Dynamic Spectrum Access. *IEEE Signal Processing Magazine, 24*(3), 79–89. doi:10.1109/MSP.2007.361604

KEY TERMS AND DEFINITIONS

Channel Acquisition: The process of securing a channel for the sole purpose of using for transmission purposes. Under DSA paradigm, the licensed owners can reclaim acquired channels at any time.

Channel Identity: An identifier mapped to a specific frequency band and can be used as an alias/reference to a channel.

Channel Size Requirement: The number of channels that a network requires to operate effectively and efficiently.

Channel Size Threshold: A threshold on the channel size requirement and used by networks implementing RENO and TAHOE mechanism in deciding the appropriate measure to take to avoid contention.

Distributed Coordination: A form of spectrum access coordination in wireless networks that requires no central authority. This is ubiquitous in decentralized networks.

Dynamic Spectrum Access: A wireless network paradigm that promotes opportunistic access to unused spectrum resources with zero tolerance to any form of interference to the activities of the licensed owners of the resources.

System Convergence: The tendency of a system of CR networks to attain a state where each network has met its channel size requirement.

Chapter 15
Asynchronous Channel Allocation in Opportunistic Cognitive Radio Networks

Sylwia Romaszko
RWTH Aachen University, Germany

Petri Mähönen
RWTH Aachen University, Germany

ABSTRACT

In the case of Opportunistic Spectrum Access (OSA), unlicensed secondary users have only limited knowledge of channel parameters or other users' information. Spectral opportunities are asymmetric due to time and space varying channels. Owing to this inherent asymmetry and uncertainty of traffic patterns, secondary users can have trouble detecting properly the real usability of unoccupied channels and as a consequence visiting channels in such a way that they can communicate with each other in a bounded period of time. Therefore, the channel service quality, and the neighborhood discovery (NB) phase are fundamental and challenging due to the dynamics of cognitive radio networks. The authors provide an analysis of these challenges, controversies, and problems, and review the state-of-the-art literature. They show that, although recently there has been a proliferation of NB protocols, there is no optimal solution meeting all required expectations of CR users. In this chapter, the reader also finds possible solutions focusing on an asynchronous channel allocation covering a channel ranking.

INTRODUCTION

A cognitive radio wireless network (CWNs) is a promising technology for solving spectrum underutilization problems thanks to the ability to operate both in unlicensed and licensed bands, so that spectrum holes can be utilized by unlicensed secondary users (SUs or cognitive radios, CRs). A spectrum hole (known as "white space") is a band of frequencies, assigned to a licensed user (incumbent or primary user, PU), but not utilized by this PU at a particular time and specific geographic location (Wyglinski, Nekovee, & Hou, 2011). Unlicensed secondary users have an opportunistic spectrum access to these

DOI: 10.4018/978-1-4666-6571-2.ch015

bands, i.e., they may utilize the spectrum only if it is empty *ad interim*, namely, only when SUs do not interfere with high-priority PUs (Liang, Chen, Li, & Mähönen, 2011). Unlicensed secondary users must promptly vacate the currently occupied band, when the presence of a licensed primary user is detected. For this reason the link recovery data and a new detected common channel cannot be announced on the previously used spectrum band on account of the incumbent activity there. However, there is a need for on-demand discovery of a new common control channel by secondary users in order to continue or to establish a communication. Owing to all these unique characteristics of CWNs, the neighborhood discovery (rendezvous) phase is a burdensome problem with which wireless researchers have recently been contending. As the existence of a common control channel (CCC) (a typical assumption in distributed wireless networks) is not a practical solution in CWNs, we concentrate on another representative technique, which is a well-known frequency hopping (FH) technique, but here in this chapter we are interested in the solutions without any synchronization or coordination between radio users (both being unfeasible and unwanted in OSA). A FH technique is a usable and appealing approach for CWNs, since the probability of interference against incumbents decreases in comparison with other possible approaches. In a neighborhood discovery (rendezvous) protocol the term channel hopping (CHH) scheme (instead of frequency hopping) is rather used as default, where a radio in a time-slotted system hops (switches) from one channel to another at a given time slot. Radios switch to (or may stay on) different channels in consecutive slots according to their channel hopping sequence (CHHS). The period of time during which a CHHS is completed is often interchangeable called a cycle or period.

Cognitive radio, as conceived by Mitola (2000), allows intelligent utilization of the spectrum, enabling unlicensed SUs to sense the channel conditions, adjust operating characteristics to the real-time environment, and access time-varying spectrum holes opportunistically. However, in contrast to a classical distributed wireless network (ad hoc network), in an asymmetric OSA the available cognitive radio resources (channel set) are different for each cognitive radio of the same network because of geographical dispersion and licensed primary user activities (heterogeneous spectrum availability). The quality of unoccupied channels can also differ for various unlicensed secondary users, although neighboring users may experience similar channel characteristics, implying a high correlation among their channel ranking data, and e.g. between their channel switching patterns. Therefore, while estimating a channel ranking one should consider both current and prior knowledge of channel availability statistics. Spectrum history knowledge is key information, since channel selection based merely on the latest data can lead to frequent disruptions of PUs and SUs, whereas sensing periodically can detect the periods where a transmission could engender possible interference to incumbents. However, in practice noise is always present and causes detection and decision problems regarding spectrum allocation. In view of the uncertainty and complexity of a distributed CWN environment and consequently the difficulty of the formulation precise (crisp) values and formulas in some algorithms, fuzzy logic (FL) theory can be a promising approach thanks to its flexibility and tolerance to imprecise data. FL allows translating qualitative and heterogeneous data into homogeneous membership values, which are further processed through a set of fuzzy inference rules. Therefore, fuzzy logic system (FLS) applications found their place in different wireless communications areas, e.g., in order to determine the channel usability in CWNs.

Channel quality information can be, or, one could even argue, should be utilized by a neighborhood discovery protocol, which considers such information. However, despite a recent proliferation of NB protocols, there is no optimal solution meeting all required expectations of unlicensed second users. This is owing to the fact that the proposed protocols mostly focus narrowly on e.g., RDV guarantee in a single cycle or improving the performance of maximum time-to-rendezvous; or the comparison to a limited number of related approaches (sometimes not a justified comparison to the proposed solution) with additional limitations or assumptions. One can find a limited number of works elaborating on channel quality issues considered explicitly in a NB protocol. In this chapter we analyze a range of possible solutions focusing on different objectives as well as utilizing various techniques in their approaches. One of the very interesting, and recently appealing techniques used in power saving protocols and neighborhood discovery is the quorum system (QS) concept, originally and widely used in the context of operating systems e.g., in achieving mutual exclusion or solving agreement problems etc. Thanks to the usability of the QS concept and relevant difference set theory, cognitive radio users can construct independently (i.e. without mutual knowledge of hopping sequences or channel availability of the neighbors) their channel hopping sequence in a way that can hop through the channels with a guarantee to meet in a bounded time period. It is already proven that such a guarantee can be obtained both in the homogeneous and heterogeneous spectrum availability case.

In view of the aforementioned issues in a CWN, we present in this chapter different techniques, each of which focuses on a particular subject matter in a CWN, but when contemplated together they form one whole approach towards neighbourhood discovery in a cognitive radio network environment. Our goal is to face the cognitive radio reality which tends more towards asynchronous than synchronous operations. However, when dealing with asynchronism many different, at first glance not directly related issues should be taken into account in order to approach the problem correctly. Therefore, in this chapter we address on the one hand channel quality estimation and ranking, but since we need to use this estimation in a cognitive radio environment, we use a technique dealing with uncertainty and imprecise information being an ordinary factor in CWNs. On the other hand we need to cope with specific asynchronous neighborhood discovery challenges, such as rendezvous guarantee, which is quite easy to deal with in synchronous mode, but in asynchronous mode, despite a proliferation of research in this area, is still an open issue. We would like to stress that channel quality ranking and asynchronous neighborhood discovery should not be consider in isolation, since they are strongly related in a real environment. One cannot perform discovery of a rendezvous channel in CWNs without knowledge of channel quality, i.e., channel rank, in the context of a cognitive radio environment. A simple meeting on a channel does not guarantee a successful rendezvous, especially in CWNs, since incumbent traffic in the sense of its frequency as well as pattern characteristics (e.g., durability of busy/idle periods), the distance or a more common factor as signal quality, has impact on the success of potential rendezvous.

Therefore, in this chapter the reader can find, in the "Neighborhood Discovery (Rendezvous) State-of-the-Art" (SoA) section, different asynchronous approaches to be found in literature, already assessed in the sense of rendezvous guarantee and asynchronous operation as well as channel ranking consideration (here we stress that in the SoA of RDV protocols, even when such consideration is envisaged, which is rare, then focus is simply on the existence of different channel qualities, but not dealing with channel estimation itself). However, before going into the basics of RDV subject matter and a comparison of the discussed schemes, for the sake of better understanding of the discussed RDV protocols, we present first the quorum system concept and properties ("Quorum Systems" section), topics born in the context of operating systems, but lately often used in wireless communications. After the reader is familiar with asynchronous neighborhood discovery, he can extend his knowledge about fuzzy logic and fuzzy logic SoA ("Fuzzy Logic in Wireless Communications" section), but directly in the CWN context, and in the following section on "Channel Quality Estimation" about channel quality basics, descriptors and SoA. Having gained sufficient understanding of FL and channel quality estimation in CWNs, the reader can learn how to estimate channel ranks by use of FL descriptors. The last subsection of this section, "FL

Channel Estimation Embodiment" discusses how FL channel rank can be applied towards a neighborhood discovery protocol. In the last section on "Neighborhood Discovery Protocols" in the first subsection, we address more closely the subject of RDV guarantee, focusing on a different channel availability perception. Finally, the second subsection on "Asynchronous Rendezvous Protocols" explains how to adapt an RDV protocol to be suitable for asynchronous mode, and at the same time, suitable for a channel ranking consideration. In other words, an induced asynchronism, through making use of an asynchronous RDV extension (ARE), allows an RDV protocol to profit more from channel quality estimation and ranking (using for example the FL estimator discussed in this chapter). At the end of this subsection a comparison of an RDV protocol using an ARE against related work is presented in order to demonstrate the potential of an asynchronous RDV approach. Finally, before giving the conclusion of this chapter, the reader can find general discussion and possible future research directions.

BACKGROUND

Quorum Systems Basics

The quorum system (QS) concept, originally used in the context of operating systems (Singhal & Shivaratri, 1994), is widely used in achieving a distributed mutual exclusion (Maekawa, 1985; Artreya, Mittal, & Peri, 2007), consistent data replication (Kumar, 1991), agreement problem (Garcia-Molina & Barbara, 1985), but also in wireless communications in gossip protocols (Zhang, Han, Ravindran, & Jensen, 2008), and power saving (PS) protocols (Tseng, Hsu, & Hsien, 2002; Jiang, Tseng, Hsu, & Lai, 2003, 2005; Kuo, 2010). In neighborhood discovery protocols the QS theory and related combinatorial theory, (Hall, 1986) including difference sets (DSs), Latin Square (LS), balanced incomplete block design (BIBD), became recently very popular (e.g., Bian, Park, & Chen, 2009, 2011; Bian & Park, 2011, 2012; Altamini, Naik, & Shen, 2010; Lee, Oh, & Gerla, 2010; Hou, Cai, Shen, & Huang, 2011; Zhang, Li, Yu, & Wang, 2011; Zhang, Li, Yu, Wang, Zhu, & Wang 2013; Romaszko, 2013; Romaszko & Mähönen, 2011, 2012a, 2012b; Romaszko, Denkovski, Pavlovska & Gavrilovska, 2013; Gu, Hua, Wang, & Lau, 2013). In this chapter the most popular (among wireless researchers) theories and properties are presented (with relevant examples); the reader is also directed to additional reading in the listed references.

Simply speaking, a *quorum* is a collection of sets that intersect with each other at least once within a certain period (cycle). The usual definition used in the context of wireless communications is as follows (Jiang et al., 2003):

A *quorum system* Q under universal set U, $U = \{0,1, \ldots, N-1\}$ with N being a cycle (also called period; symbol Z_N often used referring to $U = Z_N$) length, is a collection of non-empty subsets of U, called quorums, satisfying the intersection property $\forall A, B \in Q : A \cap B \neq \varnothing$.

The intersection property of QS is the most basic one. This property can be used in the systems where nodes are not only synchronized but also their cycles are aligned, i.e., nodes start using their cycle elements at the same time, e.g., CRs start to hop to particular channels at the same unit of time or the same slot. However, this property is not sufficient in other cases, i.e., cycle misalignment or slot misalignment. Therefore, before going into the details of different QSs, first we present an additional property, namely, the rotation closure property (RCP), which is almost always utilized along with any QS in wireless communications. If a QS satisfies this property it can be used in asynchronous communication, e.g., CRs having independent CHHSs, or in power saving protocols radios waking up at a different unit of time; in both cases, the system should have RDV guarantee in a bounded period of time.

For a quorum R in QS Q under universal set U (= Z_N) where U={0,..., N-1} and $i \in \{1,...,N-1\}$ we can define: $rotate(R,i) = (x+i) \bmod N \mid x \in R$. A QS Q has the rotation closure property *(RCP)* if and only if $\forall R', R \in Q, R' \cap rotate(R,i) \neq \varnothing$ for all $i \in \{1,...,N-1\}$.

We can say this definition also in more understandable language, i.e., if quorum Q_I and any of its shift sets (rotation) have at least one overlapping element in a cycle N that it satisfies the RCP. For example the quorum Q_1={0,1,2,3,4,8,12} under Z_{16} (N={0, ..., 15}) satisfies the RCP, since $Q_1 \cap rotate(Q_1,i) \neq \varnothing$ for each $i \in \{1,...,15\}$, however Q_2={0,4,5,10,11,14,15} does not satisfy the RCP since $Q1 \cap rotate(Q_2, i=8) = \varnothing$, because {0,4,5,10,11,14,15} ∩ {8,12,13,2,3,6,7} $= \varnothing$, but one must note that for all other i at least one common element can be found.

There are different types of QS such as, cyclic QS (Luk & Wong, 1997), grid QS (Maekawa, 1985; Luk & Wong, 1997), grid Byzantine QS (GBQS, Zhang et. al., 2008), grid diagonal QS (Romaszko & Mähönen, 2011), torus QS (Lang & Mao, 1998), mirror torus QS (Romaszko & Mähönen, 2012b), tree QS (Agrawal & Abbadi, 1991), hyperquorum system (Wu, Chen, & Chen, 2008), probabilistic quorums (Friedman, Kliot, & Avin, 2010) etc. In this chapter we present the most widely used in the ND context, which are *cyclic* QS, *grid* and *(mirror) torus* QSs.

Types of Quorum Systems

A cyclic QS is based on the cyclic block design and cyclic difference sets (DSs) of combinatorial theory (Hall, 1986), and therefore first we present the cyclic DS definition.

A subset C, where $C= \{c_1, c_2, ..., c_k\}$ modulo N of Z_N for all $c_i \in Z_N$ is called a *cyclic (N, k, λ) difference set* with k integers under Z_N (where N is the set cycle, k the set size, $|C|$ = k, and λ the time number, and k and λ are positive integers so that $2 \leq k < N$) if for a value which is smaller than N, we can find exactly λ ordered pairs (c_i, c_j) in the set, $c_i, c_j \in C$, such that $c_i - c_j \equiv d(\bmod N)$.

Moreover, if *at least* one ordered pair (c_i, c_j) can be found in *(N, k)* difference set, then such set is called a *relaxed difference set*.

Given any DS $C=\{c_1, c_2, ..., c_k\}$ under Z_N, a *cyclic quorum system* defined by C is $Q =\{Q_1, Q_2, ..., Q_n\}$, where $Q_i = \{c_1 + i, c_2 + i,..., c_k + i\} \bmod N, i \in \{0,...,N-1\}$.

For example if we take a DS under Z_8, $C = \{0, 1, 2, 4\}$, which can be generated by difference of two elements from C for every c(1, ..., 7), since $1 \equiv 1 - 0(\bmod 8); 2 \equiv 4 - 2(\bmod 8); 3 \equiv 4 - 1(\bmod 8); 4 \equiv 4 - 0(\bmod 8); 5 \equiv 1 - 4(\bmod 8); 6 \equiv 2 - 4(\bmod 8); 7 \equiv 0 - 1(\bmod 8)$, we have following cyclic QS with Q_i: Q_0={0,1,2,4}, Q_1={1,2,3,5}, Q_2={2,3,4,6}, Q_3={3,4,5,7}, Q_4={4,5,6,0}, Q_5={5,6,7,1}, Q_7={6,7,0,2}.

In a basic *grid QS* (Maekawa, 1985) elements are organized logically in a grid array, where a grid quorum is the union of a row and a column from the grid, where the cardinality of a grid quorum is $k = 2\sqrt{N} - 1$.

For example in a 4×4 grid QS ($r = \sqrt{N}$), under Z_{16} (N={0, ..., 15}), k = 7, we have the following example grid quorums (for the sake of simplicity, we denote a quorum as $Q_{row number, column number}$): (i) row 1 and column 1 is $Q_{1,1}$={0,1,2,3,4,8,12}, (ii) row 2 and column 3 is $Q_{2,3}$={2,4,5,6,7,10,14}, and (iii) row 3 and column 2 $Q_{3,2}$={1,5,8,9,10,11,13}. $Q_{1,1}$ has two common elements with $Q_{2,3}$ which are 2 and 4; $Q_{2,3}$ has two common elements with $Q_{3,2}$ which are 5, 10.

We can also have fewer consecutive elements in a grid quorum, where elements in a grid can be distributed in a *diagonal* manner (Romaszko & Mähönen, 2011):

$$f(x,y) = ((y \times r) - (r-1) \times x)) \bmod (r \times r); x = y = \{0, ..., N-1\}; r = \sqrt{N},$$

thereby forming diagonal grid quorums, e.g., $Q_{1,1}$={0,4,7,8,10,12,13}, $Q_{2,3}$={1,2,5,8,9,13,15}, and $Q_{3,2}$={1,2,4,6,10,11}, where the example $Q_{1,1}$ has 8, 13 common numbers with $Q_{2,3}$, and elements 1, 2 are common for $Q_{2,3}$ and $Q_{3,2}$. Note that this quorum maintains the RCP.

A *torus QS* (Lang & Mao, 1998) is similar to a grid QS but adopts a rectangular structure called a torus, where the last row (column) is followed by the first row (column) in a wrap-around manner, with the cycle torus size N, where $N = r \times s; s \geq r \geq 1$ (the high r represents the number of rows, and s stands for the number of columns). The cardinality of a $r \times s$ torus quorum is $r + \left\lceil \frac{s}{2} \right\rceil$, where elements are formed by selecting any column c_j (j=1...s) of r elements, plus one element out of each of the $\left\lceil \frac{s}{2} \right\rceil$ succeeding columns (called a *forward* manner) using the right wrap-around. The entire column $c_j (= r)$ is called the *head* and the rest of the elements the quorum's *tail*.

For example, a 3×6 torus depicted in Figure 1 has a cardinality of 6. Selecting the third column as the head and tail randomly from the succeeding columns we obtain a torus quorum Q_A={2,4,8,9,11,14}, or selecting the last column as the head and tail from the first columns (wrap-around next columns) we have Q_B={5,7,11,12,14,17}. Q_A and Q_B have two common elements, 11 and 14.

Romaszko & Mähönen (2012b) observed that a torus quorum's tail can be selected in a *backward* manner without loosing the cyclic property (RCP). In a backward manner a tail is selected from preceding

Figure 1. Example of a 3x6 forward torus quorum

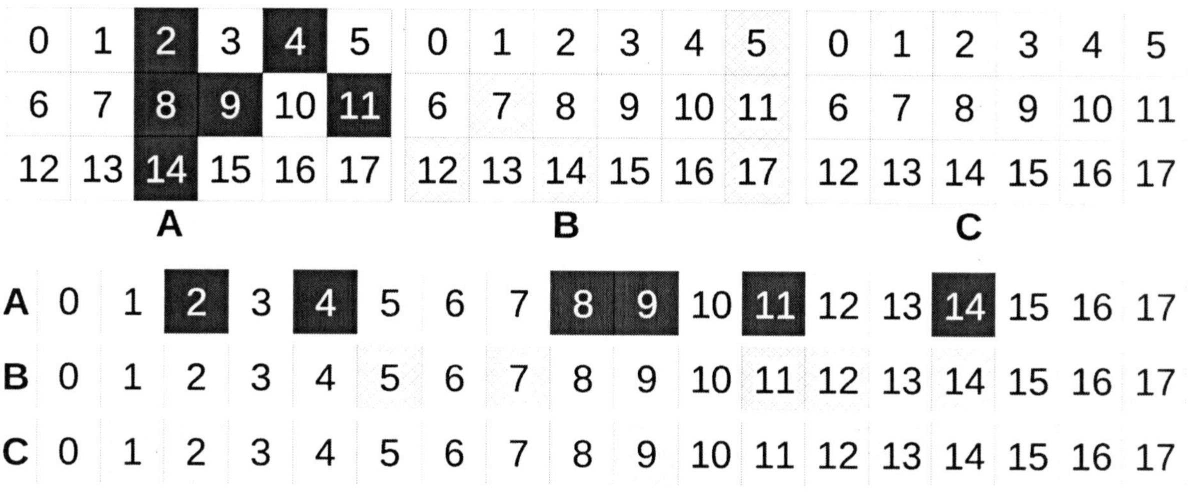

columns (instead of succeeding columns). For example, selecting the last column as the head and three elements from the preceding columns (Figure 2), we obtain the torus quorum $Q_1 = \{4,5,9,11,14,17\}$, and selecting the head from the fourth column and the tail in a backward manner we have $Q_2 = \{3,7,9,12,14,15\}$. Q_1 and Q_2 have two common elements, 9 and 14.

The reader should note that in a torus QS, elements can be selected in either a forward manner or backward manner but the manner of element selection cannot be mixed in QS; this is allowed only under additional constraints. These requirements are defined as the *mirror* torus QS extension (Romaszko & Mähönen, 2012b):

A tail of a torus quorum, $\left\lceil \dfrac{s}{2} \right\rceil$ elements can be chosen from any position of column $c_{j+k_i * i}$ (one element from a column), where $k_i \in \{1, -1\}$ and $i = \{1, ..., \left\lceil \dfrac{s}{2} \right\rceil\}$, in a wrap-around manner. For the sake of clarity, one must note that the torus quorums of the same torus QS must select their elements in the same forward (or backward) manner. In other words, if an element is selected from column c_{j+1}, the next element cannot be selected from c_{j-1}, but from the succeeding or preceding column, c_{j+2} or c_{j-2}, respectively. So k_i must be the same for all quorums of the same torus QS.

For example, in the 3×6 torus quorum illustrated in Figure 3, we select the last column as the head, the first tail's element in a backward manner, then in a forward manner and in a backward manner (which in this case is equal also to a forward manner), so $k=\{-1,1,1(-1)\}$. Therefore, we can obtain a mirror torus quorum $Q_1=\{1,4,5,11,14,17\}$. The head of the second quorum is column four, and the tail's elements are chosen with the same k order, therefore, we can have $Q_2=\{4,6,9,10,13,16\}$. Q_1 and Q_2 have one common element which is 4.

The reader should note that in a grid QS there should be at least two common elements of quorums selected from the same QS, where in a torus quorum (forward, backward or mirror) the lower bound is

Figure 2. Example of a 3x6 backward torus quorum

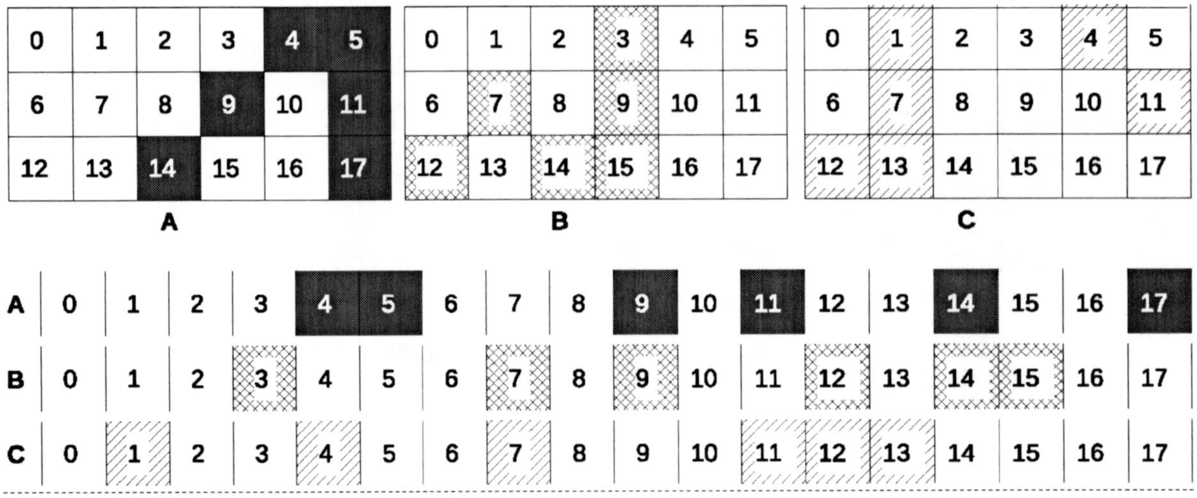

Figure 3. Example of a 3x6 mirror torus quorum

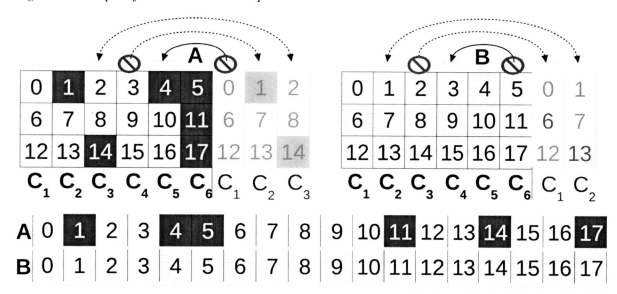

one element. All presented types of QS, namely, cyclic, (diagonal) grid, (forward, backward, mirror) torus, satisfy the RCP. Moreover, one should note that in order to verify the correctness of a quorum from cyclic QS, the quorum can also be checked according to the relaxed DS definition.

In the following we briefly present Latin square and identical-row square, since both are also used in related work (e.g., Bian et al., 2011; Bian & Park, 2012; Chao & Fu, 2013).

A *Latin square* is a $x \times x$ array filled with N different elements, such that each is occurring exactly once in each row and exactly once in each column.

A *identical-row square* (IRS) is an array where each row is a permutation of different integers.

Neighborhood Discovery (Rendezvous, RDV) Basics and State-of-the-Art

In distributed radio networks, the often used term *blind date* or *rendezvous* (RDV) refers to the ability of a pair of radios or multiple radios to establish a communication on a common channel during the so-called neighborhood discovery or rendezvous process. Considering the limited amount of time during which radios can occupy temporarily free channels and their frequent change of state (in view of PU's activity), a rendezvous must be accomplished in a bounded period of time. This *RDV guarantee* should be independent of channel availability perception, cycle misalignment, or clock drifts. The first issue refers to the asymmetric perception of channel availability, i.e., in the case of homogeneous channel availability, radios detect a uniform channel set, whereas, in the heterogeneous case, radios have a different perception of channel availability due to possible incumbents transmissions at different locations. This asymmetry and uncertainty of traffic patterns is inherent in a cognitive radio environment and therefore should be handled by a RDV protocol for CWNs. The issues of cycle misalignment and clock drifts are also natural in distributed wireless networks where radios initiate their channel hopping (starting their channel hopping sequence) at different time instances (so their cycles are not aligned) and there is a lack of clock synchronization owing to the decentralized nature of the networks.

A (bounded) period of time is measured as *time-to-rendezvous (TTR)*, which is an amount of time (in slots in theoretical analyses, sometimes in seconds in experimental analyses) within which radios can switch to a common channel (i.e. achieving a rendezvous) at an overlapping period of time (overlapping slot) with a finite delay, where the upper-bound on the delay corresponds to maximal TTR (*MTTR*) for all possible TTRs per cycle. Considering the short usability of free channels (due to possible sudden appearance of PUs) TTR should be as small as possible.

Another natural and main objective of pioneering RDV schemes (e.g., DaSilva & Guerreiro, 2008) was achieving any RDV in a single cycle. There is an evident correlation between a short (M)TTR and number RDVs per cycle, as (M)TTR decreases, the number of RDVs increases. However, as in a CWN incumbent traffic is usually unpredictable (unless using incumbents databases, e.g., radio environment map (Akyildiz et al., FARAMIR D2.1, 2010)) there is an additional goal of any ND protocol, which is a RDV guarantee on each available channel. If we are dealing with the homogeneous channel availability case, then all CRs need to be able to rendezvous on all channels (from a set) in a single period. Otherwise, RDV should occur on each common unoccupied channel in a cycle. One must note that some protocols that are analytically analyzed, regard the MTTR as the maximum time a channel hopping sequence needs to be followed until it meets its first RDV opportunity. However, this does not signify that when following the channel hopping sequence further, there are no TTR intervals that are not higher than the MTTR.

One should add that there exists also another controversy here, that some schemes provide a short (M)TTR in a cycle, however, RDVs do not occur on each channel, but only might happen on a (very) limited number of free channels (e.g., in ETCH protocol Zhang et al, 2011 in the case with different maps (CHHSs)). Moreover, it might also happen that there are a lot of RDVs encountered in a single cycle but occurring in bursts, i.e., leaving long gaps between RDVs. It is naturally obvious that a decreasing (M)TTR decreases such gaps allowing for RDVs distributed regularly in time, which is more desirable, since it avoids the case where CRs are stuck in the long no-RDV gaps before a successful rendezvous (Romaszko & Denkovski et al., 2013).

Another possible controversy might be raised regarding an *RDV balance* on the available channels. There is a great number of works arguing that the number of RDV on the channels should be balanced, which is theoretically understandable as the more balanced the RDV scheme is, the less chance of a single point of failure. On the other hand, in view of channel asymmetry and unequal channel quality, it might be more desirable to frequent more often those channels that show a better quality (although still ensuring visiting other free channels in a single cycle) because of e.g., less PU traffic there (note that the channel quality issue is discussed later in this chapter). We note that a great number of works can be found in the literature considering the former case, and a limited amount on work in the latter case. However, there also exist many schemes theoretically guaranteeing the same channel load over all the unoccupied channels, but at the same time it happens that some of the channels are unintentionally prioritized having more RDVs than others without any justification (arbitrary occurrence), e.g., in a well known RDV schemes such as (E)JS (Lin, Liu, Chu, & Leung, 2011, 2013) and SARCH (Chang, Teng, Chen, & Sheu, 2013).

Channel quality consideration in rendezvous protocols is very rarely handled. In AMRCC (Cormio & Chowdhury, 2010) CRs build a ranking table of the unoccupied channels based on PU activity detected on each channel. Then, they create their adapted CHHS, being a pseudo-random hopping sequence mapped to the ranking table, so that they can increase probability to rendezvous on a channel with a low PU activity. However, since AMRCC is a pseudo-random scheme, it cannot guarantee an RDV in a bounded time. In (Romaszko & Mähönen, 2011), (Romaszko & Denkovski et al., 2013) and (Romaszko, Torfs, Mähönen, & Blondia., 2013a, 2013b, 2014a, 2014b) CHHSs are created (based on different QS properties) attributing more slots to channels with a better quality. The scheme designed in (Romaszko & Denkovski et al., 2013) has also been implemented (ACROPOLIS D7.2 deliverable, Section 3.2.2) on USRPs (universal software radio peripheral, 2014) proving that frequenting better quality channels more often is a good approach while handling the neighborhood discovery phase in real conditions.

Synchronization establishment in distributed CWNs is a time and power consuming task, and therefore, the assumption that the unlicensed secondary users operate in a time-synchronized manner is not always justified. However, there is quite a number of works assuming synchronization for the sake of simplicity (e.g., Bahl, Chandra, & Dunagan, 2004; Balachandran & Kang, 2006; Bian et al., 2011; Chao & Fu, 2013). On the other hand, the *asynchronous operation* in distributed CWNs and its influence on the ND phase is not so well investigated, and mostly analyzed by probabilistic models to generate CHHSs e.g. the (Modified) Modular Clock ((M)MC) algorithm (Theis, Thomas, & DaSilva, 2010), Channel Rendezvous Sequence (CRSEQ) algorithm based on triangle numbers and modular operations (Shin, Yang, & Kim, 2010), Jump-Stay (JS) rendezvous algorithm using jump and stay patterns in each round (Lin et al., 2011) and many others. In (Romaszko et al., 2013a) the behavior of an ND protocol, considering channel ranking, is analyzed by a distinct approach. An asynchronism is defined by using asynchronous offsets, that can vary from -99% to 99% of a slot time unit, which should simulate the offset of a slot relative to a perfectly slot-synchronized system. The asynchronous threshold represents the minimum slot percentage required to enable communication. If an overlap of slots is higher than the threshold, a rendezvous occurs. In the analysis, different asynchronous thresholds and offsets have been considered, in order to analyze the influence of different overlap quantities. Further, an ND protocol, considering channel ranks, is evaluated in terms of RDV opportunities and TTR for different asynchronous thresholds. It is proven that asynchronism can bring noticeable benefits over the slot-synchronized case. This observation is also strengthened in Romaszko & Denkovski et al. (2013), where an asynchronous RDV protocol, benefiting from asynchronism, randomness, and gQS-based properties, is evaluated including the heterogeneity in terms of the channel rank among SUs. It is shown that the protocol performs noticeably better than the rank-random scheme (allocating rank channels the same number of slots). However, in that work asynchronism is handled differently than in all aforementioned work, namely, each cycle of CHHS has a different randomly chosen size. An RDV is considered as successful only if beacon messages are exchanged between the transmitter and receiver within a bounded slot interval. Finally, asynchronous characteristics of ND protocols have been experimentally analyzed on USRP boards only in several works (Robertson, Tran, Molnar, & Fu, 2012; ACROPOLIS D7.2 deliverable, pp. 39). Nevertheless, it has already been confirmed that added asynchronism can have a large beneficial effect in reducing the time-to-rendezvous. Further in this chapter we discuss how to introduce asynchronism in an RDV protocol in order to profit from it also by taking channel ranking into consideration.

There is only marginal study (Hsien, Lien, & Chou, 2011; Robertson et al., 2012; Romaszko & Denkovski et al., 2013) on the evaluation of ND schemes in a *real environment*. In (Hsien et al., 2011) a quorum-based rendezvous protocol has been implemented on USRP boards and evaluated in terms of the time of the first encounter between two unlicensed secondary users and in terms of the time for encounter on all channels. The experiments indicate a noticeable advantage of the sequence-based scheme over the random scheme, with and without the incumbent's activity. In (Robertson et al., 2012) on the blind neighborhood discovery for tactical networks, the performance of the Modified Modular Clock

(Theis et al., 2010) and Random Channel Access (Balachandran & Kang, 2006) approaches is compared on a testbed using USRP. It is found out that added asynchronism can have a large beneficial effect in reducing the time-to-rendezvous. The grid-based asynchronous RDV-MAC protocol (RAC^{2}E-gQS) from (Romaszko & Denkovski et al., 2013) using the QS properties of the gQS rendezvous protocol (originality designed in (Romaszko & Mähönen, 2011)) and the asynchronous character of the RAC^{2}E MAC protocol (where asynchronism is induced by using random cycle sizes and consequently various slot sizes in different cycles (Pavlovska, Denkovski, Atanasovski, & Gavrilovska, 2010)) has been implemented (ACROPOLIS D7.2 deliverable, pp. 39) on USRP2 nodes trying to achieve an RDV over 13 possible channels (but this is a case of heterogeneous spectrum availability since in experimental conditions cognitive radios locally perceive a different channel set). The proof-of-concept showed that CRs are able to establish communication (by exchanging short messages) within a single cycle.

Comparison of ND Schemes

Based on the discussion above we list the following CHH requirements: (i) RDV guarantee in a single cycle, single or multiple (*RDV: yes, no* in Table 1), (ii) RDV guarantee on each free channel N (*Degree: 0, 1, x, N*; *0* -no, *x* -on multiple channels), (iii) short (M)TTR (*sTTR: yes, no*), (iv) balanced/unbalanced or partly balanced if sometimes arbitrary unbalanced (*Bal: yes, no, par.*), (v) handling explicitly asynchronism (*Asyn: yes, no*), (vi) considering channel quality (*CHQ: yes, no*), (vii) heterogeneous channel availability consideration (*CHAv: yes, no*), (viii) proof-of-concept: analytical, experimental or both: whether an ND protocol is simulated or also implemented on a CR testbed (*PoC: yes, no*), (ix) simplicity (*Simp: yes, no*; easy or difficult to create a CHHS). Table 1 compares a number of well known protocols, often referred to in the literature, according to the nine aforementioned ND requirements. One must note that some of ND protocols can also be asymmetric, which means that each CR has a pre-assigned role as either a transmitter or receiver, where depending on its role a CR adopts different CHHS (hence, based on different algorithm); well known representatives of such an approach are ND protocols as AMOCH and ACH proposed by (Bian & Park, 2011, 2012). The reader should note that in Table 1 the degree of RDV indicates whether an RDV protocol guarantees an RDV on each channel, however, this is rather relative information, since some of the protocols, although guaranteeing optimal degree (on all free channels), might have a very large cycle (increasing with the number of available channels), e.g., SARCH by (Chang et al., 2013), EJS by (Lin et al., 2013), DRDS by (Gu et al., 2013), JS by (Lin et al., 2011), ETCH by Zhang et al. (2011), (e)MtQS-DSrdv by Romaszko (2013), DSMMAC by Hou et al. (2011), etc. (schemes listed according to the descending cycle size). Moreover, since the RAC2E-g(t)QS protocol by (Romaszko & Denkovski et al., 2013), and E2AND by (Romaszko et al., 2014) are purely asynchronous, they are indicated with capital letters (YES) in Table 1.

FUZZY LOGIC IN WIRELESS COMMUNICATIONS

Fuzzy Logic Basics

Fuzzy logic (FL) was first proposed by Zadeh (1965) as an extension of Boolean logic. In classical logic we have a crisp value, either one (1) or zero (0), expressing "true" or "false", respectively, whereas fuzzy logic permits us to assign any value in the interval [0, 1]. Fuzzy numbers are a sort of fuzzy sets (FS),

Table 1. ND protocols (legend: (-) not analyzed / not considered; (*) can tolerate clock skew; (**) there is an extension done, the slot is doubled to cope with an asynchronism, but not explicitly handled and analyzed; (***)- there is RDV guarantee on each free channel in heterogeneous channel availability case; (≈)- CHHS is not possible to create for all number of channels; only if N is a prime number; (yes/ no) -not in all possible cases of an algorithm; (*b) -it is proved that it might better)

CHH scheme	RDV	Degree	sTTR	Bal.	Asyn	CHQ	CHAv	PoC	Simp
SSCH (synchronous) Bahl et al. (2004)	yes	1	no	yes	no*	no	-	no	yes
RA (synchronous) Balachandran & Kang (2006)	no	0	no	no	-	no	-	yes	yes
SeqRDV, DaSilva & Guerreiro (2008)	yes	1	no	yes	yes	no	-	no	yes
AMRCC, Cormio & Chowdhury (2010)	no	0	-	-	no	yes	-	no	yes
CRSEQ, Shin, Yang, & Kim (2010)	≈yes	≈N	≈yes	yes	yes/ no**	no	yes	no	≈yes
(M)MC, Theis et al. (2010)	yes/no	-	yes/no	no	yes/no	no	-	yes	no
(E)JS, Lin et al. (2011, 2013)	yes	N	*b	no	yes	no	yes	no	no
AMOCH, ACH, Bian & Park (2011, 2012)	yes	N	yes	yes	yes	no	-	no	yes
Hsien et al. (2011)	yes	x	yes	-	yes	no	yes/no	yes	no
A-ETCH, Zhang et al. (2011)	≈yes	1	*b	yes		no	-	no	no
DSMMAC, Hou et al. (2011)	yes	N	*b	yes	yes	no	-	no	no
MtQS-DSrdv, Romaszko (2013)	yes	N	*b	yes	yes	no	yes	no	yes/no
enhanced MtQS-DSrdv, Romaszko & Mähönen (2012a)	yes	N	yes	yes	yes	no	yes***	no	no
ICH, 2013 Wu & Wu (2013)	yes	1	no	-	yes	no	yes/no	no	no
SARCH, Chang et al. (2013)	yes	x	yes	par.	yes	no	yes	no	no
DRDS, Gu et al. (2013)	yes	N	yes	par.	yes	no	yes	no	no
RAC2E-g(t)QS, Romaszko & Denkovski et al. (2013)	yes	x	*b	no	YES	yes	yes	yes	yes/no
E2AND, Romaszko et al. (2014b)	yes	x	yes	no	YES	yes	yes	no	yes/no

The reader should note that this table does not indicate that one protocol is better than another, but only shows the diversity of different ND protocols, and which protocols should not be omitted if referring to similar analysis of a class of a new designed protocol. It must be noted that often in a related work the performance evaluation considers a very limited number of similar works without a proper justification.

Moreover, one should note, that the table is not an exhaustive list, since some of the other works are very similar and compared to already existing ones, e.g., the Alternative Hop-and-Wait channel RDV approach (Chuang, Wu, Lee, & Kuo, 2013) is very similar to JS (Lin et al., 2011), trying to outperform JS and RW (Liu, Lin, Chu, & Leung, 2010); or an RDV approached proposed by (Wu, Han, & Kong, 2013) is similar to DSMMAC (Hoe et al., 2011), however, in (Wu et al., 2013) DSs' construction issue is partly discussed. Moreover, some of the researchers observe that already followed CHHS should be adjusted before successful rendezvous in order to avoid intolerable rendezvous delay. Liu, Pang, Wang, & Zhou (2012) propose a neighbor cooperation framework in order to allow the intended receivers to adapt their hopping sequences before RDV success with the help of neighboring nodes. On the other hand, (Gandhi, Wang, & Hu, 2012) introduce a new class of blind rendezvous algorithms, coordinated channel hopping, where CRs adjust and coordinate CHHS as they rendezvous pairwise. They propose two representatives, iterative intersection hopping and divide and conquer hopping, where the former focuses on having a new user promptly meeting the first of the existing users, and the latter addresses quickly spreading the information about a new user among existing users.

where a FS is a set without a crisp, clearly defined boundary. A FS is different from traditional set theory in the sense that an element belongs with a certain degree to a specific membership function (MSF). A predicate in fuzzy logic theory is expressed in the form *x is A*, where *x* is a variable over the universe *U*, and *A* is a FS defined over *U*. Fuzzy predicates are therefore characterized by a partial degree of truth in the interval [0, 1], since in fuzzy logic the truth of any statement is a matter of degree, because it is fuzzy. In mathematics, variables are usually numerical crisp values, but in FL non-numeric linguistic variables are allowed in order to facilitate the expression of facts and rules. For example, a linguistic variable "frequency of ON/OFF transitions" can have linguistic terms of frequency such as "low", "medium" or "high". The terms are determined by their associated MSFs, $\mu_i(u)$, which stand for the degree of membership of a fuzzy variable *u*. A fuzzy variable is utilized to be correctly interpreted without explicit knowledge of the used system model or technology, for instance, instead of saying "the SNR is 6 dB" it is simpler to express this as "link reliability is high" (Baldo & Zorzi, 2008).

A fuzzy logic system (FLS) exploits fuzzy logic reasoning, where the crisp inputs and outputs are translated to and from fuzzy representation through the *fuzzification* and *defuzzification* processes, respectively. In other words, in a FLS all measurements are translated to fuzzy sets (expressing their uncertainties) in the *fuzzification* step. Then, the inference engine determines the control rules stored in the fuzzy rule base, searching for the most appropriate rule under a certain condition. Fuzzy logic expresses and processes relationships in the form of rules composed by a set of *IF...(AND/OR)...THEN* used to determine the value of the output variables, e.g., *IF "frequency of ON/OFF transitions"=high AND...AND "opportunity"=medium THEN "Channel Usability"= veryLow*. The combination of the degrees of membership of all linguistic statements (a statement is e.g., *SINR="good"*, where *SINR="good" AND ... AND distance="short"* is a rule premise) is called the degree of rule fulfillment (or rule firing strength), $\mu_i(\underline{u})$ (Nelles, 2001). Only the rules with a degree of fulfillment greater than zero are active. In the following *act*ivation phase the output activations of the rules is estimated (Nelles, 2001), for instance $\mu_i^{act}(\underline{u}, y) = \min[\mu_i(\underline{u}), \mu_i(y)]$, where $\mu_i(\underline{u})$ stands for the degree of fulfillment of rule *i*, and $\mu_i(y)$ is the output MSF. For the conjunction of linguistic statements, *t-norm*s logic operators exist, with the most common being, *min, product,* and *bounded difference*. The rule-based decision, composed of a set of *IF-THEN* rules, determines the value of output variables (consequences). When the degree of rule fulfillment for all rules is determined, all the consequences must be aggregated (*acc*umulated), where *t-conorm*s operators are used (the most common *t-conorm*s logic operators are *max, algebraic sum* and *bounded sum*). The *max* operator is the most popular, for instance (Nelles, 2001), $\mu^{acc}(\underline{u}, y) = \max_i[\mu_i^{act}(\underline{u}, y)]$,

where $\mu^{acc}(\underline{u}, y)$ is the fuzzy output set.

The resulting fuzzy set is converted into a crisp value in the *defuzzification* step. There are a number of methods (*center of gravity (CoG), bisector, mom, lom, som*), where the most used is the center of gravity (also called "centroid") method (Nelles, 2001):

$$y_{CoG} = \frac{\displaystyle\int_{y_{\min}}^{y_{\max}} y\mu^{acc}(\underline{u}, y)dy}{\displaystyle\int_{y_{\min}}^{y_{\max}} \mu^{acc}(\underline{u}, y)dy} \, .$$

In addition, a *knowledge representation base* (Baldo & Zorzi, 2009) can be employed, which determines the relationships between crisp input (output) parameters and their fuzzy representation in order to enhance the interpretability of the knowledge possessed by the unlicensed secondary user. Therefore, purely qualitative fuzzy variables can be used, i.e., expressed in the concept of "weak", "ordinary" and "strong", or fuzzy numbers with layer-dependent landmark values. The generality of the knowledge representation based on FL (i.e., generic and technology independent knowledge) enables researchers to provide network access policies for different wireless technologies and applications. An example FLS is depicted in Figure 4 in the section on fuzzy channel quality formulation based on FL descriptors.

Fuzzy Logic State-of-the-Art

FL-based applications can be found in our day lives, e.g., an "intelligent" wash-machine, electric power steering system, antilock-braking system, or vehicle speed estimation developed for a car's control system. FL was previously applied In wireless communications, before even the concept of cognitive radio was born (Mitola, 2000), e.g., (Abdul-Haleem, Cheung, & Chuang, 1994) presented a fuzzy distributed dynamic channel assignment algorithm in, where FL is exploited as the decision making logic. (Baldo & Zorzi, 2008) designed a cognitive cross-layer architecture for cognitive radio networks, addressing the following challenges by utilizing FL: modularity, information interpretability, imprecision and uncertainty (measurements affected by errors in precision and accuracy), scalability, and complexity constraints. In

Figure 4. FLS (channel ranking estimator) example with 4 inputs, 1 output

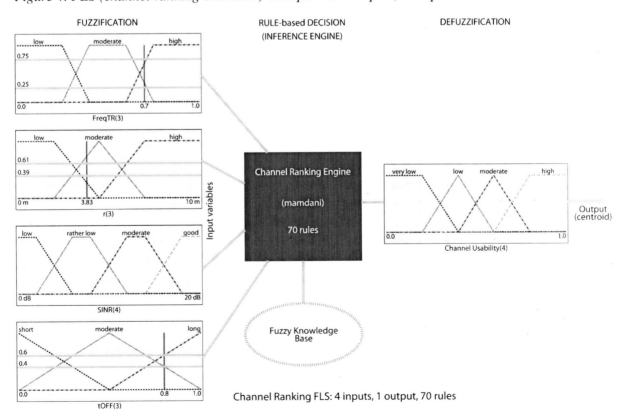

order to support easily network access policies for different wireless technologies and improve information interpretability, all variables are intended to be fuzzy variables and all parameters are intended as fuzzy control variables. (Baldo & Zorzi, 2009) also conceived a knowledge representation framework, based on FL, enabling the implementation of a cognition process which is both cross-layer and network-aware, and a distributed cognitive network access scheme based on fuzzy decision making. In (Le & Ly, 2008) the possibility of unlicensed secondary users accessing a spectrum band without interfering with licensed users is determined by using FLS with the following descriptors: spectrum utilization efficiency of the unlicensed SU, its degree of mobility, and its distance to the PU. The linguistic knowledge of spectrum access (SA) based on three descriptors was obtained from a group of five network experts. (Giupponi & Perez-Neira, 2008) built a fuzzy-based spectrum handoff algorithm using two FL controllers. The first controller estimates the distance from the PU and from the possible transmit power of the unlicensed SU based on the signal strength (SS) received at the SU from the PU, and signal-to-noise-ratio (SNR) at the PU. The second controller determines whether a spectrum handoff should be realized, or the transmit power should be adjusted. (Merentitis & Triantafyllopoulou, 2010) proposed an algorithm for cooperative dynamic spectrum access in CWNs, utilizing medium access control (MAC) layer mechanisms for message exchange between unlicensed secondary users, operating in license exempt spectrum bands, in order to achieve interference mitigation. In (Chen & Tsai, 2012) an OSA scheme was presented for CR femtocell networks, which exploits FLS for four descriptors: spectrum utilization efficiency of a secondary unlicensed user (femtocell), its degree of mobility, its SS and its distance to an incumbent (macrocell). In (Yao, Ngoga, Erman, & Popescu, 2012) a decision making algorithm was proposed for competition-based channel selection fed by two fuzzy variables, the idle time statistics (the remaining idle time on every channel is predicted) and SU competition level.

CHANNEL QUALITY ESTIMATION

Channel Quality Preliminaries and State-of-the-Art

In OSA, CRs have only limited knowledge of channel parameters or other users' information. Spectral opportunities are asymmetric due to time and space varying channels. Owing to this inherent asymmetry and uncertainty of traffic patterns, CRs can have trouble properly detecting the real usability of unoccupied channels, and as a consequence visiting channels in such a way that they can communicate with each other in a bounded period of time. However, there is a need of on-demand discovery of a control traffic channel by SUs in order to establish communication. Therefore, channel quality estimation (note that in this chapter channel quality estimation refers to terms such as frequency of transitions of ON/OFF periods (primary user activity) or level of interference based on SINR), being prerequisite knowledge for an ND protocol, is fundamental and challenging in CWNs.

Cognitive radios sense the spectrum in order to detect the presence or absence of incumbents on the licensed channels. Here for the sake of generalization, we assume that either an SU performs local sensing or the channel availability is provided by spectrum databases, and possible traffic and interference information could be available through REMs (Akyildiz et al., FARAMIR D2.1, 2010). While sensing locally, the CR maintains samples of the channel and processes them for an indication of the incumbent's activity. The spectrum sensing duration can vary with the different physical layer sensing techniques and achievable probability of correct detection of the incumbent's presence and false alarm rate for the licensed user detection. Depending on the licensed user's activity, a channel can be modeled as an ON/OFF resource alternating between busy ON (occupied by PU) and idle OFF (accessible for SUs) periods. In the channel model assumed in this chapter, an unlicensed secondary user can access a channel only when there is no incumbent activity, hence during OFF periods. In the literature researchers often exploit the channel utilization $ChU^i \in [0,1]$ which is usually defined as the average fraction of time during which channel i is in idle state (Chou, Sai Shankar, Kim, & Shin, 2007) namely: $ChU^i = \dfrac{t^i_{OFF}}{t^i_{OFF} + t^i_{ON}}$.

When a channel must be vacated by CRs, channel switching must be executed from the currently used channel to a new channel. Exploiting periodic sensing only, unlicensed secondary users are incapable of detecting *ON-OFF channel state transitions* between two consecutive spectrum measurements. Such situation should be considered when ON-OFF transitions occur frequently. That is why the frequency of ON-OFF transitions has a significant impact on the channel stability.

While determining the possibility of accessing the spectrum band, channel quality estimation can encompass different information, for example: channel utilization efficiency, frequency of transitions between ON-OFF periods, Signal strength received at an unlicensed secondary user or the distance between incumbents and SUs, SNR at the PU, level of interferences (SINR), SU's degree of mobility, or the number of users; below we elaborate on some of these.

Channel Quality Descriptors

In view of the asymmetry of channel availability and thus incumbent traffic patterns, the most common objective of wireless researchers is to use the spectrum efficiently (e.g., Le & Ly, 2008; Chen & Tsai, 2012) in their OSA algorithms. Therefore, spectrum utilization efficiency η_E is defined as the ratio of the spectrum band to be used by a cognitive radio and the available band as follows: $\eta_E = \dfrac{BW_{SU}}{BW_C} \times 100\%$, where BW_{SU} stands for the spectrum band to be used by the unlicensed user, and BW_C is available band.

Another very often used metric is the *distance* between PUs and SUs (e.g., Le & Ly, 2008; Giupponi & Perez-Neira, 2008; Chen & Tsai, 2012). There are different approaches estimating the distance between the incumbents and unlicensed secondary users, for example Hoven & Sahai (2005) noted that when the location of the licensed user is unknown, SUs can approximate their distance (r) from the primary transmitters, as locally measured SNR can be consider as an alternative metric for distance, $r = g_{12}^{-1}(\dfrac{\sigma^2}{P_{PU}} 10^{\frac{SNR_{SU} + \beta}{10}})$, where P_{PU} stands for the power of the PU, σ^2 for the noise at the primary receiver (when there is uncertainty in the noise-power, σ^2 is set to the maximum tolerable noise at that radius), $g_{12}(r)$ is the propagation path loss between the PU and SU, and β is the signal loss. One can notice that the distance is not the only variable that influences the SNR. Giupponi & Perez-Neira (2008) remarked that a low SS from an incumbent received at an unlicensed secondary user is not sufficient to decide whether the distance between them is high; therefore, additional information is needed. The authors determine a fuzzy control variable (based on nine rules) of the possible transmit power of a cognitive radio (and in this way the distance between an incumbent and secondary user) based on the

SS received at the SU and the PU's SNR assuming the possibility of determining the bit rate of the PU, e.g., if the SNR at the PU is "high" and SS at the SU is "low", then the distance is "long"; or if SNR is "low" and SS is also "low" then distance is "short".

(El Masri & Khalif, 2011) observed that the *channel stability* metric can describe the activity behavior of incumbents over the licensed channels, i.e. how the availability of these channels is distributed over time. The distribution of the channel availability can be described by the number of periods during which the channel is available for an unlicensed secondary user, and the manner in which these periods are dispersed in time, i.e. distance between successive periods and difference in their duration. In an unstable channel, transitions between busy and idle periods occur very often. A stable channel has long busy and/or idle periods. A channel can also be said to be less or more stable based on the deviation in the duration of OFF periods. With an increasing deviation, the stability of the channel also increases. For example, if we have one long OFF period and several short OFF periods (thus a high deviation) we know that we can use the long OFF period, but short OFF periods are almost useless. On the other hand, if the deviation is low, but transitions between idle and busy periods are frequent, we know that the frequent OFF periods are not sufficiently long enough.

The reader can find more information regarding a derivation of ON/OFF transition information, duration probability of ON and OFF periods, and using the history for prediction in the following works: (Kim & Shin, 2008), (Yang, Cao, & Zheng, 2008), (Lee & Akyildiz, 2008), and (Mehmeti & Spyropulos, 2013). In (Yang et al., 2008) the past channel observations are exploited to estimate the future spectrum availability, i.e. the probability that a channel will be idle in the next time slots. The channel having the longest remaining idle time is the channel deemed to be the most available. Such a proactive approach can reduce disruptions to incumbents, but it can also suffer from imperfect predictions in the case of a high randomness in the incumbent's traffic. Mehmeti & Spyropulos (2013) contest the common belief that a channel with a lower average incumbent's activity is usually a better channel (as for example assumed in Kim & Shin (2008) or in the ND domain in Cormio & Chowdhury (2010)) by showing that the impact of the variability of the channel durations can be much more significant than the duty cycle itself (utilization) while searching for a better quality spectrum hole.

The level of *interference*, estimated by the SINR is also paramount information for the channel quality estimation. How the SINR is determined depends on the network scenario and targeted interference model, but the simplified formula is as follows: $SINR = \dfrac{P}{N + I}$, where N represents the ambient noise power level, I is the total interference (sum of all power levels experienced owing to other signals currently transmitted), and P is the transmitter power (received signal strength) of the actual signal to be received. If the value is higher than a certain capture threshold (minimum SINR required for a successful reception) the signal is received correctly. The reader can find different models for controlling the transmit power of secondary users based either on the lowest permissible SINR at the primary receiver or target SINR level at the secondary receiver, or both, for example, assuming fading-free channels in (Kandukuri & Boyd, 2010) or with the fading channel environment consideration in (Huang, Liu, & Ding, 2010). (Gardellin, Lenzini, & Fontana, 2011) classify digital TV channels of IEEE 802.22 ("*IEEE 802.22 WG on WRAN*", 2011) based on their operating frequency and such that the capacity requirement of the largest number of end-users is satisfied. The channel selection is determined based on the channel quality by measuring the SINR from a transmitter to a receiver. The SINR values are estimated in relation to the performance of a communication system denoted by the bit error rate (BER). BER is the ratio between the number of wrongly received bits and the total number of transmitted bits, and it depends on the received SINR. Unlicensed secondary users select a data modulation scheme based on the received SINR and the required quality of service such that the QoS requirements can be fulfilled (the table with modulations and their respective SINRs can be found in (Gardellin et al., 2011)). (Gardellin et al., 2011) conclude that: (i) for low SINR values, channels at low frequencies are better than that at high frequencies, (ii) for high SINR values, channels are equivalent irrespective of the carrier frequency; (iii) with SINR below 2.57 dB the signal is not received correctly. Furthermore, the authors classified the channels according to the five following classes: SINR (dB) between (1) 2.57-5.06 for the capacity demand of 2.27 Mb/s, (2) 5.07-11.8 for 4.03 Mb/s, (3) 11.81- 15.31 for 8.07, (4) 15.32-19.75 for 12.61 Mb/s, and (5) SINR > 19.76 for the capacity of 18.91 Mb/s.

Furthermore, the number of users or the degree of *mobility* can also be taken into account when estimating the probability of channel access (e.g., (Giupponi & Perez-Neira, 2008), (Le & Ly, 2008), (Merentitis & Triantafyllopoulou, 2010)). Depending on the considered scenario the mobility of SUs can play a paramount role, since misdetection of a PU due to the SU's mobility leads to a false assumption of idle channel opportunity. When a CR is moving at a given velocity v (m/s), the Doppler Effect (proven by Buys-Ballot (Snellen, 1891)) needs to be taken into account: $\Delta f = v * \dfrac{f}{c} * \cos \varphi$, where Δf is Doppler frequency shift, f is the carrier frequency, c is the speed of light, and Φ is the angle between the direction of the motion and the direction of the incoming signal.

CHANNEL QUALITY ESTIMATION

Fuzzy Channel Quality Formulation Based on FL Descriptors

In this section we present a fuzzy logic system determining channel ranking based on the channel usability information (fuzzy output variable), which is derived based on the following four descriptors: (i) the frequency of ON/OFF transitions ($Freq^{TR}$), (ii) distance r from/to PU, (iii) level of interferences ($SINR$) and (iv) duration of idle time (t_{OFF}), as illustrated in Figure 4.

The first linguistic variable "frequency of ON/OFF transitions", $Freq^{TR}$, can have linguistic terms of frequency transitions between idle and busy states such as "low", "moderate" or "high" (illustrated in Figure 4). These terms are defined by their associated MSFs, $\mu_i(Freq^{TR})$, which determine the *degree of membership* of a fuzzy variable $Freq^{TR}$. For instance, if a channel is characterized by a rather high frequency of ON/OFF transitions, hence "rather high", it can be expressed as 0.7 in the range [0, 1]. Therefore, if $Freq^{TR} = 0.7$ (Figure 4), then it leads to the following degrees of membership: $\mu_{low}(Freq^{TR}) = 0$, $\mu_{moderate}(Freq^{TR}) = 0.25$, and $\mu_{high}(Freq^{TR}) = 0.75$. The reader should note that $\mu_{low}(Freq^{TR}) + \mu_{moderate}(Freq^{TR}) + \mu_{high}(Freq^{TR})$ sums up to 1.0, i.e., $\sum_{i=1}^{N} \mu_i(Freq^{Tr}) = 1$, with $N = 3$ the number of MSFs for the fuzzy variable $Freq^{TR}$.

The second variable representing the distance (r) between the PU and SU, can be described as "near", "moderate" and "far". For example a cognitive radio is 3.83 m (Figure 4) from an incumbent which can give the degrees of membership: $\mu_{near}(r) = 0.61$, $\mu_{moderate}(r) = 0.39$, $\mu_{far}(r) = 0.0$. We use meters as example unit of distance, but it can also easily be transformed to a range [0, 1] to be correctly interpreted without having explicit knowledge of the used scenario.

The third variable, *SINR*, uses the obtained SINR values from (Gardellin et al., 2011) as a reference, but with fewer classes, namely an SINR below 2.57 dB (one should note that in (Gardellin et al., 2011) it is assumed that a signal is not correctly received below an SINR of 2.57 dB, but if there is no channel which satisfies the requirements, the channel assignment scheme allows some SINR values under this lower bound) and the first class (1) we describe as "low" class. The second class (2) corresponds approximately to our "rather low" class. Our "moderate" class is similar to the third class (3). Finally, our "good" class corresponds to the fourth class (4) and an SINR above 19.76 dB (thus also 5 class). Figure 4 illustrates an example of possible MSFs of the linguistic variable SINR corresponding approximately to classes from Gardellin et al. (2011). One should note, that values of SINR can be replaced by a fuzzy representation within the range [0,1] in order to be correctly interpreted without explicit knowledge of the used coding, modulation or number of subcarrier combinations.

The fourth linguistic variable, *"opportunity"*, represented as t_{OFF}, can have three linguistic terms describing the OFF states, namely "long", "moderate" and "short", expressed in the range [0,1]. For example, if the OFF period is "long", interpreted for example as $t_{OFF} = 0.8$ (plotted in Figure 4), it leads to the following degrees of membership: $\mu_{short}(t_{OFF}) = 0.0$, $\mu_{moderate}(t_{OFF}) = 0.4$, $\mu_{long}(t_{OFF}) = 0.6$.

The output linguistic variable is the channel usability, described by four terms, "very low", "low", "medium" and "high" as can be seen in Figure 4. The depicted illustrative examples in Figure 4 show only one possible representation of MSFs, however the reader should note that MSFs of descriptors described above can have different types dependence on the objective of FLS, e.g. *triangular* or *trapezoidal*, where there is loss of information in regions where the slope of MSFs is zero (also less useful for learning). The functions can also be smooth, e.g. Gaussian, sigmoidal or generalized bell -shape, where loss of information is reduced, and hence these MSFs are better suited to the learning process.

Channel Quality Processing Using FLS

The channel ranking algorithm records the past channel activities, including the frequency of channel transitions ($Freq^{TR}$), idle period durations (t_{OFF}), the level of interferences (*SINR*), and the distance (*r*). The information is updated upon each channel sensing operation. As noted by (El Masri & Khalif, 2011) and (Mehmeti & Spyropoulos, 2013), the variability of the idle/busy time durations plays a paramount role in the channel ranking estimation process, and that is why in our FLS, $Freq^{TR}$ is selected to be the dominant factor having the greatest impact on the final decision. The reader should note that the $Freq^{TR}$ variable is correlated with the variable t_{OFF}. Figure 5 illustrates a snapshot, which has a duration of time T_i ($i=\{LOW, MODERATE, HIGH\}$ of $Freq^{TR}$), where possible relations between the frequency of ON/OFF transitions and OFF durations can be seen. Note that in this model, the longest t_{OFF} indicates which type of *"opportunity"* we can obtain.

In the T^{LOW} period, the channel is the most useful when $Freq^{TR}$ is "low" and t_{OFF} "long". Although $Freq^{TR}$ is "high" in the T^{HIGH} period, there are sometimes "long" t_{OFF} periods inbetween, which means that the channel can be visited in these durations.

The second dominant variable is the distance *r*, followed by the *SINR* variable. One must note that if the *r* is "short" and the *SINR* is "low", the channel usability is not good enough even though $Freq^{TR}$ is "low" and the OFF periods are long. There is also the relation between *SINR* and OFF periods, namely,

Figure 5. Frequency of transitions and idle/busy durations

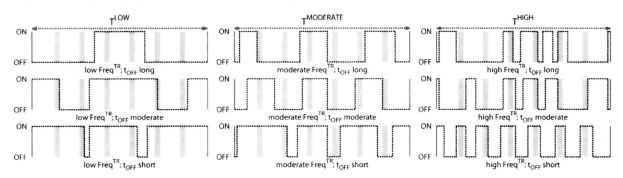

although OFF periods are short, and *SINR* is good, there is still an opportunity to use the channel because of a good rate, and while the planned data is not too big. Of course *Freq^TR* and *r* also influence the final decision.

Based on this information, the channel ranking rules are created for the fuzzy reasoner, but note that some of the rules can be reduced. For example we can have the following rule: *"IF (Freq^TR =high) AND (r=near) AND (SINR=low) THEN (Channel Usability=veryLow)"*, where the fourth variable is not mentioned, since irrespective of its value, the channel usability will be still very low. In some of the rules the logic operator NOT can also be used, e.g. the negation operator for a linguistic statement t_{OFF} = short is NOT$(\mu_i(t_{OFF})) = 1- \mu_i(t_{OFF})$. Due to the usability of the operator the number of rules can be reduced to two instead of three, when the importance of some variable can be neglected. In this chapter we will not present the rules since according to different experts the rules might be different, however, the interested reader can refer to the example channel ranking rules based on the aforementioned four descriptors in (Romaszko & Mähönen, 2014). The possible three-dimensional (3D) representation of the channel usability (described by fuzzy values in [0, 1]), as a function of the frequency of ON/OFF transitions and OFF periods, is illustrated in Figure 6 (results based on the rules presented in (Romaszko & Mähönen, 2014)). One can see the channel usability increases rather smoothly with an increasing t_{OFF} and the decreasing *Freq^TR*. While *Freq^TR* is (rather) "high" (>0.7), the channel usability is (rather) low.

FL Channel Classification Embodiment

Let us assume that a CWN network has five available channels, and secondary users exploit an ND protocol which considers channel ranking (e.g., Romaszko et al., 2013a, 2013b). With a CHHS with 5 unoccupied channels, the channels can have for example attributed 7, 6, 5, 4 and 3 slots, respectively, in a descending rank order, as in the RDV protocol in (Romaszko et al., 2013a).

Given the example data in Table 2, and according to the associated MSFs of each variable and the strongest degree of membership from (Romaszko & Mähönen, 2014), channel 1 can be described by "moderate" *Freq^TR*, "moderate" distance to PUs, "rather low" *SINR*, and "short" t_{OFF}. The reader should note that also the other degrees of membership are considered through the rules, and all rules with a degree of fulfillment greater than zero are active. Using the CoG method for defuzzification we get channel usability of 0.324. When using the CoG method, the channels are ranked in the following order: channel 2 (0.6), channel 3 (0.445), channel 5 (0.359), channel 1 (0.324), and channel 4 (0.258). The reader should note that none of the depicted channels are really good to transmit on, as there is at least one inconvenient

Figure 6. Channel usability as a function of $Freq^{TR}$ and t_{OFF}

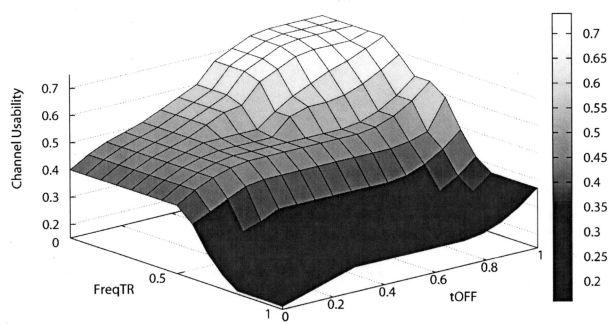

Table 2. Example channel information with five available channels

Channel x	$Freq^{TR}$	r	SINR (dB)	t_{OFF}	Channel usability
1	0.4	4	9	0.2	0.324
2	0.2	5	6	0.8	0.6
3	0.7	7	11	0.9	0.445
4	0.8	3	17	0.3	0.258
5	0.25	7	3	0.4	0.359

channel characteristic, decreasing its usability. Channel 2 is ranked as the best channel, which is rather evident, since although its *SINR* is rather low, *Freq^{TR}* and t_{OFF} give a very good opportunity to transmit, with a moderated distance. Other channels are, however, more difficult to rank, due to the presence of the worst cases of correlated data, such as, *Freq^{TR}* and t_{OFF} (channel 5 and channel 3), or distance and SINR (channel 5 and channel 4). Although rather high *Freq^{TR}*, channel 3 is the second best channel, as it has long t_{OFF} and not bad *SINR*, preceding channel 5 with a low *Freq^{TR}*, but very bad *SINR* with rather low t_{OFF}, which decreases significantly a chance for a successful transmission.

The outputs obtained by the FL channel classification estimator serve as an input for a rendezvous protocol considering a channel rank in its approach. Therefore, based on the channel ranking list (channel order from the best to the worst channel, thus according to the presented example: 2, 3, 5, 1, 4) obtained through the presented FL estimation, an ND protocol adopts an appropriate channel hopping pattern, for example by attributing more slots to better channels (Romaszko et al., 2013a).

NEIGHBORHOOD DISCOVERY PROTOCOLS

How to Achieve RDV Guarantee on Each Available Channel

As shown in the ND state-of-the-art section, the rendezvous guarantee issue on each available channel may be overcome in many different ways using either quorum system theory and properties or other methods of combinatorial theory. Based on the aforementioned description and all presented methods, we can deduct that in a neighborhood discovery algorithm every h channel hopping sequence (CHHS) for each particular channel should comply with the rotation closure property, which we can write as *cyclic quorum system channel-CHHS*:

A CHHS M^h of cycle $\Theta = \{0,...,N-1\}$, under Z_N, and $\forall i \in \Theta$, composed of h cyclic quorums (satisfying the RCP), $Q_i : \{q_1,...,q_k\}; k \in \Theta$, from the same quorum system Q, where h is the number of disjoint quorums (i.e., $Q_1 \cap Q_2 \cap ... \cap Q_i = \varnothing$) and $i=\{1,...,h\}$, is *cyclic quorum system h-CHHS*.

For example, we can have a CHHS with three free channels, hence, composed of $\{Q_1, Q_2, Q_3\} = \{\{0,1,2,5,10\}, \{3,8,11,12,14\}, \{4,6,7,9,13\}\}$ under Z_{15} is a cyclic QS 3-CHHS, since each Q_i is a cyclic QS satisfying the RCP. Moreover, each Q_i is a (relaxed) difference set, where Q_1 is also a torus quorum in a 3×5 torus array.

The above definition and example refer to the CRs that have a homogeneous channel availability perception. In this section, we focus more on RDV guarantee on all unoccupied common channels where CRs have a heterogeneous channel availability perception. This case is still a challenge in RDV protocols, since a pair of SUs or multiple SUs must have such CHHSs that satisfy the heterogeneous RCP through complying with the heterogeneous difference (map) pair properties. In the QS theory section we presented only a general case of the RCP definition. Therefore here we should adjust it for CHHSs of a pair of CRs, first for the case with homogenous channel availability perception, and then for heterogeneous case.

CHHS-Rotation Closure Property: For CHHSs *M1* and *M2* of cycle $\Theta = \{0,...,N-1\}$, under Z_N, $\forall i \in \Theta$, we can define: $\forall offset^{SLOT}, \exists i : M1_i \cap M2_{(i+offset^{SLOT}) \bmod N} \neq \varnothing$ and $offset^{SLOT} \in \Theta$. For sake of clarity, we can interpret this definition as follows: CHHS *M1* must be checked if there exists a match for each possible cycle shift, or also called cycle misalignment (i.e. slot offset, where $offset^{SLOT} \neq 0$; if $offset^{SLOT} = 0$, this is the cycle alignment case).

Some of the RDV protocols in the literature are based entirely on difference sets theory, where each sub-CHHS is a (relaxed) DS, hence it must also satisfy the CHHS-RCP, e.g. DSMMAC (Hoe et al., 2011), AACH (Wu, Han, & Kong, 2013), or DRDS (Gu et al., 2013). There exist also ND protocols composed partially of DSs and torus quorums, such as (e)MtQS-DSrdv (Romaszko & Mähönen (2012a, 2012b)). However, the construction of DSMMAC difference sets is not shown in Hoe et al. (2011), where only two CHHSs are presented as examples, for 2 and 8 available channels, hence, (7,3,1) DS and (73,9,1) DS, respectively. Nevertheless, one can notice that DSMMAC is similar to AACH, but (Wu et al., 2013) presented partially a method for the construction of DS. Based on both these works and DS theory, we can conclude that it is not easy to find disjoint DSs of the same system, especially where the number of DSs (available channels) increases. One can easily see that with 8 channels the cycle size is 73 (DSM-MAC), but with 9 channels it must be increased already to 273 to have (273, 17, 1) DS as shown in (Wu et al., 2013). For the sake of an easily implementable algorithm the cycle sizes were even increased in

(Gu et al., 2013). In that work the authors also stated that the cardinality of the maximum disjoint relaxed difference sets S_N, under Z_N, is bounded by $|S_N| \leq \sqrt{N}$ (where N is the cycle size), also presenting maximum DRDS up to $N=50$. However, the DRDS algorithm uses a larger N than is theoretically needed (according to the authors' statement and presented examples in this section), e.g. with 3 free channels DRDS has $N=27$, but it is possible to have disjoint relaxed DSs with $N=15$ ((15,5,1) DS as shown in the example above, with a cyclic QS 3-CHHS), or with 4 unoccupied channels $N=28$ ((28,7,1) DS of CHHS[CR2] later shown in an example in this section), or with 5 free channels $N=45$ ((45,9,1) DS of CHHS[CR3] later shown in an example in this section)), etc.

For multiple CRs having a heterogeneous channel availability perception we can define the following extended heterogeneous rotation closure property:

NM-*CHHS-rotation closure property*: For CHHS *M1*, being a cyclic QS, of cycle $\Theta = \{0,...,N-1\}$, under Z_N, $\forall i \in \Theta$, and *M2* of cycle $\Phi = \{0,...,M-1\}$, under Z_M, $\forall j \in \Phi$, where $N<M$, we can define: $\forall \Phi, \exists i, \exists j : M1_i \cap M2_j \neq \varnothing$.

In order to be able to find the cyclic QS CHHSs satisfying the NM-CHHS-RCP, we must find *heterogeneous difference pairs* for each CHHS pair of respective SU CHHSs. In other words, we search for each sub-CHHS of SU1 (thus *particular* channel hopping sequence for channel *r* appearing in a CHHS) matching sub-CHHS of SU2, if those sub-CHHSs match, they form a *heterogeneous (N,k,M,l) difference pair*:

Given $N \leq M$ $e = \left\lceil \dfrac{M}{N} \right\rceil$ and sets $A : \{a_1,...,a_k\}$ under Z_N and $B : \{b_1,...,b_l\}$ under Z_M, the pair (A, B) is said to be a *heterogeneous (N, k, M, l)-difference pair* if $\forall d \in \{0,...,M-1\}$ there can be found at least one ordered pair $b_j \in B, a_i^e \in A'$ in such a way that $b_j - a_i^e \equiv d(\mathrm{mod}\, M)$, where set A' is (e-extension) of set A (Lai, Ravindran, & Hyeonjoong, 2010).

(e-extension): Given two positive integers N and e, for set $A = \{a_i \mid 1 \leq i \leq k, a_i \in Z_N\}$ e-extension of A can be defined as $A' = \{a_i + j * N \mid 1 \leq i \leq k, 0 \leq j \leq e-1, a_i \in Z_N\}$. For a quorum system $Q = \{A_1, A_2,..., A_M\}$ (e-extension) of Q can be defined as $Q' = \{A_1',..., A_M'\}$.

For instance, for set $A = \{0,1,2,5,10\}$ under Z_{15} we have (e-extensions) for $e=2$ and $e=3$: $A^2 = \{0,1,2,5,10,15,16,17,20,25\}$ under Z_{30} and $A^3 = \{0,1,2,5,10,15,16,17,20,25,30,31,32,35,40\}$ under Z_{45}. Given two sets, A under Z_{15} and $B = \{0,7,9,14,15,17,21\}$ under Z_{28}, we can say that the (A, B) pair is heterogeneous (15, 5, 28, 7)-difference pair, because for A^2 and B, there can be found at least one ordered pair $b_j \in B, a_i^2 \in A^2$ such that $b_j - a_i^2 \equiv d(\mathrm{mod}\, 28)$:

$1 \equiv 21 - 20, 2 \equiv 7 - 5, 3 \equiv 0 - 25, 4 \equiv 9 - 5, 5 \equiv 7 - 2, 6 \equiv 7 - 1, 7 \equiv 9 - 2, 8 \equiv 9 - 1, 9 \equiv 14 - 5,$

$10 \equiv 15 - 5, 11 \equiv 21 - 10, 12 \equiv 14 - 2, 13 \equiv 14 - 1, 14 \equiv 15 - 1, 15 \equiv 17 - 3, 16 \equiv 17 - 1, (\mathrm{mod}\, 28)$

$17 \equiv 9 - 20, 18 \equiv 7 - 17, 19 \equiv 21 - 2, 20 \equiv 21 - 1, 21 \equiv 9 - 16, 22 \equiv 9 - 15, 23 \equiv 15 - 20,$

$24 \equiv 21 - 25, 25 \equiv 7 - 10, 26 \equiv 14 - 16, 27 \equiv 9 - 10.$

Based on all above definitions we can present the *heterogeneous difference ch-CHHS pair* for CRs having the heterogeneous channel availability perception, as follows.

Given $N \leq M$ $e = \left\lceil \dfrac{M}{N} \right\rceil$ and cyclic quorum system h-CHHSs, $M^1 : \{Q1_1, ..., Q1_w\}$ (subset of disjoint quorums of k elements under Z_N) and $M^2 : \{Q2_1, ..., Q2_u\}$ (subset of disjoint quorums of l elements under Z_M) and $w < u$, the pair (M^1, M^2) can be defined as a *heterogeneous (N, k, M, l) difference CHHS pair*, if $\forall Q1_s, \exists Q2_r$ such that $s = r; s, r \in \{1, ..., w\}$, where the sets $Q1_s : \{a_1, ..., a_k\}$ under Z_N, and $Q1_r : \{b_1, ..., b_l\}$ under Z_M, are the pairs $(Q1_s, Q2_r)$ such that $s=r$, forming a heterogeneous (N, k, M, l)-difference pair if $\forall d \in \{0, ..., M-1\}$ there can be found at least one ordered pair $b_j \in Q2_r, a_i^c \in Q1_s^c$ such that $b_j - a_i^c \equiv d (\mod M)$.

In (Romaszko & Mähönen, 2012a) it was proven that it is possible to design such an ND protocol (enhanced MtQS-DSrdv in the referred work), where CRs' CHHSs satisfy *heterogeneous (N, k, M, l) difference CHHS pair*, however it is not an easy task (as noted in Table 1) and illustrated in the example below.

Let CR1, using e MtQS-DSrdv with 3 available channels, have $CHHS^{CR1}$ composed of $S1^{CR1}, S2^{CR1}$, and $S3^{CR1}$, where sub-CHHSs are $S1^{CR1}=\{0,1,2,5,10\}$, $S2^{CR1}=\{4,6,7,9,13\}$ and $S3^{CR1}=\{3,8,11,12,14\}$ under Z_{15} in the same QS. CR2 using eMtQS-DSrdv has 4 channels, hence, $CHHS^{CR2}$ with following $S1^{CR2}, ..., S4^{CR2}$ sub-CHHSs: $S1^{CR2}=\{0,7,9,14,15,17,21\}$, $S2^{CR2}=\{4,8,11,16,18,24,25\}$, $S3^{CR2}=\{3,6,13,19,20,22,27\}$ and $S4^{CR2}=\{1,2,5,10,12,23,26\}$ under Z_{28}. And finally CR3, also using eMtQS-DSrdv has 5 channels, with following $S1^{CR3}, ..., S5^{CR3}$ sub-CHHSs: $S1^{CR3}=\{0,9,10,13,18,20,21,27,36\}$, $S2^{CR3}=\{2,5,14,23,28,31,32,34,41\}$, $S3^{CR3}=\{6,7,15,24,33,38,39,40,42\}$, $S4^{CR3}=\{8,11,12,17,26,35,37,43,44\}$, and $S5^{CR3}=\{1,3,4,16,19,22,25,29,30\}\}$ under Z_{45}. All CHHSs are *cyclic quorum system* h -CHHS: 3-CHHS 4-CHHS, and 5-CHHS, respectively.

The pair $(CHHS^{CR1}, CHHS^{CR2})$ forms a heterogeneous (15, 5, 28, 7) difference ch-CHHS pair. The pair $(CHHS^{CR1}, CHHS^{CR3})$ forms a heterogeneous (15, 5, 45, 9) difference ch-CHHS pair. However, the pair $(CHHS^{CR2}, CHHS^{CR3})$ does not form a heterogeneous (28, 7, 45, 9) difference ch-CHHS pair, since $(S1^{CR2}, S1^{CR3})$ and $(S4^{CR2}, S4^{CR3})$ are not difference pairs, as there is no ordered pair found for $d=25$, and $d=32$ (note, that this also corresponds to 25 and 32 slot offsets, cycle shift) with $S1$ respective sequences, and $d=37$ (slot offset 37) with $S4$ respective sequences. One must note that $(S2^{CR2}, S2^{CR3})$ and $(S3^{CR2}, S3^{CR3})$ are difference pairs. Proofs of the matches (or no matches) of all respective pairs can be found in (Romaszko & Mähönen, 2012a).

Asynchronous Rendezvous Protocols[1]

As already discussed, there are different approaches to dealing with the asynchronous character of an RDV protocol, with example of those particularly focusing on asynchronism being (Romaszko & Denkovski et al., 2013) or (Romaszko et al., 2013a). As already proven analytically (Romaszko et al., 2013a, 2013b) and experimentally by (Robertson et al., 2012), (Romaszko & Denkovski et al., 2013 and ACROPOLIS D7.2 deliverable, Section 3.2.2) an added asynchronism can bring benefits (reduce TTR, improve protocol efficiency) to an ND protocol. Therefore, in this section we look closer at our work presented in (Romaszko et al., 2013b, 2014b), where an asynchronous rendezvous extension is proposed, which can be added on top of any ND protocol, in order to induce asynchronism and channel ranking. The key idea of the asynchronous RDV extension (ARE) is to enable channels to have varying sizes of slots. One should note that, although differently designed, (Romaszko & Denkovski et al., 2013) also utilized a different size of slots, induced by a different size of cycles (each cycle size is randomly chosen, but the slot size is the same in a given cycle). However, although there is some asynchronous similarity, the approaches differ a lot, since in (Romaszko & Denkovski et al., 2013), sizes of cycles (and hence slots) are randomly chosen without explicit further justification (the protocol uses randomness in order to achieve asynchronism), whereas in (Romaszko et al., 2013b) the sizes of the slots depend on the rank of the channel to which the slots are attributed. In other words, certain channels have longer slots attributed to them as they are better quality channels.

According to the proposed ARE by (Romaszko et al., 2013b) the largest slot has a size of 100 time units expressed as: $AREsize^{SLOT} = 100 - SSV * priority^{CHANNEL}$; where the channel priority ranges from *0* to *N-1*, with *0* being the highest priority and *N* the number of channels, and *SSV* is the slot size variation, a parameter, which allows tuning the amount of influence the extension has on the particular channel priority. If the SSV parameter is equal to 0, ARE is not used and only the regular original protocol behavior is enforced. In (Romaszko et al., 2013b) it was illustrated that the extension in combination with certain ND protocols can improve the performance regarding TTR and number of RDVs. The asynchronous ND protocol (AND) presented in (Romaszko et al., 2014a) also shows that an ND protocol that uses ARE has a significantly better TTR and number of RDVs in comparison with well known RDV protocols.

In (Romaszko et al., 2014b) the ARE is enhanced because it does not take into account the number of channels, and hence it is suboptimal when there is a large number of channels. Therefore, three enhanced methods are proposed considering the number of channels: linear (*linea*), logarithmic (*logea*) and exponential (*expae*) ARE, where the minimum SSV (*mSSV*) is a parameter to be tuned to optimize the performance. *mSSV* describes the minimum slot size that is permitted for the lowest priority channel. All the methods support high priority channels, but with a different approach.

In the *linear* method the slot size decreases uniformly respecting a particular channel priority, i.e. the maximum slot size difference (100 − *mSSV*) divided by the number of available channels minus one:

$$LINEAsize^{SLOT} = 100 - priority^{CHANNEL} \times \frac{100 - mSSV}{N-1}.$$

In the *logarithmic* method, slot size differences are larger for a small number of high priority channels and smaller for a large number of low priority channels in comparison with the linear method:

$$LOGEAsize^{SLOT} = 100 - \frac{\ln(priority^{CHANNEL})}{\ln(N)} \times (100 - mSSV).$$ The method allows that only a few

channels have a large slot size.

In the *exponential* method, slot size differences are larger for a large number of high priority channels and smaller for a small number of low priority channels in comparison with the linear method:

$$EXPAEsize^{SLOT} = 100 - \frac{1.15^{priority^{CHANNEL}} - 1}{1.15^{N-1} - 1} \times (100 - mSSV).$$ The method has a higher number of

channels obtaining a large slot size.

In order to investigate properly the asynchronous character of ND protocols with added (enhanced) ARE, a discrete time line, the granularity of which can be tuned by passing a different time interval *t*, is defined in (Romaszko et al., 2013b, 2014b). The maximum resolution equals one; the resolution is used in the performance analyses. To perform a statistical analysis of all possible RDVs and TTRs that can occur in any random situation, a single simulation run includes a statistical analysis of all metrics for all possible offsets $offset_i$: $offset_i = i \times t$, where $i = \{0, ..., \left\lceil \dfrac{Size^{CYCLE}}{t} \right\rceil - 1\}$ and t is the time interval. This approach enables the consideration of an asynchronous threshold which defines what percentage of a full slot overlap between CHHSs is required to be considered as an RDV.

As mentioned previously, in all RDV related work the most common metric to be investigated is TTR, MTTR being its upper bound. Furthermore, also the number of RDVs in a cycle is analyzed. However, the way of estimating the number of RDVs in the synchronous case cannot be the same as in the asynchronous case, since it disregards the fact that in neighboring slots a CR can visit the same channel and, hence it has a longer time to exchange messages. Therefore, in (Romaszko et al., 2013a, 2013b, 2014b) RDV is defined as the number of RDVs on *sequentially distinct* channels (called an RDV opportunity). As already mentioned, some protocols (analytically analyzed) regard the MTTR as the maximum time a channel hopping sequence needs to be followed until it meets its first RDV opportunity. However, this does not mean that when following the channel hopping sequence further, there are no TTR intervals that are not higher than the MTTR. Works such as (Romaszko et al., 2013a, 2013b, 2014a, 2014b) take into account the complete hopping sequence and not only the first RDV occurrence from the start of the channel hopping sequence.

Furthermore, as the number of RDVs per cycle is relative to the cycle size, which is often different for every protocol, it is an unsuitable metric to compare different ND schemes and therefore the *normalized RDV per cycle* is also defined: $normalized^{RDV} = \dfrac{RDV}{time^{CYCLE}} * 100$, i.e. defining RDV per 100 time units.

Moreover, an additional interesting metric was presented and investigated in (Romaszko et al., 2013b), which is the *efficiency*, expressing the ratio of the total time that is available for a communication per cycle (total overlap time per cycle, $overlap^{RDV}$) and the time without a RDV per cycle (inter time-to-rendezvous time (total *iTTR* per cycle), t_{iTTR}): $Efficiency = \dfrac{overlap^{RDV}}{t_{iTTR}}$.

In (Romaszko et al., 2013b) the proposed ARE has been added to MtQS-DSrdv (Romaszko & Mähönen, 2012b), AMOCH (Bian et al., 2011) and DSMMAC (Hou et al., 2011), where all three protocols ensure an RDV on each available channel, but they do not support the channel ranking. It was shown that as long as the asynchronous threshold, which is the minimum required amount of overlap, is not too high (ca. 30 time units in the case of DSMMAC as can be seen in Figure 7; the form of the graph of two other protocols is similar), the proposed ARE can noticeably increase the efficiency and the percentage of RDVs also ensuring a lower TTR.

Nevertheless, it was also noted that the threshold can have a negative influence on the metrics (TTR, RDVs and efficiency) while it increases. The SSV has a positive influence on TTR, RDVs and efficiency under the condition that the threshold is not too high. At a certain threshold value, the usage of SSV can even result in a loss. The tipping point is ND scheme dependent and is different for each metric.

In (Romaszko et al., 2014b), first, the three enhancements of ARE have been evaluated as added value of the enhanced asynchronous ND (EAND; the original version, AND, was presented in (Romaszko et al., 2014a)). The protocol is intentionally designed to be used with the ARE, where in its "synchronous version" (yet without added ARE) the channel slot distribution is uniform. However, here we do not elaborate on (E)AND, the interested readers can find details in (Romaszko et al., 2014a, 2014b). Based

Figure 7. Influence of SSVs and asynchronous threshold values on the efficiency for DSMMAC

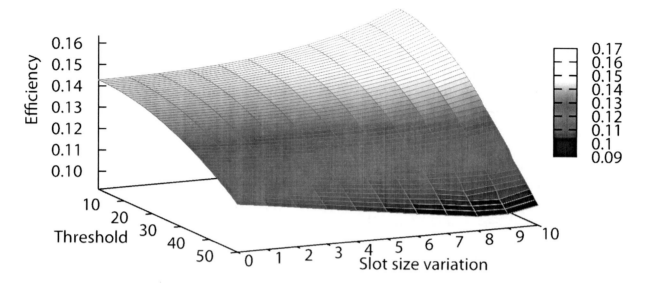

on the analysis of all three enhancements it was shown that the logarithmic approach outperforms others in terms of the normalized RDV, TTR and the cycle size. The cycle size analysis is paramount as it reveals that the smallest cycle size of EAND leads to the best RDV and TTR performance (since EAND was designed in such a way that all channels should have an equal amount of RDVs and *logae* permits only a few channels to have a large slot size and allows more channels to have smaller slot sizes where the relative difference is small, the cycle size is the smallest of all AREs). Therefore, (Romaszko et al., 2014b) compare EAND with the *logae* extension (called *eeand*) against related work, JS and EJS (Lin et al., 2011, 2013), A-ETCH (Zhang et al., 2011), mod_ETCH (modification of ETCH allowing the use of any number of channels instead of only the prime *N* channels as in the original ETCH protocol), DRDS (Gu et al., 2013), and SARCH (Chang et al., 2013) in terms of the cycle size, normalized RDV and TTR. The reader should additionally note that the full asynchronous behavior of all protocols is investigated (unlike some related work), that is, a statistical analysis of all possible RDVs and TTRs that can occur in any random situation is performed for every offset between two CHHSs with a granularity of a single time unit.

In this chapter we present (in Figure 8) only the normalized RDV (the cycle size and TTR analysis can be found in (Romaszko et al., 2014b)), but with two additional well-known related works, namely, CRSEQ (Shin et al., 2010) and asymmetric ND protocol AMOCH (Bian et al., 2011).

Figure 8 clearly shows that the EEAND protocol, exploiting the logarithmic asynchronous extension, outperforms significantly other well-known protocols (handling multiple requirements presented in Table 1), as it is explicitly designed to be used in asynchronous mode. The results are also related to the cycle size, since the EEAND has the lowest cycle size in comparison with other approaches. As the RDV performance is related to TTR, we also can deduct (see also (Romaszko et al., 2014b)) that the TTR of EEAND is significantly better than other ND protocols.

Figure 8. Normalized RDV of AMOCH, SARCH, CRSEQ, DRDS, JS, EJS, ETCH, MOD_ETCH and EEAND

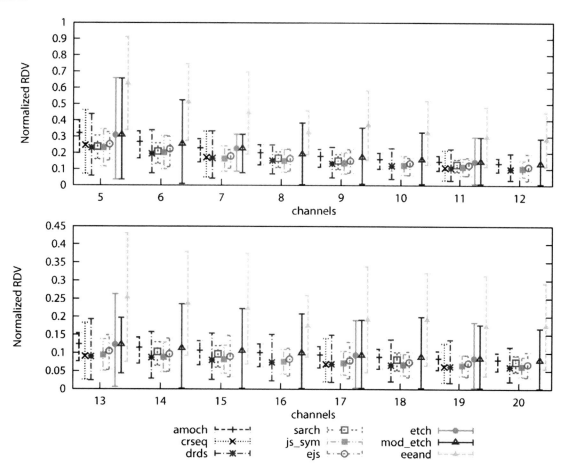

FUTURE RESEARCH DIRECTIONS

As shown in this chapter, there has recently been a great number of works on neighborhood discovery. Some of them focus on a limited number of objectives (TTR, RDV guarantee only, or RDV guarantee on all channels); others try to cover different issues (TTR, (maximum) RDV diversity, i.e. on (all) multiple channels, heterogeneous channel availability perception, asynchronism) in CWNs. However, while addressing one issue, another issue is often arises, e.g. (i) a very large cycle, which can lead to insufficient updates of CHHSs or more difficult adjustment of CHHSs; (ii) although assuring maximum RDV diversity, TTR increases along with the number of free channels as an algorithm is not scalable enough, i.e. guarantying RDV on each channel in a cycle, but the cycle also increases hence TTR performance degrades; (iii) considering channel rank in an ND algorithm but without sufficient RDV diversity. In addition, one of the major problems in the current literature is narrowing a comparison of a novel design protocol against a very limited number of other related work, making it difficult to judge whether the new proposed scheme is indeed better in general, and not just in comparison with the compared related work in the manuscript, e.g. there is an interesting tendency to almost always compare a novel RDV approach to JS (Lin et al., 2011), and also very often to RDV protocols proposed by (Bian et al., 2011), or SeqRDV by (DaSilva & Guerreiro, 2008) and of course the authors' own previously published works, while ignoring new recently proposed schemes. Note that sometimes such comparison is even not suitable because of a different nature of the new proposed scheme.

Another problem arises, when an analysis of a novel proposed approach is limited to a number of definitions and analytical proofs, without providing enough information (insufficient pseudo-code or description of an algorithm, e.g., in steps) so that other researchers cannot have the chance to learn and compare with this proposed work.

Considering the great numbers of different issues an RDV scheme must cope with, there is also a strong need to focus the current research in neighborhood discovery on experimental analyses (testbeds), instead of only handling it analytically or in a simulation environment. This new research direction could also encourage researchers to consider channel quality data in their new proposed RDV schemes, as it is an inherent issue in a real environment. Channel stability (frequency of ON/OFF transmission), duration of time a channel is really accessible, primary signal strength received at a secondary user or distance to incumbents, level of interference and other scenario-specific characteristics such as number of contending unlicensed secondary users and their mobility, should play an important role in designing a new rendezvous protocol. All of above mentioned factors can strongly influence a successful RDV, as it is one thing to be able to meet on a particular channel thanks to a proper designed CHHS, and another thing to be able to communicate on it due to good channel ranking characteristics.

Moreover, while designing a channel quality estimator algorithm, one must decide which approach is the most suitable, based on raw data information, or by interpreting this information with a degree of uncertainty (considering FL theory, neural networks or similar schemes), and also how many and which factors should be taken into account, including consideration of history, periodically collected data, or instantaneous information only. Finally it is also important to consider self-learning mechanisms, not only in channel quality estimation, but also in neighbor discovery, since depending on the environment and scenario, it can happen that some patterns are repeatable, hence, such information could be useful while designing a channel hopping pattern, e.g. which channels would be better to swap (perhaps some of them should be placed on the black list, others less frequent etc.), on which channels is it worth to stay longer, etc.

What is also worth noticing in some of RDV schemes is a consideration and handling of a great number of unoccupied channels (e.g. very often 100 channels, (JS et al., 2011), (Gu et al., 2013), (Chuang et al., 2013)) and designing and enhancing the protocols in such a way that they can handle as large a number of free channels as possible. However, one should ask whether this is needed? In a real environment, given a large number of channels, are we able to or should we visit all of them, sense and decide how to hop on them? Or would it be better to limit ourselves to a smaller number of channels, having more time for communication and more efficient RDV scheme?

CONCLUSION

In this chapter we focused on the channel service quality and the neighborhood discovery phase, two fundamental challenges in CWNs. We provided an analysis of these challenges, controversies and problems, and reviewed the state-of-the-art literature. We showed that, although recently there has been a proliferation of ND protocols, there is no optimal solution meeting all required expectations of SU users. In this chapter the reader could also find possible solutions focusing on an asynchronous channel allocation covering a channel ranking algorithm and channel hopping sequence algorithm. The reader additionally expanded his or her knowledge on methods not directly related to a wireless communication, such as fuzzy logic theory and quorum system and related theories. Moreover, a channel ranking fuzzy logic system has been presented along with a channel classification example. We also discussed neighborhood discovery requirements for the homogenous and heterogeneous channel availability perceptions with brief examples. In addition, we considered ND protocols in asynchronous mode, along with suitable metrics. Finally, we presented a comparison of eight known RDV protocols in terms of normalized RDV in asynchronous mode, but considering any possible random situation, hence, with granularity of a single time unit. Concluding this chapter we discussed controversies and proposed further research directions that might help wireless researchers in their future work.

REFERENCES

Abdul-Haleem, M., Cheung, K. F., & Chuang, J. (1994). Fuzzy logic based dynamic channel assignment. *IEEE ICCS, 2,* 773–777. doi:10.1109/ICCS.1994.474159

Agrawal, D., & Abbadi, A. (1991). An efficient and fault-tolerant solution for distributed mutual exclusion. *Journal of ACM Transactions on Database Systems, 9*(1), 1–20. doi:10.1145/103727.103728

Akyildiz, I. F. et al., (2010). Flexible and spectrum aware radio access through measurements and modelling in cognitive radio systems. *FARAMIR (ICT-248351),* Deliverable D2.1.

Altamini, M., Naik, K., & Shen, X. (2010). Parallel link rendezvous in ad hoc cognitive radio networks. In *Proceedings of IEEE Global Communications conference, Exhibition & Industry Forum (GLOBECOM).* IEEE.

Artreya, R., Mittal, N., & Peri, S. (2007). A Quorum-Based Group Mutual Exclusion Algorithm for a Distributed System with Dynamic Group Set. *IEEE Transactions on Parallel and Distributed Systems, 18*(10), 1–16.

Bahl, P., Chandra, R., & Dunagan, J. (2004). SSCH: Slotted Seeded Channel Hopping for Capacity Improvement in IEEE 802.11 Ad Hoc Wireless Networks. In *Proceedings of International Conference on Mobile Computing and Networking (MobiCom)* (pp. 216-230). ACM. doi:10.1145/1023720.1023742

Balachandran, K., & Kang, J. (2006). Neighbor discovery with dynamic spectrum access in adhoc networks. In *Proceedings of IEEE 63rd Vehicular Technology Conference,* (vol. 2, pp. 512–517). IEEE.

Baldo, N., & Zorzi, M. (2008). Fuzzy logic for cross-layer optimization in cognitive radio networks. *IEEE Communications Magazine, 46*(4), 64–71. doi:10.1109/MCOM.2008.4481342

Baldo, N., & Zorzi, M. (2009). Cognitive network access using fuzzy decision making. *IEEE Transactions on Wireless Communications, 8*(7), 3523–3535. doi:10.1109/TWC.2009.071103

Bian, K., & Park, J.-M. (2011). Asynchronous channel hopping for establishing rendezvous in cognitive radio networks. In *Proceedings of IEEE Conference on Computer Communications (INFOCOM).* IEEE. doi:10.1109/INFCOM.2011.5935056

Bian, K., & Park, J.-M. (2012). Maximizing Rendezvous Diversity in Rendezvous Protocols for Decentralized Cognitive Radio Networks. *IEEE Transactions on Mobile Computing, 12*(7), 1294–1307. doi:10.1109/TMC.2012.103

Bian, K., Park, J. M., & Chen, R. (2009). A quorum based framework for establishing control channels in dynamic spectrum access networks. In *Proceedings of ACM International Conference on Mobile Computing and Networking (MobiCom)*. ACM. doi:10.1145/1614320.1614324

Bian, K., Park, J.-M., & Chen, R. (2011). Control channel establishment in cognitive radio networks using channel hopping. *IEEE Journal on Selected Areas in Communications, 29*(4), 689–703. doi:10.1109/JSAC.2011.110403

Chang, G.-Y., Teng, W. H., Chen, H.-Y., & Sheu, J.-P. (2013). Novel Channel-Hopping Schemes for Cognitive Radio Network. *IEEE Transactions on Mobile Computing, 13*(2), 407–421. doi:10.1109/TMC.2012.260

Chao, C.-M., & Fu, H.-Y. (2013). Providing Complete Rendezvous Guarantee for Cognitive Radio Networks by Quorum Systems and Latin Squares. In *Proceedings of IEEE Wireless Communications and Networking Conference (WCNC)*. IEEE. doi:10.1109/WCNC.2013.6554545

Chen, Y. M., & Tsai, P.-S. (2012). Opportunistic spectrum access decision operation in cognitive radio femtocell networks using fuzzy logic system. *International Journal of Computational Intelligence and Information Security, 3*(3), 11.

Chou, C.-T., Sai Shankar, N., Kim, H., & Shin, K. G. (2007). What and how much to gain by spectrum agility? *IEEE Journal on Selected Areas in Communications, 25*(3), 576–588. doi:10.1109/JSAC.2007.070408

Chuang, I., Wu, H.-Y., Lee, K.-R., & Kuo, Y.-H. (2013). Alternate hop-and-wait channel rendezvous method for cognitive radio networks. In *Proceedings of IEEE International Conference on Computer Communications*. IEEE. doi:10.1109/INFCOM.2013.6566861

Cormio, C., & Chowdhury, K. R. (2010). An adaptive multiple rendezvous control channel for cognitive radio wireless ad hoc networks. In *Proceedings of IEEE International Conference on Pervasive Computing and Communications (PERCOM) WS* (pp. 346-351). IEEE. doi:10.1109/PERCOMW.2010.5470645

DaSilva, L. A., & Guerreiro, I. (2008). Sequence-Based Rendezvous for Dynamic Spectrum Access. In Proceedings of IEEE Dynamic Spectrum Access Networks Symposium (DySPAN) (pp. 1-7). IEEE.

El Masri, N. M. A., & Khalif, H. (2011). *A routing strategy for cognitive radio networks using fuzzy logic decisions. In Proceedings of Conference Cognitive Advances in Cognitive Radio*. COCORA.

Friedman, R., Kliot, G., & Avin, C. (2010). Probabilistic quorum systems in wireless ad hoc networks. *ACM Transactions on Computer Systems, 28*(3), 7. doi:10.1145/1841313.1841315

Gandhi, R., Wang, C.-C., & Hu, Y. C. (2012). Fast rendezvous for multiple clients for cognitive radios using coordinated channel hopping. In *Proceedings of IEEE International Conference on Sesing, Communication, and Networking (SECON)*, (pp. 434-442). IEEE. doi:10.1109/SECON.2012.6275809

Garcia-Molina, H., & Barbara, D. (1985). How to assign votes in distributed systems. *Journal of the ACM, 32*(4), 841–860. doi:10.1145/4221.4223

Gardellin, V. Lenzini, L. & Fontana, V. (2011). Aware channel assignment algorithm for cognitive networks. In *Proceedings of ACM Workshop on Cognitive Radio Networks (CoRoNet)*. Academic Press.

Giupponi, L., & Perez-Neira, A. I. (2008). Fuzzy-based spectrum handoff in cognitive radio networks. In *Proceedings of International Conference on Cognitive Radio Oriented Wireless Networks and Communications (CrownCom)* (pp. 1–6). Academic Press. doi:10.1109/CROWNCOM.2008.4562535

Gu, Z., Hua, Q.-S., Wang, Y., & Lau, F. C. M. (2013). Nearly Optimal Asynchronous Blind Rendezvous Algorithm for Cognitive Radio Networks. In *Proceedings of IEEE International Conference on Sensing, Communication, and Networking (SECON)*. IEEE.

Hall, J. M. (Ed.). (1986). *Combinatorial Theory*. John Wiley and Sons.

Hou, F., Cai, L. X., Shen, X., & Huang, J. (2011). Asynchronous multichannel MAC design with difference-set-based hopping sequences. *IEEE Transactions on Vehicular Technology, 60*(4), 1728–1739. doi:10.1109/TVT.2011.2119384

Hoven, N., & Sahai, A. (2005). Power scaling for cognitive radio. In *Proceedings of International Conference on Wireless Networks, Communications and Mobile Computing,* (vol. 1, pp. 250–255). Academic Press.

Hsieh, Y.-S., Lien, C.-W., & Chou, C.-T. (2011). A multi-channel testbed for dynamic spectrum access (DSA) networks. In *Proceedings of ACM Workshop on Wireless Multimedia Networking and Computing (WMUNEP)*. ACM. doi:10.1145/2069117.2069129

Huang, S., Liu, X., & Ding, Z. (2010). Distributed power control for cognitive user access based on primary link control feedback. In *Proceedings of IEEE Conference on Information Communications (INFOCOM)*. IEEE. doi:10.1109/INFCOM.2010.5461916

IEEE 802.22 WG on WRAN. (2011). Retrieved February 21, 2014, from http://www.ieee802.org/22/

Jiang, J. R., Tseng, Y. C., Hsu, C. S., & Lai, T. H. (2003). Quorum based asynchronous power-saving protocols for IEEE 802.11 ad hoc networks. In *Proceedings of IEEE International Conference on Parallel Processing*. IEEE.

Jiang, J. R., Tseng, Y. C., Hsu, C. S., & Lai, T. H. (2005). Quorum based asynchronous power-saving protocols for IEEE 802.11 ad hoc networks. *Mobile Networks and Applications, 10*(1), 169–181. doi:10.1023/B:MONE.0000048553.45798.5e

Kandukuri, S., & Boyd, S. (2010). Optimal power control in interference-limited fading wireless channels with outage-probability specifications. *IEEE Transactions on Wireless Communications, 1*(1), 46–55. doi:10.1109/7693.975444

Kim, H., & Shin, K. G. (2008). Efficient discovery of spectrum opportunities with mac-layer sensing in cognitive radio networks. *IEEE Transactions on Mobile Computing, 7*(5), 533–545. doi:10.1109/TMC.2007.70751

Kumar, A. (1991). Hierarchical quorum consensus: A new algorithm for managing replicated data. *Journal IEEE Transaction on Computers, 40*(9), 996–1004. doi:10.1109/12.83661

Kuo, Y. C. (2010). Quorum-based power-saving multicast protocols in the asynchronous ad hoc network. *Computer Networks*, *54*(11), 1911–1922. doi:10.1016/j.comnet.2010.03.001

Lai, S., Ravindran, B., & Hyeonjoong, C. (2010). Heterogeneous Quorum-Based Wake-Up Scheduling in Wireless Sensor Networks. *IEEE Transactions on Computers*, *59*(11), 1562–1575. doi:10.1109/TC.2010.20

Lang, S., & Mao, L. (1998). A torus quorum protocol for distributed mutual exclusion. In *Proceedings of International Conference on Parallel and Distributed Systems (ICPADS)*. Academic Press.

Le, H.-S. T., & Ly, H. D. (2008). Opportunistic spectrum access using fuzzy logic for cognitive radio networks. In *Proceedings of International Conference on Communications and Electronics (ICCE)* (pp. 240-245). Academic Press. doi:10.1109/CCE.2008.4578965

Lee, E. K., Oh, S. Y., & Gerla, M. (2010). *Randomized channel hopping scheme for anti jamming communication*. IFIP Wireless Days. doi:10.1109/WD.2010.5657713

Lee, W., & Akyildiz, I. F. (2008). Optimal spectrum sensing framework for cognitive radio networks. *IEEE Transactions on Wireless Communications*, *7*(10), 3845–3857. doi:10.1109/T-WC.2008.070391

Liang, Y.-C., Chen, K.-C., Li, G. Y., & Mähönen, P. (2011). Cognitive radio networking and communications: An overview. *IEEE Transactions on Vehicular Technology*, *60*(7), 3386–3407. doi:10.1109/TVT.2011.2158673

Lin, Z., Liu, H., Chu, X., & Leung, Y. W. (2011). Jump-stay based channel-hopping algorithm with guaranteed rendezvous for cognitive radio networks. In *Proceedings of IEEE International Conference on Computer Communications (INFOCOM)* (pp. 2444–2452). IEEE. doi:10.1109/INFCOM.2011.5935066

Lin, Z., Liu, H., Chu, X., & Leung, Y. W. (2013). Enhanced Jump-Stay Rendezvous Algorithm for Cognitive Radio Networks. *IEEE Communications Letters*, 1–4.

Liu, H., Lin, Z., Chu, X., & Leung, Y. W. (2010). Ring-Walk Based Channel-Hopping Algorithms with Guaranteed Rendezvous for Cognitive Radio Networks. In *Proceedings of International Workshop on Wireless Sensor, Actuator and Robot Networks (WiSARN2010-FALL), in Conjunction with IEEE/ACM International Conference on Cyber, Physical and Social Computing (CPSCom)*. IEEE. doi:10.1109/GreenCom-CPSCom.2010.30

Liu, Q., Pang, D., Wang, X., & Zhou. (2012). A Neighbor Cooperation Framework for Timeefficient Asynchronous Channel Hopping Rendezvous in Cognitive Radio Networks. In *Proceedings of IEEE Dynamic Spectrum Access Networks symposium (DySPAN)*. IEEE.

Luk, W. S., & Wong, T. T. (1997). Two new quorum based algorithms for distributed mutual exclusion. In *Proceedings of 17th International Conference on Distributed Computing Systems* (pp. 100–106). Academic Press.

Maekawa, M. (1985). A $\sqrt{N}$ algorithm for mutual exclusion in decentralized systems. *ACM Transactions on Computer Systems*, *3*(2), 145–159. doi:10.1145/214438.214445

Mehmeti, F., & Spyropoulos, T., T. (2013). Who interrupted me? Analyzing the effect of PU activity on cognitive user performance. In *Proceedings of IEEE International Conference on Communications (ICC)*. IEEE. doi:10.1109/ICC.2013.6654977

Merentitis, A., & Triantafyllopoulou, D. (n.d.). Resource allocation with mac layer node cooperation in cognitive radio networks. *Hindawi International Journal of Digital Multimedia Broadcasting, 2010*(458636), 11.

Mitola, J. (2000). *Cognitive radio: An integrated agent architecture for software defined radio.* (Ph.D. dissertation). Royal Institute of Technology, Sweden.

Nelles, O. (Ed.). (2001). *Nonlinear System Identification, From classical approaches to neural networks and fuzzy methods.* Springer-Verlag Berlin Heidelberg NY.

Pavlovska, V., Denkovski, D., Atanasovski, V., & Gavrilovska, L. (2010). RAC2E: Novel rendezvous protocol for asynchronous cognitive radios in cooperative environments. In *Proceedings of IEEE International Symposium on Personal, Indoor and Mobile Radio Communications (PIMRC)* (pp. 1848–1853). IEEE. doi:10.1109/PIMRC.2010.5671630

Robertson, A., Tran, L., Molnar, J., & Fu, E. H. F. (2012). Experimental comparison of blind rendezvous algorithms for tactical networks. In *Proceedings of IEEE International Symposium on a World of Wireless, Mobile and Multimedia Networks (WoWMoM)*. IEEE. doi:10.1109/WoWMoM.2012.6263760

Romaszko, S. (2013). A rendezvous protocol with the heterogeneous spectrum availability analysis for cognitive radio ad hoc networks. *Hindawi Journal of Electrical and Computer Engineering, 2013*(715816), 18.

Romaszko, S., Denkovski, D., Pavlovska, V., & Gavrilovska, L. (2013). Quorum system and random based asynchronous rendezvous protocol for cognitive radio ad hoc networks. *EAI Endorsed Transactions on Mobile Communications and Applications, 13*(3).

Romaszko, S., & Mähönen, P. (2011). Grid-based channel mapping in cognitive radio ad hoc networks. In *Proceedings of 22nd Annual IEEE International Symposium on Personal, Indoor and Mobile Radio Communications (PIMRC)* (pp. 438-444). IEEE. doi:10.1109/PIMRC.2011.6139999

Romaszko, S., & Mähönen, P. (2012a). Heterogeneous torus Quorum-based Rendezvous in Cognitive Radio Ad Hoc Networks. In *Proceedings of IEEE International Conference on Wireless and Mobile Computing, Networking and Communications (WiMob)*. IEEE. doi:10.1109/WiMOB.2012.6379109

Romaszko, S., & Mähönen, P. (2012b). Quorum systems towards an asynchronous communication in cognitive radio networks. *Journal of Electrical and Computer Engineering, Article ID 753541.*

Romaszko, S., & Mähönen, P. (2014). Fuzzy Channel Ranking estimation in Cognitive Wireless Networks. In *Proceedings of IEEE Wireless Communications and Networking Conference (WCNC)*. IEEE. doi:10.1109/WCNC.2014.6952964

Romaszko, S., Torfs, W., Mähönen, P., & Blondia, C. (2013a). An analysis of asynchronism of a neighborhood discovery protocol for cognitive radio networks. In *Proceedings of IEEE International Symposium on Personal, Indoor and Mobile Radio Communications (PIMRC)*. IEEE. doi:10.1109/PIMRC.2013.6666689

Romaszko, S., Torfs, W., Mähönen, P., & Blondia, C. (2013b). Benefiting from an Induced Asynchronism in Neighborhood Discovery in opportunistic Cognitive Wireless Networks. In *Proceedings of ACM International Symposium on Mobility Management and Wireless Access (MobiWac)*. ACM. doi:10.1145/2508222.2508230

Romaszko, S., Torfs, W., Mähönen, P., & Blondia, C. (2014a). AND: Asynchronous Neighborhood Discovery protocols for opportunistic Cognitive Wireless Networks. In *Proceedings of IEEE Wireless Communications and Networking Conference (WCNC)*. IEEE. doi:10.1109/WCNC.2014.6952604

Romaszko, S., Torfs, W., Mähönen, P., & Blondia, C. (2014b). *How to be an efficient Asynchronous Neighbourhood Discovery protocol in opportunistic Cognitive Wireless Networks. International Journal of Ad Hoc and Ubiquitous Computing.*

Shin, J., Yang, D., & Kim, C. (2010). A channel rendezvous scheme for cognitive radio networks. *IEEE Communications Letters, 14*(10), 954–956. doi:10.1109/LCOMM.2010.091010.100904

Singhal, M., & Shivaratri, N. G. (Eds.). (1994). *Advanced Concepts in Operating Systems, TMH*. New York: McGraw-Hill.

Snellen, M. (1891). Buys Ballot. *Meteorologische Zeitschrift, 8*, 1–6. Available from http://archive.org/stream/meteorologische00unkngoog#page/n24/mode/2up

Theis, N. C., Thomas, R. W., & DaSilva, L. A. (2010). Rendezvous for cognitive radios. *IEEE Transactions on Mobile Computing, 10*(2), 216–227. doi:10.1109/TMC.2010.60

Tseng, Y. C., Hsu, C. S., & Hsieh, T. Y. (2002). Power-saving protocols for IEEE 802.11-based multi-hop ad hoc networks. In *Proceedings of IEEE International Conference on Computer Communications (INFOCOM)* (pp. 200-209). IEEE.

Universal Software Radio Peripheral (USRP). (2014). Retrieved February 18, 2014. from http://www.ettus.com

Wu, C.-C., & Wu, S.-H. (2013). On bridging the gap between homogeneous and heterogeneous rendezvous schemes for cognitive radios. In *Proceedings of ACM International Symposium on Mobile Ad Hoc Networking and Computing (MobiHoc)*. ACM. doi:10.1145/2491288.2491295

Wu, K., Han, F., & Kong, D. (2013). Rendezvous Sequence Construction in Cognitive Radio Ad-Hoc Networks based on Difference Sets. In *Proceedings of IEEE International Symposium on Personal, Indoor and Mobile Radio Communications (PIMRC)*. IEEE. doi:10.1109/PIMRC.2013.6666442

Wu, S. H., Chen, M. S., & Chen, C. M. (2008). Fully adaptive power saving protocols for ad hoc networks using theHyper Quorum System. In *Proceedings of International Conference on Distributed Computing Systems (ICDCS)* (pp. 785–792). Academic Press.

Wyglinski, A. M., Nekovee, M., & Hou, Y. T. (Eds.). (2009). *Cognitive Radio Communications and Networks: Principles and Practice*. Elsevier Inc.

Yang, L., Cao, L., & Zheng, H. (2008). Proactive channel access in dynamic spectrum networks, *Elsevier. Physical Communication, 1*(2), 103–111. doi:10.1016/j.phycom.2008.05.001

Yao, Y., Ngoga, S. R., Erman, D., & Popescu, A. (2012). Competition-based channel selection for cognitive radio networks. In *Proceedings of IEEE Wireless Communications and Networking Conference (WCNC)*. IEEE. doi:10.1109/WCNC.2012.6214006

Yu, L., Liu, H., Leung, Y.-W., Chu, X., & Lin, Z. (2013). Multiple Radios for Effective Rendezvous in Cognitive Radio Networks. In *Proceedings of IEEE International Conference on Communication (ICC)*. IEEE. doi:10.1109/ICC.2013.6654974

Zadeh, L. A. (1965). Fuzzy Sets. *Information and Control, 8*(3), 338–353. doi:10.1016/S0019-9958(65)90241-X

Zhang, B., Han, K., Ravindran, B., & Jensen, E. D. (2008). RTQG: Real-time quorum-based gossip protocol for unreliable networks. In *Proceedings of International Conference on Availability, Security, and Reliability (ARES)*, (pp. 564–571). Academic Press. doi:10.1109/ARES.2008.139

Zhang, Y., Li, Q., Yu, G., & Wang, B. (2011). ETCH: Efficient Channel Hopping for Communication Rendezvous in Dynamic Spectrum Access Networks. In *Proceedings of IEEE Conference on Computer Communications*. IEEE. doi:10.1109/INFCOM.2011.5935070

Zhang, Y., Li, Q., Yu, G., Wang, H., Zhu, X., & Wang, B. (2013). Channel hopping based communication rendezvous in cognitive radio networks. *IEEE/ACM Transactions on Networking*.

ADDITIONAL READING

Akyildiz, I., Lee, W.-Y., & Chowdhury, K. R. (2006). CRAHNs: Cognitive radio ad hoc networks. *Ad Hoc Networks, 7*(5), 810–836. doi:10.1016/j.adhoc.2009.01.001

Chao, C.-M., & Lin, X.-H. (2007). A Fuzzy Control Quorum-Based Energy Conserving Protocol for IEEE 802.11 Ad Hoc Networks, *IEEE Wireless Communications and Networking Conference (WCNC)*. doi:10.1109/WCNC.2007.407

Chao, C.-M., Sheu, I.-C., & Chou, I.-C. (2006). An Adaptive Quorum-Based Energy Conserving Protocol for IEEE 802.11 Ad Hoc Networks. *IEEE Transactions on Mobile Computing, 5*(5), 560–570. doi:10.1109/TMC.2006.55

Chao, C.-M., & Tsai, H.-C. (2009). A Channel Hopping Multi-Channel MAC Protocol for Mobile Ad Hoc Networks, *International Conference on Wireless Communications & Signal Processing (WCSP)*.

Dai, Y., & Wu, J. (2013). Sense in Order: Channel Selection for Sensing in Cognitive Radio Networks, *International Conference on Cognitive Radio Oriented Wireless Networks (CROWNCOM)*. doi:10.4108/icst.crowncom.2013.251937

Geirhofer, S., Sun, J. Z., Tong, L., & Sadler, B. M. (2009). Cognitive Frequency Hopping Based on Interference Prediction: Theory and Experimental Results, *ACM SIGMOBILE Mobile. Computer Communications, 13*, 49–61.

Ghozzi, M, Dohler, M., Marx, F. & Palicot J. (2006). Cognitive radio: methods for the detection of free bands, *Comptes Rendus Physique, Towards reconfigurable and cognitive communications, 7(7)*, 794-804.

Htike, Z., Hong, C. S., & Lee, S. (2013). The Life Cycle of the Rendezvous Problem of Cognitive Radio Ad Hoc Networks: A Survey [JCSE]. *Journal of Computing Science and Engineering, 7*(2), 81–88. doi:10.5626/JCSE.2013.7.2.81

Htike, Z., Lee, J., & Hong, C. S. (2012). A MAC protocol for cognitive radio networks with reliable control channels assignment, *International Conference on Information Network*. doi:10.1109/ICOIN.2012.6164354

Hu, W., Willkomm, D., Abusubaih, M., Gross, J., Vlantis, G., Gerla, M., & Wolisz, A. (2009). Cognitive radios for dynamic spectrum access - Dynamic Frequency Hopping Communities for Efficient IEEE 802.22 Operation. *IEEE Communications Magazine, 45*(5), 80–87. doi:10.1109/MCOM.2007.358853

Huang, H.-J., Cao, X.-L., Jia, X.-H., & Wang, X.-L. (2006). A Bibd-Based Channal Assignment Algorithm for Multi-Radio Wireless MESH Networks, *International Conference on Machine Learning and Cybernetics*. doi:10.1109/ICMLC.2006.259095

Khatibi, S., Dehghan, M., & Poormina, M. A. (2010). Quorum-based pure directional neighbor discovery in self-organized ad hoc networks, *International Symposium on Telecommunications (IST)*. doi:10.1109/ISTEL.2010.5734073

Kumar, D., Magesh, A. & Hussain, U.M. (2012). Prediction based channel-hopping algorithm for rendezvous in cognitive radio networks, *ICTACT Journal on communication technology, 3(4)*, 619-625.

Lee, C., Lin, J., Feng, K., & Chang, C. (2011). Design and Analysis of Transmission Strategies in Channel-Hopping Cognitive Radio Networks. *IEEE Transactions on Mobile Computing, 99*, 1–30.

Liu, H., Lin, Z., Chu, X., & Leung, Y.-W. (2012). Taxonomy and Challenges of Rendezvous Algorithms in Cognitive Radio Networks, *International Conference on Computing, Networking and Communications (ICNC)*. doi:10.1109/ICCNC.2012.6167502

Luo, J., & He, Y. (2011). GeoQuorum: Load balancing and energy efficient data access in wireless sensor networks, *IEEE International Conference on Computer Communications (INFOCOM)*. doi:10.1109/INFCOM.2011.5935238

Nguyen-Thanh, N., Pham, A. T., & Nguyen, V. T. (2014). Medium Access Control Design for Cognitive Radio Networks: A Survey, *IEICE Trans. Communications. Special Section on Technologies for Effective Utilization of Spectrum White Space, E97-B*(2), 1–18.

Rahman, M.J.A., Rahbari, H. & Krunz, M. (2012). Adaptive Frequency Hopping Algorithms for Multicast Rendezvous in DSA Network, *IEEE Dynamic Spectrum Access Networks symposium (DySPAN)*.

Romaszko, S. (2012). Making a Blind Date the Guaranteed Rendezvous in Cognitive Radio Ad Hoc Networks, *European Wireless (EW)*.

Shatila, H., Khedr, M., & Reed, J. H. (2009). Opportunistic channel allocation decision making in cognitive radio communications. *International Journal of Communication Systems.*

Su, H., & Zhang, X. (2008). Channel-hopping based single transceiver MAC for cognitive radio networks, *42nd Annual Conference on Information Sciences and Systems (CISS).*

Wang, Y., Li, Y., Yuan, F., & Yang, J. (2013). A Cooperative Spectrum Sensing Scheme Based on Trust and Fuzzy Logic for Cognitive Radio Sensor Networks. *IJCSI International Journal of Computer Science Issues, 10*(1), 275–279.

Willkomm, D., Hu, W., Hollos, D., Gross, J., & Wolisz, A. (2010). Centralized and Distributed Frequency Assignment in Cognitive Radio Based Frequency Hopping Cellular Networks, *International Workshop on Cognitive Radio and Advanced Spectrum Management (CogART).* doi:10.1109/ISABEL.2010.5702860

Xiang, B., & Yu-Ming, M. (2008). *Balanced Incomplete Block Designs based scheduling scheme for multi-hop ad hoc networks, Communications, Circuits and Systems.* ICCCAS.

Yang, D., Shin, J., & Kim, C. (2010). Deterministic rendezvous scheme in multichannel access networks. *Electronics Letters, 46*(20), 1402. doi:10.1049/el.2010.1990

Zame, W., Xu, J., & van der Schaar, M. (2014). Cooperative Multi-Agent Learning and Coordination for Cognitive Radio Network. *IEEE Journal on Selected Areas in Communications, 32*(3), 464–477. doi:10.1109/JSAC.2014.140308

Zeng, R. (2006). Optimal Block Design for Asynchronous Wake-Up Schedules and its Applications in Multihop Wireless Networks. *IEEE Transactions on Mobile Computing, 5*(9), 1228–1241. doi:10.1109/TMC.2006.134

Zhang, L., Kefeng, T., Zeng, K., & Mohapatra, P. (2012). Fast Rendezvous for Cognitive Radios by Exploiting Power Leakage at Adjacent Channel, *International Symposium on Personal, Indoor and Mobile Radio Communications (PIMRC).* doi:10.1109/PIMRC.2012.6362654

KEYS TERMS AND DEFINITIONS

Asynchronous Rendezvous (RDV): Radios, not aligned in slot and cycle, have ability to rendezvous (in a bounded period of time) on a common channel (in some period of time of overlapping slots).

Asynchronous Rendezvous Extension (ARE): An extension to be added to any rendezvous protocol, in order to induce an asynchronism and channel ranking. ARE introduces varying sizes of slots.

Fuzzy Logic (FL): Expressed "true" or "false" value from the interval between 0 and 1. A fuzzy number is an element, which belongs with a certain degree ("degrees of truth") to a specific membership function.

Fuzzy Logic System (FLS): A control system based on FL, where input values are expressed in terms of logical variables from the interval between 0 and 1. FLS exploits FL reasoning, where, the crisp inputs and outputs are translated to and from fuzzy representation through the *fuzzification* and *defuzzification* processes, respectively. Fuzzy Logic expresses and processes relationships in the form of *IF...(AND/OR)...THEN* rules used to determine the value of the output variables.

Quorum System: Is a collection of sets having at least one common element.

Rendezvous (RDV): A meeting point, a blind date in distributed radio networks - ability of to establish a communication on a common channel (meeting point) during so-called neighborhood discovery or rendezvous process.

Rotation Closure Property (RCP): Allowing a quorum Q and any of its shift sets (rotation) have at least one overlapping element in a cycle N (its period).

Time-To-Rendezvous (TTR): Amount of time (in slots in theoretical analyses, in seconds in experimental analyses) within which radios can switch to a common channel (i.e. experiencing a rendezvous) at an overlapping period of time (overlapping slot) in a finite delay, where the upper-bound of the delay corresponds to Maximal TTR (*MTTR*) for all possible TTRs per cycle.

ENDNOTES

[1]   This section is based on work performed in cooperation with Wim Torfs and Chris Blondia from the University of Antwerp, Belgium.

Chapter 16

Cooperative Cognitive Radio Networking:
Towards a New Paradigm for Dynamic Spectrum Access

Bin Cao
Jingdezhen Ceramic Institute of Technology,
China & Harbin Institute of Technology
Shenzhen Graduate School, China

Hao Liang
University of Waterloo, Canada & University of
Alberta, Canada

Gang Fu
Harbin Institute of Technology Shenzhen
Graduate School, China

Qinyu Zhang
Harbin Institute of Technology Shenzhen
Graduate School, China

Jon W. Mark
University of Waterloo, Canada

ABSTRACT

Radio spectrum underutilization and energy inefficiency become urgent bottleneck problems to the sustainable development of wireless technologies. The research philosophies of wireless communications have been shifted from balancing reliability-efficiency tradeoff in the link level to seeking spectrum-energy efficiency in the network level. Global spectrum-energy efficient designs attract significant attention to improving utilization and efficiency, wherein cognitive radio networks and energy-efficient resource allocation are of particular interests. In this chapter, the authors first provide a systematic study on Cooperative Cognitive Radio Networking (CCRN). As an effort to shed light on addressing spectrum-energy inefficiency at a low complexity, an orthogonal modulation enabled two-phase cooperation framework and an Orthogonally Dual-Polarized Antenna (ODPA) based framework, as well as their resource allocation problems are given and tackled.

DOI: 10.4018/978-1-4666-6571-2.ch016

INTRODUCTION

During the past two decades, we have witnessed an explosive proliferation of applications in wireless communications and networking, e.g., vehicular ad hoc networks (VANETs), internet of things (IoTs), social networks, location-based services (LBSs), and smart grid. State-of-the-art technologies associated with products and services are changing our life styles from various aspects. While we are now enjoying the remarkable improvement in quality of service (QoS) provided by the fourth generation (4G) mobile communications networks, i.e., LTE-Advanced and WiMAX-Advanced, people from both academia and industry are seeking for the design of the next generation (xG) wireless networks, among them Wireless Vision 2020 plan towards 5G standard is notable (David, 2010).

To support sustainable development of wireless communications and networks, the demand for radio spectrum has been skyrocketing. Since the amount of usable spectrum is finite, frequency bands and their usage are strictly managed and enforced by government regulators. In the past decade, information and communication technologies (ICT) industry has turned into a highly competitive industry where companies are competing to buy valuable spectrum from governmental bodies, i.e., spectrum auction. In such a way, the winning telecommunications operators in spectrum bidding obtain the rights, also known as licenses, to transmit signals over specific bands of the electromagnetic spectrum and to conduct their communications services. With more services providers in the mobile industry, the competition during spectrum auctions has increased due to more demand from consumers and physical scarcity in available radio spectrum. In early 2013, for commercializing 4G networks, the United Kingdom government performed a spectrum auction for 250MHz of spectrum (equivalent to two-thirds of the entire 3G spectrum already in use) in the 800MHz and 2.6GHz bands. In total, five bidders have committed to pay 2.34 billion UK pounds for the right to use the frequencies for 4G services (Cao, 2013). According to a report from the Ministry of Industry and Informationization of China, China's total spectrum demand in wireless communications is about 1GHz by 2015. However, 547MHz of spectrum has already been assigned for IMT systems, leading to 420MHz of spectrum scarcity, if no new spectrum is exploited (Cao, 2013).

On the one hand, we are now facing a challenging issue that physical spectrum scarcity hinders the further evolution of wireless communications. On the other hand, investigations and experiments on spectrum utilization have clearly shown that most of the allocated spectrum is considerably underutilized. Since the spectrum is exclusively assigned to dedicated users, termed as licensed or primary users (PUs), other users cannot access to the spectrum even if it is unused. This also imposes us a conflict between spectrum scarcity and spectrum underutilization, which further exacerbates the situation of spectrum shortage. New wireless applications and services along with emerging diverse wireless networking architectures, e.g., heterogeneous networks, have been severely hampered by this inefficient fixed spectrum allocation.

In order to address spectrum underutilization, dynamic spectrum access based schemes, which allow unlicensed users (or referred as secondary users (SUs)) to share or reuse licensed bands without interfering with PUs, have attracted significant attention (Goldsmith, 2009; Wang B., 2011; Wang J., 2011; Granelli, 2010; Akyildiz, 2006. Haykin, 2005). By enabling dynamic spectrum access, SUs are allowed to (i) dynamically sense the surrounding electromagnetic environments and opportunistically access to temporally unused spectral bands, e.g., spectrum sensing based systems, (ii) concurrently and transparently transmit in licensed bands as long as PUs' transmissions are not interfered, e.g., ultra-wide band (UWB) systems, or (iii) cooperatively and trustfully negotiate with PUs for transmission opportunities by providing tangible services (Goldsmith, 2009). One promising technology to achieve dynamic spectrum access is the cognitive radio (CR) equipped by SUs (Haykin, 2005), i.e., the secondary network consisting of CRs is a cognitive radio network (CRN). In this context, these three paradigms are known as interweave CRN, underlay CRN, and overlay CRN, respectively (Goldsmith, 2009).

In an interweave CRN, communication durations of SUs are highly unstable due to the randomness in the acquisition of temporally unused spectrum, and transmission terminations when PUs reclaim the spectrum degrade SUs' transmissions performance, i.e., SUs' QoS cannot be guaranteed. Moreover, SUs need to exert more power and overhead to attain accurate spectrum sensing. Detecting temporarily unused spectrum, i.e., spectrum sensing, is one of the critical elements in CRN design, and performing an accurate spectrum sensing is a challenging task. The sensing performance highly depends on channel quality and sensing time. For example, in low signal-to-noise ratio (SNR) environments, the effect of SNR walls deteriorate and limit sensing ability, which leads to interferences to PUs from SUs because of missing detection of spectrum holes. Although some sensing schemes are with satisfied performance, e.g., cooperative sensing among SUs, they are with high complexity for implementation. Furthermore, spectrum sensing based methodology cannot fully guarantee SUs transmission performance even though SUs are with accurate sensing ability, since an SU must abandon its transmission as soon as it detects the presence of the PU, and sense another spectrum hole for transmission. If none is available, then the SU has to discontinue the transmission. The main reason leads to this issue is due to PUs are unaware of SUs demand and existence, which indicates that there is no collaborative relation between PUs and SUs.

In an underlay CRN, like the impulse radio ultra-wide band (IR-UWB) wireless communication systems, since SUs' communications cannot produce any harmful interference to PUs' communications, the interference temperature model is imposed on SUs' transmit power to avoid interference to PUs' transmissions (Goldsmith, 2009); this limits SUs to short-range communications. Such as in IR-UWB systems, the allowed power spectrum density for transmission should be strictly below -41.3 dBm/MHz. Meanwhile, cumulative interference from multiple SUs and strong interference from PUs deteriorate the transmission performance of PUs and that of SUs, respectively.

It can be concluded that in both underlay and interweave models, SUs are transparent to PUs, i.e, PUs are not aware SUs's existence around PUs. However, user cooperation is promising to provide diversity gain which can further enhance the resource utilization, like spectrum, energy, time, and space, and other communication resources.

In addition, smart PUs are willing to establish collaborative relation with SUs if PUs can obtain benefit by doing so, and SUs have the demand of using spectrum for transmission. In response to the challenging issues in spectrum sensing based CRN, an alternative promising approach for SUs to gain transmission opportunities using licensed spectrum is to provide tangible services to PUs such that the PU can transmit its traffic to its intended destination with greater reliability, e.g., faster and with satisfactory quality. If the SU can provide such service, the PU would be happy to yield a fraction of its licensed spectrum for the SU to use, so long the SU's transmission does not interfere with the PU's transmission. In this way, the PU and the SU mutually benefit from user cooperation, i.e., the PU and the SU cooperate to achieve mutual benefit.

In this regard, exploiting user cooperation between PUs and SUs, i.e., CCRN, is an interesting research issue. This framework can work when spectrum sensing cannot be accurately performed, and/or mutual interference is too strong that prohibits SUs to access the spectrum in underlay model. The focus of this chapter is on overlay CRN, also known as cooperative CRN (CCRN) (Cao, 2012; Simeone,

2008; Zhang, 2009; Xu, 2010; Hua, 2011; Su, 2012; Jia, 2009; Zhang, 2009; Li, 2010; Zou 2011; Xie, 2010; Simeone, 2007). For more details about interweave and underlay CRN, the readers are referred to (Goldsmith, 2009) and references therein.

In order to attain and implement CCRN, several open problems need to be tackled. We identify that the following issues should be addressed:

Mutual-beneficial mechanisms to motivate cooperation: To establish collaborative relationships between PUs and SUs, there must be incentives that can achieve mutual benefit, i.e., the cooperation mechanisms can create win-win situations. In this context, the design of mutual-beneficial mechanisms is not only a cross-layer technical issue, but also a consideration with human nature such as the paradox of selfishness and rationality. Stimulating cooperation motivation through introducing mutual-beneficial mechanisms is a prerequisite for the CCRN design. Moreover, performance metrics which can be accepted by both PUs and SUs need to be appropriately defined.

Cost-effective multi-user coordination to select partners: In order to fully achieve a satisfied or predefined cooperation performance or motivation, finding the most effective cooperators is crucial to both PUs and SUs in CCRN. The design of multi-user coordination protocols are supposed to be able to simply and efficiently perform the cooperator selection and/or partner matching, and management functions. Otherwise, high communication overheads or collisions during the multi-user coordination phase will degrade the cooperation performance, and lead to that PUs and/or SUs become less interesting in cooperation due to high complexity or unsatisfied performance.

Spectrum-efficient frames to attain transmissions: Besides improving spectrum utilization, high spectrum efficiency communications are essential to CCRN. The trade-off between spectrum efficiencies of transmission phases and complexities of multi-user coordination phases can be addressed by a sophisticated frame design, which is also a reflection of the trade-off between reliability and efficiency in wireless communications. In addition, the design criterion of frame structures which is capable of effective scheduling and resource allocations at a low-complexity fashion should also be taken into account. Hence spectrum-efficient frame structures to attain effective transmissions are important for CCRN.

Energy-efficient techniques to avoid interference: During the cooperative transmission phase, spectrum efficiency can be improved by merging transmission time slots assigned for PUs and SUs in CCRN, e.g., achieving two-phase transmissions for half-duplex based CCRN. In this context, fulfilling cooperation between PUs and SUs on a non-interference basis is a must. Advanced signal processing enables the ability of interference suppressions, while an ideal design should not increase the complexity and energy consumption of PUs, otherwise, PUs' incentives for cooperation will be weakened. Therefore, the burden arising from signal processing to avoid interference needs to be taken on by SUs. The trade-off between the cost and gain for SUs should be balanced to motivate cooperation, indicating that energy-efficient techniques for interference avoidance should be tackled.

Energy-sustainable policies to attain low-complexity cooperation: One feasible way to reduce the cooperation complexity for CCRN is to shift from short-term cooperation to relative long-term cooperation between PUs and SUs. In such a way, the number and time for multi-user coordination can be significantly decreased. However, enough energy is needed to perform long-term cooperation, i.e., energy should be used in sustainable manners. To address this problem, dynamic optimal or suboptimal long-term power strategies, and energy harvesting techniques to achieve energy sustainability are crucial.

Economy-efficient hardware to support applications: In the CCRN design, besides energy and spectrum efficiencies, the size- and cost-efficient hardware design for the SU-type devices should also be concerned from the commercial perspective.

For example, although spatially distributed multiple antennas can provide flexible degrees of freedom, antenna placements for MIMO systems require large physical sizes in the hardware. Moreover, the cost of MIMO systems cannot be universally accepted in hardware implementations by commercials. It is important to design an economy-efficient hardware for SUs to support future applications under the background of CCRN.

RELATED WORKS

In CCRN, unlike interweave and underlay CRN, through negotiation, an SU is fully aware of its obligation when cooperating with a PU. In this regard, SUs' transmission opportunities are dedicated rather than random in CCRN. In such a CCRN, an SU can obtain transmission opportunities from the PU in exchange for services provided to the PU, e.g., by helping relay PU's data (Simeone, 2008; Hua, 2010; Cao, 2012;Zou, 2011; Li, 2011; Zhang, 2009; Xie, 2010; Yi, 2010; Su, 2012; Attar, 2009; Lin, 2010; Chen, 2011), or by leasing a vacant spectral band from the primary network (Sodagari, 2011; Cao, 2012).

Recently, considerable research efforts have been devoted to cooperation schemes between PUs and SUs to form CCRN (Simeone, 2008; Hua, 2010; Cao, 2012). The state-of-the-art in CCRN can be generally classified into three-phase and two-phase manners (Cao, 2013).

Three-phase time division multiple access (TDMA) based CCRN (Zhang, 2009; Simeone, 2008;): In such a framework, PUs firstly broadcast their signals. In the second phase, the selected SUs forward PUs' information received in the first phase to PUs' intended destinations. In the third phase, the relaying SUs transmit to their own receivers. It is noted that the most key enabling parameters are the optimal duration length in each phase and the associated powers to achieve spectrum-energy efficient transmissions. Moreover, multi-user communications in phases two and three may result in high overhead and collisions, which would degrade the transmission performance of both PUs and SUs (Zhang, 2013).

Two-phase frequency division multiple access (FDMA) based CCRN(Su, 2010; Su, 2012; Xu, 2010): PUs use a fraction of licensed spectrum to attain two-phase cooperations with SUs, and yield remaining bandwidths to the relaying SUs for the secondary transmissions as rewards, i.e., SUs can exclusively and continuously transmit in the licensed bands. Note that the motivation of a CRN is to improve spectrum utilization by allowing SUs' transmissions based on the precondition that PUs' transmissions are guaranteed. This means SUs are imposed to achieve elastic transmission performance according to the varying wireless environment and PUs' needs. Therefore, given the bounded network capacity and performance, PUs may not have motivation for yielding dedicated bands to ensure SUs' transmissions.

Two-phase MIMO based CCRN (Hua, 2011; Zhao, 2011): By enabling SUs to exploit spatially distributed multiple antennas, SUs can utilize the capability provided by MIMO technologies, e.g., spatial beamforming, to avoid interference to PUs and that among SUs. Hence SUs use the degrees of freedom provided by MIMO systems to concurrently relay the primary traffic and transmit their own at the cost of complicated antenna operations. It is feasible for the SUs without stringent hardware and cost constraints. Since the MIMO technology is not widely and readily deployed due to the hardware and cost constraints of user devices or SU-type devices, this scheme suffers some application limitations.

Under these three popular cooperation frameworks, significant research efforts have been directed towards the performance optimization. In (Zhang, 2009), the authors propose a payment mechanism in which SUs pay charges to PUs in CCRN. The model is formulated as a non-cooperative Stackelberg game to maximize PUs' and SUs' utilities in three-phase decode-and-forward cooperation, and a unique Nash

Equilibrium is proved in SUs' payment strategies. This work assumes that both PUs and SUs transmit and relay using constant powers, and the revenue is a constant. In addition, no throughput constraints and energy issues are considered. In (Hua, 2011), an MIMO-CCRN framework is proposed to enable two-phase DF cooperation between PUs and SUs. The authors model the optimization problem as a Stackelberg game, and derive the optimal phase durations and relay selection based on NE in optimal relay powers of SUs, while SUs' transmit powers are constants. In (Xu, 2010), to address how such cooperation can be exploited in OFDMA based CCRN and the selfishness feature in resource allocation, the authors formulate an optimization framework based on Nash Bargaining Solutions to fairly and efficiently allocate resource between PUs and SUs. All the above work is based on the assumption that cooperation is always beneficial to PUs, which may be invalid under some conditions. In (Hao, 2011), by observing the effectiveness of cooperation, the authors consider the problems of when to cooperate and how to cooperate in three-phase DF CCRN, and the optimization problem is formulated as a Stackelberg game.

MAIN FOCUS OF THE CHAPTER

Basic CCRN System Modeling

In this chapter, we consider a secondary network coexisting with a primary network in a geographical area. The secondary network consisting of a secondary access point (SAP) and SUs, shares the same spectrum with the primary network consisting of a primary base station (PBS) and PUs. In such two fully operational networks, PBS and SBS are central controllers allocating and coordinating resources to the users managed by them. Transceivers in both primary and secondary networks operate in the half-duplex mode. For primary communications, if the channel condition between some sender and its receiver is too adverse to support successful transmissions, cooperation between the primary and secondary networks will be triggered. Consider downlink transmission in the primary network; in case a deep fade occurs during the propagation from PBS to a destination PU, PBS will employ appropriate SUs to improve the primary transmissions.

As show in Figure 1, the primary network and the secondary network coexist in a CCRN manner, i.e., SUs in the secondary network act as relaying nodes for PUs in the primary network. Each PU in the primary network can exclusively and continuously occupy dedicated spectrum to communicate with the PBS, i.e., in an FDMA fashion, so that there is no interference among different PUs. Take uplink communications depicted in Figure 1 as an instance, both PU_1 and PU_2 need to communicate with the primary base station (PBS), SU_1 and SU_2 want to establish communication links with secondary access point 1 (SAP_1) and SAP_2, respectively.

In Figure 1, the propagation channel quality between PU1 and the PBS cannot support ongoing service, such as the instantaneous direct throughput is incapable of providing satisfactory QoS due to severe channel shadowing. In this scenario, SU1 can help forward PU1's signals to the PBS. In such a way, the PBS can combine the signals from the direct path and the relaying path to improve the reception performance, e.g., obtaining a higher receiving signal-to-noise ratio over direct transmissions. In return, PU1 allows SU1 using PU1's licensed band to send information to SAP1. We say SU1's relaying service for PU1 is *icing on the cake*. For PU2, the propagation channel from PU2 to the PBS is blocked by buildings as shown in Figure 1. In this case, if the channel quality between SU2 and the PBS is sufficiently good to support PU1's ongoing service, SU2 provides multi-hop transmissions for PU2. As

Figure 1. System model of a classical CCRN

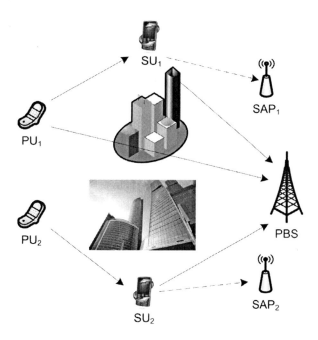

a reward, PU2 yields a fraction of time for SU2 to communicate with SAP2. We say SU2's multi-hop service for PU2 is *fuel in snowy weather*. It can be seen that user cooperation between PUs and SUs create a win-win situation.

As illustrated in Figure 1, SUs can cooperate with PUs via three-phase manner as shown in Figure 2 and two-phase manner as shown in Figure 3, respectively.

Figure 2. Three-phase CCRN

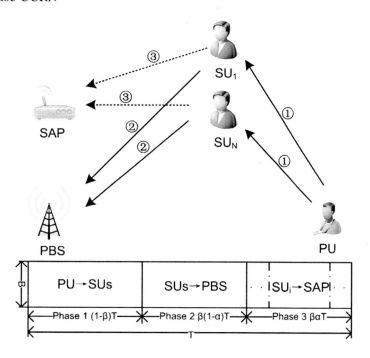

Figure 3. Two-phase CCRN

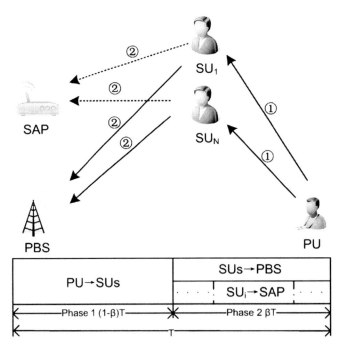

In the three-phase manner, the PU firstly broadcasts its signals to the intended SUs. After receiving PU's signal in the first phase, these SUs forward PU's signal to PBS by using either amplify-and-forward (AF) or decode-and-forward relaying protocol. Lastly, these SUs can transmit their own signals to SAP in the third phase. The duration of each phase is denoted in the figure.

Unlike three-phase manner, SUs combine the second and third phases together in the two-phase manner, i.e., SUs forward PU's signal via DF or AF to PBS, in the meanwhile SUs also transmit their signals to SAP. It can be noted that, the mutual interference among concurrent independent transmissions in the second phase would degrade the system performance, if no further multi-user processing is introduced.

The focus of this chapter is on enabling two-phase cooperative relay and transmissions between PUs and SUs in CCRN. To this end, we aim to present two cooperation frameworks for energy-efficient user cooperation between PUs and SUs to achieve two-phase CCRN.

Specifically, an orthogonal signaling based cooperation technology for leveraging the degrees of freedom in two-dimensional modulation is first discussed (Cao, 2012). In the orthogonal signaling based framework, an SU is enabled to simultaneously transmit its own data and relay PU's packets in two orthogonal channels. A cross-layer design, including user selection and signaling plan, for cooperative communications between an active PU and an SU is investigated.

Then, an orthogonally dual-polarized antenna (ODPA) based cooperation framework is presented (Cao, 2013; Cao, 2014). The use of ODPAs enables concurrent transmissions of multiple independent signals of PUs and SUs, and interference suppression via polarization zero-forcing and polarization filtering to obtain significant performance improvement. To maximize a weighted sum throughput of PUs and SUs under energy/power constraints, the problem is formulated and solved based on geometric programing and the solution algorithms are devised to attain the optimality.

Orthogonal Signaling Based Energy-Efficient CCRN Framework

As shown in Figure 4, channel time is divided into frames, each of duration T. M consecutive frames form a superframe. A prefix of duration ΔT for relay selection is appended at the start of each superframe1. Each frame is partitioned into two slots (or phases), each of duration $T/2$. The first slot is for PU transmission and the second slot is for the relay to forward the PU's message and to transmit its own data.

Note that phase shift between I channel and Q channel severely degrades the performance, since mutual interference will introduced due to phase shift. For simplicity, we assume there is no phase shift or this shift can be estimated at the receiver. There are now a lot of phase shift estimation methods (Mehrpouyan, 2012; Noels, 2005; Wang, 2010). Phase shift estimation itself is outside the scope of our work, as this technique is an independent research area.

In the first phase of each frame, the PU sends its packets to the relay and the destination using the in-phase modulation. Upon receiving the PU's packet, in the second phase, the SU forwards the PU's packets using the in-phase channel with power $(1-\alpha)P_S$ and transmits its own data using the quadrature modulation with power αP_S. The quadrature modulation framework for the PU and the SU are shown in Figures 5-7, in which Figure 5 is the orthogonal modulator at the relaying SU transmitter to relay for the PU and transmit SU's own information. Figure 6 shows the quadrature demodulation of SU's own information at the SU receiver, and Figure 7 shows the in-phase demodulation of PU's data at the PU receiver, respectively. Since these two modulations are orthogonal, there is no mutual interference between PU's and SU's concurrent transmissions.

Figure 4. Frame structure

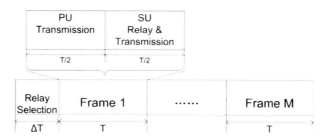

Figure 5. SU's transmitter

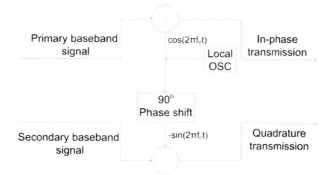

Figure 6. SU's receiver

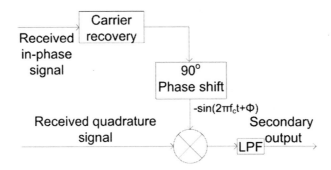

Figure 7. PU's receiver

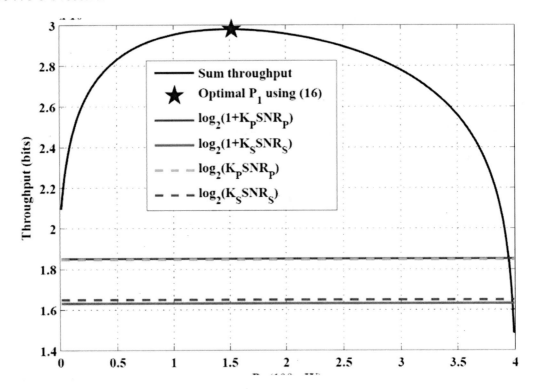

Consider a four node cooperative communication scenario in which a PU communicates with the PBS, and an SU communicates with the SBS as shown in Figure 1. The channel between any two users is assumed to follow a quasi-static Rayleigh flat fading, i.e., the channel gains are invariant during the time duration $\Delta T + T$. Let h_{PS}, h_{PB}, h_{SP}, and h_{SB} be respectively the channel gains from PU to SU, PU to PBS, SU to PBS, and SU to SBS.

The AF Relaying Mode

Consider a situation in which an SU uses the AF relaying mode and orthogonal modulations to cooperate with a PU. With the orthogonal modulations, information transmission and relaying only requires two phases. In the first phase, the PU transmits an L-bit packet x_p using in-phase modulation with power P_p to the SU and PBS, where $E\left\{x_p^2\right\} = 1$, and $E\left\{\mathbf{X}\right\}$ is the expectation operator of vector $\mathbf{X}$. After receiving the PU's signal, the SU amplifies and forwards it using the in-phase modulator with power $(1-\alpha)P_S$, and transmits its own data x_S using the quadrature modulator with power αP_S, as shown in Figure 5.

By receiving the PU's signal, allocating power for relaying the PU's signal and sending the SU's own signal, the transmitted signal of the SU in the second phase is given by

$$y_{S\_AF} = \sqrt{\frac{\left(1-\alpha\right)P_S}{\left|h_{PS}\right|^2 P_P + \sigma^2}}\left(h_{PS}\sqrt{P_P}x_P + n\right) + j\sqrt{\pm P_S}x_S \tag{1}$$

where n is the additive noise with zero mean and variance σ^2, and j is the imaginary unit denoting the quadrature component.

As shown in Figure 7, the PBS can extract PU's signal using in-phase modulator. Assuming the maximal ratio combining (MRC) technique is applied by the PBS to decode the PU's packet, the SNR of the received primary signals after combining at the PBS is obtained as

$$\mathrm{SNR}_{PB\_AF} = \gamma_{PB}P_P + \frac{\gamma_{SP}P_S\gamma_{PS}P_P\left(1-\alpha\right)}{\gamma_{SP}P_S\left(1-\alpha\right) + \gamma_{PS}P_P + 1} \tag{2}$$

where $\gamma_{ik} = \dfrac{|h_{ik}|^2}{\sigma^2}$ is the channel gain-to-noise ratio (CNR) of the link between node i and node k.

Similarly, the SNR of the received secondary signal at the SBS can be written as $\mathrm{SNR}_{SB\_AF} = \alpha\gamma_{SB}P_S$.

The DF Relaying Mode

The phase noise and other channel impairments in the propagation environment to the two different destinations may degrade the orthogonality of the channels. Compared to AF, DF relaying is more complex, but can correct errors before forwarding. We consider the user cooperation is a leader-follower mode. In this regard, we assume that SU can only access to the spectrum when and only when the SU can contribute to PU. When PU's signals cannot be correctly decoded by SU, we denote SNR=0 to represent this case.

Similarly, in the first phase, the PU broadcasts signal x_p using in-phase modulation with power P_p. If the relaying SU detects no error or the detected error is correctable, the SU relays the PU's packet and transmits its own data in the second phase using the two orthogonal channels provided by orthogonal modulations. Let $P_b = \frac{1}{2}\left(1 - \sqrt{\frac{\gamma_{PS}P_P}{\gamma_{PS}P_P + 1}}\right)$ be the probability that the detected error is not correctable

so that the PU's packet cannot be relayed. Let be the bit error rate (BER) of the PU's signal received by the SU. Assuming that bit errors are independent of each other, we have $P_e = 1 - (1 - P_b)^L \approx LP_b$ where the approximation is for small P_b and large L. In the context of using the DF relaying, the transmitted signal of the SU in the second phase is given by

$$y_{S_{DF}} = \begin{cases} \sqrt{(1-\alpha)P_S}x_P + j\sqrt{\alpha P_S}x_S, & \text{w.p. } 1\text{-}P_e; \\ 0, & \text{w.p. } P_e. \end{cases} \quad (3)$$

Therefore, the SNRs received at the PBS and SBS are given by

$$\text{SNR}_{PB\_DF} = \begin{cases} \gamma_{PB}P_P + \gamma_{SP}P_S(1-\alpha), & \text{w.p. } 1\text{-}P_e; \\ \gamma_{PB}P_P, & \text{w.p. } P_e. \end{cases} \quad (4)$$

and

$$\text{SNR}_{SB\_DF} = \begin{cases} \alpha\gamma_{SB}P_S, & \text{w.p. } 1\text{-}P_e; \\ 0, & \text{w.p. } P_e. \end{cases} \quad (5)$$

respectively, where 0 implies that no useful signal is received by the SBS, i.e., the SU does not transmit.

We formulate a throughput optimization problem under certain constraints in a CCRN. Instead of maximizing either the PU's throughput, or the SU's throughput, we propose a general metric of weighted sum throughput, namely C_W which incorporates both the PU's and SU's throughput to evaluate the co-operation performance. The optimization problem is formulated as

$$\max_{\alpha,P_P,P_S} C_W = (1-\zeta)C_{PB} + \zeta C_{SB}$$
$$\text{s.t. } C_{PB} \geq KC_{Pd} \geq C_{PT}, K \geq 1$$
$$C_{SB} \geq C_{ST} \quad (6)$$
$$P_P + P_S \leq P_M$$
$$0 < \alpha < 1$$

where ζ is a weighting parameter that strikes a balance between the PU's and SU's throughputs. In general, the parameter ζ reflects the cooperation needs of the PU. For example, if a PU is experiencing a deep fade and cannot communicate with the PBS without the help of SUs, a larger ζ value may be used to favor SUs, and vice versa. In the extreme cases when $\zeta = 0$ or $\zeta = 1$, the objective function is simplified to maximize the PU's or SU's throughput. Let $C_{Pd} = \log_2(1 + \gamma_{PB}P_P)$ be the achievable throughput of the PU to the PBS without cooperating with an SU, and K be the throughput gain that the PU wants to achieve through cooperation. The PU needs to cooperate with SUs if the achievable C_{P_d} is below its minimum throughput requirement, C_{PT}. The first constraint shows that, by cooperating with an SU, the PU can achieve a throughput greater than the threshold C_{PT}. The second constraint represents that the achievable throughout of the SU should meet its minimum throughput requirement, C_{ST}. Otherwise, the SU may not be able to establish a connection with the SBS so that the secondary transmission becomes useless. These two constraints represent the cooperation benefit that motivates the PU and SU to cooperate with each other. The third constraint indicates that the total transmission power of the PU and SU should be bounded by P_M. Lastly, the parameter α, the fraction of the power that an SU uses for transmitting its own traffic, should be selected from (0,1). In a special case when $\alpha = 1$, the SU transmits only its own data without relaying the PU's packet; alternatively, when $\alpha = 0$, the SU only helps forward the PU's packet without transmitting its own data, which is referred to as a conventional relaying scenario.

It is straightforward to verify that the above optimization problem is a convex one. We can utilize some classical optimization method to calculate the solution. In this work, we have used the Karush-Kuhn-Tucker (KKT) conditions to get the optimality.

To validate the analytical results and the effectiveness of the proposed framework, analytical and simulation results on the optimal α, and the throughput performance of the PU and SU when using the AF mode are shown in Figure 8. The SU is assumed to be able to obtain perfect CNRs. After getting the CSIs, the SU compares them with the associated cooperative conditions. In this simulation, the CSIs are within the cooperative conditions when AF is used, and the SU aims to get the optimal power allocation factor for relaying the PU's traffic and transmitting its own data. It is shown in Figure 8 that the maximum

Figure 8. Throughput comparison w/wo cooperation, AF case

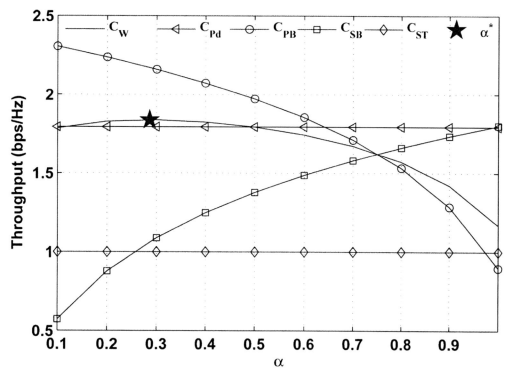

throughput is achieved when α^* is around 0.29 in simulation and the obtained value is about 0.286. It is observed that by cooperation, the PU can achieve 1.22 times the throughput compared with direct transmission while the SU can also achieve a higher throughput than its minimum throughput requirement of $C_{ST} = 1$ bps/Hz.

To show the difference in cooperative conditions of the AF and DF relaying modes, this simulation illustrates how to adapt the relaying mode according to instantaneous CSIs. In the settings, the AF mode cannot provide a feasible solution for user cooperation and the SU then checks whether the DF mode can be applied. Figure 9 plots the throughput performance of the PU and SU for the DF mode. It can be seen in the figure that the optimal α^* is about 0.341, while from the simulation it can be seen that the optimal α^* is around 0.34. In this context, the SU achieves $C_{SB\_DF} = 1.6 > C_{ST} = 1$ bps/Hz and the PU achieves $C_{PB\_DF} = 1.25 > C_{Pd} = 0.4879$ bps/Hz.

Orthogonally Dual-Polarized Antenna Based Energy-Efficient CCRN Framework

Although the MIMO technology can offer significant enhancement in throughput and link range without additional bandwidth and/or transmit power, the application of MIMO is limited by the hardware cost and physical size. In theory, typical antenna elements must be spaced at least half a wavelength at the mobile terminal and ten wavelengths at the base station or the access point to ensure fully independent spatial channel fading. This results in a large hardware size of SUs, if SUs cooperate with PUs using MIMO to attain a two-phase cooperation framework in a CCRN.

Figure 9. Throughput comparison w/wo cooperation, DF case

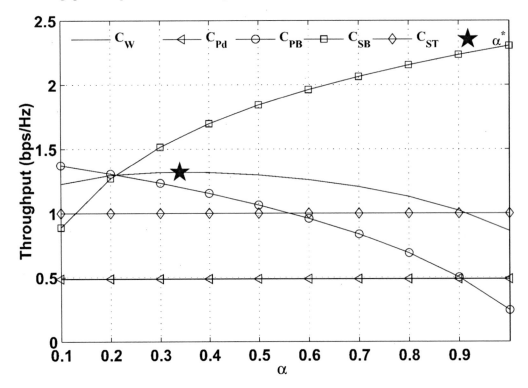

In order to reduce the stringent physical size, hardware and cost constraints for MIMO based SUs, it is feasible to incorporate dual-polarized antennas in the CCRN framework design. Recently, dual-polarized antennas have attracted a great interest for doubling or tripling the channel capacity without requiring antenna spacing by leveraging polarization at dual-polarized antennas and depolarization in wireless channels (Oestges, 2008; Andrews, 2001; Kwon, 2011.). In this regard, the use of co-located orthogonally dual-polarized antennas (ODPAs) is a space- and cost-effective alternative for SUs to employ a two-phase frame format, especially for the SU-type handset wireless devices. ODPA and its applications have been widely used in radar systems. The key issue of ODPA is the polarization information, such as channel depolarization, cross-polarization, etc. In wireless communication systems, the authors noticed that there are some research effort focused on the deployment and applications of ODPA, such as (Jacob, 2013) and the references therin. We believe that the use of ODPA will attract more attentions from academia and industry.

In this part, we present a novel polarization based two-phase CCRN framework for cooperation between PUs and SUs by exploiting ODPAs on SU-type devices. To the best of our knowledge, this is the first work that takes the use of polarization and ODPAs into account for the CCRN design. We are interested in answering the following two questions: what strategies SUs adopt to simultaneously forward PUs' traffic and transmit their own messages using ODPAs on a non-interference basis, and what are the benefits of exploiting ODPAs for the CCRN design. To this end, SUs use ODPAs, while PUs still use the legacy uni-polarized antennas and are not required to change their hardware to support polarization capability. Concurrent transmissions are enabled by carefully setting different polarization at ODPAs, i.e., SUs can simultaneously forward PUs' data and transmit their own packets by using different polarized states. In addition, the degrees of freedom provided by ODPAs can be utilized to implement interference suppression, e.g., zero-forcing in the polarization domain, and polarization filtering. To evaluate the cooperation performance of the proposed two-phase polarization based CCRN framework, sum throughput maximization under power and throughput constraints is formulated.

The system model consists of a primary network comprised of a primary base station (PBS) and a primary transmitter (PT), and a secondary network comprised of a secondary access point (SAP) and a secondary transmitter (ST), as shown in Figure 10. In a fully operational primary network, the PBS assigns licensed spectral bands to PUs. For the purpose of demonstrating the principle of polarization enabled CCRN design, the PBS plays the role of primary receiver (PR). Similarly, the SAP is the secondary receiver (SR) in the secondary network. In what follows, we use the notations PT, PR, ST and SR to describe the transceivers and the signal flows shown in Figure 10, i.e., PR instead of PBS and SR instead of SAP. The frame format of the two-phase cooperation is shown in Figure 11, where *T* represents the total frame length of the two-phase format. For the two-phase framework, a frame consists of two slots, each of duration *T/2*. It is noted that the cooperation between a PU and an SU can be easily extended to one PU cooperating with more than one SUs.

As shown in Figure 11, the frame length *T* is partitioned into two equal halves in which the PU transmits in the first *T/2* time interval, i.e., the first slot. Equipped with an ODPA, the cooperating SU relays the PU's signal to the PR using an amplify-and-forward mode, and transmits its own data in the second *T/2* interval, i.e., the second slot, to the SR.

Let D_{PS}, D_{SP}, and D_{SS} represent the 2x2 channel depolarization matrices of the links between the PT and ST, between the ST and PR, and between the ST and SR, respectively. The physical meaning of channel depolarization can be found in (Oestges, 2008). Let h_{PS}, h_{SP}, and h_{SS} be the spatial channel

Figure 10. System model

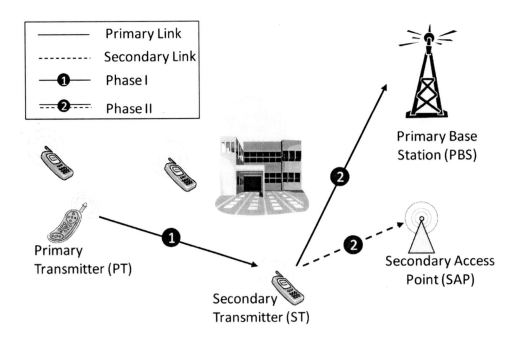

Figure 11. Frame structure

fading coefficients, assumed to follow the Rayleigh block fading model. Consider that the spatial fading coefficients are invariant during the frame duration T, and that the ambient noise in each polarization branch is Gaussian distributed with zero mean and variance $\dfrac{\sigma^2}{2}$. It is noted that polarization channel fading in terms of channel depolarization is flatter compared to spatial channel fading (Kwon, 2011); hence, it is reasonable to assume that the channel depolarization matrices are invariant during the interval T. For the analysis to follow, we assume that the spatial channel fading coefficients are known at the PU receiver and the SU transceiver. Furthermore, channel depolarization matrices are also perfectly available at the SU's transceiver, and how to model and obtain the channel depolarization matrix can be found in (Kwon, 2011).

In the first time slot of the frame, the PT sends $\sqrt{P_0}\left[0,e^{j\delta_P}\right]^T x_P$ to the chosen ST, where P_0 is the transmit power, the 2x1 vector $\left[0,e^{j\delta_P}\right]$ represents that its antenna is vertically polarized with a phase difference δ_P. The symbol x_P with unit energy is the PT's signal before radiated by ODPAs, and the superscript T denotes vector transpose.

After the PT transmitting the signal, the received signal at the ST in the first phase is expressed by

$$r_{PS} = h_{PS}\sqrt{P_0}D_{PS}\begin{bmatrix} 0 \\ \exp(j\delta_P) \end{bmatrix}x_P + n_{ST} \tag{7}$$

where n_{ST} is the two-dimensional (2D) white Gaussian noise in the dual-polarized antenna at the ST.

In the second time slot of the frame, the ST first combines the 2D signal r_{PS} and generates a 1D version, i.e., polarization combining. Then the ST forwards the PT's signal using polarization z_P with power P_1, and sends its own signal using polarization z_S with power P_2, so that the transmitted signal of the ST can be written as

$$z_P\beta(h_{PS}\sqrt{P_0}x_P + n) + z_S\sqrt{P_2}x_S \tag{8}$$

where $b = \sqrt{\dfrac{P_1}{|h_{PS}|^2 P_0 + \sigma^2}}$ is the power allocation factor, and n is a 1D additive noise with variance $\dfrac{\tilde{A}^2}{2}$.

Thus, the signals received at PR and SR can be obtained as

$$r_{SP} = \{D_{SP}z_Ph_{SP}\beta(h_{PS}\sqrt{P_0}x_P + n) + D_{SP}z_Sh_{SP}\sqrt{P_2}x_S + n_{PR}\}\,|\,v \tag{9}$$

and

$$r_{SS} = D_{SS}z_Ph_{SS}\beta(h_{PS}\sqrt{P_0}x_P + n) + D_{SS}z_Sh_{SS}\sqrt{P_2}x_S + n_{SR} \tag{10}$$

respectively, where $\{x\}|_v$ denotes the component received by the vertical polarization branch of the dual polarized signal x.

If $D_{SP}z_S = 0$ can be satisfied, then the ST causes no interference to the PR; this operation is referred to as polarization zero-forcing. Polarization zero-forcing is similar to spatial beamforming, while polarization zero-forcing utilizes the degrees of freedom provided by dual polarization at ODPAs and the depolarization effect in wireless channels. Moreover, since the variations of polarization fading are in a larger time scale than those of spatial fading, the implementation of polarization zero-forcing is less complex than spatial beamforming in terms of coefficient adaptation. In this context, the ST chooses z_S according to D_{SP} to avoid interference at the PR.

In order to maximize the PR's receiving signal-to-noise ratio (SNR), one can set z_P to render $D_{SP}z_P = [0,1]^T$. Hence the received signal at the PR is

$$r_{SP} = h_{SP} h_{PS} \sqrt{P_0} \beta x_P + h_{SP} \beta n + n_1 \tag{11}$$

where n_1, with variance $\dfrac{\sigma^2}{2}$, is the additive noise in the vertical polarization branch at the PR (Note that PR only has a single antenna with vertical polarization).

In (10), the first term is interference from the PU. By using the oblique projection polarization filtering technology (Cao, 2013), the SR is able to suppress the PU's signal. Then the received signal at the SR in the second phase is

$$r_{SS} = D_{SS} z_S h_{SS} \sqrt{P_2} x_S + n_{SR} \tag{12}$$

After the two-phase transmissions, the SNRs at the PR and SR are obtained as

$$SNR_P = \frac{2\beta^2 \mid h_{SP} \mid^2 \mid h_{PS} \mid^2 P_0}{(2\beta^2 \mid h_{SP} \mid^2 + 1)\sigma^2} \tag{13}$$

$$SNR_S = \frac{\mid h_{SS} \mid^2 P_2}{\sigma^2}$$

If the modulation scheme such as the constellation size is taken into consideration for calculating the transmission rate, e.g., M_S-ary quadrature amplitude modulation (QAM) and a pre-defined bit error rate threshold BER_{TS}, the constellation size M_S can be tightly approximated by

$$M_S = 1 - \phi_{S1} SNR_S \,/\, \ln(\phi_{S2} BER_{TS})$$

where ϕ_{S1} and ϕ_{S2} are system parameters which are constants and determined according to specific modulation type of the SU (Cao, 2012). Assume that the assigned bandwidth of the PU is B and constants $K_S = -\phi_{S1} \,/\, \ln\left(\phi_{S2} BER_{TS}\right)$ and $K_P = -\phi_{P1} \,/\, \ln\left(\phi_{P2} BER_{PS}\right)$, where BER_{PS} is the pre-defined bit error rate threshold of the PU, and these two parameters are determined by the modulation method used at the PU and SU, then the achievable throughput of the SU and PU can be written as

$$R_S = \frac{TB}{2} \log_2 (1 + K_S \gamma_{SS} P_2) \tag{14}$$

$$R_P = \frac{TB}{2} \log_2 (1 + K_P \frac{2\gamma_{SP} \gamma_{PS} P_1 P_0}{2P_1 \gamma_{SP} + \gamma_{PS} P_0 + 1}) \tag{15}$$

where $\gamma_{ik} = \dfrac{|h_{ik}|^2}{\sigma^2}$, is the channel gain-to-noise ratio (CNR) of the link between node i and node k.

We aim to maximize the sum throughout of the PU and SU, i.e., $R_P + R_S$, or equivalently the normalized sum throughput $\mathbf{T}_t = \overline{R}_P + \overline{R}_S$, where $\overline{R}_i = \dfrac{1}{2}\log_2\left(1 + K_i SNR_i\right)$. Then the objective function with power and throughput constraints can be formulated by

$$
\begin{aligned}
&\max_{P_0, P_1, P_2} && T_t = \overline{R}_P + \overline{R}_S \\
&subject\ to: && \overline{R}_P \geq \overline{R}_{PT} \\
& && \overline{R}_S \geq \overline{R}_{ST} \\
& && 0 < P_0 \leq P_P \\
& && 0 < \sum_{k=1}^{2} P_l \leq P_S \\
& && 0 < \sum_{k=0}^{2} P_k \leq P_M
\end{aligned}
\tag{16}
$$

The first constraint shows that, by cooperating with the SU, the PU can achieve a throughput greater than the minimum throughput requirement $\overline{R}_{PT}$. The second constraint indicates that the SU can achieve a throughput that is greater than the minimum throughput requirement $\overline{R}_{ST}$ to support the service. These two constraints represent the cooperation motivation of the PU and SU, i.e., mutual benefit. Only when these two constraints are both satisfied do the PU and SU cooperate with each other. The following three constraints mean that the maximum transmit power of the PU is bounded by P_P, the total transmit power of the SU is bounded by P_S, and the total transmit power of the PU and SU is bounded by P_M, respectively. The first two power constraints are due to physical hardware power limitations of the PT and ST. The rationale of the last power constraint is that, there is an allowable transmit power limit on each frequency band enforced by regulators to avoid potential interference, so that the total power of the PU and SU needs to be within this limit.

As will be shown later, the original problem in (16) is nonlinear and non-convex. It can be easily verified that the optimal problem of (16) is a geometric programming (GP) one (Cao, 2012), as SNR_p is a posynomial in P and SNR is a monomial in P, and the product of a posynomial and a monomial is still a posynomial, where $P = (P_0, P_1, P_2)$. In this context, this problem can be solved using GP algorithms.

Since T_t is increasing with respect to (w.r.t.) P_k, T_t is also increasing w.r.t. both $\sum_{k=0}^{2} P_k$ and $\sum_{l=1}^{2} P_l$. In this regard, the maximum T_t is achieved when all the power constraints are on the maximum boundaries, i.e., $\sum_{l=1}^{2} P_l^* = P_S$ and $P_0^* = \min\left(P_M - P_S, P_P\right)$, or $\sum_{l=1}^{2} P_l^* = \min\left(P_M - P_P, P_M\right)$ and $P_0^* = P_P$.

To be specific, (1) for $P_S + P_P < P_M$, the maximum T_t is achieved when $\sum_{l=1}^{2} P_l^* = \min\left(P_M - P_P, P_M\right)$

and $P_0^* = P_P$; (2) for $P_S + P_P \geq P_M$, the maximum T_t is achieved when $\sum_{l=1}^{2} P_l^* = P_S$ and $P_0^* = P_M - P_S$, or $\sum_{l=1}^{2} P_l^* = P_M - P_P$ and $P_0^* = P_P$.

Therefore, an SU needs to check the power constraints to find the optimal strategy, i.e., we have the following two cases.

$P_S + P_P < P_M$ Case

In this case, the sum of the PU and SU's maximum power is within the permitted power limit. This case will be discussed by dividing SNRs into two regimes, i.e., high and medium (or low) SNR regimes.

In the high SNR regime, e.g., $K_i SNR_i \gg 1$, equations (14) and (15) can be approximated by $\frac{TB}{2} \log_2 \left(K_i SNR_i \right)$. This approximation is reasonable when the signal strength is much higher than that of the additive noise. For example, in CDMA based wireless systems, this can be satisfied when the spreading process gain is large.

Since the optimality of (16) is achieved when $P_0^* = P_P$ and $\sum_{l=1}^{2} P_l^* = P_S$, (16) is equivalent to the following problem

$$
\begin{aligned}
&\min_{P_1, P_2} && \frac{1}{\prod_{i \in \{S,P\}} SNR_i} = \frac{2P_1 \gamma_{SP} + \gamma_{PS} P_P + 1}{2\gamma_{SP} \gamma_{SS} \gamma_{PS} P_2 P_1 P_P} \\
&subject\ to: && \frac{2\gamma_{SP} \gamma_{PS} P_1 P_P}{2P_1 \gamma_{SP} + \gamma_{PS} P_P + 1} \geq \frac{2^{2R_{PT}}}{K_P} \\
& && \gamma_{SS} P_2 \geq \frac{2^{2R_{ST}}}{K_S} \\
& && P_1 + P_2 = P_S
\end{aligned}
\tag{17}
$$

The constrained optimization problem in (18) is a GP problem and can be analytically solved by KKT conditions straightforwardly. The optimal powers thus can be obtained as

$$P_0^* = P_P$$

$$P_1^* = \frac{\sqrt{A(A + 2\gamma_{SP} P_S)} - A}{2\gamma_{SP}}$$

$$P_2^* = P_S - P_1^*$$

where $\mathbf{A} = \gamma_{PS} P_P + 1$.

In the medium to low SNR regime, i.e., when SNRs of both the PU and SU are not much larger than 1 (e.g., 0dB), the approximation does not hold. In this context, the problem for the medium or low SNR condition is then formulated by

$$
\begin{aligned}
&\min_{P_1, P_2} && \frac{1}{\prod_{i \in \{S,P\}} (1 + K_i SNR_i)} \\
&\text{subject to:} && \frac{2\gamma_{SP}\gamma_{PS}P_1 P_P}{2P_1\gamma_{SP} + \gamma_{PS}P_P + 1} \geq \frac{2^{2\bar{R}_{PT}} - 1}{K_P} \\
& && \gamma_{SS}P_2 \geq \frac{2^{2R_{ST}} - 1}{K_S} \\
& && P_1 + P_2 = P_S
\end{aligned} \tag{18}
$$

The term $1 + K_P SNR_P$ is an inverted posynomial, and it is not an inverted posynomial. However, $1/(1 + K_P SNR_P)$ is a ratio between two posynomials. To minimize the ratio or to get the upper bound of the ratio between two posynomials are proved to be a kind of non-convex problems and are also known as Complementary GP problems (Miang, 2007), which belongs to a class of intractable NP-hard problems. According to Signomial Programming (SP), this class of problems can be solved by the single or double condensation method. In this work, a single condensation approach to solving (19) below is given in Algorithm 1 as shown in Figure 12. By setting an error tolerance for exit condition, the optimal P can be obtained by iterations using Algorithm 1.

$P_S + P_P \geq P_M$ Case

By adopting the same optimization method, we can get the optimal powers in high SNR regimes as follows,

$$P_0^* = P_P$$

$$P_1^* = \frac{\sqrt{A[A + 2\gamma_{SP}(P_M - P_P)]} - A}{2\gamma_{SP}}$$

$$P_2^* = P_M - P_P - P_1^*$$

where $\mathbf{A} = \gamma_{PS} P_P + 1$.

In low to medium SNR regimes, the optimal powers can be obtained by using Algorithm 1.

To verify the correctness of the proposed schemes, some numerical results are given. With parameter values $\mathbf{K}_S = \mathbf{K}_P = 50$, $\mathbf{SNR}_S = 234.6$, $\mathbf{SNR}_P = 348.5$, $\mathbf{P}_P = 0.2\mathbf{W}$, $\mathbf{P}_S = 0.4\mathbf{W}$, $\mathbf{P}_M = 0.7\mathbf{W}$, $\mathbf{T} = 0.1\mathbf{s}$, and B=5MHz, it is shown in Figure 13 that in the high SNR regime, the solution to P_1^* is

Figure 12. Single condensation GP alogrithm

Algorithm 1 Single condensation GP Algorithm for (18

Input: A Feasible powers set $\mathcal{P} = (P_0, P_1, P_2)$;

Output: Optimal $\mathcal{P}$ that satisfies the KKT conditions:

    for $n = 1$ to N **do**

      Calculate $L^{(n)} = (1 + K_S\gamma_{SS}P_2^{(n)})(2P_1^{(n)}\gamma_{SP}$
$\gamma_{PS}P_0^{(n)} + 1 + K_P 2\gamma_{SP}\gamma_{PS}P_1^{(n)}P_0^{(\tilde{n})})$ with the giv
$\mathcal{P}$;

      Expand $L^{(n)}$ and compute each individ
term $u_m(\mathcal{P}^{(n)})$ in $L^{(n)}$ (e.g., calculati
$K_S\gamma_{SS}P_2^{(n)}K_P 2\gamma_{SP}\gamma_{PS}P_1^{(n)}$ etc.) and obt
the weight $\alpha_m^{(n)}$ as
$$\alpha_m^{(n)} = \tfrac{\text{Value of the } m_{th} \text{ term } u_m(\mathcal{P}^{(n)})}{L^{(n)}};$$

      Condense the term $L^{(n)}$ into a monomial with wei;
$\alpha_m^{(n)}$ by using

$$\tilde{g}^{(n)}(\mathcal{P}) = \prod \left(\frac{u_m(\mathcal{P})}{\alpha_m^{(n)}}\right)^{\alpha_m^{(n)}};$$

      Solve the resulting GP problem $\frac{2P_1^{(n)}\gamma_{SP}+\gamma_{PS}P_0^{(n)}}{\tilde{g}^{(n)}(\mathcal{P})}$
using an interior point method, such as the bari
method;

      if $\| \mathcal{P}^{(n)} - \mathcal{P}^{(n-1)} \| \le \varepsilon$ **then**

         Terminate and yield the $\mathcal{P}^*$ as $\mathcal{P}^{(n)}$;

      else

         Go to Step 2 and use $\mathcal{P}^{(n)}$ as the input.

      end if

   end for

   return

Figure 13. The optimal power and the associated throughput in the high SNR regime and $P_S + P_P < P_M$

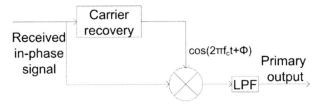

0.1518W, whereas the theoretical optimal $\mathbf{P}_1$ is 0.1512W. With $\mathbf{P}_1^*$ obtained by KKT, the associated optimal throughput approximate the theoretical optimal throughput. This illustrates that by approximating $1 + \mathbf{K}_i\mathbf{SNR}_i$ as $\mathbf{K}_i\mathbf{SNR}_i$, one can obtain a near perfect approximation.

In Figure 14, we consider a medium SNR regime with $\mathbf{K}_S = \mathbf{K}_P = 1$, $\mathbf{SNR}_S = 10.3$ and $\mathbf{SNR}_P$ =58.5, $\mathbf{P}_M$ =0.5W. By using Algorithm 1, the optimal $\mathbf{P}_1^*$ can be iteratively obtained. As shown in the figure, $\mathbf{P}_1^*$ =0.0512W achieved by the single condensation GP approaches the theoretical optimal 0.0498W, after 14 iterations. The error tolerance in this iteration is 0.002W.

Figure 14. The optimal power and the associated throughput using Algorithm 1 in the medium SNR regime and $P_S + P_P > P_M$

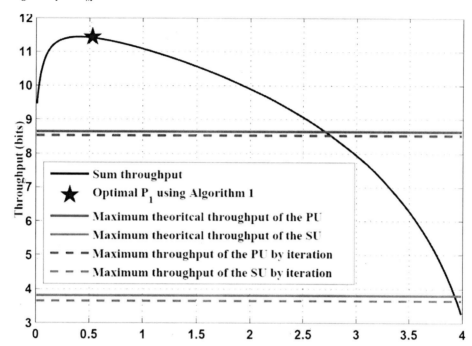

CONCLUDING REMARK

In this chapter, we have presented a promising research direction by exploiting active user cooperation between primary users and secondary users in cognitive radio networks, named CCRN. Recent research work and associated challenging issues have been outlined, and some research topics have been listed. In order to give an extensive and tutorial investigation regarding this framework, two spectrum-energy efficient paradigms to enable cooperative communications between PUs and SUs has been discussed. In the first paradigm, we have presented the use of quadrature signaling based CCRN and its key enabling techniques. In the second paradigm, we have introduced the use of polarization fading, polarization signal processing to tackle the mentioned challenges in the discussed model. Theoretic and numerical results are provided to unveil the performance and characteristics of the proposed frameworks.

REFERENCES

Akyildiz, I. F., Lee, W., Vuran, M., & Mohanty, S. (2006). Next generation/dynamic spectrum access/ cognitive radio wireless networks: A survey. *Computer Networks*, 50(13), 2127–2159. doi:10.1016/j. comnet.2006.05.001

Andrews, M., Mitra, P., & deCarvalho, R. (2001). Tripling the capacity of wireless communications using electromagnetic polarization. *Nature*, 409(1), 316–318. doi:10.1038/35053015 PMID:11201734

Attar, A., Nakhai, M. & Aghvami, A. (2009). Cognitive radio game for secondary spectrum access problem. *IEEE Trans. Wirel. Commu., 8*(4), 2121-2131.

Boyd, S., & Vandenberghe, L. (2004). *Convex optimization.* Cambridge University Press. doi:10.1017/CBO9780511804441

Cao, B. (2013). *On frameworks and techniques of cooperation between primary and secondary users enabled dynamic spectrum access.* (Unpublished doctoral dissertation). Harbin Institute of Technology, China.

Cao, B., Cai, L. X., Liang, H., Mark, J. W., Zhang, Q., Poor, H. V., & Zhuang, W. (2012). Cooperative cognitive radio networking using quadrature signaling. Paper presented in IEEE INFOCOM. Orlando, FL.

Cao, B., Liang, H., Mark, J. W., & Zhang, Q. (2013). Exploiting orthogonally dual-polarized antennas in cooperative cognitive radio networking. *IEEE Journal on Selected Areas in Communications, 31*(11), 2362–2373. doi:10.1109/JSAC.2013.13111

Cao, B., Mark, J. W., & Zhang, Q. (2012, December). *A polarization enabled cooperation framework for cognitive radio networking.* Paper presented at IEEE GLOBECOM. Anaheim, CA. doi:10.1109/GLOCOM.2012.6503262

Cao, B., Mark, J. W., Zhang, Q., Lu, R., Lin, X., & Shen, X. (2013, April). *On optimal communication strategies for cooperative cognitive radio networking.* Paper presented at IEEE INFOCOM. Turin, Italy. doi:10.1109/INFCOM.2013.6566970

Cao, B., Yu, J., Wang, Y., & Zhang, Q. (2013). Enabling polarization filtering in wireless communications: Models, algorithms, and characteristics. *IET Communications, 4*(3), 247–254. doi:10.1049/iet-com.2012.0409

Cao, B., Zhang, Q., Liang, D., Wen, S., Lin, J., & Zhang, Y. (2010, May). *Blind adaptive polarization filtering based on oblique projection.* Paper presented at IEEE ICC. Cape Town, South Africa. doi:10.1109/ICC.2010.5501769

Cao, B., Zhang, Q., Lu, R., & Mark, J. W. (2014). PEACE: Polarization enabled active cooperation scheme between primary and secondary networks. *IEEE Transactions on Vehicular Technology, 63*(8), 3677–3688. doi:10.1109/TVT.2014.2305713

Cao, B., Zhang, Q., Mark, J. W., Cai, L. X., & Poor, H. V. (2012). Toward efficient radio spectrum utilization: User cooperation in cognitive radio networking. *IEEE Network, 26*(4), 46–52. doi:10.1109/MNET.2012.6246752

Chen, Y., & Liu, K. (2011). Indirect reciprocity game modelling for cooperation stimulation in cognitive networks. *IEEE Trans. Wirel. Commu. 59*(1), 159-168.

Chiang, M., Tan, C. W., Palomar, D. D. P., O'Neill, D., & Julian, D. (2007). Power control by geometric programming. *IEEE Transactions on Wireless Communications, 6*(7), 2640–2651. doi:10.1109/TWC.2007.05960

David, K., Dixit, S., & Jefferies, N. (2010). 2020 Vision. *IEEE Vehicular Technology Magazine, 5*(3), 22–29. doi:10.1109/MVT.2010.938595

Goldsmith, A., Jafar, S., Maric, I., & Srinivasa, S. (2009). Breaking spectrum gridlock with cognitive radios: An information theoretic perspective. *Proceedings of the IEEE, 97*(5), 894–914. doi:10.1109/JPROC.2009.2015717

Granelli, F., Pawelczak, P., Prasad, R., Subbalakshmi, K., Chandramouli, R., Hoffmeyer, J., & Berger, J. (2010). Standardization and research in cognitive and dynamic spectrum access networks: IEEE SCC41 efforts and other activities. *IEEE Communications Magazine, 48*(1), 71–79. doi:10.1109/MCOM.2010.5394033

Hao, X., Cheung, M., Wong, V., & Leung, V. (2011). *A Stackelberg game for cooperative transmission and random access in cognitive radio networks.* IEEE PIMRC.

Haykin, S. (2005). Cognitive radio: Brain-empowered wireless communications. *IEEE Journal on Selected Areas in Communications, 23*(2), 201–220. doi:10.1109/JSAC.2004.839380

Hua, S., Liu, H., Wu, M., & Panwar, S. (2011, April). *Exploiting MIMO antennas in cooperative cognitive radio networks.* Paper presented at IEEE INFOCOM. Shanghai, China. doi:10.1109/INFCOM.2011.5935102

Jacob, P., Madhukumar, A., & Alphones, A. (2013). Interference Mitigation through Cross Polarized Transmission in Femto-Macro Network. *IEEE Communications Letters, 17*(10), 1940–1943. doi:10.1109/LCOMM.2013.090213.131482

Jia, J., Zhang, J., & Zhang, Q. (2009). Cooperative relay for cognitive radio networks. In *Proc. INFOCOM*, (pp. 2304-2312). IEEE.

Kwon, S., & Stuber, G. (2011). Geometrical theory of channel depolarization. *IEEE Transactions on Vehicular Technology, 60*(8), 3542–3556. doi:10.1109/TVT.2011.2163094

Li, L., Zhou, X., Xu, H., Li, G., Wang, D., & Soong, A. (2011). Simplified relay selection and power allocation in cooperative cognitive radio systems. *IEEE Trans. Wirel. Commun., 10*(1), 33-36.

Lin, P., Jia, J., Zhang, Q., & Hamdi, M. (2010). Dynamic spectrum sharing with multiple primary and secondary users. In *Proc. IEEE Int'l Conf. Commun.* IEEE. doi:10.1109/ICC.2010.5502744

Mehrpouyan, H., Nasir, A., Blostein, S., Thomas, E., Karagiannidis, G., & Svensson, T. (2012). Joint estimation of channel and oscillator phase noise in MIMO systems. *IEEE Transactions on Signal Processing, 60*(9), 4790–4807. doi:10.1109/TSP.2012.2202652

Noels, N., Steendam, H., Moeneclaey, M., & Bruneel, H. (2005). Carrier phase and frequency estimation for pilot-symbol assisted transmission: Bounds and algorithms. *IEEE Transactions on Signal Processing, 53*(12), 4578–4587. doi:10.1109/TSP.2005.859318

Oestges, C., Clerckx, B., Guillaud, M., & Debbah, M. (2008). Dual-polarized wireless communications: From propagation models to system performance evaluation. *IEEE Transactions on Wireless Communications, 7*(10), 4019–4031. doi:10.1109/T-WC.2008.070540

Simeone, O., Stanojev, I., Savazzi, S., Bar-Ness, Y., Spagnolini, U., & Pickholtz, R. (2008). Spectrum leasing to cooperating secondary ad hoc networks. *Journal on Selected Areas in Communications, 26*(1), 203–213. doi:10.1109/JSAC.2008.080118

Simeone, O., Bar-Ness, Y., & Spagnolini, U. (2007). Stable throughput of cognitive radios with and without relaying capability. *IEEE Trans. Wirel. Commun., 55*(12), 2351-2360.

Sodagari, S., Attar, A., & Bilen, A. (2011). On a trustful mechanism for expiring spectrum sharing in cognitive radio networks. *IEEE Journal on Selected Areas in Communications, 29*(4), 856–856. doi:10.1109/JSAC.2011.110416

Su, W., Matyjas, J., & Batalama, S. (2012). Active cooperation between primary users and cognitive radio users in cognitive ad-hoc networks. *IEEE Transactions on Signal Processing, 60*(4), 1796–1805. doi:10.1109/TSP.2011.2181841

Wang, B., & Liu, K. (2011). Advances in cognitive radio networks: A survey. *IEEE Journal of Selected Topics in Signal Processing, 5*(1), 5–23. doi:10.1109/JSTSP.2010.2093210

Wang, J., Ghosh, M., & Challapali, K. (2011). Emerging cognitive radio applications: A survey. *IEEE Communications Magazine, 49*(3), 74–81. doi:10.1109/MCOM.2011.5723803

Wang, Y., & Falconer, D. (2010). Phase noise estimation and suppression for single carrier SDMA. In *Proc. IEEE Wireless Commun. Netw. Conf. (WCNC)*. IEEE. doi:10.1109/WCNC.2010.5506625

Xie, M., Zhang, W., & Wong, K. (2010). A geometric approach to improve spectrum efficiency for cognitive relay networks. *IEEE Trans. Wirel. Commun., 9*(1), 268-281.

Xu, H., & Li, B. (2010, May). *Efficient resource allocation with flexible channel cooperation in OFDMA cognitive radio networks*. Paper presented at IEEE INFOCOM. San Diego, CA. doi:10.1109/INFCOM.2010.5462169

Yi, Y., Zhang, J., Zhang, Q., Jiang, T., & Zhang, J. (2010). Cooperative communication-aware spectrum leasing in cognitive radio networks. In *Proc. DySPAN*. Academic Press. doi:10.1109/DYSPAN.2010.5457883

Zhang, J., & Zhang, Q. (2009, May). *Stackelberg game for utility-based cooperative cognitive radio networks*. Paper presented at ACM MobiHoc. New Orleans, LA.

Zhang, Q., Cao, B., Wang, Y., Lin, X., Sun, L., & Zhang, N. (2013). On exploiting polarization for energy-harvesting enabled cooperative cognitive radio networking. *IEEE Wireless Communications, 20*(4), 116–124. doi:10.1109/MWC.2013.6590058

Zhang, Q., Jia, J., & Zhang, J. (2009). Cooperative relay to improve diversity in cognitive radio networks. *IEEE Commun. Magazine, 47*(2), 111-117.

Zhao, G., Yang, C., Li, D., & Soong, A. (2011). Power and channel allocation for cooperative relay in cognitive radio networks. *IEEE J. Sel. Top. Signal Proc., 5*(1), 151-158.

Zou, Y., Yao, Y., & Zheng, B. (2011). Cognitive transmissions with multiple relays in cognitive radio networks. *IEEE Trans. Wirel. Commun., 10*(2), 648-659.

KEY TERMS AND DEFINITIONS

CCRN: User cooperation enabled cognitive radio networks.

Cognitive Radios: Unlicensed wireless systems that use spectrum via dynamic fashions.

Cooperative Communications: Cooperation between wireless nodes to enhance performance.

Energy Efficiency: A factor that evaluates the power consumption performance of a system.

OPDA: An antenna that is composed by two orthogonal polarization elements.

Resource Allocation: A way to allocate resources, (e.g., power and band), in a system.

Spectrum Efficiency: A factor that evaluates the transmission rate of a given frequency band.

Compilation of References

49 CFR. (2004, May 18). *Title 49, Subtitle B – Other regulations relating to Transportation, Chapter XII – Transportation Security Administration, Department of Homeland Security, Subchapter B – Security rules for all modes of transportation, Part 1520 – Protection of sensitive security information.* Author.

49 CFR. (2011, October 1). *Title 49, Subtitle A – Office of the Secretary of Transportation, Part 15 – Protection of sensitive security information.* Author.

4G Americas. (2013). *Meeting the 1000x Challenge: The Need for Spectrum, Technology and Policy Innovation.* Author.

5 USC. (2013, August 9). *Title 5, Part I – The agencies generally, Chapter 5 – Administrative procedure, Subchapter II – Administrative procedure, Section 552 – Public information; agency rules, opinions, orders, records, and proceedings.* Author.

A Community Resource for Archiving Wireless Data at Dartmouth. (2012). *Data set of CBR and VoIP traffic measurements from the WiBro network in Seoul, Korea* [Data file]. Available from http://crawdad.cs.dartmouth.edu/~crawdad/kaist/wibro/

Aamodt, A., & Plaza, E. (1994). Case-based reasoning: Foundational issues, methodological variations, and system approaches. *AI Communications, 1,* 39–59.

Abadie, A., & Wijesekera, D. (2013). An approach for risk assessment and mitigation in cognitive radio technologies. *International Journal of Information Privacy, Security and Integrity, 1*(4), 344–359.

Abbagnale, A., & Cuomo, F. (2010). Gymkhana: A connectivity-based routing scheme for cognitive radio ad hoc networks. In *Proceedings of IEEE International Conference on Computer Communications Workshops (INFOCOM),* (pp. 1-5). IEEE. doi:10.1109/INFCOMW.2010.5466618

Abbosh, A. M., & Bialkowski, M. E. (2008). Design of Ultra wideband Planar Monopole Antennas of Circular and Elliptical Shape. *IEEE Transactions on Antennas and Propagation, 56*(1), 17–23. doi:10.1109/TAP.2007.912946

Abdelaziza, S., & ElNainay, M. (2012). *Survey of routing protocols in cognitive radio networks.* Academic Press.

Abdelmonem, M. A., Nafie, M., Ismail, M. H., & El-Soudani, M. S. (2012). Optimized spectrum sensing algorithms for cognitive LTE femtocells. *EURASIP Journal on Wireless Communications and Networking, 2012*(1), 6. doi:10.1186/1687-1499-2012-6

Abdi, Y., & Ristaniemi, T. (2013). Joint reporting and linear fusion optimization in Collaborative Spectrum Sensing for cognitive radio networks. In *Proceedings of 9th International Conference on Information, Communications and Signal Processing (ICICS).* Academic Press. doi:10.1109/ICICS.2013.6782816

Abdul-Haleem, M., Cheung, K. F., & Chuang, J. (1994). Fuzzy logic based dynamic channel assignment. *IEEE ICCS, 2,* 773–777. doi:10.1109/ICCS.1994.474159

Abidi, A. A. (2007). The path to the software-defined radio receiver. *IEEE Journal of Solid-State Circuits, 42*(5), 954–966. doi:10.1109/JSSC.2007.894307

Abidin, Z., Xiao, P., Amin, M., & Fusco, V. (2012). Circular Polarization Modulation for digital communication systems. In *Proceedings of 8th International Symposium on Communication Systems, Networks & Digital Signal Processing (CSNDSP)*, (pp. 1-6). CSNDSP. doi:10.1109/CSNDSP.2012.6292758

Aboufoul, T., & Alomainy, A. (2011). Reconfigurable printed UWB circular disc monopole antenna. In *Proceedings of Loughborough Antennas and Propagation Conference* (pp. 1-4). Academic Press. doi:10.1109/LAPC.2011.6114078

Aboufoul, T., Parini, C., Chen, X., & Alomainy, A. (2013). Pattern-Reconfigurable Planar Circular Ultra-Wideband Monopole Antenna. *IEEE Transactions on Antennas and Propagation*, *61*(10), 4973–4980. doi:10.1109/TAP.2013.2274262

Abu Tarboush, H.F., Khan, S., Nilavalan, R., Al-Raweshidy, H.S., & Budimir, D. (2009). Reconfigurable wideband patch antenna for cognitive radio. In *Proceedings of Loughborough Antennas & Propagation Conference (LAPC 2009)*, (pp. 141 – 144). LAPC.

Abutarboush, H. F., Nilavalan, R., Cheung, S. W., & Nasr, K. M. (2012). Compact printed multiband antenna with independent setting suitable for fixed and reconfigurable wireless communication systems. *IEEE Transactions on Antennas and Propagation*, *60*(8), 3867–3874. doi:10.1109/TAP.2012.2201108

Abutarboush, H. F., Nilavalan, R., Cheung, S. W., Nasr, K. M., Peter, T., Budimir, D., & Al-Raweshidy, H. (2012). A Reconfigurable Wideband and Multiband Antenna Using Dual-Patch Elements for Compact Wireless Devices. *IEEE Transactions on Antennas and Propagation*, *60*(1), 36–43. doi:10.1109/TAP.2011.2167925

Abutarboush, H. F., Nilavalan, R., Nasr, K., Cheung, S. W., Peter, T., Al-Raweshidy, H. S., & Budimir, D. (2011). Reconfigurable tri-band h-shaped antenna with frequency selectivity feature for compact wireless communication systems, *IET Microwaves. Antennas Propagation*, *5*(14), 1675–1682.

Acampora, A., Collado, A., & Georgiadis, A. (2010). Nonlinear analysis and optimization of a distributed voltage controlled oscillator for cognitive radio. In *Proceedings of IEEE International Microwave Workshop Series on RF Front-Ends for Software Defined and Cognitive Radio Solutions (IMWS)*, (pp. 1-4). IEEE. doi:10.1109/IMWS.2010.5440987

Adawi, N. S. (1988). Coverage prediction for mobile radio systems operating in the 800/900 MHz frequency range. *Transactions on Vehicular Technology*, *37*(1), 3–72. doi:10.1109/25.42678

Adhikary, A., Ntranos, V., & Caire, G. (2011). Cognitive femtocells: Breaking the spatial reuse barrier of cellular systems. In *Proceedings of Information Theory and Applications Workshop*. La Jolla, CA: Academic Press. doi:10.1109/ITA.2011.5743563

Aduwo, A., & Annamalai, A. (2004). Channel-aware inter-cluster routing protocol for wireless ad-hoc networks exploiting network diversity. In *Proceedings of IEEE 60th Vehicular Technology Conference (VTC)*, (vol. 4, pp. 2858-2862). IEEE. doi:10.1109/VETECF.2004.1400581

Advanced Television Systems Committee. (2010). *ATSC Recommended Practice: Receiver Performance Guidelines* (A/74). Author.

Agarwala, S., Rajagopal, A., Hill, A., Joshi, M., Mullinnix, S., Anderson, T., ... Ryan, M. (2007). A 65nm C64x+ multi-core DSP platform for communications infrastructure. In *Proceedings of Solid-State Circuits Conference*, (pp. 262–601). Academic Press. doi:10.1109/ISSCC.2007.373394

Agilent SystemVue. (2014). Retrieved from http://www.agilent.com/find/eesof-systemvue

Agnelli, F., Albasini, G., Bietti, I., Gnudi, A., Lacaita, A., & Manstretta, D. et al. (2006). Wireless multi-standard terminals: System analysis and design of a reconfigurable RF front-end. *IEEE Circuits and Systems Magazine*, *6*(1), 38–59. doi:10.1109/MCAS.2006.1607637

Agrawal, D., & Abbadi, A. (1991). An efficient and fault-tolerant solution for distributed mutual exclusion. *Journal of ACM Transactions on Database Systems*, *9*(1), 1–20. doi:10.1145/103727.103728

Ahmed, M. A., Seddik, K. G., Sultan, A. K., ElBatt, T., & El-Sherif, A. A. (2013). A Feedback-Soft Sensing-Based Access Scheme for Cognitive Radio Networks. *IEEE Transactions on Wireless Communications*, *12*(7), 3226–3237. doi:10.1109/TWC.2013.061413.120851

Ahmed, M. E., Kim, J. S., Mao, R., Song, J. B., & Li, H. (2012). Distributed channel allocation using kernel density estimation in cognitive radio networks. *ETRI Journal*, *34*(5), 771–774. doi:10.4218/etrij.12.0211.0537

Ahmed, M. E., Song, J. B., Han, Z., & Suh, D. Y. (2014). Sensing-transmission edifice using nonparametric traffic clustering in cognitive radio networks. *IEEE Transactions on Mobile Computing*, *13*(9), 2141–2155. doi:10.1109/TMC.2013.156

Ahmed, M. E., Song, J. B., Nguyen, T., & Han, Z. (2012). Nonparametric Bayesian identification of primary users' payloads in cognitive radio networks. In *Proceedings of International Conference on Communications ICC 2012* (pp. 1586-1591). IEEE. doi:10.1109/ICC.2012.6364306

Akkarajitsakul, K., Hossain, E., Niyato, D., & Kim, D. I. (2011). Game theoretic approaches for multiple access in wireless networks: A survey. *IEEE Communications Surveys and Tutorials*, *13*(3), 372–395. doi:10.1109/SURV.2011.122310.000119

Akter, L., Natarajan, B., & Scoglio, C. (2009). Spectrum usage modeling and forecasting in cognitive radio networks. In Y. Xiao & F. Hu (Eds.), *Cognitive Radio Networks* (pp. 37–60). Boca Raton, FL: CRC Press Taylor & Francis Group. doi:10.1201/9781420064216.ch2

Akyildiz, I. F. et al., (2010). Flexible and spectrum aware radio access through measurements and modelling in cognitive radio systems. *FARAMIR (ICT-248351)*, Deliverable D2.1.

Akyildiz, I. F., Lee, W. Y., Vuran, M. C., & Mohanty, S. (2006). Next generation/dynamic spectrum access/cognitive radio wireless networks: A survey. *Computer Networks*, *50*(13), 2127–2159. doi:10.1016/j.comnet.2006.05.001

Akyildiz, I. F., Lee, W.-Y., & Chowdhury, K. R. (2009). CRAHNs: Cognitive radio ad hoc networks. *Ad Hoc Networks*, *7*(5), 810–836. doi:10.1016/j.adhoc.2009.01.001

Akyildiz, I. F., Lee, W.-Y., Vuran, M. C., & Mohanty, S. (2008). A Survey on Spectrum Management in Cognitive Radio Networks. *IEEE Communication Magazine*, *46*(4), 40–48. doi:10.1109/MCOM.2008.4481339

Akyildiz, I. F., Lo, B. F., & Balakrishnan, R. (2011). Cooperative spectrum sensing in cognitive radio networks: A survey. *Physical Communication*, *4*(1), 40–62. doi:10.1016/j.phycom.2010.12.003

Akyildiz, I. F., Wang, X., & Wang, W. (2005). Wireless mesh networks: A survey. *Computer Networks*, *47*(4), 445–487. doi:10.1016/j.comnet.2004.12.001

Alicherry, M., Bhatia, R., & Li, L. E. (2005). Joint channel assignment and routing for throughput optimization in multi-radio wireless mesh networks. In *Proceedings of the 11th Annual International Conference on Mobile Computing and Networking*, (pp. 58-72). Academic Press. doi:10.1145/1080829.1080836

Ali, M. S., & Mehta, N. B. (2014). Modeling time-varying aggregate interference in cognitive radio systems, and application to primary exclusive zone design. *IEEE Transactions on Wireless Communications*, *13*(1), 429–439. doi:10.1109/TWC.2013.113013.130762

Aljuaid, M., & Yanikomeroglu, H. (2010). A cumulant-based characterization of the aggregate interference power in wireless networks. In *Proceedings of IEEE Vehicular Technology Conference* (pp. 1–5). IEEE. doi:10.1109/VETECS.2010.5493638

Alnwaimi, G., Arshad, K., & Moessner, K. (2011). Dynamic spectrum allocation algorithm with interference management in co-existing networks. *IEEE Communications Letters, 15*(9), 932–934. doi:10.1109/LCOMM.2011.062911.110248

Alrabaees, S., Agarwal, A., Anand, D., & Khasawneh, M. (2012c, August). Game Theory for Security in Cognitive Radio Networks. In *Proceedings of Advances in Mobile Network, Communication and its Applications* (MNCAPPS), (pp. 60-63). IEEE.

Alrabaees, S., Agarwal, A., Goel, N., Zaman, M., & Khasawneh, M. (2012a, November). Comparison of spectrum management without game theory (smwg) and with game theory (smg) for network performance in cognitive radio network. In *Proceedings of the 2012 Seventh International Conference on Broadband, Wireless Computing, Communication and Applications* (pp. 348-355). IEEE Computer Society.

Alrabaees, S., Agarwal, A., Goel, N., Zaman, M., & Khasawneh, M. (2012d, November). Routing management algorithm based on spectrum trading and spectrum competition in cognitive radio networks. In *Proceedings of the 2012 Seventh International Conference on Broadband, Wireless Computing, Communication and Applications* (pp. 614-619). IEEE Computer Society.

Alrabaees, S., Agarwal, A., Goel, N., Zaman, M., & Khasawneh, M. (201b2, October). A game theoretic approach to spectrum management in cognitive radio network. In *Proceedings of Ultra Modern Telecommunications and Control Systems and Workshops* (ICUMT), (pp. 906-913). IEEE.

Alrabaees, S., Khasawneh, M., Agarwal, A., Goel, N., & Zaman, M. (2012e, December). A game theory approach: Dynamic behaviours for spectrum management in cognitive radio network. In *Proceedings of Globecom Workshops (GC Wkshps)*, (pp. 919-924). IEEE.

Alrabaee, S., Agarwal, A., Goel, N., Zaman, M., & Khasawneh, M. (2012, August). Higher layer issues in cognitive radio network. In *Proceedings of the International Conference on Advances in Computing, Communications and Informatics* (pp. 325-330). ACM. doi:10.1145/2345396.2345450

Al-Rawi, H. A., & Yau, K.-L. A. (2013). Routing in distributed cognitive radio networks: A survey. *Wireless Personal Communications, 69*(4), 1983–2020. doi:10.1007/s11277-012-0674-7

Altamini, M., Naik, K., & Shen, X. (2010). Parallel link rendezvous in ad hoc cognitive radio networks. In *Proceedings of IEEE Global Communications conference, Exhibition & Industry Forum (GLOBECOM)*. IEEE.

Altay, C., Yilmaz, H. B., & Tugcu, T. (2012). Cooperative sensing analysis under imperfect reporting channel. In *Proceedings of IEEE Symposium on Computers and Communications (ISCC)*. IEEE. doi:10.1109/ISCC.2012.6249392

Althunibat, S., & Granelli, F. (2013, June). Novel energy-efficient reporting scheme for spectrum sensing results in cognitive radio. In *Proceedings of Communications (ICC)*, (pp. 2438-2442). IEEE. doi:10.1109/ICC.2013.6654897

Althunibat, S., Di Renzo, M., & Granelli, F. (2013, Decmebr). Optimizing the K-out-of-N Fusion rule for Cooperative Spectrum Sensing in Cognitive Radio Networks. In *Proceedings of Global Telecommunications Conference (GLOBECOM 2013)*, (pp. 1-5). IEEE.

Althunibat, S., Narayanan, S., Di Renzo, M., & Granelli, F. (2012, September). On the Energy Consumption of the Decision-Fusion Rules in Cognitive Radio Networks. In *Proceedings of Computer Aided Modeling and Design of Communication Links and Networks (CAMAD)*, (pp. 125-129). IEEE. doi:10.1109/CAMAD.2012.6335312

Althunibat, S., Renzo, M. D., & Granelli, F. (2013). Optimizing the K-out-of-N rule for cooperative spectrum sensing in cognitive radio networks. In *Proceedings of International Conference on Global Communications (Globecom'13)*. IEEE. doi:10.1109/GLOCOM.2013.6831303

Altman, E., Avrachenkov, K., & Garnaev, A. (2011). Jamming in wireless networks under uncertainty. *Mobile Networks and Applications, 16*(2), 246–254. doi:10.1007/s11036-010-0272-4

Analog Devices, Inc. (2014). *Semiconductor and Signal Processing ICs*. AD9361 Datasheet. Retrieved from http://www.analog.com

Anand, S., Sengupta, S., & Chandramouli, R. (2012). MAximum SPECTrum packing: A distributed opportunistic channel acquisition mechanism in dynamic spectrum access networks. *IET Communications*, 6(8), 872–882. doi:10.1049/iet-com.2010.0607

Andrews, M., Mitra, P., & deCarvalho, R. (2001). Tripling the capacity of wireless communications using electromagnetic polarization. *Nature*, 409(1), 316–318. doi:10.1038/35053015 PMID:11201734

Appadwedula, S., Veeravalli, V. V., & Jones, D. L. (2008). Decentralized detection with censoring sensors. *IEEE Transactions on* Signal Processing, 56(4), 1362–1373. doi:10.1109/TSP.2007.909355

Arshad, K., Imran, M. A., & Moessner, K. (2010). Collaborative Spectrum Sensing Optimisation Algorithms for Cognitive Radio Networks. *International Journal of Digital Multimedia Broadcasting*, 2010, 1–20. doi:10.1155/2010/424036

Artreya, R., Mittal, N., & Peri, S. (2007). A Quorum-Based Group Mutual Exclusion Algorithm for a Distributed System with Dynamic Group Set. *IEEE Transactions on Parallel and Distributed Systems*, 18(10), 1–16.

Atapattu, S., Tellambura, C., & Jiang, H. (2010). Performance of an Energy Detector over Channels with Both Multipath Fading and Shadowing. *IEEE Transactions on Wireless Communications*, 3662-3670.

Atapattu, S., Tellambura, C., & Jiang, H. (2011). Energy detection based cooperative spectrum sensing in cognitive radio networks. *IEEE Transactions on Communications*, 10(4), 1232–1241.

ATM Forum Technical Committee. (1999). Traffic management specification version 4.1. *af-tm-0121.000*.

Attar, A., Nakhai, M. & Aghvami, A. (2009). Cognitive radio game for secondary spectrum access problem. *IEEE Trans. Wirel. Commu.*, 8(4), 2121-2131.

Auerbach, J., Bacon, D. F., Burcea, I., Cheng, P., Fink, S. J., Rabbah, R., & Shukla, S. (2012). A compiler and runtime for heterogeneous computing. In *Proceedings of the 49th Annual Design Automation Conference on - DAC '12* (p. 271). San Francisco, CA: DAC. doi:10.1145/2228360.2228411

Auerbach, J., Bacon, D. F., Cheng, P., & Rabbah, R. (2010). Lime: a Java-Compatible and Synthesizable Language for Heterogeneous Architectures. In *Proceedings of the ACM International Conference on Object Oriented Programming Systems Languages and Applications - OOPSLA '10* (pp. 89). Reno, NV: ACM. doi:10.1145/1869459.1869469

Aumann, R. J., & Peleg, B. (1960). Von neumann-morgenstern solutions to cooperative games without side payments. *Bulletin of the American Mathematical Society*, 6(3), 173–179. doi:10.1090/S0002-9904-1960-10418-1

Avago Technologies. (2009). *Surface Mount Zero Bias Schottky Detector Diodes*. HSMS-282x Series Datasheet. Retrieved from http://www.avagotech.com

Avago. (2006). *HPND-4005 Beam Lead PIN Diode*. Retrieved January 16, 2014, from http://www.avagotech.com/docs/AV01-0593EN

Axell, E., Leus, G., & Larsson, E. G. (2010). Overview of spectrum sensing for cognitive radio. In *Proceedings of 2nd International Workshop on Cognitive Information Processing (CIP)*, (pp. 322-327). Academic Press. doi:10.1109/CIP.2010.5604136

Axell, E., Leus, G., Larsson, E., & Poor, H. (2012). Spectrum sensing for cognitive radio: State-of-the-art and recent advances. *IEEE Signal Processing Magazine*, 29(3), 101–116. doi:10.1109/MSP.2012.2183771

Babaei, A., & Jabbari, B. (2010). Interference modeling and avoidance in spectrum underlay cognitive wireless networks. In *Proceedings of IEEE International Conference on Communications* (pp. 1–5). IEEE. doi:10.1109/ICC.2010.5501850

Badarneh, O. S., & Salameh, H. B. (2011). Opportunistic routing in cognitive radio networks: exploiting spectrum availability and rich channel diversity. In *Proceedings of IEEE Global Telecommunications Conference (GLOBECOM)*, (pp. 1-5). IEEE. doi:10.1109/GLOCOM.2011.6134241

Badoi, C.-I., Croitoru, V., & Prasad, R. (2010). IPSAG: an IP spectrum aware geographic routing algorithm proposal for multi-hop cognitive radio networks. In *Proceedings of IEEE 8th International Conference on Communications (COMM)*, (pp. 491-496). IEEE. doi:10.1109/ICCOMM.2010.5509020

Bae, Y. H., Alfa, A. S., & Choi, B. D. (2009). Performance analysis of modified IEEE 802.11-based cognitive radio networks. *IEEE Communications Letters*, *14*(10), 975–977. doi:10.1109/LCOMM.2010.082310.100322

Bagheri, H., Pakravan, M. R., & Khalaj, B. H. (2004). Iterative multi-user power allocation: performance evaluation & modification. In Proceedings of IEEE Region 10 (pp. 216–219). TENCON. doi:10.1109/TEN-CON.2004.1414746

Bahl, P., Chandra, R., & Dunagan, J. (2004). SSCH: Slotted Seeded Channel Hopping for Capacity Improvement in IEEE 802.11 Ad Hoc Wireless Networks. In *Proceedings of International Conference on Mobile Computing and Networking (MobiCom)* (pp. 216-230). ACM. doi:10.1145/1023720.1023742

Balachandran, K., & Kang, J. (2006). Neighbor discovery with dynamic spectrum access in adhoc networks. In *Proceedings of IEEE 63rd Vehicular Technology Conference,* (vol. 2, pp. 512–517). IEEE.

Balasubramanian, S., Boumaiza, S., Sarbishaei, H., Quach, T., Orlando, P., & Volakis, J. et al. (2012). Ultimate transmission. *IEEE Microwave Magazine*, *13*(1), 64–82. doi:10.1109/MMM.2011.2173983

Balasubramanyn, V. B., Thamilarasu, G., & Sridhar, R. (2007, June). Security solution for data integrity in wireless biosensor networks. In *Proceedings of Distributed Computing Systems Workshops,* (pp. 79-79). IEEE.

Baldo, N., & Zorzi, M. (2008). Fuzzy logic for cross-layer optimization in cognitive radio networks. *IEEE Communications Magazine*, *46*(4), 64–71. doi:10.1109/MCOM.2008.4481342

Baldo, N., & Zorzi, M. (2009). Cognitive network access using fuzzy decision making. *IEEE Transactions on Wireless Communications*, *8*(7), 3523–3535. doi:10.1109/TWC.2009.071103

Banerjee, J.S., & Karmakar, K. (2012). A Comparative Study on Cognitive Radio Implementation Issues. *International Journal of Computer Applications*, *45*(15), 44-51.

Banerjee, J. S., & Chakraborty, A. (2014). Modeling of Software Defined Radio Architecture & Cognitive Radio, the Next Generation Dynamic and Smart Spectrum Access Technology. In M. H. Rehmani & Y. Faheem (Eds.), *Cognitive Radio Sensor Networks: Applications, Architectures, and Challenges* (pp. 127–158). Hershey, PA: IGI Global. doi:10.4018/978-1-4666-6212-4.ch006

Banerjee, J. S., Chakraborty, A., & Karmakar, K. (2013). Architecture of Cognitive Radio Networks. In N. Meghanathan & Y. B. Reddy (Eds.), *Cognitive Radio Technology Applications for Wireless and Mobile Ad Hoc Networks* (pp. 125–152). Hershey, PA: IGI Global. doi:10.4018/978-1-4666-4221-8.ch007

Banković, Z., Stepanović, D., Bojanić, S., & Nieto-Taladriz, O. (2007). Improving network security using genetic algorithm approach. *Computers & Electrical Engineering*, *33*(5), 438–451. doi:10.1016/j.compeleceng.2007.05.010

Ban, T. W., Choi, W., Jung, B. C., & Sung, D. K. (2009). Multi-user diversity in a spectrum sharing system. *IEEE Transactions on Wireless Communications*, *8*(1), 102–106. doi:10.1109/T-WC.2009.080326

Bantouna, A., Stavroulaki, V., Kritikou, Y., Tsagkaris, K., Demestichas, P., & Moessner, K. (2012). An overview of learning mechanisms for cognitive systems. *EURASIP Journal on Wireless Communications and Networking, 22.*

Baraniuk, R. G. (2007, July). Compressive Sensing. *Signal Processing Magazine, IEEE, 24*(4), 118–121. doi:10.1109/MSP.2007.4286571

Baraniuk, R. G., Cevher, V., Duarte, M. F., & Hegde, C. (2010, April). Model-Based Compressive Sensing. *IEEE Transactions on* Information Theory, *56*, 1982–2001.

Barcelo, J., Bellalta, B., & Oliver, M. (2011). Learning-BEB: Avoiding collisions in WLAN. *Other IFIP Publications,* (1).

Baron, D., Sarvotham, S., & Baraniuk, R. G. (2010, January). Bayesian Compressive Sensing Via Belief Propagation. *IEEE Transactions on* Signal Processing, *58*(1), 269–280. doi:10.1109/TSP.2009.2027773

Bartia, P., & Bahl, I. (1982). Frequency Agile Microstrip Antennas. *Microwave Journal, 37*, 1136–1139.

Bartoli, G., Tassi, A., Marabissi, D., Tarchi, D., & Fantacci, R. (2011). An Optimized Resource Allocation Scheme Based on a Multidimensional Multiple-Choice Approach with Reduced Complexity. *Proceedings of the IEEE, ICC*, 2011.

Basar, T., & Olsder, G. J. (1999). *Dynamic Noncooperative Game Theory (Series in Classics in Applied Mathematics)*. Philadelphia, PA: SIAM.

Bayat, S., Louie, R. H. Y., Li, Y., & Vucetic, B. (2011). Cognitive Radio Relay Networks with Multiple Primary and Secondary Users: Distributed Stable Matching Algorithms for Spectrum Access. In *Proceedings of IEEE ICC 2011*, (pp. 1-6). IEEE.

Bayhan, S., & Alagöz, F. (2014). A Markovian approach for best-fit channel selection in cognitive radio networks. *Ad Hoc Networks, 12*, 165–177. doi:10.1016/j.adhoc.2011.08.007

Bazerque, J. A., & Giannakis, G. B. (2010). Distributed spectrum sensing for cognitive radio networks by exploiting sparsity. *IEEE Transactions on* Signal Processing, *58*(3), 1847–1862. doi:10.1109/TSP.2009.2038417

Behdad, N., & Sarabandi, K. (2006). Dual-band Reconfigurable Antenna with a very wide Tunability Range. *IEEE Transactions on Antennas and Propagation, 54*(2), 409–416. doi:10.1109/TAP.2005.863412

Bellasi, D. E., Bettini, L., Benkeser, C., Burger, T., Qiuting, H., & Studer, C. (2013). VLSI Design of a Monolithic Compressive-Sensing Wideband Analog-to-Information Converter. *IEEE Journal on* Emerging and Selected Topics in Circuits and Systems, *3*(4), 552–565.

Ben Abdallah, R., Risset, T., Fraboulet, A., & Martin, J. (2010). Virtual Machine for Software Defined Radio: Evaluating the Software VM Approach. In *Proceedings of 2010 10th IEEE International Conference on Computer and Information Technology* (pp. 1970–1977). Bradford, UK: IEEE.

Ben Letaief, K., & Wei Zhang,. (2009). Cooperative Communications for Cognitive Radio Networks. *Proceedings of the IEEE, 97*(5), 878–893. doi:10.1109/JPROC.2009.2015716

Benidris, F. Z., Benmammar, B., & Bendimerad, F. T. (2012). Comparative studies of artificial intelligence techniques in the context of cognitive radio. In *Proceedings of the International Conference on Multimedia Information Processing*. Mascara, Algeria: Academic Press.

Berry, R. A. (2012). Network market design part II: Spectrum markets. *IEEE Communications Magazine, 50*(11), 84–90. doi:10.1109/MCOM.2012.6353687

Berry, R., Honig, M. L., & Vohra, R. (2010). Spectrum markets: Motivation, challenges and implications. *IEEE Communications Magazine, 48*(11), 146–155. doi:10.1109/MCOM.2010.5621982

Bhargavaand, V., & Hussain, E. (Eds.). (2007). *Cognitive Radio Networks*. Springer-Verlag.

Bhatnagar, M. R., Hjørungnes, A., & Debbah, M. (2010). Delay-tolerant decode-and-forward based cooperative communication over Ricean channels. *IEEE Transactions on Wireless Communications, 9*(4), 1277–1282. doi:10.1109/TWC.2010.04.090343

Bhatti, F. A., Rowe, G. B., & Sowerby, K. W. (2012). Spectrum sensing using principal component analysis. In *Proceedings of IEEE Wireless Communications and Networking Conference*, (pp. 725 –730). IEEE.

Bian, K., & Park, J.-M. (2011). Asynchronous channel hopping for establishing rendezvous in cognitive radio networks. In *Proceedings of IEEE Conference on Computer Communications (INFOCOM)*. IEEE. doi:10.1109/INFCOM.2011.5935056

Bian, K., Park, J. M., & Chen, R. (2009). A quorum based framework for establishing control channels in dynamic spectrum access networks. In *Proceedings of ACM International Conference on Mobile Computing and Networking (MobiCom)*. ACM. doi:10.1145/1614320.1614324

Bian, K., Park, J. M., & Chen, R. (2011). Control channel establishment in cognitive radio networks using channel hopping. *IEEE Journal on* Selected Areas in Communications, *29*(4), 689–703.

Bian, K., & Park, J.-M. (2012). Maximizing Rendezvous Diversity in Rendezvous Protocols for Decentralized Cognitive Radio Networks. *IEEE Transactions on Mobile Computing, 12*(7), 1294–1307. doi:10.1109/TMC.2012.103

Bian, K., Park, J.-M., & Chen, R. (2011). Control channel establishment in cognitive radio networks using channel hopping. *IEEE Journal on Selected Areas in Communications, 29*(4), 689–703. doi:10.1109/JSAC.2011.110403

Bilsen, G., Engels, M., Lauwereins, R., & Peperstraete, J. (1996). Cyclo-Static Dataflow. *IEEE Transactions on Signal Processing, 44*(2), 397–408. doi:10.1109/78.485935

Bkassiny, M., Jayaweera, S. K., Li, Y., & Avery, K. A. (2012). Wideband spectrum sensing and non-parametric signal classification for autonomous self-learning cognitive radios. *IEEE Transactions on Wireless Communications, 11*(7), 2596–2605. doi:10.1109/TWC.2012.051512.111504

Blanquero, R., & Carrizosa, E. (2000). On covering methods for d.c. optimization. *Journal of Global Optimization, 18*(3), 265–274. doi:10.1023/A:1008366808825

Blei, D., & Jordan, M. (2006). Variational inference for dirichlet process mixtures. *Bayesian Analysis, 1*(1), 121–144. doi:10.1214/06-BA104

Bogomolnaia, A., & Jackson, M. O. (2002). The stability of hedonic coalition structures. *Games and Economic Behavior, 38*(2), 201–230. doi:10.1006/game.2001.0877

Bonfiglio, D., Mellia, M., Meo, M., Rossi, D., & Tofanelli, P. (2007, August). Revealing skype traffic: When randomness plays with you. *ACM Computer Communication Review, 37*(4), 37–48. doi:10.1145/1282427.1282386

Bougard, B., De Sutter, B., Verkest, D., Van der Perre, L., & Lauwereins, R. (2008). A Coarse-Grained Array Accelerator for Software-Defined Radio Baseband Processing. *IEEE Micro, 28*(4), 41–50. doi:10.1109/MM.2008.49

Boulet, P. (2007). *Array-OL revisited, multidimensional intensive signal processing specification* (INRIA research report num 6113). Academic Press.

Boyd, S., & Vandenberghe, L. (2004). *Convex Optimization*. Academic Press.

Boyd, S., & Vandenberghe, L. (2004). *Convex optimization*. Cambridge University Press. doi:10.1017/CBO9780511804441

Broadband Commission, ITU. (2012). *The State of Broadband 2012: Achieving Digital Inclusion for All*. Author.

Buchwald, G. J., Kuffner, S. L., Ecklund, L. M., Brown, M., & Callaway, E. H. (2008). The design and operation of the IEEE 802.22.1 disabling beacon for the protection of TV whitespace incumbents. In *Proceedings of IEEE Symposium on New Frontiers in Dynamic Spectrum Access Networks* (pp. 1–6). IEEE. doi:10.1109/DYSPAN.2008.55

Buck, J. T. (1994). A dynamic dataflow model suitable for efficient mixed hardware and software implementations of DSP applications. In *Proceedings of Third International Workshop on Hardware/Software Codesign* (pp. 165–172). Grenoble, France: Academic Press. doi:10.1109/HSC.1994.336710

Buck, I., Foley, T., Horn, D., Sugerman, J., Fatahalian, K., Houston, M., & Hanrahan, P. (2006). Sequoia: Programming the Memory Hierarchy. In *Proceedings of the 2006 ACM/IEEE conference on Supercomputing* (p. 83). Tampa, FL: ACM.

Buddhikot, M. M. (2007, April). Understanding Dynamic Spectrum Access: Models, Taxonomy and Challenges. In *Proceedings of 2007 IEEE International Symposium on Dynamic Spectrum Access Networks*. Dublin, Ireland: IEEE. doi:10.1109/DYSPAN.2007.88

Buddhikot, M. M., Kolodzy, P., Miller, S., Ryan, K., & Evans, J. (2005, June). DIMSUMnet: New Directions in Wireless Networking Using Coordinated Dynamic Spectrum Access. In *Proceedings of 2005 IEEE International Symposium on a World of Wireless Mobile and Multimedia Networks (WOWMOM)*. IEEE. doi:10.1109/WOWMOM.2005.36

Buddhikot, M. M., & Ryan, K. (2005, November). Spectrum Management in Coordinated Dynamic Spectrum Access Based Cellular Networks. In *Proceedings of IEEE Dynamic Spectrum Access Networks (DYSPAN 2005)*. Baltimore, MD: IEEE. doi:10.1109/DYSPAN.2005.1542646

Burbank, J. L. (2008, May). Security in cognitive radio networks: The required evolution in approaches to wireless network security. In *Proceedings of Cognitive Radio Oriented Wireless Networks and Communications*, (pp. 1-7). IEEE.

Butt, M. A. (2013). Cognitive radio network: Security enhancements. *Journal of Global Research in Computer Science, 4*(2), 36–41.

Cabric, D., & Brodersen, R. W. (2005). Physical layer design issues unique to cognitive radio systems. In *Proceedings of IEEE 16th International Symposium on Personal, Indoor and Mobile Radio Communications (PIMRC),* (vol. 2, pp. 759-763). IEEE.

Cabric, D., Mishra, S. M., Willkomm, D., Brodersen, R., & Wolisz, A. (2005). A Cognitive Radio Approach for Usage of Virtual Unlicensed Spectrum. In Proceedings of 14th IST Mobile Wireless Communications Summit. Dresden, Germany: IST.

Cabric, D., Tkachenko, A., & Brodersen, R. W. (2006). Spectrum sensing measurements of pilot, energy, and collaborative detection. In *Proceedings of IEEE Military Communications Conference,* (pp. 1-7). IEEE. doi:10.1109/MILCOM.2006.301994

Cabric, S. D., Mishra, S. M., & Brodersen, R. W. (2004). Implementation issues in spectrum sensing for cognitive radios. In *Proceedings of the Asilomar Conference on Signals, Systems, and Computers,* (Vol. 1, pp. 772-776). Academic Press. doi:10.1109/ACSSC.2004.1399240

Cabric, D., Tkachenko, A., & Brodersen, R. W. (2006). Experimental study of spectrum sensing based on energy detection and network. In *Proc. of ACM TAPAS.* Boston, MA: ACM. doi:10.1145/1234388.1234400

Cagatay Talay, A., & Altilar, D. T. (2009). ROPCORN: Routing protocol for cognitive radio ad hoc networks. In *Proceedings of IEEE International Conference on Ultra Modern Telecommunications & Workshops (ICUMT),* (pp. 1-6). IEEE.

Cao, B. (2013). *On frameworks and techniques of cooperation between primary and secondary users enabled dynamic spectrum access.* (Unpublished doctoral dissertation). Harbin Institute of Technology, China.

Cao, B., Cai, L. X., Liang, H., Mark, J. W., Zhang, Q., Poor, H. V., & Zhuang, W. (2012). Cooperative cognitive radio networking using quadrature signaling. Paper presented in IEEE INFOCOM. Orlando, FL.

Cao, B., Mark, J. W., & Zhang, Q. (2012, December). *A polarization enabled cooperation framework for cognitive radio networking.* Paper presented at IEEE GLOBECOM. Anaheim, CA. doi:10.1109/GLOCOM.2012.6503262

Cao, B., Mark, J. W., Zhang, Q., Lu, R., Lin, X., & Shen, X. (2013, April). *On optimal communication strategies for cooperative cognitive radio networking.* Paper presented at IEEE INFOCOM. Turin, Italy. doi:10.1109/INFCOM.2013.6566970

Cao, B., Zhang, Q., Liang, D., Wen, S., Lin, J., & Zhang, Y. (2010, May). *Blind adaptive polarization filtering based on oblique projection.* Paper presented at IEEE ICC. Cape Town, South Africa. doi:10.1109/ICC.2010.5501769

Cao, L., & Zheng, H. (2007). On the Efficiency and Complexity of Distributed Spectrum Allocation. In *Proceedings of Cognitive Radio Oriented Wireless Networks and Communications,* (pp. 357-366). CROWNCOM.

Cao, B., Liang, H., Mark, J. W., & Zhang, Q. (2013). Exploiting orthogonally dual-polarized antennas in cooperative cognitive radio networking. *IEEE Journal on Selected Areas in Communications, 31*(11), 2362–2373. doi:10.1109/JSAC.2013.13111

Cao, B., Yu, J., Wang, Y., & Zhang, Q. (2013). Enabling polarization filtering in wireless communications: Models, algorithms, and characteristics. *IET Communications, 4*(3), 247–254. doi:10.1049/iet-com.2012.0409

Cao, B., Zhang, Q., Lu, R., & Mark, J. W. (2014). PEACE: Polarization enabled active cooperation scheme between primary and secondary networks. *IEEE Transactions on Vehicular Technology, 63*(8), 3677–3688. doi:10.1109/TVT.2014.2305713

Cao, B., Zhang, Q., Mark, J. W., Cai, L. X., & Poor, H. V. (2012). Toward efficient radio spectrum utilization: User cooperation in cognitive radio networking. *IEEE Network, 26*(4), 46–52. doi:10.1109/MNET.2012.6246752

Cao, D. Y., & Cheng, J. X. (2010). A genetic algorithm based on modified selection operator and crossover operator. *Computer Technology and Development, 20*(2), 44–47.

Cao, L., & Zheng, H. (2005). Distributed spectrum allocation via local bargaining. In *Proc. IEEE SECON 2005,* (pp. 475-486). IEEE.

Cao, W., Zhang, B., Liu, A., Yu, T., Guo, D., & Pan, K. (2012). A reconfigurable microstrip antenna with radiation pattern selectivity and polarization diversity. *IEEE Antennas and Wireless Propagation Letters, 11,* 453–456. doi:10.1109/LAWP.2012.2193549

Cao, Y., Jiang, T., Wang, C., & Zhang, L. (2012, October). CRAC: Cognitive radio assisted cooperation for downlink transmissions in OFDMA-based cellular networks. *IEEE Journal on Selected Areas in Communications, 30*(9), 1614–1622. doi:10.1109/JSAC.2012.121004

Cao, Y., Li, Y., & Ye, F. (2011, December). Improved Fair Spectrum Allocation Algorithms Based on Graph Coloring Theory in Cognitive Radio Networks. *Journal of Computer Information Systems, 7*(13), 4694–4701.

Cardoso, L. S., Debbah, M., Bianchi, P., & Najim, J. (2008, May). Cooperative spectrum sensing using random matrix theory. In *Proceedings of Wireless Pervasive Computing,* (pp. 334-338). IEEE. doi:10.1109/ISWPC.2008.4556225

Carey-Smith, B. E., Warr, P. A., Rogers, P. R., Beach, M. A., & Hilton, G. S. (2005). Flexible frequency discrimination subsystems for reconfigurable radio front-ends. *EURASIP Journal on Wireless Communications and Networking, 2005*(3), 372196. doi:10.1155/WCN.2005.354

Cassandras, C., & Li, W. (2005). Sensor Networks and Cooperative Control. *European Journal of Control, 11*(4-5), 436–463. doi:10.3166/ejc.11.436-463

Castrillon, J., Leupers, R., & Ascheid, G. (2013). MAPS: Mapping Concurrent Dataflow Applications to Heterogeneous MPSoCs. *IEEE Transactions on Industrial Informatics, 9*(1), 527–545. doi:10.1109/TII.2011.2173941

Castrillon, J., Schürmans, S., Stulova, A., Sheng, W., Kempf, T., & Leupers, R. et al. (2011). Component-based waveform development: The Nucleus tool flow for efficient and portable software defined radio. *Analog Integrated Circuits and Signal Processing, 69*(2-3), 173–190. doi:10.1007/s10470-011-9670-1

Cattell, R. (1966). The scree test for the number of factors. *Multivariate Behavioral Research, 1*(2), 245–276. doi:10.1207/s15327906mbr0102_10

Cendrillon, R., & Moonen, M. (2005). Iterative spectrum balancing for digital subscriber lines. In *Proceedings of IEEE International Conference on Communications* (pp. 1937–1941). IEEE. doi:10.1109/ICC.2005.1494677

Cendrillon, R., Huang, J., Chiang, M., & Moonen, M. (2007). Autonomous spectrum balancing for digital subscriber lines. *IEEE Transactions on Signal Processing, 55*(8), 4241–4257. doi:10.1109/TSP.2007.895989

Cendrillon, R., Yu, W., Moonen, M., Verlinden, J., & Bostoen, T. (2006). Optimal multiuser spectrum balancing for digital subscriber lines. *IEEE Transactions on Communications, 54*(5), 922–933. doi:10.1109/TCOMM.2006.873096

Chakraborty, A., & Banerjee, J. S. (2013). An Advance Q Learning (AQL) Approach for Path Planning and Obstacle Avoidance of a Mobile Robot. *International Journal of Intelligent Mechatronics and Robotics, 3*(1), 53–73. doi:10.4018/ijimr.2013010105

Chang, C. S. (2000). *Performance guarantees in communication networks.* London: Springer. doi:10.1007/978-1-4471-0459-9

Chang, G.-Y., Teng, W. H., Chen, H.-Y., & Sheu, J.-P. (2013). Novel Channel-Hopping Schemes for Cognitive Radio Network. *IEEE Transactions on Mobile Computing, 13*(2), 407–421. doi:10.1109/TMC.2012.260

Chao, C.-M., & Fu, H.-Y. (2013). Providing Complete Rendezvous Guarantee for Cognitive Radio Networks by Quorum Systems and Latin Squares. In *Proceedings of IEEE Wireless Communications and Networking Conference (WCNC).* IEEE. doi:10.1109/WCNC.2013.6554545

Chapin, J. M., & Lehr, W. H. (2007, May). The Path to Market Success for Dynamic Spectrum Access Technology. *IEEE Communications Magazine, 45*(5), 96–103. doi:10.1109/MCOM.2007.358855

Chatzinotas, S., Christopolous, D., & Ottersten, B. (2011). Coordinated multipoint decoding with dual-polarized antennas. In *Proceedings of Wireless Communications and Mobile Computing Conference (IWCMC),* (pp. 157-161). IWCMC.

Chaudhari, S., & Koivunen, V. (2013). Impact of reporting-channel coding on the performance of distributed sequential sensing. In *Proceeding of IEEE 14th Workshop on Signal Processing Advances in Wireless Communications (SPAWC).* IEEE.

Chaudhari, S., Lunden, J., & Koivunen, V. (2012). Effects of quantization and channel errors on sequential detection in cognitive radios. In *Proceeding of 46th Annual Conference on Information Sciences and Systems (CISS)*. CISS.

Chaudhari, S., & Lunden, J., & Koivunen, V. (2012). Cooperative Sensing with Imperfect Reporting Channels: Hard Decisions or Soft Decisions. *IEEE Transactions on Signal Processing*. doi:10.1109/TSP.2011.2170978

Chaudhari, S., Lunden, J., & Koivunen, V. (2011). BEP walls for collaborative spectrum sensing. In *Proceedings of International Conference on Acoustic, Speech, Signal Processing (ICASSP'11)*. Prague, Czech Republic: IEEE.

Chehata, A., Ajib, W., & Elbiaze, H. (2011). An on-demand routing protocol for multi-hop multi-radio multi-channel cognitive radio networks. In Proceedings of IEEE Wireless Days (WD) (pp. 1–5). IEEE. doi:10.1109/WD.2011.6098215

Chen, P.-Y., Cheng, S.-M., & Ao, W. C. (2011). Multi-path routing with end-to-end statistical QoS provisioning in underlay cognitive radio networks. In *Proceedings of IEEE International Conference on Computer Communications Workshops (INFOCOM WKSHPS)*, (pp. 7-12). IEEE.

Chen, S., Newman, T. R., Evans, J. B., & Wyglinski, A. M. (2010, April). Genetic algorithm-based optimization for cognitive radio networks. In *Proceedings of Sarnoff Symposium*, (pp. 1-6). IEEE. doi:10.1109/SARNOF.2010.5469780

Chen, Y., & Liu, K. (2011). Indirect reciprocity game modelling for cooperation stimulation in cognitive networks. *IEEE Trans. Wirel. Commu. 59*(1), 159-168.

Chen, Z., Wang, C.-X., Hong, X., Thompson, J., Vorobyov, S., & Ge, X. (2010). Interference modeling for cognitive radio networks with power or contention control. In *Proceedings of IEEE Wireless Communications and Networking Conference* (pp. 1–6). IEEE. doi:10.1109/WCNC.2010.5506443

Chen, C., Cheng, H., & Yao, Y. D. (2011). Cooperative spectrum sensing in cognitive radio networks in the presence of the primary user emulation attack. *IEEE Transactions on Wireless Communications, 10*(7), 2135–2141.

Cheng, G., Liu, W., Li, Y., & Cheng, W. (2007). Joint on-demand routing and spectrum assignment in cognitive radio networks. In *Proceedings of IEEE International Conference on Communications (ICC)*, (pp. 6499-6503). IEEE. doi:10.1109/ICC.2007.1075

Cheng, H. T., & Zhuang, W. (2011). Simple channel sensing order in cognitive radio networks. *IEEE Journal on Selected Areas in Communications, 29*(4), 676–688.

Cheng, N., Zhang, N., Lu, N., Shen, X. W., Mark, J., & Liu, F. (2014). Opportunistic spectrum access for CR-VANETs: A game-theoretic approach. *IEEE Transactions on Vehicular Technology, 63*(1), 237–251. doi:10.1109/TVT.2013.2274201

Cheng, P., Deng, R., & Chen, J. (2012). Energy-efficient cooperative spectrum sensing in sensor-aided cognitive radio networks. *IEEE Wireless Communications, 19*(6), 100–105. doi:10.1109/MWC.2012.6393524

Chen, J., Li, Z., Si, J., Huang, H., & Gao, R. (2012). Outage probability analysis of partial AF relay selection in spectrum-sharing scenario with multiple primary users. *Electronics Letters, 48*(19), 1211–1212. doi:10.1049/el.2012.2217

Chen, R. R., Teo, K. H., & Farhang-Boroujeny, B. (2011). Random access protocols for collaborative spectrum sensing in multi-band cognitive radio networks. *IEEE Journal on Selected Areas in Communications, 5*(1), 124–136.

Chen, R., Park, J. M., Hou, Y. T., & Reed, J. H. (2008). Toward secure distributed spectrum sensing in cognitive radio networks. *IEEE Communications Magazine, 46*(4), 50–55. doi:10.1109/MCOM.2008.4481340

Chen, R., Park, J. M., & Reed, J. H. (2008). Defense against primary user emulation attacks in cognitive radio networks. *IEEE Journal on* Selected Areas in Communications, 26(1), 25–37.

Chen, S., & Wyglinski, A. (2009, December). Efficient spectrum utilization via cross-layer optimization in distributed cognitive radio networks. *Computer Communications, 18*(18), 1931–1943. doi:10.1016/j.comcom.2009.05.016

Chen, Y. (2010). Improved energy detector for random signals in Gaussian noise. *IEEE Transactions on Wireless Communications*, 9(2), 558–563. doi:10.1109/TWC.2010.5403535

Chen, Y. M., & Tsai, P.-S. (2012). Opportunistic spectrum access decision operation in cognitive radio femtocell networks using fuzzy logic system. *International Journal of Computational Intelligence and Information Security*, 3(3), 11.

Chen, Y. S., & Hong, J. S. (2013). A relay-assisted protocol for spectrum mobility and handover in cognitive LTE networks. *IEEE Systems Journal*, 7(1), 77–91. doi:10.1109/JSYST.2012.2205089

Chiang, M., Tan, C. W., Palomar, D. D. P., O'Neill, D., & Julian, D. (2007). Power control by geometric programming. *IEEE Transactions on Wireless Communications*, 6(7), 2640–2651. doi:10.1109/TWC.2007.05960

Chittur, A. (2001). *Model generation for an intrusion detection system using genetic algorithms*. (High School Honors Thesis). Ossining High School. In cooperation with Columbia Univ.

Chiu, H. S., Yeung, K. L., & Lui, K.-S. (2009). J-CAR: An efficient joint channel assignment and routing protocol for IEEE 802.11-based multi-channel multi-interface mobile ad hoc networks. *IEEE Transactions on Wireless Communications*, 8(4), 1706–1715. doi:10.1109/TWC.2009.080174

Choe, S. (2010). OFDMA cognitive radio medium access control using multichannel ALOHA. In *Proc. of ISWCS 2010*, (p. 244-249). York, UK: ISWCS. doi:10.1109/ISWCS.2010.5624407

Choi, K. W., & Hossain, E. (2011). Opportunistic access to spectrum holes between packet bursts: A learning-based approach. *IEEE Transactions on Wireless Communications*, 10(8), 2497–2509. doi:10.1109/TWC.2011.060711.100154

Choi, K. W., Jeon, W. S., & Jeong, D. G. (2009). Sequential detection of cyclostationary signal for cognitive radio systems. *IEEE Transactions on Wireless Communications*, 8(9), 4480–4485. doi:10.1109/TWC.2009.090288

Chou, C.-T., Sai Shankar, N., Kim, H., & Shin, K. G. (2007). What and how much to gain by spectrum agility? *IEEE Journal on Selected Areas in Communications*, 25(3), 576–588. doi:10.1109/JSAC.2007.070408

Chouhan, L., & Trivedi, A. (2012, September). Priority based MAC scheme for cognitive radio network: A queuing theory modelling. In *Proceedings of Ninth International Conference on Wireless and Optical Communications Networks* (pp. 1-5). Piscataway, NJ: IEEE. doi:10.1109/WOCN.2012.6331886

Christian, I., Sangman, M., Ilyong, C., & Lee, J. (2012). Spectrum mobility in cognitive radio networks. *IEEE Communications Magazine*, 50(6), 114–121. doi:10.1109/MCOM.2012.6211495

Christian, X., Lu, D., Li, P., & Fang, Y. (2011). Coolest path: spectrum mobility aware routing metrics in cognitive ad hoc networks. In *Proceedings of the 2011 31st International Conference on Distributed Computing Systems (ICDCS)*, (pp. 182-191). IEEE.

Chuang, I., Wu, H.-Y., Lee, K.-R., & Kuo, Y.-H. (2013). Alternate hop-and-wait channel rendezvous method for cognitive radio networks. In *Proceedings of IEEE International Conference on Computer Communications*. IEEE. doi:10.1109/INFCOM.2013.6566861

Clancy, T. C., & Goergen, N. (2008, May). Security in cognitive radio networks: Threats and mitigation. In *Proceedings of Cognitive Radio Oriented Wireless Networks and Communications*, (pp. 1-8). IEEE.

Clancy, C., Hecker, J., Stuntebeck, E., & O'Shea, T. (2007). Applications of machine learning to cognitive radio networks. *IEEE Wireless Communications*, 4(4), 47–52. doi:10.1109/MWC.2007.4300983

Clermidy, F., Lemaire, R., Popon, X., Ktenas, D., & Thonnart, Y. (2009). An Open and Reconfigurable Platform for 4G Telecommunication: Concepts and Application. In *Proceedings of Euromicro Conference on Digital System Design, Architectures, Methods and Tools* (pp. 449–456). Patras, Greece: Academic Press. doi:10.1109/DSD.2009.200

Cohen, A., & Rohou, E. (2010). Processor virtualization and split compilation for heterogeneous multicore embedded systems. In *Proceedings of the 47th Design Automation Conference* (pp. 102 - 107). Anaheim, CA: Academic Press. doi:10.1145/1837274.1837303

Coilcraft. (2012). *0402HP Ceramic Chip Inductors.* Retrieved January 16, 2014, from http://www.coilcraft.com/0402hp.cfm

Collins, B. (2000). Polarization-diversity antennas for compact base stations. *Microwave Journal*, 76-88.

Cordeiro, C., Challapali, K., & Ghosh, M. (2006). Cognitive PHY and MAC layers for dynamic spectrum access and sharing of TV bands. In *Proceedings of the first international workshop on Technology and policy for accessing spectrum* (p. 3). ACM. doi:10.1145/1234388.1234391

Cordeiro, C., Challapali, K., & Sai Shankar, N. (2005, November). IEEE 802.22: The First Worldwide Wireless Standard Based On Cognitive Radios. *New Frontier in Dynamic Spectrum Access Networks, 2005*, 328–337.

Cordeiro, C., Philips, R., Briarcliff, N. M., Challapali, K., Birru, D., & Sai, N. S. (2005, November). IEEE 802.22: The First Worldwide Wireless Standard Based on Cognitive Radios. In *Proceedings of IEEE Dynamic Spectrum Access Networks (DYSPAN 2005)*. Baltimore, MD: IEEE. doi:10.1109/DYSPAN.2005.1542649

Cormio, C., & Chowdhury, K. R. (2010). An adaptive multiple rendezvous control channel for cognitive radio wireless ad hoc networks. In *Proceedings of IEEE International Conference on Pervasive Computing and Communications (PERCOM) WS* (pp. 346-351). IEEE. doi:10.1109/PERCOMW.2010.5470645

Cormio, C., & Chowdhury, K. R. (2009). A survey on MAC protocols for cognitive radio networks. *Ad Hoc Networks, 7*(7), 1315–1329. doi:10.1016/j.adhoc.2009.01.002

Cover, T. M., & Thomas, J. A. (1991). *Elements of Information Theory*. John Wiley & Sons, Inc. doi:10.1002/0471200611

Craninckx, J., Liu, M., Hauspie, D., Giannini, V., Kim, T., Lee, J., et al. (2007). A Fully Reconfigurable Software-Defined Radio Transceiver in 0.13μm CMOS. In *Proceedings of IEEE Solid-State Circuits Conference*. IEEE.

Cruz, R. L. (1991). A calculus for network delay I: Network elements in isolation. *IEEE Transactions on Information Theory, 37*(1), 114–131. doi:10.1109/18.61109

Cruz, R. L. (1991). A calculus for network delay II: Network analysis. *IEEE Transactions on Information Theory, 37*(1), 132–141. doi:10.1109/18.61110

CST. (2013). *Computer Simulation Technology.* Retrieved January 16, 2014, from https://www.cst.com/Products

Cui, C., & Wang, Y. (2013, October). Analysis and Optimization of Sensing Reliability for Relay-Based Dual-Stage Collaborative Spectrum Sensing in Cognitive Radio Networks. *Wireless Personal Communications, 72*(4), 2321–2337. doi:10.1007/s11277-013-1152-6

Cui, S., Goldsmith, A. J., & Bahai, A. (2004). Energy-efficiency of MIMO and cooperative MIMO techniques in sensor networks. *IEEE Journal on* Selected Areas in Communications, 22(6), 1089–1098.

Cullum, J. K., & Willoughby, R. A. (1985). Lanczos Algorithms for Large Symmetric Eigenvalue Computations: *Programms (vol. 2)*. Birkhäuser Boston.

Dandawate, A. V., & Giannakis, G. B. (1994). Statistical tests for presence of cyclostationarity. *IEEE Transactions on Signal Processing, 42*(9), 2355–2369. doi:10.1109/78.317857

Darabi, H. (2010). Highly integrated and tunable RF front-ends for reconfigurable multi-band transceivers. In *Proceedings of IEEE Custom Integrated Circuits Conference (CICC)*, (pp. 1-8). IEEE. doi:10.1109/CICC.2010.5617622

Dardaillon, M., Marquet, K., Risset, T., Martin, J., & Charles, H. (2014). Compilation for heterogeneous SoCs : bridging the gap between software and target-specific mechanisms. In Proceedings of Workshop on High Performance Energy Efficient Embedded Systems - HIPEAC. Vienna, Austria: Academic Press.

DaSilva, L. A., & Guerreiro, I. (2008). Sequence-Based Rendezvous for Dynamic Spectrum Access. In Proceedings of IEEE Dynamic Spectrum Access Networks Symposium (DySPAN) (pp. 1-7). IEEE.

David, K., Dixit, S., & Jefferies, N. (2010). 2020 Vision. *IEEE Vehicular Technology Magazine*, *5*(3), 22–29. doi:10.1109/MVT.2010.938595

de Baynast, A., Mohnen, P., & Petrova, M. (2008, March). Arq based cross-layer optimization for wireless multicarrier transmission on cognitive radio networks. *Computer Networks*, *4*(4), 778–794. doi:10.1016/j.comnet.2007.11.009

De Domenico, A., Strinati, E. C., & Di Benedetto, M. (2012). A survey on MAC strategies for cognitive radio networks. *IEEE Communications Surveys and Tutorials*, *14*(1), 21–44. doi:10.1109/SURV.2011.111510.00108

de Oliveira Castro, P., Louise, S., & Barthou, D. (2010). A Multidimensional Array Slicing DSL for Stream Programming. In *Proceedings of International Conference on Complex, Intelligent and Software Intensive Systems* (pp. 913–918). Krakow, Poland: Academic Press. doi:10.1109/CISIS.2010.135

Defence Advanced Research Project Agency (DARPA). (n.d.). DARPA Next Generation (XG) Communication Program.

Delahaye, J., Palicot, J., Moy, C., & Leray, P. (2007). *Partial Reconfiguration of FPGAs for Dynamical Reconfiguration of a Software Radio Platform*. In *Proceedings of 16th IST Mobile and Wireless Communications Summit* (pp. 1–5). Budapest, Hungary: IST.

Demmel, J. W. (1997). *Applied Numerical Linear Algebra*. SIAM. doi:10.1137/1.9781611971446

Deng, S., Chen, J., He, H., & Tang, W. (2007). Collaborative strategy for route and spectrum selection in cognitive radio networks. *IEEE Future Generation Communication and Networking*, *2*, 168–172. doi:10.1109/FGCN.2007.88

Deng, S., Du, L., Wang, B., Wang, W., Zheng, Y., & Chai, K. K. (2013). Dynamic channel reservation based on forecast in cognitive radio. In *Proceedings of IEEE Vehicular Technology Conference*. Dresden, Germany: IEEE. doi:10.1109/VTCSpring.2013.6692522

Derakhshani, M., & Le-Ngoc, T. (2012). Aggregate interference and capacity-outage analysis in a cognitive radio network. *IEEE Transactions on Vehicular Technology*, *61*(1), 196–207. doi:10.1109/TVT.2011.2174464

Derakhshani, M., Le-Ngoc, T., & Nasiri-Kenari, M. (2011). Efficient cooperative cyclostationary spectrum sensing in cognitive radios at low SNR regimes. *IEEE Transactions on Wireless Communications*, *10*(11), 3754–3764. doi:10.1109/TWC.2011.080611.101580

Derudder, V., Bougard, B., Couvreur, A., Dewilde, A., Dupont, S., Folens, L., ... Van der Perre, L. (2009). A 200Mbps+ 2.14 nJ/b digital baseband multi processor system-on-chip for SDRs. In *Proceedings of Symposium on VLSI Circuits* (pp. 292–293). Kyoto, Japan: VLSI.

Dhar, R., George, G., Malani, A., & Steenkiste, P. (2006, September). Supporting integrated MAC and PHY software development for the USRP SDR. In *Proceedings of Networking Technologies for Software Defined Radio Networks*, (pp. 68-77). IEEE. doi:10.1109/SDR.2006.4286328

Di Renzo, M., Graziosi, F., & Santucci, F. (2009, April). Cooperative spectrum sensing in cognitive radio networks over correlated log-normal shadowing. In *Proceedings of Vehicular Technology Conference*, (pp. 1-5). IEEE. doi:10.1109/VETECS.2009.5073470

Di Renzo, M., Imbriglio, L., Graziosi, F., & Santucci, F. (2009). Distributed data fusion over correlated log-normal sensing and reporting channels: Application to cognitive radio networks. *IEEE Transactions on* Wireless Communications, *8*(12), 5813–5821.

Dietrich, F., Dietze, K., Nealy, J., & Stutzman, W. (2001). Spatial, polarization and patern density for wireless handheld terminals. *IEEE Trans. Antennas and Propagation*, 1271-1281.

Digham, F. F., Alouini, M. S., & Simon, M. K. (2003). On the energy detection of unknown signals over fading channels. In *Proceedings of IEEE International Conference on Communications*, (Vol. 5, pp. 3575–3579). IEEE. doi:10.1109/ICC.2003.1204119

Digham, F. F., Alouini, M.-S., & Simon, M. K. (2007). On the Energy Detection of Unknown Signals Over Fading Channels. *IEEE Transactions on Communications, 55*(1), 21–24. doi:10.1109/TCOMM.2006.887483

Dikmese, S., Wong, J. L., Gokceoglu, A., Guzzon, E., Valkama, M., & Renfors, M. (2013, July). Reducing computational complexity of eigenvalue based spectrum sensing for cognitive radio. In *Proceedings of Cognitive Radio Oriented Wireless Networks (CROWNCOM)*, (pp. 61-67). IEEE. doi:10.4108/icst.crowncom.2013.252014

Ding, L., Melodia, T., Batalama, S. N., Matyjas, J. D., & Medley, M. J. (2010). Cross-layer routing and dynamic spectrum allocation in cognitive radio ad hoc networks. *IEEE Transactions on Vehicular Technology, 59*(4), 1969–1979. doi:10.1109/TVT.2010.2045403

Djoumessi, E. E., & Ke Wu. (2010). Reconfigurable RF front-end for frequency-agile direct conversion receivers and cognitive radio system applications. In *Proceedings of IEEE Radio and Wireless Symposium (RWS)*, (pp. 272-275). IEEE. doi:10.1109/RWS.2010.5434205

Dong, W., Chen, C., Liu, X., He, Y., Liu, Y., Bu, J., & Xu, X. (2010). Dplc: Dynamic packet length control in wireless sensor networks. In *Proceedings of IEEE INFOCOM*. IEEE.

Donoho, D. L. (2006, April). Compressed sensing. *IEEE Transactions on* Information Theory, 52, 1289–1306.

Doppler, K., Rinne, M., Wijting, C., Ribeiro, C., & Hugl, K. (2009, December). Device-to-device communication as an underlay to LTE-advanced networks. *IEEE Communications Magazine, 47*(12), 42–49. doi:10.1109/MCOM.2009.5350367

Dorsey, W. M., Zaghloul, A. I., & Parent, M. G. (2009). Perturbed Square-Ring Slot Antenna With Reconfigurable Polarization. *IEEE Antennas and Wireless Propagation Letters, 8*(1), 603–606. doi:10.1109/LAWP.2009.2022286

Duan, B. S. L., Huang, J., & Shou, B. (2011). Investment and pricing with spectrum uncertainty: A cognitive operator's perspective. *IEEE Transactions on Mobile Computing, 10*(11), 1590–1604. doi:10.1109/TMC.2011.78

Duan, L., Huang, J., & Shou, B. (2010a, 6-9 April). Competition with Dynamic Spectrum Leasing. In *Proceedings of 2010 IEEE Symposium on New Frontiers in Dynamic Spectrum (DYSPAN)*. Singapore: IEEE. doi:10.1109/DYSPAN.2010.5457903

Duan, L., Huang, J., & Shou, B. (2010b, April). Competition with Dynamic Spectrum Leasing. In *Proceedings of 2010 IEEE International Symposium on Dynamic Spectrum Access Networks*. Singapore: IEEE.

Duan, L., Huang, J., & Shou, B. (2012). Duopoly Competition in Dynamic Spectrum Leasing and Pricing. *IEEE Transactions on Mobile Computing, 11*(11), 1706–1719. doi:10.1109/TMC.2011.213

Duarte, M. F., Sarvotham, S., Baron, D., Wakin, M. B., & Baraniuk, R. G. (2005). Distributed Compressed Sensing of Jointly Sparse Signals. In *Proceedings of the Asilomar Conference on Signals, Systems and Computers*. Academic Press. doi:10.1109/ACSSC.2005.1600024

Duarte, M. F., & Baraniuk, R. G. (2012, February). Kronecker Compressive Sensing. *IEEE Transactions on Image Processing, 21*(2), 494–504.

Duffy, K. R. (2013). Article. *Decentralized Constraint Satisfaction., 21*(4), 1298–1308.

Dutta, A., Saha, D., Grunwald, D., & Sicker, D. (2010). An architecture for software defined cognitive radio. In *Proceedings of ACM/IEEE Symposium on Architectures for Networking and Communications Systems (ANCS)*, (pp. 1-12). ACM/IEEE. doi:10.1145/1872007.1872014

Eberhart, R. C., & Shi, Y. (1998). Comparison between genetic algorithms and particle swarm optimization. In *Proceedings of Evolutionary Programming*. Academic Press.

Ebrahimi, E., Kelly, J., & Hall, P. (2011). Integrated wide-narrowband antenna for multi-standard radio. *IEEE Transactions on Antennas and Propagation, 59*(7), 2628–2635. doi:10.1109/TAP.2011.2152353

Eker, J., Janneck, J. W., Lee, E. A., Liu, J., Liu, X., Ludvig, J., ... Xiong, Y. (2003). Taming heterogeneity - the Ptolemy approach. *Proceedings of the IEEE, 91*(1), 127–144.

Ekin, S., Yilmaz, F., Celebi, H., Qaraqe, K., Alouini, M.-S., & Serpedin, E. (2009). Achievable capacity of a spectrum sharing system over hyper fading channels. In *Proceedings of IEEE Global Communications Conference (GLOBECOM)*. IEEE. doi:10.1109/GLOCOM.2009.5425715

El Masri, N. M. A., & Khalif, H. (2011). *A routing strategy for cognitive radio networks using fuzzy logic decisions. In Proceedings of Conference Cognitive Advances in Cognitive Radio*. COCORA.

Electromagnetic compatibility and Radio spectrum Matters (ERM); System Reference document (SRdoc): Spectrum Requirements for Short Range Device, Metropolitan Mesh Machine Networks (M3N) and Smart Metering (SM) applications. (2011). TR 103 055 v1.1.1, ETSI.

El-Hajj, W., Safa, H., & Guizani, M. (2011). Survey of security issues in cognitive radio networks. *Journal of Internet Technology, 12*(2), 181–198.

Elleuch, I., Abdelkefi, F., & Siala, M. (2013, July). Reduced complexity blind beamforming for multiple antenna spectrum sensing. In *Proceedings of Wireless Communications and Mobile Computing Conference (IWCMC)*, (pp. 1769-1773). IEEE. doi:10.1109/IWCMC.2013.6583824

Elzanati, A. M., Abdelkader, M. F., Seddik, K. G., & Ghuniem, A. M. (2013). Collaborative compressive spectrum sensing using kronecker sparsifying basis. In *Proceedings of Wireless Communications and Networking Conference (WCNC)* (pp. 2902-2907). IEEE. doi:10.1109/WCNC.2013.6555022

Epiq Solutions. (2014). *Software Defined Radio Solutions.* Maveriq Multichannel Reconfigurable RF Transceiver Datasheet. Retrieved from http//www.epiqsolutions.com

Erman, J., Arlitt, M., & Mahanti, A. (2006, September). Traffic classification using clustering algorithms. In *Proceedings of ACM SIGCOMM Workshop on Mining Network Data* (pp. 281-286). Pisa, Italy: ACM.

Erman, J., Mahanti, A., Arlitt, M., Cohen, I., & Williamson, C. (2007, October). Offline/realtime traffic classification using semi-supervised learning. In *Proceedings of International Symposium on Computer Performance, Modeling, Measurements, and Evaluation*, (vol. 64, pp. 1194-1213). Cologne, Germany: Academic Press.

Eronen, P., & Arkko, J. (2004, November). Role of authorization in wireless network security. In Proceedings of DIMACS Workshop. Academic Press.

Etkin, R., Parekh, A., & Tse, D. (2005, November). Spectrum Sharing for Unlicensed Bands. In Proceedings of IEEE Dynamic Spectrum Access Networks (DYSPAN 2005). Baltimore, MD: IEEE.

Etkin, R., Parekh, A., & Tse, D. (2007). Spectrum sharing for unlicensed bands. *IEEE JSAC, 25*(3), 517–528.

Ettus Research. (2013, Nov). Retrieved from Ettus Research: http://www.ettus.com/

Ettus Research. (2014). *USRP Networked Series.* Retrieved January 16, 2014, from https://www.ettus.com/product

Ezirim, K., Sengupta, S., & Troja, E. (2013, January). (Multiple) Channel acquisition and contention handling mechanisms for dynamic spectrum access in a distributed system of cognitive radio networks. In *Proceedings of the 2013 International Conference on Computing, Networking and Communications (ICNC) (ICNC '13)*, (pp. 252-256). ICNC.

Fang, M., Malone, D., Duffy, K. R., & Leith, D. J. (2013). Decentralised learning MACs for collision-free access in WLANs. *Wireless Networks, 19*(1), 83–98. doi:10.1007/s11276-012-0452-1

Fano, R. (1950). Theoretical Limitations on the Broadband Matching of Arbitrary Impedances. *Journal of the Franklin Institute, 249*, 57–83, 139–154.

Fan, R., & Jiang, H. (2009). Channel sensing-order setting in cognitive radio networks: A two-user case. *IEEE Transactions on Vehicular Technology, 58*(9), 4997–5008.

Fantacci, R., Marabissi, D., & Tarchi, D. (2009, September). Adaptive scheduling algorithms for multimedia traffic in wireless OFDMA systems. *Physical Communication, 2*(3), 228–234. doi:10.1016/j.phycom.2009.08.003

Fantacci, R., & Tarchi, D. (2009). A Novel Cognitive Networking Scenario for IEEE 802.16 Networks. In *Proc. of IEEE GLOBECOM 2009*. Honolulu, HI: IEEE. doi:10.1109/GLOCOM.2009.5425637

Farhang-Boroujeny, B. (2008, May). Filter bank spectrum sensing for cognitive radios. *IEEE Transactions on* Signal Processing, *56*(5), 1801–1811. doi:10.1109/TSP.2007.911490

Farraj, A. K., Miller, S. L., & Qaraqe, K. A. (2011). Queue performance measures for cognitive radios in spectrum sharing systems. In *Proceedings of IEEE International Workshop on Recent Advances in Cognitive Communications and Networking (RACCN) – Global Telecommunications Conference (GLOBECOM) Workshop.* IEEE. doi:10.1109/GLOCOMW.2011.6162606

Farraj, A. K. (2013). Impact of cognitive communications on the performance of the primary users. *Wireless Personal Communications, 71*(2), 975–985. doi:10.1007/s11277-012-0855-4

Farraj, A. K. (2013). Queue model analysis for spectrum-sharing cognitive systems under outage probability constraint. *Wireless Personal Communications, 73*(3), 1021–1035. doi:10.1007/s11277-013-1245-2

Farraj, A. K. (2014). Analysis of primary users' queueing behavior in a spectrum-sharing cognitive environment. *Wireless Personal Communications, 75*(2), 1283–1293. doi:10.1007/s11277-013-1423-2

Farraj, A. K. (2014). Switched-diversity approach for cognitive scheduling. *Wireless Personal Communications, 74*(2), 933–952. doi:10.1007/s11277-013-1331-5

Farraj, A. K., & Ekin, S. (2013). Performance of cognitive radios in dynamic fading channels under primary outage constraint. *Wireless Personal Communications, 73*(3), 637–649. doi:10.1007/s11277-013-1207-8

Farraj, A. K., & Hammad, E. M. (2013). Impact of quality of service constraints on the performance of spectrum sharing cognitive users. *Wireless Personal Communications, 69*(2), 673–688. doi:10.1007/s11277-012-0606-6

Farraj, A. K., & Hammad, E. M. (2013). Performance of primary users in spectrum sharing cognitive radio environment. *Wireless Personal Communications, 68*(3), 575–585. doi:10.1007/s11277-011-0469-2

Farraj, A. K., & Miller, S. L. (2013). Scheduling in a spectrum-sharing cognitive environment under outage probability constraint. *Wireless Personal Communications, 70*(2), 785–805. doi:10.1007/s11277-012-0722-3

Fathi, Y., & Tawfik, M. H. (2012). Versatile performance expression for energy detector over á-μ generalized fading channels. *Electronics Letters, 48*(17), 1081–1082. doi:10.1049/el.2012.1744

Faulhaber, G. R., & Farber, D. (2002). Spectrum management: property rights, markets, and the commons. *AEI-Brookings Joint Center for Regulatory Studies Working Paper,* (02-12), 6.

Fayrouz, H., Wenceslas, R., Lakhdar, Z., & Oussama, F. (2010). Design of fully-integrated RF front-end for large image rejection and wireless communication applications. In *Proceedings of 17ʰ IEEE International Conference on Electronics, Circuits, and Systems (ICECS),* (pp. 902-905). IEEE.

FCC Spectrum Policy Task Force. (2002, November). *Report of the Spectrum Efficiency Working Group* (FCC Spectrum Policy Task Force, Technical Report). Author.

FCC. (2002). *ET Docket No. 02-135 Notice of proposed rule making and order.* FCC.

FCC. (2002). *Report of the spectrum efficiency working* (FCC spectrum policy task force, Tech. Rep., Nov. 2002). Washington, DC: FCC.

FCC. (2002). *Spectrum Policy Task Force* (ET Docket 02-135). Washington, DC: FCC.

FCC. (2003). *ET Docket No 03-222 Notice of proposed rule making and order.* Washington, DC: FCC.

FCC. (2004). *FCC adopted rules for unlicensed use of television white spaces. FCC adopted rules for unlicensed use of television white spaces* (Docket no. 04-186, Second Report and Order and Memorandum Opinion and Order). FCC.

FCC. (2010). *ET Docket No. 10-174 2nd memo opinion and order.* FCC.

Federal Communication Commission (FCC). (2004, May). *Notice of Proposed Rule Making* (ET Docket no. 04-113). Washington, DC: FCC.

Federal Communication Commuication (FCC). (2003, December). *Notice of Proposed Rule Making and Order* (ET Docket no. 03-322). Washington, DC: FCC.

Federal Communications Commission. (2002). *Spectrum Policy Task Force Report* (ET Docket No. 02-135). Washington, DC: FCC.

Federal Communications Commission. (2010). *Mobile broadband: The benefits of additional spectrum Tech report 6*. Washington, DC: FCC.

Feng, H., Wang, Y., & Li, S. L. S. (2008). Statistical test based on finding the optimum lag in cyclic autocorrelation for detecting free bands in cognitive radios. In *Proceedings of International Conference on Cognitive Radio Oriented Wireless Networks and Communications* (pp. 1–6). Academic Press. doi:10.1109/CROWNCOM.2008.4562528

Feng, X., Gan, X., & Wang, X. (2011). Energy-constrained cooperative spectrum sensing in cognitive radio networks. In *Proceedings of Global Telecommunications Conference (GLOBECOM 2011)*. IEEE.

Feng, Z., Li, W., Li, Q., Le, V., & Gulliver, T. A. (2011). Dynamic spectrum management for wcdma/dvb heterogeneous systems. *IEEE Transactions on Wireless Communications*, 10(5), 1582–1593. doi:10.1109/TWC.2011.030911.100915

Ferrari, G., & Pagliari, R. (2006). Decentralized binary detection with noisy communication links. *Transactions on Aerospace Electronic Systems*, 42(4), 1554–1563. doi:10.1109/TAES.2006.314597

Fette, B. (2009). Cognitive Radio Technology (2nd ed.). Elsevier.

Fette, B. A. (Ed.). (2009). Cognitive radio technology. Academic Press.

Fettweis, G., & Zimmermann, E. (2008, September). ICT energy consumption-trends and challenges. In *Proceedings of the 11th International Symposium on Wireless Personal Multimedia Communications* (Vol. 2, p. 6). Academic Press.

Fettweis, G., Lohning, M., Petrovic, D., Windisch, M., Zillmann, P., & Tave, W. (2005). Dirty RF: A New Paradigm. In *Proceedings of IEEE 16th International Symposium on Personal, Indoor and Mobile Radio Communications (PIMRC)*, (pp. 2347–2355). IEEE.

Filippini, I., Ekici, E., & Cesana, M. (2009). Minimum maintenance cost routing in cognitive radio networks. In *Proceedings of IEEE 6th International Conference on Mobile Ad hoc and Sensor Systems (MASS)*, (pp. 284-293). IEEE. doi:10.1109/MOBHOC.2009.5336987

First IEEE Symposium. (November 2005). *Proceedings of the First IEEE Symposium on New Frontiers in Dynamic Spectrum Access Network*. IEEE.

Folino, G., Pizzuti, C., & Spezzano, G. (2005). GP ensemble for distributed intrusion detection systems. In *Pattern Recognition and Data Mining* (pp. 54–62). Springer Berlin Heidelberg. doi:10.1007/11551188_6

Forouzan, A. R., & Garth, L. M. (2005). Generalized iterative spectrum balancing and grouped vectoring for maximal throughput of digital subscriber lines. In *Proceedings of IEEE Global Communications Conference* (pp. 2359–2363). IEEE. doi:10.1109/GLOCOM.2005.1578085

Fradet, P., Girault, A., & Poplavko, P. (2012). SPDF: A schedulable parametric data-flow MoC. In Proceedings of 2012 Design, Automation & Test in Europe Conference & Exhibition (DATE) (pp. 769–774). Dresden, Germany: Academic Press.

Fragkiadakis, A. G., Tragos, E. Z., Tryfonas, T., & Askoxylakis, I. G. (2012). Design and performance evaluation of a lightweight wireless early warning intrusion detection prototype. *EURASIP Journal on Wireless Communications and Networking*, (1): 1–18.

Friedman, R., Kliot, G., & Avin, C. (2010). Probabilistic quorum systems in wireless ad hoc networks. *ACM Transactions on Computer Systems*, 28(3), 7. doi:10.1145/1841313.1841315

Friis, H. T. (1946). A note on a simple transmission formula. *Proc. IRE*, 34(5), 254–256. doi:10.1109/JRPROC.1946.234568

Fruth, M. (2006, November). Probabilistic model checking of contention resolution in the IEEE 802.15.4 low-rate wireless personal area network protocol. In *Proceedings of Leveraging Applications of Formal Methods, Verification and Validation,* (pp. 290-297). IEEE.

Fujii, T., & Yamao, Y. (2006). *Multi-band routing for ad-hoc cognitive radio networks.* IEEE SDR.

Fukunaga, K. (1990). *Introduction to Statistical Pattern Recognition.* Academic Press.

Gandhi, R., Wang, C.-C., & Hu, Y. C. (2012). Fast rendezvous for multiple clients for cognitive radios using coordinated channel hopping. In *Proceedings of IEEE International Conference on Sesing, Communication, and Networking (SECON),* (pp. 434-442). IEEE. doi:10.1109/SECON.2012.6275809

Ganesan, G., & Li, Y. (2007). Cooperative spectrum sensing in cognitive radio, part I: Two user networks. *IEEE Transactions on Communications, 6*(6), 2204–2213.

Ganesan, G., & Li, Y. (2007). Cooperative spectrum sensing in cognitive radio, part II: Multiuser networks. *IEEE Transactions on Communications, 6*(6), 2214–2222.

Gann, D., Dodgson, M., & Bhardwaj, D. (2011). Physical-digital Integration in City Infrastructure. *IBM Journal of Research and Development, 55,* 8:1-8:10.

Gao, Y., Jiang, Y., Lin, T., & Zhang, X. (2010, September). Performance bounds for a cognitive radio network with network calculus analysis. In *Proceedings of 2nd IEEE International Conference on Network Infrastructure and Digital Content* (pp. 11-15). Piscataway, NJ: IEEE. doi:10.1109/ICNIDC.2010.5659456

Gao, Y., Xu, W., Yang, K., Niu, K., & Lin, J. (2013). Energy-efficient transmission with cooperative spectrum sensing in cognitive radio networks. In *Proceedings of Wireless Communications and Networking Conference (WCNC),* (pp. 7-12). IEEE.

Gao, Y., Yang, J., Zhang, X., & Jiang, Y. (2011, May). Capacity limits for a cognitive radio network under fading channel. In *Proceedings of IFIP Networking 2011 Workshops* (LNCS) (pp. 42-51). Berlin: Springer.

Gao, L., Wang, X., Xu, Y., & Zhang, Q. (2011). Spectrum trading in cognitive radio networks: A contract-theoretic modeling approach. *IEEE Journal on Selected Areas in Communications, 29*(4), 843–855. doi:10.1109/JSAC.2011.110415

Gao, X., Wu, G., & Miki, T. (2004). End-to-end QoS provisioning in mobile heterogeneous networks. *Wireless Communications, IEEE, 11*(3), 24–34. doi:10.1109/MWC.2004.1308940

Gao, Y., & Jiang, Y. (2012). Analysis on the capacity of a cognitive radio network under delay constraints. *IEICE Transactions on Communications, 95*(4), 1180–1189. doi:10.1587/transcom.E95.B.1180

Garcia-Molina, H., & Barbara, D. (1985). How to assign votes in distributed systems. *Journal of the ACM, 32*(4), 841–860. doi:10.1145/4221.4223

Gardellin, V. Lenzini, L. & Fontana, V. (2011). Aware channel assignment algorithm for cognitive networks. In *Proceedings of ACM Workshop on Cognitive Radio Networks (CoRoNet).* Academic Press.

Gardner, W. A. (1986). Measurement of spectral correlation. *IEEE Transactions on Acoustics, Speech, and Signal Processing, 34*(5), 1111–1123. doi:10.1109/TASSP.1986.1164951

Gardner, W. A. (1988). Signal interception: A unifying theoretical framework for feature detection. *IEEE Transactions on Communications, 36*(8), 897–906. doi:10.1109/26.3769

Gardner, W. A. (1991). Exploitation of spectral redundancy in cyclostationary signals. *IEEE Signal Processing Magazine, 8*(2), 14–36. doi:10.1109/79.81007

Gardner, W. A., Napolitano, A., & Paura, L. (2006). Cyclostationarity: Half a century of research. *Signal Processing, 86*(4), 639–697. doi:10.1016/j.sigpro.2005.06.016

Gardner, W. A., & Spooner, C. M. (1992). Signal interception: Performance advantages of cyclic-feature detectors. *IEEE Transactions on Communications, 40*(1), 149–159. doi:10.1109/26.126716

Gatsis, N., Marques, A. G., & Giannakis, G. B. (2010). Power control for cooperative dynamic spectrum access networks with diverse QoS constraints. *IEEE Transactions on Communications, 58*(3), 933–944. doi:10.1109/TCOMM.2010.03.080491

Geirhofer, S., Tong, L., & Sadler, B. M. (2006, October). A measurement-based model for dynamic spectrum access in WLAN channels. In *Proceedings of IEEE Military Communications Conference*. Washington, DC: IEEE. doi:10.1109/MILCOM.2006.302405

Geirhofer, S., Tong, L., & Sadler, B. M. (2007). Cognitive radios for dynamic spectrum access-dynamic spectrum access in the time domain: Modeling and exploiting white space. *IEEE Communications Magazine, 45*(5), 66–72. doi:10.1109/MCOM.2007.358851

Gelabert, X., Sallent, O., Pérez-Romero, J., & Agustí, R. (2010). Spectrum sharing in cognitive radio networks with imperfect sensing: A discrete-time Markov model. *Computer Networks, 54*(14), 2519–2536. doi:10.1016/j.comnet.2010.04.005

Gelman, A., Carlin, J. B., Stern, H. S., & Rubin, D. B. (2003). *Bayesian data analysis*. Chapman & Hall/CRC Press.

Gevorgian, S. (2009). *Ferroelectrics in Microwave Devices, Circuits and Systems: Physics, Modelling, Fabrication and Measurements*. London: Springer-Verlag. doi:10.1007/978-1-84882-507-9

Ghahremani, S., Khokhar, R. H., Noor, R. M., Naebi, A., & Kheyrihassankandi, J. (2012). On QoS routing in Mobile WiMAX cognitive radio networks. In *Proceedings of IEEE International Conference on Computer and Communication Engineering (ICCCE)*, (pp. 467-471). IEEE. doi:10.1109/ICCCE.2012.6271231

Gharavol, E., Liang, Y., & Mouthaan, K. (2010, December). *Collaborative nonlinear transceiver optimization in multi-tier MIMO cognitive radio networks with deterministically imperfect CSI*. Paper presented at the IEEE Global Telecommunications Conference (IEEE GLOBECOM). Miami, FL. doi:10.1109/GLOCOM.2010.5683394

Ghasemi, A., & Sousa, E. S. (2005). Collaborative spectrum sensing for opportunistic access in fading environments. In *Proceedings of First IEEE International Symposium on New Frontiers in Dynamic Spectrum Access Networks*, (pp. 131–136). IEEE. doi:10.1109/DYSPAN.2005.1542627

Ghasemi, A., & Sousa, E. S. (2007). Opportunistic spectrum access in fading channels through collaborative sensing. *Journal on Selected Areas in Communications, 2*(2), 71-82.

Ghasemi, A., E. S. (2007). Asymptotic performance of collaborative spectrum sensing under correlated log-normal shadowing. *IEEE Communications Letters,* 34–36.

Ghasemi, A., & Sousa, E. (2008). Spectrum sensing in cognitive radio networks: Requirements, challenges and design trade-offs. *IEEE Communications Magazine, 46*(4), 32–39. doi:10.1109/MCOM.2008.4481338

Ghasemi, A., & Sousa, E. S. (2007, January). Opportunistic spectrum access in fading channels through collaborative sensing. *Journal of Communication, 2*(2), 71–82.

Ghasemi, A., & Sousa, E. S. (2008). Interference aggregation in spectrum-sensing cognitive wireless networks. *IEEE Journal of Selected Topics in Signal Processing, 2*(1), 41–56. doi:10.1109/JSTSP.2007.914897

Ghosh, A., Ratasuk, R., Mondal, B., Mangalvedhe, N., & Thomas, T. (2010, June). LTE-advanced: Next-generation wireless broadband technology. *IEEE Wireless Communications, 17*(3), 10–22. doi:10.1109/MWC.2010.5490974

Ghosh, J., & Ramamoorthi, R. (2003). *Bayesian nonparametrics*. New York, NY: Springer-Verlag.

Giupponi, L., & Perez-Neira, A. I. (2008). Fuzzy-based spectrum handoff in cognitive radio networks. In *Proceedings of International Conference on Cognitive Radio Oriented Wireless Networks and Communications (CrownCom)* (pp. 1–6). Academic Press. doi:10.1109/CROWNCOM.2008.4562535

Glossner, J., Iancu, D., Moudgill, M., Nacer, G., Jinturkar, S., Stanley, S., & Schulte, M. (2007). The Sandbridge SB3011 Platform. *EURASIP Journal on Embedded Systems, 2007*(1), 1–16. doi:10.1155/2007/56467

GNU Radio Framework. (n.d.). Retrieved from http://gnuradio.org

Goldfarb, D., & Liu, S. (1991). An o(n3l) primal interior point algorithm for convex quadratic programming. *Mathematical Programming*, 49(3), 325–340.

Goldsmith, A. (2005). *Wireless Communications*. Cambridge University Press. doi:10.1017/CBO9780511841224

Goldsmith, A., Jafar, S. A., Maric, I., & Srinivasa, S. (2009). Breaking spectrum gridlock with cognitive radios: An information theoretic perspective. *Proceedings of the IEEE*, 97(5), 894–914. doi:10.1109/JPROC.2009.2015717

Gong, X., Ishaque, A., Dartmann, G., & Ascheid, G. (2011, March). *MSE-based stochastic transceiver optimization in downlink cognitive radio networks*. Paper presented at the IEEE Wireless Communications and Networking Conference (WCNC). Quintana-roo, Mexico. doi:10.1109/WCNC.2011.5779369

Gonzalez, C. R. A., Dietrich, C. B., Sayed, S., Volos, H. I., Gaeddert, J. D., & Robert, P. M. et al. (2009). Open-source SCA-based core framework and rapid development tools enable software-defined radio education and research. *IEEE Communications Magazine*, 47(10), 48–55. doi:10.1109/MCOM.2009.5273808

Gonzalez-Pina, J., Ameur-Boulifa, R., & Pacalet, R. (2012). DiplodocusDF, a Domain-Specific Modelling Language for Software Defined Radio Applications. In *Proceedings of 38th Euromicro Conference on Software Engineering and Advanced Applications* (pp. 1–8). Cesme, Izmir: Academic Press. doi:10.1109/SEAA.2012.36

Goodrich, M. T., & Tamassia, R. (2001). *Algorithm Design: Foundations*. Analysis, and Internet Examples.

Google Maps. (n.d.). Retrieved from maps.google.com

Gorin, J., Wipliez, M., Preteux, F., & Raulet, M. (2010). A portable Video Tool Library for MPEG Reconfigurable Video Coding using LLVM representation. In *Proceedings of Conference on Design and Architectures for Signal and Image Processing (DASIP)* (pp. 183–190). Edinburgh, UK: Academic Press. doi:10.1109/DASIP.2010.5706263

Goubier, T., Sirdey, R., Louise, S., & David, V. (2011). ∑C A Programming Model and Language for Embedded Manycores. In *Proceedings of Algorithms and Architectures for Parallel Processing - 11th International Conference, ICA3PP* (pp. 385–394). Melbourne, Australia: Academic Press.

Granelli, F., Pawelczak, P., Prasad, R., Subbalakshmi, K., Chandramouli, R., Hoffmeyer, J., & Berger, J. (2010). Standardization and research in cognitive and dynamic spectrum access networks: IEEE SCC41 efforts and other activities. *IEEE Communications Magazine*, 48(1), 71–79. doi:10.1109/MCOM.2010.5394033

Grau, A., Romeu, J., Lee, M.-J., Blanch, S., Jofre, J., & De Flaviis, F. (2010). A dual-linearly-polarized MEMS-reconfigurable antenna for narrow-band MIMO communication systems. *IEEE Transactions on Antennas and Propagation*, 58(1), 4–17. doi:10.1109/TAP.2009.2036197

Grimm, M., Allen, M., Marttila, J., Valkama, M., & Thoma, R. (2014). Joint Mitigation of Nonlinear RF and Baseband Distortions in Wideband Direct-Conversion Receivers. *IEEE Transactions on Microwave Theory and Techniques*, 62(1), 166–182. doi:10.1109/TMTT.2013.2292603

Grimm, M., Sharma, R. K., Hein, M. A., & Thomä, R. S. (2012). DSP-based Mitigation of RF Front-end Nonlinearity in Cognitive Wideband Receivers. *Frequenz.*, 6(9-10), 303–310.

Gronsund, P., Pham, H. N., & Engelstad, P. (2009). Towards dynamic spectrum access in primary OFDMA systems. In *Proc. of IEEE PIMRC 2009*, (pp. 848-852). Tokyo, Japan: IEEE. doi:10.1109/PIMRC.2009.5450079

Gu, Z., Hua, Q.-S., Wang, Y., & Lau, F. C. M. (2013). Nearly Optimal Asynchronous Blind Rendezvous Algorithm for Cognitive Radio Networks. In *Proceedings of IEEE International Conference on Sensing, Communication, and Networking (SECON)*. IEEE.

Guan, Q., Yu, F. R., Jiang, S., & Wei, G. (2010). Prediction-based topology control and routing in cognitive radio mobile ad hoc networks. *IEEE Transactions on Vehicular Technology*, 59(9), 4443–4452. doi:10.1109/TVT.2010.2069105

Gu, C. S., Zhang, X. X., & Huang, H. (2011). Online wireless mesh network traffic classification using machine learning. *Journal of Computer Information Systems, 7*(5), 1524–1532.

Guibene, W., Moussavinik, H., & Hayar, A. (2011). Combined compressive sampling and distribution discontinuities detection approach to wideband spectrum sensing for cognitive radios. In *Proceedings of Ultra Modern Telecommunications and Control Systems and Workshops (ICUMT),* (pp. 1--7). IEEE.

Gummaraju, J., Coburn, J., Turner, Y., & Rosenblum, M. (2008). Streamware: Programming General-Purpose Multicore Processors Using Streams. In *Proceedings of the 13th International Conference on Architectural Support for Programming Languages and Operating Systems - ASPLOS XIII* (p. 297). Seattle, WA: Academic Press. doi:10.1145/1346281.1346319

Guo, C., Wu, X., Feng, C., & Zeng, Z. (2013). Spectrum Sensing for Cognitive Radios Based on Directional Statistics of Polarization Vectors. *IEEE Journal on Selected Areas in Communications, 31*(3), 379–393. doi:10.1109/JSAC.2013.130305

Haffner, P., Sen, S., Spatscheck, O., & Wang, D. (2005, August). ACAS: automated construction of application signatures. In *Proceedings of ACM SIGCOMM Workshop on Mining Network Data*, (pp.197-202). Philadhelpia, PA: ACM.

Hall, P., Gardner, P., Kelly, J., Ebrahimi, E., Hamid, M., Ghanem, F., et al. (2009). Reconfigurable antenna challenges for future radio systems. In *Proceedings of Third European Conference on Antennas and Propagation* (pp. 949-955). Academic Press.

Hall, P., Kapoulas, S., Chauhan, R., & Kalialakis, C. (1997). Microstrip Patch Antenna with Integrated Adaptive Tuning. In *Proceedings of IEEE Conference on Antennas and Propagation*. IEEE. doi:10.1049/cp:19970306

Hall, J. M. (Ed.). (1986). *Combinatorial Theory*. John Wiley and Sons.

Hamdi, K., Zhang, W., & Letaief, K. B. (2007). Power control in cognitive radio systems based on spectrum sensing side information. In *Proceedings of IEEE International Conference on Communications (ICC)*. IEEE. doi:10.1109/ICC.2007.853

Hamid, M. R., Gardner, P., Hall, P. S., & Ghanem, F. (2011). Switched-Band Vivaldi Antenna. *IEEE Transactions on Antennas and Propagation, 59*(5), 1472–1480. doi:10.1109/TAP.2011.2122293

Hamza Mohamed, D., Aissa, S., & Aniba, G. (2014). Equal Gain Combining for Cooperative Spectrum Sensing in Cognitive Radio Networks. *IEEE Transactions on Wireless Communications*. doi:10.1109/TWC.2014.2317788

Hamza, D., & Aissa, S. (2013). Wideband Spectrum Sensing Order for Cognitive Radios with Sensing Errors and Channel SNR Probing Uncertainty. *IEEE Wireless Communications Letters, 2*(2), 151–154. doi:10.1109/WCL.2012.121812.120809

Han, W., Li, J., Li, Z., Si, J., Zhang, Y. (2013). Efficient Soft Decision Fusion Rule in Cooperative Spectrum Sensing. *IEEE Transactions on Signal Processing, 61*(8), 1931-1943. doi:10.1109/TSP.2013.2245659

Hanif, M. F., Shafi, M., Smith, P. J., & Dmochowski, P. (2009). Interference and deployment issues for cognitive radio systems in shadowing environments. In *Proceedings of IEEE International Conference on Communications* (pp. 1–6). IEEE. doi:10.1109/ICC.2009.5199089

Han, Y., Pandharipande, A., & Ting, S. H. (2009). Cooperative Decode-and- Forward Relaying for Secondary Spectrum Access. *IEEE Transactions on Wireless Communications, 8*(10), 4945–4950. doi:10.1109/TWC.2009.081484

Han, Z., & Liu, K. R. (2008). *Resource allocation for wireless networks.* Cambridge, UK: Cambridge University Press. doi:10.1017/CBO9780511619748

Hao, X., Cheung, M. H., Wong, V. W. S., & Leung, V. C. M. (2012). Hedonic coalition formation game for cooperative spectrum sensing and channel access in cognitive radio networks. *IEEE Transactions on Wireless Communications, 11*(11), 3968–3979. doi:10.1109/TWC.2012.092412.111833

Hao, X., Cheung, M., Wong, V., & Leung, V. (2011). *A Stackelberg game for cooperative transmission and random access in cognitive radio networks.* IEEE PIMRC.

Harada, H. (2008). A feasibility study on software defined cognitive radio equipment. In *Proceedings of 3rd IEEE Symposium on New Frontiers in Dynamic Spectrum Access Networks (DySPAN),* (pp. 1-12). IEEE. doi:10.1109/DYSPAN.2008.8

Harada, H. (2008). A small-size software defined cognitive radio prototype. In *Proceedings of IEEE 19th International Symposium on Personal, Indoor and Mobile Radio Communications (PIMRC)* (pp. 1-5). IEEE.

Harrison, K., Mishra, S. M., & Sahai, A. (2010). How much white-space capacity is there? In *Proceedings of IEEE Symposium on New Frontiers in Dynamic Spectrum* (pp. 1–10). IEEE. doi:10.1109/DYSPAN.2010.5457914

Hashemi, H. (1993). The indoor radio propagation channel. *Proceedings of the IEEE, 81*(7), 943–968. doi:10.1109/5.231342

Hassan, R., Cohanim, B. D., & Venter, G. (2005). A comparison of particle swarm optimization and the genetic algorithm. In *Proceedings of 1st AIAA Multidisciplinary Design Optimization Specialist Conference.* AIAA.

Haupt, J., Baraniuk, R., Castro, R., & Nowak, R. (2012). Sequentially designed compressed sensing. In *Proceedings of Statistical Signal Processing Workshop (SSP),* (pp. 401-404). IEEE.

Havary-Nassab, V., Hassan, S., & Valaee, S. (2010). Compressive detection for wide-band spectrum sensing. In *Proceedings of Acoustics Speech and Signal Processing (ICASSP),* (pp. 3094--3097). IEEE.

Hayar, A. M., Pacalet, R., & Knopp, R. (2007). Cognitive radio research and implementation challenges. In *Proceedings of Conference Record of the Forty-First Asilomar Conference on Signals, Systems and Computers (ACSSC),* (pp. 782-786). ACSSC.

Haykin, S. (2005). Cognitive radio: Brain-empowered wireless communications. *IEEE Journal on Selected Areas in Communications, 23*(2), 201–220.

He, A., Bae, K. K., Newman, R. T., & Gaeddert, J. K. (2010, May). A survey of artificial intelligence for cognitive radios. *IEEE Transactions on Vehicular Technology, 4*(4), 1578–1592. doi:10.1109/TVT.2010.2043968

He, A., Gaeddert, J., Bae, K. K., Newman, T. R., Reed, J., Morales, L., & Park, C. H. (2009). Development of a case-based reasoning cognitive engine for ieee 802.22 wran applications. *SIGMOBILE Mob. Comput. Commun. Rev., 13*(2), 37–48. doi:10.1145/1621076.1621081

Heinen, S., & Wunderlich, R. (2011). *High dynamic range RF frontends from multiband multistandard to cognitive radio. In Proceedings of Semiconductor Conference Dresden* (pp. 1–8). SCD.

Herath, S. P., Rajatheva, N., & Tellambura, C. (2011). Energy detection of unknown signals in fading and diversity reception. *IEEE Transactions on Communications, 59*(9), 2443–2453. doi:10.1109/TCOMM.2011.071111.090349

Hernandez-Serrano, J., León, O., & Soriano, M. (2011). Modeling the lion attack in cognitive radio networks. *EURASIP Journal on Wireless Communications and Networking, 2,* 2011.

Hofmann, K., Shen, L., González-Rodríguez, E., Maune, H., Shah, I., Dahlhaus, D., & Jakoby, R. (2012). Fully Integrated High Voltage Charge Pump for Energy-Efficient Reconfigurable Multi-band RF-Transceivers. In *Proceedings of International Telecommunications Energy Conference.* Academic Press. doi:10.1109/INTLEC.2012.6374483

Hong, S. (2010). Multi-resolution bayesian compressive sensing for cognitive radio primary user detection. In *Proceedings of Global Telecommunications Conference (GLOBECOM 2010),* (pp. 1-6). IEEE. doi:10.1109/GLOCOM.2010.5683291

Hong, X. H. X., Wang, C.-X. W. C.-X., & Thompson, J. (2008). Interference modeling of cognitive radio networks. In *Proceedings of IEEE Vehicular Technology Conference* (pp. 1851–1855). IEEE.

Horrein, P.-H., Hennebert, C., & Pétrot, F. (2011). Integration of GPU Computing in a Software Radio Environment. *Journal of Signal Processing Systems for Signal, Image, and Video Technology, 69*(1), 55–65. doi:10.1007/s11265-011-0639-1

Horst, R., Phong, T. Q., Thoai, N. V., & Vries, J. (1991). On solving a d.c. programming problem by a sequence of linear programs. *Journal of Global Optimization, 1*(2), 183–293. doi:10.1007/BF00119991

Hossain, E., Niyato, D., & Han, Z. (2009). *Dynamic spectrum access and management in cognitive radio networks*. Cambridge University Press. doi:10.1017/CBO9780511609909

Hotelling, H. (1933). Analysis of a complex of statistical variables into principal components. *Journal of Educational Psychology, 24*(6), 417–441. doi:10.1037/h0071325

Hou, F., Cai, L. X., Shen, X., & Huang, J. (2011). Asynchronous multichannel MAC design with difference-set-based hopping sequences. *IEEE Transactions on Vehicular Technology, 60*(4), 1728–1739. doi:10.1109/TVT.2011.2119384

Hoven, N., & Sahai, A. (2005). Power scaling for cognitive radio. In *Proceedings of International Conference on Wireless Networks, Communications and Mobile Computing,* (vol. 1, pp. 250–255). Academic Press.

Hsieh, Y.-S., Lien, C.-W., & Chou, C.-T. (2011). A multichannel testbed for dynamic spectrum access (DSA) networks. In *Proceedings of ACM Workshop on Wireless Multimedia Networking and Computing (WMUNEP)*. ACM. doi:10.1145/2069117.2069129

Hsu, C.-J., Pino, J. L., & Hu, F.-J. (2010). A mixed-mode vector-based dataflow approach for modeling and simulating LTE physical layer. In *Proceedings of Design Automation Conference (DAC)* (pp. 18–23). Anaheim, CA: DAC. doi:10.1145/1837274.1837282

Hsu, S.-H., & Chang, K. (2007). A novel reconfigurable microstrip antenna with switchable circular polarization. *IEEE Antennas and Wireless Propagation Letters, 6*(11), 160–162. doi:10.1109/LAWP.2007.894150

Hu, Z., Ranganathan, R., Zhang, C., Qiu, R., Bryant, M., Wick, M., et al. (2012). Robust non-negative matrix factorization for joint spectrum sensing and primary user localization in cognitive radio networks. In *Proceedings of IEEE Waveform Diversity and Design Conference*. IEEE.

Hua, S., Liu, H., Wu, M., & Panwar, S. (2011, April). *Exploiting MIMO antennas in cooperative cognitive radio networks*. Paper presented at IEEE INFOCOM. Shanghai, China. doi:10.1109/INFCOM.2011.5935102

Huang, C., Buisman, K., Nanver, L., Zampardi, P., Larson, L., & de Vreede, L. (2010). Design concepts for semiconductor based ultra-linear varactor circuits. In *Proceedings of IEEE Bipolar/BiCMOS Circuits and Technology Meeting*. IEEE.

Huang, C., Buisman, K., Zampardi, P., Nanver, L., Larson, L., & de Vreede, L. (2010). Low Distortion Tunable RF Components, a Compound Semiconductor Opportunity. In *Proceedings of Compund Semiconductor Manufacturing Technology Conference*. Academic Press.

Huang, D., Wu, S., & Wang, P. (2010). Cooperative spectrum sensing and locationing: A sparse Bayesian learning approach. In *Proceedings of Global Telecommunications Conference (GLOBECOM 2010),* (pp. 1-5). IEEE. doi:10.1109/GLOCOM.2010.5684081

Huang, K., Lin, C., & Hsiung, P. (2010). A space efficient and multi-objective case-based reasoning in cognitive radio. In *Proceedings of IET International Conference on Frontier Computing Theory, Technologies and Applications*. IET. doi:10.1049/cp.2010.0532

Huang, S., Liu, X., & Ding, Z. (2009, April). Optimal sensing-transmission structure for dynamic spectrum access. In *Proceedings of IEEE International Conference on Computer Communications IEEE INFOCOM 2009*. (pp. 2295-2303). Rio, Brazil: IEEE. doi:10.1109/INFCOM.2009.5062155

Huang, S., Liu, X., & Ding, Z. (2010). Distributed power control for cognitive user access based on primary link control feedback. In *Proceedings of IEEE Conference on Information Communications (INFOCOM)*. IEEE. doi:10.1109/INFCOM.2010.5461916

Huang, X.-L., Wang, G., Hu, F., Kumar, S., & Zhang, Y. (2011). Optimal packet size design for multimedia transmissions in cognitive radio networks. In *Proceedings of 6th International ICST Conference on Communications and Networking in China (CHINACOM)*. ICST.

Huang, C. C., & Wang, L. C. (2012). Dynamic sampling rate adjustment for compressive spectrum sensing over cognitive radio network. *IEEE Wireless Communications Letters*, *1*(2), 57–60. doi:10.1109/WCL.2012.010912.110136

Huang, J., Berry, R., & Honig, M. (2005, November). Spectrum Sharing with Distributed Interference Compensation. In *Proceedings of IEEE Dynamic Spectrum Access Networks (DYSPAN)*. Baltimore, MD: IEEE.

Huang, K.-C., Jing, X., & Raychaudhuri, D. (2009). MAC protocol adaptation in cognitive radio networks: An experimental study. In *Proceedings of Computer Communications and Networks*, (pp. 1-6). IEEE.

Huang, Y.-K., Pang, A.-C., Hsiu, P.-C., Zhuang, W., & Liu, P. (2012, March). Distributed Throughput Optimization for ZigBee Cluster-Tree Networks. *IEEE Transactions on Parallel and Distributed Systems*, *23*(3), 513–520. doi:10.1109/TPDS.2011.192

Huberman, S., Leung, C., & Le-Ngoc, T. (2011). A clustering approach to autonomous spectrum balancing using multiple reference lines for dsl. In *Proceedings of IEEE Global Telecommunications Conference* (pp. 1–5). IEEE. doi:10.1109/GLOCOM.2011.6133937

Huberman, S., Leung, C., & Le-Ngoc, T. (2012). Constant offset autonomous spectrum balancing using multiple reference lines for vdsl. *IEEE Transactions on Signal Processing*, *60*(12), 6719–6723. doi:10.1109/TSP.2012.2215606

Huberman, S., Leung, C., & Le-Ngoc, T. (2012). Dynamic spectrum management (dsm) algorithms for multi-user xdsl. *IEEE Communications Surveys and Tutorials*, *14*(1), 109–130. doi:10.1109/SURV.2011.092110.00090

Hulbert, A. P. (2005). Spectrum sharing through beacons. In *Proceedings of IEEE International Symposium on Personal, Indoor and Mobile Radio Communications* (pp. 989–993). IEEE.

IEEE 802.22 WG on WRAN. (2011). Retrieved February 21, 2014, from http://www.ieee802.org/22/

IEEE P802.22/D0.5. (2008). *Draft Standard for Wireless Regional Area Networks Part 22: Cognitive Wireless RAN Medium Access Control (MAC) and Physical Layer (PHY) specifications: Policies and procedures for operation in the TV Bands*. IEEE.

IEEE Std 802.22-2011. (2011). *IEEE Standard for Information Technology Telecom-munications and information exchange between systems Wireless Regional Area Networks (WRAN) Specific requirements — Part 22: Cognitive Wireless RAN Medium Access Control (MAC) and Physical Layer (PHY) Specifications: Policies and Procedures for Operation in the TV Bands*. IEEE.

Ikram, A., & Rashdi, A. (2012, October). Complexity analysis of eigenvalue based spectrum sensing techniques in cognitive radio networks. In *Proceedings of Communications (APCC)*, (pp. 290-294). IEEE.

Ileri, O., Samardzija, D., & Mandayam, N. B. (2005). Demand responsive pricing and competitive spectrum allocation via a spectrum server. In *Proceedings of the 1st IEEE International Symposium on New Frontiers in Dynamic Spectrum Access Networks* (pp. 194-202). Baltimore, MD: IEEE. doi:10.1109/DYSPAN.2005.1542635

Im, S., Kang, Y., Kim, W., Kim, S., Kim, J., & Lee, H. (2011). Dynamic spectrum allocation with efficient sinr-based interference management. In *Proceedings of IEEE Vehicular Technology Conference* (pp. 1–5). IEEE. doi:10.1109/VETECF.2011.6092824

Ireson-Paine, J. (2013). *Expert systems*. Retrieved February 10, 2014, from http://www.j-paine.org/students/lectures/lect3/lect3.html.abgerufen

Ismail, M. H., & Matalgah, M. M. (2006). BER analysis of BPSK modulation over the Weibull fading channel with CCI. In *Proceedings of the Vehicular Technology Conference* (pp. 1-5). Montreal, Canada: Academic Press. doi:10.1109/VTCF.2006.334

Izzo, L., Paura, L., & Tanda, M. (1992). Signal interception in non-Gaussian noise. *IEEE Transactions on Communications*, *40*(6), 1030–1037. doi:10.1109/26.142793

Jaafar, W., Ajib, W., & Haccoun, D. (2011). A novel relay-aided transmission scheme in cognitive radio networks. In *Proceedings of the IEEE Global Telecommunications Conference* (pp. 1-6). Houston, TX: IEEE. doi:10.1109/GLOCOM.2011.6133927

Jääskeläinen, P. O., de La Lama, C. S., Huerta, P., & Takala, J. H. (2010). OpenCL-based design methodology for application-specific processors. In *Proceedings of International Conference on Embedded Computer Systems: Architectures, Modeling and Simulation* (pp. 223–230). Samos, Greece: Academic Press.

Jackson, D. A. (1993). Stopping rules in principal components analysis: A comparison of heuristical and statistical approaches. *Ecology*, *74*(8), 2204–2214. doi:10.2307/1939574

Jacob, P., Madhukumar, A., & Alphones, A. (2013). Interference Mitigation through Cross Polarized Transmission in Femto-Macro Network. *IEEE Communications Letters*, *17*(10), 1940–1943. doi:10.1109/LCOMM.2013.090213.131482

Jain, R. (2012). Network market design part I: Bandwidth markets. *IEEE Communications Magazine*, *50*(11), 78–83. doi:10.1109/MCOM.2012.6353686

Jalier, C., Lattard, D., Jerraya, A., Sassatelli, G., Benoit, P., & Torres, L. (2010). Heterogeneous vs homogeneous MPSoC approaches for a mobile LTE modem. In *Proceedings of Conference on Design, Automation and Test in Europe* (pp. 184–189). Dresden, Germany: Academic Press. doi:10.1109/DATE.2010.5457213

Jashni, B., Tadaion, A. A., & Ashtiani, F. (2010). Dynamic link/frequency selection in multi-hop cognitive radio networks for delay sensitive applications. In *Proceedings of IEEE 17th International Conference on Telecommunications (ICT)*, (pp. 128-132). IEEE.

Jayaweera, S., & Christodoulou, C. (2011). *Radiobots: Architecture, Algorithms and Realtime Reconfigurable Antenna Designs for Autonomous, Self-learning Future Cognitive Radios*. UNM Technical Report: EECE-TR-11-0001. Retrieved from http://repository.unm.edu

Jayaweera, S. K., Bkassiny, M., & Avery, K. A. (2011, August). Asymmetric Cooperative Communications Based Spectrum Leasing via Auctions in Cognitive Radio Networks. *IEEE Transactions on Wireless Communications*, *10*(8), 2716–2724. doi:10.1109/TWC.2011.061311.102044

Jayaweera, S. K., & Li, M. T. (2009). Dynamic Spectrum Leasing in Cognitive Radio Networks via Primary-Secondary User Power Control Games. *IEEE Transactions on Wireless Communications*, *8*(6), 3300–3310. doi:10.1109/TWC.2009.081230

Jayaweera, S. K., Vazquez-Vilar, C., & Mosquera, C. (2010). Dynamic spectrum leasing: A new paradigm for spectrum sharing in cognitive radio networks. *IEEE Transactions on Vehicular Technology*, *59*(5), 2328–2339. doi:10.1109/TVT.2010.2042741

Jelenkovi, P., & Tan. (2008). Dynamic packet fragmentation for wireless channels with failures. In *Proceedings of ACM MobiHoc*. ACM.

Jeong, S. S., Jeong, D. G., & Jeon, W. S. (2009, October). Nonquiet primary user detection for OFDMA-based cognitive radio systems. *IEEE Transactions on Wireless Communications*, *8*(10), 5112–5123. doi:10.1109/TWC.2009.080970

Jia, Y. Y., Tan, Z. H., Huang, Q., & Zhou, Z. Y. (2011). Study on the Measurement, Analysis and Modeling of Primary User Spectrum Occupation in GSM Band. Institute of Broadband Wireless Mobile Communication.

Jia, J., Zhang, J., & Zhang, Q. (2009). Cooperative relay for cognitive radio networks. In *Proc. INFOCOM*, (pp. 2304-2312). IEEE.

Jiang, J. R., Tseng, Y. C., Hsu, C. S., & Lai, T. H. (2003). Quorum based asynchronous power-saving protocols for IEEE 802.11 ad hoc networks. In *Proceedings of IEEE International Conference on Parallel Processing*. IEEE.

Jiang, Y. (2010). *A note on applying stochastic network calculus*. Retrieved May 1, 2010, from http://q2s,ntnu.no/-iiang/publications/note-on-applying-snetcal-v20100501

Jiang, H., Lai, L., Fan, R., & Poor, H. V. (2009). Optimal selection of channel sensing order in cognitive radio. *IEEE Transactions on* Wireless Communications, *8*(1), 297–307.

Jiang, J. R., Tseng, Y. C., Hsu, C. S., & Lai, T. H. (2005). Quorum based asynchronous power-saving protocols for IEEE 802.11 ad hoc networks. *Mobile Networks and Applications, 10*(1), 169–181. doi:10.1023/B:MONE.0000048553.45798.5e

Jiang, Y., & Liu, Y. (2008). *Stochastic network calculus.* Berlin: Springer.

Jiao, L., Li, F. Y., & Pla, V. (2011). Dynamic channel aggregation strategies in cognitive radio networks with spectrum adaptation. In *Proceedings of IEEE Global Telecommunications Conference* (GLOBECOM), (pp. 1-6). IEEE.

Jiao, L., Li, F. Y., & Pla, V. (2012). Modeling and performance analysis of channel assembling in multichannel cognitive radio networks with spectrum adaptation. *IEEE Transactions on Vehicular Technology, 61*(6), 2686–2697.

Jin, Z., Anand, S., & Subbalakshmi, K. P. (2009, June). Detecting primary user emulation attacks in dynamic spectrum access networks. In *Proceedings of Communications*, (pp. 1-5). IEEE. doi:10.1109/ICC.2009.5198911

Jin, Z., Anand, S., & Subbalakshmi, K. P. (2012). Impact of primary user emulation attacks on dynamic spectrum access networks. *IEEE Transactions on Communications, 60*(9), 2635–2643. doi:10.1109/TCOMM.2012.071812.100729

Ji, Z., & Liu, K. J. R. (2008). Dynamic spectrum sharing: A game theoretical overview. *IEEE Transactions on Wireless Communications, 7*(11), 4273–4283.

Jolliffe, I. (1986). *Principal Component Analysis.* Springer-Verlag. doi:10.1007/978-1-4757-1904-8

Jondral, F., Wiesler, A., & Machauer, R. (2000). A Software Defined Radio Structure for 2nd and 3rd Generation Mobile Communication Standards. In *Proceedings of IEEE International Symposium on Spread Spectrum Techniques and Applications.* IEEE. doi:10.1109/ISSSTA.2000.876511

Jondral, F. K. (2005). Software-Defined Radio: Basics and Evolution to Cognitive Radio. *EURASIP Journal on Wireless Communications and Networking*, (3): 275–283.

Jondral, F. K. (2005, August). Software-Defined Radio - Basics and Evolution to Cognitive Radio. *EURASIP Journal on Wireless Communications and Networking*, (3): 275–283.

Jorge, C., Wu, M.-Y., & Shu, W. (2008). Protocols and architectures for channel assignment in wireless mesh networks. *Ad Hoc Networks, 6*(7), 1051–1077. doi:10.1016/j.adhoc.2007.10.002

Joseph, P. M., Mwangoka, W., & Rodriguez, J. (2011). Broker based secondary spectrum trading. In *Proceedings of 2011 6th International ICST Conference on Cognitive Radio Oriented Wireless Networks and Communications* (CROWNCOM), (pp. 186-190). ICST.

Joung, J., & Sayed, A. H. (2010, March). Multiuser two-way amplify-and-forward relay processing and power control methods for beamforming systems. *IEEE Transactions on Signal Processing, 58*(3), 1833–1846. doi:10.1109/TSP.2009.2038668

Jung, S. M., Kim, T. K., Eom, J. H., & Chung, T. M. (2013). The Quantitative Overhead Analysis for Effective Task Migration in Biosensor Networks. In Proceedings of BioMed Research International. Biosensor Networks.

Kahn, G. (1974). The semantics of a simple language for parallel programming. In *Proceedings of Information Processing* (pp. 471 – 475). Stockholm, Sweden: IFIP.

Kalil, M. A., Puschmann, A., & Mitschele-Thiel, A. (2012). Switch: A multichannel mac protocol for cognitive radio ad hoc networks. In *Proceedings of IEEE Vehicular Technology Conference (VTC Fall).* IEEE. doi:10.1109/VTCFall.2012.6399238

Kaltiokallio, M., Saari, V., Kallioinen, S., Parssinen, A., & Ryynnen, J. (2012). Wideband 2 to 6 GHz RF front-end with blocker filtering. *IEEE Journal of Solid-State Circuits, 47*(7), 1636–1645. doi:10.1109/JSSC.2012.2191348

Kamruzzaman, S., Kim, E., Jeong, D. G., & Jeon, W. S. (2012). Energy-aware routing protocol for cognitive radio ad hoc networks. *IET Communications, 6*(14), 2159–2168. doi:10.1049/iet-com.2011.0698

Kandeepan, S., De Nardis, L., Di Benedetto, M., Guidotti, A., & Corazza, G. E. (2010). Cognitive Satellite Terrestrial Radios. In *Proceedings of IEEE Global Telecommunications Conference (GLOBECOM 2010)*, (pp. 1-6). Miami, FL: IEEE.

Kandukuri, S., & Boyd, S. (2010). Optimal power control in interference-limited fading wireless channels with outage-probability specifications. *IEEE Transactions on Wireless Communications*, *1*(1), 46–55. doi:10.1109/7693.975444

Kang, B.-J. (2009, September). Spectrum sensing issues in cognitive radio networks. In *Proceedings of Communications and Information Technology*, (pp. 824-828). IEEE.

Kang, X., Zhang, R., Liang, Y.-C., & Garg, H. K. (2009). Optimal power allocation for cognitive radio under primary user's outage loss constraint. In *Proceedings of IEEE International Conference on Communications (ICC)*. IEEE.

Kang, W., Lee, S., & Kim, K. (2010). Design of symmetric beam pattern reconfigurable antenna. *Electronics Letters*, *46*(23), 1536–1537. doi:10.1049/el.2010.2527

Kang, W., Park, J., & Yoon, Y. (2008). Simple reconfigurable antenna with radiation pattern. *Electronics Letters*, *44*(3), 182–183. doi:10.1049/el:20082994

Karlof, C., & Wagner, D. (2003). Secure routing in wireless sensor networks: Attacks and countermeasures. *Ad Hoc Networks*, *1*(2), 293–315. doi:10.1016/S1570-8705(03)00008-8

Kartheek, M., Misra, R., & Sharma, V. (2011, January). Performance analysis of data and voice connections in a cognitive radio network. In *Proceedings of Communications (NCC)*, (pp. 1-5). IEEE. doi:10.1109/NCC.2011.5734764

Kauffmann, B., Baccelli, F., Chaintreau, A., Papagiannaki, K., & Diot, C. (2005). *Self organization of interfering 802.11 wireless access networks*. Academic Press.

Kay, S. (1998). *Fundamentals of Statistical Signal Processing: Detection Theory*. Academic Press.

Kelly, F. (1996). *Notes on effective bandwidths. In Stochastic networks: Theory and applications* (pp. 141–168). Oxford, UK: Oxford University Press.

Kelly, F. P. (1997). Charging and Rate Control for Elastic Traffic. *Euro. Trans. Telecommun.*, *8*(1), 33–37. doi:10.1002/ett.4460080106

Kennedy, J., & Eberhart, R. (1995). Particle swarm optimization. In *Proceedings of IEEE International Conference on Neural Networks*. IEEE.

Kennedy, J., & Eberhart, R. (1995). Particle swarm optimization. In *Proceedings of ICNN'95 - International Conference on Neural Networks* (Vol. 4, pp. 1942–1948). IEEE. doi:10.1109/ICNN.1995.488968

Kennedy, J., & Eberhart, R. C. (2001). *Swarm Intelligence*. Morgan Kauffman Publishers.

Khalaf, Z., Nafkha, A., & Palicot, J. (2012, June). Blind spectrum detector for cognitive radio using compressed sensing and symmetry property of the second order cyclic autocorrelation. In *Proceedings of 7th International ICST Conference on Cognitive Radio Oriented Wireless Networks and Communications (CROWNCOM)*, (pp. 291-296). IEEE.

Khalife, H., Ahuja, S., Malouch, N., & Krunz, M. (2008). Probabilistic path selection in opportunistic cognitive radio networks. In *Proceedings of IEEE Global Telecommunications Conference (GlobeCom)*, (pp. 1-5). IEEE. doi:10.1109/GLOCOM.2008.ECP.931

Khan, Z., Lehtomaki, J. J., Mustonen, M., & Matinmikko, M. (2011, June). Sensing order dispersion for autonomous cognitive radios. In *Proceedings of Cognitive Radio Oriented Wireless Networks and Communications (CROWNCOM)*, (pp. 191-195). IEEE. doi:10.4108/icst.crowncom.2011.246247

Khan, Z., Glisic, S., DaSilva, L., & Lehtomaki, J. (2011). Modeling the dynamics of coalition formation games for cooperative spectrum sharing in an interference channel. *Transactions on Computational Intelligence and Artificial Intelligence in Games*, *3*(1), 17–30. doi:10.1109/TCIAIG.2010.2080358

Khan, Z., Lehtomaeki, J., Codreanu, M., Latva-aho, M., & DaSilva, L. A. (2010). Throughput efficient dynamic coalition formation in distributed cognitive radio networks. *EURASIP Journal on Wireless Communications and Networking*, *2010*(87).

Khasawneh, M., & Agarwal, A. (2014, March). A survey on security in Cognitive Radio networks. In *Proceedings of Computer Science and Information Technology* (CSIT), (pp. 64-70). IEEE. doi:10.1109/CSIT.2014.6805980

Khasawneh, M., Agarwal, A., Goel, N., Zaman, M., & Alrabaee, S. (2012, July). Sureness efficient energy technique for cooperative spectrum sensing in cognitive radios. In *Proceedings of Telecommunications and Multimedia (TEMU)*, (pp. 25-30). IEEE. doi:10.1109/TEMU.2012.6294727

Khasawneh, M., Agarwal, A., Goel, N., Zaman, M., & Alrabaee, S. (2012a, September). A game theoretic approach to power trading in cognitive radio systems. In *Proceedings of Software, Telecommunications and Computer Networks* (SoftCOM), (pp. 1-5). IEEE.

Khasawneh, M., Agarwal, A., Goel, N., Zaman, M., & Alrabaee, S. (2012b, October). A price setting approach to power trading in cognitive radio networks. In *Proceedings of Ultra Modern Telecommunications and Control Systems and Workshops* (ICUMT), (pp. 878-883). IEEE.

Khasawneh, M., Agarwal, A., Goel, N., Zaman, M., & Alrabaee, S. (2012c, July). Sureness efficient energy technique for cooperative spectrum sensing in cognitive radios. In *Proceedings of Telecommunications and Multimedia* (TEMU), (pp. 25-30). IEEE.

Khedr, M., & Shatila, H. (2009). Cogmax- A cognitive radio approach for wimax systems. In *Proceedings of IEEE/ACS International Conference on Computer Systems and Applications (AICCSA 2009)*. IEEE. doi:10.1109/AICCSA.2009.5069379

Kim, H., & Shin, K. G. (2008, October). Fast discovery of spectrum opportunities in cognitive radio networks. In *Proceedings of New Frontiers in Dynamic Spectrum Access Networks*, (pp. 1-12). IEEE. doi:10.1109/DYSPAN.2008.30

Kim, J., & Shin, H. (2011). Design considerations for cognitive radio based CMOS TV white space transceivers. In *Proceedings of International SoC Design Conference (ISOCC)*, (pp. 238-241). ISOCC.

Kim, M., Kimtho, P., & Takada, J.-i. (2010). Performance enhancement of cyclostationarity detector by utilizing multiple cyclic frequencies of OFDM signals. In *Proceedings of IEEE Symposium on New Frontiers in Dynamic Spectrum* (pp. 1–8). IEEE. doi:10.1109/DYSPAN.2010.5457876

Kim, B., Pan, B., Nikolaou, S., Kim, Y.-S., Papapolymerou, J., & Tentzeris, M. (2008). A novel single-feed circular microstrip antenna with reconfigurable polarization capability. *IEEE Transactions on Antennas and Propagation*, 56(3), 630–638. doi:10.1109/TAP.2008.916894

Kim, H., & Shin, K. G. (2008). Efficient discovery of spectrum opportunities with MAC-layer sensing in cognitive radio networks. *IEEE Transactions on Mobile Computing*, 7(5), 533–545. doi:10.1109/TMC.2007.70751

Kim, H., & Shin, K. G. (2013). Optimal online sensing sequence in multichannel cognitive radio networks. *IEEE Transactions on* Mobile Computing, 12(7), 1349–1362.

Kim, J. O. (1978). *Factor Analysis: Statistical Methods and Practical Issues*. Beverly Hills, CA: Sage.

Kirolos, S., Ragheb, T., Laska, J., Duarte, M. E., Massoud, Y., & Baraniuk, R. G. (2006). Practical Issues in Implementing Analog-to-Information Converters. In *Proceedings of the 6th International Workshop on System-on-Chip for Real-Time Applications*, (pp. 141-146). Cairo: Academic Press.

Kitsunezuka, M., Kodama, H., Oshima, N., Kunihiro, K., Maeda, T., & Fukaishi, M. (2012). A 30-MHz–2.4-GHz CMOS receiver with integrated RF filter and dynamic-range-scalable energy detector for cognitive radio systems. *IEEE Journal of Solid-State Circuits*, 47(5), 1084–1093. doi:10.1109/JSSC.2012.2185531

Klozar, L., & Prokopec, J. (2011, April). Propagation path loss models for mobile communication. In *Proceedings of Radioelektronika (RADIOELEKTRONIKA)*, (pp. 1-4). IEEE. doi:10.1109/RADIOELEK.2011.5936478

Kondareddy, Y. R., & Agrawal, P. (2008). Selective broadcasting in multi-hop cognitive radio networks. In *Proceedings of IEEE Sarnoff Symposium*, (pp. 1-5). IEEE. doi:10.1109/SARNOF.2008.4520042

Kortun, A., Ratnarajah, T., Sellathurai, M., & Zhong, C. (2010). On the Performance of Eigenvalue-Based Spectrum Sensing for Cognitive Radio. In *Proceedings of IEEE Symposium on New Frontiers in Dynamic Spectrum*, (pp. 1-6). IEEE. doi:10.1109/DYSPAN.2010.5457844

Kostylev, V. I. (2002). Energy detection of a signal with random amplitude. In *Proceedings of the IEEE International Conference on Communications* (pp.1606-1610). New York, NY: IEEE. doi:10.1109/ICC.2002.997120

Krishnamurthy, S., Thoppian, M., Venkatesan, S., & Prakash, R. (2005). Control channel based MAC-layer configuration, routing and situation awareness for cognitive radio networks. In *Proceedings of IEEE Military Communications Conference (MILCOM)*, (pp. 455-460). IEEE. doi:10.1109/MILCOM.2005.1605725

Kumar, A. (1991). Hierarchical quorum consensus: A new algorithm for managing replicated data. *Journal IEEE Transaction on Computers*, *40*(9), 996–1004. doi:10.1109/12.83661

Kuo, Y. C. (2010). Quorum-based power-saving multicast protocols in the asynchronous ad hoc network. *Computer Networks*, *54*(11), 1911–1922. doi:10.1016/j.comnet.2010.03.001

Kusaladharma, S., & Tellambura, C. (2012). Aggregate interference analysis for underlay cognitive radio networks. *IEEE Wireless Communications Letters*, *1*(6), 641–644. doi:10.1109/WCL.2012.091312.120600

Kwan, A., Bassam, S. A., & Ghannouchi, F. M. (2012). Sub-sampling technique for spectrum sensing in cognitive radio systems. In *Proceedings of IEEE Radio and Wireless Symposium (RWS)*, (pp. 347-350). IEEE. doi:10.1109/RWS.2012.6175350

Kwon, S., & Stuber, G. (2011). Geometrical theory of channel depolarization. *IEEE Transactions on Vehicular Technology*, *60*(8), 3542–3556. doi:10.1109/TVT.2011.2163094

Labonte, F., Mattson, P., Thies, W., Buck, I., Kozyrakis, C., & Horowitz, M. (2004). The stream virtual machine. In *Proceedings of 13th International Conference on Parallel Architecture and Compilation Techniques (PACT)* (pp. 267–277). Stanford, CA: Academic Press.

LaGrone, A. H. (1960). Forecasting television service fields. *Proceedings of the IRE*, *48*(6), 1009–1015. doi:10.1109/JRPROC.1960.287501

Lai, L., Jiang, H., & Poor, H. V. (2008, October). Medium access in cognitive radio networks: A competitive multi-armed bandit framework. In *Proceedings of Signals, Systems and Computers*, (pp. 98-102). IEEE. doi:10.1109/ACSSC.2008.5074370

Lai, L., El Gamal, H., Jiang, H., & Poor, H. V. (2011). Cognitive medium access: Exploration, exploitation, and competition. *IEEE Transactions on* Mobile Computing, *10*(2), 239–253.

Lai, S., Ravindran, B., & Hyeonjoong, C. (2010). Heterogeneous Quorum-Based Wake-Up Scheduling in Wireless Sensor Networks. *IEEE Transactions on Computers*, *59*(11), 1562–1575. doi:10.1109/TC.2010.20

Lang, S., & Mao, L. (1998). A torus quorum protocol for distributed mutual exclusion. In *Proceedings of International Conference on Parallel and Distributed Systems (ICPADS)*. Academic Press.

La, R., & Anantharam, V. (2002). A game-theoretic look at the Gaussian multiaccess channel. *DIMACS Series in Discrete Mathematics and Theoretical Computer Science*, *66*, 87–106.

Larson, L., Abdelhalem, S., Thomas, C., & Gudem, P. (2012). High-performance silicon-based RF front-end design techniques for adaptive and cognitive radios. In *Proceedings of IEEE Compound Semiconductor Integrated Circuit Symposium (CSICS)*, (pp. 1-4). IEEE. doi:10.1109/CSICS.2012.6340060

Lasaulce, S., Hayel, Y., Azouzi, R., & El,. (2009). Introduce hierarchy in energy games. *IEEE Transactions on Wireless Communications*, *8*(7), 3833–3843. doi:10.1109/TWC.2009.081443

Lathi, B. P., & Ding, Z. (2009). *Modern Digital and Analog Communication Systems*. Oxford University Series.

Lazrak, O., Leray, P., & Moy, C. (2012). HDCRAM Proof-of-Concept for Opportunistic Spectrum Access. In *Proceedings of 15th Euromicro Conference on Digital System Design* (pp. 453–458). Izmir, Turkey: Academic Press.

Le Boudec, J. Y., & Thiran, P. (2001). *Network calculus: A theory of deterministic queuing systems for the internet.* Berlin: Springer. doi:10.1007/3-540-45318-0

Le, H. S. T., & Liang, Q. (2007). An efficient power control scheme for cognitive radios. In *Proceedings of IEEE Wireless Communications and Networking Conference (WCNC).* IEEE. doi:10.1109/WCNC.2007.476

Le, H.-S. T., & Ly, H. D. (2008). Opportunistic spectrum access using fuzzy logic for cognitive radio networks. In *Proceedings of International Conference on Communications and Electronics (ICCE)* (pp. 240-245). Academic Press. doi:10.1109/CCE.2008.4578965

Leake, D. B. (n.d.). Case-based reasoning. In *Encyclopedia of Computer Science.* Retrieved January 10, 2013, from http://dl.acm.org/citation.cfm?id=1074100.1074199

Leaves, P., Moessner, K., Tafazolli, R., Grandblaise, D., Bourse, D., Tonjes, R., & Breveglieri, M. (2004). Dynamic spectrum allocation in composite reconfigurable wireless networks. *IEEE Communications Magazine, 42*(5), 72–81. doi:10.1109/MCOM.2004.1299346

Lee, C. H., & Wolf, W. (2008). Energy efficient techniques for cooperative spectrum sensing in cognitive radios. In *Proceedings of Consumer Communications and Networking Conference,* (pp. 968-972). IEEE. doi:10.1109/ccnc08.2007.223

Lee, K., Sung, G., & Lee, I. (2011, June). *Linear precoder designs for cognitive radio multiuser MIMO downlink systems.* Paper presented at the 2011 IEEE International Conference on Communications (ICC). Kyoto, Japan. doi:10.1109/icc.2011.5963176

Lee, W., Kim, Y., Brady, M. H., & Cioffi, J. M. (2006). Band preference in dynamic spectrum management. In *Proceedings of IEEE Global Communications Conference* (*Vol. 2,* pp. 1–5). IEEE.

Lee, C. (1990). Fuzzy logic in control systems: Fuzzy logic controller. *IEEE Transactions on Systems, Man, and Cybernetics, 2*(2), 404–418. doi:10.1109/21.52551

Lee, C., & Haenggi, M. (2012). Interferenec and outage in Poisson cognitive networks. *IEEE Transactions on Wireless Communications, 11*(4), 1392–1401. doi:10.1109/TWC.2012.021512.110131

Lee, E. A., & Messerschmitt, D. G. (1987). Synchronous Data Flow. *Proceedings of the IEEE, 75*(9), 1235–1245. doi:10.1109/PROC.1987.13876

Lee, E. K., Oh, S. Y., & Gerla, M. (2010). *Randomized channel hopping scheme for anti jamming communication.* IFIP Wireless Days. doi:10.1109/WD.2010.5657713

Lee, H.-, Kim, D. M., Hwang, Y., Yu, S. M., & Kim, S.-L. (2013). Feasibility of cognitive machine-to-machine communication using cellular bands. *IEEE Wireless Communications, 20*(2), 97–103. doi:10.1109/MWC.2013.6507400

Lee, K., Sung, H., Park, E., & Lee, I. (2010, December). Joint optimization for one and two-way MIMO AF multiple-relay systems. *IEEE Transactions on Wireless Communications, 9*(12), 3671–3681. doi:10.1109/TWC.2010.102210.091021

Lee, M. K., Lim, S. C., & Yang, K. (2012). Blind Compensation for Phase Noise in OFDM Systems over Constant Modulus Modulation. *IEEE Transactions on Communications, 60*(3), 620–625. doi:10.1109/TCOMM.2011.121511.110088

Lee, W., & Akyildiz, I. F. (2008). Optimal spectrum sensing framework for cognitive radio networks. *IEEE Transactions on Wireless Communications, 7*(10), 3845–3857. doi:10.1109/T-WC.2008.070391

Lee, W.-Y., & Akyildiz, I. (2008, October). Optimal spectrum sensing framework for cognitive radio networks. *IEEE Transactions on Wireless Communications, 10,* 3845–3857.

Le, L. B., & Hossain, E. (2008). Resource allocation for spectrum underlay in cognitive radio networks. *IEEE Transactions on Wireless Communications, 7*(12), 5306–5315. doi:10.1109/T-WC.2008.070890

Leon, O., Hernandez-Serrano, J., & Soriano, M. (2009). A new cross-layer attack to TCP in cognitive radio networks. In *Proceedings of Second International Workshop on Cross Layer Design 2009. IWCLD 2009.* Academic Press. doi:10.1109/IWCLD.2009.5156526

Leon, O., Hernandez-Serrano, J., & Soriano, M. (2010). Securing cognitive radio networks. *International Journal of Communication Systems, 23,* 633–652.

Leschhorn, R., & Buchin, B. (2004). Military software defined radios-rohde and schwarz status and perspective. In *Proceedings of the SDR 04 Technical Conference and Product Exposition.* Academic Press.

Letaief, K.B., & Zhang, W. (n.d.). Cooperative communications for cognitive radio networks. *Proceedings of IEEE, 97*(5), 878-893.

Leung, C., Huberman, S., & Le-Ngoc, T. (2010). Autonomous spectrum balancing using multiple reference lines for digital subscriber lines. In *Proceedings of IEEE Global Communications Conference* (pp. 1–6). IEEE. doi:10.1109/GLOCOM.2010.5683960

Li, L., Zhou, X., Xu, H., Li, G., Wang, D., & Soong, A. (2011). Simplified relay selection and power allocation in cooperative cognitive radio systems. *IEEE Trans. Wirel. Commun., 10*(1), 33-36.

Li, S., Hong, S., Han, Z., & Wu, Z. (2011). Bayesian Compressed Sensing based Dynamic Joint Spectrum Sensing and Primary User Localization for Dynamic Spectrum Access. In *Proceedings of Global Telecommunications Conference (GLOBECOM 2011)*, (pp. 1-5). IEEE.

Li, X., Han, Q., Chakravarthy, V., & Wu, Z. (2010). Joint spectrum sensing and primary user localization for cognitive radio via compressed sensing. In Proceedings of Military Communications Conference, (pp. 329--334). IEEE.

Li, Z., Shi, P., Chen, W., & Yan, Y. (2011). Square-Law Combining Double-threshold Energy Detection in Nakagami Channel. *International Journal of Digital Content Technology & its Applications, 5*(12), 307-315.

Liang, J., Chiau, C. C., Chen, X., & Parini, C. (2005). Study of a printed circular disc monopole antenna for UWB systems. *IEEE Transactions on Antennas and Propagation, 53*(11), 3500–3504. doi:10.1109/TAP.2005.858598

Liang, Y. C., Zeng, Y., Peh, E. C., & Hoang, A. T. (2008). Sensing-throughput tradeoff for cognitive radio networks. *IEEE Transactions on Wireless Communications, 7*(4), 1326–1337. doi:10.1109/TWC.2008.060869

Liang, Y.-C., Chen, K.-C., Li, G. Y., & Mähönen, P. (2011). Cognitive radio networking and communications: An overview. *IEEE Transactions on Vehicular Technology, 60*(7), 3386–3407. doi:10.1109/TVT.2011.2158673

Liang, Y., Somekh-Baruch, A., Poor, H. V., Shamai, S., & Verdú, S. (2009). Capacity of cognitive interference channels with and without secrecy. *IEEE Transactions on* Information Theory, *55*(2), 604–619.

Li, C., Yang, L., & Zhu, W. (2010, December). Two-way MIMO relay precoder design with channel state information. *IEEE Transactions on Communications, 58*(12), 3358–3363. doi:10.1109/TCOMM.2010.12.100002

Li, D. (2010). Performance analysis of uplink cognitive cellular networks with opportunistic scheduling. *IEEE Communications Letters, 14*(9), 827–829. doi:10.1109/LCOMM.2010.072910.100962

Li, D. (2011). Efficient power allocation for multiuser cognitive radio networks. *Wireless Personal Communications, 59*(4), 589–597. doi:10.1007/s11277-010-9926-6

Lien, S. Y., Tseng, C. C., & Chen, K. C. (2008). Carrier sensing based multiple access protocols for cognitive radio networks. In *Proceedings of the IEEE International Conference on Communications* (pp.3208-3214). Beijing, China: IEEE. doi:10.1109/ICC.2008.604

Lien, S.-Y., Cheng, S.-M., Shih, S.-Y., & Chen, K.-C. (2012). Radio Resource Management for QoS Guarantees in Cyber-Physical Systems. *IEEE Transactions on Parallel and Distributed Systems, 23*(9), 1752–1761. doi:10.1109/TPDS.2012.151

Lim, K., & Laskar, J. (2008). Emerging opportunities of RF IC/system for future cognitive radio wireless communications. In *Proceedings of IEEE Radio and Wireless Symposium*, (pp. 703-706). IEEE. doi:10.1109/RWS.2008.4463589

Limberg, T., Winter, M., Bimberg, M., Klemm, R., Matus, E., Tavares, M. B. S., ... Robelly, P. (2008). A fully programmable 40 GOPS SDR single chip baseband for LTE/WiMAX terminals. In *Proceedings of Solid-State Circuits Conference*, (pp. 466–469). Edinburgh, UK: Academic Press. doi:10.1109/ESSCIRC.2008.4681893

Lime Microsystems. (2014). *Ultra flexible FPRF solutions*. LMS6002D and LMS7002D Configurable Broadband Transceiver IC Datasheets. Retrieved from http://www.limemicro.com

Lin, D. T., Hyungil Chae, L. L., & Flynn, M. P. (2012). A 600MHz to 3.4GHz flexible spectrum-sensing receiver with spectrum-adaptive reconfigurable DT filtering. In *Proceedings of IEEE Radio Frequency Integrated Circuits Symposium (RFIC)*, (pp. 269-272). IEEE. doi:10.1109/RFIC.2012.6242279

Lin, F., Hu, Z., Hou, S., Yu, J., Zhang, C., Guo, N., et al. (2011, July). Cognitive radio network as wireless sensor network (ii): Security consideration. In *Proceedings of Aerospace and Electronics Conference* (NAECON), (pp. 324-328). IEEE.

Lin, L., Guo, C., Feng, C., & Wu, X. (2012). A dual-polarized antenna based sensing algorithm for OFDM signals in LTE-Advanced systems. In Proceedings of Globecom Workshops (GC Wkshps), 2012 IEEE, (pp. 697-702). IEEE.

Lin, L., Guo, C., Feng, C., & Wu, X. (2012). Degree of polarization spectrum sensing algorithm for cognitive radios. In *Proceedings of 2012 15th International Symposium on Wireless Personal Multimedia Communications (WPMC)*, (pp. 35-39). WPMC.

Lin, X., & Rasool, S. (2007). A distributed joint channel-assignment, scheduling and routing algorithm for multi-channel ad-hoc wireless networks. In *Proceedings of IEEE 26th International Conference on Computer Communications (INFOCOM)*, (pp. 1118-1126). IEEE. doi:10.1109/INFCOM.2007.134

Lin, Y., Mullenix, R., Woh, M., Mahlke, S., Mudge, T., Reid, A., & Flautner, K. (2006). SPEX: A programming language for software defined radio. In *Proceedings of SDR Forum Technical Conference* (pp. 13 – 17). Orlando, FL: SDR.

Lin, Z., Liu, H., Chu, X., & Leung, Y. W. (2011). Jump-stay based channel-hopping algorithm with guaranteed rendezvous for cognitive radio networks. In *Proceedings of IEEE International Conference on Computer Communications (INFOCOM)* (pp. 2444–2452). IEEE. doi:10.1109/INFCOM.2011.5935066

Linderman, M. D., Collins, J. D., Wang, H., & Meng, T. H. Y. (2008). Merge : A Programming Model for Heterogeneous Multi-core Systems. In *Proceedings of the 13th International Conference on Architectural Support for Programming Languages and Operating Systems* (pp. 287–296). Seattle, WA: Academic Press. doi:10.1145/1346281.1346318

Lindmark, B., & Nilsson, M. (2001). On the available diversity gain from different dual-polarized antennas. *IEEE Journal on Selected Areas in Communications, 19*(2), 287–294. doi:10.1109/49.914506

Lin, P., Jia, J., Zhang, Q., & Hamdi, M. (2010). Dynamic spectrum sharing with multiple primary and secondary users. In *Proc. IEEE Int'l Conf. Commun.* IEEE. doi:10.1109/ICC.2010.5502744

Lin, Z., Liu, H., Chu, X., & Leung, Y. W. (2013). Enhanced Jump-Stay Rendezvous Algorithm for Cognitive Radio Networks. *IEEE Communications Letters*, 1–4.

Liu, F., Chunyan, F., Caili, G., Wang, Y., & Dong, W. (2010). Virtual Polarization Detection: A Vector Signal Sensing Method for Cognitive Radios. In *Proceedings of IEEE 71st Vehicular Technology Conference (VTC 2010-Spring)*, (pp. 1-5). IEEE.

Liu, F., Feng, C., Guo, C., Wang, Y., & Dong, W. (2009). Polarization Spectrum Sensing Scheme for Cognitive Radios. In *Proceedings of Int. Conf. on Wireless Comm. Netwokring and Mobile Computing*, (pp. 1-4). Beijing: Academic Press. doi:10.1109/WICOM.2009.5301939

Liu, H., Lin, Z., Chu, X., & Leung, Y. W. (2010). Ring-Walk Based Channel-Hopping Algorithms with Guaranteed Rendezvous for Cognitive Radio Networks. In *Proceedings of International Workshop on Wireless Sensor, Actuator and Robot Networks (WiSARN2010-FALL), in Conjunction with IEEE/ACM International Conference on Cyber, Physical and Social Computing (CPSCom)*. IEEE. doi:10.1109/GreenCom-CPSCom.2010.30

Liu, L., Han, Z., Wu, Z., & Qian, L. (2011). Collaborative Compressive Sensing based Dynamic Spectrum Sensing and Mobile Primary User Localization in Cognitive Radio Networks. In *Proceedings of Global Telecommunications Conference (GLOBECOM 2011)*, (pp. 1-5). IEEE.

Liu, Q., Pang, D., Wang, X., & Zhou. (2012). A Neighbor Cooperation Framework for Timeefficient Asynchronous Channel Hopping Rendezvous in Cognitive Radio Networks. In *Proceedings of IEEE Dynamic Spectrum Access Networks symposium (DySPAN)*. IEEE.

Liu, R., Wu, Y., Wang, D., Yang, Y., & Kang, S. (2013, October). A class of low complexity spectrum sensing algorithms based on statistical covariances. In *Proceedings of Wireless Communications & Signal Processing (WCSP)*, (pp. 1-5). IEEE.

Liu, J., & Annamalai, A. (2005). Channel-Aware Routing Protocol for Wireless Ad-Hoc Networks: Generalized Multiple-Route Path Selection Diversity. *IEEE Vehicular Technology Conference*, 62, 2258.

Liu, L., Hu, G., Huang, Z., & Peng, Y. (2012). White Cloud or Black Cloud-Opportunity and Challenge of Spectrum Sharing on Cloud Computing. *Advanced Materials Research*, 430-432(1), 1290–1293.

Liu, Y.-S., & Su, Z.-J. (2007). Distributed dynamic spectrum management for digital subscriber lines. *IEICE Transactions on Communications*, E90-B(3), 491–498. doi:10.1093/ietcom/e90-b.3.491

Li, Z., Yu, F. R., & Huang, M. (2009). A cooperative spectrum sensing consensus scheme in cognitive radios. In *Proceedings of the IEEE INFOCOM* (pp.2546-2550). Rio de Janeiro, Brazil: IEEE. doi:10.1109/INFCOM.2009.5062184

Loctronix. (2013). *Advanced Positioning for Challenging Environments*. ASR-2300 MIMO SDR Motion Sensing Module Datasheet. Retrieved from http//www.loctronix.com

Lodi, A., Cappelli, A., Bocchi, M., Mucci, C., Innocenti, M., & DeBartolomeis, C. et al. (2006). XiSystem: A XiRisc-Based SoC With Reconfigurable IO Module. *IEEE Journal of Solid-State Circuits*, 41(1), 85–96. doi:10.1109/JSSC.2005.859319

Lombardo, P., Pastina, D., & Bucciarelli, T. (2001). Adaptive polarimetric target detection with coherent radar. II. Detection against non-Gaussian background. *IEEE Transactions on Aerospace and Electronic Systems*, 37(4), 1207–1220. doi:10.1109/7.976960

Long, T., & Juebo, W. (2012). Research and analysis on cognitive radio network security. In *Proceedings of Wireless Sensor Network*. Academic Press.

López-Benítez, M., & Casadevall, F. (2011, June). Modeling and simulation of time-correlation properties of spectrum use in cognitive radio. In *Proceedings of Cognitive Radio Oriented Wireless Networks and Communications (CROWNCOM)*, (pp. 326-330). IEEE. doi:10.4108/icst.crowncom.2011.246158

Lourandakis, E., Weigel, R., Mextorf, H., & Knoechel, R. (2012). Circuit agility. *IEEE Microwave Magazine*, 13(1), 111–121. doi:10.1109/MMM.2011.2173987

Lugaric, L., Krajcar, S., & Simic, Z. (2010). Smart City Platform for Emergent Phenomena Power System Testbed Simulator. In Proceedings of Innovative Smart Grid Technologies Conference Europe. Academic Press. doi:10.1109/ISGTEUROPE.2010.5638890

Lui, R., & Yu, W. (2005). Low-complexity near-optimal spectrum balancing for digital subscriber lines. In *Proceedings of IEEE International Conference on Communications* (pp. 1947–1951). IEEE. doi:10.1109/ICC.2005.1494679

Luk, W. S., & Wong, T. T. (1997). Two new quorum based algorithms for distributed mutual exclusion. In *Proceedings of 17th International Conference on Distributed Computing Systems* (pp. 100–106). Academic Press.

Lunden, J., Koivunen, V., Huttunen, A., & Poor, H. V. (2007). Spectrum sensing in cognitive radios based on multiple cyclic frequencies. In *Proceedings of International Conference on Cognitive Radio Oriented Wireless Networks and Communications* (pp. 37–43). IEEE. doi:10.1109/CROWNCOM.2007.4549769

Lunden, J., Koivunen, V., Huttunen, A., & Poor, H. V. (2007, November). Censoring for collaborative spectrum sensing in cognitive radios. In *Proceedings of Signals, Systems and Computers*, (pp. 772-776). IEEE. doi:10.1109/ACSSC.2007.4487321

Lunden, J., Koivunen, V., Huttunen, A., & Poor, H. V. (2009). Collaborative cyclostationary spectrum sensing for cognitive radio systems. *IEEE Transactions on Signal Processing*, 57(11), 4182–4195. doi:10.1109/TSP.2009.2025152

Luo, L., & Roy, S. (2008). *A two-stage sensing technique for dynamic spectrum access.* Paper presented at the IEEE International Conference on Communications (ICC '08). Beijing, China. doi:10.1109/ICC.2008.785

Luu, L., & Daneshrad, B. (2007). An adaptive weaver architecture radio with spectrum sensing capabilities to relax RF component requirements. *IEEE Journal on Selected Areas in Communications, 25*(3), 538–545. doi:10.1109/JSAC.2007.070404

Lu, W., & Traore, I. (2004). Detecting new forms of network intrusion using genetic programming. *Computational Intelligence, 20*(3), 475–494. doi:10.1111/j.0824-7935.2004.00247.x

Ma, M., & Tsang, D. H. (2008). Joint spectrum sharing and fair routing in cognitive radio networks. In *Proceedings of IEEE 5th Consumer Communications and Networking Conference (CCNC)*, (pp. 978-982). IEEE. doi:10.1109/ccnc08.2007.225

Maekawa, M. (1985). A $\sqrt{N}$ algorithm for mutual exclusion in decentralized systems. *ACM Transactions on Computer Systems, 3*(2), 145–159. doi:10.1145/214438.214445

Maharjan, S., Zhang, Y., & Gjessing, S. (2011, March). Economic Approaches for Cognitive Radio Networks: A Survey. *Wireless Personal Communications, 57*(1), 33–51. doi:10.1007/s11277-010-0005-9

Mahdi, A., Mohanan, J., & Kalil, M. A.-T. (2012). Adaptive discrete particle swarm optimization for cognitive radios. In *Proceedings of IEEE International Conference on Communication (ICC)*. IEEE. doi:10.1109/ICC.2012.6364817

Ma, J., Li, G. Y., & Juang, B. H. (2009). Signal processing in cognitive radio. *Proceedings of the IEEE, 97*(5), 805–823. doi:10.1109/JPROC.2009.2015707

Ma, J., Zhao, G., & Li, Y. (2008). Soft combination and detection for cooperative spectrum sensing in cognitive radio networks. *Transactions on Wireless Communications, 7*(11), 4502–4507. doi:10.1109/T-WC.2008.070941

Majeed, P. G., & Kumar, S. (2014). Genetic Algorithms in Intrusion Detection Systems: A Survey. *International Journal of Innovation and Applied Studies, 5*(3), 233–240.

Maldonado, D., Le, B., Hugine, A., Rondeau, T. W., & Bostian, C. W. (2005). Cognitive radio applications to dynamic spectrum allocation: a discussion and an illustrative example. In *Proceedings of IEEE International Symposium on New Frontiers in Dynamic Spectrum Access Networks (DySPAN)*, (pp. 597-600). IEEE. doi:10.1109/DYSPAN.2005.1542677

Maleki, S., Chepuri, S. P., & Leus, G. (2011, June). Energy and throughput efficient strategies for cooperative spectrum sensing in cognitive radios. In *Proceedings of Signal Processing Advances in Wireless Communications (SPAWC)*, (pp. 71-75). IEEE. doi:10.1109/SPAWC.2011.5990482

Maleki, S., Pandharipande, A., & Leus, G. (2010). Two-stage spectrum sensing for cognitive radios. In *Proceedings of 2010 IEEE International Conference on Acoustics, Speech and Signal Processing* (pp. 2946–2949). IEEE. doi:10.1109/ICASSP.2010.5496149

Maleki, S., Chepuri, S. P., & Leus, G. (2013). Optimization of hard fusion based spectrum sensing for energy-constrained cognitive radio networks. *Physical Communication, 9*, 193–198. doi:10.1016/j.phycom.2012.07.003

Maleki, S., & Leus, G. (2013). Censored truncated sequential spectrum sensing for cognitive radio networks. *IEEE Journal on Selected Areas in Communications, 31*(3), 364–378.

Maleki, S., Pandharipande, A., & Leus, G. (2011). Energy-efficient distributed spectrum sensing for cognitive sensor networks. *IEEE Sensors Journal, 11*(3), 565–573. doi:10.1109/JSEN.2010.2051327

Mallat, S. G., & Zhifeng, Z. (1993). Matching pursuits with time-frequency dictionaries. *IEEE Transactions on Signal Processing, 41*(12), 3397–3415. doi:10.1109/78.258082

Maple, C., Williams, G., & Yue, Y. (2007). *Reliability, availability, and security of wireless networks in the community.* Academic Press.

Mardia, K. V., Kent, J. T., & Bibby, J. M. (1979). *Multivariate Analysis.* Academic Press.

Martin, J., Bernard, C., Clermidy, F., & Durand, Y. (2009). A Microprogrammable Memory Controller for high-performance dataflow applications. In *Proceedings of ESSCIRC* (pp. 348–351). Athens, Greece: Academic Press. doi:10.1109/ESSCIRC.2009.5325981

Maskery, M., Krishnamurthy, V., & Zhao, Q. (2009). Decentralized dynamic spectrum access for cognitive radios: Cooperative design of a non-cooperative game. *IEEE Transactions on* Communications, *57*(2), 459–469. doi:10.1109/TCOMM.2009.02.070158

Masonta, M. T., Mzyece, M., & Ntlatlapa, N. (2013). Spectrum decision in cognitive radio networks: A survey. *IEEE Communications Surveys and Tutorials*, *15*(3), 1088–1107. doi:10.1109/SURV.2012.111412.00160

Mathur, C. N., & Subbalakshmi, K. P. (2007). Security issues in cognitive radio networks. In *Cognitive networks: Towards self-aware networks*, (pp. 284-293). Academic Press.

Matsui, M., Nakagawa, T., Kudo, R., Ishihara, K., & Mizoguchi, M. (2011). Blind frequency-dependent IQ imbalance compensation scheme using CMA for OFDM system. In *Proceedings of IEEE 22nd International Symposium on Personal, Indoor and Mobile Radio Communications (PIMRC)*, (pp. 1386-1390). IEEE.

Mayer, A., Maurer, L., Hueber, G., Dellsperger, T., Christen, T., Burger, T., & Zhiheng, C. (2007). RF front-end architecture for cognitive radios. In *Proceedings of IEEE 18th International Symposium on Personal, Indoor and Mobile Radio Communications (PIMRC)*, (pp. 1-5). IEEE.

McClaning, K. (2012). *Wireless Receiver Design for Digital Communications*. Scitech Publishing.

McHenry, M. A., McCloskey, D., Roberson, D., & MacDonald, J. T. (2005). *Spectrum Occupancy Measurements Chicago*. Shared Spectrum Company.

Mehmeti, F., & Spyropoulos, T., T. (2013). Who interrupted me? Analyzing the effect of PU activity on cognitive user performance. In *Proceedings of IEEE International Conference on Communications (ICC)*. IEEE. doi:10.1109/ICC.2013.6654977

Mehrpouyan, H., Nasir, A., Blostein, S., Thomas, E., Karagiannidis, G., & Svensson, T. (2012). Joint estimation of channel and oscillator phase noise in MIMO systems. *IEEE Transactions on Signal Processing*, *60*(9), 4790–4807. doi:10.1109/TSP.2012.2202652

Mei, B., Vernalde, S., Verkest, D., De Man, H., & Lauwereins, R. (2002). DRESC: A retargetable compiler for coarse-grained reconfigurable architectures. In *Proceedings of IEEE International Conference on Field-Programmable Technology (FPT)* (pp. 166–173). Hong Kong: IEEE.

Mellia, M., Pescape, A., & Salgarelli, L. (2009). Traffic classification and its applications to modern networks, Computer Networks. *The International Journal of Computer and Telecommunications Networking*, *53*(5), 759–760.

Melo, M. A. K., Araujo dos Santos, J. C., & Dias, M. H. C. (2011, October). On the use of Tamir's model for site-specific path loss prediction of HF/VHF systems in forests. In Proceedings of Microwave & Optoelectronics Conference (IMOC), (pp. 147-151). IEEE.

Merentitis, A., & Triantafyllopoulou, D. (n.d.). Resource allocation with mac layer node cooperation in cognitive radio networks. *Hindawi International Journal of Digital Multimedia Broadcasting, 2010*(458636), 11.

Metrolink. (2012, September). *Methodology for Initial Assessment of Spectrum Requirements and Required Numbers of Base Stations in a Multi-railroad, Dense Traffic Area. Vers. No. 1.0.* Author.

Metrolink. (2013, April). *All Lines Timetable.* Author.

Miller, S. L., & Childers, D. G. (2004). *Probability and Random Processes: With Applications to Signal Processing and Communications.* Elsevier Academic Press.

Min, A. W., & Shin, K. G. (2008). Exploiting multi-channel diversity in spectrum-agile networks. In *Proceedings of IEEE 27th International Conference on Computer Communications (INFOCOM)*. IEEE. doi:10.1109/INFOCOM.2008.256

Minden, G. J., Evans, J. B., Searl, L., DePardo, D., Petty, V. R., Rajbanshi, R., … Agah, A. (2007). KUAR: A Flexible Software-Defined Radio Development Platform. In *Proceedings of 2nd IEEE International Symposium on New Frontiers in Dynamic Spectrum Access Networks* (pp. 428–439). Dublin, Ireland: IEEE. doi:10.1109/DYSPAN.2007.62

Mishali, M., & Eldar, Y. C. (2009). Blind multiband signal reconstruction: Compressed sensing for analog signals. *IEEE Transactions on* Signal Processing, *57*(3), 993–1009. doi:10.1109/TSP.2009.2012791

Mishra, S. M., Sahai, A., & Brodersen, R. W. (2006, June). Cooperative sensing among cognitive radios. In *Proceedings of Communications,* (Vol. 4, pp. 1658-1663). IEEE. doi:10.1109/ICC.2006.254957

Mishra, S., Sahai, A., & Brodersen, R. (2006). Cooperative sensing among cognitive radios. In *Proceedings of IEEE International Conference on Communications* (pp. 1658–1663). IEEE.

Mishra, A., Banerjee, S., & Arbaugh, W. (2005). Weighted coloring based channel assignment for WLANs. *SIGMOBILE Mob. Comput. Commun. Rev., 9*(3), 19–31. doi:10.1145/1094549.1094554

Mishra, S. M., Sahai, A., & Brodersen, R. W. (2006). Cooperative sensing among cognitive radios. In *Proceedings of the IEEE International Conference on Communications* (pp. 1658-1663).Istanbul, Turkey: IEEE.

Misra, R., & Kannu, A. P. (2012a, June). Optimal sensing-order in cognitive radio networks with cooperative centralized sensing. In *Proceedings of Communications (ICC),* (pp. 1566-1570). IEEE. doi:10.1109/ICC.2012.6363863

Misra, R., & Kannu, A. P. (2012b, December). Optimal decentralized sensing-orders in multi-user cognitive radio networks. In *Proceedings of Global Communications Conference (GLOBECOM),* (pp. 1447-1452). IEEE. doi:10.1109/GLOCOM.2012.6503317

Mitola Iii, J. (2000). *An Integrated Agent Architecture for Software Defined Radio.* Academic Press.

Mitola, J. (1992). Software radios-survey, critical evaluation and future directions. In *Proceedings of National Telesystems Conference* (pp. 13/15–13/23). Washington, DC: Academic Press. doi:10.1109/NTC.1992.267870

Mitola, J. (1999). Cognitive radio for flexible mobile multimedia communications. In *Proceedings of Mobile Multimedia Communications,* (pp. 3-10). IEEE. doi:10.1109/MOMUC.1999.819467

Mitola, J. (2000). *Cognitive radio: An integrated agent architecture for software defined radio.* (Ph.D. dissertation). KTH Royal Inst. Technol.

Mitola, J., & Maguire, G. Q. (1999). Cognitive radio: Making software radios more personal. *IEEE Personal Communication, 6,* 13–18.

Mitola, J. (1993). Software Radios: Survey, Critical Evaluation and Future Directions. *IEEE Aerospace and Electronic Systems Magazine, 8*(4), 25–36. doi:10.1109/62.210638

Mitola, J. (1995). The software radio architecture. *IEEE Communications Magazine, 33*(5), 26–38. doi:10.1109/35.393001

Mitola, J. (2000). *Cognitive radio: an integrated agent architecture for software defined radio. KTH.* Stockholm: The Royal Institute of Technology.

Mitola, J. III. (2006). Cognitive radio architecture. In F. H. Fitzek & M. D. Katz (Eds.), *Cooperation in Wireless Networks: Principles and Applications* (pp. 243–311). Springer. doi:10.1007/1-4020-4711-8_9

Mitola, J., & Maguire, G. Q. (1999). Cognitive radio: Making software radios more personal. *IEEE Personal Communications, 6*(4), 13–18. doi:10.1109/98.788210

Moore, A. W., & Papagiannaki, K. (2005, March). Toward the accurate identification of network application. In *Proceedings of International Workshop Passive and Active Meassurement* (pp. 41-54). Boston, MA: Academic Press. doi:10.1007/978-3-540-31966-5_4

Moraes, R. B., Dortschy, B., Klautau, A., & Rius i Riuy, J. (2010). Semiblind spectrum balancing for dsl. *IEEE Transactions on Signal Processing, 58*(7), 3717–3727. doi:10.1109/TSP.2010.2045553

Morvaj, B., Lugaric, L., & Krajcar, S. (2011). Demonstrating Smart Buildings and Smart Grid Features in a Smart Energy City. In *Proceedings of International Youth Conference on Energetics*. Academic Press.

Mu, H., & Tugnait, J. K. (2013, April). MSE-based source and relay precoder design for cognitive multiuser multi-way relay systems. *IEEE Transactions on Signal Processing*, 58, 1833–1846.

Mukhopadhyay, R., Yunseo Park, , Sen, P., Srirattana, N., Jongsoo Lee, , & Chang-Ho Lee, et al. (2005). Reconfigurable RFICs in si-based technologies for a compact intelligent RF front-end. *IEEE Transactions on Microwave Theory and Techniques*, 53(1), 81–93. doi:10.1109/TMTT.2004.839352

Mumey, B., Tang, J., Judson, I. R., & Stevens, D. (2012). On routing and channel selection in cognitive radio mesh networks. *IEEE Transactions on Vehicular Technology*, 61(9), 4118–4128. doi:10.1109/TVT.2012.2213310

Myerson, R. B. (1991). *Game Theory, Analysis of Conflict*. Cambridge, MA: Harvard Univ. Press.

Nabar, R.U., Bolcskei, H., Erceg, V., Gesbert, D., & Paulraj, A.J. (2002). Performance of multiantenna signaling techniques in the presence of polarization diversity. *IEEE Transactions on Signal Processing*, 2553-2562.

Najimi, M., Ebrahimzadeh, A., Andargoli, S. M. H., & Fallahi, A. (2013). A Novel Sensing Nodes and Decision Node Selection Method for Energy Efficiency of Cooperative Spectrum Sensing in Cognitive Sensor Networks. *IEEE Sensors Journal*, 1(5), 1610–1621. doi:10.1109/JSEN.2013.2240900

Nallagonda, S., Roy, S. D., & Kundu, S. (2011). Performance of energy detection based spectrum sensing in fading channels. In *Proceedings of the Computer & Communication Technology* (pp. 575-580). Allahabad, India: Academic Press. doi:10.1109/ICCCT.2011.6075107

Nallagonda, S., Roy, S. D., & Kundu, S. (2011). Performance of Cooperative Spectrum Sensing in Rician and Weibull Fading Channels. In *Proceedings of the India Conference* (pp. 1-5). Hyderabad, India: Academic Press.

Nallagonda, S., Roy, S. D., & Kundu, S. (2012). Performance of cooperative spectrum sensing in Fading Channels. In *Proceedings of the Recent Advances in Information Technology* (pp. 1-6). ISM Dhanbad. doi:10.1109/RAIT.2012.6194506

Nallagonda, S., Roy, S. D., Kundu, S., Ferrari, G., & Raheli, R. (2013c). Cooperative spectrum sensing with censoring of cognitive radios in Rayleigh fading under majority logic fusion. In *Proceedings of the National Conference on Communications* (pp. 1-5). IIT Delhi. doi:10.1109/NCC.2013.6487926

Nallagonda, S., Roy, S. D., Kundu, S., Ferrari, G., & Raheli, R. (2013). Performance of MRC fusion based cooperative spectrum sensing with censoring of cognitive radios in Rayleigh fading channel. In *Proceedings of the Wireless Communications and Mobile Computing Conference* (pp. 30-35). Sardinia, Italy: Academic Press. doi:10.1109/IWCMC.2013.6583530

Nallagonda, S., Roy, S. D., Kundu, S., Ferrari, G., & Raheli, R. (2014). Cooperative spectrum sensing with censoring of cognitive radios in fading channel under majority logic fusion. In F. Bader & M. G. Di Benedetto (Eds.), Cognitive Communications and Cooperative HetNet Coexistence (pp. 133-161), Springer International Publishing. doi:10.1007/978-3-319-01402-9_7

Nallagonda, S., Shravan kumar, B., Roy, S. D., & Kundu, S. (2013). Performance of cooperative spectrum sensing with soft data fusion schemes in fading channels. In *Proceedings of the India Conference* (pp. 1-5). IIT Kanpur. doi:10.1109/INDCON.2013.6725924

Nallagonda, S., Roy, S. D., & Kundu, S. (2013). Performance evaluation of cooperative spectrum sensing with censoring of cognitive radios in Rayleigh fading channel. *Wireless Personal Communications*, 70(4), 1409–1424. doi:10.1007/s11277-012-0756-6

Nandiraju, D. S., Nandiraju, N. S., & Agrawal, D. P. (2009). Adaptive state-based multi-radio multi-channel multi-path routing in Wireless Mesh Networks. *Pervasive and Mobile Computing*, 5(1), 93–109. doi:10.1016/j.pmcj.2008.11.003

Naphade, M., Banavar, G., Harrison, C., Paraszczak, J., & Morris, R. (2011). Smarter Cities and their Innovation Challenges. *IEEE Computer Society, 44*(6), 32–39. doi:10.1109/MC.2011.187

National Atlas of the United States. (n.d.). Retrieved from www.nationalatlas.gov

Neal, R. (2000, March). Markov chain sampling methods for Dirichlet process mixture models. *Journal of Computational and Graphical Statistics*, 249–265.

Nekovee, M. (2006). Dynamic spectrum access with cognitive radios: Future architectures and research challenges. In *Proceedings of 1st International Conference on Cognitive Radio Oriented Wireless Networks and Communications*, (pp. 1-5). Academic Press. doi:10.1109/CROWNCOM.2006.363464

Nelles, O. (Ed.). (2001). *Nonlinear System Identification, From classical approaches to neural networks and fuzzy methods.* Springer-Verlag Berlin Heidelberg NY.

New, M. F., Sarvotham, S., Baron, D., Wakin, M. B., & Baraniuk, R. G. (2005). Distributed Compressed Sensing of Jointly Sparse Signals. In *Proceedings of Signals, Systems and Computers,* (pp. 1537-1541). Academic Press.

Newman, R. T. (2008). *Multiple Objective Fitness Functions for Cognitive Radio Adaptation.* (PhD thesis). Department of Electrical Engineering & Computer Science.

Newman, T. R., Rajbanshi, R., Wyglinski, A. M., Evans, J. B., & Minden, G. J. (2007). Population adaptation for genetic algorithm-based cognitive radios. In *Proceedings of 2nd International Conference on Cognitive Radio Oriented Wireless Networks and Communication (CROWNCOM 2007).* Academic Press. doi:10.1109/CROWNCOM.2007.4549811

Newman, T. R., Barker, B. A., Wyglinski, A. M., Agah, A., Evans, J. B., & Minden, G. J. (2007, November). Cognitive engine implementation for wireless multicarrier transceivers. *Wireless Communications and Mobile Computing, 9*(9), 1129–1142. doi:10.1002/wcm.486

Ngo, H. H., Wu, X., Le, P. D., & Srinivasan, B. (2010, April). An Authentication Model for Wireless Network Services. In *Proceedings of Advanced Information Networking and Applications* (AINA), (pp. 996-1003). IEEE. doi:10.1109/AINA.2010.40

Ngo, D. T., Khakurel, S., & Le-Ngoc, T. (2014). Joint subchannel assignment and power allocation for ofdma femtocell networks. *IEEE Transactions on Wireless Communications, 13*(1), 342–355. doi:10.1109/TWC.2013.111313.130645

Nguyen, G. D., Kompella, S., Wieselthier, J. E., & Ephremides, A. (2010). Channel sharing in cognitive radio networks. In *Proceedings of IEEE Military Communications Conference (MILCOM),* (pp. 2268-2273). IEEE.

Nguyen, N. T., Zheng, G., Han, Z., & Zheng, R. (2011, April). Device fingerprinting to enhance wireless security using nonparametric Bayesian method. In *Proceedings of IEEE International Conference on Computer Communications (IEEE INFOCOM 2011),* (pp. 1404-1412). Shanghai, China: IEEE. doi:10.1109/INFCOM.2011.5934926

Nguyen, N. T., & Armitage, G. (2008). A survey of techniques for internet traffic classification using machine learning. *IEEE Communications Surveys and Tutorials, 10*(4), 56–76. doi:10.1109/SURV.2008.080406

Nguyen, V.-T., Villaine, F., & Le Guillou, Y. (2012). Cognitive Radio RF: Overview and Challenges. *VLSI Design, 12*, 1–13.

Nickalls, R. (1993). A new approach to solving the cubic: Cardan's solution revealed. *Mathematical Gazette, 77*(480), 354–359. doi:10.2307/3619777

Nie, N., & Comaniciu, C. (2006). Adaptive Channel Allocation Spectrum Etiquette for Cognitive Radio Networks. *Mobile Network Application*, 779-797.

Ning, J., & Hofmann, K. (2013). *An Integrated High Voltage Controller for a Reconfigurable Antenna Array.* IEEE Norchip. doi:10.1109/NORCHIP.2013.6702004

Niu, R., Chen, B., & Varshney, P. K. (2003). Decision fusion rules in wireless sensor networks using fading statistics. In *Proceedings of the Annual Conference on Information Sciences and Systems* (pp. 1-6). The Johns Hopkins University.

Niyato, D., & Hossain, E. (2007). A game-theoretic approach to competitive spectrum sharing in cognitive radio networks. *Proceedings of the IEEE*, 16–20.

Niyato, D., & Hossain, E. (2007). Equilibrium and disequilibrium pricing for spectrum trading in cognitive radio: A control-theoretic approach. *GLOBECOM, 07*, 4852–4856.

Niyato, D., & Hossain, E. (2008). Market-equilibrium, competitive, and cooperative pricing for spectrum sharing in cognitive radio networks: Analysis and comparison. *IEEE Transactions on Wireless Communications, 7*(11), 4273–4283. doi:10.1109/T-WC.2008.070546

Niyato, D., & Hossain, E. (2008). Spectrum trading in cognitive radio networks: A market-equilibrium-based approach. *IEEE Wireless Communications, 15*(6), 71–80. doi:10.1109/MWC.2008.4749750

Niyato, D., & Hossain, E. (2008, January). Competitive Pricing for Spectrum Sharing in Cognitive Radio Networks: Dynamic game, Inefficiency of Nash Equilibrium, and Collusion. *IEEE Journal on Selected Areas in Communications, 26*(1), 192–202. doi:10.1109/JSAC.2008.080117

Niyato, D., & Hossain, E. (2009). Cognitive radio for next-generation wireless networks: an approach to opportunistic channel selection in IEEE 802.11-based wireless mesh. *IEEE Wireless Communications, 1*, 46–54.

Niyato, D., Hossain, E., & Han, Z. (2009). Dynamic spectrum access in IEEE 802.22-based cognitive wireless networks: A game theoretic model for competitive spectrum bidding and pricing. *IEEE Wireless Communications, 16*(2), 16–23. doi:10.1109/MWC.2009.4907555

Niyato, D., Hossain, E., & Zhu Han,. (2009). Dynamics of multiple-seller and multiple-buyer spectrum trading in cognitive radio networks: A game-theoretic modeling approach. *IEEE Transactions on Mobile Computing, 8*(8), 1009–1022. doi:10.1109/TMC.2008.157

Noam, E. M. (2012). The economists' contribution to radio spectrum access: The past, the present, and the future. *Proceedings of the IEEE, 100*, 1692–1697. doi:10.1109/JPROC.2012.2187133

Noam, Y., & Leshem, A. (2009). Iterative power pricing for distributed spectrum coordination in dsl. *IEEE Transactions on Communications, 57*(4), 948–953. doi:10.1109/TCOMM.2009.4814362

Noels, N., Steendam, H., Moeneclaey, M., & Bruneel, H. (2005). Carrier phase and frequency estimation for pilot-symbol assisted transmission: Bounds and algorithms. *IEEE Transactions on Signal Processing, 53*(12), 4578–4587. doi:10.1109/TSP.2005.859318

Novak, L. M., Sechtin, M. B., & Cardullo, M. J. (1989). Studies of target detection algorithms that use polarimetric radar data. *IEEE Transactions on Aerospace and Electronic Systems, 25*(2), 150–165. doi:10.1109/7.18677

Nussbaum, D., et al. (2009). Open Platform for Prototyping of Advanced Software Defined Radio and Cognitive Radio Techniques. In *Proceedings of 12th Euromicro Conference on Digital System Design, Architectures, Methods and Tools* (pp. 435–440). Patras, Greece: Academic Press. doi:10.1109/DSD.2009.123

Nutaq. (2013). *Design, Testing & Deployment*. Radio 420X Multimode SDR FMC RF Transceiver Datasheet. Retrieved from http://www.nutaq.com

Nutaq. (2014). Retrieved from http://www.nutaq.com

Nuttall, A. (1975). Some integrals involving the function (corresp.). *IEEE Transactions on Information Theory, 21*(1), 95–96. doi:10.1109/TIT.1975.1055327

Oestges, C., Clerckx, B., Guillaud, M., & Debbah, M. (2008). Dual-polarized wireless communications: From propagation models to system performance evaluation. *IEEE Transactions on Wireless Communications, 7*(10), 4019–4031. doi:10.1109/T-WC.2008.070540

Ohring, M. (2002). *Materials Science of Thin Films: Deposition & Structure* (2nd ed.). San Diego, CA: Academic Press.

Oh, S. W., Le, T. P., Zhang, W., Ahmed, S. N., Zeng, Y., & Kua, K. J. (2008). *TV white-space sensing prototype.* Wireless Communications and Mobile Computing.

Open Air Interface. (2014). Retrieved from http://www.openairinterface.org

Oto, M., & Akan, O. (2012, April). Energy-efficient packet size optimization for cognitive radio sensor networks. *IEEE Transactions on Wireless Communications, 4*(4), 1544–1553. doi:10.1109/TWC.2012.021412.021512.111398

Pan, M., Huang, R., & Fang, Y. (2008). Cost design for opportunistic multi-hop routing in cognitive radio networks. In *Proceedings of IEEE Military Communications Conference (MILCOM)*, (pp. 1-7). IEEE.

Pandharipande, A., & Linnartz, J. P. (2007, June). Performance analysis of primary user detection in a multiple antenna cognitive radio. In *Proceedings of Communications*, (pp. 6482-6486). IEEE. doi:10.1109/ICC.2007.1072

Pan, M., Li, P., Song, Y., Fang, Y., Lin, P., & Glisic, S. (2013). When spectrum meets clouds: Optimal session based spectrum trading under spectrum uncertainty. *IEEE Journal on Selected Areas in Communications, 32*(3), 615–627. doi:10.1109/JSAC.2014.140320

Pan, M., Zhang, C., Li, P., & Fang, Y. (2012). Spectrum harvesting and sharing in multi-hop cognitive radio networks under uncertain spectrum supply. *IEEE Journal on Selected Areas in Communications, 30*(2), 369–378. doi:10.1109/JSAC.2012.120216

Papadimitratos, P., Sanjaranaryanan, S., & Mishra, A. (2005). A bandwidth sharing approach to improve licensed spectrum utilization. *IEEE Communications Magazine, 43*(12), 10–14. doi:10.1109/MCOM.2005.1561918

Papandriopoulos, J., & Evans, J. S. (2006). Low-complexity distributed algorithms for spectrum balancing in multi-user dsl networks. In *Proceedings of International Conference on Communications* (pp. 3270–3275). Academic Press. doi:10.1109/ICC.2006.255311

Papandriopoulos, J., & Evans, J. S. (2009). Scale: A low-complexity distributed protocol for spectrum balancing in multiuser dsl networks. *IEEE Transactions on Information Theory, 55*(8), 3711–3724. doi:10.1109/TIT.2009.2023751

Papapanagiotou, I., Toumpakaris, D., Lee, J., & Devetsikiotis, M. (2009). Fourth Quarter). A survey on next generation mobile WiMAX networks: Objectives, features and technical challenges. *IEEE Communications Surveys and Tutorials, 11*(4), 3–18. doi:10.1109/SURV.2009.090402

Papoulis, A., & Pillai, S. U. (2002). *Probability, Random Variables, and Stochastic Processes* (4th ed.). McGraw-Hill.

Park, B., Lee, K., Choi, S., & Hong, S. (2010). A 3.1–10.6 GHz RF receiver front-end in 0.18 µm CMOS for ultra-wideband applications. In *Proceedings of IEEE MTT-S International Microwave Symposium Digest (MTT)*, (pp. 1616-1619). IEEE.

Park, H., Li, J., & Wang, H. (1995). Polarization-space-time domain generalized likelihood ratio detection of radar targets. *Signal Processing, 41*(2), 153–164. doi:10.1016/0165-1684(94)00097-J

Parvin, S., & Fujii, T. (2012). Radio environment aware stable routing scheme for multi-hop cognitive radio network. *IEEE IET Networks, 1*(4), 207–216. doi:10.1049/iet-net.2012.0103

Pastina, D., Lombardo, P., & Bucciarelli, T. (2001). Adaptive polarimetric target detection with coherent radar. Part I: Detection against Gaussian. *IEEE Transactions on Aerospace and Electronic Systems, 37*(4), 1194–1206. doi:10.1109/7.976959

Paula, A. D., & Panazio, C. (2014). Cooperative spectrum sensing under unreliable reporting channels. Elsevier.

Pavlovska, V., Denkovski, D., Atanasovski, V., & Gavrilovska, L. (2010). RAC2E: Novel rendezvous protocol for asynchronous cognitive radios in cooperative environments. In *Proceedings of IEEE International Symposium on Personal, Indoor and Mobile Radio Communications (PIMRC)* (pp. 1848–1853). IEEE. doi:10.1109/PIMRC.2010.5671630

Pearson, K. (1901). Llll. on lines and planes of closest fit to systems of points in space. Retrieved from Philosophical Magazine Series.

Peh, E. C. Y., Liang, Y. C., Guan, Y. L., & Pei, Y. (2011, Dec.). Energy-Efficient Cooperative Spectrum Sensing in Cognitive Radio Networks. In *Proceedings of Global Telecommunications Conference (GLOBECOM 2011)*, (pp. 1-5). IEEE. doi:10.1109/GLOCOM.2011.6134342

Penna, F., Garello, R., Figlioli, D., & Spirito, M. A. (2009). Exact non-asymptotic threshold for eigenvalue-based spectrum sensing. In *Proceedings of 4th International Conference on Cognitive Radio Oriented Wireless Networks and Communications*, (pp. 1-5). Academic Press. doi:10.1109/CROWNCOM.2009.5189008

Pentek. (2014). Retrieved from http://www.pentek.com

Perron, E., Diggavi, S., & Telatar, I. E. (2009, April). On cooperative wireless network secrecy. In Proceedings of INFOCOM 2009, (pp. 1935-1943). IEEE.

Perruisseau-Carrier, J., Pardo-Carrera, P., & Miskovsky, P. (2010). Modeling, design and characterization of a very wideband slot antenna with reconfigurable band rejection. *IEEE Transactions on Antennas and Propagation*, *58*(7), 2218–2226. doi:10.1109/TAP.2010.2048872

Petit, W. A. (2009). Interoperable Positive Train Control (PTC). In *Proceedings of AREMA 2009 Annual Conference, Communications and Signals Track, September* (pp. 20-23). AREMA.

Pham, H. N., Zhang, Y., Engelstad, P. E., Skeie, T., & Eliassen, F. (2010, March). Energy minimization approach for optimal cooperative spectrum sensing in sensor-aided cognitive radio networks. In *Proceedings of Wireless Internet Conference (WICON)*, (pp. 1-9). IEEE. doi:10.4108/ICST.WICON2010.8531

Pham, D. T., & Hoai An, L. T. (1997). Convex analysis approach to d.c. programming: Theory, algorithms, and applications. *Acta Mathematica Vietnamica*, *22*(1), 289–355.

Pham, H. N., Gronsund, P., Engelstad, P., & Grondalen, O. (2010). A Dynamic Spectrum Access Scheme for Unlicensed Systems Coexisting with Primary OFDMA Systems. In *Proc. of IEEE CCNC 2010*. Las Vegas, NV: IEEE. doi:10.1109/CCNC.2010.5421723

Pleskachev, V., & Vendik, I. (2004). Tunable Microwave Filters based on Ferroelectric Capacitors. In *Proceedings of Conference on Microwaves, Radar and Wireless Communications (MIKON)*. Academic Press. doi:10.1109/MIKON.2004.1358556

Poelman, A. (1981). Virtual polarization adaptation a method of increasing the detection capability of a radar system through polarization-vector processing. In *Proc. Communications, Radar and Signal Processing*, (pp. 261-270). IEEE.

Polo, Y. L., Wang, &., Andharipande, A., & Leus, G. (2009). Compressive wide-band spectrum sensing. In *Proceedings of IEEE International Conference on Acoustics, Speech and Signal Processing (ICASSP)*, (pp. 2337-2340). Taipei, Taiwan: IEEE.

Poor, H. V. (1994). *An introduction to signal detection and estimation. Book*. Springer-Verlag. doi:10.1007/978-1-4757-2341-0

Positive Train Control White Paper. (2012, May). *Joint Council on Transit Wireless Communications Transit Technology Series*. Author.

Pradhan, S. S., Kusuma, J., & Ramchandran, K. (2002). Distributed compression in a dense microsensor network. *IEEE Signal Processing Magazine*, *19*(2), 51–60. doi:10.1109/79.985684

Pratt, T., Nguyen, S., & Walkenhorst, B. T. (2008). Dual-polarized architectures for sensing with wireless communications signals. In *Proceedings of IEEE Military Communications Conference*, (pp. 1-6). IEEE. doi:10.1109/MILCOM.2008.4753072

Proakis, J. G., & Salehi, M. (2008). *Digital Communications*. New York: McGraw-Hill.

Pulley, D., & Baines, R. (2003). Software defined baseband processing for 3G base stations. In *Proceedings of Fourth International Conference on 3G Mobile Communication Technologies (Vol. 2003*, pp. 123–127). London, UK: Academic Press. doi:10.1049/cp:20030350

Qian, L. P., & Jun, Y. (2009). Monotonic optimization for non-concave power control in multiuser multicarrier network systems. In *Proceedings of IEEE International Conference on Computer Communications* (pp. 172–180). IEEE. doi:10.1109/INFCOM.2009.5061919

Qian, L., Ye, F., Gao, L., Gan, X., Chu, T., & Tian, X. et al. (2011). Spectrum trading in cognitive radio networks: An agent-based mode under demand uncertainty. *IEEE Transactions on Communications*, 59(11), 3192–3203. doi:10.1109/TCOMM.2011.100411.100446

QinetiQ. (2007). *Cognitive radio technology, a study for Ofcom* (QinetiQ, Tech. Rep., Feb. 2007). QinetiQ.

Qin, P.-Y., Guo, Y., Cai, Y., Dutkiewicz, E., & Liang, C.-H. (2011). A reconfigurable antenna with frequency and polarization agility. *IEEE Antennas and Wireless Propagation Letters*, 10, 1373–1376. doi:10.1109/LAWP.2011.2178226

Qin, P.-Y., Guo, Y., Weily, A., & Liang, C.-H. (2012). A pattern reconfigurable U-slot antenna and its applications in MIMO systems. *IEEE Transactions on Antennas and Propagation*, 60(2), 516–528. doi:10.1109/TAP.2011.2173439

Quan, Z. Q. Z., Cui, S. C. S., Poor, H., & Sayed, A. (2008). Collaborative wideband sensing for cognitive radios. *IEEE Signal Processing Magazine*, 25(6), 60–73. doi:10.1109/MSP.2008.929296

Quan, Z., Cui, S., & Sayed, A. H. (2008). Optimal Linear Cooperation for Spectrum Sensing in Cognitive Radio Networks. *IEEE Journal of Selected Topics in Signal Processing*, 2(1), 28–40. doi:10.1109/JSTSP.2007.914882

Qusay, H. M., & Mahmou, D. (2007). *Cognitive networks: Towards self-aware networks*. John Wiley & Sons Ltd.

Rabbachin, A., Quek, T. Q. S., & Win, M. Z. (2011). Cognitive network interference. *IEEE Journal on Selected Areas in Communications*, 29(2), 480–493.

Rahimzadeh, P., Moradian, M., & Ashtiani, F. (2012). Saturation throughput analysis of a cognitive IEEE 802.11-based WLAN overlaid on an IEEE 802.16e WiMAX. In *Proc. of IEEE PIMRC 2012*, (p. 1515–1520). Sydney, Australia: IEEE. doi:10.1109/PIMRC.2012.6362588

Ramacher, U., Raab, W., Hachmann, J. A. U., Langen, D., Berthold, J., Kramer, R., ... Harrington, J. (2011). Architecture and implementation of a Software-Defined Radio baseband processor. In *Proceedings of International Symposium on Circuits and Systems (ISCAS)* (pp. 2193–2196). Rio de Janeiro, Brazil: ISCAS. doi:10.1109/ISCAS.2011.5938035

Ramacher, U. (2007). Software-Defined Radio Prospects for Multistandard Mobile Phones. *Computer*, 40(10), 62–69. doi:10.1109/MC.2007.362

Rao, A., & Alouini, M. S. (2011). Performance of cooperative spectrum sensing over non-identical fading environments. *IEEE Transactions on Wireless Communications*, 59(1), 3249–3253. doi:10.1109/TCOMM.2011.082911.100222

Rappaport, T. S. (2002). Wireless communications: Principles and practice (Vol. 2). Prentice Hall PTR.

Rashid, R. A., Baguda, Y. S., Fisal, N., Sarijari, M. A., Yusof, S. K. S., Ariffin, S. H. S., & Mohd, A. (2011). Optimizing Achievable Throughput for Cognitive Radio Network using. *Swarm Intelligence*, (October): 354–359.

Rawat, A. S., Anand, P., Chen, H., & Varshney, P. K. (2011). Collaborative spectrum sensing in the presence of Byzantine attacks in cognitive radio networks. *IEEE Transactions on* Signal Processing, 59(2), 774–786. doi:10.1109/TSP.2010.2091277

Raychaudhuri, D., Mandayam, N. B., Evans, J. B., Ewy, B. J., Seshan, S., & Steenkiste, P. (2006). CogNet: an Architectural Foundation for Experimental Cognitive Radio Networks within the Future Internet. In *Proceedings of First ACM/IEEE International Workshop on Mobility in the Evolving Internet Architecture* (pp. 11-16). ACM. doi:10.1145/1186699.1186707

Ray, D. (2007). *A Game-Theoretic Perspective on Coalition Formation*. New York: Oxford Univ. Press. doi:10.1093/acprof:oso/9780199207954.001.0001

Razavi, B. (2009). Challenges in the design of cognitive radios. In *Proceedings of IEEE Custom Integrated Circuits Conference (CICC)*, (pp. 391-398). IEEE.

Reddy, Y. (2010). Efficient spectrum allocation using case-based reasoning and collaborative filtering approaches. In *Proceedings of Fourth International Conference on Sensor Technologies and Applications (SENSORCOMM)*. Academic Press. doi:10.1109/SENSORCOMM.2010.62

Reddy, Y. (2013, June). Security Issues and Threats in Cognitive Radio Networks. In *Proceedings of AICT 2013,the Ninth Advanced International Conference on Telecommunications* (pp. 84-89). Academic Press.

Rehmani, M. H. (2011). *Opportunistic Data Dissemination in Ad-Hoc Cognitive Radio Networks*. (Ph.D. Dissertation). Universit'e Pierre et Marie Curie-Paris VI.

Rehmani, M. H., Viana, A. C., Khalife, H., & Fdida, S. (2010). A cognitive radio based internet access framework for disaster response network deployment. In *Proceedings of International Symposium on Applied Sciences in Biomedical and Communication Technologies (ISABEL)*. Academic Press. doi:10.1109/ISABEL.2010.5702851

Rehmani, M. H., Viana, A. C., Khalife, H., & Fdida, S. (2013). Surf: A distributed channel selection strategy for data dissemination in multi-hop cognitive radio networks. *Computer Communications, 36*(10), 1172–1185. doi:10.1016/j.comcom.2013.03.005

Reisenfeld, S. (2009). *Performance bounds for detect and avoid signal sensing*. Paper presented at the Second International Workshop on Cognitive Radio and Advanced Spectrum Management (CogART 2009). Aalborg, Germany. doi:10.1109/COGART.2009.5167248

Rieser, C. (2004). *Biologically inspired cognitive radio engine model utilizing distributed genetic algorithms for secure and robust wireless communications and networking*. (Ph.D. dissertation). Virginia Polytechnic Institute.

Robertson, A., Tran, L., Molnar, J., & Fu, E. H. F. (2012). Experimental comparison of blind rendezvous algorithms for tactical networks. In *Proceedings of IEEE International Symposium on a World of Wireless, Mobile and Multimedia Networks (WoWMoM)*. IEEE. doi:10.1109/WoWMoM.2012.6263760

Roberts, R. S., Brown, W. A., & Loomis, H. H. Jr. (1991). Computationally efficient algorithms for cyclic spectral analysis. *IEEE Signal Processing Magazine, 8*(2), 38–49. doi:10.1109/79.81008

Rodenbeck, C. T., Chang, K., & Aubin, J. (2004). Automated pattern measurement for circularly-polarized antennas using the phase-amplitude method. *Microwave Journal, 47*(7), 68–78.

Rodriguez, K. M. V., & Tafazolli, R. (2005). Auction driven dynamic spectrum allocation: Optimal bidding, pricing and service priorities for multi-rate, multi-class cdma. In *Proc. of IEEE PIMRC05*, (vol. 3, pp. 1850-1854). IEEE. doi:10.1109/PIMRC.2005.1651761

Romaszko, S. (2013). A rendezvous protocol with the heterogeneous spectrum availability analysis for cognitive radio ad hoc networks. *Hindawi Journal of Electrical and Computer Engineering, 2013*(715816), 18.

Romaszko, S., & Mähönen, P. (2011). Grid-based channel mapping in cognitive radio ad hoc networks. In *Proceedings of 22nd Annual IEEE International Symposium on Personal, Indoor and Mobile Radio Communications (PIMRC)* (pp. 438-444). IEEE. doi:10.1109/PIMRC.2011.6139999

Romaszko, S., & Mähönen, P. (2012). Heterogeneous torus Quorum-based Rendezvous in Cognitive Radio Ad Hoc Networks. In *Proceedings of IEEE International Conference on Wireless and Mobile Computing, Networking and Communications (WiMob)*. IEEE. doi:10.1109/WiMOB.2012.6379109

Romaszko, S., & Mähönen, P. (2014). Fuzzy Channel Ranking estimation in Cognitive Wireless Networks. In *Proceedings of IEEE Wireless Communications and Networking Conference (WCNC)*. IEEE. doi:10.1109/WCNC.2014.6952964

Romaszko, S., Denkovski, D., Pavlovska, V., & Gavrilovska, L. (2013). Quorum system and random based asynchronous rendezvous protocol for cognitive radio ad hoc networks. *EAI Endorsed Transactions on Mobile Communications and Applications, 13*(3).

Romaszko, S., Torfs, W., Mähönen, P., & Blondia, C. (2013). An analysis of asynchronism of a neighborhood discovery protocol for cognitive radio networks. In *Proceedings of IEEE International Symposium on Personal, Indoor and Mobile Radio Communications (PIMRC)*. IEEE. doi:10.1109/PIMRC.2013.6666689

Romaszko, S., Torfs, W., Mähönen, P., & Blondia, C. (2013). Benefiting from an Induced Asynchronism in Neighborhood Discovery in opportunistic Cognitive Wireless Networks. In *Proceedings of ACM International Symposium on Mobility Management and Wireless Access (MobiWac)*. ACM. doi:10.1145/2508222.2508230

Romaszko, S., Torfs, W., Mähönen, P., & Blondia, C. (2014). AND: Asynchronous Neighborhood Discovery protocols for opportunistic Cognitive Wireless Networks. In *Proceedings of IEEE Wireless Communications and Networking Conference (WCNC)*. IEEE. doi:10.1109/WCNC.2014.6952604

Romaszko, S., & Mähönen, P. (2012). Quorum systems towards an asynchronous communication in cognitive radio networks. *Journal of Electrical and Computer Engineering, Article ID 753541*.

Romaszko, S., Torfs, W., Mähönen, P., & Blondia, C. (2014). *How to be an efficient Asynchronous Neighbourhood Discovery protocol in opportunistic Cognitive Wireless Networks. International Journal of Ad Hoc and Ubiquitous Computing*.

Rondeau, T. W., Le, B., Rieser, C. J., & Bostian, C. W. (2004). Cognitive radios with genetic algorithms: Intelligent control of software defined radios. In *Proceedings of Software Defined Radio Forum Technical Conference*. Academic Press.

Ru, Z., Klumperink, E. A. M., Saavedra, C. E., & Nauta, B. (2010). A 300–800 MHz tunable filter and linearized LNA applied in a low-noise harmonic-rejection RF-sampling receiver. *IEEE Journal of Solid-State Circuits*, *45*(5), 967–978. doi:10.1109/JSSC.2010.2041403

Ryzhik, I. S. (1994). Table of Integrals, Series, and Products (5th ed.). Academic Press.

Saad, S., Ismail, M., & Nordin, R. (2013). A survey on power control techniques in femtocell networks. *Journal of Communication*, *8*(12), 845–854. doi:10.12720/jcm.8.12.845-854

Saad, W., Han, Z., Basar, T., Debbah, M., & Hjorungnes, A. (2011). Coalition Formation Games for Collaborative Spectrum Sensing. *IEEE Transactions on Vehicular Technology*, *60*(1), 276–297. doi:10.1109/TVT.2010.2089477

Saad, W., Han, Z., Debbah, M., Hjorungnes, A., & Basar, T. (2009). Coalitional game theory for communication networks. *IEEE Signal Processing Magazine*, *26*(5), 77–97. doi:10.1109/MSP.2009.000000

Saad, W., Han, Z., Zheng, R., Hjorungnes, A., Basar, T., & Poor, H. V. (2012). Coalitional games in partition form for joint spectrum sensing and access in cognitive radio networks. *IEEE Journal of Selected Topics in Signal Processing*, *6*(2), 195–209. doi:10.1109/JSTSP.2011.2175699

Saaty, T. L. (1978). Exploring the interface between hierarchies, multiplied objectives and fuzzy sets. *Fuzzy Sets and Systems Journal (Elsevier)*, *1*(1), 57–68. doi:10.1016/0165-0114(78)90032-5

Sadeghi, H., & Azmi, P. (2008). Cyclostationarity-based cooperative spectrum sensing for cognitive radio networks. In *Proceedings of International Symposium on Telecommunications* (pp. 429–434). Academic Press. doi:10.1109/ISTEL.2008.4651341

Sadeghi, H., Azmi, P., & Arezumand, H. (2011). Optimal multi-cycle cyclostationarity-based spectrum sensing for cognitive radio networks. In *Proceedings of Iranian Conference on Electrical Engineering* (pp. 1–6). IEEE.

Sadek, A. K., Han, Z., & Liu, K. R. (2010). Distributed relay-assignment protocols for coverage expansion in cooperative wireless networks. *IEEE Transactions on Mobile Computing*, *9*(4), 505–515.

Sadr, S., Anpalagan, A., & Raahemifar, K. (2009). Radio resource allocation algorithms for the downlink of multiuser ofdm communication systems. *IEEE Communications Surveys and Tutorials*, *11*(3), 92–106. doi:10.1109/SURV.2009.090307

Safatly, L., Aziz, B., Nafkha, A., Louet, Y., Nasser, Y., El-Hajj, A., & Kabalan, K. Y. (2014). A Blind Spectrum Sensing Using Symmetry Property of Cyclic Autocorrelation Function: From Theory to Practice. *EURASIP Journal on Wireless Communications and Networking*, *2014*(26).

Sagias, N. C., Karagiannidis, G. K., & Tombras, G. S. (2004). Error-rate analysis of switched diversity receivers in Weibull fading. *Electronics Letters*, *40*(11), 681–682. doi:10.1049/el:20040479

Sahai, A., Hoven, N., & Tandra, R. (2004). Some Fundamental Limits on Cognitive Radio. In *Proceedings of the 42nd Allerton Conference*. Champaign, IL: Academic Press.

Sahai, A., Tandra, R., Mishra, S. M., & Hoven, N. (2006). Fundamental design tradeoffs in cognitive radio systems. In *Proceedings of the first international workshop on Technology and policy for accessing spectrum* (p. 2). ACM. doi:10.1145/1234388.1234390

Saito, V. S. N. V. M., & Sugiyama, V. I. (2006). Single-Chip Baseband Signal Processor for Software-Defined Radio. *FUJITSU Sci. Tech. J, 42*(2), 240–247.

Salami, G., Durowoju, O., Attar, A., Holland, O., Tafazolli, R., & Aghvami, H. (2011). A Comparison Between the Centralized and Distributed Approaches for Spectrum Management. *IEEE Communications Surveys and Tutorials, 13*(2), 274–289. doi:10.1109/SURV.2011.041110.00018

Saleem, Y., Bashir, A., Ahmed, E., Qadir, J., & Baig, A. (2012). Spectrum-aware dynamic channel assignment in cognitive radio networks. In *Proceedings of IEEE International Conference on Emerging Technologies (ICET)*, (pp. 1-6). IEEE. doi:10.1109/ICET.2012.6375468

Saleem, Y., & Husain Rehmani, M. (2014). Primary radio user activity models for cognitive radio networks: A survey. *Journal of Network and Computer Applications, 43*, 1–16. doi:10.1016/j.jnca.2014.04.001

Sampath, A., Yang, L., Cao, L., Zheng, H., & Zhao, B. Y. (2008). High throughput spectrum-aware routing for cognitive radio networks. In *Proceedings of IEEE International Conference on Cognitive Radio Oriented Wireless Networks (CrownCom)*. IEEE.

Sandholm, T., Larson, K. S., Andersson, M. R., Shehory, O., & Tohmé, F. (1999). Coalition structure generation with worst case guarantees. *Elsevier Artificial Intelligence Journal, 111*(1-2), 209–238. doi:10.1016/S0004-3702(99)00036-3

Santos Filho, J. C. S., & Yacoub, M. D. (2006). Simple precise approximations to Weibull sums. *IEEE Communications Letters, 10*(8), 614–616. doi:10.1109/LCOMM.2006.1665128

Satarkar, S. (2009). *Performance analysis of the WiNC2R platform*. (PhD Thesis). Rutgers, The State University of New Jersey.

Sawyer, S., & Tapia, A. (2005). The sociotechnical nature of mobile computing work: Evidence from a study of policing in the United States. *International Journal of Technology and Human Interaction, 1*(3), 1–14. doi:10.4018/jthi.2005070101

Schmidt-Knorreck, C., Pacalet, R., Minwegen, A., Deidersen, U., Kempf, T., Knopp, R., & Ascheid, G. (2012). Flexible front-end processing for software defined radio applications using application specific instruction-set processors. In *Proceedings of Conference on Design and Architectures for Signal and Image Processing (DASIP)*, (pp. 1 – 8). Karlsruhe, Germany: DASIP.

Schulte, M. J., Glossner, J., Mamidi, S., Moudgill, M., & Vassiliadis, S. (2004). A low-power multithreaded processor for baseband communication systems. In *Proceedings of Computer Systems: Architectures, Modeling, and Simulation* (LNCS), (vol. 3133, pp. 333–346). Berlin: Springer.

Sciacca, L., & Evans, R. (2002). *Cooperative Sensor Networks with Bandwidth Constraints*. Battlespace Digitization and Network-Centric Warfare II.

Sengupta, S., Chandramouli, R. a., & Chatterjee, M. (2008). A Game Theoretic Framework for Distributed Self-Coexistence Among IEEE 802.22 Networks. In *Proceedings of Global Telecommunications Conference*, (pp. 1-6). IEEE. doi:10.1109/GLOCOM.2008.ECP.598

Sengupta, S., Brahma, S., Chatterjee, M., & Sai Shankar, N. (2013). Self-coexistence among interference-aware IEEE 802.22 networks with enhanced air-interface. *Pervasive and Mobile Computing, 9*(4), 454–471. doi:10.1016/j.pmcj.2011.08.003

Sengupta, S., & Chatterjee, M. (2009, August). An Economic Framework for Dynamic Spectrum Access and Service Pricing. *IEEE/ACM Transactions on Networking, 17*(4), 1200–1213. doi:10.1109/TNET.2008.2007758

Shanmugam, K., & Breipohl, A. M. (1971). An error correcting procedure for learning with an imperfect teacher. *IEEE Transactions on Systems, Man, and Cybernetics, 1*(3), 223–229. doi:10.1109/TSMC.1971.4308289

Sharma, A., & Jain, P. (2010). Effects of rain on radio propagation in GSM. *Int. Journal of Advanced Engineering & Applications*.

Sharma, R. K., & Wallace, J. W. (2009). Analysis of fusion and combining for wireless source detection. In Proceedings of Int. ITG Workshop on Smart Antennas-WSA. Berlin: ITG.

Sharma, S. K. Chatzinotas, S. & Ottersten, B. (Sept. 2012). Satellite Cognitive Communications: Interference modeling and techniques selection. In *Proceedings of 6th ASMS/SPSC Conference*. Baiona, Spain: IEEE. doi:10.1109/ASMS-SPSC.2012.6333061

Sharma, S. K., Chatzinotas, S., & Ottersten, B. (2012). Exploiting Polarization for Spectrum Sensing in Cognitive SatComs. In *Proceedings of 7th International ICST Conference on Cognitive Radio Oriented Wireless Networks and Communications (CROWNCOM)*, (pp. 36-41). Stockholm: ICST. doi:10.4108/icst.crowncom.2012.248473

Sharma, S. K., Chatzinotas, S., & Ottersten, B. (2012). Spectrum sensing in dual polarized fading channels for cognitive SatComs. In *Proceedings of IEEE Global Communications Conference (GLOBECOM)*, (pp. 287-294). Anaheim, CA: IEEE. doi:10.1109/GLOCOM.2012.6503643

Sharma, S. K., Chatzinotas, S., & Ottersten, B. (2013). Spatial filtering for underlay cognitive satcoms. In *Proceedings of 5th International Conference on Personal Satellite Services*. Toulouse, France: Academic Press. doi:10.1007/978-3-319-02762-3_17

Sharma, S. K., Chatzinotas, S., & Ottersten, B. (2013). Transmit beamforming for spectral coexistence of satellite and terrestrial networks. In *Proceedings of 8th International Conference on Cognitive Radio Oriented Wireless Networks (CROWNCOM)*, (pp. 275-281). Washington, DC: Academic Press. doi:10.1109/CROWNCom.2013.6636830

Sharma, S. K., Chatzinotas, S., & Ottersten, B. (Sept. 2013). Cognitive Radio Techniques for Satellite Communication Systems. In *Proceedings of IEEE 78th Vehicular Technology Conference (VTC Fall)* (pp. 1-5). Las Vegas, NV: IEEE.

Sharma, S. K., Chatzinotas, S., & Ottersten, B. (2013). *Cognitive beamhopping for spectral coexistence of multibeam satellites. Paper presented at the Future Network and Mobile Summit (FutureNetworkSummit)*. Lisbon.

Sharma, S. K., Chatzinotas, S., & Ottersten, B. (2013). Interference alignment for spectral coexistence of multibeam satellites. *EURASIP Journal on Wireless Communications and Networking*. doi:10.1186/1687-1499-2013-46

Sharma, S. K., Chatzinotas, S., & Ottersten, B. (2014). *Cogniitve beamhopping for spectral coexistence of cognitive SatComs. International J. Sat. Commun. and Networking*.

Shaw, T., Suo, Z., Huang, M., Liniger, E., Laibowitz, R., & Baniecki, J. (1999). The Effect of Stress on the Dielectric Properties of Barium Strontium Titanate Thin Films. *Applied Physics Letters*, *75*(14), 2129–2131. doi:10.1063/1.124939

Shellhammer, S. J. (2008). Spectrum sensing in IEEE 802.22. In *Proceedings of Cognitive Information Processing Workshop*. IEEE.

Shellhammer, S., & Tandra, R. (2006). Performance of the Power Detector with Noise Uncertainty. *IEEE 802.22-06/0134r0 Std.*

Shen, B., Ullah, S., & Kwak, K. (2010). Deflection coefficient maximization criterion based optimal cooperative spectrum sensing. *AEÜ: International Journal of Electronics and Communications*, *64*(9), 819–827. doi:10.1016/j.aeue.2009.06.006

Shen, J., & Alsusa, E. (2013). Joint cycle frequencies and lags utilization in cyclostationary feature spectrum sensing. *IEEE Transactions on Signal Processing*, *61*(21), 5337–5346. doi:10.1109/TSP.2013.2278810

Shenker, S. J. (1995, September). Fundamental Design Issues for the Future Internet. *IEEE Selected Areas in Communications*, *13*(7), 1176–1188. doi:10.1109/49.414637

Shi, Y., & Eberhart, R. C. (1999). Empirical study of particle swarm optimization. In *Proceedings of IEEE Congress on Evolutionary Computation (CEC 99)*. IEEE. doi:10.1109/CEC.1999.785511

Shi, H., Li, J., Ma, Y., & Li, Z. (2013). Coverage probability driven dynamic spectrum allocation in heterogeneous wireless networks. *Science China Information Sciences*, *56*(8), 1–7. doi:10.1007/s11432-013-4920-8

Shin, J., Yang, D., & Kim, C. (2010). A channel rendezvous scheme for cognitive radio networks. *IEEE Communications Letters*, *14*(10), 954–956. doi:10.1109/LCOMM.2010.091010.100904

Shokri-Ghadikolaei, H., Sheikholeslami, F., & Nasiri-Kenari, M. (2013). Distributed Multiuser Sequential Channel Sensing Schemes in Multichannel Cognitive Radio Networks. *IEEE Transactions on Wireless Communications*, *12*(5), 2055–2067.

Siegel, C., Ziegler, V., von Wächter, C., Schönlinner, B., Prechtel, U., & Schumacher, H. (2006). Switching Speed Analysis of Low Complexity RF-MEMS Switches. In *Proceedings of IEEE German Microwave Conference*. IEEE.

Simeone, O., Bar-Ness, Y., & Spagnolini, U. (2007). Stable throughput of cognitive radios with and without relaying capability. *IEEE Trans. Wirel. Commun.*, *55*(12), 2351-2360.

Simeone, O., Stanojev, I., Savazzi, S., Bar-Ness, Y., Spagnolini, U., & Pickholtz, R. (2008). Spectrum leasing to cooperating secondary ad hoc networks. *Journal on Selected Areas in Communications*, *26*(1), 203–213. doi:10.1109/JSAC.2008.080118

Simon, M. K., & Alouini, M. S. (2004). *Digital Communication over Fading Channels* (2nd ed.). John Wiley and Sons. doi:10.1002/0471715220

Simon, M. K., & Alouini, M. S. (2005). *Digital communication over fading channels* (Vol. 95). Hoboken, NJ: John Wiley & Sons.

Singh, A., Bhatnagar, M. R., & Mallik, R. K. (2011). Cooperative spectrum sensing with an improved energy detector in cognitive radio network. In *Proceedings of National Conference on communications (NCC'11)*. Bangalore, India: IEEE. doi:10.1109/NCC.2011.5734777

Singh, A., Bhatnagar, M. R., & Mallik, R. K. (2012). Cooperative spectrum sensing in multiple antenna based cognitive radio network using an improved energy detector. *IEEE Communications Letters*, *16*(1), 64–67. doi:10.1109/LCOMM.2011.103111.111884

Singhal, M., & Shivaratri, N. G. (Eds.). (1994). *Advanced Concepts in Operating Systems, TMH*. New York: McGraw-Hill.

Singh, J., & Rana, A. (2013). Exploring the Big Data Spectrum. *International Journal of Emerging Technology and Advanced Engineering*, *3*(4), 73–76.

Siyoum, F., Geilen, M., Moreira, O., Nas, R., & Corporaal, H. (2011). Analyzing synchronous dataflow scenarios for dynamic software-defined radio applications. In *Proceedings of 2011 International Symposium on System on Chip (SoC)* (pp. 14–21). Eindhoven, Netherlands: Academic Press. doi:10.1109/ISSOC.2011.6089222

Sklar, B. (1988). *Digital Communications: Fundamentals and Applications*. Prentice Hall.

Skyworks Solutions. (2008). *SKY13298-360LF: GaAs SP2T Switch for Ultra Wideband (UWB) 3–8 GHz*. Retrieved January 16, 2014, from www.skyworksinc.com/uploads/documents/200797C.pdf

Skyworks. (2013). *Surface Mount Mixer and Detector Schottky Diodes*. SMS76xx Datasheet. Retrieved from http://skyworksinc.com

Sloane, N. J. A. (2014). *Bell or exponential numbers: Ways of placing n labeled balls into n indistinguishable boxes. In The On-Line Encyclopedia of Integer Sequence*. OEIS.

Snellen, M. (1891). Buys Ballot. *Meteorologische Zeitschrift*, *8*, 1–6. Available from http://archive.org/stream/meteorologische00unkngoog#page/n24/mode/2up

Sodagari, S., Attar, A., & Bilen, A. (2011). On a trustful mechanism for expiring spectrum sharing in cognitive radio networks. *IEEE Journal on Selected Areas in Communications*, *29*(4), 856–856. doi:10.1109/JSAC.2011.110416

So, J., & Vaidya, N. H. (2004). Multi-channel mac for ad hoc networks: handling multi-channel hidden terminals using a single transceiver. In *Proceedings of the 5th ACM International Symposium on Mobile Ad Hoc Networking and Computing* (pp. 222-233). ACM. doi:10.1145/989459.989487

Sonnenschein, A., & Fishman, P. M. (1992). Radiometric detection of spread-spectrum signals in noise of uncertain power. *IEEE Transactions on Aerospace and Electronic Systems, 28*(3), 654–660. doi:10.1109/7.256287

Southwell, R., Huang, J., & Liu, X. (2012). Spectrum mobility games. In *Proceedings of IEEE Conference on Computer Communications*, (pp. 37-45). IEEE.

Sree Harsha, K. (2003). *Principles of Physical Vapor Deposition of Thin Films*. Oxford, UK: Elsevier Ltd.

Srivastava, M. S., & Khatri, C. G. (1979). *An Introduction to Multivariate Statistics*. Elsevier North Holland.

Stadius, K., Kaltiokallio, M., Ollikainen, J., Parnanen, T., Saari, V., & Ryynanen, J. (2011). A 0.7 – 2.6 GHz high-linearity rf front-end for cognitive radio spectrum sensing. In *Proceedings of IEEE International Symposium on Circuits and Systems (ISCAS)*, (pp. 2181-2184). IEEE. doi:10.1109/ISCAS.2011.5938032

Stanojev, Simeone, Bar-Ness, & Yu. (2008). Spectrum Leasing via Distributed Cooperation in Cognitive Radio. In *Proceedings of IEEE ICC 2008*, (pp. 3427-3431). IEEE.

Stapor, D. (1995). Optimal receive antenna polarization in the presence of interference and noise. *IEEE Transactions on Antennas and Propagation, 43*(5), 473–477. doi:10.1109/8.384191

Starr, T., Sorbara, M., Cioffi, J. M., & Silverman, P. J. (2003). *DSL Advances*. Prentice-Hall.

Statovci, D., Nordstrom, T., & Nilsson, R. (2006). The normalized-rate iterative algorithm: A practical dynamic spectrum management method for dsl. *EURASIP Journal on Applied Signal Processing, 2006*, 1–17. doi:10.1155/ASP/2006/95175 PMID:16758000

Steenkiste, P., Sicker, D., Minden, G., & Raychaudhuri, D. (2009, March). Future directions in cognitive radio network research. NSF Workshop Report, 4(1), 1-2.

Stegun, M. A. (1972). Handbook of Mathematical Functions with Formulas, Graphs, and Mathematical Tables (9th ed. ed.). New York: Dover Press.

Stevenson, C., Chouinard, G., Lei, Z., Hu, W., Shellhammer, S., & Caldwell, W. (2009). Gennaio). IEEE 802.22: The first cognitive radio wireless regional area network standard. *IEEE Communications Magazine, 47*(1), 130–138. doi:10.1109/MCOM.2009.4752688

Storch, F. W. (1999). *Statistical Analysis in Climate Research*. Cambridge University Press.

Stuijk, S., Geilen, M., Theelen, B., & Basten, T. (2011). Scenario-aware dataflow: Modeling, analysis and implementation of dynamic applications. In *Proceedings of International Conference on Embedded Computer Systems: Architectures, Modeling and Simulation* (pp. 404–411). Samos, Greece: Academic Press. doi:10.1109/SAMOS.2011.6045491

Su, W., Matyjas, J.D., & Batalama, S. (2010). Active Cooperation between Primary Users and Cognitive Radio Users in Cognitive Ad-Hoc Networks. In *Proceedings of ICASSP*, (pp. 3174-3177). ICASSP.

Su, Y., & Mihaela, V. D. (2008). A new look at multi-user power control games. In *Proceedings of IEEE International Conference on Communications*, (pp. 1072-1076). IEEE. doi:10.1109/ICC.2008.209

Subhedar, M., & Birajdar, G. (2011). Spectrum sensing techniques in cognitive radio networks: A survey. *International Journal of Next-Generation Networks, 3*(2), 37–51. doi:10.5121/ijngn.2011.3203

Subramanian, A. P., Al-Ayyoub, M., Gupta, H., Das, S. R., & Buddhikot, M. M. (2008). Near-optimal dynamic spectrum allocation in cellular networks. In *Proceedings of IEEE Symposium on New Frontiers in Dynamic Spectrum Access Networks*, (pp. 1–11). IEEE. doi:10.1109/DYSPAN.2008.41

Subramanian, A. P., Gupta, H., Das, S. R., & Buddhikot, M. M. (2007, April). Fast Spectrum Allocation in Coordinated Dynamic Spectrum Access Based Cellular Networks. In *Proceedings of 2007 IEEE International Symposium on Dynamic Spectrum Access Networks*. Dublin, Ireland: IEEE. doi:10.1109/DYSPAN.2007.50

Subramanian, A. P., Gupta, H., Das, S. R., & Cao, J. (2008). Minimum interference channel assignment in multiradio wireless mesh networks. *IEEE Transactions on Mobile Computing, 7*(12), 1459–1473.

Sugeno, M. (1985). An introductory survey of fuzzy control. *Information Sciences, 1*(1-2), 59–83. doi:10.1016/0020-0255(85)90026-X

Sun, C., Zhang, W., & Ben Letaief, K. (2007). Cooperative Spectrum Sensing for Cognitive Radios under Bandwidth Constraints. In *Proceedings of 2007 IEEE Wireless Communications and Networking Conference*, (pp. 1–5). IEEE. doi:10.1109/WCNC.2007.6

Sun, C., Zhang, W., & Letaief, K. B. (2007). Cluster-Based Cooperative Spectrum Sensing in Cognitive Radio Systems. In *Proceedings of 2007 IEEE International Conference on Communications*, (pp. 2511–2515). IEEE. doi:10.1109/ICC.2007.415

Sun, H., Nallan, A., Wang, C-X., & Chen, Y. (2013, April). Wideband spectrum sensing for cognitive radio networks: A survey. *IEEE Wireless Communications, 20*(2), 74-81.

Sun, H., Nallanathan, A., Jiang, J., & Wang, C. X. (2011). Cooperative spectrum sensing with diversity reception in cognitive radios. In *Proceedings of the International Conference on Communications and Networking in China* (216-220). Harbin, China: Academic Press. doi:10.1109/ChinaCom.2011.6158151

Sun, W. J. (2013). *Research on Key Techniques of Spectrum Resources Management in Cognitive Radio Networks*. The PLA Information Engineering University.

Sundance. (2014). Retrieved from http://www.sundance.com

Sun, H., Chiu, W., & Nallanathan, A. (2012, November). Adaptive Compressive Spectrum Sensing for Wideband Cognitive Radios. *IEEE Communications Letters, 16*(11), 1812–1815. doi:10.1109/LCOMM.2012.092812.121648

Suris, J. E., Dasilva, L. A., Han, Z., & Mackenzie, A. B. (2007). Cooperative game theory for distributed spectrum sharing. In *Proceedings of IEEE International Conference on Communications*, (pp. 5282-5287). IEEE. doi:10.1109/ICC.2007.874

Sutton, P. D., Lotze, J., Lahlou, H., Fahmy, S. A., Nolan, K. E., & Ozgul, B. et al. (2010). Iris: An architecture for cognitive radio networking testbeds. *IEEE Communications Magazine, 48*(9), 114–122. doi:10.1109/MCOM.2010.5560595

Sutton, P. D., Nolan, K. E., & Doyle, L. E. (2008). Cyclostationary signatures in practical cognitive radio applications. *IEEE Journal on Selected Areas in Communications, 26*(1), 13–24. doi:10.1109/JSAC.2008.080103

Su, W., Matyjas, J., & Batalama, S. (2012). Active cooperation between primary users and cognitive radio users in cognitive ad-hoc networks. *IEEE Transactions on Signal Processing, 60*(4), 1796–1805. doi:10.1109/TSP.2011.2181841

Svenson, P., Berg, T., Horling, P., Malm, M., & Martenson, C. (2007, July). Using the impact matrix for predictive situational awareness. In *Proceedings of Information Fusion*, (pp. 1-7). IEEE. doi:10.1109/ICIF.2007.4408072

Taherpour, A., Nasiri-Kenari, M., & Jamshidi, A. (2007). Efficient cooperative spectrum sensing in cognitive radio networks. In *Proceedings of IEEE International Symposium on Personal, Indoor and Mobile Radio Communications* (pp. 1–6). IEEE. doi:10.1109/PIMRC.2007.4394424

Taherpour, A., Nasiri-Kenari, M., & Gazor, S. (2010). Multiple antenna spectrum sensing in cognitive radios. *IEEE Transactions on Wireless Communications, 9*(2), 814–823. doi:10.1109/TWC.2009.02.090385

Taherpour, A., Norouzi, Y., Nasiri-Kenari, M., Jamshidi, A., & Zeinalpour-Yazdi, Z. (2007). Asymptotically optimum detection of primary user in cognitive radio networks. *IET Communications, 1*(6), 1138–1145. doi:10.1049/iet-com:20060645

Talbot, J., & Jakeman, M. (2011). *Security risk management body of knowledge* (Vol. 69). John Wiley & Sons.

Tandel, P., Valiveti, S., Agrawal, K. P., & Kotecha, K. (2010). Non-Repudiation in Ad Hoc Networks. In *Proceedings of Communication and Networking* (pp. 405-415). Springer Berlin Heidelberg.

Tandra, R., & Sahai, A. (2005). Fundamental limits on detection in low SNR under noise uncertainty. In *Proceedings of International Conference on Wireless Networks, Communications and Mobile Computing* (Vol. 1, pp. 464–469). IEEE. doi:10.1109/WIRLES.2005.1549453

Tandur, D., & Moonen, M. (2007). Joint adaptive compensation of transmitter and receiver IQ imbalance under carrier frequency offset in OFDM-based systems. *IEEE Transactions on Signal Processing*, 55(11), 5246–5252. doi:10.1109/TSP.2007.898788

Tanenbaum, A. S. (2003). *Computer Networks*. Pearson Education.

Tani, A., & Fantacci, R. (2010). A Low-Complexity Cyclostationary-Based Spectrum Sensing for UWB and WiMAX Coexistence With Noise Uncertainty. *IEEE Transactions on Vehicular Technology*, 59(6), 2940–2950. doi:10.1109/TVT.2010.2049511

Tan, K., Liu, H., Zhang, J., Zhang, Y., Fang, J., & Voelker, G. M. (2011). Sora: High-performance software radio using general-purpose multi-core processors. *Communications of the ACM*, 54(1), 99–107. doi:10.1145/1866739.1866760

Tarchi, D., Fantacci, R., & Bonciani, E. (2010). Analysis and comparison of scheduling techniques for a BWA OFDMA mobile system. *Wireless Communications and Mobile Computing*, 10(7), 888–898.

Tawk, Y., Constantine, J., Hemmady, S., Balakrishnan, G., Avery, K., & Christodoulou, C. G. (2012). Demonstration of a Cognitive Radio Front End Using an Optically Pumped Reconfigurable Antenna System (OPRAS). *IEEE Transactions on Antennas and Propagation*, 60(2), 1075–1083. doi:10.1109/TAP.2011.2173139

Tawk, Y., Costantine, J., Avery, K., & Christodoulou, C. G. (2011). Implementation of a Cognitive Radio Front-End Using Rotatable Controlled Reconfigurable Antennas. *IEEE Transactions on Antennas and Propagation*, 59(5), 1773–1778. doi:10.1109/TAP.2011.2122239

Tawk, Y., Costantine, J., & Christodolou, C. (2014). Cognitive-Radio and Antenna Functionalities: A Tutorial. *IEEE Antennas and Propagation Magazine*, 56(1), 231–243. doi:10.1109/MAP.2014.6821791

Teguig, D., Scheers, B., & Nir, V. L. (2012). Data fusion schemes for cooperative spectrum sensing in cognitive radio networks. In *Proceedings of the Military Communications and Information Systems Conference* (pp. 1-7). Gdansk, Poland: Academic Press.

Teh, Y., Jordan, M., Beal, M., & Blei, D. (2006, December). Hierarchical Dirichlet processes. *Journal of the American Statistical Association*, 101(476), 1566–1581. doi:10.1198/016214506000000302

Thai, Q., & Reisenfeld, S. (2011). *Adaptive exploitation of cyclostationarity features for detecting modulated signals*. Paper presented at the International Conference on Advanced Technologies for Communications (ATC). New York, NY. doi:10.1109/ATC.2011.6027450

Theis, N. C., Thomas, R. W., & DaSilva, L. A. (2010). Rendezvous for cognitive radios. *IEEE Transactions on Mobile Computing*, 10(2), 216–227. doi:10.1109/TMC.2010.60

Thies, W. (2009). *Language and Compiler Support for Stream Programs*. (PhD Thesis). Massachusetts Institute of Technology, Cambridge, MA.

Thrall, R., & Lucas, W. (1963). N-person games in partition function form. *Naval Res. Logistics Quart.*, 10(1), 281–298. doi:10.1002/nav.3800100126

Tian, Z. (2008). Compressed Wideband Sensing in Cooperative Cognitive Radio Networks. In *Proceedings of IEEE Global Telecommunications Conference (GLOBECOM)*. IEEE. doi:10.1109/GLOCOM.2008.ECP.721

Tian, Z., & Giannakis, G. G. (2007). Compressed Sensing for Wideband Cognitive Radios. In *Proceedings of Acoustics, Speech and Signal Processing*. Honolulu, HI: IEEE.

Tian, Z., Tafesse, Y., & Sadler, B. M. (2012). Cyclic feature detection with sub-Nyquist sampling for wideband spectrum sensing. *IEEE Journal of Selected Topics in Signal Processing*, 58-69.

Timmers, M., Pollin, S., Dejonghe, A., Bahai, A., Van der Perre, L., & Catthoor, F. (2008). Accumulative interference modeling for cognitive radios with distributed channel access. In *Proceedings of International Conference on Cognitive Radio Oriented Wireless Networks and Communications* (pp. 1–7). IEEE. doi:10.1109/CROWNCOM.2008.4562479

Tosh, D., & Sengupta, S. (2013, September). Self-Coexistence in Cognitive Radio Networks using Multi-Stage Perception Learning. In *Proceedings of Vehicular Technology Conference*, (pp. 1-5). Academic Press. doi:10.1109/VTCFall.2013.6692404

Tragos, E., Zeadally, S., Fragkiadakis, A., & Siris, V. (2013). Spectrum assignment in cognitive radio networks: A comprehensive survey. *IEEE Communications Surveys and Tutorials*, *15*(3), 1108–1135. doi:10.1109/SURV.2012.121112.00047

Trelea, I. C. (2003). The particle swarm optimization algorithm: Convergence analysis and parameter selection. *Information Processing Letters*, *85*(6), 317–325. doi:10.1016/S0020-0190(02)00447-7

Truong, D. N., Cheng, W. H., Mohsenin, T., Yu, Z., Jacobson, A. T., & Landge, G. et al. (2009). A 167-processor computational platform in 65 nm CMOS. *IEEE Journal of Solid-State Circuits*, *44*(4), 123–127.

Tse, D., & Viswanath, P. (2005). *Fundamentals of Wireless Communication*. Cambridge University Press. doi:10.1017/CBO9780511807213

Tseng, Y. C., Hsu, C. S., & Hsieh, T. Y. (2002). Power-saving protocols for IEEE 802.11-based multi-hop ad hoc networks. In *Proceedings of IEEE International Conference on Computer Communications (INFOCOM)* (pp. 200-209). IEEE.

Tsiaflakis, P., Vangorp, J., Moonen, M., Verlinden, J., & Van Acker, K. (2006). An efficient search algorithm for the lagrange multipliers of optimal spectrum balancing in multi-user xdsl systems. In *Proceedings of IEEE International Conference on Acoustics, Speech and Signal Processing* (pp. 101–104). IEEE. doi:10.1109/ICASSP.2006.1660915

Tsiaflakis, P., Diehl, M., & Moonen, M. (2008). Distributed spectrum management algorithms for multiuser dsl networks. *IEEE Transactions on Signal Processing*, *56*(10), 4825–4843. doi:10.1109/TSP.2008.927460

Turk, M., & Pentland, A. (1991). Eigenfaces for recognition. *Journal of Cognitive Neuroscience*, *3*(1), 71–86. doi:10.1162/jocn.1991.3.1.71 PMID:23964806

Tu, S.-Y., Chen, K.-C., & Prasad, R. (2009, September). Spectrum Sensing of OFDMA Systems for Cognitive Radio Networks. *IEEE Transactions on Vehicular Technology*, *58*(7), 3410–3425. doi:10.1109/TVT.2009.2014775

Tyagi, A. K., & Rajakumar, R. V. (2010). A wideband RF frontend architecture for software defined radio. In *Proceedings of 2010 International Conference on Industrial and Information Systems (ICIIS)*, (pp. 25-30). Academic Press. doi:10.1109/ICIINFS.2010.5578742

Tychogiorgos, G., Gkelias, A., & Leung, K. K. (2012). Utility-Proportional Fairness in Wireless Networks. In *Proceedings of 2012 IEEE 23rd International Symposium on Personal, Indoor and Mobile Radio Communications, PIMRC, 2012*. IEEE.

U.S. Weather Bureau. (1961, May). *Technical Paper No. 40, Chart 14: 100-Year 60-Minute Rainfall (inches)*. Washington, DC: U.S. Government Printing Office.

Ueberhuber, C. W. (1997). Numerical Computation: *Methods, Software, and Analysis (vol. 1)*. Springer.

Umar, R., & Sheikh, A. U. H. (2012). Spectrum Access and Sharing for Cognitive Radio. In Developments in Wireless Network Prototyping, Design, and Deployment: Future Generations (pp. 241–271). Academic Press.

Umar, R., & Sheikh, A. U. H. (2012). Cognitive Radio oriented wireless networks: Challenges and solutions. In *Proceedings of 2012 International Conference on Multimedia Computing and Systems* (pp. 992–997). IEEE. doi:10.1109/ICMCS.2012.6320105

Umar, R., & Sheikh, A. U. H. (2012). A comparative study of spectrum awareness techniques for cognitive radio oriented wireless networks. *Physical Communication*. doi:10.1016/j.phycom.2012.07.005

Umashankar, G., & Kannu, A. P. (2013). Throughput Optimal Multi-Slot Sensing Procedure for a Cognitive Radio. *IEEE Communications Letters*, *17*(12), 2292–2295. doi:10.1109/LCOMM.2013.102613.131825

UMTS Forum Report 44. (2011). *Mobile traffic forecasts 2010-2020, report January 2011*. Author.

Universal Software Radio Peripheral (USRP). (2014). Retrieved February 18, 2014. from http://www.ettus.com

Unnikrishnan, J., & Veeravalli, V. V. (2008). Cooperative sensing for primary detection in cognitive radio. *IEEE Journal of Selected Topics in Signal Processing, 2*(1), 18–27. doi:10.1109/JSTSP.2007.914880

Urgaonkar, R., & Neely, M. (2009, June). Opportunistic scheduling with reliability guarantees in cognitive radio networks. *IEEE Transactions on Mobile Computing, 6*(6), 766–777. doi:10.1109/TMC.2009.38

Urkowitz, H. (1967). Energy detection of unknown deterministic signals. *Proceedings of the IEEE, 55*(4), 523–531. doi:10.1109/PROC.1967.5573

Valkama, M., Shahed Hagh Ghadam, A., Anttila, L., & Renfors, M. (2006). Advanced Digital Signal Processing Techniques for Compensation of Nonlinear Distortion in Wideband Multicarrier Radio Receivers. *IEEE Transactions on Microwave Theory and Techniques, 54*(6), 2356–2366. doi:10.1109/TMTT.2006.875274

Van Berkel, K., Heinle, F., Meuwissen, P. P. E., Moerman, K., & Weiss, M. (2005). Vector Processing as an Enabler for Software-Defined Radio in Handheld Devices. *EURASIP Journal on Advances in Signal Processing, 2005*(16), 2613–2625. doi:10.1155/ASP.2005.2613

Van de Beek, J., Riihija, J., Achtzehn, A., & Mahonen, P. (2012). TV white space in Europe. *IEEE Transactions on Mobile Computing, 11*(2), 178–188. doi:10.1109/TMC.2011.203

Van der Schaar, M., & Fu, F. (2009). Spectrum access games and strategic learning in cognitive radio networks for delay-critical applications. *Proceedings of the IEEE, 97*(4), 720–740.

Van Trees, H. L. (1968). *Detection, estimation, and modulation theory part-1*. New York: John Wiley and Sons.

Vandebril, R., Barel, M. V., & Mastronardi, N. (2008). *Matrix Computations and Semiseparable Matrices: Eigenvalue and Singular Value Methods*. JHU Press.

Vaughan, R. (1990). Polarization diversity in mobile communications. *IEEE Transactions on Vehicular Technology, 39*(3), 177–186. doi:10.1109/25.130998

Vazquez-Vilar, G., Mosquera, C., & Jayaweera, S. K. (2010). Primary user enters the game: Performance of dynamic spectrum leasing in cognitive radio networks. *IEEE Transactions on Wireless Communications, 9*(12), 1–5. doi:10.1109/TWC.2010.101310.101056

Vijayandran, L., Dharmawansa, P., Ekman, T., & Tellambura, C. (2012). Analysis of aggregate interference and primary system performance in finite area cognitive radio networks. *IEEE Transactions on Communications, 60*(7), 1811–1822. doi:10.1109/TCOMM.2012.051412.100739

Vishay. (2012). *Thin Film Surface Mounted RF Capacitor*. Retrieved January 16, 2014, from http://www.vishay.com/docs/61090/rfcs.pdf

Visotsky, E., Kuffner, S., & Peterson, R. (2005). On collaborative detection of TV transmissions in support of dynamic spectrum sharing. In *Proceedings of the IEEE International Symposium on New Frontiers in Dynamic Spectrum Access Networks* (pp. 338-345). Baltimore, MD: IEEE. doi:10.1109/DYSPAN.2005.1542650

Viswanathan, R., & Varshney, P. K. (1997). Distributed detection with multiple sensors I. Fundamentals. *Proceedings of the IEEE, 85*(1), 54–63. doi:10.1109/5.554208

von Neumann, J., & Morgenstern, O. (1944). *Theory of Games and Economic Behavior*. Princeton, NJ: Princeton Univ. Press.

Vu, M., Ghassemzadeh, S. S., & Tarokh, V. (2008). Interference in a cognitive network with beacon. In *Proceedings of IEEE Wireless Communications and Networking Conference* (pp. 876–881). IEEE.

Wang, F., Krunz, M., & Cui, S. G. (2008). Spectrum sharing in cognitive radio networks. In *Proceedings of IEEE INFOCOM*, (pp. 36-40). IEEE.

Wang, Q., & Zheng, H. (2006). Route and spectrum selection in dynamic spectrum networks. In *Proceedings of IEEE Consumer Communications and Networking Conference (CNCC)*, (pp. 342-346). IEEE.

Wang, W. (2009). Spectrum sensing for cognitive radio. In *Proceedings of Third International Symposium on Intelligent Information Technology Application Workshops (IITAW)*, (pp. 410-412). Academic Press. doi:10.1109/IITAW.2009.49

Wang, X., Guo, W., Lu, W., & Wang, W. (2011). Adaptive compressive sampling for wideband signals. In *Proceedings of Vehicular Technology Conference (VTC Spring)*, (pp. 1-5). IEEE.

Wang, Y., Pandharipande, A., Polo, Y. L., & Leus, G. (2009). Distributed compressive wide-band spectrum sensing. In *Proceedings of Information Theory and Applications Workshop*. Academic Press.

Wang, Y., Tian, Z., & Feng, C. (2010). A two-step compressed spectrum sensing scheme for wideband cognitive radios. In *Proceedings of Global Telecommunications Conference (GLOBECOM 2010)*, (pp. 1-5). IEEE. doi:10.1109/GLOCOM.2010.5683246

Wang, Y., Tian, Z., & Feng, C. (2011). Cooperative spectrum sensing based on matrix rank minimization. In *Proceedings of Acoustics, Speech and Signal Processing (ICASSP)*, (pp. 3000-3003). IEEE.

Wang, B., & Liu, K. (2011). Advances in cognitive radio networks: A survey. *IEEE Journal of Selected Topics in Signal Processing*, *5*(1), 5–23. doi:10.1109/JSTSP.2010.2093210

Wang, H. S., & Moayeri, N. (1995). Finite-state Markov channel-a useful model for radio communication channels. *IEEE Transactions on Vehicular Technology*, *44*(1), 163–171. doi:10.1109/25.350282

Wang, J., Ghosh, M., & Challapali, K. (2011). Emerging cognitive radio applications: A survey. *IEEE Communications Magazine*, *49*(3), 74–81. doi:10.1109/MCOM.2011.5723803

Wang, P. H., Collins, J. D., Chinya, G. N., Jiang, H., Tian, X., & Girkar, M., … Wang, H. (2007). EXOCHI: Architecture and Programming Environment for A Heterogeneous Multi-core Multithreaded System. In *Proceedings of the ACM SIGPLAN Conference on Programming Language Design and Implementation* (pp. 156 – 166). San Diego, CA: ACM. doi:10.1145/1250734.1250753

Wang, R., Lau, V. K., Lv, L., & Chen, B. (2009). Joint cross-layer scheduling and spectrum sensing for OFDMA cognitive. *IEEE Transactions on Wireless Communications*, *8*(5), 2410–2416. doi:10.1109/TWC.2009.071147

Wang, R., & Tao, M. (2012, March). Joint source and relay precoding designs for MIMO two-way relaying based on MSE criterion. *IEEE Transactions on Signal Processing*, *61*(7), 1770–1785.

Wang, X. Y., Wong, A., & Ho, P. H. (2010). Extended knowledge-based reasoning approach to spectrum sensing for cognitive radio. *IEEE Transactions on Mobile Computing*, *9*(4), 465–478. doi:10.1109/TMC.2009.148

Wang, Y., & Falconer, D. (2010). Phase noise estimation and suppression for single carrier SDMA. In *Proc. IEEE Wireless Commun. Netw. Conf. (WCNC)*. IEEE. doi:10.1109/WCNC.2010.5506625

WARP. (2014). Retrieved from http://warp.rice.edu

Watkins, D. S. (2004). *Fundamentals of matrix computations* (Vol. 64). John Wiley & Sons.

Webb, W. (1998). The role of economic techniques in spectrum management. *IEEE Communications Magazine*, *36*(3), 102–107. doi:10.1109/35.663334

Wei, D., Guo, C., Liu, F., & Zeng, Z. (2009). A SINR improving scheme based on optimal polarization receiving for the cognitive radios. In *Proceedings of IEEE Int. Conf. on Network Infrastructure and Digital Content*, (pp. 100-104). Beijing: IEEE.

Wei, J., & Zhang, X. (2010, March). Energy-efficient distributed spectrum sensing for wireless cognitive radio networks. In *Proceedings of INFOCOM IEEE Conference on Computer Communications Workshops*, (pp. 1-6). IEEE. doi:10.1109/INFCOMW.2010.5466680

Wei, S., Youming, L., & Miaoliang, Y. (2009). Low-complexity grouping spectrum management in multi-user dsl networks. In *Proceedings of International Conference on Communications and Mobile Computing* (pp. 381–385). Academic Press. doi:10.1109/CMC.2009.125

Weidling, F., Datla, D., Petty, V., Krishnan, P., & Minden, G. (2005). A Framework for RF Spectrum Measurements and Analysis. In *Proceedings of the 1st International Symposium on New Frontiers in Dynamic Spectrum Access Networks* (pp. 573-576). Baltimore, MD: IEEE.

Weifang, W. (2010). Denial of service attacks in cognitive radio networks. In *Proceedings of Environmental Science and Information Application Technology* (ESIAT), (pp. 530-533). ESIAT.

Weingart, T., Yee, G. V., Sicker, D. C., & Grunwald, D. (2007). Implementation of a reconfiguration algorithm for cognitive radio. In *Proceedings of 2nd International Conference on Cognitive Radio Oriented Wireless Networks and Communications (CrownCom)*, (pp. 171-180). Academic Press. doi:10.1109/CROWNCOM.2007.4549792

Weingart, T., Sicker, D. C., & Grunwald, D. (2007). A statistical method for reconfiguration of cognitive radios. *IEEE Wireless Communications*, 4(4), 34–40. doi:10.1109/MWC.2007.4300981

Weiss, T. A., & Jondral, F. K. (2004). Spectrum pooling: An innovative strategy for the enhancement of spectrum efficiency. *IEEE Communications Magazine*, 42(3), S8–S14. doi:10.1109/MCOM.2004.1273768

Westerhagen, A. P. (2014). *Cognitive radio networks: Elements and architectures.* (Doctoral dissertation). Blekinge Institute of Technology, Sweden.

Wiens, A., Arnous, M., Maune, H., Sazegar, M., Nikfalazar, M., Kohler, C., et al. (2013). Load Modulation for High Power Applications based on Printed Ceramics. In *Proceedings of IEEE MTT-S International Microwave Symposium.* IEEE. doi:10.1109/MWSYM.2013.6697548

Wireless Innovation Forum. (n.d.). *WinnForum Top Ten Innovations.* Retrieved from http://groups.winnforum.org/winnforum_top_ten

Woh, M., Harel, Y., Mahlke, S., Mudge, T., Chakrabarti, C., & Flautner, K. (2006). SODA: A Low-power Architecture For Software Radio. In *Proceedings of 33rd International Symposium on Computer Architecture, ISCA* (pp. 89–101). Boston, MA: Academic Press.

Woh, M., Lin, Y., Seo, S., Mahlke, S., Mudge, T., Chakrabarti, C., … Flautner, K. (2008). From SODA to scotch: The evolution of a wireless baseband processor. In *Proceedings of 41st IEEE/ACM International Symposium on Microarchitecture* (pp. 152–163). Como, Italy: IEEE. doi:10.1109/MICRO.2008.4771787

Wolkowicz, H., & Styan, G. P. (1980). Bounds for eigenvalues using traces. *Linear Algebra and Its Applications*, 29, 471–506. doi:10.1016/0024-3795(80)90258-X

Wood, F., & Black, M. J. (2008, August). A nonparametric Bayesian alternative to spike sorting. *Journal of Neuroscience Methods*, 173(1), 1–12. doi:10.1016/j.jneumeth.2008.04.030 PMID:18602697

Wu, C.-C., & Wu, S.-H. (2013). On bridging the gap between homogeneous and heterogeneous rendezvous schemes for cognitive radios. In *Proceedings of ACM International Symposium on Mobile Ad Hoc Networking and Computing (MobiHoc).* ACM. doi:10.1145/2491288.2491295

Wu, K., Han, F., & Kong, D. (2013). Rendezvous Sequence Construction in Cognitive Radio Ad-Hoc Networks based on Difference Sets. In *Proceedings of IEEE International Symposium on Personal, Indoor and Mobile Radio Communications (PIMRC).* IEEE. doi:10.1109/PIMRC.2013.6666442

Wu, S. H., Chen, M. S., & Chen, C. M. (2008). Fully adaptive power saving protocols for ad hoc networks using the Hyper Quorum System. In *Proceedings of International Conference on Distributed Computing Systems (ICDCS)* (pp. 785–792). Academic Press.

Wu, S.-L., Lin, C.-Y., Tseng, Y.-C., & Sheu, J.-P. (2000). A new multi-channel MAC protocol with on-demand channel assignment for multi-hop mobile ad hoc networks. In *Proceedings of Parallel Architectures, Algorithms and Networks,* (pp. 232-237). IEEE.

Wu, G., Talwar, S., Johnsson, K., Himayat, N., & Johnson, K. (2011, April). M2M: From mobile to embedded internet. *IEEE Communications Magazine*, 49(4), 36–43. doi:10.1109/MCOM.2011.5741144

Wu, H., Yang, F., Tan, K., Chen, J., Zhang, Q., & Zhang, Z. (2006). Distributed Channel Assignment and Routing in MultiRadio Multichannel Multihop Wireless Network. *Selected Areas in Communications*, 24(11), 1972–1983. doi:10.1109/JSAC.2006.881638

Wyglinski, A. M., Nekovee, M., & Hou, Y. T. (Eds.). (2009). *Cognitive Radio Communications and Networks: Principles and Practice*. Elsevier Inc.

Xia, B., Wahab, M. H., Yang, Y., Fan, Z., & Sooriyabandara, M. (2009). Reinforcement learning based spectrum-aware routing in multi-hop cognitive radio networks. In *Proceedings of IEEE 4th International Conference on Cognitive Radio Oriented Wireless Networks and Communications (CrownCom)*, (pp. 1-5). IEEE. doi:10.1109/CROWNCOM.2009.5189189

Xia, W., Wang, S., Liu, W., & Chen, W. (2009, September). Cluster-based energy efficient cooperative spectrum sensing in cognitive radios. In *Proceedings of Wireless Communications, Networking and Mobile Computing*, (pp. 1-4). Academic Press. doi:10.1109/WICOM.2009.5301709

Xiangzhen, L., & Xuping, Z. (2008). Feature Detection Based on Multiple Cyclic Frequencies in Cognitive Radios. In *Proceedings of China-Japan Joint Microwave Conference* (pp. 290 – 293). IEEE. doi:10.1109/CJMW.2008.4772428

Xie, L., Heegaard, P. E., Zhang, Y., & Xiang, J. (2012). Reliable Channel Selection and Routing for Real-Time Services over Cognitive Radio Mesh Networks. In Quality, Reliability, Security and Robustness in Heterogeneous Networks (pp. 41-57). Springer. doi:10.1007/978-3-642-29222-4_4

Xie, M., Zhang, W., & Wong, K. (2010). A geometric approach to improve spectrum efficiency for cognitive relay networks. *IEEE Trans. Wirel. Commun.*, 9(1), 268-281.

Xin, Q., Wang, X., Cao, J., & Feng, W. (2011). Joint Admission Control, Channel Assignment and QoS Routing for Coverage Optimization in Multi-hop Cognitive Radio Cellular Networks. In *Proceedings of IEEE 8th International Conference on Mobile Adhoc and Sensor Systems (MASS)*, (pp. 55-62). IEEE.

Xing-Peng Mao., & Mark J. W. (2006). On Polarization Diversity in Mobile Communications. In *Proceedings of International Conference on Communication Technology*, (pp. 1-4). Academic Press.

Xing, Y., Chandramouli, R., & Cordeiro, C. (2007, April). Price Dynamics in Competitive Agile Spectrum Access Markets. *IEEE Journal on Selected Areas in Communications*, 25(3), 613–621. doi:10.1109/JSAC.2007.070411

Xu, H., & Li, B. (2010, May). *Efficient resource allocation with flexible channel cooperation in OFDMA cognitive radio networks*. Paper presented at IEEE INFOCOM. San Diego, CA. doi:10.1109/INFCOM.2010.5462169

Xu, S., Zhang, Q., & Lin, W. (2009). Pso-based ofdm adaptive power and bit allocation for multiuser cognitive radio system. In *Proceedings of 5th International Conference on Wireless Communications, Networking and Mobile Computing (WiCom '09)*. Academic Press.

Xu, Y., Panigrahi, S., & Le-Ngoc, T. (2005). Selective iterative water-filling for digital subscriber loops. In *Proceedings of International Conference on Information, Communications and Signal Processing* (pp. 101–105). Academic Press.

Xu, D., Liu, X., & Han, Z. (2013). Decentralized bargaining: A two-tier market for efficient and flexible dynamic spectrum access. *IEEE Transactions on Mobile Computing*, 12(9), 1697–1711. doi:10.1109/TMC.2012.130

Xu, W., Trappe, W., Zhang, Y., & Wood, T. (2005, May). The feasibility of launching and detecting jamming attacks in wireless networks. In *Proceedings of the 6th ACM International Symposium on Mobile Ad Hoc Networking and Computing* (pp. 46-57). ACM. doi:10.1145/1062689.1062697

Xu, Y., Anpalagan, A., Wu, Q., Shen, L., Gao, Z., & Wang, J. (2013). Decision-Theoretic Distributed Channel Selection for Opportunistic Spectrum Access: Strategies, Challenges and Solutions. *IEEE Communications Surveys and Tutorials*, 15(4), 1689–1713. doi:10.1109/SURV.2013.030713.00189

Xu, Y., Le-Ngoc, T., & Panigrahi, S. (2008). Global concave minimization for optimal spectrum balancing in multi-user dsl networks. *IEEE Transactions on Signal Processing*, 56(7), 2875–2885. doi:10.1109/TSP.2008.917378

Yang, C., & Li, J. (2010). Joint economical and technical consideration of dynamic spectrum sharing: A multi-stage Stackelberg game perspective. *VTC*, 1-5.

Yang, C., & Li, J. (2013). Pricing-based dynamic spectrum leasing: A hierarchical multi-stage Stackelberg game perspective. *IEICE Transactions on Communications*, 96(6), 1511–1521. doi:10.1587/transcom.E96.B.1511

Yang, L., Cao, L., & Zheng, H. (2007). Proactive channel access in dynamic spectrum networks. In *Proceedings of IEEE International Conference on Cognitive Radio Oriented Wireless Networks and Communications*. Orlando, FL: IEEE.

Yang, L., Cao, L., & Zheng, H. (2008). Proactive channel access in dynamic spectrum networks. *Physical Communication*, 1(2), 103–111. doi:10.1016/j.phycom.2008.05.001

Yang, S., & Hajek, B. (2007). VCG-Kelly Mechanisms for Divisible Goods: Adapting VCG Mechanisms to One-Dimensional Signals. *IEEE JSAC*, 25(6), 1237–1243.

Yang, Z., Cheng, G., Liu, W., Yuan, W., & Cheng, W. (2008). Local coordination based routing and spectrum assignment in multi-hop cognitive radio networks. *Mobile Networks and Applications*, 13(1-2), 67–81. doi:10.1007/s11036-008-0025-9

Yao, Y. (2014). *A software framework for prioritized spectrum access in heterogeneous cognitive radio networks.* (Doctoral dissertation). Blekinge Institute of Technology, Sweden.

Yao, Y., Ngoga, S. R., Erman, D., & Popescu, A. P. (2012). Competition-based channel selection for cognitive radio networks, In *Proceedings of IEEE Wireless Communications and Networking Conference*. Paris, France: IEEE. doi:10.1109/WCNC.2012.6214006

Ye, Z., Memik, G., & Grosspietsch, J. (2008). Energy Detection Using Estimated Noise Variance for Spectrum Sensing in Cognitive Radio Networks. In *Proceedings of 2008 IEEE Wireless Communications and Networking Conference* (pp. 711–716). IEEE. doi:10.1109/WCNC.2008.131

Yenumula, B. R. (2013). Security Issues and Threats in Cognitive Radio Networks. In *Proceedings of AICT:The Ninth Advanced International Conference on Telecommunications*, (pp. 85-90). AICT.

Yeung, G. K., & Gardner, W. A. (1996). Search-efficient methods of detection of cyclostationary signals. *IEEE Transactions on Signal Processing*, 44(5), 1214–1223. doi:10.1109/78.502333

Yi Tan, K., & Shamil Sengupta. (2010). Competitive spectrum trading in dynamic Spectrum access markets: A price war. In *Proceedings of GLOBECOM'10*, (pp. 1-5). IEEE.

Yi, Y., Zhang, J., Zhang, Q., Jiang, T., & Zhang, J. (2010). Cooperative communication-aware spectrum leasing in cognitive radio networks. In *Proc. DySPAN*. Academic Press. doi:10.1109/DYSPAN.2010.5457883

Yin, W., Wen, Z., Li, S., Meng, J., & Han, Z. (2011). Dynamic compressive spectrum sensing for cognitive radio networks. In *Proceedings of Information Sciences and Systems (CISS)*, (pp. 1-6). IEEE.

Younis, O., & Fahmy, S. (2004). *Distributed clustering in ad-hoc sensor networks: a hybrid, energy-efficient approach. In Proceedings of IEEE INFOCOM 2004* (Vol. 1, pp. 629–640). IEEE. doi:10.1109/INFCOM.2004.1354534

Youssef, M., Ibrahim, M., Abdelatif, M., Chen, L., & Vasilakos, A. (2014). Routing Metrics of Cognitive Radio Networks: A Survey. *IEEE Communications Surveys and Tutorials*, 16(1), 92–109. doi:10.1109/SURV.2013.082713.00184

Yu, L., Liu, H., Leung, Y.-W., Chu, X., & Lin, Z. (2013). Multiple Radios for Effective Rendezvous in Cognitive Radio Networks. In *Proceedings of IEEE International Conference on Communication (ICC)*. IEEE. doi:10.1109/ICC.2013.6654974

Yuan, Y., Bahl, P., Chandra, R., Chou, P. A., Ferrell, J. I., Moscibroda, T., et al. (2007). KNOWS: Cognitive Radio Networks Over White Spaces. In *Proceedings of IEEE Dynamic Spectrum Access Networks Conference*, (pp. 416-427). IEEE. doi:10.1109/DYSPAN.2007.61

Yuan, Z., Niyato, D., Li, H., Song, J. B., & Han, Z. (2012). Defeating primary user emulation attacks using belief propagation in cognitive radio networks. *IEEE Journal on Selected Areas in Communications*, 30(10), 1850–1860.

Yucek, T., & Arslan, H. (2009). A survey of spectrum sensing algorithms for cognitive radio applications. *IEEE Communications Surveys and Tutorials, 11*(1), 116–130. doi:10.1109/SURV.2009.090109

Yu-Chun, W., Haiguang, W., & Zhang, P. (2009). Protection of wireless microphones in IEEE 802.22 cognitive radio network. In *Proceedings of IEEE International Conference on Communications Workshops* (pp. 1–5). IEEE.

Yu, H., Tang, W., & Li, S. (2012). Joint optimal sensing and power allocation for cooperative relay in cognitive radio networks. In *Proceedings of the IEEE International Conference on Communications* (pp. 1635-1640). Ottawa, Canada: IEEE. doi:10.1109/ICC.2012.6363718

Yun, Y. H., & Cho, J. H. (2009). An orthogonal cognitive radio for a satellite communication link. In *Proceedings of IEEE 20th International Symposium on Personal, Indoor and Mobile Radio Communications*, (pp. 3154-3158). Tokyo: IEEE.

Yu, W., Ginis, G., & Cioffi, J. M. (2002). Distributed multiuser power control for digital subscriber lines. *IEEE Journal on Selected Areas in Communications, 20*(5), 1105–1115. doi:10.1109/JSAC.2002.1007390

Zadeh, L. A. (1965). Fuzzy Sets. *Information and Control, 8*(3), 338–353. doi:10.1016/S0019-9958(65)90241-X

Zander, S., Nguyen, T., & Armitage, G. (2005, November). Automated traffic classification and application identification using machine learning. In *Proceedings of the IEEE Conference on Local Computer Networks*. Sydney, Australia: IEEE.

Zeeshan, M., Manzoor, M. F., & Qadir, J. (2010). Backup channel and cooperative channel switching on-demand routing protocol for multi-hop cognitive radio ad hoc networks (BCCCS). In *Proceedings of IEEE 6th International Conference on Emerging Technologies (ICET)*, (pp. 394-399). IEEE.

Zeng, F., Li, C., & Tian, Z. (2011). Distributed compressive spectrum sensing in cooperative multihop cognitive networks. *IEEE Journal of Selected Topics in Signal Processing*, 37-48.

Zeng, Y., & Liang, Y. C. (2007). Maximum-Minimum Eigenvalue Detection for Cognitive Radio. In *Proceedings of IEEE 18th International Symposium on Personal, Indoor and Mobile Radio Communications*, (pp. 1-5). IEEE.

Zeng, Y., & Liang, Y. C. (2007). Covariance Based Signal Detections for Cognitive Radio. In *Proceedings of 2nd IEEE International Symposium on New Frontiers in Dynamic Spectrum Access Networks*, (pp. 202-207). IEEE. doi:10.1109/DYSPAN.2007.33

Zeng, Y., & Liang, Y. C. (2010). Robust spectrum sensing in cognitive radio. In *Proceedings of IEEE 21st International Symposium on Personal, Indoor and Mobile Radio Communications Workshops*, (pp. 1-8). IEEE.

Zeng, Y., Koh, L. C., & Liang, Y. C. (2008). Maximum Eigenvalue Detection: Theory and Application. Academic Press.

Zeng, Y., Liang, Y.-C., Hoang, A. T., & Peh, E. C. Y. (2009). Reliability of Spectrum Sensing Under Noise and Interference Uncertainty. In *Proceedings of 2009 IEEE International Conference on Communications Workshops* (pp. 1–5). IEEE. doi:10.1109/ICC.2009.5199287

Zeng, Y., & Liang, Y. C. (2009). Eigenvalue-based spectrum sensing algorithms for cognitive radio. *IEEE Transactions on Communications, 57*(6), 1784–1793. doi:10.1109/TCOMM.2009.06.070402

Zeng, Y., & Liang, Y. C. (2009). Spectrum sensing algorithms for cognitive radio based on statistical covariances. *IEEE Transactions on* Vehicular Technology, *58*(4), 1804–1815.

Zeng, Y., & Liang, Y. C. (2009). Spectrum-Sensing Algorithms for Cognitive Radio Based on Statistical Covariances. *IEEE Transactions on Vehicular Technology, 58*(4), 1804–1815. doi:10.1109/TVT.2008.2005267

Zeng, Y., Liang, Y.-C., Hoang, A. T., & Zhang, R. (2010). A Review on Spectrum Sensing for Cognitive Radio: Challenges and Solutions. *EURASIP Journal on Advances in Signal Processing, 2010*, 1–16. doi:10.1155/2010/381465

Zhang, B., Han, K., Ravindran, B., & Jensen, E. D. (2008). RTQG: Real-time quorum-based gossip protocol for unreliable networks. In *Proceedings of International Conference on Availability, Security, and Reliability (ARES)*, (pp. 564–571). Academic Press. doi:10.1109/ARES.2008.139

Zhang, J., & Zhang, Q. (2009, May). *Stackelberg game for utility-based cooperative cognitive radio networks.* Paper presented at ACM MobiHoc. New Orleans, LA.

Zhang, J., Yao, F., Liu, Y., & Cao, L. (2012). Robust route and channel selection in cognitive radio networks. In *Proceedings of IEEE 14th International Conference on Communication Technology (ICCT)*, (pp. 202-208). IEEE.

Zhang, Q., Jia, J., & Zhang, J. (2009). Cooperative relay to improve diversity in cognitive radio networks. *IEEE Commun. Magazine, 47*(2), 111-117.

Zhang, R., Cui, S., & Liang, Y.-C. (2008). On ergodic sum capacity of fading cognitive multiple-access channel. In *Proceedings of Forty-Sixth Annual Allerton Conference*. Academic Press. doi:10.1109/ALLERTON.2008.4797650

Zhang, W., & Sanada, Y. (2010). *Dual-stage detection scheme for detect and avoid*. Paper presented at the IEEE International Conference on Ultra-Wideband (ICUWB). Nanjing, China. doi:10.1109/ICUWB.2010.5615282

Zhang, W., Mallik, R. K., & Letaief, K. B. (2008). Cooperative spectrum sensing optimization in cognitive radio networks. In *Proceedings of the International conference on Communications* (pp. 3411-3415). Beijing, China: Academic Press. doi:10.1109/ICC.2008.641

Zhang, Y., Li, Q., Yu, G., & Wang, B. (2011). ETCH: Efficient Channel Hopping for Communication Rendezvous in Dynamic Spectrum Access Networks. In *Proceedings of IEEE Conference on Computer Communications*. IEEE. doi:10.1109/INFCOM.2011.5935070

Zhang, Z., & Xie, X. (2008, June). Application research of evolution in cognitive radio based on GA. In *Proceedings of Industrial Electronics and Applications*, (pp. 1575-1579). IEEE.

Zhang, Q., Cao, B., Wang, Y., Lin, X., Sun, L., & Zhang, N. (2013). On exploiting polarization for energy-harvesting enabled cooperative cognitive radio networking. *IEEE Wireless Communications, 20*(4), 116–124. doi:10.1109/MWC.2013.6590058

Zhang, Q., Jia, J., & Zhang, J. (2009). Cooperative relay to improve diversity in cognitive radio networks. *IEEE Communications Magazine, 47*(2), 111–117. doi:10.1109/MCOM.2009.4785388

Zhang, Q., Kokkeler, B. J., Smit, G. J. M., & Walters, K. H. G. (2009). Cognitive Radio baseband processing on a reconfigurable platform. *Physical Communication, 2*(1-2), 33–46. doi:10.1016/j.phycom.2009.02.008

Zhang, R. (2009). On peak versus average interference power constraints for protecting primary users in cognitive radio networks. *IEEE Transactions on Wireless Communications, 8*(4), 2112–2120. doi:10.1109/TWC.2009.080714

Zhang, R., Liang, Y., Chai, C., & Cui, S. (2009, June). Optimal beamforming for two-way multi-antenna relay channel with analogue network coding. *IEEE Journal on Selected Areas in Communications, 27*(5), 699–712. doi:10.1109/JSAC.2009.090611

Zhang, R., Lim, T., Liang, Y. C., & Zeng, Y. (2010). Multi-antenna based spectrum sensing for cognitive radios: A GLRT approach. *IEEE Transactions on Communications, 58*(1), 84–88. doi:10.1109/TCOMM.2010.01.080158

Zhang, R., Yang, C., Yu, A., Wang, B., Tang, H., Chen, H., & Zhang, J. (2008). Wet Chemical Etching Method for BST Thin Films Annealed at High Temperature. *Applied Surface Science, 254*(21), 6697–6700. doi:10.1016/j.apsusc.2008.05.233

Zhang, W., & Letaief, K. B. (2008). Cooperative spectrum sensing with transmit and relay diversity in cognitive radio networks. *IEEE Transactions on Wireless Communications, 7*(12), 4761–4766. doi:10.1109/T-WC.2008.060857

Zhang, W., Mallik, R. K., & Letaief, K. B. (2009). Cooperative spectrum sensing optimization in cognitive radio networks. In *Proceedings of International Conference on Communications (ICC'08)*. Beijing, China: IEEE.

Zhang, W., Mallik, R. K., & Letaief, K. B. (2009). Optimization of cooperative spectrum sensing with energy detection in cognitive radio networks. *IEEE Transactions on Wireless Communications, 8*(12), 5761–5766. doi:10.1109/TWC.2009.12.081710

Zhang, X., & Li, C. (2009, June). The security in cognitive radio networks: A survey. In *Proceedings of the 2009 International Conference on Wireless Communications and Mobile Computing: Connecting the World Wirelessly* (pp. 309-313). ACM. doi:10.1145/1582379.1582447

Zhang, X., & Su, H. (2011). Opportunistic spectrum sharing schemes for CDMA-based uplink MAC in cognitive radio networks. *IEEE Journal on Selected Areas in Communications, 29*(4), 716–730. doi:10.1109/JSAC.2011.110405

Zhang, X., Sun, Y., Chen, X., Zhou, S., Wang, J., & Shroff, N. B. (2013). Distributed power allocation for coordinated multipoint transmissions in distributed antenna systems. *IEEE Transactions on Wireless Communications, 12*(5), 2281–2291. doi:10.1109/TWC.2013.040213.120863

Zhang, Y., Li, Q., Yu, G., Wang, H., Zhu, X., & Wang, B. (2013). Channel hopping based communication rendezvous in cognitive radio networks. *IEEE/ACM Transactions on Networking.*

Zhang, Y., Yu, R., Nekovee, M., Liu, Y., Xie, S., & Gjessing, S. (2012). Cognitive machine-to-machine communications: Visions and potentials for the smart grid. *IEEE Network, 26*(3), 6–13. doi:10.1109/MNET.2012.6201210

Zhang, Z., Jiang, H., Tan, P., & Slevinsky, J. (2013). Channel exploration and exploitation with imperfect spectrum sensing in cognitive radio networks. *IEEE Journal on Selected Areas in Communications, 31*(3), 429–441.

Zhan, Z., Zhang, J., Li, Y., & Chung, H. S.-H. (2009). Adaptive particle swarm optimization. *IEEE Transactions on Systems, Man, and Cybernetics, 6*(6), 1362–1381. doi:10.1109/TSMCB.2009.2015956 PMID:19362911

Zhao, G., Yang, C., Li, D., & Soong, A. (2011). Power and channel allocation for cooperative relay in cognitive radio networks. *IEEE J. Sel. Top. Signal Proc., 5*(1), 151-158.

Zhao, J., & Wang, X. (2012, October). Channel sensing order in multi-user cognitive radio networks. In *Proceedings of Dynamic Spectrum Access Networks (DYSPAN),* (pp. 397-407). IEEE.

Zhao, J., Zheng, H., & Yang, G.-H. (2005). Distributed coordination in dynamic spectrum allocation networks. In *Proceedings of New Frontiers in Dynamic Spectrum Access Networks,* (pp. 259-268). IEEE.

Zhao, N., Yu, F. R., Sun, H., & Nallanathan, A. (2012). An energy-efficient cooperative spectrum sensing scheme for cognitive radio networks. In *Proceedings of Global Communications Conference (GLOBECOM),* (pp. 3600-3604). IEEE. doi:10.1109/GLOCOM.2012.6503674

Zhao, Y., Gaeddert, J., Morales, L., Bae, K., Um, J., & Reed, J. H. (2007). Development of radio environment map enabled case- and knowledge-based learning algorithms for ieee 802.22 wran cognitive engines. In *Proceedings of 2nd International Conference on Cognitive Radio Oriented Wireless Networks and Communications (CrownCom 07).* Academic Press.

Zhao, Y., Kang, G., Wang, J., Liang, X., & Liu, Y. (2013). A soft fusion scheme for cooperative spectrum sensing based on the log-likelihood ratio. In *Proceedings of IEEE 24th International Symposium on Personal Indoor and Mobile Radio Communications (PIMRC).* IEEE. doi:10.1109/PIMRC.2013.6666688

Zhao, J. H., Li, F., & Zhang, X.-. (2012). Parameter adjustment based on improved genetic algorithm for cognitive radio networks. *Journal of China Universities of Posts and Telecommunications, 19*(3), 22–26. doi:10.1016/S1005-8885(11)60260-4

Zhao, N., Pu, F., Xu, X., & Chen, N. (2013). Optimization of multi-channel cooperative sensing in cognitive radio networks. *IET Communications, 7*(12), 1177–1190. doi:10.1049/iet-com.2012.0748

Zhao, Q., Geirhofer, S., Tong, L. T. L., & Sadler, B. M. (2008). Opportunistic spectrum access via periodic channel sensing. *IEEE Transactions on Signal Processing, 56*(2), 785–796. doi:10.1109/TSP.2007.907867

Zhao, Q., & Sadler, B. M. (2007). A survey of dynamic spectrum access: Signal processing, networking, and regulatory Policy. *IEEE Signal Processing Magazine, 24*(3), 79–89. doi:10.1109/MSP.2007.361604

Zhao, Q., Tong, L., & Swami, A. (2005, November). Decentralized cognitive MAC for dynamic spectrum access. In *Proceedings of 1st IEEE International Symposium on New Frontiers in Dynamic Spectrum Access Networks.* (pp. 224-232). Baltimore, MD: IEEE.

Zhao, Q., Tong, L., Swami, A., & Chen, Y. (2007). Decentralized cognitive MAC for opportunistic spectrum access in ad hoc networks: A POMDP framework. *IEEE Journal on Selected Areas in Communications, 25*(3), 589–600.

Zhao, Z., Xu, S., Zheng, S., & Shang, J. (2009, July). Cognitive radio adaptation using particle swarm optimization. *Wireless Communications & Mobile Computing*, 7(7), 875–881. doi:10.1002/wcm.633

Zheng, X. (2003). *Network Algorithm and Complexity Theory*. National University of Defense Technology.

Zheng, Y., Hristov, A., Giere, A., & Jakoby, R. (2008). Suppression of Harmonic Radiation of Tunable Planar Inverted-F Antenna by Ferroelectric Varactor Loading. In *Proceedings of IEEE MTT-S International Microwave Symposium*. IEEE.

Zheng, H., & Cao, L. (2005). Device-centric spectrum management. In *Proceedings of the 1st IEEE International Symposium on New Frontiers in Dynamic Spectrum Access Networks* (pp. 56-65). Baltimore, MD: IEEE.

Zheng, S., Lou, C., & Yang, X. (2010). Cooperative spectrum sensing using particle swarm optimisation. *Electronics Letters*, 46(22), 1525. doi:10.1049/el.2010.2115

Zheng, Y., Sazegar, M., Maune, H., Zhou, X., Binder, J., & Jakoby, R. (2011). Compact Substrate Integrated Waveguide Tunable Filter based on Ferroelectric Ceramics. *IEEE Microwave and Wireless Components Letters*, 21(9), 477–479. doi:10.1109/LMWC.2011.2162615

Zhou, D. (2003, January). Security issues in ad hoc networks. In *The handbook of ad hoc wireless networks* (pp. 569–582). CRC Press, Inc.

Zhou, H., Berry, R., Honig, M. L., & Vohra, R. (2013). Complexity of allocation problems in spectrum markets with interference complementarities. *IEEE Journal on Selected Areas in Communications*, 31(3), 489–499. doi:10.1109/JSAC.2013.130314

Zhou, X., Lin, L., Wang, J., & Zhang, X. (2009). Cross-layer routing design in cognitive radio networks by colored multigraph model. *Wireless Personal Communications*, 49(1), 123–131. doi:10.1007/s11277-008-9561-7

Zhu, L., & Mao, H. (2011, March). An efficient authentication mechanism for cognitive radio networks. In *Proceedings of Power and Energy Engineering Conference* (APPEEC), (pp. 1-5). IEEE. doi:10.1109/APPEEC.2011.5748783

Zhu, L., & Zhou, H. (2008, December). Two types of attacks against cognitive radio network MAC protocols. In *Proceedings of Computer Science and Software Engineering*, (Vol. 4, pp. 1110-1113). IEEE. doi:10.1109/CSSE.2008.1536

Zhu, J., & Liu, K. (2007). Cognitive Radios for Dynamic Spectrum Access-Dynamic Spectrum Sharing: A Game Theoretical Overview. *IEEE Communications Magazine*, 45(5), 88–94. doi:10.1109/MCOM.2007.358854

Zhu, Q., Yuan, Z., Song, J. B., Han, Z., & Basar, T. (2012). Interference aware routing game for cognitive radio multi-hop networks. *IEEE Journal on Selected Areas in Communications*, 30(10), 2006–2015. doi:10.1109/JSAC.2012.121115

Zorba, N., Realp, M., Lagunas, M., & Pérez-Neira, A. I. (2008). Dual polarization for MIMO processing in multibeam satellite systems. In *Proceedings of 10th International Workshop on Signal Processing for Space Communications*, (pp. 1-7). Academic Press. doi:10.1109/SPSC.2008.4686702

Zou, Y., Yao, Y., & Zheng, B. (2011). Cognitive transmissions with multiple relays in cognitive radio networks. *IEEE Trans. Wirel. Commun.*, 10(2), 648-659.

Zou, Y., Yao, Y. D., & Zheng, B. (2011). A selective-relay based cooperative spectrum sensing scheme without dedicated reporting channels in cognitive radio networks. *Transactions on Wireless Communications*, 10(4), 1188–1198. doi:10.1109/TWC.2011.021611.100913

Zukerman, M. (2013). Introduction to queueing theory and stochastic teletraffic models. *arXiv preprint arXiv:1307.2968*. Retrieved July 11, 2013, from http://arxiv.org/abs/1307.2968

About the Contributors

Naima Kaabouch is currently an associate professor in the Electrical Engineering Department at the University of North Dakota, Grand Forks, USA. She received her BS and MS degrees in Electrical Engineering from the University of Paris 11 as well as and a PhD in Electrical Engineering from the University of Paris 6, France. Her research interests include signal/image processing, sensing, smart systems, wireless communications, and cognitive radio.

Wen-Chen Hu received a BE, an ME, an MS, and a PhD, all in Computer Science, from Tamkang University, Taiwan, the National Central University, Taiwan, the University of Iowa, Iowa City, and the University of Florida, Gainesville, in 1984, 1986, 1993, and 1998, respectively. He is currently an associate professor in the Department of Computer Science of the University of North Dakota, Grand Forks. He was an assistant professor in the Department of Computer Science and Software Engineering at the Auburn University, Alabama, for years. He is the Editor-in-Chief of the *International Journal of Handheld Computing Research* (IJHCR) and an associate editor of the *Journal of Information Technology Research* (JITR), and has acted in more than 100 positions as editors and editorial advisory/review board members of international journals/books and track/session chairs and program committee members of international conferences. He has also won a couple of awards of best papers, best reviewers, and community services. Dr. Hu has been teaching more than 10 years at US universities in over 10 different computer/IT-related courses, and advising/consulting more than 100 graduate students. He has published over 100 articles in refereed journals, conference proceedings, books, and encyclopedias, edited more than 10 books and conference proceedings, and solely authored a book titled *Internet-Enabled Handheld Devices, Computing, and Programming: Mobile Commerce and Personal Data Applications*. His current research interests include handheld/mobile/smartphone/tablet computing, location-based services, Web-enabled information system such as search engines and Web mining, electronic and mobile commerce systems, and Web technologies. He is a member of the IEEE Computer Society and ACM (Association for Computing Machinery).

* * *

Andre' Abadie conducts research in the management of secure information systems and information assurance for government and military operational environments. He received his Doctorate in Information Technology with a concentration in Information Security and Assurance from George Mason University,

Fairfax, VA. He received his Master of Science in Government Information Leadership from the US National Defense University iCollege, Master in Military Art and Science from the US Army Command and General Staff College, Master of Science in Information Assurance from the University of Maryland University College, and Bachelor of Science in Computer Engineering from the US Military Academy.

Fatma Abdelkefi (M'99) received the Engineer Degree in Electrical Engineering from ENIT, Tunisia, in 1998, the MS degree in Digital Communications Systems in 1999, and PhD degree with highest honours in 2002, both from TELECOM Paris. In 2002, she joined the Signal Processing Group of ENSEA, France, as a Research Associate. Then, she spent two years with the Communication Technology Laboratory, ETHZ, Switzerland. From 2006 to 2008, she was a Scientific Collaborator with the EPFL, Switzerland. Since 2009, she joined Sup'Com where she is an Associate Professor. Her research interests include digital communications, information theory, equalization, and multicarrier system.

Mohamed F. Abdelkader received his BSc (with highest honor) degree in Electrical Engineering from Suez Canal University, Port Said, Egypt, in 2002. He received his MSc and PhD degrees in Electrical Engineering from University of Maryland, College Park, in 2005 and 2010, respectively. Since 2010, Dr. Abdelkader has been an assistant professor in the Electrical Engineering Department at Port Said University, Egypt. He is also the founder and current director of the Machine learning and Signal Processing (MLSP) research center at Port Said University. His current research interests include image and video processing, computer vision, signal processing for wireless communication, cognitive radio, compressive sensing. Dr. Abdelkader is the recipient of the Egyptian President certificate of honor for Egyptian university graduates 2002.

Tamer Aboufoul received the MSc degree in Radio Frequency Communication Systems from the University of Southampton, Southampton, UK, in December 2007, and the PhD degree specialized in Reconfigurable Antennas and Radio Systems from Queen Mary University of London in January 2014. From 2008 to 2010, he was an Electronic Engineer in Harwell Science and Innovation Campus, Oxfordshire, UK. In September 2010, he joined the School of Electronic Engineering and Computer Science, Queen Mary, University of London, London, UK, to pursue his research studies. His main research interests include cognitive radio, compact Ultra-Wideband (UWB) antennas, reconfigurable antennas, multi-beam antenna arrays, and RF/microwave circuit design. Mr. Aboufoul was selected as a finalist for the prestigious IET Postgraduate Awards 2013.

Anjali Agarwal received her PhD (Electrical Engineering) in 1996 from Concordia University, Montreal, MSc (Electrical Engineering) in 1986 from University of Calgary, Alberta, and BE (Electronics and Communication Engineering) in 1983 from Delhi College of Engineering, India. She is currently a Professor in the Department of Electrical and Computer Engineering at Concordia University, Montreal. Prior to joining faculty in Concordia, she has worked as a Protocol Design Engineer and a Software Engineer in industry. Her current research interests are in the various aspects and types of wireless networks including Cognitive Radio Networks, Heterogeneous Wireless Access Networks, Sensor Networks, and in Cloud Computing. Dr. Agarwal is a senior member of the IEEE.

Akram Alomainy received the Master of Engineering in Communication Engineering and PhD degree in Electrical and Electronic Engineering (specialised in antennas and radio propagation) from Queen Mary University of London (QMUL), United Kingdom, in July 2003 and July 2007, respectively. He joined the School of Electronic Engineering and Computer Science, QMUL, in 2007, where he is now a Senior Lecturer (Associate Professor) in the Antennas and Electromagnetics Research Group. His current research interests include small and compact antennas for wireless body area networks, radio propagation, antenna interactions with human body, localisation, reconfigurable structures and radio front-ends and advanced algorithm for smart and intelligent systems. He has authored and co-authored 5 book chapters and more than 130 technical papers in leading journals and peer-reviewed conferences with H-Index of 22. Dr. Alomainy won the Isambard Brunel Kingdom Award in 2011. He is a member of the IET, UK, and Senior Member of the IEEE, USA.

Saed Alrabaee received his BSc degree in Electrical and Computer Engineering from Hashemite University, Jordan, 2006, and received his Master degree in Electrical and Computer Engineering form Concordia University, Canada, 2012. He is currently a PhD student in Information System Engineering.

Saud G. Althunibat, University of Trento (UNITN), Trento (Italy), was born in Alkarak, Jordan, in 1982. He received the BSc in Electrical Engineering Communication in 2004 from Mutah University, Jordan, and the MSc Degree in Electrical Engineering Communication in 2010 from the University of Jordan, Jordan. Currently, Saud is a Marie Curie Early stage researcher working within the GREENET project at University of Trento and pursuing a PhD degree under the supervision of Dr. Fabrizio Granelli. He is a reviewer in many international journals and a TPC member in many international conferences. He is the recipient of the best-paper award in IEEE CAMAD 2012, and was selected as exemplary reviewer in *IEEE Communication Letters* 2013. His research interests include Cognitive Radio Networks, Physical-Layer Security, Resource Allocation, and Heterogeneous Networks.

He Bai received his BEng degree from Beijing University of Posts and Telecommunications (BUPT) in 2004 and PhD degree from Chinese Academy of Science (CAS) in 2009. Now he works in Academy of Broadcasting Science (ABS) as an associate professor. He mainly engages in the research of Digital TV Transmission technology, which includes terrestrial digital TV transmission, IPv6 network evolution, 700MHz wireless frequency application and TCP/IP data network planning related topics. He took responsibility for multiple projects including "Research on NGB-based IPv6-related Technology", "Trial Commercial of IPv6-based content distribution business" and "Research on Key-technology for NGB-based Smart Home IOT system". He is the main drafter for national standard for China Digital Terrestrial TV Single Frequency Network technology requirements.

Damindra Bandara is a PhD student in Information Technology at George Mason University. She works as a Graduate Teaching Assistant at Department of Computer Science in the same university. She received her Bachelor degree in Electrical and Electronic Engineering from University of Peradeniya, Sri Lanka. Her research interests are security of software defined radios, safety and security of wireless controlled trains, and formal verification of protocols.

Jyoti Sekhar Banerjee has joined Bengal Institute of Technology under Techno India Group, Kolkata, India, in 2005, as Lecturer, and presently, he is working as Assistant Professor, Dept. of ECE. Previously, he was with Narula Institute of Technology, Kolkata, India, and also served an Indian IT company. He also served many other reputed educational Institutes in India in various positions. He received his BTech in Electronics and Communication Engineering from University of Kalyani, India, in 2003, and ME in Information Technology having specialization in Network Security from West Bengal University of Technology, India, in 2005. His areas of research interests include wireless communication, data communication and computer network, network security, databases, microprocessors, and microcontrollers. He is the author of many research papers published in international conferences and journals. Mr. Banerjee has also served as a review committee member of IEEE sponsored ICCCCM-13.

Manav R. Bhatnagar received his PhD in Wireless Communications from Department of Informatics, University of Oslo, Norway, in 2008, and MTech in Communications Engineering from Indian Institute of Technology Delhi, India, in 2005. From 2008 to 2009, he was Postdoctoral Research Fellow with University Graduate Center (UNIK), Kjeller, Norway. He held visiting appointments with SPiNCOM Group, University of Minnesota Twin Cities; SUPÉLEC, France; UNIK, Kjeller, Norway; Department of Communications and Networking, Aalto University, Finland; and INRIA/IRISA Lab, Lannion, France. He is currently an Associate Professor with the Department of Electrical Engineering, Indian Institute of Technology Delhi, India. His research interests include signal processing for MIMO systems, cooperative communications, FSO communication, cognitive radio, software defined radio, power line communications, and satellite communications. He was selected "Exemplary Reviewer" of *IEEE Communications Letters* for 2010 and 2012. He has served as Editor for *IEEE Transactions on Wireless Communications* during 2011-2014.

Farrukh A. Bhatti received his BE and MS degrees in Avionics Engineering and Electrical Engineering from the National University of Sciences and Technology, Pakistan, in 2004 and 2008, respectively. He received his PhD degree in Electrical and Electronic Engineering from The University of Auckland, New Zealand, in 2014. During his PhD, he undertook research field trips to the Industrial Research Ltd Wellington and the Mobile and Portable Radio Research Group, Virginia Polytechnic Institute and State University, Blacksburg, VA, USA. Presently, he is an Assistant Professor in the Department of Electrical Engineering at the Institute of Space Technology, Islamabad, Pakistan. His research interests include advanced cellular networks, multiple-antenna systems, software-defined radios, and cognitive radio networks.

Bin Cao received his PhD, Master, and Bachelor degrees from Harbin Institute of Technology, China, in 2013, 2009, and 2007, respectively, all in Information and Communication Engineering. From 2010 to 2012, he has been a visiting PhD Student in University of Waterloo, Canada, financially supported by the Chinese Scholarship Council. Since May 2014, he is an associate professor in Jingdezhen Ceramic Institute, Jingdezhen, Jiangxi, China. He also serves as the director of Center for Communications Engineering in Jingdezhen Ceramic Institute. His research interests include signal processing for wireless communications, cognitive radio networking, and resource allocation for wireless networks.

Arpita Chakraborty has joined Bengal Institute of Technology under Techno India Group, Kolkata, India, in 2008, as Lecturer, and presently, she is working as Assistant Professor, Dept. of ECE. She received her BTech degree in Electronics and Communication Engineering (ECE) from BPPIMT, West Bengal University of Technology, India, and MTech Degree in Intelligent Automation and Robotics (IAR) from Jadavpur University, India. She was awarded university medal for standing first in order of merit at MTech examination. She is the author of many research papers published in international conferences and journals. Her current research interests include cognitive radio networks, next generation wireless communications, digital signal processing, and advance robotics. She has also served as a review committee member of IEEE sponsored ICCCCM-13.

Henri-Pierre Charles is research director at CEA-List since September 2010. He was previously assistant professor in Versailles University for 17 years, authorized to supervise PhD students (HDR) since 2008. During this period, he participated in many European and national projects, mainly in the high performance computing domain. He has supervised many PhD theses on code optimizations and high performance computing. Since his arrival in CEA, he applied his knowledge on Multi-Processor System on Chip (MPSoC) and embedded systems. His research interests are in dynamic compilation, low-level code optimization, and new methods for code generation for high performances and multimedia applications. He has developed the concept of "compilettes" that can be defined as tiny code generators embedded into applications, able to generate optimal code depending on the data characteristics.

Symeon Chatzinotas received the MEng degree in Telecommunications from Aristotle University of Thessaloniki, Thessaloniki, Greece, in 2003, and the MSc and the PhD degrees in Electronic Engineering from the University of Surrey, Surrey, UK, in 2009. He is currently a Research Scientist with the Interdisciplinary Centre for Security, Reliability, and Trust, University of Luxembourg. In the past, he has worked on numerous research and development projects for the Institute of Informatics and Telecommunications, National Center for Scientific Research Demokritos, Athens, Greece; the Institute of Telematics and Informatics, Center of Research and Technology Hellas, Thessaloniki, Greece; and the Mobile Communications Research Group, Center of Communication Systems Research, University of Surrey. He is the author of more than 110 technical papers in refereed international journals, conferences, and scientific books, and he is currently crediting a book on *Cooperative and Cognitive Satellite Systems*. His research interests include multiuser information theory, cooperative and cognitive communications, and transceiver optimization for terrestrial and satellite networks.

Mickaël Dardaillon is a PhD student at INSA Lyon since October 2011, in the Inria Socrate team of the CITI Laboratory, in collaboration with the CEA Leti. He received an Engineering degree in Electronics, Signal, and Image in 2011 from Polytech'Orléans, France, as well as a Master degree in Electronics, Signal, and Microsystems from Université d'Orléans, France, in a double degree program. His current research efforts concentrate on compilation for heterogeneous Multi-Processor System on Chip (MPSoC) in the context of software-defined radio. He is interested in embedded systems, compilation, and hardware architecture.

Mahsa Derakhshani received the BSc and MSc degrees from Sharif University of Technology, Tehran, Iran, in 2006 and 2008, respectively, and the PhD degree from McGill University, Montreal, QC, Canada, in 2013, all in Electrical Engineering. She is currently a Postdoctoral Research Fellow with the Department of Electrical and Computer Engineering, University of Toronto, Toronto, ON, Canada, and a Research Assistant with the Department of Electrical and Computer Engineering, McGill University. Her research interests include radio resource allocation for wireless networks, applications of convex optimization and game theory for communication systems, and spectrum sensing techniques in cognitive radio networks. Dr. Derakhshani received the John Bonsall Porter Prize, the McGill Engineering Doctoral Award, and the Fonds de Recherche du Québec–Nature et Technologies Postdoctoral Fellowship.

Mohamed Deriche received his undergraduate degree from the National Polytechnic School of Algeria. He then joined the University of Minnesota, USA, where he completed his MS and PhD in 1988 and 1992, respectively. He worked as a Post Doctorate Fellow with the University of Minnesota Radiology Department in the area of MRI. He then joined the Queensland University of Technology, Australia, as a Lecturer in 1994, then Associate Professor in 2001. In 2001, he joined the EE Department at King Fahd University of Petroleum and Minerals, Saudi Arabia, where he is currently leading the signal processing group. He has published over 150 refereed papers. He delivered a number of tutorial and invited talks at international conferences. He is a recipient of the IEEE third Millennium Medal for 2000. In 2006, he received the Shauman award from best researcher in the Arab world in the area of engineering sciences, and in 2009, he received the excellence in research award at KFUPM and the excellence in teaching award in 2013. Dr. Deriche supervised more than 20 PhD and Master students and more than 60 BS theses in the areas of signal and image processing. He completed more than 20 Major Funded Research Projects including grants from the Australian Research Council (ARC) and King Abdulaziz City for Science and Technology (KACST), Saudi Arabia.

Marco Di Renzo received the Laurea (cum laude) and the PhD degrees in Electrical and Information Engineering from the University of L'Aquila, Italy, in 2003 and in 2007, respectively. In October 2013, he received the Habilitation à Diriger des Recherches degree in Wireless Communications Theory, from the University of Paris-Sud XI, France. Currently, he is a Chargé de Recherche Titulaire with the French National Center for Scientific Research (CNRS), as well as a faculty member of the Laboratory of Signals and Systems, a joint research laboratory of CNRS, SUPELEC, and the University of Paris-Sud XI, Paris, France. His main research interests are in the area of wireless communications theory. Dr. Di Renzo is the recipient of several awards, which include the 2013 IEEE/COMSOC Best Young Researcher Award for the EMEA Region. He currently serves as an Editor of the *IEEE Communications Letters* and the *IEEE Transactions on Communications*.

Ali El-Hajj was born in Lebanon. He received the Lisense degree in Physics from the Lebanese University, Lebanon, in 1979, the degree of "Ingenieur" from L'Ecole Superieure d'Electricite, France, in 1981, and the "Docteur Ingenieur" degree from the University of Rennes I, France, in 1983. From 1983 to 1987, he was with the Electrical Engineering Department at the Lebanese University. In 1987, he joined the American University of Beirut, where he is currently Professor of Electrical and Computer Engineering. His research interests are numerical solution of electromagnetic field problems, antennas and propagation, and spreadsheet applications to engineering problems.

Ines Elleuch was born in Tunisia in 1983. She received the Diplôme d'Ingénieur in Signals and Systems (valedictorian) in 2006 and the Master of Science degree (with high Honors) in Electronic Systems and Communication Networks in 2007, both from Ecole Polytechnique de Tunisie, Tunisia. She is currently working towards the PhD degree in Telecommunications at Ecole Supérieure des Communications de Tunis, Tunisia. Her research interests include statistical signal processing, array processing, spectrum sensing for cognitive radio, and compressed sensing.

Ahmed M. Elzanati is a teaching assistant in the Electronics and Communications Engineering Department at Sinai University, Egypt, since 2009. Eng. Elzanati received the BS (with highest honors) and MS degrees in Electrical Engineering from Port Said University (Formerly Suez Canal University), Port Said, Egypt, in 2008 and 2013, respectively. He is currently pursuing his PhD degree in Electronics, Informatics, and Telecommunications Engineering at University of Bologna, Italy. His research interests include cognitive radio, compressive sensing, cooperative communications, and wireless sensor network. Eng. Elzanati is a recipient of the European Commission Scholarship Erasmus Mundus EU-METALIC 2 project in 2014.

Kenneth Ezirim is a PhD candidate in the Department of Computer Science, Graduate Center City University of New York. He received his BS and MS degrees in Information Technology and Computer Engineering (First Class Hons.) from South-West State University, Russia, in 2006 and 2008, respectively. In 2013, he was awarded a Master of Philosophy degree in Computer Science by the Graduate Center, City University of New York. He has held lecturing positions at Brooklyn College and John Jay College of the City University of New York since 2011. His research interests include wireless networking, cognitive radio networks, machine learning, data mining, and game theory. He is a member of IEEE Computer Society.

Romano Fantacci is a full Professor of Computer Networks at the University of Florence, Florence, Italy. He is the author of more than 300 papers published in prestigious communication science journal and holds some patents. He guest edited special issues in IEEE journals and magazines and served as symposium chair of several IEEE conferences. He received the AEI/IEEE "Renato Mariani" Award in 1982, IEE IERE Benefactor premium in 1990, and IEEE COMSOC Award Distinguished Contributions to Satellite Communications in 2002. Romano was Associate Editor for *Telecommunication Systems, IEEE Trans. Commun., IEEE Trans. Wireless Commun.*, and funder Area Editor for *IEEE Transactions on Wireless Communications*. Currently, he is serving as associate editor for *Int. J. Commun. Systems* and *IEEE Networks*. He is member of the Editorial Board of *Security Commun. Networks* and Regional Editor for *IET Commun.* Since 2005, he is an IEEE Fellow.

Abdallah K. Farraj is a Fulbright Scholar. He received the BSc and MSc degrees in Electrical Engineering from the University of Jordan in 2000 and 2005, respectively, and the PhD degree in Electrical Engineering from Texas A&M University in 2012. He is currently a Postdoctoral Fellow at the University of Toronto. His research interests include cognitive communications, cyber security of smart grids, downhole telemetry systems, wireless communications, and signal processing applications.

Gianluigi Ferrari (http://www.tlc.unipr.it/ferrari) is an Associate Professor of Telecommunications at the Department of Information Engineering of the University of Parma, where he coordinates the activities of the Wireless Ad-hoc and Sensor Networks (WASN) Lab (http://wasnlab.tlc.unipr.it/). As of today, he has published more than 200 papers in leading international journals and conferences, 21 book chapters, 9 patents, and 7 books. He edited the book *Sensor Networks: Where Theory Meets Practice* (Springer 2010). His research interests include wireless ad hoc and sensor networking, Internet of Things, adaptive signal processing, and communication theory. Prof. Ferrari is a corecipient of scientific awards at the following conferences: IWWAN'06; EMERGING'10; BSN 2011; ITST-2011; SENSORNETS 2012; EvoCOMNET 2013; BSN 2014. He currently serves on the editorial boards of several international journals and is an IEEE Senior Member.

Hendrik C. Ferreira received the MSc (Eng.) Electronical and DSc (Eng.) from the University of Pretoria, South Africa, in 1977 and 1980, respectively. In 1983, he joined the University of Johannesburg, South Africa, where he was promoted to professor in 1989. His field of specialization is Digital Communications and Communications emphasizing Emerging Technologies. His fundamental research interest is in Information Theory, especially in Coding Techniques. Prof Ferreira was a past chairman (1991-1993) of the Communications and Signal Processing Chapter of the IEEE South Africa section. He has served as chairman of several conferences, including the International 1999 IEEE Information Theory Workshop in the Kruger National Park, South Africa, and the 2013 IEEE ISPLC conference hosted at the University of Johannesburg. He was the founding chairman (2010-2012) of the Information Theory Chapter of the IEEE South Africa Section. The South African science funding foundations awarded him in 1989 with the FRD Presidential Award for Young Investigators and in 2012 with the NRF A-Rating for Researchers recognized by their peers as international leaders.

Gang Fu received his BASc degree from Harbin Institute of Technology in 2012, in Communication Engineering. Currently, he is with his Master degree in Information and Communication Engineering in Harbin Institute of Technology Shenzhen Graduate. His research interests include cognitive radio and network protocol design.

Yuehong Gao received her BEng degree from Beijing University of Posts and Telecommunications (BUPT) in 2004 and PhD degree from BUPT in 2010. She was a joint-PhD in the Centre for Quantifiable Quality of Service in Communication Systems (Q2S), Norwegian University of Science and Technology (NTNU) under the support of China Scholarship Council and NTNU during 2008 to 2011, and she received her PhD degree from NTNU in 2012. Now she works in the School of Information and Communication Engineering at BUPT as an associate professor. Her research interests include quality of service guarantees in communication networks, performance evaluation of wireless communication systems and system level simulation methodologies, with special emphasis on cognitive radio networks, 3G/4G/5G networks and heterogeneous networks related topics.

Erick Gonzalez-Rodriguez was born in Mexico City, Mexico, in 1984. He graduated as Electronic and Communications Engineer in 2006 from Tecnológico de Monterrey, campus Mexico City. In 2006, he joined Nortel Networks, Mexico, where he was involved in the design of wireline networks. He received the M.Sc. degree in Information and Communication Engineering in 2010 from Technische Universität Darmstadt, Germany. Currently is working towards the Dr.-Ing. (Ph.D.) degree at the Institute for Microwave Engineering and Photonics, Technische Universität Darmstadt, Germany. His research activities focus on tunable microwave components for reconfigurable wireless communications, software-defined-radio and cognitive radio.

Fabrizio Granelli is IEEE ComSoc Distinguished Lecturer for 2012-13, and Associate Professor at the Dept. of Information Engineering and Computer Science (DISI) of the University of Trento (Italy). Since 2008, he is deputy head of the academic council in Information Engineering. He received the Laurea (MSc) degree in Electronic Engineering and the PhD in Telecommunications Engineering from the University of Genoa, Italy, in 1997 and 2001, respectively. In August 2004 and August 2010, he was visiting professor at the State University of Campinas (Brasil). He is author or co-author of more than 130 papers with topics related to networking, with focus on performance modeling, wireless communications and networks, cognitive radios and networks, green networking and smart grid communications. Dr. Granelli was guest-editor of *ACM Journal on Mobile Networks and Applications, ACM Transactions on Modeling and Computer Simulation,* and *Hindawi Journal of Computer Systems, Networks, and Communications.* He is Founder and General Vice-Chair of the First International Conference on Wireless Internet (WICON'05) and General Chair of the 11th and 15th IEEE Workshop on Computer-Aided Modeling, Analysis, and Design of Communication Links and Networks (CAMAD'06 and IEEE CAMAD'10). He is TPC Co-Chair of IEEE GLOBECOM Symposium on Communications QoS, Reliability and Performance Modeling in the years 2007, 2008, 2009, and 2012. He was officer (Secretary 2005-2006, Vice-Chair 2007-2008, Chair 2009-2010) of the IEEE ComSoc Technical Committee on Communication Systems Integration and Modeling (CSIM) and Associate Editor of *IEEE Communications Letters* (2007-2011).

Eman M. Hammad received her undergraduate degree in Electrical Engineering from the University of Jordan in 2000 and the MS degree in Electrical and Computer Engineering from Texas A&M University in 2011, and she is currently a PhD student at the University of Toronto. Her research interests include wireless communication systems, smart grid, cyber security, dynamical systems, and game theory.

Zhu Han (S'01–M'04–SM'09–F'14) received the BS degree in Electronic Engineering from Tsinghua University, in 1997, and the MS and PhD degrees in Electrical Engineering from the University of Maryland, College Park, in 1999 and 2003, respectively. From 2000 to 2002, he was an R&D Engineer of JDSU, Germantown, Maryland. From 2003 to 2006, he was a Research Associate at the University of Maryland. From 2006 to 2008, he was an assistant professor in Boise State University, Idaho. Currently, he is an Associate Professor in Electrical and Computer Engineering Department at the University of Houston, Texas. His research interests include wireless resource allocation and management, wireless communications and networking, game theory, wireless multimedia, security, and smart grid communication. Dr. Han is an Associate Editor of *IEEE Transactions on Wireless Communications* since 2010. Dr. Han is the winner of IEEE Fred W. Ellersick Prize 2011. Dr. Han is an NSF CAREER award recipient 2010.

Gang Hu is an Assistant Professor in the Department of Networking, Computer College of National University of Defense Technology (NUDT), Changsha, China. He received his PhD in Computer Science and Technology from NUDT in 2010. He has been a visiting scholar at Hong Kong University of Science and Technology for 6 months in 2007. His current research interests include the spectrum management of cognitive radio networks, wireless security, and mobile computing. Dr. Hu has published more than 30 papers in the area of wireless networks, most of the papers are focused on the cognitive radio networks, these papers are published in ICC, CROWNCOM, DYSPAN, GLOBECOM, CHINACOM, etc.

Sean Huberman received his B.Sc. Engineering (first class honours) from Queen's University (Kingston, Canada) in 2008. As of November 2014, he completed his Ph.D. in Electrical Engineering at McGill University (Montreal, Canada). His research interests include techniques of dynamic resource allocation and mathematical optimization with applications to wireless and wireline half- and full-duplex systems.

Mohammed (Al-)Husseini received his BE and ME degrees in Computer and Communications Engineering in 1999 and 2001, respectively, and his PhD degree in Electrical and Computer Engineering in January 2012, from the American University of Beirut (AUB), Beirut, Lebanon. In 2013, he was a visiting researcher at the University of New Mexico, Albuquerque, NM, USA, where he also was an exchange research scholar from 2009 to 2011. From 2001 to 2007, he worked as a research assistant at the ECE Department, AUB, and between 2004 and 2007 as a lecturer at the Arab Open University, Beirut, Lebanon. Currently, he is a senior researcher at the Beirut Research and Innovation Center, Beirut, Lebanon. His research interests include antennas and sources for high-power electromagnetics and the design and applications of antenna arrays, reconfigurable antennas, UWB antennas, antennas for Cognitive Radio, wearable antennas, metamaterials, RF energy harvesting, and RF circuits. He has over 90 publications in international refereed journals and conference proceedings.

Rolf Jakoby was born in Kinheim, Germany, in 1958. He received the Dipl.-Ing. and Dr.-Ing. degrees in electrical engineering from the University of Siegen, Germany, in 1985 and 1990, respectively. In Jan. 1991, he joined the Research Center of Deutsche Telekom in Darmstadt, Germany. Since April 1997 he has a full professorship at TU Darmstadt, Germany. Its interdisciplinary research is focused on RFID, micro- and millimeterwave sensors and on reconfigurable RF passive devices by using novel approaches with metamaterial structures, liquid crystal and ferroelectric thick/thin film technologies. He is editor-in-chief of "FREQUENZ", member of the German Society for Information Technology (ITG) and of the IEEE societies MTT and AP. In 1992, he received an award from the CCI Siegen and in 1997, the ITG-Prize for an excellent publication in the IEEE AP Transactions. He is participating on fifteen patents and on sixteen scientific awards.

Ping Ji received her BA degree in Computer Science and Technology from Tsinghua University, Beijing, China, and her PhD degree in Computer Science from the University of Massachusetts at Amherst. Since fall 2003, she has been a faculty member, and currently a Full Professor, of the Mathematics and Computer Science Department of John Jay College of Criminal Justice of the City University of New York (CUNY), and the Deputy Executive Officer of the Computer Science PhD Program of CUNY Graduate Center. Prof. Ji is an awardee of a number of PSC CUNY Research Awards, NSF research grants, and US army research grants. Her research interests include the broad aspects of Computer Networks, Network Security and Digital Forensics, Wireless and Sensor Networks, Distributed Network Signaling Architectures, and Performance Evaluation.

Karim Y. Kabalan was born in Jbeil, Lebanon. He received the BS degree in Physics from the Lebanese University in 1979, and the MS and PhD degrees in Electrical Engineering from Syracuse University, in 1983 and 1985, respectively. During the 1986 fall semester, he was a visiting assistant professor of Electrical Engineering at Syracuse University, Currently, he is a Professor of Electrical and Computer Engineering Department, Faculty of Engineering and Architecture, American University of Beirut. His research interests are antenna theory, numerical solution of electromagnetic field problems, and software development.

Mohamed A. Kalil is an assistant professor at the Faculty of Science, Suez University, Suez, Egypt. He received the BSc degree in Computing Science and MSc degree in Computer Science from the Faculty of Science, Suez Canal University, Ismailia, Egypt, in 1996 and 2001, respectively. He received a Doctoral degree in Computer Science in 2011 from Ilmenau University of Technology, Ilmenau, Germany. From 2006 to 2011, he served as research associate in the Integrated Communication Systems Group, Ilmenau University of Technology, Ilmenau, Germany. From 2011 to 2013, he served as a team leader for the cognitive radio network group in the same university. His research interests include Dynamic Spectrum Access, Opportunistic Spectrum Access, Cognitive Radio Networks, especially MAC protocols and spectrum handoff. He has many papers, mostly in the area of wireless networks and performance evaluation.

Arun Pachai Kannu received the MS and PhD degrees in Electrical Engineering from The Ohio State University in 2004 and 2007, respectively. From 2007 to 2009, he was a Senior Engineer in Qualcomm Inc., San Diego, USA. He is currently an Assistant Professor in the department of Electrical Engineering, Indian Institute of Technology Madras, Chennai, India. He works on detection, estimation, information theoretic problems in wireless communications.

Mahmoud Khasawneh is currently a PhD student in the department of electrical and computer engineering at Concordia University. He received his MSc in Electrical and Computer Engineering from Concordia University in 2012. He received the Bachelor`s degree in Computer Engineering from Jordan University of Science and Technology (JUST) in 2010. During his studies at Concordia University, he has published many conference papers, journals, and a book chapter. All of them are about cognitive radio networks. He received many awards due to his outstanding research work such as Concordia University International Tuition Fee Remission Award and Conference & Exposition Award.

Sumit Kundu received Bachelor of Engineering (BE) with First class Honors in Electronics and Communication Engg from National Institute of Technology (NIT), Durgapur, India (previously Regional Engineering College, Durgapur under the University of Burdwan, India) in 1991, Master of Technology (M.Tech) in Telecommunication Systems Engg from Indian Institute of Technology, Kharagpur, in 1993-94, and PhD in Wireless Communications from Indian Institute of Technology, Kharagpur, in April 2004. He has been a faculty in the Dept. of Electronics and Communication Engg, National Institute of Technology, Durgapur, since 1995, where he is currently a Professor. His research interests include Radio Resource Management in Wireless Networks, Wireless Sensor Networks, Cognitive Radio Networks, and Cooperative Communications. He has published extensively in several international journals and conferences. He is a senior member of IEEE.

Tho Le-Ngoc received the BEng (with Distinction) in 1976, the MEng in 1978 from McGill University, Montréal, QC, Canada, and the PhD degree in 1983 from the University of Ottawa, Canada. He was with Spar Aerospace Limited during 1977-1982 and with SRTelecom Inc during 1982-1985. From 1985 to 2000, he was a Professor in the Department of Electrical and Computer Engineering of Concordia University during 1985-2000, and of McGill University since 2000. His research interest is in the area of broadband access communications. Dr. Le-Ngoc is a Fellow of the IEEE, the EIC, the CAE, and the RSC. He is the recipient of the 2004 Canadian Award in Telecommunications Research and the 2005 IEEE Canada Fessenden Award. He is the Canada Research Chair (Tier I) on Broadband Access Communications.

Andre Leon Nel (BSc(Mech) '79, MIng '87, PhD '93) is a professor in Mechanical Engineering Science at the University of Johannesburg, South Africa. Early in his career, he became interested in Telecommunications research – originally in the packing of codes in higher dimensional space and later in the application of Game Theory to problems in allocation of resources to competing users in a variety of areas. He is presently working in fields of Game Theory applied to Risk Management and the optimal use of limited resources in a competitive environment. He is interested in work that transcends disciplinary boundaries but that use similar tools and techniques. He has graduated more than 25 doctoral candidates during the past 15 years in the aforementioned fields of research.

Husheng Li received the BS and MS degrees in Electronic Engineering from Tsinghua University, Beijing, China, in 1998 and 2000, respectively, and the PhD degree in Electrical Engineering from Princeton University, Princeton, NJ, in 2005. From 2005 to 2007, he worked as a senior engineer at Qualcomm Inc., San Diego, CA. In 2007, he joined the EECS department of the University of Tennessee, Knoxville, TN, as an assistant professor. He was promoted to associate professor in 2013. His research is mainly focused on statistical signal processing, wireless communications, networking, smart grid, and game theory. Dr. Li is the recipient of the Best Paper Awards of *EURASIP Journal of Wireless Communications and Networks* 2005 (together with his PhD advisor, Prof. H. V. Poor), IEEE ICC 2011, IEEE SmartGridComm 2012, and the Best Demo Award of IEEE Globecom 2010.

Jiandong Li graduated from Xidian University with Bachelor Degree, Master Degree, and PhD in Communications and Electronic Systems in 1982, 1985, and 1991, respectively. He is with Xidian University from 1985, an associate professor from 1990 to 1994, professor from 1994, respectively. He was a visiting professor of Cornell University (Jan. 2002-Jan. 2003), awarded National Science Fund for Distinguished Young Scholars by NSFC, and is the distinguished professor of Changjiang Scholar. He is a senior member of IEEE, the Chair of IEEE ComSoc Xi'an Chapter and the fellow of China Institute of Communication (CIC). He was the member of PCN specialist group for China 863 Communication high technology program between Jan. 1993 and Oct. 1994 and from 1999 to 2000. He is also the member of the specialist group of the next generation of wireless mobile communication networks for The Ministry of Industry and Information Technology of the People's Republic of China. His current research interests include cognitive wireless networks, interference management for heterogeneous wireless networks, next generation mobile communications.

Zhidu Li received his BEng from Beijing University of Posts and Telecommunications (BUPT), Beijing, China, in July 2012. He is currently working toward a PhD in the Wireless Theories and Technologies Laboratory as a master-doctor combined program graduate student in BUPT. His research interests include quality of service guarantees in communication networks and performance evaluation of wireless communication systems.

Hao Liang received his PhD degree in Electrical and Computer Engineering from the University of Waterloo, Canada, in 2013. From 2013 to 2014, he was a postdoctoral research fellow in the Broadband Communications Research (BBCR) Lab and Electricity Market Simulation and Optimization Lab (EM-SOL) at the University of Waterloo. Since 2014, he has been an Assistant Professor in the Department of Electrical and Computer Engineering at the University of Alberta, Canada. His research interests are in the areas of smart grid, wireless communications, and wireless networking. He is a recipient of the Best Student Paper Award from IEEE 72nd Vehicular Technology Conference (VTC Fall-2010), Ottawa, ON, Canada. He was the System Administrator of *IEEE Transactions on Vehicular Technology* (2009-2013).

Lixia Liu is an Assistant Engineer in the People's Liberation Army of China. She received her PhD in Computer Science and Technology from NUDT in 2012. The title of her PhD thesis is "Research on Efficient Resource Utilization-Oriented Spectrum Access Technique for Wireless Cognitive Radio Networks." Her current research interests include the spectrum management of cognitive radio networks, wireless sensor networks, and parallel and distributed computing. Dr. Liu has published more than 10 papers in the area of cognitive radio networks, most of the papers are indexed by EI. After her graduation, she is interested in the spectrum optimization for 4G system.

Ali H. Mahdi received his BSc degree in Information Engineering from University of Baghdad, Baghdad, Iraq, in 2003, and his MSc degree in Computer Engineering from University of Technology, Baghdad, Iraq, in 2006. He is currently working toward his PhD degree with Integrated Communication Systems Group, Technische Universität Ilmenau, Ilmenau, Germany, where he is working on device to device communication using cognitive radio aspects. In 2004, he worked as an IT/communication specialist in PARSONS company for infrastructures in Iraq, project of security and justice. In 2005, he joined Lucent Technologies in Iraq, as Network Operation Center Manager at the project of Advance First Response Network (AFRN) for ministry of interior. In 2006, he moved to the academic field as a lecturer at Al-Khwarizmi College of Engineering, University of Baghdad, Baghdad, Iraq. He has many publications in different international conferences and journals. He got a scholarship from German Academic Exchange Service (DAAD) in 2010 for doctoral studies in Germany. His research interests include the application of artificial intelligence techniques for cognitive radios and radio resources optimization.

Petri Mähönen is a professor of networked systems at the RWTH Aachen University, Germany. He is a chair and founding director of Institute for Networked Systems (iNETS) at the RWTH Aachen University. He is also a member of steering board and research area coordinator of the UMIC Research Centre in Aachen, one of the research clusters funded by German Federal and State Government excellence initiative. He serves also at the scientific advisory board of KTH@Wireless in Stockholm, Sweden, and is an associated editor for IEEE TMC. In 2006, he was awarded Telenor research prize, and in 2011, he was serving as a general chair of IEEE DySPAN conference. His research interests and background include work on wireless Internet, software defined radios, cognitive radio networking, spectral efficiency, spatial statistics, and large-scale mm-wave systems.

Ranjan K. Mallik is a Professor in the Department of Electrical Engineering, Indian Institute of Technology (IIT) Delhi. He received the BTech degree from IIT Kanpur and the MS and PhD degrees from the University of Southern California, Los Angeles, all in Electrical Engineering. He has worked as a scientist in the Defence Electronics Research Laboratory, Hyderabad, India, and as a faculty member in IIT Kharagpur and IIT Guwahati. His research interests are in diversity combining and channel modeling for wireless communications, space-time systems, cooperative communications, multiple-access systems, power line communications, difference equations, and linear algebra. He is a recipient of the Shanti Swarup Bhatnagar Prize and the Hari Om Ashram Prerit Dr. Vikram Sarabhai Research Award. He is a member of Eta Kappa Nu, and a fellow of IEEE, the Indian National Academies INAE, INSA, NASI, and IASc, TWAS, IET(UK), IETE(India), and IE(I).

Dania Marabissi was born in Chianciano, Italy. She received her degree in Telecommunications Engineering and her PhD degree in Informatics and Telecommunications Engineering from the University of Florence in 2000 and 2004, respectively. She joined the Department of Information Engineering at the University of Florence in 2000, where she now works as an assistant professor. She conducts research on physical and MAC layers design for broadband wireless systems. Dr. Marabissi is responsible for the Information and Communication Technology (TiCOM) Consortium between the University of Florence and Selex Elsag SpA, a Finmeccanica Company. She has been involved in several national and European research projects and is the author of technical papers published in international journals and conferences. She was the associate editor for *IEEE Transaction on Vehicular Technology* and she is an IEEE senior member.

Jon W. Mark received the PhD degree in Electrical Engineering from McMaster University, Canada, in 1970. Upon graduation, he joined the Department of Electrical and Computer Engineering at the University of Waterloo, became a full Professor in 1978, and served as Department Chair from July 1984 to June 1990. In 1996, he established the Centre for Wireless Communications (CWC) at the University of Waterloo and has since been serving as the founding Director. A Life Fellow of the IEEE, Dr. Mark is the recipient of the 2000 Canadian Award in Telecommunications Research and the 2000 Award of Merit by the Educational Foundation of the Association of Chinese Canadian Professionals. He has served as a member of several Editorial Boards.

Kevin Marquet is associate professor at INSA Lyon since 2010. His research topics focus on the design and evaluation of embedded systems, especially memory management and energy consumption.

Jérôme Martin graduated from the Ecole Polytechnique in 2000 and Telecom ParisTech in 2002, with specializations in computer science, embedded software, signal processing, and integrated circuits. After working as embedded software developer for industry, he became research engineer at CEA-Leti in Grenoble in 2005, where he contributed to several advanced systems-on-chip prototypes in the fields of telecommunications and multicore processing. His main topics of interest include digital integrated circuit architectures, DSPs, hardware-software interactions, multicore programming techniques, as well as SoC design and modeling methodologies.

Holger Maune was born in Cologne, Germany, in 1981. He received the Dipl.-Ing. and Dr.-Ing. degree in communications engineering from the Technische Universität Darmstadt, Darmstadt, Germany, in 2006 and 2011 respectively. His research is currently focused on advanced RF devices employing tunable dielectrics on the system level and for high power applications.

Wessam Mesbah received his PhD in 2008 from McMaster University, Hamilton, ON, Canada, and his MSc and BSc degrees (with highest honors) in Electrical Engineering from Alexandria University, Alexandria, Egypt, in 2003 and 2000, respectively. He joined the Electrical Engineering Department of King Fahd University For Petroleum and Minerals in February 2010, where he is currently an assistant professor. Between January 2009 and February 2010, he held the position of postdoctoral fellow at Texas A&M University, Qatar, Doha, Qatar. His research interests are in the areas of cooperative communications and relay channels, layered multimedia transmission, wireless sensor networks, multiuser MIMO/OFDM systems, cognitive radio, optimization, game theory, and advanced signal processing techniques.

Scott L. Miller received the BS, MS, and PhD degrees in Electrical Engineering from the University of California at San Diego (UCSD). In 1988, he joined the Department of Electrical and Computer Engineering at the University of Florida, where he was an Assistant and Associate Professor from 1988 through 1998. In August 1998, he joined the Electrical and Computer Engineering Department at Texas A&M University, where he is currently the Debbie and Dennis Segers '75 Professor of Electrical Engineering. He has also held visiting positions at Motorola Inc., University of Utah, and UCSD. His current research interests are in the general area of wireless communications and signal processing. He is a fellow of the IEEE and is a past chair of the IEEE Communications Theory Technical Committee. His hobbies include swimming, tennis, and cycling.

Rakesh Misra received the BTech and MTech degrees in Electrical Engineering (with specialization in Communication Engineering) from Indian Institute of Technology (IIT) Madras, Chennai, India, in 2011. He was a Summer Research Fellow at Indian Institute of Science, Bangalore, in 2010, and at Bell Labs India in 2009. He is currently a PhD Candidate in Electrical Engineering at Stanford University. His research interests include the design and analysis of networked systems, with a focus on wireless systems. He is a recipient of the Stanford Graduate Fellowship at Stanford University, and has previously received the Institute Merit Prize from IIT Madras in recognition of his academic achievements.

Fahham Mohammed received his BE degree with distinction in Electronics and Communication from Osmania University (OU), Hyderabad, India, in 2008. He received his MS degree from the King Fahd University of Petroleum and Minerals (KFUPM), Kingdom of Saudi Arabia, in 2013. He was a Research Assistant during his Master's degree program in KFUPM (2010-2013). He completed his MS Thesis

under the supervision of Dr. Mohamed Deriche, with Dr. Asrar U. H. Sheikh as one of the committee members. Before joining KFUPM, he worked as a Project Engineer in Wipro Technologies, a leading MNC in Bangalore, India (2009-2010), where he was part of a Testing team in DVR technology, which handled the TiVo Inc. project. Currently, he is working as an Electrical Engineer in the Environment, Health, and Safety (EHS) Department in KFUPM. His research interests include Wireless Communications and Cognitive Radio Networks.

S. Hossein Mousavinezhad, Electrical Engineering Professor, Idaho State University, is the principal investigator of the National Science Foundation's research grant, National Wireless Research Collaboration Symposium 2014. He has published a book (with Dr. Hu of University of North Dakota) on mobile computing in 2013. Professor Mousavinezhad is an active member of IEEE and ASEE having chaired sessions in national and regional conferences. He is an ABET Program Evaluator for Electrical Engineering and Computer Engineering programs. He is the Founding General Chair of the IEEE International Electro Information Technology Conferences. Professor Mousavinezhad serves as MGA Vice President and Van Valkenburg Early Career Teaching Award Chair for IEEE Education Society. He is a member of the Editorial Advisory Board of the international research journal *Integrated Computer-Aided Engineering*.

Hua Mu (S'11-M'13) received the BS degree in Applied Physics for Northeastern University, China, in 2005, the MS degree in Electrical Engineering from Shanghai Jiao Tong University, Shanghai, China, in 2008, and the PhD degree in Electrical and Computer Engineering from Auburn University, Auburn, AL, in 2012. Since Aug. 2013, she has been with Florida Institute of Technology as an Assistant Professor. Her research interests lie in the areas of wireless communication and signal processing, in particular cognitive radio, relay networks, and MIMO techniques.

Sandeep Narayanan received the BTech degree in Electronics and Communication Engineering from Amrita University, India, in 2010, and the MSc degree in Signal Processing and Communications from the University of Edinburgh, UK, in 2011. Currently, Sandeep is a Marie-Curie Early-Stage-Researcher (ESR) with Wireless Embedded Systems Technologies (WEST) Aquila, Italy, and he is currently working towards his PhD degree in Wireless Communication Systems at the Center of Excellence for Research DEWS, University of L'Aquila, Italy. His research interests include wireless communication theory and signal processing, with main emphasis on Spatial Modulation for MIMO wireless systems, cooperative communications, and cognitive radio networks.

Masoumeh Nasiri-Kenari received her BS and MS degrees in Electrical Engineering from Isfahan University of Technology, Isfahan, Iran, in 1986 and 1987, respectively, and her PhD degree in Electrical Engineering from University of Utah, Salt Lake City, in 1993. From 1987 to 1988, she was a Technical Instructor and Research Assistant at Isfahan University of Technology. Since 1994, she has been with the Department of Electrical Engineering, Sharif University of Technology, Tehran, Iran, where she is now a Professor. Prof. Nasiri-Kenari founded Wireless Research Laboratory (WRL) of the Electrical Engineering Department in 2001 to coordinate the research activities in the field of wireless communication. Current main activities of WRL are Cooperative and Cognitive Communications, Network Coding, Molecular Communications, and WBAN. From 1999-2001, she was a Co-Director of Advanced Communication Research Laboratory, Iran Telecommunication Research Center, Tehran, Iran. She is a recipient of Distinguished Researcher Award and Distinguished Lecturer Award of EE department at Sharif University of Technology in years 2005 and 2007, respectively. She is now a visiting professor at USC, where she spends her sabbatical leave.

Youssef Nasser obtained his Master's and PhD degrees in Signal Processing and Communications from the National Polytechnic Institute of Grenoble, France in 2003 and 2006, respectively. He is currently with the ECE Department at the American University of Beirut. From 2006 to 2010, he worked as a Senior Research Engineer/Assistant Professor with IETR, Rennes, France. From 2003 to 2006, he worked as R&D Engineer with CEA-Leti Grenoble, France. Since 2007, Y. Nasser has been involved in the standardization of European DVB broadcasting systems. He has coordinated different tasks in many European projects. Y. Nasser is currently coordinating two projects: "Localization in HetNets" and "Convergence between DVB and LTE." Y. Nasser has published more than 90 papers in international peer-reviewed journals and conferences. He has been and is still involved in the organization of different conferences, an active TPC member of IEEE conferences. He is a Senior Member of the IEEE.

Björn Ottersten received his PhD from Stanford University in 1989. In 1991, he was appointed Professor at the Royal Institute of Technology, Stockholm. Currently, Dr. Ottersten is Director for the Interdisciplinary Centre for Security, Reliability, and Trust at the University of Luxembourg. Dr. Ottersten was Director of Research at ArrayComm, San Jose, a start-up company based on Ottersten's patented technology; he is a board member of the Swedish Research Council, and as Digital Champion of Luxembourg, he advises the European Commission. Dr. Ottersten acts on several editorial boards and is currently editor-in-chief of *EURASIP Signal Processing*. Dr. Ottersten is a Fellow of IEEE and EURASIP and on the IEEE Signal Processing Society Board of Governors. He has co-authored 7 papers receiving IEEE paper awards, and in 2011, he received the IEEE Signal Processing Society Technical Achievement Award. He is a first recipient of the European Research Council advanced research grant.

Adrian Popescu is a Full Professor in the Faculty of Computing, Blekinge Institute of Technology, Karlskrona, Sweden. The main research interests are on green networking with particular focus on video networks, cognitive radio networks, network virtualization, cloud computing, performance modeling, and analysis. He has a strong research activity as a result of his involvement in diverse Swedish and European research projects like CELTIC-PLUS CONVINcE, EUREKA MOBICOME, EU FP7 PERIMETER, Euro-NGI Routing in Overlay Networks, Euro-NG Networking over Cognitive Radio Networks, Swedish VINNOVA Performance Analysis of Internet Applications. He is the author and co-author of about 150 publications in different journals, books, and conferences. He is a senior member of ACM and also member of IEEE and IEEE CS. He has served as an area editor of the *Computer Networks* journal, Elsevier.

Alexandru Popescu's main research topics include Cognitive Radio Network architectures, wireless architectures, Peer-to-Peer systems, and routing and optimization algorithms. Alexandru has experience from international projects, interacting and collaborating with other researchers, as well as teaching and supervision experience. He is the author of more than 20 scientific publications at conferences and journals around the world and holds a PhD degree in Telecommunications Systems together with a MSc degree in Electrical Engineering, both from the Blekinge Institute of Technology, in Karlskrona, Sweden. Alexandru has a particular interested in cutting-edge science and technology and a background in the performance modelling field from the University of Bradford in the UK.

Riccardo Raheli is a Professor of Communication Engineering at the University of Parma, Italy, which he joined in 1991. Previously, he was with the Scuola Superiore S. Anna, Pisa, from 1988 to 1991, and Siemens Telecomunicazioni, Milan, from 1986 to 1988. In 1990 and 1993, he was a Visiting Assistant Professor at the University of Southern California, Los Angeles, USA. His scientific interests are in the general area of systems for communication, processing and storage of information, in which he has published extensively in quality journals and conference proceedings, as well as in scientific monographs and industrial patents. He has served as Editorial Board Member and Technical Program Committee Co-Chair of prestigious international journals and conferences.

Ali H. Ramadan was born in Blat, Lebanon. He received his PhD degree in Electrical and Computer Engineering from the American University of Beirut (AUB), Beirut, Lebanon, in January 2014. From 2008 to 2009, he worked as a research assistant, with research duties in antenna design, at the Electrical and Computer Engineering Department, AUB. From October 2010 to October 2013, he visited the University of New Mexico, Albuquerque, NM, USA, as an exchange research scholar. He is currently a research associate at the Electrical and Computer Engineering Department, AUB. His research interests include the design and applications of antenna arrays, UWB antennas, fractal antennas, pattern/polarization diversity antennas for MIMO applications, reconfigurable/tunable antennas, and RF front ends for cognitive radio applications.

Bushra Rashid is a Lab Engineer at Wah Engineering College, Wah Cantt, Pakistan. She did B.S. in Electrical Engineering from Wah Engineering College, Wah Cantt, Pakistan in 2013. She is currently enrolled in M.S Electrical Engineering program at COMSATS Institute of Information Technology, Wah Cantt, Pakistan.

Mubashir Husain Rehmani is an Assistant Professor at COMSATS Institute of Information Technology, Wah Cantt, Pakistan. He did Post Doctorate, PhD, and M.S. from University of Paris Est, University Pierre and Marie Curie and University of Paris XI, France in 2012, 2011 and 2008, respectively. He is an Associate Editor at IEEE Communications Magazine and Elsevier CAEE journal and serving as a Guest Editor in Elsevier Pervasive and Mobile Computing Journal and Elsevier COMPELECENG Journal. He served in the TPC of IEEE WoWMoM 2014, IEEE ICC 2014, and ACM CoNext 2013 SW conferences.

Sam Reisenfeld received the BS in Information Engineering from the University of Illinois, Chicago, and MS and PhD degrees in Communication Systems Engineering from the University of California at Los Angeles (UCLA), in 1969, 1971, and 1979, respectively. From June 1969 until September 1988, he was a Communication Systems Engineer at the Hughes Aircraft Company, Space and Communications Group, El Segundo, California. He was responsible for the design and development of numerous NASA, defense, and commercial satellite communication systems. From September 1988 until June 2009, he was an Associate Professor of Communication Systems Engineering at the University of Technology, Sydney, located at Sydney, NSW, Australia. Since June 2009, he has been an Associate Professor in the Department of Engineering, Faculty of Science, at Macquarie University, North Ryde, NSW, Australia. His current research interests are cognitive radio networks, wireless communication systems, and satellite communication systems.

Tanguy Risset is professor at Insa-Lyon. He is head of the Inria Socrate Team (Software and Cogntive radio for telecommunication). He obtained his PhD working on Systolic Arrays in ENS-Lyon in 1994. Until 2001, he worked as an Inria researcher with Patrice Quinton in Rennes on high level synthesis within the MMAlpha framework. Then, he started the Inria team Compsys (compilation for embedded systems) at ENS-Lyon. He moved to Insa-Lyon as a professor in 2005 and created the Socrate Team in 2011.

Sylwia Romaszko obtained her PhD degree at the University of Antwerp from Belgium. She has worked as a postdoctoral fellow at POPs at the University of Lille in France (July 2009 - July 2010), a developer in OMP partners in Belgium (2010), a researcher at the University of Antwerp in Belgium (2003-2009), a research assistant at the Technical University of Szczecin in Poland (2002-2003), and administrator and Webmaster in X-Comp Computer Service of Companies in Szczecin in Poland (2001-2002). She is currently working as a postdoctoral researcher at iNETs at the RWTH Aachen University (August 2010 - March 2014). Her research interests are focused on cognitive wireless networks and resource optimization problems in wireless networks. She was a project manager for RWTH in EU FP7 ACROPOLIS Network of Excellence project.

Gerard Rowe completed the degrees of BE, ME, and PhD (in Electrical and Electronic Engineering) at the University of Auckland in 1978, 1980, and 1984, respectively. He joined the Department of Electrical and Computer Engineering at the University of Auckland in 1984, where he is currently an Associate Professor, and serves as Associate Dean (Teaching and Learning) within the Faculty of Engineering. He is a member of the Department's Radio Systems Group, and his (disciplinary) research interests lie in the areas of radio systems, electromagnetics, and bioelectromagnetics.

Sanjay Dhar Roy received his BE (Hons.) degree in Electronics and Telecommunication Engineering in 1997 from Jadavpur University, Kolkata, India, and MTech degree in Telecommunication Engineering in 2008 from NIT Durgapur. He received his PhD degree from NIT Durgapur in 2011. He worked for Koshika Telecom Ltd. from 1997 to 2000. After that, he joined the Department of Electronics and Communication Engineering, National Institute of Technology Durgapur, as a Lecturer in 2000, and is currently an Assistant Professor there. His research interests include Radio Resource Management, Handoff, and Cognitive Radio Networks. As of today, he has published 60 research papers in various journals and conferences. Dr. Dhar Roy is a member of IEEE (Communication Society) and is a reviewer of *IET Communications, Electronics Letters and Journal of PIER*, IJCS, Wiley, *International Journal of Electronics*, Taylor & Francis, *Wireless Personal Communication*, Springer. Dr. Dhar Roy has also reviewed for *IEEE GlobeCom, IEEE WCNC, IEEE PIMRC, IEEE VTC*, etc.

Lise Safatly received her PhD in Electrical and Computer Engineering from the American University of Beirut (AUB), Beirut, Lebanon. Her dissertation, titled "Robust Signal Processing Algorithms for RF Complexity Reduction in Cognitive Radio Systems," was supervised by Prof. Ali El-Hajj and Prof. Karim Y. Kabalan. In 2010, Lise obtained her "Master de Recherche" in Electronic Systems from Ecole Polytechnique de Nantes and her Engineering Diploma in Telecommunications and Information Technology Engineering from the Lebanese University, faculty of engineering, with high distinction. Lise authored or co-authored several publications in refereed journals and international conferences. Her research has focused on the implementation of several algorithms and techniques for cognitive radio transceivers to assist and relax the hardware limitations and complexity.

Yasir Saleem is currently a Master's student under a joint Programme of Sunway University, Malaysia and Lancaster University, United Kingdom. He received his BS degree in Information Technology in 2012 from National University of Sciences and Technology, Islamabad, Pakistan. He is also a reviewer of journals and conferences such as Elsevier Computers & Electrical Engineering, International Journal of Communication Networks and Information Security (IJCNIS), IEEE ICC 2013 and IEEE Globecom 2014. His research interests include cognitive radio networks, cognitive radio sensor networks, cloud computing, wireless sensor networks and unmanned aerial vehicles.

Farrukh Salim is currently a PhD candidate at Technische Universität Ilmenau, Germany. He was a lecturer in NED University, Pakistan from June 2010 – August 2014. He completed his MS in Computer Engineering from Hanyang University, South Korea in 2010. He received his Bachelors of Engineering (BE) degree in Computer Systems Engineering from Mehran University of Engineering and Technology (MUET), Pakistan in 2005. His current research interests include wireless communications, specifically, cognitive radio networks, wireless sensor networks and wireless ad-hoc networks.

Karim G. Seddik is an assistant professor in the Electronics and Communications Engineering Department at the American University in Cairo (AUC), Egypt. Before joining AUC, he was an assistant professor in the Electrical Engineering Department at Alexandria University, Egypt. Dr. Seddik received the BS (with highest honors) and MS degrees in Electrical Engineering from Alexandria University, Alexandria, Egypt, in 2001 and 2004, respectively. He received his PhD degree at the Electrical and Computer Engineering Department, University of Maryland, College Park, 2008. His current research interests include cooperative communications and networking, MIMO–OFDM systems, cognitive radio, layered channel coding, and distributed detection in wireless sensor networks. Dr. Seddik is a recipient of the Certificate of Honor from the Egyptian President for being ranked first among all departments in College of Engineering, Alexandria University, in 2002, the Graduate School Fellowship from the University of Maryland in 2004 and 2005, and the Future Faculty Program Fellowship from the University of Maryland in 2007.

Shamik Sengupta is an Assistant Professor in the Department of Computer Science and Engineering at University of Nevada, Reno (UNR). He received his PhD degree from the School of Electrical Engineering and Computer Science, University of Central Florida, in 2007. Before joining University of Nevada, Reno, he worked as a Post-Doctorate Researcher at Stevens Institute of Technology and an Assistant Professor at City University of New York (John Jay College of Criminal Justice). His research interests include cognitive radio and DSA networks, game theory, cybersecurity, network economics, and self-configuring wireless mesh networks. He has authored over 50 international conferences and journal publications. He is the recipient of an IEEE GLOBECOM 2008 best paper award and a recipient of NSF CAREER award in 2012. Shamik Sengupta serves on the organizing and technical program committee of several IEEE conferences. He served as Guest Editor in *Eurasip Journal on Wireless Communications and Networking*, Special Issue on Advances in 4G Wireless, and is now serving as an editorial board member of *Springer Computing Journal*. For more information visit http://www.cse.unr.edu/~shamik/.

Shree Krishna Sharma received the BE degree in Electronics and Communication from Birla Institute of Technology, Mesra, India, in 2004; the MSc degree in Information and Communication Engineering from the Institute of Engineering, Pulchowk, Nepal, in 2010; the MA degree in Economics from Tribhuvan University Patan Multiple Campus, Patan, Nepal; and the MRes degree in Computing Science from Staffordshire University, Staffordshire, UK, in 2011. He is currently working toward the PhD degree with the Interdisciplinary Centre for Security, Reliability, and Trust, University of Luxembourg, Luxembourg City, Luxembourg. In the past, he was also involved with Kathmandu University, Dhulikhel, Nepal, as a Teaching Assistant, and he served as a Part-Time Lecturer for eight engineering colleges in Nepal. He was with Nepal Telecom for more than four years as a Telecom Engineer in the field of information technology and telecommunication. His research interests include cognitive wireless communications, resource allocation, and interference mitigation in heterogeneous wireless networks. Mr. Sharma received an Indian Embassy Scholarship for his BE study, an Erasmus Mundus Scholarship for his MRes study, and an Aids Training-Research PhD grant from the National Research Fund of Luxembourg.

Asrar U. H. Sheikh graduated from the University of Birmingham with degrees of MSc and PhD in 1966 and 1969, respectively. He is currently Professor and Head of Electrical Engineering Department at the University of Lahore. Previous to this, he has been the Rector, Foundation University Islamabad, Pakistan (2006-2007) and Bugshan-Bell Labs Chair Professor at KFUPM (2000-2013). Dr. Sheikh has published over 300 papers in technical journals, conference proceedings, chapters in books. He is a Life Fellow of the IEEE and a Fellow of the IET. His current interests are in several aspects of communication systems, particularly in wireless communications.

Mohamed Siala received his General Engineering degree from Ecole Polytechnique, Palaiseau, France, in 1988, his specialization Engineering Degree in Telecommunications from Telecom ParisTech, Paris, France, in 1990, and his PhD in Digital Communications from Telecom ParisTech, Paris, France, in 1995. From 1990 to 1992, he was with Alcatel Radio-Telephones, Colombes, France, working on the GSM physical layer. In 1995, he joined Wavecom, Issy-les-Moulineaux, France, where he worked on advanced multicarrier modulations and channel estimation for low-orbit mobile satellite communications. From 1997 to 2001, he worked at Orange Labs, Issy-les-Moulineaux, France, on the physical layer of 3G systems and participated actively on the standardization of the physical layer of the UMTS system. In 2001, he joined Sup'Com, Tunis, Tunisia, where he is now a full professor. His research interests are in the areas of digital and wireless communications with special emphasis on multicarrier systems, channel estimation, synchronization, adaptive modulation and coding, ARQ, MIMO systems, space-time coding, relaying, cooperative networks, and cognitive radio.

Ajay Singh received the BTech degree from the Kurukshetra University, Haryana, India, in 2003, and the ME degree from Panjab University, Chandigarh, in 2007, both in Electronics and Communication Engineering. He obtained the PhD degree in from the Department of Electrical Engineering, Indian Institute of Technology Delhi, New Delhi, India, in 2012. Since 2013, he has been with the Department of Electronics and Telecommunication Engineering, National Institute of Technology Raipur, Chhattisgarh, India, where he is currently an Assistant Professor. He is an IEEE member. He is serving as a reviewer in a number of international journals including *IEEE Transactions on Wireless Communications, IEEE Transactions on Vehicular Technology, IEEE Transactions on Communications, IEEE Transactions on Selected Area in Communications, IEEE Communications Letters*, and *IEEE Wireless Communications Letters*. His current research interests are in the areas of cognitive radio and cooperative communications.

Ju Bin Song received the PhD degree from the Department of Electronic and Electrical Engineering, University College London (UCL), London, UK, in 2001, as well as the BS and MS degrees in 1987 and 1989, respectively. From 1992 to 1997, he was a Senior Researcher in the Electronics and Telecommunications Research Institute (ETRI), South Korea. He was a Research Fellow in the Department of Electronic and Electrical Engineering at UCL in 2001. During 2002-2003, he was an Assistant Professor in the School of Information and Computer Engineering at Hanbat National University, South Korea. He is currently a Professor in the Department of Electronics and Radio Engineering, Kyung Hee University, South Korea, since 2003. From 2009 to 2010, he was a Visiting Professor in the Department of Electrical Engineering and Computer Science at the University of Houston, Texas, USA. He is currently Head of the Department of Electronics and Radio Engineering, Kyung Hee University. His current research interests include resource allocation in communication systems and networks, cooperative communications, game theory, optimization, cognitive radio networks, and smart grid. Dr. Song serves as a member of the technical program committee for international conferences on communications and networks and is an editor for international journals on communications and networks. Dr. Song was a Kyung Hee University Best Teaching Award recipient in 2004 and 2012, respectively.

Kevin W. Sowerby received the BE and PhD degrees in Electrical and Electronic Engineering from the University of Auckland in 1986 and 1989, respectively. During 1989 and 1990, he was a Leverhulme Visiting Fellow at the University of Liverpool (UK). In 1990, he returned to the University of Auckland to take up an academic post in what is now the Department of Electrical and Computer Engineering, where he is now a professor. He served as Deputy Head of Department (Research) from 2008 to 2012. Dr. Sowerby has spent extended periods of research leave at Virginia Tech (1997), Simon Fraser University (1997 and 2005), Motorola Research Labs (2005), Georgia Tech (2012), TNO in The Netherlands (2012), and Hong Kong University of Science and Technology (2012-13). His research interests include wireless communications systems and, in particular, methods for designing reliable high-capacity networks. Within the IEEE, he was the founding Chair of New Zealand's first Communications Society Chapter, and he has served as Chair of the New Zealand North Section and the New Zealand Council. He is currently the Region 10 representative on the IEEE Admission and Advancement Committee.